Stochastic Integration with Jumps

Stochastic processes with jumps and random measures are gaining importance as drivers in applications like financial mathematics and signal processing. This book develops the stochastic integration theory for both integrators (semimartingales) and random measures from a common point of view.

Highlights feature the DCT and Egoroff's theorem, as well as comprehensive analogs to results from ordinary integration theory, for instance, previsible envelopes and an algorithm computing the stochastic integrals of càglàd integrands pathwise.

An integrator under P is continuous as a map into L^q for any finite q, provided that P is replaced with a suitable probability, and there is control of the transition; this extends to random measures when $q = 2$. This has the consequence that every integrator is controlled by some previsible process in much the same way a Wiener process is controlled by time t. The previsible controller furnishes Picard norms that reduce SDEs to simple (global) fixed-point problems with an easy stability theory and numerical pathwise approximation schemes.

Full proofs are given for all results, and motivation is stressed throughout. A large appendix contains most of the analysis that readers will need as a prerequisite. A comprehensive reference list and an index of notation are also provided. Extra material is available from the book's Web site at http://www.ma.utexas.edu/users/cup.

This will be an invaluable reference for graduate students and researchers in mathematics, physics, electrical engineering, and finance who need to use stochastic differential equations.

Klaus Bichteler is a Professor of Mathematics at the University of Texas at Austin. He received his Ph.D. in physics from Hamburg University in 1965. He has written extensively on general relativity, representation theory, integration, probability, and Malliavin calculus.

Volume 89
Stochastic Integration with Jumps

27 N. H. Bingham, C. M. Goldie and J. L. Teugels *Regular Variation*
29 N. White (ed.) *Combinatorial Geometries*
30 M. Pohst and H. Zassenhaus *Algorithmic Algebraic Number Theory*
31 J. Aczel and J. Dhombres *Functional Equations in Several Variables*
32 M. Kuczma, B. Choczewski and R. Ger *Iterative Functional Equations*
33 R. V. Ambartzumian *Factorization Calculus and Geometric Probability*
34 G. Gripenberg, S.-O. Londen and O. Staffans *Volterra Integral and Functional Equations*
35 G. Gasper and M. Rahman *Basic Hypergeometric Series*
36 E. Torgersen *Comparison of Statistical Experiments*
38 N. Korneichuk *Exact Constants in Approximation Theory*
39 R. Brualdi and H. Ryser *Combinatorial Matric Theory*
40 N. White (ed.) *Matroid Applications*
41 S. Sakai *Operator Algebras in Dynamical Systems*
42 W. Hodges *Basic Model Theory*
43 H. Stahl and V. Totik *General Orthogonal Polynomials*
45 G. Da Prato and J. Zabczyk *Stochastic Equations in Infinite Dimensions*
46 A. Björner et al. *Oriented Matroids*
47 G. Edgar and L. Sucheston *Stopping Times and Directed Processes*
48 C. Sims *Computation with Finitely Presented Groups*
49 T. Palmer *Banach Algebras and the General Theory of *-Algebras I*
50 F. Borceux *Handbook of Categorical Algebra I*
51 F. Borceux *Handbook of Categorical Algebra II*
52 F. Borceux *Handbook of Categorical Algebra III*
53 V. F. Kolchin *Random Graphics*
54 A. Katok and B. Hasselblatt *Introduction to the Modern Theory of Dynamical Systems*
55 V. N. Sachkov *Combinatorial Methods in Discrete Mathematics*
56 V. N. Sachkov *Probabilistic Methods in Discrete Mathematics*
57 M. Cohn *Skew Fields*
58 R. Gardner *Geometric Topography*
59 G. A. Baker Jr. and P. Graves-Morris *Pade Approximants, 2nd Edition*
60 J. Krajicek *Bounded Arithmetic Propositional Logic and Complexity Theory*
61 H. Groemer *Geometric Applications of Fourier Series and Spherical Harmonics*
62 H. O. Fattorini *Infinite Dimensional Optimization and Control Theory*
63 A. C. Thompson *Minkowski Geometry*
64 R. B. Bapat and T. E. S. Raghavan *Nonnegative Matrices with Applications*
65 K. Engel *Sperner Theory*
66 D. Cvetkovic, P. Rowlinson and S. Simic *Eigenspaces of Graphs*
67 F. Bergeron, G. Labelle and P. Leroux *Combinatorial Species and Tree-Like Structures*
68 R. Goodman and N. Wallach *Representations and Invariants of the Classical Groups*
69 T. Beth, D. Jungnickel and H. Lenz *Design Theory I, 2nd Edition*
70 A. Pietsch and J. Wenzel *Orthonormal Systems for Banach Space Geometry*
71 G. E. Andrews, R. Askey and R. Roy *Special Functions*
72 R. Ticciati *Quantum Field Theory for Mathematicians*
73 M. Stern *Semimodular Lattices*
74 I. Lasiecka and R. Triggiani *Control Theory for Partial Differential Equations I*
75 I. Lasiecka and R. Triggiani *Control Theory for Partial Differential Equations II*
76 A. A. Ivanov *Geometry of Sporadic Groups 1*
77 A. Schinzel *Polynomials with Special Regard to Reducibility*
78 H. Lenz, T. Beth and D. Jungnickel *Design Theory II, 2nd Edition*
79 T. Palmer *Banach Algebras and the General Theory of *-Algebras II*
80 O. Stormark *Lie's Structural Approach to PDE Systems*
81 C. F. Dunkl and Y. Xu *Orthogonal Polynomials of Several Variables*
82 J. P. Mayberry *The Foundations of Mathematics in the Theory of Sets*

ENCYCLOPEDIA OF MATHEMATICS AND ITS APPLICATIONS

Stochastic Integration with Jumps

KLAUS BICHTELER

University of Texas at Austin

CAMBRIDGE
UNIVERSITY PRESS

CAMBRIDGE UNIVERSITY PRESS
Cambridge, New York, Melbourne, Madrid, Cape Town, Singapore,
São Paulo, Delhi, Dubai, Tokyo

Cambridge University Press
The Edinburgh Building, Cambridge CB2 8RU, UK

Published in the United States of America by Cambridge University Press, New York

www.cambridge.org
Information on this title: www.cambridge.org/9780521142144

First published 2002
This digitally printed version 2010

A catalogue record for this publication is available from the British Library

Library of Congress Cataloguing in Publication data

Bichteler, Klaus
Stochastic integration with jumps / Klaus Bichteler.
p. cm – (Encyclopedia of mathematics and its applications)
Includes bibliographical references and indexes.
ISBN 0-521-81129-5
1. Stochastic Integrals. 2. Jump processes. I. Title. II. Series.

QA274.22 .B53 2002
519.2—dc21 2001043017

ISBN 978-0-521-81129-3 Hardback
ISBN 978-0-521-14214-4 Paperback

Contents

Preface

This book originated with several courses given at the University of Texas. The audience consisted of graduate students of mathematics, physics, electrical engineering, and finance. Most had met some stochastic analysis during work in their field; the course was meant to provide the mathematical underpinning. To satisfy the economists, driving processes other than Wiener process had to be treated; to give the mathematicians a chance to connect with the literature and discrete-time martingales, I chose to include driving terms with jumps. This plus a predilection for generality for simplicity's sake led directly to the most general stochastic Lebesgue–Stieltjes integral.

The spirit of the exposition is as follows: just as having finite variation and being right-continuous identifies the useful Lebesgue–Stieltjes distribution functions among all functions on the line, are there criteria for processes to be useful as "random distribution functions." They turn out to be straightforward generalizations of those on the line. A process that meets these criteria is called an *integrator*, and its integration theory is just as easy as that of a deterministic distribution function on the line – provided Daniell's method is used. (This proviso has to do with the lack of convexity in some of the target spaces of the stochastic integral.)

For the purpose of error estimates in approximations both to the stochastic integral and to solutions of stochastic differential equations we define various *numerical sizes of an integrator* Z and analyze rather carefully how they propagate through many operations done on and with Z, for instance, solving a stochastic differential equation driven by Z. These size-measurements arise as generalizations to integrators of the famed Burkholder–Davis–Gundy inequalities for martingales. The present exposition differs in the ubiquitous use of numerical estimates from the many fine books on the market, where convergence arguments are usually done in probability or every once in a while in Hilbert space L^2. For reasons that unfold with the story we employ the L^p-norms in the whole range $0 \le p < \infty$. An effort is made to furnish reasonable estimates for the universal constants that occur in this context.

Such attention to estimates, unusual as it may be for a book on this subject, pays handsomely with some new results that may be edifying even to the expert. For instance, it turns out that every integrator Z can be controlled by an increasing previsible process much like a Wiener process is controlled by time t; and if not with respect to the given probability, then at least with respect to an equivalent one that lets one view the given integrator as a map into Hilbert space, where computation is comparatively facile. This *previsible controller* obviates prelocal arguments [91] and can be used to construct Picard norms for the solution of stochastic differential equations driven by Z that allow growth estimates, easy treatment of stability theory, and even *pathwise algorithms* for the solution. These schemes extend without ado to *random measures*, including the previsible control and its application to stochastic differential equations driven by them.

All this would seem to lead necessarily to an enormous number of technicalities. A strenuous effort is made to keep them to a minimum, by these devices: everything not directly needed in stochastic integration theory and its application to the solution of stochastic differential equations is either omitted or relegated to the Supplements or to the Appendices. A short survey of the beautiful "General Theory of Processes" developed by the French school can be found there.

A warning concerning the usual conditions is appropriate at this point. They have been replaced throughout with what I call the *natural conditions*. This will no doubt arouse the ire of experts who think one should not "tamper with a mature field." However, many fine books contain erroneous statements of the important Girsanov theorem – in fact, it is hard to find a correct statement in unbounded time – and this is traceable directly to the employ of the usual conditions (see example 3.9.14 on page 164 and 3.9.20). In mathematics, correctness trumps conformity. The natural conditions confer the same benefits as do the usual ones: path regularity (section 2.3), section theorems (page 437 ff.), and an ample supply of stopping times (*ibidem*), without setting a trap in Girsanov's theorem.

The students were expected to know the basics of point set topology up to Tychonoff's theorem, general integration theory, and enough functional analysis to recognize the Hahn–Banach theorem. If a fact fancier than that is needed, it is provided in appendix A, or at least a reference is given.

The exercises are sprinkled throughout the text and form an integral part. They have the following appearance:

Exercise 4.3.2 This is an exercise. It is set in a smaller font. It requires no novel argument to solve it, only arguments and results that have appeared earlier. Answers to some of the exercises can be found in appendix B. Answers to most of them can be found in appendix C, which is available on the web *via* `http://www.ma.utexas.edu/users/cup/Answers`.

I made an effort to index every technical term that appears (page 489), and to make an index of notation that gives a short explanation of every symbol and lists the page where it is defined in full (page 483). Both indexes appear in expanded form at `http://www.ma.utexas.edu/users/cup/Indexes`.

http://www.ma.utexas.edu/users/cup/Errata contains the errata. I plead with the gentle reader to send me the errors he/she found via email to kbi@math.utexas.edu, so that I may include them, with proper credit of course, in these errata.

At this point I recommend reading the conventions on page 363.

1

Introduction

1.1 Motivation: Stochastic Differential Equations

Stochastic Integration and Stochastic Differential Equations (SDEs) appear in analysis in various guises. An example from physics will perhaps best illuminate the need for this field and give an inkling of its particularities. Consider a physical system whose state at time t is described by a vector X_t in $\mathbb{R}^n$. In fact, for concreteness' sake imagine that the system is a space probe on the way to the moon. The pertinent quantities are its location and momentum. If x_t is its location at time t and p_t its momentum at that instant, then X_t is the 6-vector (x_t, p_t) in the phase space $\mathbb{R}^6$. In an ideal world the evolution of the state is governed by a differential equation:

$$\frac{dX_t}{dt} = \begin{pmatrix} dx_t/dt \\ dp_t/dt \end{pmatrix} = \begin{pmatrix} p_t/m \\ F(x_t, p_t) \end{pmatrix} .$$

Here m is the mass of the probe. The first line is merely the definition of p: momentum = mass $\times$ velocity. The second line is Newton's second law: the rate of change of the momentum is the force F. For simplicity of reading we rewrite this in the form

$$dX_t = a(X_t)\, dt , \tag{1.1.1}$$

which expresses the idea that the change of X_t during the time-interval dt is proportional to the time dt elapsed, with a proportionality constant or *coupling coefficient* a that depends on the state of the system and is provided by a model for the forces acting. In the present case $a(X)$ is the 6-vector $(p/m, F(X))$. Given the initial state X_0, there will be a unique solution to (1.1.1). The usual way to show the existence of this solution is Picard's iterative scheme: first one observes that (1.1.1) can be rewritten in the form of an *integral equation*:

$$X_t = X_0 + \int_0^t a(X_s)\, ds . \tag{1.1.2}$$

Then one starts Picard's scheme with $X_t^0 = X_0$ or a better guess and defines the iterates inductively by

$$X_t^{n+1} = X^0 + \int_0^t a(X_s^n)\, ds .$$

If the coupling coefficient a is a Lipschitz function of its argument, then the *Picard iterates* X^n will converge uniformly on every bounded time-interval and the limit X^∞ is a solution of (1.1.2), and thus of (1.1.1), and the only one. The reader who has forgotten how this works can find details on pages 274–281. Even if the solution of (1.1.1) cannot be written as an analytical expression in t, there exist extremely fast numerical methods that compute it to very high accuracy. Things look rosy.

In the less-than-ideal real world our system is subject to unknown forces, *noise*. Our rocket will travel through gullies in the gravitational field that are due to unknown inhomogeneities in the mass distribution of the earth; it will meet gusts of wind that cannot be foreseen; it might even run into a gaggle of geese that deflect it. The evolution of the system is better modeled by an equation

$$dX_t = a(X_t)\,dt + dG_t\ , \tag{1.1.3}$$

where G_t is a noise that contributes its differential dG_t to the change dX_t of X_t during the interval dt. To accommodate the idea that the noise comes from without the system one assumes that there is a *background noise* Z_t – consisting of gravitational gullies, gusts, and geese in our example – and that its effect on the state during the time-interval dt is proportional to the difference dZ_t of the *cumulative noise* Z_t during the time-interval dt, with a proportionality constant or *coupling coefficient* b that depends on the state of the system:

$$dG_t = b(X_t)\,dZ_t\ .$$

For instance, if our probe is at time t halfway to the moon, then the effect of the gaggle of geese at that instant should be considered negligible, and the effect of the gravitational gullies is small. Equation (1.1.3) turns into

$$dX_t = a(X_t)\,dt + b(X_t)\,dZ_t\ , \tag{1.1.4}$$

in integrated form $$X_t = X_t^0 + \int_0^t a(X_s)\,ds + \int_0^t b(X_s)\,dZ_s\ . \tag{1.1.5}$$

What is the meaning of this equation in practical terms? Since the background noise Z_t is not known one cannot solve (1.1.5), and nothing seems to be gained. Let us not give up too easily, though. Physical intuition tells us that the rocket, though deflected by gullies, gusts, and geese, will probably not turn all the way around but will rather still head somewhere in the vicinity of the moon. In fact, for all we know the various noises might just cancel each other and permit a perfect landing.

What are the chances of this happening? They seem remote, perhaps, yet it is obviously important to find out how likely it is that our vehicle will at least hit the moon or, better, hit it reasonably closely to the intended landing site. The smaller the noise dZ_t, or at least its *effect* $b(X_t)\,dZ_t$, the better we feel the chances will be. In other words, our intuition tells us to look for a statistical inference: from some reasonable or measurable assumptions on the background noise Z or its effect $b(X)dZ$ we hope to conclude about the likelihood of a successful landing.

This is all a bit vague. We must cast the preceding contemplations in a mathematical framework in order to talk about them with precision and, if possible, to obtain quantitative answers. To this end let us introduce the set Ω of all possible *evolutions of the world*. The idea is this: at the beginning $t = 0$ of the reckoning of time we may or may not know the *state-of-the-world* ω_0, but thereafter the course that the history $\omega : t \mapsto \omega_t$ of the world actually will take has the vast collection Ω of evolutions to choose from. For any two possible courses-of-history[1] $\omega : t \mapsto \omega_t$ and $\omega' : t \mapsto \omega'_t$ the state-of-the-world might take there will generally correspond different cumulative background noises $t \mapsto Z_t(\omega)$ and $t \mapsto Z_t(\omega')$. We stipulate further that there is a function $\mathbb{P}$ that assigns to certain subsets E of Ω, the *events*, a *probability* $\mathbb{P}[E]$ that they will occur, i.e., that the actual evolution lies in E. It is known that no reasonable probability $\mathbb{P}$ can be defined on *all* subsets of Ω. We assume therefore that the collection of all events that can ever be observed or are ever pertinent form a σ-algebra $\mathcal{F}$ of subsets of Ω and that the function $\mathbb{P}$ is a probability measure on $\mathcal{F}$. It is not altogether easy to defend these assumptions. Why should the observable events form a σ-algebra? Why should $\mathbb{P}$ be σ-additive? We content ourselves with this answer: there is a well-developed theory of such triples $(\Omega, \mathcal{F}, \mathbb{P})$; it comprises a rich calculus, and we want to make use of it. Kolmogorov [57] has a better answer:

Project 1.1.1 *Make a mathematical model for the analysis of random phenomena that does not require σ-additivity at the outset but furnishes it instead.*

So, for every possible course-of-history[1] $\omega \in \Omega$ there is a *background noise* $Z_. : t \mapsto Z_t(\omega)$, and with it comes the *effective noise* $b(X_t)\, dZ_t(\omega)$ that our system is subject to during dt. Evidently the state X_t of the system depends on ω as well. The obvious thing to do here is to compute, for every $\omega \in \Omega$, the solution of equation (1.1.5), to wit,

$$X_t(\omega) = X_t^0 + \int_0^t a(X_s(\omega))\, ds + \int_0^t b(X_s(\omega))\, dZ_s(\omega) , \qquad (1.1.6)$$

as the limit of the Picard iterates $X_t^0 \stackrel{\text{def}}{=} X_0$,

$$X_t^{n+1}(\omega) \stackrel{\text{def}}{=} X_t^0 + \int_0^t a(X_s^n(\omega))\, ds + \int_0^t b(X_s^n(\omega))\, dZ_s(\omega) . \qquad (1.1.7)$$

Let T be the time when the probe hits the moon. This depends on chance, of course: $T = T(\omega)$. Recall that x_t are the three spatial components of X_t.

---

[1] The redundancy in these words is for emphasis. [Note how repeated references to a footnote like this one are handled. Also read the last line of the chapter on page 41 to see how to find a repeated footnote.]

Our interest is in the function $\omega \mapsto x_T(\omega) = x_{T(\omega)}(\omega)$, the location of the probe at the time T. Suppose we consider a landing successful if our probe lands within F feet of the ideal landing site s at the time T it does land. We are then most interested in the probability

$$p_F \stackrel{\text{def}}{=} \mathbb{P}(\{\omega \in \Omega : \|x_T(\omega) - s\| < F\})$$

of a successful landing – its value should influence strongly our decision to launch. Now x_T is just a function on Ω, albeit defined in a circuitous way. We should be able to compute the set $\{\omega \in \Omega : \|x_T(\omega) - s\| < F\}$, and if we have enough information about $\mathbb{P}$, we should be able to compute its probability p_F and to make a decision. This is all classical ordinary differential equations (ODE), complicated by the presence of a parameter ω: straightforward in principle, if possibly hard in execution.

The Obstacle

As long as the *paths* $Z_.(\omega) : s \mapsto Z_s(\omega)$ of the background noise are right-continuous and have finite variation, the integrals $\int \cdots_s dZ_s$ appearing in equations (1.1.6) and (1.1.7) have a perfectly clear classical meaning as Lebesgue–Stieltjes integrals, and Picard's scheme works as usual, under the assumption that the coupling coefficients a, b are Lipschitz functions (see pages 274–281).

Now, since we do not know the background noise Z precisely, we must make a model about its statistical behavior. And here a formidable obstacle rears its head: the simplest and most plausible statistical assumptions about Z force it to be so irregular that the integrals of (1.1.6) and (1.1.7) cannot be interpreted in terms of the usual integration theory. The moment we stipulate some symmetry that merely expresses the idea that we don't know it all, obstacles arise that cause the paths of Z to have infinite variation and thus prevent the use of the Lebesgue–Stieltjes integral in giving a meaning to expressions like $\int X_s \, dZ_s(\omega)$.

Here are two assumptions on the random *driving term* Z that are eminently plausible:

(a) The expectation of the increment $dZ_t \approx Z_{t+h} - Z_t$ should be zero; otherwise there is a *drift* part to the noise, which should be subsumed in the first driving term $\int \cdot \, ds$ of equation (1.1.6). We may want to assume a bit more, namely, that if everything of interest, including the noise $Z_.(\omega)$, was actually observed up to time t, then the future increment $Z_{t+h} - Z_t$ still averages to zero. Again, if this is not so, then a part of Z can be shifted into a driving term of finite variation so that the remainder satisfies this condition – see theorem 4.3.1 on page 221 and proposition 4.4.1 on page 233. The mathematical formulation of this idea is as follows: let $\mathcal{F}_t$ be the σ-algebra generated by the collection of all observations that can be made before and at time t; $\mathcal{F}_t$ is commonly and with intuitive appeal called the *history* or *past* at time t. In these terms our assumption is that the *conditional expectation*

$$\mathbb{E}\left[Z_{t+h} - Z_t \middle| \mathcal{F}_t\right]$$

of the future differential noise given the past vanishes. This makes Z a *martingale on the filtration* $\mathcal{F}_. = \{\mathcal{F}_t\}_{0 \le t < \infty}$ – these notions are discussed in detail in sections 1.3 and 2.5.

(b) We may want to assume further that Z does not change too wildly with time, say, that the paths $s \mapsto Z_s(\omega)$ are continuous. In the example of our space probe this reflects the idea that it will not blow up or be hit by lightning; these would be huge and sudden disturbances that we avoid by careful engineering and by not launching during a thunderstorm.

A background noise Z satisfying (a) and (b) has the property that *almost none of its paths* $Z_.(\omega)$ *is differentiable at any instant* – see exercise 3.8.13 on page 152. By a well-known theorem of real analysis,[2] the path $s \mapsto Z_s(\omega)$ does not have finite variation on any time-interval; and this irregularity happens for almost every $\omega \in \Omega$!

We are stumped: since $s \mapsto Z_s$ does not have finite variation, the integrals $\int \cdots dZ_s$ appearing in equations (1.1.6) and (1.1.7) do not make sense in any way we know, and then neither do the equations themselves.

Historically, the situation stalled at this juncture for quite a while. Wiener made an attempt to define the integrals in question in the sense of distribution theory, but the resulting *Wiener integral* is unsuitable for the iteration scheme (1.1.7), for lack of decent limit theorems.

Itô's Way Out of the Quandary

The problem is evidently to give a meaning to the integrals appearing in (1.1.6) and (1.1.7). Not only that, any prospective integral must have rather good properties: to show that the iterates X^n of (1.1.7) form a Cauchy sequence and thus converge there must be estimates available; to show that their limit is the solution of (1.1.6) there must be a limit theorem that permits the interchange of limit and integral, to wit,

$$\int_0^t \lim_n b(X_s^n)\, dZ_s = \lim_n \int_0^t b(X_s^n)\, dZ_s\, .$$

In other words, what is needed is an integral satisfying the Dominated Convergence Theorem, say. Convinced that an integral with this property cannot be defined *pathwise*, i.e., ω for ω, the Japanese mathematician Itô decided to try for an integral in the sense of the L^2-mean. His idea was this: while the sums

$$S_{\mathcal{P}}(\omega) \stackrel{\text{def}}{=} \sum_{k=1}^{K} b(X_{\sigma_k}(\omega)) \left(Z_{s_{k+1}}(\omega) - Z_{s_k}(\omega)\right)\, , \quad s_k \le \sigma_k \le s_{k+1}\, , \qquad (1.1.8)$$

[2] See for example [96, pages 94–100] or [9, page 157 ff.].

which appear in the usual definition of the integral, do not converge for any $\omega \in \Omega$, there may obtain *convergence in mean* as the partition $\mathcal{P} = \{s_0 < s_1 < \ldots < s_{K+1}\}$ is refined. In other words, there may be a random variable I such that

$$\| S_{\mathcal{P}} - I \|_{L^2} \to 0 \quad \text{as} \quad \text{mesh}[\mathcal{P}] \to 0 .$$

And if $S_{\mathcal{P}}$ should not converge in L^2-mean, it may converge in L^p-mean for some other $p \in (0, \infty)$, or at least in measure ($p = 0$).

In fact, this approach succeeds, but not without another observation that Itô made: for the purpose of Picard's scheme it is not necessary to integrate all processes.[3] *An integral defined for non-anticipating integrands suffices.* In order to describe this notion with a modicum of precision, we must refer again to the σ-algebras $\mathcal{F}_t$ comprising the history known at time t. The integrals $\int_0^t a(X_0)\,ds = a(X_0)\cdot t$ and $\int_0^t b(X_0)\,dZ_s(\omega) = b(X_0)\cdot\big(Z_t(\omega) - Z_0(\omega)\big)$ are at any time measurable on $\mathcal{F}_t$ because Z_t is; then so is the first Picard iterate X^1_t. Suppose it is true that the iterate X^n of Picard's scheme is at all times t measurable on $\mathcal{F}_t$; then so are $a(X^n_t)$ and $b(X^n_t)$. Their integrals, being limits of sums as in (1.1.8), will again be measurable on $\mathcal{F}_t$ at all instants t; then so will be the next Picard iterate X^{n+1}_t and with it $a(X^{n+1}_t)$ and $b(X^{n+1}_t)$, and so on. In other words, the integrands that have to be dealt with *do not anticipate the future*; rather, they are at any instant t measurable on the past $\mathcal{F}_t$. If this is to hold for the approximation of (1.1.8) as well, we are forced to choose for the point σ_i at which $b(X)$ is evaluated the left endpoint s_{i-1}. We shall see in theorem 2.5.24 that the choice $\sigma_i = s_{i-1}$ permits martingale[4] drivers Z – recall that it is the martingales that are causing the problems.

Since our object is to obtain statistical information, evaluating integrals and solving stochastic differential equations in the sense of a mean would pose no philosophical obstacle. It is, however, now not quite clear what it is that equation (1.1.5) models, if the integral is understood in the sense of the mean. Namely, what is the mechanism by which the random variable dZ_t affects the change dX_t in mean but not through its actual realization $dZ_t(\omega)$? Do the possible but not actually realized courses-of-history[1] somehow influence the behavior of our system? We shall return to this question in remarks 3.7.27 on page 141 and give a rather satisfactory answer in section 5.4 on page 310.

Summary: The Task Ahead

It is now clear what has to be done. First, the stochastic integral in the L^p-mean sense for non-anticipating integrands has to be developed. This

---

[3] A process is simply a function $Y : (s, \omega) \mapsto Y_s(\omega)$ on $\mathbb{R}_+ \times \Omega$. Think of $Y_s(\omega) = b(X_s(\omega))$.
[4] See page 5 and section 2.5, where this notion is discussed in detail.

is surprisingly easy. As in the case of integrals on the line, the integral is defined first in a non-controversial way on a collection $\mathcal{E}$ of *elementary integrands*. These are the analogs of the familiar step functions. Then that *elementary integral* is extended to a large class of processes in such a way that it features the Dominated Convergence Theorem. This is not possible for arbitrary driving terms Z, just as not every function z on the line is the distribution function of a σ-additive measure – to earn that distinction z must be right-continuous and have finite variation. The stochastic driving terms Z for which an extension with the desired properties has a chance to exist are identified by conditions completely analogous to these two and are called *integrators*.

For the extension proper we employ Daniell's method. The arguments are so similar to the usual ones that it would suffice to state the theorems, were it not for the deplorable fact that Daniell's procedure is generally not too well known, is even being resisted. Its efficacy is unsurpassed, in particular in the stochastic case.

Then it has to be shown that the integral found can, in fact, be used to solve the stochastic differential equation (1.1.5). Again, the arguments are straightforward adaptations of the classical ones outlined in the beginning of section 5.1, jazzed up a bit in the manner well known from the theory of ordinary differential equations in Banach spaces (e.g., [22, page 279 ff.] – the reader need not be familiar with it, as the details are developed in chapter 5). A pleasant surprise waits in the wings. Although the integrals appearing in (1.1.6) cannot be understood pathwise in the ordinary sense, there is an algorithm that solves (1.1.6) pathwise, i.e., ω–by–ω. This answers satisfactorily the question raised above concerning the meaning of solving a stochastic differential equation "in mean."

Indeed, why not let the cat out of the bag: the algorithm is simply the method of Euler–Peano. Recall how this works in the case of the deterministic differential equation $dX_t = a(X_t)\,dt$. One gives oneself a threshold δ and defines inductively an approximate solution $X_t^{(\delta)}$ at the points $t_k \stackrel{\text{def}}{=} k\delta$, $k \in \mathbb{N}$, as follows: if $X_{t_k}^{(\delta)}$ is constructed, wait until the driving term t has changed by δ, and let $t_{k+1} \stackrel{\text{def}}{=} t_k + \delta$ and

$$X_{t_{k+1}}^{(\delta)} = X_{t_k}^{(\delta)} + a(X_{t_k}^{(\delta)}) \times (t_{k+1} - t_k) ;$$

between t_k and t_{k+1} define $X_t^{(\delta)}$ by linearity. The compactness criterion A.2.38 of Ascoli–Arzelà allows the conclusion that the polygonal paths $X^{(\delta)}$ have a limit point as $\delta \to 0$, which is a solution. This scheme actually expresses more intuitively the meaning of the equation $dX_t = a(X_t)\,dt$ than does Picard's. If one can show that it converges, one should be satisfied that the limit is for all intents and purposes a solution of the differential equation.

In fact, the adaptive version of this scheme, where one waits until the *effect of the driving term* $a(X_{t_k}^{(\delta)}) \times (t - t_k)$ is sufficiently large to define t_{k+1}

and $X^{(\delta)}_{t_{k+1}}$, does converge for almost all $\omega \in \Omega$ in the stochastic case, when the deterministic driving term $t \mapsto t$ is replaced by the stochastic driver $t \mapsto Z_t(\omega)$ (see section 5.4).

So now the reader might well ask why we should go through all the labor of stochastic integration: integrals do not even appear in this scheme! And the question of what it means to solve a stochastic differential equation "in mean" does not arise. The answer is that there seems to be no way to prove the almost sure convergence of the Euler–Peano scheme directly, due to the absence of compactness. One has to show[5] that the Picard scheme works before the Euler–Peano scheme can be proved to converge.

So here is a new perspective: what we mean by a solution of equation (1.1.4),

$$dX_t(\omega) = a(X_t(\omega))\,dt + b(X_t(\omega))\,dZ_t(\omega)\,,$$

is a limit to the Euler–Peano scheme. Much of the labor in these notes is expended just to establish via stochastic integration and Picard's method that this scheme does, in fact, converge almost surely.

Two further points. First, even if the model for the background noise Z is simple, say, is a Wiener process, the stochastic integration theory must be developed for integrators more general than that. The reason is that the solution of a stochastic differential equation is itself an integrator, and in this capacity it can best be analyzed. Moreover, in mathematical finance and in filtering and control theory, the solution of one stochastic differential equation is often used to drive another.

Next, in most applications the state of the system will have many components and there will be several background noises; the stochastic differential equation (1.1.5) then becomes[6]

$$X^\nu_t = C^\nu_t + \sum_{1\le\eta\le d}\int_0^t F^\nu_\eta[X^1,\dots,X^n]\,dZ^\eta\,, \qquad \nu = 1,\dots,n\,.$$

The state of the system is a vector $X = (X^\nu)^{\nu=1\dots n}$ in $\mathbb{R}^n$ whose evolution is driven by a collection $\{Z^\eta : 1 \le \eta \le d\}$ of scalar integrators. The d vector fields $F_\eta = (F^\nu_\eta)^{\nu=1\dots n}$ are the *coupling coefficients*, which describe the effect of the background noises Z^η on the change of X. $C_t = (C^\nu_t)^{\nu=1\dots n}$ is the *initial condition* – it is convenient to abandon the idea that it be constant. It eases the reading to rewrite the previous equation in vector notation as[7]

$$X_t = C_t + \int_0^t F_\eta[X]\,dZ^\eta\,. \tag{1.1.9}$$

[5] So far – here is a challenge for the reader!
[6] See equation (5.2.1) on page 282 for a more precise discussion.
[7] We shall use the ***Einstein convention*** throughout: summation over repeated indices *in opposite positions* (the η in (1.1.9)) is implied.

The form (1.1.9) offers an intuitive way of reading the stochastic differential equation: the noise Z^η *drives* the state X in the direction $F_\eta[X]$. In our example we had four driving terms: $Z^1_t = t$ is time and F_1 is the systemic force; Z^2 describes the gravitational gullies and F_2 their effect; and Z^3 and Z^4 describe the gusts of wind and the gaggle of geese, respectively. The need for several noises will occasionally call for estimates involving whole slews $\{Z^1, ..., Z^d\}$ of integrators.

1.2 Wiener Process

Wiener process[8] is the model most frequently used for a background noise. It can perhaps best be motivated by looking at Brownian motion, for which it was an early model. Brownian motion is an example not far removed from our space probe, in that it concerns the motion of a particle moving under the influence of noise. It is simple enough to allow a good stab at the background noise.

Example 1.2.1 (Brownian Motion) Soon after the invention of the microscope in the 17th century it was observed that pollen immersed in a fluid of its own specific weight does not stay calmly suspended but rather moves about in a highly irregular fashion, and never stops. The English physicist Brown studied this phenomenon extensively in the early part of the 19th century and found some systematic behavior: the motion is the more pronounced the smaller the pollen and the higher the temperature; the pollen does not aim for any goal – rather, during any time-interval its path appears much the same as it does during any other interval of like duration, and it also looks the same if the direction of time is reversed. There was speculation that the pollen, being live matter, is propelling itself through the fluid. This, however, runs into the objection that it must have infinite energy to do so (jars of fluid with pollen in it were stored for up to 20 years in dark, cool places, after which the pollen was observed to jitter about with undiminished enthusiasm); worse, ground-up granite instead of pollen showed the same behavior.

In 1905 Einstein wrote three Nobel-prize–worthy papers. One offered the Special Theory of Relativity, another explained the Photoeffect (for this he got the Nobel prize), and the third gave an explanation of Brownian motion. It rests on the idea that the pollen is kicked by the much smaller fluid molecules, which are in constant thermal motion. The idea is not, as one might think at first, that the little jittery movements one observes are due to kicks received from particularly energetic molecules; estimates of the distribution of the kinetic energy of the fluid molecules rule this out. Rather, it is this: the pollen suffers an enormous number of collisions with the molecules of the surrounding fluid, each trying to propel it in a different direction, but mostly canceling each other; *the motion observed is due to*

[8] "Wiener process" is sometimes used without an article, in the way "Hilbert space" is.

statistical fluctuations. Formulating this in mathematical terms leads to a stochastic differential equation[9]

$$\begin{pmatrix} dx_t \\ dp_t \end{pmatrix} = \begin{pmatrix} p_t/m\, dt \\ -\alpha\, p_t\, dt + d\boldsymbol{W}_t \end{pmatrix} \tag{1.2.1}$$

for the location (x,p) of the pollen *in its phase space.* The first line expresses merely the definition of the momentum p; namely, the rate of change of the location x in $\mathbb{R}^3$ is the velocity $v = p/m$, m being the mass of the pollen. The second line attributes the change of p during dt to two causes: $-\alpha p\, dt$ describes the resistance to motion due to the viscosity of the fluid, and $d\boldsymbol{W}_t$ is the sum of the very small momenta that the enormous number of collisions impart to the pollen during dt. The random driving term is denoted $\boldsymbol{W}$ here rather than Z as in section 1.1, since the model for it will be a Wiener process.

This explanation leads to a plausible model for the background noise $\boldsymbol{W}$: $d\boldsymbol{W}_t = \boldsymbol{W}_{t+dt} - \boldsymbol{W}_t$ is the sum of a huge number of exceedingly small momenta, so by the Central Limit Theorem A.4.4 we expect $d\boldsymbol{W}_t$ to have a normal law. (For the notion of a law or distribution see section A.3 on page 391. We won't discuss here Lindeberg's or other conditions that would make this argument more rigorous; let us just assume that whatever condition on the distribution of the momenta of the molecules needed for the CLT is satisfied. We are, after all, doing heuristics here.)

We do not see any reason why kicks in one direction should, on the average, be more likely than in any other, so this normal law should have expectation zero and a multiple of the identity for its covariance matrix. In other words, it is plausible to stipulate that $d\boldsymbol{W}$ be a 3-vector of identically distributed independent normal random variables. It suffices to analyze one of its three scalar components; let us denote it by dW.

Next, there is no reason to believe that the total momenta imparted during non-overlapping time-intervals should have anything to do with one another. In terms of W this means that for consecutive instants $0 = t_0 < t_1 < t_2 < \ldots < t_K$ the corresponding family of consecutive ***increments***

$$\left\{ W_{t_1} - W_{t_0}\,,\, W_{t_2} - W_{t_1}\,,\, \ldots\,,\, W_{t_K} - W_{t_{K-1}} \right\}$$

should be independent. In self-explanatory terminology: we stipulate that W have ***independent increments.***

The background noise that we visualize does not change its character with time (except when the temperature changes). Therefore the law of $W_t - W_s$ should not depend on the times s,t individually but only on their difference, the elapsed time $t - s$. In self-explanatory terminology: we stipulate that W be ***stationary.***

---

[9] Edward Nelson's book, *Dynamical Theories of Brownian Motion* [82], offers a most enjoyable and thorough treatment and opens vistas to higher things.

Subtracting W_0 does not change the differential noises dW_t, so we simplify the situation further by stipulating that $W_0 = 0$.

Let $\delta = var(W_1) = \mathbb{E}[W_1^2]$. The variances of $W_{(k+1)/n} - W_{k/n}$ then must be δ/n, since they are all equal by stationarity and add up to δ by the independence of the increments. Thus the variance of W_q is δq for a rational $q = k/n$. By continuity the variance of W_t is δt, and the stationarity forces the variance of $W_t - W_s$ to be $\delta(t-s)$. ∎

Our heuristics about the cause of the Brownian jitter have led us to a stochastic differential equation, (1.2.1), including a model for the driving term W with rather specific properties: it should have stationary independent increments dW_t distributed as $N(0, \delta \cdot dt)$ and have $W_0 = 0$.

Does such a background noise exist? Yes; see theorem 1.2.2 below. If so, what further properties does it have? Volumes; see, e.g., [47]. How many such noises are there? Essentially one for every *diffusion coefficient* δ (see lemma 1.2.7 on page 16 and exercise 1.2.14 on page 19). They are called *Wiener processes.*

Existence of Wiener Process

What is meant by "Wiener process[8] exists"? It means that there is a probability space $(\Omega, \mathcal{F}, \mathbb{P})$ on which there lives a family $\{W_t : t \geq 0\}$ of random variables with the properties specified above. The quadruple $(\Omega, \mathcal{F}, \mathbb{P}, \{W_t : t \geq 0\})$ is a ***mathematical model*** for the noise envisaged. The case $\delta = 1$ is representative (exercise 1.2.14), so we concentrate on it:

Theorem 1.2.2 (Existence and Continuity of Wiener Process) *(i) There exist a probability space $(\Omega, \mathcal{F}, \mathbb{P})$ and on it a family $\{W_t : 0 \leq t < \infty\}$ of random variables that has stationary independent increments, and such that $W_0 = 0$ and the law of the increment $W_t - W_s$ is $N(0, t-s)$.*

(ii) Given such a family, one may change every W_t on a negligible set in such a way that for every $\omega \in \Omega$ the ***path*** *$t \mapsto W_t(\omega)$ is a continuous function.*

Definition 1.2.3 *Any family $\{W_t : t \in [0, \infty)\}$ of random variables (defined on some probability space) that has continuous paths and stationary independent increments $W_t - W_s$ with law $N(0, t-s)$, and that is normalized to $W_0 = 0$, is called a* ***standard Wiener process.***

A standard Wiener process can be characterized more simply as a continuous martingale W scaled by $W_0 = 0$ and $\mathbb{E}[W_t^2] = t$ (see corollary 3.9.5). In view of the discussion on page 4 it is thus not surprising that it serves as a background noise in the majority of stochastic models for physical, genetic, economic, and other phenomena and plays an important role in harmonic analysis and other branches of mathematics. For example, three-dimensional Wiener process[8] "knows" the zeroes of the ζ-function, and thus the distribution of the prime numbers – alas, so far it is reluctant to part with this knowledge. Wiener process is frequently called Brownian motion in the literature. We prefer to reserve the name "***Brownian motion***" for the physical phenomenon described in example 1.2.1 and capable of being described to various degrees of accuracy by different mathematical models [82].

Proof of Theorem 1.2.2 (i). To get an idea how we might construct the probability space $(\Omega, \mathcal{F}, \mathbb{P})$ and the W_t, consider dW as a map that associates with any interval $(s,t]$ the random variable $W_t - W_s$ on Ω, i.e., as a measure on $[0,\infty)$ with values in $L^2(\mathbb{P})$. It is after all in this capacity that the noise W will be used in a stochastic differential equation (see page 5). Eventually we shall need to integrate functions with dW, so we are tempted to extend this measure by linearity to a map $\int \cdot \, dW$ from step functions

$$\phi = \sum_k r_k \cdot 1_{(t_k, t_{k+1}]}$$

on the half-line to $L^2(\mathbb{P})$ to random variables via

$$\int \phi \, dW = \sum_k r_k \cdot (W_{t_{k+1}} - W_{t_k}) \, .$$

Suppose that the family $\{W_t : 0 \le t < \infty\}$ has the properties listed in (i). It is then rather easy to check that $\int \cdot \, dW$ extends to a linear isometry U from $L^2[0,\infty)$ to $L^2(\mathbb{P})$ with the property that $U(\phi)$ has a normal law $N(0,\sigma^2)$ with mean zero and variance $\sigma^2 = \int_0^\infty \phi^2(x)\,dx$, and so that functions perpendicular in $L^2[0,\infty)$ have independent images in $L^2(\mathbb{P})$. If we apply U to a basis of $L^2[0,\infty)$, we shall get a sequence (ξ_n) of independent $N(0,1)$ random variables. The verification of these claims is left as an exercise.

We now stand these heuristics on their head and arrive at the

Construction of Wiener Process Let $(\Omega, \mathcal{F}, \mathbb{P})$ be a probability space that admits a sequence (ξ_n) of independent identically distributed random variables, each having law $N(0,1)$. This can be had by the following simple construction: prepare countably many copies of $(\mathbb{R}, \mathcal{B}^\bullet(\mathbb{R}), \gamma_1)$ [10] and let $(\Omega, \mathcal{F}, \mathbb{P})$ be their product; for ξ_n take the nth coordinate function. Now pick any orthonormal basis (ϕ_n) of the Hilbert space $L^2[0,\infty)$. Any element f of $L^2[0,\infty)$ can be written uniquely in the form

$$f = \textstyle\sum_{n=1}^\infty a_n \phi_n \, ,$$

with $\|f\|^2_{L^2[0,\infty)} = \sum_{n=1}^\infty a_n^2 < \infty$. So we may define a map Φ by

$$\Phi(f) = \textstyle\sum_{n=1}^\infty a_n \xi_n \, .$$

[10] $\mathcal{B}^\bullet(\mathbb{R})$ is the σ-algebra of Borel sets on the line, and $\gamma_1(dx) = (1/\sqrt{2\pi}) \cdot e^{-x^2/2}\,dx$ is the normalized Gaussian measure, see page 419. For the infinite product see lemma A.3.20.

Φ evidently associates with every class in $L^2[0,\infty)$ an *equivalence class* of square integrable functions in $L^2(\mathbb{P}) = L^2(\Omega, \mathcal{F}, \mathbb{P})$. Recall the argument: the finite sums $\sum_{n=1}^N a_n\xi_n$ form a Cauchy sequence in the space $L^2(\mathbb{P})$, because

$$\mathbb{E}\Big[\Big(\sum_{n=M}^N a_n\xi_n\Big)^2\Big] = \sum_{n=M}^N a_n^2 \le \sum_{n=M}^\infty a_n^2 \xrightarrow[M\to\infty]{} 0 .$$

Since the space $L^2(\mathbb{P})$ is complete there is a limit in 2-mean; since $L^2(\mathbb{P})$, the space of equivalence classes, is Hausdorff, this limit is unique. Φ is clearly a linear isometry from $L^2[0,\infty)$ *into* $L^2(\mathbb{P})$. It is worth noting here that our recipe Φ does not produce a function but merely an equivalence class modulo $\mathbb{P}$-negligible functions. It is necessary to make some hard estimates to pick a suitable representative from each class, so as to obtain actual random variables (see lemma A.2.37).

Let us establish next that the law of $\Phi(f)$ is $N(0, \|f\|^2_{L^2[0,\infty)})$. To this end note that $f = \sum_n a_n\phi_n$ has the same norm as $\Phi(f)$:

$$\int_0^\infty f^2(t)\,dt = \sum a_n^2 = \mathbb{E}[(\Phi(f))^2] .$$

The simple computation

$$\mathbb{E}\Big[e^{i\alpha\Phi(f)}\Big] = \mathbb{E}\Big[e^{i\alpha\sum_n a_n\xi_n}\Big] = \prod_n \mathbb{E}\Big[e^{i\alpha a_n\xi_n}\Big] = e^{-\alpha^2\sum_n a_n^2/2}$$

shows that the characteristic function of $\Phi(f)$ is that of a $N(0, \sum a_n^2)$ random variable (see exercise A.3.45 on page 419). Since the characteristic function determines the law (page 410), the claim follows.

A similar argument shows that if $f_1, f_2, \ldots$ are orthogonal in $L^2[0,\infty)$, then $\Phi(f_1), \Phi(f_2), \ldots$ are not only also orthogonal in $L^2(\mathbb{P})$ but are actually independent:

clearly $\quad \Big\|\sum_k \alpha_k f_k\Big\|^2_{L^2[0,\infty)} = \sum_k \alpha_k^2 \cdot \|f_k\|^2_{L^2[0,\infty)}$,

whence $\quad \mathbb{E}\Big[e^{i\sum_k \alpha_k\Phi(f_k)}\Big] = \mathbb{E}\Big[e^{i\Phi(\sum_k \alpha_k f_k)}\Big] = e^{-\|\sum_k \alpha_k f_k\|^2/2}$

$$= \prod_k e^{-\alpha_k^2\cdot\|f_k\|^2/2} = \prod_k \mathbb{E}\Big[e^{i\alpha_k\Phi(f_k)}\Big] .$$

This says that the joint characteristic function is the product of the marginal characteristic functions, so the random variables are independent (see exercise A.3.36).

For any $t \ge 0$ let $\dot{W}_t$ be the class $\Phi\big(1_{[0,t]}\big)$ and simply pick a member W_t of $\dot{W}_t$. If $0 \le s < t$, then $\dot{W}_t - \dot{W}_s = \Phi\big(1_{(s,t]}\big)$ is distributed $N(0, t-s)$ and our family $\{W_t\}$ is stationary. With disjoint intervals being orthogonal functions of $L^2[0,\infty)$, our family has independent increments. ∎

Proof of Theorem 1.2.2 (ii). We start with the following observation: due to exercise A.3.47, the curve $t \mapsto \dot{W}_t$ is continuous from $\mathbb{R}_+$ to the space $L^p(\mathbb{P})$, for any $p < \infty$. In particular, for $p = 4$

$$\mathbb{E}\big[|W_t - W_s|^4\big] = 4 \cdot |t - s|^2 . \tag{1.2.2}$$

Next, in order to have the parameter domain open let us extend the process $\dot{W}_t$ constructed in part (i) of the proof to negative times by $\dot{W}_{-t} = \dot{W}_t$ for $t > 0$. Equality (1.2.2) is valid for any family $\{W_t : t \geq 0\}$ as in theorem 1.2.2 (i). Lemma A.2.37 applies, with $(E, \rho) = (\mathbb{R}, |\ |)$, $p = 4$, $\beta = 1$, $C = 4$: there is a selection $W_t \in \dot{W}_t$ such that the path $t \to W_t(\omega)$ is continuous for all $\omega \in \Omega$. We modify this by setting $W_.(\omega) \equiv 0$ in the negligible set of those points ω where $W_0(\omega) \neq 0$ and then forget about negative times. ∎

Uniqueness of Wiener Measure

A standard Wiener process is, of course, not unique: given the one we constructed above, we paint every element of Ω purple and get a new Wiener process that differs from the old one simply because its domain Ω is different. Less facetious examples are given in exercises 1.2.14 and 1.2.16. What is unique about a Wiener process is its *law* or *distribution.*

Recall – or consult section A.3 for – the notion of the law of a real-valued random variable $f : \Omega \to \mathbb{R}$. It is the measure $f[\mathbb{P}]$ on the codomain of f, $\mathbb{R}$ in this case, that is given by $f[\mathbb{P}](B) \stackrel{\text{def}}{=} \mathbb{P}[f^{-1}(B)]$ on Borels $B \in \mathcal{B}^\bullet(\mathbb{R})$. Now any standard Wiener process $W_.$ on some probability space $(\Omega, \mathcal{F}, \mathbb{P})$ can be identified in a natural way with a random variable $\underline{W}$ that has values in the space $\mathscr{C} = C[0, \infty)$ of continuous real-valued functions on the half-line. Namely, $\underline{W}$ is the map that associates with every $\omega \in \Omega$ the function or *path* $w = \underline{W}(\omega)$ whose value at t is $w_t = \overline{W}_t(w) \stackrel{\text{def}}{=} W_t(\omega)$, $t \geq 0$. We also call $\underline{W}$ a *representation* of $W_.$ on path space.[11] It is determined by the equation

$$\overline{W}_t \circ \underline{W}(\omega) = W_t(\omega) , \qquad t \geq 0 , \ \omega \in \Omega .$$

Wiener measure is the law or distribution of this $\mathscr{C}$-valued random variable $\underline{W}$, and this will turn out to be unique.

Before we can talk about this law, we have to identify the equivalent of the Borel sets $B \subset \mathbb{R}$ above. To do this a little analysis of ***path space*** $\mathscr{C} = C[0, \infty)$ is required. $\mathscr{C}$ has a natural topology, to wit, the topology of uniform convergence on compact sets. It can be described by a metric, for instance,[12]

$$d(w, w') = \sum_{n \in \mathbb{N}} \sup_{0 \leq s \leq n} \big| w_s - w'_s \big| \wedge 2^{-n} \qquad \text{for } w, w' \in \mathscr{C} . \tag{1.2.3}$$

[11] "Path space," like "frequency space" or "outer space," may be used without an article.
[12] $a \vee b$ $(a \wedge b)$ is the larger (smaller) of a and b.

Exercise 1.2.4 (i) A sequence $(w^{(n)})$ in $\mathscr{C}$ converges uniformly on compact sets to $w \in \mathscr{C}$ if and only if $d(w^{(n)}, w) \to 0$. $\mathscr{C}$ is complete under the metric d.

(ii) $\mathscr{C}$ is Hausdorff, and is separable, i.e., it contains a countable dense subset.

(iii) Let $\{w^{(1)}, w^{(2)}, \ldots\}$ be a countable dense subset of $\mathscr{C}$. Every open subset of $\mathscr{C}$ is the union of balls in the countable collection

$$B_q(w^{(n)}) \stackrel{\text{def}}{=} \left\{ w : d(w, w^{(n)}) < q \right\}, \qquad n \in \mathbb{N},\ 0 < q \in \mathbb{Q}.$$

Being separable and complete under a metric that defines the topology makes $\mathscr{C}$ a ***polish space.*** The Borel σ-algebra $\mathcal{B}^\bullet(\mathscr{C})$ on $\mathscr{C}$ is, of course, the σ-algebra generated by this topology (see section A.3 on page 391). As to our standard Wiener process W, defined on the probability space $(\Omega, \mathcal{F}, \mathbb{P})$ and identified with a $\mathscr{C}$-valued map $\underline{W}$ on Ω, it is not altogether obvious that inverse images $\underline{W}^{-1}(B)$ of Borel sets $B \subset \mathscr{C}$ belong to $\mathcal{F}$; yet this is precisely what is needed if the law $\underline{W}[\mathbb{P}]$ of $\underline{W}$ is to be defined, in analogy with the real-valued case, by

$$\underline{W}[\mathbb{P}](B) \stackrel{\text{def}}{=} \mathbb{P}[\underline{W}^{-1}(B)], \qquad B \in \mathcal{B}^\bullet(\mathscr{C}).$$

Let us show that they do. To this end denote by $\mathcal{F}^0_\infty[\mathscr{C}]$ the σ-algebra on $\mathscr{C}$ generated by the real-valued functions $\overline{W}_t : w \mapsto w_t$, $t \in [0, \infty)$, the ***evaluation maps***. Since $\overline{W}_t \circ W = W_t$ is measurable on $\mathcal{F}_t$, clearly

$$\underline{W}^{-1}(E) \in \mathcal{F}, \qquad \forall E \in \mathcal{F}^0_\infty[\mathscr{C}]. \tag{1.2.4}$$

Let us show next that every ball $B_r(w^{(0)}) \stackrel{\text{def}}{=} \left\{ w : d(w, w^{(0)}) < r \right\}$ belongs to $\mathcal{F}^0_\infty[\mathscr{C}]$. To prove this it evidently suffices to show that for fixed $w^{(0)} \in \mathscr{C}$ the map $w \mapsto d(w, w^{(0)})$ is measurable on $\mathcal{F}^0_\infty[\mathscr{C}]$. A glance at equation (1.2.3) reveals that this will be true if for every $n \in \mathbb{N}$ the map $w \mapsto \sup_{0 \le s \le n} |w_s - w^{(0)}_s|$ is measurable on $\mathcal{F}^0_\infty[\mathscr{C}]$. This, however, is clear, since the previous supremum equals the countable supremum of the functions

$$w \mapsto \left| w_q - w^{(0)}_q \right|, \qquad q \in \mathbb{Q},\ q \le n,$$

each of which is measurable on $\mathcal{F}^0_\infty[\mathscr{C}]$. We conclude with exercise 1.2.4 (iii) that every open set belongs to $\mathcal{F}^0_\infty[\mathscr{C}]$, and that therefore

$$\mathcal{F}^0_\infty[\mathscr{C}] = \mathcal{B}^\bullet(\mathscr{C}). \tag{1.2.5}$$

In view of equation (1.2.4) we now know that the inverse image under $\underline{W} : \Omega \to \mathscr{C}$ of a Borel set in $\mathscr{C}$ belongs to $\mathcal{F}$. We are now in position to talk about the image $\underline{W}[\mathbb{P}]$:

$$\underline{W}[\mathbb{P}](B) \stackrel{\text{def}}{=} \mathbb{P}[\underline{W}^{-1}(B)], \qquad B \in \mathcal{B}^\bullet(\mathscr{C}).$$

of $\mathbb{P}$ under $\underline{W}$ (see page 405) and to define Wiener measure:

Definition 1.2.5 *The law of a standard Wiener process* $(\Omega, \mathcal{F}, \mathbb{P}, W.)$, *that is to say the probability* $\mathbb{W} = \underline{W}[\mathbb{P}]$ *on* $\mathscr{C}$ *given by*

$$\mathbb{W}(B) \stackrel{\text{def}}{=} \underline{W}[\mathbb{P}](B) = \mathbb{P}[\underline{W}^{-1}(B)]\,, \qquad B \in \mathcal{B}^\bullet(\mathscr{C})\,,$$

is called ***Wiener measure****. The topological space* $\mathscr{C}$ *equipped with Wiener measure* $\mathbb{W}$ *on its Borel sets is called* ***Wiener space****. The real-valued random variables on* $\mathscr{C}$ *that map a path* $w \in \mathscr{C}$ *to its value at* t *and that are denoted by* $\overline{W}_t$ *above, and often simply by* w_t, *constitute the* ***canonical Wiener process****.*[8]

Exercise 1.2.6 The name is justified by the observation that the quadruple $(\mathscr{C}, \mathcal{B}^\bullet(\mathscr{C}), \mathbb{W}, \{\overline{W}_t\}_{0 \le t < \infty})$ is a standard Wiener process.

Definition 1.2.5 makes sense only if any two standard Wiener processes have the same distribution on $\mathscr{C}$. Indeed they do:

Lemma 1.2.7 *Any two standard Wiener processes have the same law.*

Proof. Let $(\Omega, \mathcal{F}, \mathbb{P}, W.)$ and $(\Omega', \mathcal{F}', \mathbb{P}', W'_.)$ be two standard Wiener processes and let $\mathbb{W}$ denote the law of $W.$. Consider a complex-valued function on $\mathscr{C} = C[0, \infty)$ of the form

$$\phi(w) = \exp\Big(i \sum_{k=1}^{K} r_k (w_{t_k} - w_{t_{k-1}})\Big) = \exp\Big(i \sum_{k=1}^{K} r_k (\overline{W}_{t_k}(w) - \overline{W}_{t_{k-1}}(w))\Big)\,, \tag{1.2.6}$$

$r_k \in \mathbb{R}$, $0 = t_0 < t_1 < \ldots < t_K$. Its $\mathbb{W}$-integral can be computed:

$$\int \phi(w)\,\mathbb{W}(dw) = \int \exp\Big(i \sum_{k=1}^{K} r_k (\overline{W}_{t_k} \circ \underline{W} - \overline{W}_{t_{k-1}} \circ \underline{W})\Big)\, d\mathbb{P}$$

by independence:
$$= \prod_{k=1}^{K} \int \exp\big(i r_k (W_{t_k} - W_{t_{k-1}})\big)\, d\mathbb{P}$$

$$= \prod_{k}^{K} \int_{-\infty}^{\infty} e^{i r_k x} \cdot \frac{e^{-x^2/2(t_k - t_{k-1})}}{\sqrt{2\pi(t_k - t_{k-1})}}\, dx$$

by exercise A.3.45:
$$= \prod_{k}^{K} e^{-r_k^2 (t_k - t_{k-1})/2}\,.$$

The same calculation can be done for $\mathbb{P}'$ and shows that its distribution $\mathbb{W}'$ under $\underline{W}'$ coincides with $\mathbb{W}$ on functions of the form (1.2.6). Now note that these functions are bounded, and that their collection $\mathcal{M}$ is closed under multiplication and complex conjugation and generates the same σ-algebra as the collection $\{\overline{W}_t : t \ge 0\}$, to wit $\mathcal{F}^0_\infty[\mathscr{C}] = \mathcal{B}^\bullet(C[0, \infty))$. An application of the Monotone Class Theorem in the form of exercise A.3.5 finishes the proof.

Namely, the vector space $\mathcal{V}$ of bounded complex-valued functions on $\mathscr{C}$ on which $\mathbb{W}$ and $\mathbb{W}'$ agree is sequentially closed and contains $\mathcal{M}$, so it contains every bounded $\mathcal{B}^\bullet(C[0,\infty))$-measurable function. ∎

Non-Differentiability of the Wiener Path

The main point of the introduction was that a novel integration theory is needed because the driving term of stochastic differential equations occurring most frequently, Wiener process, has paths of infinite variation. We show this now. In fact,[2] since a function that has finite variation on some interval is differentiable at almost every point of it, the claim is immediate from the following result:

Theorem 1.2.8 (Wiener) *Let W be a standard Wiener process on some probability space $(\Omega,\mathcal{F},\mathbb{P})$. Except for the points ω in a negligible subset $\mathcal{N}$ of Ω, the path $t \mapsto W_t(\omega)$ is nowhere differentiable.*

Proof [27]. Suppose that $t \mapsto W_t(\omega)$ is differentiable at some instant s. There exists a $K \in \mathbb{N}$ with $s < K-1$. There exist $M, N \in \mathbb{N}$ such that for all $n \ge N$ and all $t \in (s-5/n, s+5/n)$, $|W_t(\omega) - W_s(\omega)| \le M \cdot |t-s|$. Consider the first three consecutive points of the form j/n, $j \in \mathbb{N}$, in the interval $(s, s+5/n)$. The triangle inequality produces

$$|W_{\frac{j+1}{n}}(\omega) - W_{\frac{j}{n}}(\omega)| \le |W_{\frac{j+1}{n}}(\omega) - W_s(\omega)| + |W_{\frac{j}{n}}(\omega) - W_s(\omega)| \le 7M/n$$

for each of them. The point ω therefore lies in the set

$$\mathcal{N} = \bigcup_K \bigcup_M \bigcup_N \bigcap_{n\ge N} \bigcup_{k\le K\cdot n} \bigcap_{j=k}^{k+2} \Big[|W_{\frac{j+1}{n}} - W_{\frac{j}{n}}| \le 7M/n\Big] .$$

To prove that $\mathcal{N}$ is negligible it suffices to show that the quantity

$$\begin{aligned} Q &\stackrel{\text{def}}{=} \mathbb{P}\Big[\bigcap_{n\ge N} \bigcup_{k\le K\cdot n} \bigcap_{j=k}^{k+2} \Big[|W_{\frac{j+1}{n}} - W_{\frac{j}{n}}| \le 7M/n\Big]\Big] \\ &\le \liminf_{n\to\infty} \mathbb{P}\Big[\bigcup_{k\le K\cdot n} \bigcap_{j=k}^{k+2} \Big[|W_{\frac{j+1}{n}} - W_{\frac{j}{n}}| \le 7M/n\Big]\Big] \end{aligned}$$

vanishes. To see this note that the events

$$\Big[|W_{\frac{j+1}{n}} - W_{\frac{j}{n}}| \le 7M/n\Big] , \qquad j = k, k+1, k+2,$$

are independent and have probability

$$\begin{aligned} \mathbb{P}\Big[|W_{\frac{1}{n}}| \le 7M/n\Big] &= \frac{1}{\sqrt{2\pi/n}} \int_{-7M/n}^{+7M/n} e^{-x^2 n/2}\, dx \\ &= \frac{1}{\sqrt{2\pi}} \int_{-7M/\sqrt{n}}^{+7M/\sqrt{n}} e^{-\xi^2/2}\, d\xi \le \frac{14M}{\sqrt{2\pi n}} . \end{aligned}$$

Thus

$$Q \le \liminf_{n\to\infty} K\cdot n \cdot \left(\frac{const}{\sqrt{n}}\right)^3 = 0 .$$

∎

Remark 1.2.9 In the beginning of this section Wiener process[8] was motivated as a driving term for a stochastic differential equation describing physical Brownian motion. One could argue that the non-differentiability of the paths was a result of overly much idealization. Namely, the total momentum imparted to the pollen (in our billiard ball model) during the time-interval $[0, t]$ by collisions with the gas molecules is in reality a function of finite variation in t. In fact, it is constant between kicks and jumps at a kick by the momentum imparted; it is, in particular, not continuous. If the interval dt is small enough, there will not be any kicks at all. So the assumption that the differential of the driving term is distributed $N(0, dt)$ is just too idealistic. It seems that one should therefore look for a better model for the driver, one that takes the microscopic aspects of the interaction between pollen and gas molecules into account.

Alas, no one has succeeded so far, and there is little hope: first, the total variation of a momentum transfer during $[0, t]$ turns out to be huge, since it does not take into account the cancellation of kicks in opposite directions. This rules out any reasonable estimates for the convergence of any scheme for the solution of the stochastic differential equation driven by a more accurately modeled noise, in terms of this variation. Also, it would be rather cumbersome to keep track of the statistics of such a process of finite variation if its structure between any two of the huge number of kicks is taken into account.

We shall therefore stick to Wiener process as a model for the driver in the model for Brownian motion and show that the statistics of the solution of equation (1.2.1) on page 10 are close to the statistics of the solution of the corresponding equation driven by a finite variation model for the driver, provided the number of kicks is sufficiently large (exercise A.4.14). We shall return to this circle of problems several times, next in example 2.5.26 on page 79.

Supplements and Additional Exercises

Fix a standard Wiener process $W_.$ on some probability space $(\Omega, \mathcal{F}, \mathbb{P})$. For any s let $\mathcal{F}_s^0[W_.]$ denote the σ-algebra generated by the collection $\{W_r : 0 \le r \le s\}$. That is to say, $\mathcal{F}_s^0[W_.]$ is the smallest σ-algebra on which the $W_r : r \le s$ are all measurable. Intuitively, $\mathcal{F}_t^0[W_.]$ contains all information about the random variables W_s that can be observed up to and including time t. The collection

$$\mathcal{F}_.^0[W_.] = \{\mathcal{F}_s^0[W_.] : 0 \le s < \infty\}$$

of σ-algebras is called the ***basic filtration of the Wiener process*** $W_.$.

Exercise 1.2.10 $\mathcal{F}_s^0[W_.]$ increases with s and is the σ-algebra generated by the increments $\{W_r - W_{r'} : 0 \le r, r' \le s\}$. For $s < t$, $W_t - W_s$ is independent of $\mathcal{F}_s^0[W_.]$. Also, for $0 \le s < t < \infty$,

$$\mathbb{E}[W_t | \mathcal{F}_s^0[W_.]] = W_s \quad \text{and} \quad \mathbb{E}\left[W_t^2 - W_s^2 | \mathcal{F}_s^0[W_.]\right] = t - s\,. \tag{1.2.7}$$

Equations (1.2.7) say that both W_t and $W_t^2 - t$ are *martingales*[4] on $\mathcal{F}^0_.[W.]$. Together with the continuity of the path they determine the law of the whole process W uniquely. This fact, Lévy's characterization of Wiener process,[8] is proven most easily using stochastic integration, so we defer the proof until corollary 3.9.5. In the meantime here is a characterization that is just as useful:

Exercise 1.2.11 Let $X. = (X_t)_{t\ge 0}$ be a real-valued process with continuous paths and $X_0 = 0$, and denote by $\mathcal{F}^0_.[X.]$ its basic filtration – $\mathcal{F}^0_s[X.]$ is the σ-algebra generated by the random variables $\{X_r : 0 \le r \le s\}$. Note that it contains the basic filtration $\mathcal{F}^0_.[M^z_.]$ of the process $M^z_. : t \mapsto M^z_t \overset{\text{def}}{=} e^{zX_t - z^2 t/2}$ whenever $0 \ne z \in \mathbb{C}$. The following are equivalent:
(i) X is a standard Wiener process; (ii) the M^z are martingales[4] on $\mathcal{F}^0_.[X.]$; (iii) $M^\alpha : t \mapsto e^{i\alpha X_t + \alpha^2 t/2}$ is an $\mathcal{F}^0_.[M^\alpha_.]$-martingale for every real α.

Exercise 1.2.12 For any bounded Borel function ϕ and $s < t$

$$\mathbb{E}\left[\phi(W_t)|\mathcal{F}^0_s[W.]\right] = \frac{1}{\sqrt{2\pi(t-s)}}\int_{-\infty}^{+\infty} \phi(y)\cdot e^{-(y-W_s)^2/2(t-s)}\,dy\,. \tag{1.2.8}$$

Exercise 1.2.13 For any bounded Borel function ϕ on $\mathbb{R}$ and $t > 0$ define the function $T_t\phi$ by $T_0\phi = \phi$ if $t = 0$, and for $t > 0$ by

$$(T_t\phi)(x) = \frac{1}{\sqrt{2\pi t}}\int_{-\infty}^{+\infty} \phi(x+y)e^{-y^2/2t}\,dy\,.$$

Then T_t is a semigroup (i.e., $T_t \circ T_s = T_{t+s}$) of positive (i.e., $\phi \ge 0 \implies T_t\phi \ge 0$) linear operators with $T_0 = I$ and $T_t 1 = 1$, whose restriction to the space $C_0(\mathbb{R})$ of bounded continuous functions that vanish at infinity is continuous in the sup-norm topology. Rewrite equation (1.2.8) as

$$\mathbb{E}\left[\phi(W_t)|\mathcal{F}^0_s[W.]\right] = (T_{t-s}\phi)(W_s)\,.$$

Exercise 1.2.14 Let $(\Omega, \mathcal{F}, \mathbb{P}, W.)$ be a standard Wiener process. (i) For every $a > 0$, $\sqrt{a}\cdot W_{t/a}$ is a standard Wiener process. (ii) $t \mapsto t\cdot W_{1/t}$ is a standard Wiener process. (iii) For $\delta > 0$, the family $\{\sqrt{\delta}W_t : t \ge 0\}$ is a background noise as in example 1.2.1, but with diffusion coefficient δ.

Exercise 1.2.15 (d-Dimensional Wiener Process) (i) Let $1 \le n \in \mathbb{N}$. There exist a probability space $(\Omega, \mathcal{F}, \mathbb{P})$ and a family $(\boldsymbol{W}_t : 0 \le t < \infty)$ of $\mathbb{R}^d$-valued random variables on it with the following properties:

(a) $\boldsymbol{W}_0 = 0$.

(b) $\boldsymbol{W}.$ has independent increments. That is to say, if $0 = t_0 < t_1 < \ldots < t_K$ are consecutive instants, then the corresponding family of consecutive increments

$$\left\{\boldsymbol{W}_{t_1} - \boldsymbol{W}_{t_0}\,,\, \boldsymbol{W}_{t_2} - \boldsymbol{W}_{t_1}\,,\,\ldots\,,\, \boldsymbol{W}_{t_K} - \boldsymbol{W}_{t_{K-1}}\right\}$$

is independent.

(c) The increments $\boldsymbol{W}_t - \boldsymbol{W}_s$ are stationary and have normal law with covariance matrix

$$\int (W^\eta_t - W^\eta_s)(W^\theta_t - W^\theta_s)\,d\mathbb{P} = (t-s)\cdot\delta^{\eta\theta}\,.$$

Here $\delta^{\eta\theta} \overset{\text{def}}{=} \begin{cases} 1 & \text{if } \eta = \theta \\ 0 & \text{if } \eta \ne \theta \end{cases}$ is the ***Kronecker delta.***

(ii) Given such a family, one may change every $\boldsymbol{W}_t$ on a negligible set in such a way that for every $\omega \in \boldsymbol{W}$ the path $t \mapsto \boldsymbol{W}_t(\omega)$ is a continuous function from $[0, \infty)$

to $\mathbb{R}^d$. Any family $\{\boldsymbol{W}_t : t \in [0,\infty)\}$ of $\mathbb{R}^d$-valued random variables (defined on some probability space) that has the three properties (a)–(c) and also has continuous paths is called a ***standard d-dimensional Wiener process.***

(iii) The law of a standard d-dimensional Wiener process is a measure defined on the Borel subsets of the topological space

$$\mathscr{C}^d = C_{\mathbb{R}^d}[0,\infty)$$

of continuous paths $w : [0,\infty) \to \mathbb{R}^d$ and is unique. It is again called Wiener measure and is also denoted by $\mathbb{W}$.

(iv) An $\mathbb{R}^d$-valued process $(\Omega, \mathcal{F}, (Z_t)_{0\le t<\infty})$ with continuous paths whose law is Wiener measure is a standard d-dimensional Wiener process.

(v) Define the basic filtration $\mathcal{F}^0_s[\boldsymbol{W}.]$ and redo exercises 1.2.10–1.2.13 after proper reformulation.

Exercise 1.2.16 (The Brownian Sheet) A ***random sheet*** is a family $S_{\eta,t}$ of random variables on some common probability space $(\Omega, \mathcal{F}, \mathbb{P})$ indexed by the points of some domain in $\mathbb{R}^2$, say of $\check{\boldsymbol{H}} \stackrel{\text{def}}{=} \{(\eta,t) : \eta \in \mathbb{R}\ ,\ 0 \le t < \infty\}$. Any two points $z_1 = (\eta_1, t_1)$ and $z_2 = (\eta_2, t_2)$ in $\check{\boldsymbol{H}}$ with $\eta_1 \le \eta_2$ and $0 \le t_1 \le t_2$ determine a rectangle $(z_1, z_2] = (\eta_1, \eta_2] \times (t_1, t_2]$, and with it goes the "increment"

$$dS((z_1, z_2]) = S_{\eta_2,t_2} - S_{\eta_2,t_1} - S_{\eta_1,t_2} + S_{\eta_1,t_1}\ .$$

A ***Brownian sheet*** or ***Wiener sheet*** on $\check{\boldsymbol{H}}$ is a random sheet with the following properties:

$W_{0,0} = 0$; if $R_1, \ldots R_K$ are disjoint rectangles, then the corresponding family

$$\{dS(R_1), \ldots, dS(R_K)\}$$

of random variables is independent; for any rectangle $\check{\boldsymbol{H}}$, the law of $dS(R)$ is $N(0, \lambda(R))$, $\lambda(R)$ being the Lebesgue measure of R.

Show: there exists a Brownian sheet; its paths, or better, *sheets*, $(\eta, t) \mapsto S_{\eta,t}(\omega)$ can be chosen to be continuous for every $\omega \in \Omega$; the law of a Brownian sheet is a probability defined on all Borel subsets of the polish space $C(\check{\boldsymbol{H}})$ of continuous functions from $\check{\boldsymbol{H}}$ to the reals and is unique; for fixed η, $t \mapsto \eta^{-1/2} S_{\eta,t}$ is a standard Wiener process.

Exercise 1.2.17 Define the Brownian box and show that it is continuous.

1.3 The General Model

Wiener process is not the only driver for stochastic differential equations, albeit the most frequent one. For instance, the solution of a stochastic differential equation can be used to drive yet another one; even if it is not used for this purpose, it can best be analyzed in its capacity as a driver. We are thus automatically led to consider the class of all *drivers* or *integrators.*

As long as the integrators are Wiener processes or solutions of stochastic differential equations driven by Wiener processes, or are at least continuous, we can take for the underlying probability space Ω the path space $\mathscr{C}$ of the previous section (exercise 1.2.6). Recall how the uniqueness proof of the law of a Wiener process was facilitated greatly by the polish topology on $\mathscr{C}$. Now there are systems that should be modeled by drivers having jumps, for instance, the signal from a Geiger counter or a stock price. The corresponding space of trajectories does not consist of continuous paths anymore. After some analysis we shall see in section 2.3 that the appropriate path space Ω is the space $\mathscr{D}$ of right-continuous paths with left limits. The probabilistic analysis leads to estimates involving the so-called maximal process, which means that the naturally arising topology on $\mathscr{D}$ is again the topology of uniform convergence on compacta. However, under this topology $\mathscr{D}$ fails to be polish because it is not separable, and the relation between measurability and topology is not so nice and "tight" as in the case $\mathscr{C}$. Skorohod has given a useful polish topology on $\mathscr{D}$, which we shall describe later (section A.7). However, this topology is not compatible with the vector space structure of $\mathscr{D}$ and thus does not permit the use of arguments from Fourier analysis, as in the uniqueness proof of Wiener measure.

These difficulties can, of course, only be sketched here, lest we never reach our goal of solving stochastic differential equations. Identifying them has taken probabilists many years, and they might at this point not be too clear in the reader's mind. So we shall from now on follow the French School and mostly disregard topology. To identify and analyze general integrators we shall distill a general mathematical model directly from the heuristic arguments of section 1.1. It should be noted here that when a specific physical, financial, etc., system is to be modeled by specific assumptions about a driver, a model for the driver has to be constructed (as we did for Wiener process,[8] the driver of Brownian motion) and shown to fit this general mathematical model. We shall give some examples of this later (page 267).

Before starting on the general model it is well to *get acquainted with some notations and conventions* laid out in the beginning of appendix A on page 363 that are fairly but not altogether standard and are designed to cut down on verbiage and to enhance legibility.

Filtrations on Measurable Spaces

Now to the general probabilistic model suggested by the heuristics of section 1.1. First we need a probability space on which the random variables X_t, Z_t, etc., of section 1.1 are realized as functions – so we can apply functional calculus – and a notion of *past* or *history* (see page 6). Accordingly, we stipulate that we are given a ***filtered measurable space*** on which everything of interest lives. This is a pair $(\Omega, \mathcal{F}_.)$ consisting of a set Ω and an increasing family

$$\mathcal{F}_. = \{\mathcal{F}_t\}_{0 \le t < \infty}$$

of σ-algebras on Ω. It is convenient to begin the reckoning of time at $t = 0$; if the starting time is another finite time, a linear scaling will reduce the situation to this case. It is also convenient to end the reckoning of time at ∞. The reader interested in only a finite time-interval $[0, u)$ can use everything said here simply by reading the symbol ∞ as another name for his ultimate time u of interest.

To say that $\mathcal{F}_.$ is increasing means of course that $\mathcal{F}_s \subseteq \mathcal{F}_t$ for $0 \le s \le t$. The family $\mathcal{F}_.$ is called a ***filtration*** or ***stochastic basis*** on Ω. The intuitive meaning of it is this: Ω is the set of all evolutions that the world or the system under consideration might take, and $\mathcal{F}_t$ models the collection of "all events that will be observable by the time t," the "history at time t." We close the filtration at ∞ with three objects: first there are

the *algebra* of sets $\qquad \mathcal{A}_\infty \overset{\text{def}}{=} \bigcup_{0 \le t < \infty} \mathcal{F}_t$

and the σ-*algebra* $\qquad \mathcal{F}_\infty \overset{\text{def}}{=} \bigvee_{0 \le t < \infty} \mathcal{F}_t$

that it generates. Lastly there is the universal completion $\mathcal{F}^*_\infty$ of $\mathcal{F}_\infty$ – see page 407 of appendix A.

A ***random variable*** is simply a universally (i.e., $\mathcal{F}^*_\infty$-) measurable function on Ω.

The filtration $\mathcal{F}_.$ is ***universally complete*** if $\mathcal{F}_t$ is universally complete at any instant $t < \infty$. We shall eventually require that $\mathcal{F}_.$ have this and further properties.

The Base Space

The noises and other processes of interest are functions on the ***base space***

$$\boldsymbol{B} \overset{\text{def}}{=} [0, \infty) \times \Omega\,.$$

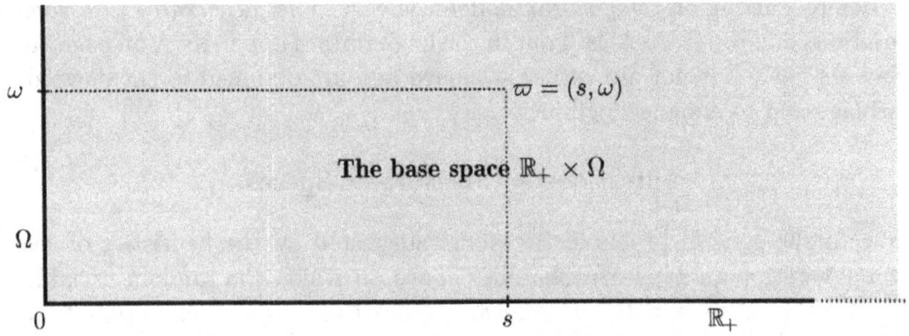

Figure 1.1 The base space

Its typical point is a pair (s, ω), which will frequently be denoted by ϖ. The spirit of this exposition is to reduce stochastic analysis to the analysis of real-valued functions on $\boldsymbol{B}$. The base space has a rather rich structure, being a product whose fibers $\{s\} \times \Omega$ carry finer and finer σ-algebras $\mathcal{F}_s$ as time s increases. This structure gives rise to quite a bit of terminology, which we will be discussing for a while. Fortunately, most notions attached to a filtration are quite intuitive.

Processes

Processes are simply functions[13] on the base space $\boldsymbol{B}$. We are mostly concerned with processes that take their values in the reals $\mathbb{R}$ or in the extended reals $\overline{\mathbb{R}}$ (see item A.1.2 on page 363). So unless the range is explicitly specified differently, ***a process is numerical.*** A process is ***measurable*** if it is measurable on the product σ-algebra $\mathcal{B}^\bullet[0,\infty) \otimes \mathcal{F}^*_\infty$ on $\mathbb{R}_+ \times \Omega$.

It is customary to write $Z_s(\omega)$ for the value of the process Z at the point $\varpi = (s, \omega) \in \boldsymbol{B}$, and to denote by Z_s the function $\omega \mapsto Z_s(\omega)$:

$$Z_s : \omega \mapsto Z_s(\omega) , \qquad s \in \mathbb{R}_+ .$$

The process Z is ***adapted*** to the filtration $\mathcal{F}_.$ if at every instant t the random variable Z_t is measurable on $\mathcal{F}_t$; one then writes

$$Z_t \in \mathcal{F}_t .$$

In other words, the symbol $\in$ is not only shorthand for "is an element of" but also for "is measurable on" (see page 391).

The ***path*** or ***trajectory*** of the process Z, on the other hand, is the function

$$Z_.(\omega) : s \mapsto Z_s(\omega) , \qquad 0 \le s < \infty ,$$

one for each $\omega \in \Omega$. A statement such as "Z is ***continuous*** (***left-continuous***, ***right-continuous***, ***increasing***, ***of finite variation***, etc.)" means that every path of Z has this property.

Stopping a process is a useful and ubiquitous tool. The process Z stopped at time t is the function[12] $(s, \omega) \mapsto Z_{s \wedge t}(\omega)$ and is denoted by Z^t. After time t its path is constant with value $Z_t(\omega)$.

Remark 1.3.1 Frequently the only randomness of interest is the one introduced by some given process Z of interest. Then one appropriate filtration is the ***basic filtration*** $\mathcal{F}^0_.[Z] = \{\mathcal{F}^0_t[Z] : 0 \le t < \infty\}$ of Z. $\mathcal{F}^0_t[Z]$ is the σ-algebra generated by the random variables $\{Z_s : 0 \le s \le t\}$. An instance of this was considered in exercise 1.2.10. We shall see soon (pages 37–40) that there are more convenient filtrations, even in this simple case.

Exercise 1.3.2 The projection on Ω of a measurable subset of the base space is universally measurable. A measurable process has Borel-measurable paths.

Exercise 1.3.3 A process Z is adapted to its basic filtration $\mathcal{F}^0_.[Z]$. Conversely, if Z is adapted to the filtration $\mathcal{F}_.$, then $\mathcal{F}^0_t[Z] \subseteq \mathcal{F}_t$ for all t.

---

[13] The reader has no doubt met before the propensity of probabilists to give new names to everyday mathematical objects – for instance, calling the elements of Ω outcomes, the subsets events, the functions on Ω random variables, etc. This is meant to help intuition but sometimes obscures the distinction between a physical system and its mathematical model.

Wiener Process Revisited On the other hand, it occurs that a Wiener process W is forced to live together with other processes on a filtration $\mathcal{F}_.$ larger than its own basic one (see, e.g., example 5.5.1). A modicum of compatibility is usually required:

Definition 1.3.4 *W is **standard Wiener process on the filtration** $\mathcal{F}_.$ if it is adapted to $\mathcal{F}_.$ and $W_t - W_s$ is independent of $\mathcal{F}_s$ for $0 \le s < t < \infty$.*

See corollary 3.9.5 on page 160 for Lévy's characterization of standard Wiener process on a filtration.

Right- and Left-Continuous Processes Let $\mathscr{D}$ denote the collection of all paths $[0,\infty) \to \mathbb{R}$ that are right-continuous and have finite left limits at all instants $t \in \mathbb{R}_+$, and $\mathscr{L}$ the collection of paths that are left-continuous and have right limits in $\mathbb{R}$ at all instants. A path in $\mathscr{D}$ is also called[14] ***càdlàg*** and a path in $\mathscr{L}$ ***càglàd***. The paths of $\mathscr{D}$ and $\mathscr{L}$ have discontinuities only where they jump; they do not oscillate. Most of the processes that we have occasion to consider are adapted and have paths in one or the other of these classes. They deserve their own symbols: the family of adapted processes whose paths are right-continuous and have left limits is denoted by $\mathfrak{D} = \mathfrak{D}[\mathcal{F}_.]$, and the family of adapted processes whose paths are left-continuous and have right limits is denoted by $\mathfrak{L} = \mathfrak{L}[\mathcal{F}_.]$. Clearly $\mathscr{C} = \mathscr{L} \cap \mathscr{D}$, and $\mathfrak{C} = \mathfrak{L} \cap \mathfrak{D}$ is the collection of continuous adapted processes.

The Left-Continuous Version $X_{.-}$ of a right-continuous process X with left limits has at the instant t the value

$$X_{t-} \stackrel{\text{def}}{=} \begin{cases} 0 & \text{for } t = 0\,, \\ \lim\limits_{t>s\to t} X_s & \text{for } 0 < t \le \infty\,. \end{cases}$$

Clearly $X_{.-} \in \mathfrak{L}$ whenever $X \in \mathfrak{D}$. Note that the left-continuous version is forced to have the value zero at the instant zero. Given an $X \in \mathfrak{L}$ we might – but seldom will – consider its ***right-continuous version*** $X_{.+}$:

$$X_{t+} \stackrel{\text{def}}{=} \lim_{t<u\downarrow t} X_u\,, \qquad 0 \le t < \infty\,.$$

If $X \in \mathfrak{D}$, then taking the right-continuous version of $X_{.-}$ leads back to X. But if $X \in \mathfrak{L}$, then the left-continuous version of $X_{.+}$ differs from X at $t = 0$, unless X happens to vanish at that instant. This slightly unsatisfactory lack of symmetry is outweighed by the simplification of the bookkeeping it affords in Itô's formula and related topics (section 4.2). Here is a mnemonic device: imagine that all processes have the value 0 for strictly negative times; this forces $X_{0-} = 0$.

---

[14] càdlàg is an acronym for the French "continu à droite, limites à gauche" and càglàd for "continu à gauche, limites à droite." Some authors write "*corlol*" and "*collor*" and others write "*rcll*" and "*lcrl*." "càglàd," though of French origin, is pronounceable.

The Jump Process ΔX of a right-continuous process X with left limits is the difference between itself and its left-continuous version:

$$\Delta X_t \stackrel{\text{def}}{=} (X_t - X_{t-}) = \begin{cases} X_0 & \text{for } t = 0, \\ X_t - X_{t-} & \text{for } t > 0. \end{cases}$$

$\Delta X_.$ is evidently adapted to the basic filtration $\mathcal{F}^0_.[X]$ of X.

Progressive Measurability The adaptedness of a process Z reflects the idea that at no instant t should a look into the future be necessary to determine the value of the random variable Z_t. It is still thinkable that some property of the whole path of Z up to time t depends on more than the information in $\mathcal{F}_t$. (Such a property would have to involve the values of Z at uncountably many instants, of course.) Progressive measurability rules out such a contingency: Z is ***progressively measurable*** if for every $t \in \mathbb{R}$ the stopped process Z^t is measurable on $\mathcal{B}^\bullet[0,\infty) \otimes \mathcal{F}_t$. This is the the same as saying that the restriction of Z to $[0,t] \times \Omega$ is measurable on $\mathcal{B}^\bullet[0,t] \otimes \mathcal{F}_t$ and means that any measurable information about the whole path up to time t is contained in $\mathcal{F}_t$.

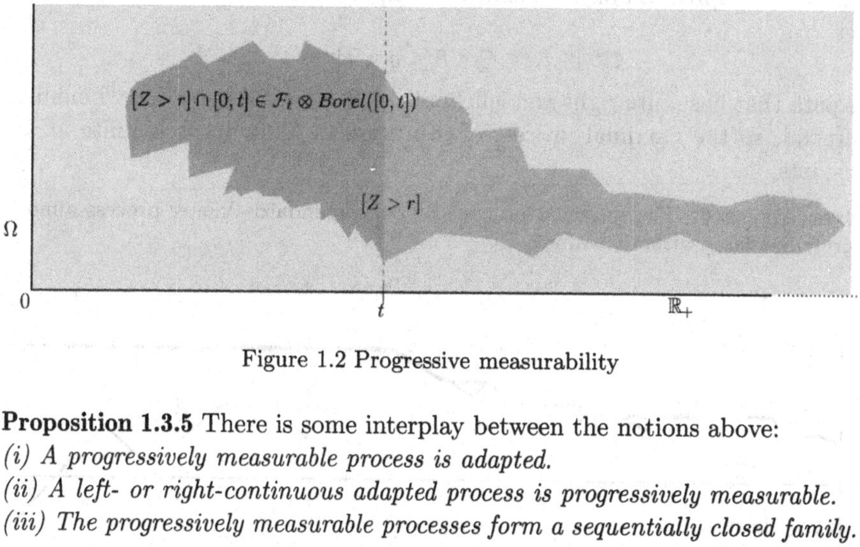

Figure 1.2 Progressive measurability

Proposition 1.3.5 There is some interplay between the notions above:
(i) A progressively measurable process is adapted.
(ii) A left- or right-continuous adapted process is progressively measurable.
(iii) The progressively measurable processes form a sequentially closed family.

Proof. (i): Z_t is the composition of $\omega \mapsto (t,\omega)$ with Z. (ii): If Z is left-continuous and adapted, set

$$Z^{(n)}_s(\omega) = Z_{\frac{k}{n}}(\omega) \ \text{ for } \ \frac{k}{n} \le s < \frac{k+1}{n} .$$

Clearly $Z^{(n)}$ is progressively measurable. Also, $Z^{(n)}(\varpi) \xrightarrow[n\to\infty]{} Z(\varpi)$ at every point $\varpi = (s,\omega) \in \boldsymbol{B}$. To see this, let s_n denote the largest rational of the form k/n less than or equal to s. Clearly $Z^{(n)}(\varpi) = Z_{s_n}(\omega)$ converges to $Z_s(\omega) = Z(\varpi)$.

Suppose now that Z is right-continuous and fix an instant t. The stopped process Z^t is evidently the pointwise limit of the functions

$$Z_s^{(n)}(\omega) = \sum_{k=0}^{+\infty} Z_{\frac{k+1}{n}\wedge t}(\omega) \cdot 1_{\left(\frac{k}{n},\frac{k+1}{n}\right]}(s) ,$$

which are measurable on $\mathcal{B}^\bullet[0,\infty) \otimes \mathcal{F}_t$. (iii) is left to the reader. ∎

The Maximal Process $Z^\star$ of a process $Z : \boldsymbol{B} \to \overline{\mathbb{R}}$ is defined by

$$Z_t^\star = \sup_{0 \le s \le t} |Z_s| , \qquad 0 \le t \le \infty .$$

This is a supremum over uncountably many indices and so is not in general expected to be measurable. However, when Z is progressively measurable and the filtration is universally complete, then $Z^\star$ is again progressively measurable. This is shown in corollary A.5.13 with the help of some capacity theory. We shall deal mostly with processes Z that are left- or right-continuous, and then we don't need this big cannon. $Z^\star$ is then also left- or right-continuous, respectively, and if Z is adapted, then so is $Z^\star$, inasmuch as it suffices to extend the supremum in the definition of $Z_t^\star$ over instants s in the countable set

$$\mathbb{Q}^t \stackrel{\text{def}}{=} \{q \in \mathbb{Q} : 0 \le q < t\} \cup \{t\} .$$

A path that has finite right and left limits in $\mathbb{R}$ is bounded on every bounded interval, so the maximal process of a process in $\mathfrak{D}$ or in $\mathfrak{L}$ is finite at all instants.

Exercise 1.3.6 The maximal process $W^\star$ of a standard Wiener process almost surely increases without bound.

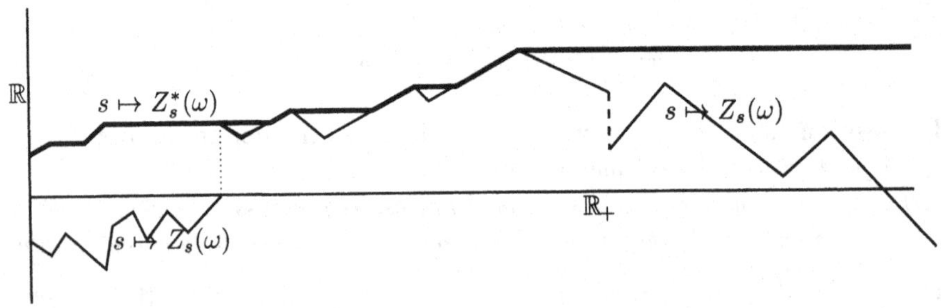

Figure 1.3 A path and its maximal path

The Limit at Infinity of a process Z is the random variable $\lim_{t\to\infty} Z_t(\omega)$, provided this limit exists almost surely. For consistency's sake it should be and is denoted by $Z_{\infty-}$. It is convenient and unambiguous to use also the notation Z_∞:

$$Z_\infty = Z_{\infty-} \stackrel{\text{def}}{=} \lim_{t\to\infty} Z_t .$$

The maximal process $Z^\star$ always has a limit $Z^\star_\infty$, possibly equal to $+\infty$ on a large set. If Z is adapted and right-continuous, say, then Z_∞ is evidently measurable on $\mathcal{F}_\infty$.

Stopping Times and Stochastic Intervals

Definition 1.3.7 *A **random time** is a universally measurable function on Ω with values in $[0,\infty]$. A random time T is a **stopping time** if*

$$[T \le t] \in \mathcal{F}_t \qquad \forall\, t \in \mathbb{R}_+ . \tag{$*$}$$

*This notion depends on the filtration $\mathcal{F}_.$; and if this dependence must be stressed, then T is called an $\mathcal{F}_.$**-stopping time**. The collection of all stopping times is denoted by $\mathfrak{T}$, or by $\mathfrak{T}[\mathcal{F}_.]$ if the filtration needs to be made explicit.*

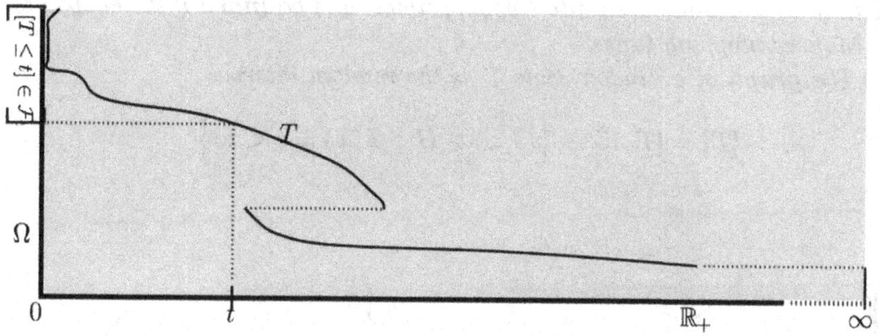

Figure 1.4 Graph of a stopping time

Condition $(*)$ expresses the idea that at the instant t no look into the future is necessary in order to determine whether the time T has arrived. In section 1.1, for example, we were led to consider the first time T our space probe hit the moon. This time evidently depends on chance and is thus a random time. Moreover, the event $[T \le t]$ that the probe hits the moon at or before the instant t can certainly be determined if the history $\mathcal{F}_t$ of the universe up to this instant is known: in our stochastic model, T should turn out to be a stopping time. If the probe never hits the moon, then the time T should be $+\infty$, as $+\infty$ is by general convention the infimum of the void set of numbers. This explains why a stopping time is permitted to have the value $+\infty$. Here are a few natural notions attached to random and stopping times:

Definition 1.3.8 *(i) If T is a stopping time, then the collection*

$$\mathcal{F}_T \stackrel{\text{def}}{=} \Big\{ A \in \mathcal{F}_\infty : A \cap [T \le t] \in \mathcal{F}_t \quad \forall\, t \in [0,\infty] \Big\}$$

is easily seen to be a σ-algebra on Ω. It is called the ***past*** *at time T or the past of T.* To paraphrase: an event A occurs in the past of T if at any instant t at which T has arrived no look into the future is necessary to determine whether the event A has occurred.
(ii) The value of a process Z at a random time T is the random variable

$$Z_T : \omega \mapsto Z_{T(\omega)}(\omega) .$$

(iii) Let S,T be two random times. The random interval $(\!(S,T]\!]$ is the set

$$(\!(S,T]\!] \stackrel{\text{def}}{=} \Big\{(s,\omega) \in \boldsymbol{B} \,:\, S(\omega) < s \le T(\omega),\ s < \infty\Big\} ;$$

and $(\!(S,T)\!)$, $[\![S,T]\!]$, and $[\![S,T)\!)$ are defined similarly. Note that the point (∞,ω) does not belong to any of these intervals, even if $T(\omega) = \infty$. If both S and T are stopping times, then the random intervals $(\!(S,T]\!]$, $(\!(S,T)\!)$, $[\![S,T]\!]$, and $[\![S,T)\!)$ are called ***stochastic intervals****. A stochastic interval is* ***finite*** *if its endpoints are finite stopping times, and* ***bounded*** *if the endpoints are bounded stopping times.*
(iv) The ***graph*** *of a random time T is the random interval*

$$[\![T]\!] = [\![T,T]\!] \stackrel{\text{def}}{=} \Big\{(s,\omega) \in \boldsymbol{B} \,:\, T(\omega) = s < \infty\Big\} .$$

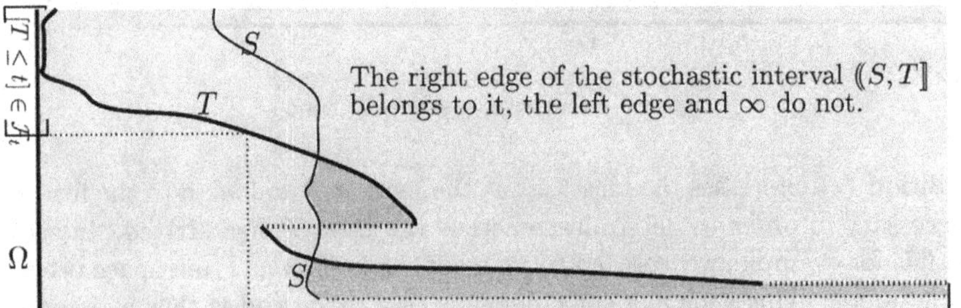

Figure 1.5 A stochastic interval

Proposition 1.3.9 *Suppose that Z is a progressively measurable process and T is a stopping time. The* ***stopped process*** *Z^T, defined by $Z^T_s(\omega) = Z_{T\wedge s}(\omega)$, is progressively measurable, and Z_T is measurable on $\mathcal{F}_T$.*

We can paraphrase the last statement by saying "a progressively measurable process is adapted to the 'expanded filtration' $\{\mathcal{F}_T : T \text{ a stopping time}\}$."

Proof. For fixed t let $\mathcal{F}^t = \mathcal{B}^\bullet[0,\infty) \otimes \mathcal{F}_t$. The map[12] that sends (s,ω) to $(T(\omega) \wedge t \wedge s, \omega)$ from $\boldsymbol{B}$ to itself is easily seen to be $\mathcal{F}^t/\mathcal{F}^t$-measurable.

$(Z^T)^t = Z^{T\wedge t}$ is the composition of Z^t with this map and is therefore $\mathcal{F}^t$-measurable. This holds for all t, so Z^T is progressively measurable. For $a \in \mathbb{R}$, $[Z_T > a] \cap [T \le t] = [Z^T_t > a] \cap [T \le t]$ belongs to $\mathcal{F}_t$ because Z^T is adapted. This holds for all t, so $[Z_T > a] \in \mathcal{F}_T$ for all a: $Z_T \in \mathcal{F}_T$, as claimed. ∎

Exercise 1.3.10 If the process Z is progressively measurable, then so is the process $t \mapsto Z_{T\vee t} - Z_T$.

Let $T_1 \le T_2 \le \ldots \le T_\infty = \infty$ be an increasing sequence of stopping times and X a progressively measurable process. For $r \in \mathbb{R}$ define $K = \inf\{k \in \mathbb{N} : X_{T_k} > r\}$. Then $T_K : \omega \mapsto T_{K(\omega)}(\omega)$ is a stopping time.

Some Examples of Stopping Times

Stopping times occur most frequently as first hitting times – of the moon in our example of section 1.1, or of sets of bad behavior in much of the analysis below. First hitting times are stopping times, provided that the filtration $\mathcal{F}_.$ satisfies some natural conditions – see figure 1.6 on page 40. This is shown with the help of a little capacity theory in appendix A, section A.5. A few elementary results, established with rather simple arguments, will go a long way:

Proposition 1.3.11 *Let I be an adapted process with increasing right-continuous paths and let $\lambda \in \mathbb{R}$. Then*

$$T^\lambda \stackrel{\text{def}}{=} \inf\{t : I_t \ge \lambda\}$$

is a stopping time, and $I_{T^\lambda} \ge \lambda$ on the set $[T^\lambda < \infty]$. Moreover, the functions $\lambda \mapsto T^\lambda(\omega)$ are increasing and left-continuous.

Proof. $T^\lambda(\omega) \le t$ if and only if $I_t(\omega) \ge \lambda$. In other words, $[T^\lambda \le t] = [I_t \ge \lambda] \in \mathcal{F}_t$, so T^λ is a stopping time. If $T^\lambda(\omega) < \infty$, then there is a sequence (t_n) of instants that decreases to $T^\lambda(\omega)$ and has $I_{t_n}(\omega) \ge \lambda$. The right-continuity of I produces $I_{T^\lambda(\omega)}(\omega) \ge \lambda$.

That $T^\lambda \le T^\mu$ when $\lambda \le \mu$ is obvious: $T^{\boldsymbol{\cdot}}$ is indeed increasing. If $T^\lambda \le t$ for all $\lambda < \mu$, then $I_t \ge \lambda$ for all $\lambda < \mu$, and thus $I_t \ge \mu$ and $T^\mu \le t$. That is to say, $\sup_{\lambda<\mu} T^\lambda = T^\mu$: $\lambda \mapsto T^\lambda$ is left-continuous. ∎

The main application of this near-trivial result is to the maximal process of some process I. Proposition 1.3.11 applied to $(Z - Z^S)^\star$ yields the

Corollary 1.3.12 *Let S be a finite stopping time and $\lambda > 0$. Suppose that Z is adapted and has right-continuous paths. Then the first time the maximal gain of Z after S exceeds λ,*

$$T = \inf\{t > S : \sup_{S<s\le t} |Z_s - Z_S| \ge \lambda\} = \inf\{t : |Z - Z^S|^\star_t \ge \lambda\} ,$$

is a stopping time strictly greater than S, and $|Z - Z_S|^\star_T \ge \lambda$ on $[T < \infty]$.

Proposition 1.3.13 *Let Z be an adapted process, T a stopping time, X a random variable measurable on $\mathcal{F}_T$, and $S = \{s_0 < s_1 < \ldots < s_N\} \subset [0, u)$ a finite set of instants. Define*

$$T' = \inf\{s \in S : s > T\ ,\ Z_s \diamond X\} \wedge u\ ,$$

where $\diamond$ stands for any of the relations $>, \geq, =, \leq, <$. Then T' is a stopping time, and $Z_{T'} \in \mathcal{F}_{T'}$ satisfies $Z_{T'} \diamond X$ on $[T' < u]$.

Proof. If $t \geq u$, then $[T' \leq t] = \Omega \in \mathcal{F}_t$. Let then $t < u$. Then

$$[T' \leq t] = \bigcup\{[T < s] \cap [Z_s \diamond X] : S \ni s \leq t\}\ .$$

Now $[T < s] \in \mathcal{F}_s$ and so $[T < s]Z_s \in \mathcal{F}_s$. Also, $[X > x] \cap [T < s] \in \mathcal{F}_s$ for all x, so that $[T < s]X \in \mathcal{F}_s$ as well. Hence $[T < s] \cap [Z_s \diamond X] \in \mathcal{F}_s$ for $s \leq t$ and so $[T' \leq t] \in \mathcal{F}_t$. Clearly $Z_{T'}[T' \leq t] = \bigcup_{S \ni s \leq t} Z_s[T'{=}s] \in \mathcal{F}_t$ for all $t \in S$, and so $Z_{T'} \in \mathcal{F}_{T'}$. ∎

Proposition 1.3.14 *Let S be a stopping time, let $c > 0$, and let $X \in \mathfrak{D}$. Then*

$$T = \inf\{t > S :\ |\Delta X_t| \geq c\}$$

is a stopping time that is strictly later than S on the set $[S < \infty]$, and $|\Delta X_T| \geq c$ on $[T < \infty]$.

Proof. Let us prove the last point first. Let $t_n \geq T$ decrease to T and $|\Delta X_{t_n}| \geq c$. Then (t_n) must be ultimately constant. For if it is not, then it can be replaced by a strictly decreasing subsequence, in which case both X_{t_n} and X_{t_n-} converge to the same value, to wit, X_T. This forces $\Delta X_{t_n} \xrightarrow[n\to\infty]{} 0$, which is impossible since $|\Delta X_{t_n}| \geq c > 0$. Thus $T > S$ and $\Delta X_T \geq c$.

Next observe that $T \leq t$ precisely if for every $n \in \mathbb{N}$ there are numbers q, q' in the countable set

$$\mathbb{Q}^t = (\mathbb{Q} \cap [0, t]) \cup \{t\}$$

with $S < q < q'$ and $q' - q < 1/n$, and such that $|X_{q'} - X_q| \geq c - 1/n$. This condition is clearly necessary. To see that it is sufficient note that in its presence there are rationals $S < q_n < q'_n \leq t$ with $q'_n - q_n \to 0$ and $|X_{q'_n} - X_{q_n}| \geq c - 1/n$. Extracting a subsequence we may assume that both (q_n) and (q'_n) converge to some point $s \in [S, t]$. (q_n) can clearly not contain a constant subsequence; if (q'_n) does, then $|\Delta X_s| \geq c$ and $T \leq t$. If (q'_n) has no constant subsequence, it can be replaced by a strictly monotone subsequence. We may thus assume that both (q_n) and (q'_n) are strictly monotone. Recalling the first part of the proof we see that this is possible only if (q_n) is increasing and (q'_n) decreasing, in which case $T \leq t$ again. The upshot of all this is that

$$[T \leq t] = \bigcap_{n \in \mathbb{N}} \bigcup_{\substack{q, q' \in \mathbb{Q}^t \\ q < q' < q+1/n}} [S < q] \cap \big[|X_{q'} - X_q| \geq c - 1/n\big]\ ,$$

a set that is easily recognizable as belonging to $\mathcal{F}_t$. ∎

Further elementary but ubiquitous facts about stopping times are developed in the next exercises. Most are clear upon inspection, and they are used freely in the sequel.

Exercise 1.3.15 (i) An instant $t \in \mathbb{R}_+$ is a stopping time, and its past equals $\mathcal{F}_t$. (ii) The infimum of a finite number and the supremum of a countable number of stopping times are stopping times.

Exercise 1.3.16 Let S, T be any two stopping times. (i) If $S \le T$, then $\mathcal{F}_S \subseteq \mathcal{F}_T$. (ii) In general, the sets $[S < T]$, $[S \le T]$, $[S = T]$ belong to $\mathcal{F}_{S\wedge T} = \mathcal{F}_S \cap \mathcal{F}_T$.

Exercise 1.3.17 A random time T is a stopping time precisely if the (indicator function of the) random interval $[\![0, T)\!)$ is an adapted process.[15] If S, T are stopping times, then any stochastic interval with left endpoint S and right endpoint T is an adapted process.

Exercise 1.3.18 Let T be a stopping time and $A \in \mathcal{F}_T$. Setting

$$T_A(\omega) \stackrel{\text{def}}{=} \begin{cases} T(\omega) & \text{if } \omega \in A \\ \infty & \text{if } \omega \notin A \end{cases} = T + \infty \cdot A^c$$

defines a new stopping time T_A, the ***reduction*** of the stopping time T to A.

Exercise 1.3.19 Let T be a stopping time and $A \in \mathcal{F}_\infty$. Then $A \in \mathcal{F}_T$ if and only if the reduction T_A is a stopping time. A random variable f is measurable on $\mathcal{F}_T$ if and only if $f \cdot [T \le t] \in \mathcal{F}_t$ at all instants t.

Exercise 1.3.20 The following "discrete approximation from above" of a stopping time T will come in handy on several occasions. For $n = 1, 2, \ldots$ set

$$T^{(n)} \stackrel{\text{def}}{=} \begin{cases} 0 & \text{on } [T = 0]; \\ \dfrac{k+1}{n} & \text{on } \left[\dfrac{k}{n} < T \le \dfrac{k+1}{n}\right], \qquad k = 0, 1, \ldots; \\ \infty & \text{on } [T = \infty]. \end{cases}$$

Using convention A.1.5 on page 364, we can rewrite this as

$$T^{(n)} \stackrel{\text{def}}{=} \sum_{k=0}^{\infty} \frac{k+1}{n} \cdot \left[\frac{k}{n} < T \le \frac{k+1}{n}\right] + \infty \cdot [T = \infty].$$

Then $T^{(n)}$ is a stopping time that takes only countably many values, and T is the pointwise infimum of the decreasing sequence $T^{(n)}$.

Exercise 1.3.21 Let $X \in \mathfrak{D}$. (i) The set $\{s \in \mathbb{R}_+ : |\Delta X_s(\omega)| \ge \epsilon\}$ is discrete (has no accumulation point in $\mathbb{R}_+$) for every $\omega \in \Omega$ and $\epsilon > 0$. (ii) There exists a countable family $\{T_n\}$ of stopping times with bounded disjoint graphs $[\![T_n]\!]$ at which the jumps of X occur:

$$[\Delta X \ne 0] \subseteq \bigcup_n [\![T_n]\!] .$$

(iii) Let h be a Borel function on $\mathbb{R}$ and assume that for all $t < \infty$ the sum $J_t \stackrel{\text{def}}{=} \sum_{0 \le s \le t} h(\Delta X_s)$ converges absolutely. Then $J_.$ is adapted.

[15] See convention A.1.5 and figure A.14 on page 365.

Probabilities

A probabilistic model of a system requires, of course, a probability measure $\mathbb{P}$ on the pertinent σ-algebra $\mathcal{F}_\infty$, the idea being that a priori assumptions on, or measurements of, $\mathbb{P}$ plus mathematical analysis will lead to estimates of the random variables of interest.

The need to consider a family $\mathfrak{P}$ of pertinent probabilities does arise: first, there is often not enough information to specify one particular probability as the right one, merely enough to narrow the class. Second, in the context of stochastic differential equations and Markov processes, whole slews of probabilities appear that may depend on a starting point or other parameter (see theorem 5.7.3). Third, it is possible and often desirable to replace a given probability by an equivalent one with respect to which the stochastic integral has superior properties (this is done in section 4.1 and is put to frequent use thereafter). Nevertheless, we shall mostly develop the theory for a fixed probability $\mathbb{P}$ and simply apply the results to each $\mathbb{P} \in \mathfrak{P}$ separately. The pair $(\mathcal{F}_., \mathbb{P})$ or $(\mathcal{F}_., \mathfrak{P})$, as the case may be, is termed a ***measured filtration***.

Let $\mathbb{P} \in \mathfrak{P}$. It is customary to denote the integral with respect to $\mathbb{P}$ by $\mathbb{E}^\mathbb{P}$ and to call it the ***expectation***; that is to say, for $f : \Omega \to \mathbb{R}$ measurable on $\mathcal{F}_\infty$,

$$\mathbb{E}^\mathbb{P}[f] = \int f \, d\mathbb{P} = \int f(\omega) \, \mathbb{P}(d\omega) \,, \qquad \mathbb{P} \in \mathfrak{P} \,.$$

If there is no doubt which probability $\mathbb{P} \in \mathfrak{P}$ is meant, we write simply $\mathbb{E}$.

A subset $N \subset \Omega$ is commonly called $\mathbb{P}$-negligible, or simply ***negligible*** when there is no doubt about the probability, if its outer measure $\mathbb{P}^*[N]$ equals zero. This is the same as saying that it is contained in a set of $\mathcal{F}_\infty$ that has measure zero. A function on Ω is negligible if it vanishes off a negligible set; this is the same as saying that the upper integral[16] of its absolute value vanishes. The functions that differ negligibly, i.e., only in a negligible set, from f constitute the ***equivalence class*** $\dot{f}$. We have seen in the proof of theorem 1.2.2 (ii) that in the present business we sometimes have to make the distinction between a random variable and its class, boring as this is. We write $f \doteq g$ if f and g differ negligibly and also $\dot{f} \doteq \dot{g}$ if f and g belong to the same equivalence class, etc.

A property P of the points of Ω is said to hold $\mathbb{P}$-almost surely or simply ***almost surely***, if the set N of points of Ω where it does not hold is negligible. The abbreviation $\mathbb{P}$-a.s. or simply ***a.s.*** is common. The terminology "almost everywhere" and its short form "a.e." will be avoided in context with $\mathbb{P}$ since it is employed with a different meaning in chapter 3.

---

[16] See page 396.

The Sizes of Random Variables

With every probability $\mathbb{P}$ on $\mathcal{F}_\infty$ there come many different ways of measuring the size of a random variable. We shall review a few that have proved particularly useful in many branches of mathematics and that continue to be so in the context of stochastic integration and of stochastic differential equations.

For a function f measurable on the universal completion $\mathcal{F}^*_\infty$ and $0 < p < \infty$, set

$$\| f \|_p \stackrel{\text{def}}{=} \| f \|_{L^p} \stackrel{\text{def}}{=} \Big(\int |f|^p \, d\mathbb{P} \Big)^{1/p} .$$

If there is need to stress which probability $\mathbb{P} \in \mathfrak{P}$ is meant, we write $\| f \|_{L^p(\mathbb{P})}$. The ***p-mean*** $\| \; \|_p$ is

absolute-homogeneous: $\quad \|r \cdot f\|_p = |r| \cdot \|f\|_p$

and ***subadditive***: $\quad \|f + g\|_p \le \|f\|_p + \|g\|_p$

in the range $1 \le p < \infty$, but not for $0 < p < 1$. Since it is often more convenient to have subadditivity at one's disposal rather than homogeneity, we shall mostly employ the subadditive versions

$$\lceil\!\lceil f \rceil\!\rceil_p \stackrel{\text{def}}{=} \begin{cases} \|f\|_{L^p(\mathbb{P})} = \Big(\int |f|^p \, d\mathbb{P} \Big)^{1/p} & \text{for } 1 \le p < \infty, \\ \|f\|^p_{L^p(\mathbb{P})} = \int |f|^p \, d\mathbb{P} & \text{for } 0 < p \le 1 . \end{cases} \tag{1.3.1}$$

L^p or $L^p(\mathbb{P})$ denotes the collection of measurable functions f with $\lceil\!\lceil f \rceil\!\rceil_p < \infty$, the p-integrable functions. The collection of $\mathcal{F}_t$-measurable functions in L^p is $L^p(\mathcal{F}_t)$ or $L^p(\mathcal{F}_t, \mathbb{P})$. It is well known that L^p is a complete pseudometric space under the distance $dist_p(f, g) = \lceil\!\lceil f - g \rceil\!\rceil_p$ – it is to make $dist_p$ a metric that we generally prefer the subadditive size measurement $\lceil\!\lceil \; \rceil\!\rceil_p$ over its homogeneous cousin $\| \; \|_p$.

Two random variables in the same class have the same p-means, so we shall also talk about $\lceil\!\lceil \dot{f} \rceil\!\rceil_p$, etc.

The prominence of the p-means $\| \; \|_p$ and $\lceil\!\lceil \; \rceil\!\rceil_p$ among other size measurements that one might think up is due to Hölder's inequality A.8.4, which provides a partial alleviation of the fact that L^1 is not an algebra, and to the method of interpolation (see proposition A.8.24). Section A.8 contains further information about the p-means and the L^p-spaces. A process Z is called ***p-integrable*** if the random variables Z_t are all p-integrable, and ***L^p-bounded*** if

$$\sup_t \| Z_t \|_p < \infty , \qquad 0 < p \le \infty.$$

The largest class of useful random variables is that of measurable a.s. finite ones. It is denoted by L^0, $L^0(\mathbb{P})$, or $L^0(\mathcal{F}_t, \mathbb{P})$, as the context requires. It extends the slew of the L^p-spaces at $p = 0$. It plays a major role in stochastic analysis due to the fact that it forms an algebra and does not change when $\mathbb{P}$ is replaced by an equivalent probability (exercise A.8.11). There are several ways to attach a numerical size to a function $f \in L^0$, the most common[17] being

$$\lceil\!\lceil f \rceil\!\rceil_0 = \lceil\!\lceil f \rceil\!\rceil_{0;\mathbb{P}} \stackrel{\text{def}}{=} \inf\Big\{\lambda : \mathbb{P}[|f| > \lambda] \le \lambda\Big\}.$$

It measures ***convergence in probability***, also called ***convergence in measure***; namely, $f_n \to f$ in probability if

$$dist_0(f_n, f) \stackrel{\text{def}}{=} \lceil\!\lceil f_n - f \rceil\!\rceil_0 \xrightarrow[n\to\infty]{} 0.$$

$\lceil\!\lceil \ \rceil\!\rceil_0$ is subadditive but not homogeneous (exercise A.8.1). There is also a whole slew of absolute-homogeneous but non-subadditive functionals, one for every $\alpha \in \mathbb{R}$, that can be used to describe the topology of $L^0(\mathbb{P})$:

$$\| f \|_{[\alpha]} = \| f \|_{[\alpha;\mathbb{P}]} \stackrel{\text{def}}{=} \inf\{\lambda > 0 : \mathbb{P}[|f| > \lambda] \le \alpha\}.$$

Further information about these various size measurements and their relation to each other can be found in section A.8. The reader not familiar with L^0 or the basic notions concerning topological vector spaces is strongly advised to peruse that section. In the meantime, here is a little mnemonic device: functionals with "straight sides" $\| \ \|$ are homogeneous, and those with a little "crossbar," like $\lceil\!\lceil \ \rceil\!\rceil$, are subadditive. Of course, if $1 \le p < \infty$, then $\| \ \|_p = \lceil\!\lceil \ \rceil\!\rceil_p$ has both properties.

Exercise 1.3.22 While some of the functionals $\lceil\!\lceil \ \rceil\!\rceil_p$ are not homogeneous – the $\lceil\!\lceil \ \rceil\!\rceil_p$ for $0 \le p < 1$ – and some are not subadditive – the $\| \ \|_p$ for $0 < p < 1$ and the $\| \ \|_{[\alpha]}$ – all of them respect the order: $|f| \le |g| \implies \| f \|_. \le \| g \|_.$. Functionals with this property are termed ***solid***.

Two Notions of Equality for Processes

Modifications Let $\mathbb{P}$ be a probability in the pertinent class $\mathfrak{P}$. Two processes X, Y are $\mathbb{P}$-***modifications*** of each other if, at every instant t, $\mathbb{P}$-almost surely $X_t = Y_t$. We also say "X is a modification of Y," suppressing as usual any mention of $\mathbb{P}$ if it is clear from the context. In fact, it may happen that $[X_t \ne Y_t]$ is so wild that the only set of $\mathcal{F}_t$ containing it is the whole space Ω. It may, in particular, occur that X is adapted but a modification Y of it is not.

Indistinguishability Even if X and a modification Y are adapted, the sets $[X_t \ne Y_t]$ may vary wildly with t. They may even cover all of Ω. In other words, while the values of X and Y might at no finite instant be distinguishable with $\mathbb{P}$, an apparatus rigged to ring the first time X and

[17] It is commonly attributed to Ky–Fan.

Y differ may ring for sure, even immediately. There is evidently a need for a more restrictive notion of equality for processes than merely being almost surely equal at every instant.

To approach this notion let us assume that X, Y are progressively measurable, as respectable processes without ESP are supposed to be. It seems reasonable to say that X, Y are indistinguishable if their entire paths $X_., Y_.$ agree almost surely, that is to say, if the event

$$N \stackrel{\text{def}}{=} [X_. \neq Y_.] = \bigcup_{k\in\mathbb{N}} [|X - Y|_k^\star > 0]$$

has no chance of occurring. Now $N^k = [X_.^k \neq Y_.^k] = [|X - Y|_k^\star > 0]$ is the uncountable union of the sets $[X_s \neq Y_s]$, $s \le k$, and looks at first sight nonmeasurable. By corollary A.5.13, though, N^k belongs to the universal completion of $\mathcal{F}_k$; there is no problem attaching a measure to it. There is still a little conceptual difficulty in that N^k may not belong to $\mathcal{F}_k$ itself, meaning that it is not observable at time k, but this seems like splitting hairs. Anyway, the filtration will soon be enlarged so as to become regular; this implies its universal completeness, and our little trouble goes away. We see that no apparatus will ever be able to detect any difference between X and Y if they differ only on a set like $\bigcup_{k\in\mathbb{N}} N^k$, which is the countable union of negligible sets in $\mathcal{A}_\infty$. Such sets should be declared inconsequential. Letting $\mathcal{A}_{\infty\sigma}$ denote the collection of sets that are countable unions of sets in $\mathcal{A}_\infty$, we are led to the following definition of indistinguishability. Note that it makes sense without any measurability assumption on X, Y.

Definition 1.3.23 *(i) A subset of Ω is **nearly empty** if it is contained in a negligible set of $\mathcal{A}_{\infty\sigma}$. A random variable is nearly zero if it vanishes outside a nearly empty set.*

*(ii) A property P of the points $\omega \in \Omega$ is said to hold **nearly** if the set N of points of Ω where it does not hold is nearly empty. Writing $f = g$ for two random variables f, g generally means that f and g nearly agree.*

*(iii) Two processes X, Y are **indistinguishable** if $[X_. \neq Y_.]$ is nearly empty. A process or subset of the base space $\boldsymbol{B}$ that cannot be distinguished from zero is called **evanescent**. When we write $X = Y$ for two processes X, Y we mean generally that X and Y are indistinguishable.*

When the probability $\mathbb{P} \in \mathfrak{P}$ must be specified, then we talk about $\mathbb{P}$-nearly empty sets or $\mathbb{P}$-nearly vanishing random variables, properties holding $\mathbb{P}$-nearly, processes indistinguishable with $\mathbb{P}$ or $\mathbb{P}$-indistinguishable, and $\mathbb{P}$-evanescent processes.

A set N is nearly empty if someone with a finite if possibly very long life span t can measure it ($N \in \mathcal{F}_t$) and find it to be negligible, or if it is the countable union of such sets. If he and his offspring must wait past the expiration of time (check whether $N \in \mathcal{F}_\infty$) to ascertain that N is negligible – in other words, if this must be left to God – then N is not nearly empty even though it be negligible. Think of nearly empty sets as sets whose negligibility can be detected before the expiration of time.

There is an apologia for the introduction of this class of sets in warnings 1.3.39 on page 39 and 3.9.20 on page 167.

Example 1.3.24 Take for Ω the unit interval $[0,1]$. For $n \in \mathbb{N}$ let $\mathcal{F}_n$ be the σ-algebra generated by the closed intervals $[k2^{-n},(k+1)2^{-n}]$, $0 \le k \le 2^n$. To obtain a filtration indexed by $[0,\infty)$ set $\mathcal{F}_t = \mathcal{F}_n$ for $n \le t < n+1$. For $\mathbb{P}$ take Lebesgue measure λ. The negligible sets in $\mathcal{F}_n$ are the sets of ***dyadic rationals*** of the form $k2^{-n}$, $0 < k < 2^n$. In this case $\mathcal{A}_\infty$ is the algebra of finite unions of intervals with dyadic-rational endpoints, and its span $\mathcal{F}_\infty$ is the σ-algebra of all Borel sets on $[0,1]$. A set is nearly empty if and only if it is a subset of the dyadic rationals in $(0,1)$. There are many more negligible sets than these. Here is a striking phenomenon: consider a countable set $\mathcal{I}$ of irrational numbers dense in $[0,1]$. It is Lebesgue negligible but has outer measure 1 for any of the measured triples $\big([0,1],\mathcal{F}_t,\mathbb{P}_{|\mathcal{F}_t}\big)$. The upshot: the notion of a nearly empty set is rather more restrictive than that of a negligible set. In the present example there are $2^{\aleph_0}$ of the former and $\ge 2^{\aleph_1}$ of the latter. For more on this see example 1.3.32.

Exercise 1.3.25 $N \subset \Omega$ is negligible if and only if there is, for every $\epsilon > 0$, a set of $\mathcal{A}_{\infty\sigma}$ that has measure less than ϵ and contains N. It is nearly empty if and only if there exists a set of $\mathcal{A}_{\infty\sigma}$ that has measure equal to zero and contains N. $N \subset \Omega$ is nearly empty if and only if there exist instants $t_n < \infty$ and negligible sets $N_n \in \mathcal{F}_{t_n}$ whose union contains N.

Exercise 1.3.26 A subset of a nearly empty set is nearly empty; so is the countable union of nearly empty sets. A subset of an evanescent set is evanescent; so is the countable union of evanescent sets.

Near-emptyness and evanescence are "solid" notions: if f,g are random variables, g nearly zero and $|f| \le |g|$, then f is nearly zero; if X,Y are processes, Y evanescent and $|X| \le |Y|$, then X is evanescent. The pointwise limit of a sequence of nearly zero random variables is nearly zero. The pointwise limit of a sequence of evanescent processes is evanescent.

A process X is evanescent if and only if the projection $\pi_\Omega[X \ne 0]$ is nearly empty.

Exercise 1.3.27 (i) Two stopping times that agree almost surely agree nearly.

(ii) If T is a nearly finite stopping time and $N \in \mathcal{F}_T$ is negligible, then it is nearly empty.

Exercise 1.3.28 Indistinguishable processes are modifications of each other. Two adapted left- or right-continuous processes that are modifications of each other are indistinguishable.

The Natural Conditions

We shall now enlarge the given filtration slightly, and carefully. The purpose of this is to gain regularity results for paths of integrators (theorem 2.3.4) and to increase the supply of stopping times (exercise 1.3.30 and appendix A, pages 436–438).

Right-Continuity of a Filtration Many arguments are simplified or possible only when the filtration $\mathcal{F}_.$ is right-continuous:

Definition 1.3.29 *The **right-continuous version** $\mathcal{F}_{.+}$ of a filtration $\mathcal{F}_.$ is defined by*

$$\mathcal{F}_{t+} \stackrel{\text{def}}{=} \bigcap_{u>t} \mathcal{F}_u \qquad \forall\, t \geq 0 .$$

*The given filtration $\mathcal{F}_.$ is termed **right-continuous** if $\mathcal{F}_. = \mathcal{F}_{.+}$.*

The following exercise develops some of the benefits of having the filtration right-continuous. We shall see soon (proposition 2.2.11) that it costs nothing to replace any given filtration by its right-continuous version, so that we can easily avail ourselves of these benefits.

Exercise 1.3.30 The right-continuity of the filtration implies all of this: (i) A random time T is a stopping time if and only if $[T < t] \in \mathcal{F}_t$ for all $t > 0$. This is often easier to check than that $[T \leq t] \in \mathcal{F}_t$ for all t. For instance (compare with proposition 1.3.11): (ii) If Z is an adapted process with right- or left-continuous paths, then for any $\lambda \in \mathbb{R}$

$$T^{\lambda+} \stackrel{\text{def}}{=} \inf\{t : Z_t > \lambda\}$$

is a stopping time. Moreover, the functions $\lambda \mapsto T^{\lambda+}(\omega)$ are increasing and right-continuous. (iii) If T is a stopping time, then $A \in \mathcal{F}_T$ iff $A \cap [T < t] \in \mathcal{F}_t \quad \forall\, t$.

(iv) The infimum T of a countable collection $\{T_n\}$ of stopping times is a stopping time, and its past is $\mathcal{F}_T = \bigcap_n \mathcal{F}_{T_n}$ (cf. exercise 1.3.15).

(v) $\mathcal{F}_.$ and $\mathcal{F}_{.+}$ have the same adapted left-continuous processes. A process adapted to the filtration $\mathcal{F}_.$ and progressively measurable on $\mathcal{F}_{.+}$ is progressively measurable on $\mathcal{F}_.$.

Regularity of a Measured Filtration It is still possible that there exist measurable indistinguishable processes X, Y of which one is adapted, the other not. This unsatisfactory state of affairs is ruled out if the filtration is regular. For motivation consider a subset $N \subset \Omega$ that is not measurable on $\mathcal{F}_t$ (too wild to be observable now, at time t) but that is measurable on $\mathcal{F}_u$ for some $u > t$ (observable then) and turns out to have probability $\mathbb{P}[N] = 0$ of occurring. Or N might merely be a subset of such a set. The class of such N and their countable unions is precisely the class of nearly empty sets. It does no harm but confers great technical advantage to declare such an event N to be both observable and impossible now. Precisely:

Definition 1.3.31 *(i) Given a measured filtration $(\mathcal{F}_., \mathfrak{P})$ on Ω and a probability $\mathbb{P}$ in the pertinent class $\mathfrak{P}$, set*

$$\mathcal{F}_t^{\mathbb{P}} \stackrel{\text{def}}{=} \left\{A \subset \Omega : \exists A_{\mathbb{P}} \in \mathcal{F}_t \ \text{ so that } \ |A - A_{\mathbb{P}}| \ \text{ is } \mathbb{P}\text{-nearly empty}\right\} .$$

Here $|A - A_{\mathbb{P}}|$ is the symmetric difference $(A \setminus A_{\mathbb{P}}) \cup (A_{\mathbb{P}} \setminus A)$ (see convention A.1.5). $\mathcal{F}_t^{\mathbb{P}}$ is easily seen to be a σ-algebra; in fact, it is the σ-algebra generated by $\mathcal{F}_t$ and the $\mathbb{P}$-nearly empty sets. The collection

$$\mathcal{F}_.^{\mathbb{P}} \stackrel{\text{def}}{=} \left\{\mathcal{F}_t^{\mathbb{P}}\right\}_{0 \leq t \leq \infty}$$

is the $\mathbb{P}$***-regularization*** *of* $\mathcal{F}_.$. *The filtration* $\mathcal{F}^{\mathfrak{P}}$ *composed of the* σ*-algebras*

$$\mathcal{F}_t^{\mathfrak{P}} \stackrel{\text{def}}{=} \bigcap_{\mathbb{P}\in\mathfrak{P}} \mathcal{F}_t^{\mathbb{P}} , \qquad t \geq 0 ,$$

is the $\mathfrak{P}$***-regularization****, or simply the* ***regularization****, when* $\mathfrak{P}$ *is clear from the context.*

(ii) The measured filtration $(\mathcal{F}_., \mathfrak{P})$ *is* ***regular*** *if* $\mathcal{F}_. = \mathcal{F}_.^{\mathfrak{P}}$. *We then also write "*$\mathcal{F}_.$ *is* $\mathfrak{P}$***-regular****," or simply "*$\mathcal{F}_.$ *is* ***regular****" when* $\mathfrak{P}$ *is understood.*

Let us paraphrase the regularity of a filtration in intuitive terms: "an event that proves in the long run to be indistinguishable, whatever the probability in the admissible class $\mathfrak{P}$, from some event observable now is considered to be observable now."

$\mathcal{F}_t^{\mathbb{P}}$ contains the completion of $\mathcal{F}_t$ under the restriction $\mathbb{P}_{|\mathcal{F}_t}$, which in turn contains the universal completion. The regularization of $\mathcal{F}$ is thus universally complete. If $\mathcal{F}$ is regular, then the maximal process of a progressively measurable process is again progressively measurable (corollary A.5.13). This is nice. The main point of regularity is, though, that it allows us to prove the path regularity of integrators (section 2.3 and definition 3.7.6). The following exercises show how much – or rather how little – is changed by such a replacement and develop some of the benefits of having the filtration regular. We shall see soon (proposition 2.2.11) that it costs nothing to replace a given filtration by its regularization, so that we can easily avail ourselves of these benefits.

Example 1.3.32 In the right-continuous measured filtration $(\Omega = [0,1], \mathcal{F}, \mathbb{P} = \lambda)$ of example 1.3.24 the $\mathcal{F}_t$ are all universally complete, and the couples $(\mathcal{F}_t, \mathbb{P})$ are complete. Nevertheless, the regularization differs from $\mathcal{F}$: $\mathcal{F}_t^{\mathbb{P}}$ is the σ-algebra generated by $\mathcal{F}_t$ and the dyadic-rational points in $(0,1)$. For more on this see example 1.3.45.

Exercise 1.3.33 A random variable f is measurable on $\mathcal{F}_t^{\mathbb{P}}$ if and only if there exists an $\mathcal{F}_t$-measurable random variable $f_{\mathbb{P}}$ that $\mathbb{P}$-nearly equals f.

Exercise 1.3.34 (i) $\mathcal{F}_.^{\mathfrak{P}}$ is regular. (ii) A random variable f is measurable on $\mathcal{F}_t^{\mathfrak{P}}$ if and only if for every $\mathbb{P} \in \mathfrak{P}$ there is an $\mathcal{F}_t$-measurable random variable $\mathbb{P}$-nearly equal to f.

Exercise 1.3.35 Assume that $\mathcal{F}_.$ is right-continuous and let $\mathbb{P}$ be a probability on $\mathcal{F}_\infty$.

(i) Let X be a right-continuous process adapted to $\mathcal{F}_.^{\mathbb{P}}$. There exists a process X' that is $\mathbb{P}$-nearly right-continuous and adapted to $\mathcal{F}_.$ and cannot be distinguished from X with $\mathbb{P}$. If X is a set, then X' can be chosen to be a set; and if X is increasing, then X' can be chosen increasing and right-continuous everywhere.

(ii) A random time T is a stopping time on $\mathcal{F}_.^{\mathbb{P}}$ if and only if there exists an $\mathcal{F}_.$-stopping time $T_{\mathbb{P}}$ that nearly equals T. A set A belongs to $\mathcal{F}_T$ if and only if there exists a set $A_{\mathbb{P}} \in \mathcal{F}_{T_{\mathbb{P}}}$ that is nearly equal to A.

Exercise 1.3.36 (i) The right-continuous version of the regularization equals the regularization of the right-continuous version; if $\mathcal{F}_.$ is regular, then so is $\mathcal{F}_{.+}$.

(ii) Substituting $\mathcal{F}_{.+}$ for $\mathcal{F}_.$ will increase the supply of adapted and of progressively measurable processes, and of stopping times, and will sometimes enlarge the spaces $L^p[\mathcal{F}_t, \mathbb{P}]$ of equivalence classes (sometimes it will not – see exercise 1.3.47.).

Exercise 1.3.37 $\mathcal{F}_t^{\mathfrak{P}}$ contains the σ-algebra generated by $\mathcal{F}_t$ and the nearly empty sets, and coincides with that σ-algebra if there happens to exist a probability with respect to which every probability in $\mathfrak{P}$ is absolutely continuous.

Definition 1.3.38 (The Natural Conditions) *Let $(\mathcal{F}_., \mathfrak{P})$ be a measured filtration. The* ***natural enlargement*** *of $\mathcal{F}_.$ is the filtration $\mathcal{F}_{.+}^{\mathfrak{P}}$ obtained by regularizing the right-continuous version of $\mathcal{F}_.$ (or, equivalently, by taking the right-continuous version of the regularization — see exercise 1.3.36).*

Suppose that Z is a process and the pertinent class $\mathfrak{P}$ of probabilities is understood; then the natural enlargement of the basic filtration $\mathcal{F}_.^0[Z]$ is called the ***natural filtration*** *of Z and is denoted by $\mathcal{F}_.[Z]$. If $\mathfrak{P}$ must be mentioned, we write $\mathcal{F}_.^{\mathfrak{P}}[Z]$.*

A measured filtration is said to satisfy the ***natural conditions*** *if it equals its natural enlargement.*

Warning 1.3.39 The reader will find the term ***usual conditions*** at this juncture in most textbooks, instead of "natural conditions." The usual conditions require that $\mathcal{F}_.$ equal its ***usual enlargement***, which is effected by replacing $\mathcal{F}_.$ with its right-continuous version and throwing into every $\mathcal{F}_{t+}$, $t < \infty$, ***all*** $\mathbb{P}$-negligible sets of $\mathcal{F}_\infty$ and their subsets, i.e., all sets that are negligible for the outer measure $\mathbb{P}^*$ constructed from $(\mathcal{F}_\infty, \mathbb{P}^*)$. The latter class is generally cardinalities bigger than the class of nearly empty sets (see example 1.3.24). Doing the regularization (frequently called ***completion***) of the filtration this way evidently has the consequence that a probability absolutely continuous with respect to $\mathbb{P}$ on $\mathcal{F}_0$ is already absolutely continuous with respect to $\mathbb{P}$ on $\mathcal{F}_\infty$. Failure to observe this has occasionally led to vacuous investigations of the local equivalence of probabilities and to erroneous statements of Girsanov's theorem (see example 3.9.14 on page 164 and warning 3.9.20 on page 167). The term "usual conditions" was coined by the French School and is now in universal use.

We shall see in due course that definition 1.3.38 of the enlargement furnishes the advantages one expects: path regularity of integrators and a plentiful supply of stopping times, without incurring some of the disadvantages that come with too liberal an enlargement. Here is a mnemonic device: the natural conditions are obtained by adjoining the nearly empty (instead of the negligible) sets to the right-continuous version of the filtration; and they are nearly the usual conditions, but not quite: ***The natural enlargement does not in general contain every negligible set of $\mathcal{F}_\infty$!*** ▮

The natural conditions can of course be had by the simple expedient of replacing the given filtration with its natural enlargement – and, according to proposition 2.2.11, doing this costs nothing so far as the stochastic integral is concerned. Here is one pretty consequence of doing such a replacement. Consider a progressively measurable subset B of the base space $\boldsymbol{B}$. The ***debut*** of B is the time (see figure 1.6)

$$D_B(\omega) \stackrel{\text{def}}{=} \inf\{t : (t,\omega) \in B\} .$$

It is shown in corollary A.5.12 that under the natural conditions D_B is a stopping time. The proof uses some capacity theory, which can be found in appendix A. Our elementary analysis of integrators won't need to employ this big result, but we shall make use of the larger supply of stopping times provided by the regularity and right-continuity and established in exercises 1.3.35 and 1.3.30.

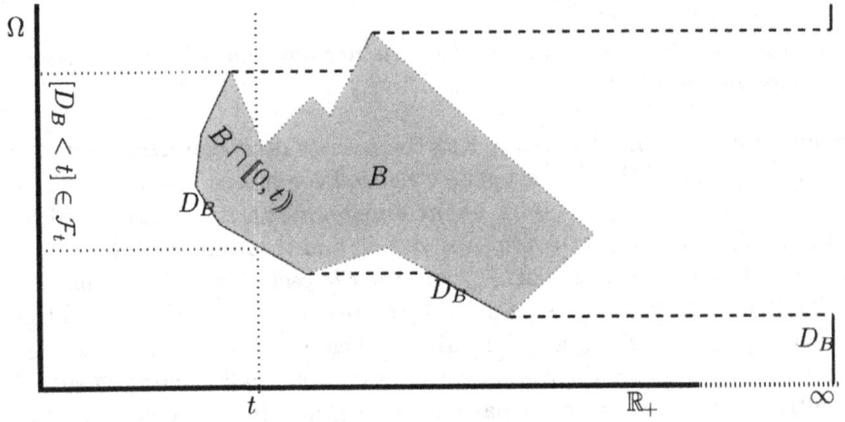

Figure 1.6 The debut of A

Exercise 1.3.40 The natural enlargement has the same nearly empty sets and evanescent processes as the original measured filtration.

Local Absolute Continuity A probability $\mathbb{P}'$ on $\mathcal{F}_\infty$ is ***locally absolutely continuous*** with respect to $\mathbb{P}$ if for all finite $t < \infty$ its restriction $\mathbb{P}'_t$ to $\mathcal{F}_t$ is absolutely continuous with respect to the restriction $\mathbb{P}_t$ of $\mathbb{P}$ to $\mathcal{F}_t$. This is evidently the same as saying that a $\mathbb{P}$-nearly empty set is $\mathbb{P}'$-nearly empty and is written $\mathbb{P}' \ll\!. \ \mathbb{P}$. This can very well happen without $\mathbb{P}'$ being absolutely continuous with respect to $\mathbb{P}$ on $\mathcal{F}_\infty$! If both $\mathbb{P}' \ll\!. \ \mathbb{P}$ and $\mathbb{P} \ll\!. \ \mathbb{P}'$, we say that $\mathbb{P}$ and $\mathbb{P}'$ are ***locally equivalent*** and write $\mathbb{P} \approx\!. \ \mathbb{P}'$; it simply means that $\mathbb{P}$ and $\mathbb{P}'$ have the same nearly empty sets. For more on the subject see pages 162–167.

Exercise 1.3.41 Let $\mathbb{P}' \ll\!. \ \mathbb{P}$. (i) A $\mathbb{P}$-evanescent process is $\mathbb{P}'$-evanescent. (ii) $(\mathcal{F}^{\mathbb{P}}_{.}, \mathbb{P}')$ is $\mathbb{P}'$-regular. (iii) If T is a $\mathbb{P}$-nearly finite stopping time, then a $\mathbb{P}$-negligible set in $\mathcal{F}_T$ is $\mathbb{P}'$-negligible.

Exercise 1.3.42 In order to see that the definition of local absolute continuity conforms with our usual use of the word "local" (page 51), show that $\mathbb{P}' \ll\!. \, \mathbb{P}$ if and only if there are arbitrarily large finite stopping times T so that $\mathbb{P}' \ll \mathbb{P}$ on $\mathcal{F}_T$. If $\mathfrak{P}$ is enlarged by adding every measure locally absolutely continuous with respect to some probability in $\mathfrak{P}$, then the regularization does not change. In particular, if there exists a probability $\mathbb{P}$ in $\mathfrak{P}$ with respect to which all of the others are locally absolutely continuous, then $\mathcal{F}^{\mathfrak{P}}_{.} = \mathcal{F}^{\mathbb{P}}_{.}$.

Exercise 1.3.43 Replacing $\mathcal{F}_t$ by $\mathcal{F}^{\mathfrak{P}}_t$ is harmless in this sense: it will increase the supply of adapted and of progressively measurable processes, but it will not change the spaces $L^p[\mathcal{F}_t, \mathbb{P}]$ of equivalence classes $0 \le p \le \infty, 0 \le t \le \infty, \mathbb{P} \in \mathfrak{P}$.

Exercise 1.3.44 Construct a measured filtration $(\mathcal{F}_., \mathbb{P})$ that is not regular yet has the property that the pairs $(\mathcal{F}_t, \mathbb{P})$ all are complete measure spaces.

Example 1.3.45 Recall the measured filtration $(\Omega = [0,1], \mathcal{F}^{\mathbb{P}}_{.}, \mathbb{P} = \lambda)$ of example 1.3.32. It is right-continuous and regular. On $\mathcal{F}_\infty = \mathcal{B}^\bullet[0,1]$ let $\mathbb{P}'$ be Dirac measure at 0.[18] Its restriction to $\mathcal{F}^{\mathbb{P}}_t$ is absolutely continuous with respect to $\mathbb{P}$; in fact, for $n \le t < n+1$ a Radon–Nikodym derivative is $M_t \stackrel{\text{def}}{=} 2^n \cdot [0, 2^{-n}]$. So $\mathbb{P}'$ is locally absolutely continuous with respect to $\mathbb{P}$, evidently without being absolutely continuous with respect to $\mathbb{P}$ on $\mathcal{F}_\infty$. For another example see theorem 3.9.19 on page 167.

Exercise 1.3.46 The set $\mathcal{N}$ of theorem 1.2.8 where the Wiener path is differentiable at at least one instant was actually nearly empty.

Exercise 1.3.47 (A Zero-One Law) Let W be a standard Wiener process on $(\Omega, \mathbb{P})$. (i) The $\mathbb{P}$-regularization of the basic filtration $\mathcal{F}^0_{.}[W]$ is right-continuous. (ii) Set $T^{\pm} \stackrel{\text{def}}{=} \inf\{t > 0 : W_t \gtrless 0\}$. Then $\mathbb{P}[T^+ = 0] = \mathbb{P}[T^- = 0] = 1$; to paraphrase "$W$ starts off by oscillating about 0."

Exercise 1.3.48 A standard Wiener process[8] is ***recurrent***. That is to say, for every $s \in [0, \infty)$ and $x \in \mathbb{R}$ and almost all $\omega \in \Omega$ there is an instant $t > s$ at which $W_t(\omega) = x$.

Repeated Footnotes: 3[1] 5[2] 6[4] 8[7] 9[8] 14[12]

---

[18] Dirac measure at ω is the measure $A \mapsto A(\omega)$ – see convention A.1.5.

2

Integrators and Martingales

Now that the basic notions of filtration, process, and stopping time are at our disposal, it is time to develop the stochastic integral $\int X\, dZ$, as per Itô's ideas explained on page 5. We shall call X the ***integrand*** and Z the ***integrator***. Both are now processes.

For a guide let us review the construction of the ordinary Lebesgue–Stieltjes integral $\int x\, dz$ on the half-line; the stochastic integral $\int X\, dZ$ that we are aiming for is but a straightforward generalization of it. The Lebesgue–Stieltjes integral is constructed in two steps. First, it is defined on step functions x. This can be done whatever the integrator z. If, however, the Dominated Convergence Theorem is to hold, even on as small a class as the step functions themselves, restrictions must be placed on the integrator: z must be right-continuous and must have finite variation. This chapter discusses the stochastic analog of these restrictions, identifying the processes that have a chance of being useful stochastic integrators.

Given that a distribution function z on the line is right-continuous and has finite variation, the second step is one of a variety of procedures that extend the integral from step functions to a much larger class of integrands. The most efficient extension procedure is that of Daniell; it is also the only one that has a straightforward generalization to the stochastic case. This is discussed in chapter 3.

Step Functions and Lebesgue–Stieltjes Integrators on the Line

By way of motivation for this chapter let us go through the arguments in the second paragraph above in "abbreviated detail." A function $x : s \mapsto x_s$ on $[0, \infty)$ is a *step function* if there are a partition

$$\mathcal{P} = \{0 = t_1 < t_2 < \ldots < t_{N+1} < \infty\}$$

and constants $r_n \in \mathbb{R}$, $n = 0, 1, \ldots, N$, such that

$$x_s = \begin{cases} r_0 & \text{if} \quad s = 0 \\ r_n & \text{for} \quad t_n < s \le t_{n+1}\,,\ n = 1, 2, \ldots, N, \\ 0 & \text{for} \quad s > t_{N+1}. \end{cases} \tag{2.1}$$

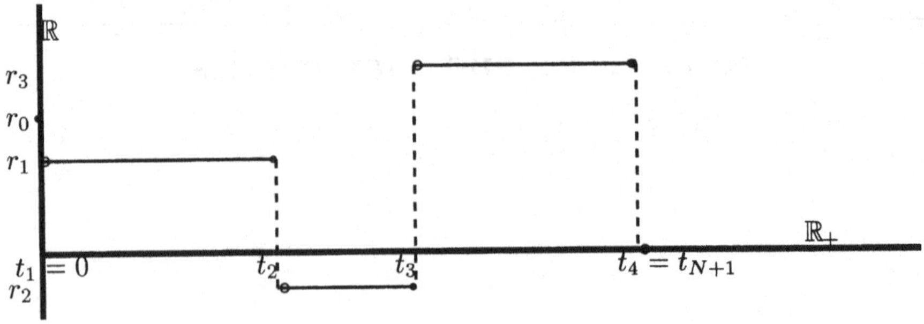

Figure 2.7 A step function on the half-line

The point $t = 0$ receives special treatment inasmuch as the measure $\mu = dz$ might charge the singleton $\{0\}$. The integral of such an *elementary integrand* x against a distribution function or *integrator* $z : [0, \infty) \to \mathbb{R}$ is

$$\int x\, dz = \int x_s\, dz_s \stackrel{\text{def}}{=} r_0 \cdot z_0 + \sum_{n=1}^{N} r_n \cdot \left(z_{t_{n+1}} - z_{t_n}\right) . \tag{2.2}$$

The collection $\mathfrak{e}$ of step functions is a vector space, and the map $x \mapsto \int x\, dz$ is a linear functional on it. It is called the *elementary integral.*

If z is just any function, nothing more of use can be said. We are after an extension satisfying the Dominated Convergence Theorem, though. If there is to be one, then z must be right-continuous; for if (t_n) is any sequence decreasing to t, then

$$z_{t_n} - z_t = \int 1_{(t,t_n]}\, dz \xrightarrow[n\to\infty]{} 0 ,$$

because the sequence $\left(1_{(t,t_n]}\right)$ of elementary integrands decreases pointwise to zero. Also, for every t the set

$$\int \mathfrak{e}_1\, dz^t \stackrel{\text{def}}{=} \left\{\int x\, dz^t \ : x \in \mathfrak{e},\ |x| \le 1\right\}$$

must be bounded.[1] For if it were not, then there would exist elementary integrands $x^{(n)}$ with $|x^{(n)}| \le 1_{[0,t]}$ and $\int x^{(n)}\, dz > n$; the functions $x^{(n)}/n \in \mathfrak{e}$ would converge pointwise to zero, being dominated by $1_{[0,t]} \in \mathfrak{e}$, and yet their integrals would all exceed 1. The condition can be rewritten quantitatively

as
$$\lceil z \rceil_t \stackrel{\text{def}}{=} \sup\left\{\left|\int x\, dz\right| : |x| \le 1_{[0,t]}\right\} < \infty \quad \forall\, t < \infty , \tag{2.3}$$

or as
$$\|y\|_z \stackrel{\text{def}}{=} \sup\left\{\left|\int x\, dz\right| : |x| \le y\right\} < \infty \quad \forall\, y \in \mathfrak{e}_+ ,$$

[1] Recall from page 23 that z^t is z stopped at t.

or again thus: the image under $\int \cdot\, dz$ of any ***order interval***

$$[-y, y] \stackrel{\text{def}}{=} \{x \in \mathfrak{e} : -y \le x \le y\}$$

is a bounded subset of the range $\mathbb{R}$, $y \in \mathfrak{e}_+$. If (2.3) is satisfied, we say that z has *finite variation.*

In summary, if there is to exist an extension satisfying the Dominated Convergence Theorem, then z must be right-continuous and have finite variation. As is well known, these two conditions are also sufficient for the existence of such an extension.

The present chapter defines and analyzes the stochastic analogs of these notions and conditions; the elementary integrands are certain step functions on the half-line that depend on chance $\omega \in \Omega$; z is replaced by a process Z that plays the role of a "random distribution function"; and the conditions of right-continuity and finite variation have their straightforward analogs in the stochastic case. Discussing these and drawing first conclusions occupies the present chapter; the next one contains the extension theory via Daniell's procedure, which works just as simply and efficiently here as it does on the half-line.

Exercise 2.1 According to most textbooks, a distribution function $z : [0, \infty) \to \mathbb{R}$ has finite variation if for all $t < \infty$ the number

$$\,\vert z \vert_t = \sup\Big\{ |z_0| + \sum_i |z_{t_{i+1}} - z_{t_i}| : \ 0 = t_1 \le t_2 \le \ldots \le t_{I+1} = t \Big\},$$

called the variation of z on $[0, t]$, is finite. The supremum is taken over all finite partitions $0 = t_1 \le t_2 \le \ldots \le t_{I+1} = t$ of $[0, t]$. To reconcile this with the definition given above, observe that the sum is nothing but the integral of a step function, to wit, the function that takes the value $\text{sgn}(z_0)$ on $\{0\}$ and $\text{sgn}(z_{t_{i+1}} - z_{t_i})$ on the interval $(z_{t_i}, z_{t_{i+1}}]$. Show that

$$\vert z \vert_t = \sup\Big\{ \Big| \int x_s \, dz_s \Big| : |x| \le [0, t] \Big\} = \| [0, t] \|_z \, .$$

Exercise 2.2 The map $y \mapsto \|y\|_z$ is additive and extends to a positive measure on step functions. The latter is called the *variation measure* $\vert\mu\vert = d\vert z\vert = |dz|$ of $\mu = dz$. Suppose that z has finite variation. Then z is right-continuous if and only if $\mu = dz$ is σ-additive. If z is right-continuous, then so is $\vert z\vert$. $\vert z\vert$ is increasing and its limit at ∞ equals

$$\vert z \vert_\infty = \sup\Big\{ \Big| \int x_s \, dz_s \Big| : |x| \le 1 \Big\} .$$

If this number is finite, then z is said to have *bounded* or *totally finite variation.*

Exercise 2.3 A function on the half-line is a step function if and only if it is left-continuous, takes only finite many values, and vanishes after some instant. Their collection $\mathfrak{e}$ forms both an algebra and a vector lattice closed under chopping. The uniform closure of $\mathfrak{e}$ contains all continuous functions that vanish at infinity. The confined uniform closure of $\mathfrak{e}$ contains all continuous functions of compact support.

2.1 The Elementary Stochastic Integral

Elementary Stochastic Integrands

The first task is to identify the stochastic analog of the step functions in equation (2.1). The simplest thing coming to mind is this: a process X is an ***elementary stochastic integrand*** if there are a finite partition

$$\mathcal{P} = \{0 = t_0 = t_1 < t_2 \ldots < t_{N+1} < \infty\}$$

of the half-line and simple random variables $f_0 \in \mathcal{F}_0, f_n \in \mathcal{F}_{t_n}, n = 1, 2, \ldots, N$ such that

$$X_s(\omega) = \begin{cases} f_0(\omega) & \text{for} \quad s = 0 \\ f_n(\omega) & \text{for} \quad t_n < s \le t_{n+1}\,, \; n = 1, 2, \ldots, N, \\ 0 & \text{for} \quad s > t_{N+1}\,. \end{cases}$$

In other words, for $t_n < s \le t \le t_{n+1}$, the random variables $X_s = X_t$ are simple and measurable on the σ-algebra $\mathcal{F}_{t_n}$ that goes with the left endpoint t_n of this interval. If we fix $\omega \in \Omega$ and consider the path $t \mapsto X_t(\omega)$, then we see an ordinary step function as in figure 2.7 on page 44. If we fix t and let ω vary, we see a simple random variable measurable on a σ-algebra strictly prior to t. Convention A.1.5 on page 364 produces this compact notation for X:

$$X = f_0 \cdot [\![0]\!] + \sum_{n=1}^{N} f_n \cdot (\!(t_n, t_{n+1}]\!] \,. \tag{2.1.1}$$

The collection of elementary integrands will be denoted by $\mathcal{E}$, or by $\mathcal{E}[\mathcal{F}_.]$ if we want to stress the fact that the notion depends – through the measurability assumption on the f_n – on the filtration.

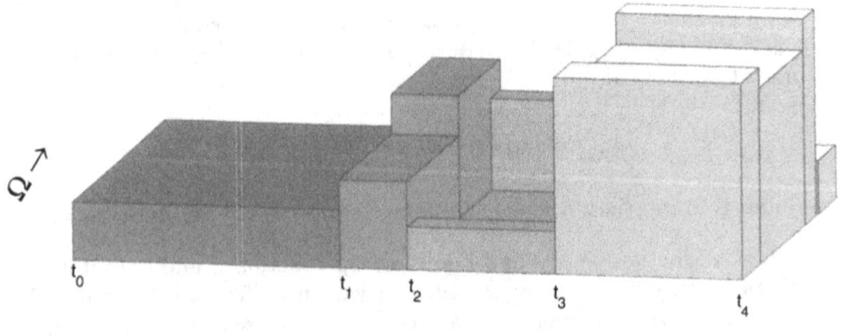

Figure 2.8 An elementary stochastic integrand

Exercise 2.1.1 An elementary integrand is an adapted left-continuous process.

Exercise 2.1.2 If X, Y are elementary integrands, then so are any linear combination, their product, their pointwise infimum $X \wedge Y$, their pointwise maximum $X \vee Y$, and the "chopped function" $X \wedge 1$. In other words, $\mathcal{E}$ is an algebra and vector lattice of bounded functions on $\boldsymbol{B}$ closed under chopping. (For the proof of proposition 3.3.2 it is worth noting that this is the sole information about $\mathcal{E}$ used in the extension theory of the next chapter.)

Exercise 2.1.3 Let $\mathcal{A}$ denote the collection of idempotent functions, i.e., sets,[2] in $\mathcal{E}$. Then $\mathcal{A}$ is a ring of subsets of $\boldsymbol{B}$ and $\mathcal{E}$ is the linear span of $\mathcal{A}$. $\mathcal{A}$ is the ring generated by the collection $\{\{0\} \times A : A \in \mathcal{F}_0\} \cup \{(s,t] \times A : s < t,\ A \in \mathcal{F}_s\}$ of rectangles, and $\mathcal{E}$ is the linear span of these rectangles.

The Elementary Stochastic Integral

Let Z be an adapted process. The integral against dZ of an elementary integrand $X \in \mathcal{E}$ as in (2.1.1) is, in complete analogy with the deterministic case (2.2), defined by

$$\int X\, dZ = f_0 \cdot Z_0 + \sum_{n=1}^{N} f_n \cdot (Z_{t_{n+1}} - Z_{t_n}) . \tag{2.1.2}$$

This is a random variable: for $\omega \in \Omega$

$$\Big(\int X\, dZ\Big)(\omega) = f_0(\omega) \cdot Z_0(\omega) + \sum_{n=1}^{N} f_n(\omega) \cdot \big(Z_{t_{n+1}}(\omega) - Z_{t_n}(\omega)\big) .$$

However, although stochastic analysis is about dependence on chance ω, it is considered babyish to mention the ω; so mostly we shan't after this. The path of X is an ordinary step function as in (2.1). The present definition agrees ω-for-ω with the classical definition (2.2). The linear map $X \mapsto \int X\, dZ$ of (2.1.2) is called the ***elementary stochastic integral***.

Exercise 2.1.4 $\int X\, dZ$ does not depend on the representation (2.1.1) of X and is linear in both X and Z.

The Elementary Integral and Stopping Times

A description in terms of stopping times and stochastic intervals of both the elementary integrands and their integrals is natural and most useful. Let us call a stopping time ***elementary*** if it takes only finitely many values, all of them finite.

Let $S \le T$ be two elementary stopping times. The elementary stochastic interval $(\!(S, T]\!]$ is then an elementary integrand.[2] To see this let

$$\{0 \le t_1 < t_2 < \ldots < t_{N+1} < \infty\}$$

[2] See convention A.1.5 and figure A.14 on page 365.

be the values that S and T take, written in order. If $s \in (t_n, t_{n+1}]$, then the random variable $(\!(S,T]\!]_s$ takes only the values 0 or 1; in fact, $(\!(S,T]\!]_s(\omega) = 1$ precisely if $S(\omega) \le t_n$ and $T(\omega) \ge t_{n+1}$. In other words, for $t_n < s \le t_{n+1}$

$$\begin{aligned}(\!(S,T]\!]_s &= [S \le t_n] \cap [T \ge t_{n+1}] \\ &= [S \le t_n] \setminus [T \le t_n] \in \mathcal{F}_{t_n} ,\end{aligned}$$

so that $$(\!(S,T]\!] = \sum_{n=1}^{N} (t_n, t_{n+1}] \times \Big([S \le t_n] \cap [T \ge t_{n+1}]\Big) :$$

$(\!(S,T]\!]$ is a set in $\mathcal{E}$. Let us compute its integral against the integrator Z:

$$\begin{aligned}\int (\!(S,T]\!] \, dZ &= \sum_{n=1}^{N} ([S \le t_n][T \ge t_{n+1}])(Z_{t_{n+1}} - Z_{t_n}) \\ &= \sum_{1 \le m < n \le N+1} ([S = t_m][T = t_n])(Z_{t_n} - Z_{t_m}) \\ &= \sum_{1 \le m < n \le N+1} ([S = t_m][T = t_n])(Z_T - Z_S) \\ &= Z_T - Z_S . \qquad (2.1.3)\end{aligned}$$

This is just as it should be.

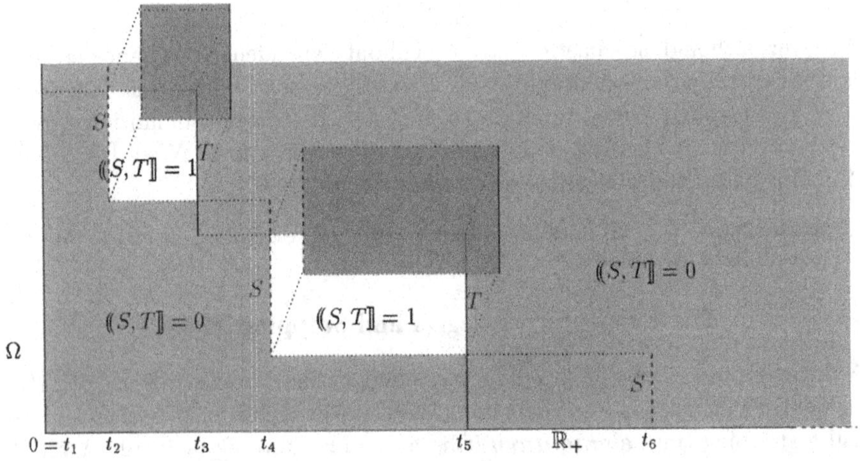

Figure 2.9 The indicator function of the stochastic interval $(\!(S,T]\!]$

Next let $A \in \mathcal{F}_0$. The stopping time 0 can be reduced by A to produce the stopping time 0_A (see exercise 1.3.18 on page 31). Its graph $[\![0_A]\!] = \{0\} \times A$ is evidently an elementary integrand with integral $A \cdot Z_0$. Finally, let $0 = T_1 < \ldots < T_{N+1}$ be elementary stopping times and $r_1, \ldots, r_N$ real numbers, and let f_0 be a simple random variable measurable on $\mathcal{F}_0$. Since f_0 can be written as $f_0 = \sum_k \rho_k \cdot A_k$, $A_k \in \mathcal{F}_0$, the process $f_0 \cdot [\![0]\!]$ is again an elementary integrand with integral $f_0 \cdot Z_0$. The linear combination

$$X = f_0 \cdot [\![0]\!] + \textstyle\sum_{n=1}^N r_n \cdot (\!(T_n, T_{n+1}]\!] \tag{2.1.4}$$

is then also an elementary integrand and its integral against dZ is

$$\int X\, dZ = f_0 \cdot Z_0 + \textstyle\sum_{n=1}^N r_n \cdot (Z_{T_{n+1}} - Z_{T_n}) .$$

Exercise 2.1.5 Let $0 = T_1 \le T_2 \le \ldots \le T_{N+1}$ be elementary stopping times and let $f_0 \in \mathcal{F}_0, f_1 \in \mathcal{F}_{T_1}, \ldots, f_N \in \mathcal{F}_{T_N}$ be simple functions. Then

$$X = f_0 \cdot [\![0]\!] + \textstyle\sum_{n=1}^N f_n \cdot (\!(T_n, T_{n+1}]\!]$$

is an elementary integrand, and its integral is

$$\int X\, dZ = f_0 \cdot Z_0 + \textstyle\sum_{n=1}^N f_n \cdot (Z_{T_{n+1}} - Z_{T_n}) .$$

Exercise 2.1.6 Every elementary integrand is of the form (2.1.4).

L^p-Integrators

Formula (2.1.2) associates with every elementary integrand $X : \boldsymbol{B} \to \mathbb{R}$ a random variable $\int X\, dZ$. The linear map $X \mapsto \int X\, dZ$ from $\mathcal{E}$ to L^0 is just like a signed measure, except that its values are random variables instead of numbers – the technical term is that the elementary integral defined by (2.1.2) is a ***vector measure***. Measures with values in topological vector spaces like L^p, $0 \le p < \infty$, turn out to have just as simple an extension theory as do measures with real values, provided they satisfy some simple conditions. Recall from the introduction to this chapter that a distribution function z on the half-line must be right-continuous, and its associated elementary integral must map order-bounded sets of step functions to bounded sets of reals, if there is to be a satisfactory extension. Precisely this is required of our random distribution function Z, too:

Definition 2.1.7 (Integrators) *Let Z be a numerical process adapted to $\mathcal{F}_.$, $\mathbb{P}$ a probability on $\mathcal{F}_\infty$, and $0 \le p < \infty$.*

*(i) Let T be any stopping time, possibly $T = \infty$. We say that Z is **$\mathcal{I}^p$-bounded on the stochastic interval** $[\![0, T]\!]$ if the family of random variables*

$$\int \mathcal{E}_1\, dZ^T = \left\{ \int X\, dZ^T : X \in \mathcal{E}, |X| \le 1 \right\}$$

if T is elementary:
$$= \left\{ \int X\, dZ : X \in \mathcal{E}, |X| \le [\![0, T]\!] \right\}$$

is a bounded subset of L^p.

(ii) Z is an ***L^p-integrator*** *if it satisfies the following two conditions:*

$$Z \text{ is right-continuous in probability;} \tag{RC-0}$$

$$Z \text{ is } \mathcal{I}^p\text{-bounded on every bounded interval } [\![0,t]\!] . \tag{B-p}$$

(B-p) simply says that the image under $\int \cdot\, dZ$ of any ***order interval***

$$[-Y,Y] \stackrel{\text{def}}{=} \{X \in \mathcal{E} : -Y \le X \le Y\}\,, \qquad Y \in \mathcal{E}_+\,,$$

is a bounded subset of the range L^p, or again that $\int \cdot\, dZ$ is continuous in the topology of confined uniform convergence (see item A.2.5 on page 370).

(iii) Z is a ***global L^p-integrator*** *if it is right-continuous in probability and $\mathcal{I}^p$-bounded on $[\![0,\infty)\!)$.*

If there is a need to specify the probability, then we talk about $\mathcal{I}^p[\mathbb{P}]$-boundedness and (global) $L^p(\mathbb{P})$-integrators.

The reader might have wondered why in (2.1.1) the values f_n that X takes on the interval $(t_n, t_{n+1}]$ were chosen to be measurable on the smallest possible σ-algebra, the one attached to the left endpoint t_n. The way the question is phrased points to the answer: had f_n been allowed to be measurable on the σ-algebra that goes with the right endpoint, or the midpoint, of that interval, then we would have ended up with a larger space $\mathcal{E}$ of elementary integrands. A process Z would have a harder time satisfying the boundedness condition (B-p), and the class of L^p-integrators would be smaller. We shall see soon (theorem 2.5.24) that it is precisely the choice made in equation (2.1.1) that permits martingales to be integrators.

The reader might also be intimidated by the parameter p. Why consider all exponents $0 \le p < \infty$ instead of picking one, say $p = 2$, to compute in Hilbert space, and be done with it? There are several reasons. First, a given integrator Z might not be an L^2-integrator but merely an L^1-integrator or an L^0-integrator. One could argue here that every integrator is an L^0-integrator, so that it would suffice to consider only these. In fact, L^0-integrators are very flexible (see proposition 2.1.9 and proposition 3.7.4); almost every reasonable process can be integrated in the sense L^0 (theorem 3.7.17); neither the feature of being an integrator nor the integral change when $\mathbb{P}$ is replaced by an equivalent measure (proposition 2.1.9 and proposition 3.6.20), which is of principal interest for statistical analysis; and finally L^0 is an algebra. On the other hand, the topological vector space L^0 is not locally convex, and the absence of a single homogeneous gauge measuring the size of its functions makes for cumbersome arguments – this problem can be overcome by replacing in a controlled way the given probability $\mathbb{P}$ by an equivalent one for which the driving term is an L^2-integrator or better – see theorem 4.1.2 on page 191. Second and more importantly, in the stability theory of stochastic differential equations Kolmogoroff's lemma A.2.37 will be used. The exponent p in inequality (A.2.4) will generally have to be strictly greater than the dimension of some parameter space (theorem 5.3.10) or of the state space (example 5.6.2).

The notion of an L^∞-integrator could be defined along the lines of definition 2.1.7, but this would be useless; there is no satisfactory extension theory for L^∞-valued vector measures. Replacing L^p with an Orlicz space whose defining Young function satisfies a so-called Δ_2-condition leads to a satisfactory integration theory, as does replacing it with a Lorentz space $L^{p,\infty}$, $p < \infty$. The most reasonable generalization is touched upon in exercise 3.6.19. We shall not pursue these possibilities.

Local Properties

A word about global *versus* "plain" L^p-integrators. The former are evidently the analogs of distribution functions with *totally finite* or *bounded* variation $\mathchoice{}{}{}{}|z|_\infty$, while the latter are the analogs of distribution functions z on $\mathbb{R}_+$ with just plain finite variation: $|z|_t < \infty \quad \forall t < \infty$. $|z|_t$ may well tend to ∞ as $t \to \infty$, as witness the distribution function $z_t = t$ of Lebesgue measure.

Note that a global integrator is defined in terms of the sup-norm on $\mathcal{E}$: the image of the ***unit ball***

$$\mathcal{E}_1 \overset{\text{def}}{=} \{X \in \mathcal{E} : |X| \le 1\} = \{X \in \mathcal{E} : -1 \le X \le 1\}$$

under the elementary integral must be a bounded subset of L^p. It is not good enough to consider only global integrators – a Wiener process, for instance, is not one. Yet it is frequently sufficient to prove a general result for them; given a "plain" integrator Z, the result in question will apply to every one of the stopped processes Z^t, $0 \le t < \infty$, these being evidently global L^p-integrators. In fact, in the stochastic case it is natural to consider an even more local notion:

Definition 2.1.8 *Let P be a property of processes – P might be the property of being a (global) L^p-integrator or of having continuous paths, for example. A stopping time T is said to* **reduce** *Z to a process having the property P if the stopped process Z^T has P.*

The process Z is said to have the property P ***locally*** *if there are* ***arbitrarily large stopping times*** *that reduce Z to processes having P, that is to say, if for every $\epsilon > 0$ and $t \in (0,\infty)$ there is a stopping time T with $\mathbb{P}[T < t] < \epsilon$ such that the stopped process Z^T has the property P.*

A local L^p-integrator is generally not an L^p-integrator. If $p = 0$, though, it is; this is a first indication of the flexibility of L^0-integrators. A second indication is the fact that being an L^0-integrator depends on the probability only up to local equivalence:

Proposition 2.1.9 *(i) A local L^0-integrator is an L^0-integrator; in fact, it is $\mathcal{I}^0$-bounded on every finite stochastic interval.*

(ii) If Z is a global $L^0(\mathbb{P})$-integrator, then it is a global $L^0(\mathbb{P}')$-integrator for any measure $\mathbb{P}'$ absolutely continuous with respect to $\mathbb{P}$.

(iii) Suppose that Z is an $L^0(\mathbb{P})$-integrator and $\mathbb{P}'$ is a probability on $\mathcal{F}_\infty$ locally absolutely continuous with respect to $\mathbb{P}$. Then Z is an $L^0(\mathbb{P}')$-integrator.

Proof. (i) To say that Z is a local $L^0(\mathbb{P})$-integrator means that, given an instant t and an $\epsilon > 0$, we can find a stopping time T with $\mathbb{P}[T \le t] < \epsilon$ such that the set of classes

$$\mathcal{B}^\epsilon \stackrel{\text{def}}{=} \left\{ \int X \, dZ^{T\wedge t} : X \in \mathcal{E}\,, |X| \le 1 \right\}$$

is bounded in $L^0(\mathbb{P})$. Every random variable $\int X \, dZ$ in the set

$$\mathcal{B} \stackrel{\text{def}}{=} \left\{ \int X \, dZ^t : X \in \mathcal{E}\,, |X| \le 1 \right\}$$

differs from the random variable $\int X \, dZ^{T\wedge t} \in \mathcal{B}^\epsilon$ only in the set $[T \le t]$. That is, the distance of these two random variables is less than ϵ if measured with $[\![\ ]\!]_0$[3]. Thus $\mathcal{B} \subset L^0$ is a set with the property that for every $\epsilon > 0$ there exists a bounded set $\mathcal{B}^\epsilon \subset L^0$ with $\sup_{f\in\mathcal{B}} \inf_{f'\in\mathcal{B}'} [\![f - f']\!]_0 \le \epsilon$. Such a set is itself bounded in L^0. The second half of the statement follows from the observation that the instant t above can be replaced by an almost surely finite stopping time without damaging the argument. For the right-continuity in probability see exercise 2.1.11.

(iii) If the set $\{\int X \, dZ : X \in \mathcal{E}, |X| \le [\![0, t]\!]\}$ is bounded in $L^0(\mathbb{P})$, then it is bounded in $L^0(\mathcal{F}_t, \mathbb{P})$. Since the injection of $L^0(\mathcal{F}_t, \mathbb{P})$ into $L^0(\mathcal{F}_t, \mathbb{P}')$ is continuous (exercise A.8.19), this set is also bounded in the latter space. Since it is known that $t_n \downarrow t$ implies $Z_{t_n} \to Z_t$ in $L^0(\mathcal{F}_{t_1}, \mathbb{P})$, it also implies $Z_{t_n} \to Z_t$ in $L^0(\mathcal{F}_{t_1}, \mathbb{P}')$ and then in $L^0(\mathcal{F}_\infty, \mathbb{P}')$. (ii) is even simpler. ∎

Exercise 2.1.10 (i) If Z is an L^p-integrator, then for any stopping time T so is the stopped process Z^T. A local L^p-integrator is locally a global L^p-integrator.
(ii) If the stopped processes Z^S and Z^T are plain or global L^p-integrators, then so is the stopped process $Z^{S\vee T}$. If Z is a local L^p-integrator, then there is a sequence of stopping times reducing Z to global L^p-integrators and increasing a.s. to ∞.

Exercise 2.1.11 A locally right-continuous process is right-continuous. A process locally right-continuous in probability is right-continuous in probability. An adapted process locally of finite variation nearly has finite variation.

Exercise 2.1.12 The sum of two (local, plain, global) L^p-integrators is a (local, plain, global) L^p-integrator. If Z is a (local, plain, global) L^q-integrator and $0 \le p \le q < \infty$, then Z is a (local, plain, global) L^p-integrator.

Exercise 2.1.13 Argue along the lines on page 43 that both conditions (B-p) and (RC-0) are necessary for the existence of an extension that satisfies the Dominated Convergence Theorem.

---

[3] The topology of L^0 is discussed briefly on page 33 ff., and in detail in section A.8.

Exercise 2.1.14 The map $X \mapsto \int X\, dZ$ is evidently a measure (that is to say a linear map on a space of functions) that has values in a vector space (L^p). Not every vector measure $\mathcal{I} : \mathcal{E} \to L^p$ is of the form $\mathcal{I}[X] = \int X\, dZ$. In fact, the stochastic integrals are exactly the vector measures $\mathcal{I} : \mathcal{E} \to L^0$ that satisfy $\mathcal{I}[f \cdot [\![0,0]\!]] \in \mathcal{F}_0$ for $f \in \mathcal{F}_0$ and

$$\mathcal{I}[f \cdot (\!(s,t]\!]] = f \cdot \mathcal{I}[(\!(s,t]\!]] \in \mathcal{F}_t$$

for $0 \le s \le t$ and simple functions $f \in L^\infty(\mathcal{F}_s)$.

2.2 The Semivariations

Numerical expressions for the boundedness condition (B-p) of definition 2.1.7 are desirable, in fact are necessary, to do the estimates we should expect, for instance, in Picard's scheme sketched on page 5. Now, the only difference with the classical situation discussed on page 44 is that the range $\mathbb{R}$ of the measure has been replaced by $L^p(\mathbb{P})$. It is tempting to emulate the definition (2.3) of the ordinary variation on page 44. To do that we have to agree on a substitute for the absolute value, which measures the size of elements of $\mathbb{R}$, by some device that measures the size of the elements of $L^p(\mathbb{P})$.

The obvious choice is the subadditive p-mean of equation (1.3.1) on page 33. With it the analog of inequality (2.3) becomes

$$\lceil\!\lceil Y \rceil\!\rceil_{Z\text{-}p} \stackrel{\text{def}}{=} \sup\Big\{ \Big\lceil\!\Big\lceil \int X\, dZ \Big\rceil\!\Big\rceil_p : X \in \mathcal{E},\, |X| \le Y \Big\}, \tag{2.2.1}$$

$0 \le p < \infty$. The functional $Y \mapsto \lceil\!\lceil Y \rceil\!\rceil_{Z\text{-}p}$ is called ***the*** $\lceil\!\lceil\ \rceil\!\rceil_p$***-semivariation*** of Z. Recall our little mnemonic device: functionals with "straight sides" like $\|\ \|$ are homogeneous, and those with a little "crossbar" like $\lceil\!\lceil\ \rceil\!\rceil$ are subadditive. Of course, for $1 \le p < \infty$, $\|\ \|_p = \lceil\!\lceil\ \rceil\!\rceil_p$ is both; we then also write $\|Y\|_{Z\text{-}p}$ for $\lceil\!\lceil Y \rceil\!\rceil_{Z\text{-}p}$. In the case $p = 0$, the homogeneous gauges $\|\ \|_{[\alpha]}$ occasionally come in handy; the corresponding semivariation is

$$\|Y\|_{Z\text{-}[\alpha]} \stackrel{\text{def}}{=} \sup\Big\{ \Big\| \int X\, dZ \Big\|_{[\alpha]} : X \in \mathcal{E},\, |X| \le Y \Big\}, \qquad p = 0,\ \alpha \in \mathbb{R}.$$

If there is need to mention the measure $\mathbb{P}$, we shall write $\|\ \|_{Z\text{-}p;\mathbb{P}}$, $\lceil\!\lceil\ \rceil\!\rceil_{Z\text{-}p;\mathbb{P}}$, and $\|\ \|_{Z\text{-}[\alpha;\mathbb{P}]}$. It is clear that we could define a Z-semivariation for any other functional on measurable functions that strikes the fancy. We shall refrain from that.

In view of exercise A.8.18 on page 451 the boundedness condition (B-p) can be rewritten in terms of the semivariation as

$$\lceil\!\lceil \lambda \cdot Y \rceil\!\rceil_{Z\text{-}p} \xrightarrow[\lambda \to 0]{} 0 \qquad \forall\, Y \in \mathcal{E}_+ . \tag{B-p}$$

When $0 < p < \infty$, this reads simply: $\lceil\!\lceil Y \rceil\!\rceil_{Z\text{-}p} < \infty \quad \forall\, Y \in \mathcal{E}_+$.

Proposition 2.2.1 *The semivariation* $\lceil\!\lceil\ \rceil\!\rceil_{Z\text{-}p}$ *is subadditive.*

Proof. Let $Y_1, Y_2 \in \mathcal{E}_+$ and let $r < \lceil\!\lceil Y_1 + Y_2 \rceil\!\rceil_{Z\text{-}p}$. There exists an integrand $X \in \mathcal{E}$ with $|X| \le Y_1 + Y_2$ and $r < \lceil\!\lceil \int X\, dZ \rceil\!\rceil_p$. Set $Y_1' \stackrel{\text{def}}{=} |X| \wedge Y_1$, $Y_2' \stackrel{\text{def}}{=} |X| - |X| \wedge Y_1 = Y_2$, and

$$X_{1+} \stackrel{\text{def}}{=} Y_1' \wedge X_+ , \qquad X_{2+} \stackrel{\text{def}}{=} X_+ - Y_1' \wedge X_+ ,$$
$$X_{1-} \stackrel{\text{def}}{=} Y_1' - Y_1' \wedge X_+ , \quad X_{2-} \stackrel{\text{def}}{=} X_- + Y_1' \wedge X_+ - Y_1' .$$

The columns of this matrix add up to Y_1' and Y_2', the rows to X_+ and X_-. The entries are positive elementary integrands. This is evident, except possibly for the positivity of X_{2-}. But on $[X_- = 0]$ we have $Y_1' = X_+ \wedge Y_1$ and with it $X_{2-} = 0$, and on $[X_+ = 0]$ we have instead $Y_1' = X_- \wedge Y_1$ and therefore $X_{2-} = X_- - X_- \wedge Y_1 \ge 0$. We estimate

$$\begin{aligned}
r &< \left\lceil\!\!\left\lceil \int X\, dZ \right\rceil\!\!\right\rceil_p = \left\lceil\!\!\left\lceil \int (X_{1+} - X_{1-})\, dZ + \int (X_{2+} - X_{2-})\, dZ \right\rceil\!\!\right\rceil_p \\
&\le \left\lceil\!\!\left\lceil \int (X_{1+} - X_{1-})\, dZ \right\rceil\!\!\right\rceil_p + \left\lceil\!\!\left\lceil \int (X_{2+} - X_{2-})\, dZ \right\rceil\!\!\right\rceil_p \qquad (*) \\
&\le \lceil\!\lceil X_{1+} + X_{1-} \rceil\!\rceil_{Z\text{-}p} + \lceil\!\lceil X_{2+} + X_{2-} \rceil\!\rceil_{Z\text{-}p} \\
\text{as } Y_i' \le Y_i: \quad &= \lceil\!\lceil Y_1' \rceil\!\rceil_{Z\text{-}p} + \lceil\!\lceil Y_2' \rceil\!\rceil_{Z\text{-}p} \le \lceil\!\lceil Y_1 \rceil\!\rceil_{Z\text{-}p} + \lceil\!\lceil Y_2 \rceil\!\rceil_{Z\text{-}p} .
\end{aligned}$$

The subadditivity of $\lceil\!\lceil\ \rceil\!\rceil_{Z\text{-}p}$ is established. Note that the subadditivity of $\lceil\!\lceil\ \rceil\!\rceil_p$ was used at $(*)$. ∎

At this stage the case $p = 0$ seems complicated, what with the boundedness condition (B-p) looking so clumsy. As the story unfolds we shall see that L^0-integrators are actually rather flexible and easy to handle. Proposition 2.1.9 gave a first indication of this; in theorem 3.7.17 it is shown in addition that every halfway decent process is integrable in the sense L^0 on every almost surely finite stochastic interval.

Exercise 2.2.2 The semivariations $\lceil\!\lceil\ \rceil\!\rceil_{Z\text{-}p}$, $\|\ \|_{Z\text{-}p}$, and $\|\ \|_{Z\text{-}[\alpha]}$ are ***solid***; that is to say, $|Y| \le |Y'| \implies \|Y\|_{\bullet} \le \|Y'\|_{\bullet}$. The last two are absolute-homogeneous.

Exercise 2.2.3 Suppose that V is an adapted increasing process. Then for $X \in \mathcal{E}$ and $0 \le p < \infty$, $\lceil\!\lceil X \rceil\!\rceil_{V\text{-}p}$ equals the p-mean of the Lebesgue–Stieltjes integral $\int |X|\, dV$.

The Size of an Integrator

Saying that Z is a global L^p-integrator simply means that the elementary stochastic integral with respect to it is a continuous linear operator from one topological vector space, $\mathcal{E}$, to another, L^p; the size of such is customarily measured by its operator norm. In the case of the Lebesgue–Stieltjes integral this was the total variation $\vert z \vert_{\infty}$ (see exercise 2.2). By analogy we are led to set

$$\vert Z \vert_{\mathcal{I}^p} \stackrel{\text{def}}{=} \sup\Big\{ \Big\| \int X\, dZ \Big\|_p : X \in \mathcal{E},\ |X| \le 1 \Big\}, \qquad 0 < p < \infty\,,$$

$$\lceil Z \rceil_{\mathcal{I}^p} \stackrel{\text{def}}{=} \sup\Big\{ \Big\lceil\!\Big\lceil \int X\, dZ \Big\rceil\!\Big\rceil_p : X \in \mathcal{E},\ |X| \le 1 \Big\}, \qquad 0 \le p < \infty\,,$$

$$\vert Z \vert_{[\alpha]} \stackrel{\text{def}}{=} \sup\Big\{ \Big\| \int X\, dZ \Big\|_{[\alpha]} : X \in \mathcal{E},\ |X| \le 1 \Big\}, \qquad p = 0,\ \alpha \in \mathbb{R}\,,$$

depending on our current predilection for the size-measuring functional. If Z is merely an L^p-integrator, not a global one, then these numbers are generally infinite, and the quantities of interest are their finite-time versions

$$\lceil Z^t \rceil_{\mathcal{I}^p} \stackrel{\text{def}}{=} \sup\Big\{ \Big\lceil\!\Big\lceil \int X\, dZ \Big\rceil\!\Big\rceil_p : X \in \mathcal{E}\,,\ |X| \le [\![0,t]\!] \Big\}\,,$$

$0 \le t < \infty$, $0 < p < \infty$, etc.

Exercise 2.2.4

(i)
$$\lceil Z + Z' \rceil_{\mathcal{I}^p} \le \lceil Z \rceil_{\mathcal{I}^p} + \lceil Z' \rceil_{\mathcal{I}^p}\,,$$
$$\lceil Z \rceil_{\mathcal{I}^p} = \vert Z \vert_{\mathcal{I}^p}^{p\wedge 1}\,, \quad \text{and} \quad \lceil \lambda Z \rceil_{\mathcal{I}^p} = |\lambda|^{1\wedge p} \lceil Z \rceil_{\mathcal{I}^p}$$

for $p > 0$. (ii) $\mathcal{I}^p$ forms a vector space on which $Z \mapsto \lceil Z \rceil_{\mathcal{I}^p}$ is subbadditive. (iii) If $0 < p < \infty$, then (B-p) is equivalent with $\lceil Z \rceil_{\mathcal{I}^p} < \infty$ or $\vert Z \vert_{\mathcal{I}^p} < \infty$. (iv) If $p = 0$, then (B-p) is equivalent with $\vert Z \vert_{[\alpha]} < \infty \quad \forall \alpha > 0$.

Exercise 2.2.5 If Z is an L^p-integrator and T is an elementary stopping time, then the stopped process Z^T is a global L^p-integrator and

$$\lceil \lambda \cdot Z^T \rceil_{\mathcal{I}^p} = \lceil\!\lceil \lambda \cdot [\![0,T]\!] \rceil\!\rceil_{Z\text{-}p} \qquad \forall \lambda \in \mathbb{R}\,.$$

Also,
$$\lceil\!\lceil [\![0,T]\!] \rceil\!\rceil_{Z\text{-}p} = \lceil Z^T \rceil_{\mathcal{I}^p}\,, \quad \| [\![0,T]\!] \|_{Z\text{-}p} = \vert Z^T \vert_{\mathcal{I}^p}\,,$$

and
$$\| [\![0,T]\!] \|_{Z\text{-}[\alpha]} = \vert Z^T \vert_{[\alpha]}\,.$$

Exercise 2.2.6 Let $0 \le p \le q < \infty$. An L^q-integrator is an L^p-integrator. Give an inequality between $\lceil\!\lceil Z \rceil\!\rceil_{\mathcal{I}^p}$ and $\lceil\!\lceil Z \rceil\!\rceil_{\mathcal{I}^q}$ and between $\| Z^T \|_{\mathcal{I}^p}$ and $\| Z^T \|_{\mathcal{I}^q}$ in case p is strictly positive.

Exercise 2.2.7 If Z is an L^p-integrator and $X \in \mathcal{E}$, then $Y_t \stackrel{\text{def}}{=} \int X\, dZ^t = \int_0^t X\, dZ$ defines a global L^p-integrator Y. For any $X' \in \mathcal{E}$,

$$\int X'\, dY = \int X' X\, dZ\,.$$

Exercise 2.2.8 $1/p \mapsto \log \vert Z \vert_{\mathcal{I}^p}$ is convex for $0 < p < \infty$.

Vectors of Integrators

A stochastic differential equation frequently is driven by not one or two but a whole slew $\boldsymbol{Z} = (Z^1, Z^2, \ldots, Z^d)$ of integrators, even an infinity of them.[4] It eases the notation to set,[5] for $\boldsymbol{X} = (X_1, \ldots, X_d) \in \mathcal{E}^d$,

$$\int \boldsymbol{X}\, d\boldsymbol{Z} \stackrel{\text{def}}{=} \int X_\eta\, dZ^\eta \tag{2.2.2}$$

and to define the integrator size of the d-tuple $\boldsymbol{Z}$ by

$$\lceil \boldsymbol{Z} \rceil_{\mathcal{I}^p} = \sup\left\{ \left\| \int \boldsymbol{X}\, d\boldsymbol{Z} \right\|_{L^p} : \boldsymbol{X} \in \mathcal{E}_1^d \right\}, \qquad p > 0\,;$$

$$\lceil \boldsymbol{Z} \rceil_{[\alpha]} = \sup\left\{ \left\| \int \boldsymbol{X}\, d\boldsymbol{Z} \right\|_{[\alpha]} : \boldsymbol{X} \in \mathcal{E}_1^d \right\}, \qquad p = 0, \alpha > 0\,,$$

and so on. These definitions take advantage of possible cancellations among the Z^η. For instance, if $\boldsymbol{W} = (W^1, W^2, \ldots, W^d)$ are independent standard Wiener processes stopped at the instant t, then $\lceil \boldsymbol{W} \rceil_{\mathcal{I}^2}$ equals $\sqrt{d{\cdot}t}$ rather than the first-impulse estimate $d\sqrt{t}$. Lest the gentle reader think us too nitpicking, let us point out that this definition of the integrator size is instrumental in establishing previsible control of random measures in theorem 4.5.25 on page 251, control which in turn greatly facilitates the solution of differential equations driven by random measures (page 296).

Definition 2.2.9 *A vector $\boldsymbol{Z}$ of adapted processes is an L^p-integrator if its components are right-continuous in probability and its $\mathcal{I}^p$-size $\lceil \boldsymbol{Z}^t \rceil_{\mathcal{I}^p}$ is finite for all $t < \infty$.*

Exercise 2.2.10 $\mathcal{E}^d$ is a self-confined algebra and vector lattice closed under chopping of bounded functions, and the vector $\boldsymbol{Z}$ of càdlàg adapted processes is an L^p-integrator if and only if the map $\boldsymbol{X} \mapsto \int \boldsymbol{X}\, d\boldsymbol{Z}$ is continuous from $\mathcal{E}^d$ equipped with the topology of confined uniform convergence (see item A.2.5) to L^p.

The Natural Conditions

The notion of an L^p-integrator depends on the filtration. If Z is an L^p-integrator with respect to the given filtration $\mathcal{F}_.$ and we change every $\mathcal{F}_t$ to a larger σ-algebra $\mathcal{G}_t$, then Z will still be adapted and right-continuous in probability – these features do not mention the filtration. But doing so will generally increase the supply of elementary integrands, so that now Z

[4] See equation (1.1.9) on page 8 or equation (5.1.3) on page 271 and section 3.10 on page 171.

[5] We shall use the ***Einstein convention*** throughout: summation over repeated indices *in opposite positions* (the η in (2.2.2)) is implied.

has a harder time satisfying the boundedness condition (B-p). Namely, since $\mathcal{E}[\mathcal{F}_.] \subset \mathcal{E}[\mathcal{G}_.]$,

the collection $$\left\{\int X\, dZ : X \in \mathcal{E}[\mathcal{G}_.],\ |X| \le 1\right\}$$

is larger than $$\left\{\int X\, dZ : X \in \mathcal{E}[\mathcal{F}_.],\ |X| \le 1\right\};$$

and while the latter is bounded in L^p, the former need not be. However, a slight enlargement is innocuous:

Proposition 2.2.11 *Suppose that Z is an $L^p(\mathbb{P})$-integrator on $\mathcal{F}_.$ for some $p \in [0, \infty)$. Then Z is an $L^p(\mathbb{P})$-integrator on the natural enlargement $\mathcal{F}^{\mathbb{P}}_{.+}$, and the sizes $\lceil Z^t \rceil_{\mathcal{I}^p}$ computed on $\mathcal{F}^{\mathbb{P}}_{.+}$ are at most twice what they are computed on $\mathcal{F}_.$ – if $Z_0 = 0$, they are the same.*

Proof. Let $\mathcal{E}^{\mathbb{P}} = \mathcal{E}[\mathcal{F}^{\mathbb{P}}_{.+}]$ denote the elementary integrands for the natural enlargement and set

$$\mathcal{B} \stackrel{\text{def}}{=} \left\{\int X\, dZ^t : X \in \mathcal{E}_1\right\} \quad \text{and} \quad \mathcal{B}^{\mathbb{P}} \stackrel{\text{def}}{=} \left\{\int X\, dZ^t : X \in \mathcal{E}^{\mathbb{P}}_1\right\}.$$

$\mathcal{B}$ is a bounded subset of L^p, and so is its "solid closure"

$$\mathcal{B}^{\diamond} \stackrel{\text{def}}{=} \{f \in L^p : |f| \le |g| \ \text{ for some } \ g \in \mathcal{B}\}.$$

We shall show that $\mathcal{B}^{\mathbb{P}}$ is contained in $\overline{\mathcal{B}^{\diamond}} + \overline{\mathcal{B}^{\diamond}}$, where $\overline{\mathcal{B}^{\diamond}}$ is the closure of $\mathcal{B}^{\diamond}$ in the topology of convergence in measure; the claim is then immediate from this consequence of solidity and Fatou's lemma A.8.7:

$$\sup\left\{\lceil\!\lceil f \rceil\!\rceil_p : f \in \overline{\mathcal{B}^{\diamond}} + \overline{\mathcal{B}^{\diamond}}\right\} \le 2 \sup\left\{\lceil\!\lceil f \rceil\!\rceil_p : f \in \mathcal{B}\right\}.$$

Let then $X \in \mathcal{E}^{\mathbb{P}}_1$, writing it as in equation (2.1.1):

$$X = f_0 \cdot [\![0]\!] + \sum_{n=1}^{N} f_n \cdot (\!(t_n, t_{n+1}]\!], \qquad f_n \in \mathcal{F}^{\mathbb{P}}_{t_n+}.$$

For every $n \in \mathbb{N}$ there is a simple random variable $f'_n \in \mathcal{F}_{t_n+}$ that differs negligibly from f_n. Let k be so large that $t_n + 1/k < t_{n+1}$ for all n and set

$$X^{(k)} \stackrel{\text{def}}{=} f'_0 \cdot [\![0]\!] + \sum_{n=1}^{N} f'_n \cdot (\!(t_n + 1/k, t_{n+1}]\!], \qquad k \in \mathbb{N}.$$

The sum on the right clearly belongs to $\mathcal{E}_1$, so its stochastic integral

$$f'_0 \cdot Z_0 + \sum f'_n \cdot \left(Z_{t_{n+1}} - Z_{t_n+1/k}\right)$$

belongs to $\mathcal{B}$. The first random variable $f'_0 Z_0$ is majorized in absolute value by $|Z_0| = \left|\int [\![0]\!]\, dZ\right|$ and thus belongs to $\mathcal{B}^{\diamond}$. Therefore $\int X^{(k)}\, dZ$ lies in the sum $\mathcal{B}^\diamond + \mathcal{B} \subset \overline{\mathcal{B}^\diamond} + \overline{\mathcal{B}^\diamond}$. As $k \to \infty$ these stochastic integrals on $\mathcal{E}$ converge in probability to

$$f_0' \cdot Z_0 + \sum f_n' \cdot (Z_{t_{n+1}} - Z_{t_n}) = \int X\, dZ ,$$

which therefore belongs to $\overline{\mathcal{B}^\diamond} + \overline{\mathcal{B}^\diamond}$.

Recall from exercise 1.3.30 that a regular and right-continuous filtration has more stopping times than just a plain filtration. We shall therefore make our life easy and replace the given measured filtration by its natural enlargement:

Assumption 2.2.12 ***The given measured filtration*** $(\mathcal{F}_., \mathbb{P})$ ***is henceforth assumed to be both right-continuous and regular.***

Exercise 2.2.13 On Wiener space $(\mathscr{C}, \mathcal{B}^\bullet(\mathscr{C}), \mathbb{W})$ consider the canonical Wiener process $w_.$ (w_t takes a path $w \in \mathscr{C}$ to its value at t). The $\mathbb{W}$-regularization of the basic filtration $\mathcal{F}_.^0[w]$ is right-continuous (see exercise 1.3.47): it is the natural filtration $\mathcal{F}_.[w]$ of w. Then the triple $(\mathscr{C}, \mathcal{F}_.[w], \mathbb{W})$ is an instance of a measured filtration that is right-continuous and regular. $w : (t, w) \mapsto w_t$ is a continuous process, adapted to $\mathcal{F}_.[w]$ and p-integrable for all $p \geq 0$, but not L^p-bounded for any $p \geq 0$.

Exercise 2.2.14 Let $\mathcal{A}'$ denote the ring on $\boldsymbol{B}$ generated by $\{[\![0_A]\!] : A \in \mathcal{F}_0\}$ and the collection $\{(\!(S, T]\!] : S, T \text{ bounded stopping times}\}$ of stochastic intervals, and let $\mathcal{E}'$ denote the step functions over $\mathcal{A}'$. Clearly $\mathcal{A} \subset \mathcal{A}'$ and $\mathcal{E} \subset \mathcal{E}'$. Every $X \in \mathcal{E}'$ can be written in the form

$$X = f_0 \cdot [\![0]\!] + \textstyle\sum_{n=0}^N f_n \cdot (\!(T_n, T_{n+1}]\!] ,$$

where $0 = T_0 \leq T_1 \leq \ldots \leq T_{N+1}$ are bounded stopping times and $f_n \in \mathcal{F}_{T_n}$ are simple. If Z is a global L^p-integrator, then the definition

$$\int X\, dZ \overset{\text{def}}{=} f_0 \cdot Z_0 + \textstyle\sum_n f_n \cdot (Z_{T_{n+1}} - Z_{T_n}) \qquad (*)$$

provides an extension of the elementary integral that has the same modulus of continuity. Any extension of the elementary integral that satisfies the Dominated Convergence Theorem must have a domain containing $\mathcal{E}'$ and coincide there with $(*)$.

2.3 Path Regularity of Integrators

Suppose Z, Z' are modifications of each other, that is to say, $Z_t = Z_t'$ almost surely at every instant t. An inspection of (2.1.2) then shows that for every elementary integrand X the random variables $\int X\, dZ$ and $\int X\, dZ'$ nearly coincide: as integrators, Z and Z' are the same and should be identified. It is shown in this section that from all of the modifications one can be chosen that has rather regular paths, namely, càdlàg ones.

Right-Continuity and Left Limits

Lemma 2.3.1 *Suppose Z is a process adapted to $\mathcal{F}$ that is $\mathcal{I}^0[\mathbb{P}]$-bounded on bounded intervals. Then the paths whose restrictions to the positive rationals have an oscillatory discontinuity occur in a $\mathbb{P}$-nearly empty set.*

Proof. Fix two rationals a, b with $a < b$, an instant $u < \infty$, and a finite set

$$\mathcal{S} = \{s_0 < s_1 < \ldots < s_N\}$$

of consecutive rationals in $[0, u)$. Next set $T_0 \overset{\text{def}}{=} \min\{s \in \mathcal{S} : Z_s < a\} \wedge u$ and continue by induction:

$$T_{2k+1} = \inf\{s \in \mathcal{S} : s > T_{2k}, Z_s > b\} \wedge u$$
$$T_{2k} = \inf\{s \in \mathcal{S} : s > T_{2k-1}, Z_s < a\} \wedge u .$$

It was shown in proposition 1.3.13 that these are stopping times, evidently elementary. ($T_n(\omega)$ will be equal to u for some index $n(\omega)$ and all higher ones, but let that not bother us.) Let us now estimate the number $U_{\mathcal{S}}^{[a,b]}$ of upcrossings of the interval $[a, b]$ that the path of Z performs on $\mathcal{S}$. (We say that $\mathcal{S} \ni s \mapsto Z_s(\omega)$ *upcrosses* the interval $[a, b]$ on $\mathcal{S}$ if there are points $s < t$ in $\mathcal{S}$ with $Z_s(\omega) < a$ and $Z_t(\omega) > b$. To say that this path has n ***upcrossings*** means that there are n pairs: $s_1 < t_1 < s_2 < t_2 < \ldots < s_n < t_n$ in $\mathcal{S}$ with $Z_{s_\nu} < a$ and $Z_{t_\nu} > b$.) If $\mathcal{S} \ni s \mapsto Z_s(\omega)$ upcrosses the interval $[a, b]$ n times or more on $\mathcal{S}$, then $T_{2n-1}(\omega)$ is strictly less than u, and vice versa:

$$\left[U_{\mathcal{S}}^{[a,b]} \geq n\right] = [T_{2n-1} < u] \in \mathcal{F}_u . \tag{2.3.1}$$

This observation produces the inequality[2]

$$\left[U_{\mathcal{S}}^{[a,b]} \geq n\right] \leq \frac{1}{n(b-a)} \left(\sum_{k=0}^{\infty} (Z_{T_{2k+1}} - Z_{T_{2k}}) + |Z_u - a| \right), \tag{2.3.2}$$

for if $U_{\mathcal{S}}^{[a,b]} \geq n$, then the (finite!) sum on the right contributes more than n times a number greater than $b - a$. The last term of the sum might be negative, however. This occurs when $T_{2k}(\omega) < s_N$ and thus $Z_{T_{2k}}(\omega) < a$, and $T_{2k+1}(\omega) = u$ because there is no more $s \in \mathcal{S}$ exceeding $T_{2k}(\omega)$ with $Z_s(\omega) > b$. The last term of the sum is then $Z_u(\omega) - Z_{T_{2k}}(\omega)$. This number might well be negative. However, it will not be less than $Z_u(\omega) - a$: the last term $\left|Z_u(\omega) - a\right|$ of (2.3.2) added to the last non-zero term of the sum will always be positive.

The stochastic intervals $(\!(T_{2k}, T_{2k+1}]\!]$ are elementary integrands, and their integrals are $Z_{T_{2k+1}} - Z_{T_{2k}}$. This observation permits us to rewrite (2.3.2) as

$$\left[U_{\mathcal{S}}^{[a,b]} \geq n\right] \leq \frac{1}{n(b-a)} \left(\int \sum_{k=0}^{\infty} (\!(T_{2k}, T_{2k+1}]\!] \, dZ^u + |Z_u - a| \right) . \tag{2.3.3}$$

This inequality holds for any adapted process Z. To continue the estimate observe now that the integrand $\sum_{k=1}^{\infty} (\!(T_{2k}, T_{2k+1}]\!]$ is majorized in absolute value by 1. Measuring both sides of (2.3.3) with $\lceil\!\lceil \ \rceil\!\rceil_0$ yields the inequality

$$\begin{aligned} \mathbb{P}\left[U_{\mathcal{S}}^{[a,b]} \geq n\right] &= \left\lceil\!\left\lceil \left[U_{\mathcal{S}}^{[a,b]} \geq n\right] \right\rceil\!\right\rceil_{L^0(\mathbb{P})} \\ &\leq \lceil\!\lceil 1/n(b-a) \rceil\!\rceil_{Z^u\text{–}0;\mathbb{P}} + \left\lceil\!\left\lceil (a - Z_u)/n(b-a) \right\rceil\!\right\rceil_{L^0(\mathbb{P})} \\ &\leq 2 \cdot \lceil\!\lceil 1/n(b-a) \rceil\!\rceil_{Z^u\text{–}0;\mathbb{P}} + |a|/n(b-a) . \end{aligned}$$

Now let $\mathbb{Q}_+^{u-}$ denote the set of positive rationals less than u. The right-hand side of the previous inequality does not depend on $\mathcal{S} \subset \mathbb{Q}_+^{u-}$. Taking the supremum over all finite subsets $\mathcal{S}$ of $\mathbb{Q}_+^{u-}$ results, in obvious notation, in the inequality

$$\mathbb{P}\Big[U_{\mathbb{Q}_+^{u-}}^{[a,b]} \geq n\Big] \leq 2 \cdot \lceil\!\lceil 1/n(b-a) \rceil\!\rceil_{Z^u\text{–}0;\mathbb{P}} + a/n(b-a)\ .$$

Note that the set on the left belongs to $\mathcal{F}_u$ (equation (2.3.1)). Since Z is assumed $\mathcal{I}^0$-bounded on $[0,u]$, taking the limit as $n \to \infty$ gives

$$\mathbb{P}\Big[U_{\mathbb{Q}_+^{u-}}^{[a,b]} = \infty\Big] = 0\ .$$

That is to say, the restriction to $\mathbb{Q}_+^{u-}$ of nearly no path upcrosses the interval $[a,b]$ infinitely often. The set

$$Osc = \bigcup_{u \in \mathbb{N}} \ \bigcup_{a,b \in \mathbb{Q},\, a<b} \Big[U_{\mathbb{Q}_+^{u}}^{[a,b]} = \infty\Big]$$

therefore belongs to $\mathcal{A}_{\infty\sigma}$ and is $\mathbb{P}$-negligible: it is a $\mathbb{P}$-nearly empty set. If ω is in its complement, then the path $t \mapsto Z_t(\omega)$ restricted to the rationals has no oscillatory discontinuity. ∎

The upcrossing argument above is due to Doob, who used it to show the regularity of martingales (see proposition 2.5.13).

Let Ω_0 be the complement of Osc. By our standing regularity assumption 2.2.12 on the filtration, Ω_0 belongs to $\mathcal{F}_0$. The path $Z_.(\omega)$ of the process Z of lemma 2.3.1 has for every $\omega \in \Omega_0$ left and right limits through the rationals at any time t. We may define

$$Z'_t(\omega) = \begin{cases} \lim_{\mathbb{Q} \ni q \downarrow t} Z_q(\omega) & \text{for } \omega \in \Omega_0, \\ 0 & \text{for } \omega \in Osc. \end{cases}$$

The limit is understood in the extended reals $\overline{\mathbb{R}}$, since nothing said so far prevents it from being $\pm\infty$. The process Z' is right-continuous and has left limits at any finite instant.

Assume now that Z is an L^0-integrator. Since then Z is right-continuous in probability, Z and Z' are modifications of one another. Indeed, for fixed t let q_n be a sequence of rationals decreasing to t. Then $Z_t = \lim_n Z_{q_n}$ in measure, in fact nearly, since $\big[Z_t \neq \lim_n Z_{q_n}\big] \in \mathcal{F}_{q_1}$. On the other hand, $Z'_t = \lim_n Z_{q_n}$ nearly, by definition. Thus $Z_t = Z'_t$ nearly for all $t < \infty$. Since the filtration satisfies the natural conditions, Z' is adapted. Any other right-continuous modification of Z is indistinguishable from Z' (exercise 1.3.28).

Note that these arguments apply to any probability with respect to which Z is an L^p-integrator: Osc is nearly empty for every one of them. Let $\mathfrak{P}[Z]$

denote their collection. The version Z' that we found is thus "universally regular" in the sense that it is adapted to the "small enlargement"

$$\mathcal{F}_{.+}^{\mathfrak{P}[Z]} \stackrel{\text{def}}{=} \bigcap \{\mathcal{F}_{.+}^{\mathbb{P}} : \mathbb{P} \in \mathfrak{P}[Z]\} .$$

Denote by $\mathfrak{P}_0[Z]$ the class of probabilities under which Z is actually a global $L^0(\mathbb{P})$-integrator. If $\mathbb{P} \in \mathfrak{P}_0[Z]$, then we may take ∞ for the time u of the proof and see that the paths of Z that have an oscillatory discontinuity anywhere, including at ∞, are negligible. In other words, then $Z'_\infty \stackrel{\text{def}}{=} \lim_{t\uparrow\infty} Z'_t$ exists, except possibly in a set that is $\mathbb{P}$-negligible simultaneously for all $\mathbb{P} \in \mathfrak{P}_0[Z]$.

Boundedness of the Paths

For the remainder of the section we shall assume that a modification of the L^0-integrator Z has been chosen that is right-continuous and has left limits at all finite times and is adapted to $\mathcal{F}_{.+}^{\mathfrak{P}[Z]}$. So far it is still possible that this modification takes the values $\pm\infty$ frequently. The following ***maximal inequality*** of weak type rules out this contingency, however:

Lemma 2.3.2 *Let T be any stopping time and $\lambda > 0$. The maximal process $Z^\star$ of Z satisfies, for every $\mathbb{P} \in \mathfrak{P}[Z]$,*

$$\mathbb{P}[Z_T^\star \geq \lambda] \leq \lceil Z^T/\lambda \rceil_{\mathcal{I}^0[\mathbb{P}]} , \qquad p = 0;$$

$$\|Z_T^\star\|_{[\alpha]} \leq \lceil Z^T \rceil_{[\alpha;\mathbb{P}]} , \qquad p = 0, \alpha \in \mathbb{R};$$

$$\mathbb{P}[Z_T^\star \geq \lambda] \leq \lambda^{-p} \cdot \lceil Z^T \rceil_{\mathcal{I}^p[\mathbb{P}]}^p , \qquad 0 < p < \infty.$$

Proof. We resurrect our finite set $\mathcal{S} = \{s_0 < s_1 < \ldots < s_N\}$ of consecutive positive rationals strictly less than u and define

$$U = \inf\{s \in \mathcal{S} : |Z_s^T| > \lambda\} \wedge u .$$

This is an elementary stopping time (proposition 1.3.13). Now

$$\Big[\sup_{s\in\mathcal{S}} |Z_s^T| > \lambda\Big] = [U < u] \in \mathcal{F}_u ,$$

on which set $$|Z_U^T| = \Big|\int [\![0,U]\!]\, dZ^T\Big| > \lambda .$$

Applying $[\![\ ]\!]_p$ to the resulting inequality

$$\Big[\sup_{s\in\mathcal{S}} |Z_s^T| > \lambda\Big] \leq \lambda^{-1}\Big|\int [\![0,U]\!]\, dZ^T\Big|$$

gives $$\Big[\!\Big[\Big[\sup_{s\in\mathcal{S}} |Z_s^T| > \lambda\Big]\Big]\!\Big]_p \leq \Big[\!\Big[\lambda^{-1}\int [\![0,U]\!]\, dZ^T\Big]\!\Big]_p \leq \lceil \lambda^{-1} Z^T \rceil_{\mathcal{I}^p} .$$

We observe that the ultimate right does not depend on $\mathcal{S} \subset \mathbb{Q}_+ \cap [0,u)$. Taking the supremum over $\mathcal{S} \subset \mathbb{Q}_+ \cap [0,u)$ therefore gives

$$\left\| \left[\sup_{s<u} |Z_s^T| > \lambda \right] \right\|_p = \left\| \left[\sup_{s\in\mathbb{Q}_+, s<u} |Z_s^T| > \lambda \right] \right\|_p \leq \lceil \lambda^{-1} Z^T \rceil_{\mathcal{I}^p} . \tag{2.3.4}$$

Letting $u \to \infty$ yields the stated inequalities (see exercise A.8.3). ∎

Exercise 2.3.3 Let $\boldsymbol{Z} = (Z^1, \ldots, Z^d)$ be a vector of L^0-integrators. The maximal process of its euclidean length

$$|\boldsymbol{Z}|_t \stackrel{\text{def}}{=} \Big(\sum_{1\leq\eta\leq d} (Z_t^\eta)^2 \Big)^{1/2}$$

satisfies
$$\big\| |\boldsymbol{Z}|_t^\star \big\|_{[\alpha]} \leq K_0^{(A.8.6)} \cdot \lceil \boldsymbol{Z} \rceil_{[\alpha\kappa_0]} , \qquad 0 < \alpha < 1 .$$

(See theorem 2.3.6 on page 63 for the case $p > 0$ and a hint.)

Redefinition of Integrators

Note that the set $[Z_u^{\star T} > \lambda]$ on the left in inequality (2.3.4) belongs to $\mathcal{F}_u \in \mathcal{A}_\infty$. Therefore

$$N \stackrel{\text{def}}{=} [Z_T^\star = \infty] \cap [T < \infty] = \bigcup_{u\in\mathbb{N}} [Z_u^{\star T} = \infty]$$

is a $\mathbb{P}$-negligible set of $\mathcal{A}_{\infty\sigma}$; it is $\mathbb{P}$-nearly empty. This is true for all $\mathbb{P} \in \mathfrak{P}[Z]$. We now alter Z by setting it equal to zero on N. Since $\mathcal{F}_.$ is assumed to be right-continuous and regular, we obtain an adapted right-continuous modification of Z whose paths are real-valued, in fact bounded on bounded intervals. The upshot:

Theorem 2.3.4 *Every L^0-integrator Z has a modification all of whose paths are right-continuous, have left limits at every finite instant, and are bounded on every finite interval. Any two such modifications are indistinguishable. Furthermore, this modification can be chosen adapted to $\mathcal{F}_{.+}^{\mathfrak{P}[Z]}$. Its limit at infinity exists and is $\mathbb{P}$-almost surely finite for all $\mathbb{P}$ under which Z is a global L^0-integrator.*

Convention 2.3.5 *Whenever an L^p-integrator Z on a regular right-continuous filtration appears it will henceforth be understood that a right-continuous real-valued modification with left limits has been chosen, adapted to $\mathcal{F}_{.+}^{\mathfrak{P}[Z]}$ as it can be.*

Since a local L^p-integrator Z is an L^0-integrator (proposition 2.1.9), it is also understood to have càdlàg paths and to be adapted to $\mathcal{F}_{.+}^{\mathfrak{P}[Z]}$.

In remark 3.8.5 we shall meet a further regularity property of the paths of an integrator Z; namely, while the sums $\sum_k |Z_{T_k} - Z_{T_{k-1}}|$ may diverge as the random partition $\{0 = T_1 \leq T_2 \leq \ldots \leq T_K = t\}$ of $[\![0,t]\!]$ is refined, the sums $\sum_k |Z_{T_k} - Z_{T_{k-1}}|^2$ of squares stay bounded, even converge.

The Maximal Inequality

The last "weak type" inequality in lemma 2.3.2 can be replaced by one of "strong type," which holds even for a whole vector

$$\boldsymbol{Z} = (Z^1, \ldots, Z^d)$$

of L^p-integrators and extends the result of exercise 2.3.3 for $p = 0$ to strictly positive p. The ***maximal process*** of $\boldsymbol{Z}$ is the d-tuple of increasing processes

$$\boldsymbol{Z}^\star = (Z^{\eta\star})_{\eta=1}^d \stackrel{\text{def}}{=} (|Z^\eta|^\star)_{\eta=1}^d .$$

Theorem 2.3.6 *Let $0 < p < \infty$ and let $\boldsymbol{Z}$ be an L^p-integrator. The euclidean length $|\boldsymbol{Z}^\star|$ of its maximal process satisfies $|\boldsymbol{Z}|^\star \le |\boldsymbol{Z}^\star|$ and*

$$\big\| |\boldsymbol{Z}^\star|_t \big\|_{L^p} = \big\lceil |\boldsymbol{Z}^\star|^t \big\rceil_{\mathcal{I}^p} \le C_p^\star \cdot \big\lceil \boldsymbol{Z}^t \big\rceil_{\mathcal{I}^p} , \tag{2.3.5}$$

with universal constant $C_p^\star \le \dfrac{10}{3} \cdot K_p^{(A.8.5)} \le 3.35 \cdot 2^{\frac{2-p}{2p} \vee 0}$.

Proof. Let $\mathcal{S} = \{0 = s_0 < s_1 < \ldots < t\}$ be a finite partition of $[0, t]$ and pick a $q > 1$. For $\eta = 1 \ldots d$, set $T_0^\eta = -1$, $Z_{-1}^\eta = 0$, and define inductively $T_1^\eta = 0$ and

$$T_{n+1}^\eta = \inf\{s \in \mathcal{S} : s > T_n^\eta \text{ and } |Z_s^\eta| > q \big| Z_{T_n^\eta}^\eta \big| \} \wedge t .$$

These are elementary stopping times, only finitely many distinct. Let N^η be the *last* index n such that $|Z_{T_n^\eta}^\eta| > |Z_{T_{n-1}^\eta}^\eta|$. Clearly $\sup_{s \in \mathcal{S}} |Z_s^\eta| \le q |Z_{T_{N^\eta}^\eta}^\eta|$. Now $\omega \mapsto T_{N^\eta(\omega)}^\eta$ is not a stopping time, inasmuch as one has to check $\boldsymbol{Z}_t$ at instants t later than $T_{N^\eta}^\eta$ in order to determine whether $T_{N^\eta}^\eta$ has arrived. This unfortunate fact necessitates a slightly circuitous argument.

Set $\zeta_0^\eta = 0$ and $\zeta_n^\eta = \big| Z_{T_n^\eta}^\eta \big|$ for $n = 1, \ldots, N^\eta$. Since $\zeta_n^\eta \ge q \zeta_{n-1}^\eta$ for $1 \le n \le N^\eta$,

$$\zeta_{N^\eta} \le L_q \left(\sum_{n=1}^{N^\eta} (\zeta_n^\eta - \zeta_{n-1}^\eta)^2 \right)^{1/2} ;$$

we leave it to the reader to show by induction that this holds when the choice $L_q^2 \stackrel{\text{def}}{=} (q+1)/(q-1)$ is made.

Since
$$\sup_{s \in \mathcal{S}} |Z_s^\eta| \le q \Big| Z_{T_{N^\eta}^\eta}^\eta \Big| = q\, \zeta_{N^\eta}^\eta \le q L_q \left(\sum_{n=1}^{N^\eta} (\zeta_n^\eta - \zeta_{n-1}^\eta)^2 \right)^{1/2} \tag{2.3.6}$$

$$\le q L_q \left(\sum_{n=1}^{\infty} \left(Z_{T_n^\eta}^\eta - Z_{T_{n-1}^\eta}^\eta \right)^2 \right)^{1/2} ,$$

the quantity $\zeta^{\mathcal{S}} \stackrel{\text{def}}{=} \left\| \left(\sum_{\eta=1}^d \big(\sup_{s \in \mathcal{S}} |Z_s^\eta| \big)^2 \right)^{1/2} \right\|_{L^p}$

satisfies, thanks to the Khintchine inequality of theorem A.8.26,

$$\zeta^S \le qL_q \left\| \left(\sum_{\eta=1}^{d} \sum_{n=1}^{\infty} \left| Z^\eta_{T^\eta_n} - Z^\eta_{T^\eta_{n-1}} \right|^2 \right)^{1/2} \right\|_{L^p(\mathbb{P})}$$

$$\le qL_q K_p^{(A.8.5)} \left\| \left\| \sum_{n,\eta} \left(Z^\eta_{T^\eta_n} - Z^\eta_{T^\eta_{n-1}} \right) \epsilon_{n,\eta}(\tau) \right\|_{L^p(d\tau)} \right\|_{L^p(\mathbb{P})}$$

by Fubini:
$$= qL_q K_p \left\| \left\| \sum_{n,\eta} \left(Z^\eta_{T^\eta_n} - Z^\eta_{T^\eta_{n-1}} \right) \epsilon_{n,\eta}(\tau) \right\|_{L^p(\mathbb{P})} \right\|_{L^p(d\tau)}$$

$$= qL_q K_p \left\| \left\| \int \sum_{\eta} \Big(\sum_{n} (\!(T^\eta_{n-1}, T^\eta_n]\!]\, \epsilon_{n,\eta}(\tau) \Big)\, dZ^\eta \right\|_{L^p(\mathbb{P})} \right\|_{L^p(d\tau)}$$

$$\le qL_q K_p \left\| \lceil \boldsymbol{Z}^t \rceil_{\mathcal{I}^p} \right\|_{L^p(d\tau)} = qL_q K_p \lceil \boldsymbol{Z}^t \rceil_{\mathcal{I}^p} .$$

In the penultimate line $(\!(T^\eta_0, T^\eta_1]\!]$ stands for $[\![0]\!]$. The sums are of course really finite, since no more summands can be non-zero than $\mathcal{S}$ has members. Taking now the supremum over all finite partitions of $[0,t]$ results, in view of the right-continuity of Z, in $\| \, |\boldsymbol{Z}^\star|_t \|_{L^p} \le qL_q K_p \lceil \boldsymbol{Z}^t \rceil_{\mathcal{I}^p}$. The constant qL_q is minimal for the choice $q = (1+\sqrt{5})/2$, where it equals $(q+1)/\sqrt{q-1} \le 10/3$. Lastly, observe that for a positive increasing process I, $I = |\boldsymbol{Z}^\star|$ in this case, the supremum in the definition of $\lceil I^t \rceil_{\mathcal{I}^p}$ on page 55 is assumed at the elementary integrand $[\![0,t]\!]$, where it equals $\| I^t \|_{L^p}$. This proves the equality in (2.3.5); since $|\boldsymbol{Z}^\star|$ is plainly right-continuous, it is an L^p-integrator. ❏

Exercise 2.3.7 The absolute value $|Z|$ of an L^p-integrator Z is an L^p-integrator, and

$$\lceil |Z|^t \rceil_{\mathcal{I}^p} \le 3 \lceil Z^t \rceil_{\mathcal{I}^p} , \qquad 0 \le p < \infty, 0 \le t \le \infty .$$

Consequently, $\mathcal{I}^p$ forms a vector lattice under pointwise operations.

Law and Canonical Representation

2.3.8 Adapted Maps between Filtered Spaces Let $(\Omega, \mathcal{F}_.)$ and $(\overline{\Omega}, \overline{\mathcal{F}}_.)$ be filtered probability spaces. We shall say that a map $\underline{R} : \Omega \to \overline{\Omega}$ is ***adapted to*** $\mathcal{F}_.$ ***and*** $\overline{\mathcal{F}}_.$ if $\underline{R}$ is $\mathcal{F}_t/\overline{\mathcal{F}}_t$-measurable at all instants t. This amounts to saying that for all t

$$\underline{\mathcal{F}}_t \stackrel{\text{def}}{=} \underline{R}^{-1}(\overline{\mathcal{F}}_t) = \overline{\mathcal{F}}_t \circ \underline{R}\ ^2 \tag{2.3.7}$$

is a sub-σ-algebra of $\mathcal{F}_t$. Occasionally we call such a map $\underline{R}$ a ***morphism of filtered spaces*** or a ***representation*** of $(\Omega, \mathcal{F}_.)$ on $(\overline{\Omega}, \overline{\mathcal{F}}_.)$, the idea being that it forgets unwanted information and leaves only the "aspect of interest" $(\overline{\Omega}, \overline{\mathcal{F}}_.)$. With such $\underline{R}$ comes naturally the map $(t,\omega) \mapsto (t, \underline{R}(\omega))$ of the base space of Ω to the base space of $\overline{\Omega}$. We shall denote this map by $\underline{R}$ as well; this won't lead to confusion.

The following facts are obvious or provide easy exercises:

(i) If the process $\overline{X}$ on $\overline{\Omega}$ is left-continuous (right-continuous, càdlàg, continuous, of finite variation), then $X \stackrel{\text{def}}{=} \overline{X} \circ \underline{R}$ has the same property on Ω.

(ii) If $\overline{T}$ is an $\overline{\mathcal{F}}_.$-stopping time, then $T \stackrel{\text{def}}{=} \overline{T} \circ \underline{R}$ is an $\underline{\mathcal{F}}_.$-stopping time. If the process $\overline{X}$ is adapted (progressively measurable, an elementary integrand) on $(\overline{\Omega}, \overline{\mathcal{F}}_.)$, then $X \stackrel{\text{def}}{=} \overline{X} \circ \underline{R}$ is adapted (progressively measurable, an elementary integrand) on $(\Omega, \underline{\mathcal{F}}_.)$ and on $(\Omega, \mathcal{F}_.)$. $\overline{X}$ is predictable[6] on $(\overline{\Omega}, \overline{\mathcal{F}}_.)$ if and only if X is predictable on $(\Omega, \underline{\mathcal{F}}_.)$; it is then predictable on $(\Omega, \mathcal{F}_.)$.

(iii) If a probability $\mathbb{P}$ on $\underline{\mathcal{F}}_\infty \subset \mathcal{F}_\infty$ is given, then the image of $\mathbb{P}$ under $\underline{R}$ provides a probability $\overline{\mathbb{P}}$ on $\overline{\mathcal{F}}_\infty$. In this way the whole slew $\mathfrak{P}$ of pertinent probabilities gives rise to the pertinent probabilities $\overline{\mathfrak{P}}$ on $(\overline{\Omega}, \overline{\mathcal{F}}_\infty)$.

Suppose $\overline{Z}_.$ is a càdlàg process on $\overline{\Omega}$. Then $\overline{Z}$ is an $L^p(\overline{\mathbb{P}})$-integrator on $(\overline{\Omega}, \overline{\mathcal{F}}_.)$ if and only if $X \stackrel{\text{def}}{=} \overline{Z} \circ \underline{R}$ is an $L^p(\mathbb{P})$-integrator on $(\Omega, \underline{\mathcal{F}}_.)$.[7] To see this let $\underline{\mathcal{E}}$ denote the elementary integrands for the filtration $\underline{\mathcal{F}}_. \stackrel{\text{def}}{=} \underline{R}^{-1}(\overline{\mathcal{F}}_.)$. It is easily seen that $\underline{\mathcal{E}} = \overline{\mathcal{E}} \circ \underline{R}$, in obvious notation, and that the collections of random variables

$$\left\{ \int \overline{X}\, d\overline{Z} : \overline{X} \in \overline{\mathcal{E}}, |\overline{X}| \le \overline{Y} \right\} \quad \text{and} \quad \left\{ \int X\, dZ : X \in \underline{\mathcal{E}}, |X| \le Y \right\}$$

upon being measured with $\lceil\!\lceil \ \rceil\!\rceil^*_{L^p(\overline{\mathbb{P}})}$ and $\lceil\!\lceil \ \rceil\!\rceil^*_{L^p(\mathbb{P})}$, respectively, produce the same sets of numbers when $Y = \overline{Y} \circ R$. The equality of the suprema reads

$$\left\lceil\!\left\lceil \overline{Y} \right\rceil\!\right\rceil_{\overline{Z}\text{–}p;\overline{\mathbb{P}}} = \lceil\!\lceil Y \rceil\!\rceil_{Z\text{–}p;\mathbb{P}} \tag{2.3.8}$$

for $Y = \overline{Y} \circ \underline{R}$, $\overline{\mathbb{P}} = \underline{R}[\mathbb{P}]$, and $Z = \overline{Z} \circ \underline{R}$ considered as an integrator on $(\Omega, \underline{\mathcal{F}}_.)$.[7]

Let us then henceforth forget information that may be present in $\mathcal{F}_.$ but not in $\underline{\mathcal{F}}_.$, by replacing the former filtration with the latter. That is to say,

$$\mathcal{F}_t = \underline{R}^{-1}(\overline{\mathcal{F}}_t) = \overline{\mathcal{F}}_t \circ \underline{R} \quad \forall t \ge 0\,, \quad \text{and then } \mathcal{E} = \overline{\mathcal{E}} \circ \underline{R}\,.$$

Once the integration theory of Z and $\overline{Z}$ is established in chapter 3, the following further facts concerning a process X of the form $X = \overline{X} \circ \underline{R}$ will be obvious:

(iv) X is previsible with $\mathbb{P}$ if and only if $\overline{X}$ is previsible with $\overline{\mathbb{P}}$.

(v) X is Z–p;$\mathbb{P}$-integrable if and only if $\overline{X}$ is $\overline{Z}$–p;$\overline{\mathbb{P}}$-integrable, and then

$$(\overline{\boldsymbol{X}}{*}\overline{Z})_. \circ \underline{R} = (\boldsymbol{X}{*}\boldsymbol{Z})_.\,. \tag{2.3.9}$$

(vi) X is Z-measurable if and only if $\overline{X}$ is $\overline{Z}$-measurable. Any Z-measurable process differs Z-negligibly from a process of this form.

---

[6] A process is predictable if it belongs to the sequential closure of the elementary integrands – see section 3.5.

[7] Note the underscore! One cannot expect in general that Z be an $L^p(\mathbb{P})$-integrator, i.e., be bounded on the potentially much larger space $\mathcal{E}$ of elementary integrands for $\mathcal{F}_.$.

2.3.9 Canonical Path Space In algebra one tries to get insight into the structure of an object by representing it with morphisms on objects of the same category that have additional structure. For example, groups get represented on matrices or linear operators, which one can also add, multiply with scalars, and measure by size. In a similar vein[8] the typical target space of a representation is a space of paths, which usually carries a topology and may even have a linear structure:

Let (E,ρ) be some polish space. $\mathscr{D}_E$ denotes the set of all càdlàg paths $x_. : [0,\infty) \to E$. If $E = \mathbb{R}$, we simply write $\mathscr{D}$; if $E = \mathbb{R}^d$, we write $\mathscr{D}^d$. A path in $\mathscr{D}^d$ is identified with a path on $(-\infty,\infty)$ that vanishes on $(-\infty,0)$.

A natural topology on $\mathscr{D}_E$ is the topology τ of uniform convergence on bounded time-intervals; it is given by the complete metric

$$d(x_.,y_.) \stackrel{\text{def}}{=} \sum_{n\in\mathbb{N}} 2^{-n} \wedge \rho(x_.,y_.)^\star_n \qquad x_.,y_. \in \mathscr{D}_E ,$$

where

$$\rho(x_.,y_.)^\star_t \stackrel{\text{def}}{=} \sup_{0\le s\le t} \rho(x_s,y_s) .$$

The maximal theorem 2.3.6 shows that this topology is pertinent. Yet its Borel σ-algebra is rarely useful; it is too fine. Rather, it is the ***basic filtration*** $\mathcal{F}^0_.[\mathscr{D}_E]$, generated by the right-continuous ***evaluation process***

$$\overline{R}_s : x_. \mapsto x_s , \qquad 0 \le s < \infty , x_. \in \mathscr{D}_E ,$$

and its right-continuous version $\mathcal{F}^0_{.+}[\mathscr{D}_E]$ that play a major role. The final σ-algebra $\mathcal{F}^0_\infty[\mathscr{D}_E]$ of the basic filtration coincides with the Baire σ-algebra of the topology σ of pointwise convergence on $\mathscr{D}_E$. On the space $\mathscr{C}_E$ of continuous paths the σ-algebras generated by σ and τ coincide (generalize equation (1.2.5)).

The right-continuous version $\mathcal{F}^0_{.+}[\mathscr{D}_E]$ of the basic filtration will also be called the ***canonical filtration***. The space $\mathscr{D}_E$ equipped with the topology τ [9] and its canonical filtration $\mathcal{F}^0_{.+}[\mathscr{D}_E]$ is ***canonical path space***.[10]

Consider now a càdlàg adapted E-valued process R on $(\Omega,\mathcal{F}_.)$. Just as a Wiener process was considered as a random variable with values in canonical path space $\mathscr{C}$ (page 14), so can now our process R be regarded as a map $\underline{R}$ from Ω to path space $\mathscr{D}_E$, the image of an $\omega \in \Omega$ under $\underline{R}$ being the path $R_.(\omega) : t \mapsto R_t(\omega)$. Since R is assumed adapted, $\underline{R}$ represents $(\Omega,\mathcal{F}_.)$ on path space $(\mathscr{D}_E,\mathcal{F}^0_.[\mathscr{D}_E])$ in the sense of item 2.3.8. If $\mathcal{F}_.$ is right-continuous, then $\underline{R}$ represents $(\Omega,\mathcal{F}_.)$ on canonical path space $(\mathscr{D}_E,\mathcal{F}^0_{.+}[\mathscr{D}_E])$. We call $\underline{R}$ the ***canonical representation*** of R on path space.

---

[8] I hope that the reader will find a little farfetchedness more amusing than offensive.

[9] A glance at theorems 2.3.6, 4.5.1, and A.4.9 will convince the reader that τ is most pertinent, despite the fact that it is not polish and that its Borels *properly* contain the pertinent σ-algebra $\mathcal{F}_\infty$.

[10] "Path space", like "frequency space" or "outer space," may be used without an article.

If $(\Omega, \mathcal{F}_.)$ carries a distinguished probability $\mathbb{P}$, then the ***law*** of the process R is of course nothing but the image $\overline{\mathbb{P}} \stackrel{\text{def}}{=} \underline{R}[\mathbb{P}]$ of $\mathbb{P}$ under $\underline{R}$. The triple $(\mathscr{D}_E, \mathcal{F}^0_{.+}[\mathscr{D}_E], \overline{\mathbb{P}})$ carries all statistical information about the process R – which now "is" the evaluation process $\overline{R}_.$ – and has forgotten all other information that might have been available on $(\Omega, \mathcal{F}_., \mathbb{P})$.

2.3.10 Integrators on Canonical Path Space Suppose that E comes equipped with a distinguished slew $\boldsymbol{z} = (z^1, \ldots, z^d)$ of continuous functions. Then $t \mapsto \overline{Z}_t \stackrel{\text{def}}{=} \boldsymbol{z} \circ \overline{R}_t$ is a distinguished adapted $\mathbb{R}^d$-valued process on the path space $(\mathscr{D}_E, \mathcal{F}^0_.[\mathscr{D}_E], \overline{\mathbb{P}})$. These data give rise to the collection $\mathfrak{P}[\overline{Z}]$ of all probabilities on path space for which $\overline{Z}$ is an integrator. We may then define the ***natural filtration*** on $\mathscr{D}_E$: it is the regularization of $\mathcal{F}^0_{.+}[\mathscr{D}_E]$, taken for the collection $\mathfrak{P}[\overline{Z}]$, and it is denoted by $\mathcal{F}_.[\mathscr{D}_E]$ or $\mathcal{F}_.[\mathscr{D}_E; \boldsymbol{z}]$.

2.3.11 Canonical Representation of an Integrator Suppose that we face an integrator $\boldsymbol{Z} = (Z^1, \ldots, Z^d)$ on $(\Omega, \mathcal{F}_., \mathfrak{P})$ and a collection $C = (C^1, C^2, \ldots)$ of real-valued processes, certain functions f_η of which we might wish to integrate with $\boldsymbol{Z}$, say. We glob the data together in the obvious way into a process $R_t \stackrel{\text{def}}{=} (C_t, \boldsymbol{Z}_t) : \Omega \to E \stackrel{\text{def}}{=} \mathbb{R}^{\mathbb{N}} \times \mathbb{R}^d$, which we identify with a map $\underline{R} : \Omega \to \mathscr{D}_E$. "$\underline{R}$ forgets all information except the aspect of interest $(C, \boldsymbol{Z})$." Let us write $\overline{\omega}_. = (c^\nu_., \boldsymbol{z}^\eta_.)$ for the generic point of $\overline{\Omega} = \mathscr{D}_E$. On E there are the distiguished last d coordinate functions $z^1, \ldots, z^d$. They give rise to the distinguished process $\overline{Z} : t \mapsto \big(z^1(\overline{\omega}_t), \ldots, z^d(\overline{\omega}_t)\big)$. Clearly the image under $\underline{R}$ of any probability in $\mathfrak{P} \subset \mathfrak{P}[\boldsymbol{Z}]$ makes $\overline{Z}$ into an integrator on path space.[10] The integral $\int_0^t f_\eta[\overline{\omega}_.]_s \, d\overline{Z}^\eta_s(\overline{\omega}_.)$, which is frequently and with intuitive appeal written as

$$\int_0^t f_\eta[c_., \boldsymbol{z}_.]_s \, dz^\eta , \tag{2.3.10}$$

then equals $\int_0^t f_\eta[C_., \boldsymbol{Z}_.]_s \, dZ^\eta_s$, after composition with $\underline{R}$, that is, and after information beyond $\underline{\mathcal{F}}_.$ has been discarded.[7] In other words,

$$\boldsymbol{X}*\boldsymbol{Z} = (\overline{\boldsymbol{X}*\boldsymbol{Z}}) \circ \underline{R} . \tag{2.3.11}$$

In this way we arrive at the ***canonical representation*** $\underline{R}$ of $(C, \boldsymbol{Z})$ on $(\mathscr{D}_E, \mathcal{F}_.[\mathscr{D}_E])$ with pertinent probabilities $\overline{\mathfrak{P}} \stackrel{\text{def}}{=} \underline{R}[\mathfrak{P}]$. For an application see page 316.

2.4 Processes of Finite Variation

Recall that a process V has ***bounded variation*** if its paths are functions of bounded variation on the half-line, i.e., if the number

$$\vert V \vert_\infty(\omega) = |V_0| + \sup\Big\{\sum_{i=1}^{I} |V_{t_{i+1}}(\omega) - V_{t_i}(\omega)|\Big\}$$

is finite for every $\omega \in \Omega$. Here the supremum is taken over all finite partitions $\mathcal{T} = \{t_1 < t_2 < \ldots < t_{I+1}\}$ of $\mathbb{R}_+$. V has ***finite variation*** if the stopped processes V^t have bounded variation, at every instant t. In this case the variation process $\vert V\vert$ of V is defined by

$$\vert V\vert_t(\omega) = |V_0(\omega)| + \sup_{\mathcal{T}} \left\{ \sum_{i=1}^{I} |V_{t\wedge t_{i+1}}(\omega) - V_{t\wedge t_i}(\omega)| \right\} . \tag{2.4.1}$$

The integration theory of processes of finite variation can of course be handled path-by-path. Yet it is well to see how they fit in the general framework.

Proposition 2.4.1 *Suppose V is an adapted right-continuous process of finite variation. Then $\vert V\vert$ is adapted, increasing, and right-continuous with left limits. Both V and $\vert V\vert$ are L^0-integrators.*

If $\vert V\vert_t \in L^p$ at all instants t, then V is an L^p-integrator. In fact, for $0 \le p < \infty$ and $0 \le t \le \infty$

$$\lceil\!\lceil [\![0,t]\!] \rceil\!\rceil_{V\text{-}p} = \lceil V^t \rceil_{\mathcal{I}^p} \le \left\lceil\!\left\lceil \vert V\vert_t \right\rceil\!\right\rceil_p . \tag{2.4.2}$$

Proof. Due to the right-continuity of V, taking the partition points t_i of equation (2.4.1) in the set $\mathbb{Q}^t = (\mathbb{Q} \cap [0,t]) \cup \{t\}$ will result in the same path $t \mapsto \vert V\vert_t(\omega)$; and since the collection of finite subsets of $\mathbb{Q}^t$ is countable, the process $\vert V\vert$ is adapted. For every $\omega \in \Omega$, $t \mapsto \vert V\vert_t(\omega)$ is the cumulative distribution function of the variation $\vert dV\vert(\omega)$ of the scalar measure $dV_.(\omega)$ on the half-line. It is therefore right-continuous (exercise 2.2). Next, for $X \in \mathcal{E}_1$ as in equation (2.1.1) we have

$$\begin{aligned} \left|\int X\, dV^t\right| &= \left|f_0 \cdot V_0 + \sum\nolimits_n f_n \cdot (V^t_{t_{n+1}} - V^t_{t_n})\right| \\ &\le |V_0| + \sum\nolimits_n \left|V^t_{t_{n+1}} - V^t_{t_n}\right| \le \vert V\vert_t . \end{aligned}$$

We apply $\lceil\!\lceil\ \rceil\!\rceil_p$ to this and obtain inequality (2.4.2). ∎

Our adapted right-continuous process of finite variation therefore can be written as the difference of two adapted increasing right-continuous processes $V^\pm$ of finite variation: $V = V^+ - V^-$ with

$$V_t^+ = 1/2\left(\vert V\vert_t + V_t\right) , \; V_t^- = 1/2\left(\vert V\vert_t - V_t\right) .$$

It suffices to analyze *increasing* adapted right-continuous processes I.

Remark 2.4.2 The reverse of inequality (2.4.2) is not true in general, nor is it even true that $\vert V\vert_t \in L^p$ if V is an L^p-integrator, except if $p = 0$. The reason is that the collection $\mathcal{E}$ is too small; testing V against its members is not enough to determine the variation of V, which can be written as

$$\vert V\vert_t = |V_0| + \sup \int_0^t \operatorname{sgn}(V_{t_{i+1}} - V_{t_i})\, dV .$$

Note that the integrand here is not elementary inasmuch as $(V_{t_{i+1}} - V_{t_i}) \notin \mathcal{F}_{t_i}$. However, in (2.4.2) equality holds if V is previsible (exercise 4.3.13) or increasing. Example 2.5.26 on page 79 exhibits a sequence of processes whose variation grows beyond all bounds yet whose $\mathcal{I}^2$-norms stay bounded.

Exercise 2.4.3 Prove the right-continuity of $\vert V\vert$ directly.

Decomposition into Continuous and Jump Parts

A measure μ on $[0,\infty)$ is the sum of a measure ${}^c\mu$ that does not charge points and an atomic measure ${}^j\mu$ that is carried by a countable collection $\{t_1, t_2, \ldots\}$ of points. The cumulative distribution function[11] of ${}^c\mu$ is continuous and that of ${}^j\mu$ constant, except for jumps at the times t_n, and the cumulative distribution function of μ is the sum of these two. All of this is classical, and every path of an increasing right-continuous process can be decomposed in this way. In the stochastic case we hope that the continuous and jump components are again adapted, and this is indeed so; also, the times of the jumps of the discontinuous part are not too wildly scattered:

Theorem 2.4.4 *A positive increasing adapted right-continuous process I can be written uniquely as the sum of a continuous increasing adapted process ${}^cI$ that vanishes at 0 and a right-continuous increasing adapted process ${}^jI$ of the following form: there exist a countable collection $\{T_n\}$ of stopping times with bounded disjoint graphs,*[12] *and bounded positive $\mathcal{F}_{T_n}$-measurable functions f_n, such that*

$$ {}^jI = \sum_n f_n \cdot [\![T_n, \infty)\!) \ . $$

Proof. For every $i \in \mathbb{N}$ define inductively $T^{i,0} = 0$ and

$$ T^{i,j+1} = \inf\left\{t > T^{i,j} : \Delta I_t \geq 1/i\right\} \ . $$

From proposition 1.3.14 we know that the $T^{i,j}$ are stopping times. They increase a.s. strictly to ∞ as $j \to \infty$; for if $T = \sup_j T^{i,j} < \infty$, then $I_t = \infty$ after T. Next let $T^{i,j}_k$ denote the reduction of $T^{i,j}$ to the set

$$ [\Delta I_{T^{i,j}} \leq k+1] \cap [T^{i,j} \leq k] \in \mathcal{F}_{T^{i,j}} \ . $$

(See exercises 1.3.18 and 1.3.16.) Every one of the $T^{i,j}_k$ is a stopping time with a bounded graph. The jump of I at time $T^{i,j}_k$ is bounded, and the set $[\Delta I \neq 0]$ is contained in the union of the graphs of the $T^{i,j}_k$. Moreover, the collection $\{T^{i,j}_k\}$ is countable; so let us count it: $\{T^{i,j}_k\} = \{T'_1, T'_2, \ldots\}$. The T'_n do not have disjoint graphs, of course. We force the issue by letting T_n be the reduction of T'_n to the set $\bigcup_{m<n}[T'_n \neq T'_m] \in \mathcal{F}_{T'_n}$ (exercise 1.3.16). It is plain upon inspection that with $f_n = \Delta I_{T_n}$ and ${}^cI = I - {}^jI$ the statement is met. ∎

Exercise 2.4.5 ${}^jI_t = \sum_{s\leq t} \Delta I_s = \sum_n \sum_{s\leq t} f_n \cdot [T_n = s]$.

Exercise 2.4.6 Call a subset S of the base space ***sparse*** if it is contained in the union of the graphs of countably many stopping times. Such stopping times can be chosen to have disjoint graphs; and if S is measurable, then it actually equals the union of the disjoint graphs of countably many stopping times (use theorem A.5.10).

---

[11] See page 406.

[12] We say that T has a ***bounded graph*** if $[\![T]\!] \subset [\![0, t]\!]$ for some finite instant t.

Next let V be an adapted càdlàg process of finite variation. Then V is the sum $V = {}^cV + {}^jV$ of two adapted càdlàg processes of finite variation, of which ${}^cV$ has continuous paths and $d^jV = S \cdot d^jV$ with $S \stackrel{\text{def}}{=} [\Delta V \neq 0] = [\Delta^j V \neq 0]$ sparse. For more see exercise 4.3.4.

The Change-of-Variable Formula

Theorem 2.4.7 *Let I be an adapted positive increasing right-continuous process and $\Phi : [0, \infty) \to \mathbb{R}_+$ a continuously differentiable function. Set*

$$T^\lambda = \inf\{t : I_t \geq \lambda\} \quad \textit{and} \quad T^{\lambda+} = \inf\{t : I_t > \lambda\}\,, \qquad \lambda \in \mathbb{R}\,.$$

Both $\{T^\lambda\}$ and $\{T^{\lambda+}\}$ form increasing families of stopping times, $\{T^\lambda\}$ left-continuous and $\{T^{\lambda+}\}$ right-continuous. For every bounded measurable process X [13]

$$\int_{[0}^{\infty} X_s \, d\Phi(I_s) = \int_0^\infty X_{T^\lambda} \cdot \Phi'(\lambda) \cdot [T^\lambda < \infty] \, d\lambda \tag{2.4.3}$$

$$= \int_0^\infty X_{T^{\lambda+}} \cdot \Phi'(\lambda) \cdot [T^{\lambda+} < \infty] \, d\lambda\,. \tag{2.4.4}$$

Proof. Thanks to proposition 1.3.11 the T^λ are stopping times and are increasing and left-continuous in λ. Exercise 1.3.30 yields the corresponding claims for $T^{\lambda+}$. $T^\lambda < T^{\lambda+}$ signifies that $I = \lambda$ on an interval of strictly positive length. This can happen only for countably many different λ. Therefore the right-hand sides of (2.4.3) and (2.4.4) coincide.

To prove (2.4.3), say, consider the family $\mathcal{M}$ of bounded measurable processes X such that for all finite instants u

$$\int [\![0, u]\!] \cdot X \, d\Phi(I) = \int_0^\infty X_{T^\lambda} \cdot \Phi'(\lambda) \cdot [T^\lambda \leq u] \, d\lambda\,. \tag{?}$$

$\mathcal{M}$ is clearly a vector space closed under pointwise limits of bounded sequences. For processes X of the special form

$$X = f \cdot [0, t]\,, \qquad f \in L^\infty(\mathcal{F}_\infty)\,, \tag{$*$}$$

the left-hand side of (?) is simply

$$f \cdot \big(\Phi(I_{t\wedge u}) - \Phi(I_{0-})\big) = f \cdot \big(\Phi(I_{t\wedge u}) - \Phi(0)\big)$$

---

[13] Recall from convention A.1.5 that $[T^\lambda < \infty]$ equals 1 if $T^\lambda < \infty$ and 0 otherwise. Indicator function aficionados read these integrals as $\int_0^\infty X_{T^\lambda} \cdot \Phi'(\lambda) \cdot 1_{[T^\cdot < \infty]}(\lambda) \, d\lambda$, etc.

and the right-hand side is[13]

$$
\begin{aligned}
& f \cdot \int_0^\infty [0,t](T^\lambda) \cdot \Phi'(\lambda) \cdot [T^\lambda \le u] \, d\lambda \\
=& f \cdot \int_0^\infty [T^\lambda \le t] \cdot \Phi'(\lambda) \cdot [T^\lambda \le u] \, d\lambda \\
=& f \cdot \int_0^\infty [T^\lambda \le t \wedge u] \cdot \Phi'(\lambda) \, d\lambda \\
=& f \cdot \int_0^\infty [\lambda \le I_{t\wedge u}] \cdot \Phi'(\lambda) \, d\lambda = f \cdot \big(\Phi(I_{t\wedge u}) - \Phi(0)\big)
\end{aligned}
$$

as well. That is to say, $\mathcal{M}$ contains the processes of the form $(*)$, and also the constant process 1 (choose $f \equiv 1$ and $t \ge u$). The processes of the form $(*)$ generate the measurable processes and so, in view of theorem A.3.4 on page 393, $(*)$ holds for all bounded measurable processes. Equation (2.4.3) follows upon taking u to ∞. ∎

Exercise 2.4.8 $I_t = \inf\{\lambda : T^\lambda > t\} = \int [\![T^\lambda, t]\!] \, d\lambda = \int [\![T^\lambda, \infty)\!)_t \, d\lambda$ (see convention A.1.5). A stochastic interval $[\![T, \infty)\!)$ is an increasing adapted process (ibidem). Equation (2.4.3) can thus be read as saying that $\Phi(I)$ is a "continuous superposition" of such simple processes:

$$
\Phi(I) = \int_0^\infty \Phi'(\lambda) [\![T^\lambda, \infty)\!) \, d\lambda \, .
$$

Exercise 2.4.9 (i) If the right-continuous adapted process I is strictly increasing, then $T^\lambda = T^{\lambda+}$ for every $\lambda \ge 0$; in general, $\{\lambda : T^\lambda < T^{\lambda+}\}$ is countable.

(ii) Suppose that $T^{\lambda+}$ is nearly finite for all λ and $\mathcal{F}_.$ meets the natural conditions. Then $(\mathcal{F}_{T^{\lambda+}})_{\lambda\ge 0}$ inherits the natural conditions; if Λ is an $\mathcal{F}_{T^{.+}}$-stopping time, then $T^{\Lambda+}$ is an $\mathcal{F}_.$-stopping time.

Exercise 2.4.10 Equations (2.4.3) and (2.4.4) hold for measurable processes X whenever one or the other side is finite.

Exercise 2.4.11 If $T^{\lambda+} < \infty$ almost surely for all λ, then the filtration $(\mathcal{F}_{T^{\lambda+}})_\lambda$ inherits the natural conditions from $\mathcal{F}_.$.

2.5 Martingales

Definition 2.5.1 *An integrable process M is an $(\mathcal{F}_., \mathbb{P})$-**martingale** if*[14]

$$
\mathbb{E}^{\mathbb{P}}[M_t | \mathcal{F}_s] = M_s
$$

*for $0 \le s < t < \infty$. We also say that M is a $\mathbb{P}$-**martingale on** $\mathcal{F}_.$, or simply a **martingale** if the filtration $\mathcal{F}_.$ and probability $\mathbb{P}$ meant are clear from the context.*

Since the conditional expectation above is unique only up to $\mathbb{P}$-negligible and $\mathcal{F}_s$-measurable functions, the equation should be read "M_s is a (one of very many) conditional expectation of M_t given $\mathcal{F}_s$."

[14] $\mathbb{E}^{\mathbb{P}}[M_t | \mathcal{F}_s]$ is the conditional expectation of M_t given $\mathcal{F}_s$ – see theorem A.3.24 on page 407.

A martingale on $\mathcal{F}_.$ is clearly adapted to $\mathcal{F}_.$. The martingales form a class of integrators that is complementary to the class of finite variation processes – in a sense that will become clearer as the story unfolds – and that is much more challenging. The name "martingale" seems to derive from the part of a horse's harness that keeps the beast from throwing up its head and thus from rearing up; the term has also been used in gambling for centuries. The defining equality for a martingale says this: given the whole history $\mathcal{F}_s$ of the game up to time s, the gambler's fortune at time $t > s$, $\$M_t$, is expected to be just what she has at time s, namely, $\$M_s$; in other words, she is engaged in a fair game. Roughly, martingales are processes that show, on the average, no drift (see the discussion on page 4).

The class of L^0-integrators is rather stable under changes of the probability (proposition 2.1.9), but the class of martingales is not. It is rare that a process that is a martingale with respect to one probability is a martingale with respect to an equivalent or otherwise pertinent measure. For instance, if the dice in a fair game are replaced by loaded ones, the game will most likely cease to be fair, that being no doubt the object of the replacement. Therefore we will fix a probability $\mathbb{P}$ on $\mathcal{F}_\infty$ throughout this section. $\mathbb{E}$ is understood to be the expectation $\mathbb{E}^{\mathbb{P}}$ with respect to $\mathbb{P}$.

Example 2.5.2 Here is a frequent construction of martingales. Let g be an integrable random variable, and set $M_t^g = \mathbb{E}[g|\mathcal{F}_t]$, the conditional expectation of g given $\mathcal{F}_t$. Then M^g is a uniformly integrable martingale – it is shown in exercise 2.5.14 that all uniformly integrable martingales are of this form. It is an easy exercise to establish that the collection

$$\{\mathbb{E}[g|\mathcal{G}] : \mathcal{G} \text{ a sub-}\sigma\text{-algebra of } \mathcal{F}_\infty\}$$

of random variables is uniformly integrable.

Exercise 2.5.3 Suppose M is a martingale. Then $\mathbb{E}[f \cdot (M_t - M_s)] = 0$ for $s < t$ and any $f \in L^\infty(\mathcal{F}_s)$. Next assume M is ***square integrable***: $M_t \in L^2(\mathcal{F}_t, \mathbb{P})$ $\forall t$. Then

$$\mathbb{E}[(M_t - M_s)^2|\mathcal{F}_s] = \mathbb{E}[M_t^2 - M_s^2|\mathcal{F}_s], \qquad 0 \le s < t < \infty.$$

Exercise 2.5.4 If W is a Wiener process on the filtration $\mathcal{F}_.$, then it is a martingale on $\mathcal{F}_.$ and on the natural enlargement of $\mathcal{F}_.$, and so are $W_t^2 - t$, and $e^{zW_t - z^2t/2}$ for any $z \in \mathbb{C}$.

Exercise 2.5.5 Let $(\Omega, \mathcal{F}, \mathbb{P})$ be a probability space and $\mathfrak{F}$ a collection of sub-σ-algebras of $\mathcal{F}$ that is increasingly directed. That is to say, for any two $\mathcal{F}_1, \mathcal{F}_2 \in \mathfrak{F}$ there is a σ-algebra $\mathcal{G} \in \mathfrak{F}$ containing both $\mathcal{F}_1$ and $\mathcal{F}_2$. Let $g \in L^1(\mathcal{F}, \mathbb{P})$ and for $\mathcal{G} \in \mathfrak{F}$ set $g^{\mathcal{G}} \overset{\text{def}}{=} \mathbb{E}[f|\mathcal{G}]$. The collection $\{g^{\mathcal{G}} : \mathcal{G} \in \mathfrak{F}\}$ is uniformly integrable and converges in L^1-mean to the conditional expectation of g with respect to the σ-algebra $\bigvee \mathfrak{F} \subset \mathcal{F}$ generated by $\mathfrak{F}$.

Exercise 2.5.6 Let $(\Omega, \mathcal{F}, \mathbb{P})$ be a probability space and $f, f' : \Omega \to \overline{\mathbb{R}}_+$ two $\mathcal{F}$-measurable functions such that $[f \le q] = [f' \le q]$ $\mathbb{P}$-almost surely for all rationals q. Then $f = f'$ $\mathbb{P}$-almost surely.

Submartingales and Supermartingales

The martingales are the processes of primary interest in the sequel. It eases their analysis though to introduce the following generalizations. An integrable process Z adapted to $\mathcal{F}_.$ is a ***submartingale*** (***supermartingale***) if

$$\mathbb{E}\left[Z_t|\mathcal{F}_s\right] \geq Z_s \text{ a.s. } (\leq Z_s \text{ a.s., respectively}), \qquad 0 \leq s \leq t < \infty .$$

Exercise 2.5.7 The fortune of a gambler in Las Vegas is a supermartingale, that of the casino is a submartingale.

Since the absolute-value function $|\cdot|$ is convex, it follows immediately from Jensen's inequality in A.3.24 that the absolute value of a martingale M is a submartingale:

$$\left|M_s\right| = \left|\mathbb{E}[M_t|\mathcal{F}_s]\right| \leq \mathbb{E}\Big[\left|M_t\right| |\mathcal{F}_s\Big] \text{ a.s.}, \quad 0 \leq s < t < \infty .$$

Taking expectations, $\qquad \mathbb{E}[|M_s|] \leq \mathbb{E}[|M_t|] \qquad 0 \leq s < t < \infty$

follows. This argument lends itself to some generalizations:

Exercise 2.5.8 Let $\boldsymbol{M} = (M^1, \ldots, M^d)$ be a vector of martingales and $\Phi : \mathbb{R}^d \to \mathbb{R}$ convex and so that $\Phi(\boldsymbol{M}_t)$ is integrable for all t. Then $\Phi(\boldsymbol{M})$ is a submartingale. Apply this with $\Phi(\boldsymbol{x}) = |\boldsymbol{x}|_p$ to conclude that $t \mapsto |\boldsymbol{M}_t|_p$ is a submartingale for $1 \leq p \leq \infty$ and that $t \mapsto \|M_t^1\|_{L^p}$ is increasing if M^1 is p-integrable (bounded if $p = \infty$).

Exercise 2.5.9 A martingale M on $\mathcal{F}_.$ is a martingale on its own basic filtration $\mathcal{F}^0_.[M]$. Similar statements hold for sub– and supermartingales.

Here is a characterization of martingales that gives a first indication of the special role they play in stochastic integration:

Proposition 2.5.10 *The adapted integrable process M is a martingale (submartingale, supermartingale) on $\mathcal{F}_.$ if and only if*

$$\mathbb{E}\Big[\int X\, dM\Big] = 0 \quad (\geq 0,\ \leq 0,\ \textit{respectively})$$

for every positive elementary integrand X that vanishes on $[\![0]\!]$.

Proof. $\Longrightarrow$: A glance at equation (2.1.1) shows that we may take X to be of the form $X = f \cdot (\!(s,t]\!]$, with $0 \leq s < t < \infty$ and $f \geq 0$ in $L^\infty(\mathcal{F}_s)$. Then

$$\begin{aligned}
\mathbb{E}\Big[\int X\, dM\Big] &= \mathbb{E}[f \cdot (M_t - M_s)] = \mathbb{E}[f \cdot M_t] - \mathbb{E}[f \cdot M_s] \\
&= \mathbb{E}\Big[f \cdot \Big(\mathbb{E}[M_t|\mathcal{F}_s] - M_s\Big)\Big] \\
&= (\geq, \leq)\ \mathbb{E}\Big[f \cdot (M_s - M_s)\Big] = 0 .
\end{aligned}$$

$\Longleftarrow$: If, on the other hand,

$$\mathbb{E}\Big[\int X\, dM\Big] = \mathbb{E}\Big[f \cdot \Big(\mathbb{E}[M_t|\mathcal{F}_s] - M_s\Big)\Big] = (\geq, \leq)\ 0$$

for all $f \geq 0$ in $L^\infty(\mathcal{F}_s)$, then $\mathbb{E}\big[M_t|\mathcal{F}_s\big] - M_s = (\geq, \leq)\ 0$ almost surely, and M is a martingale (submartingale, supermartingale). ∎

Corollary 2.5.11 *Let M be a martingale (submartingale, supermartingale). Then for any two elementary stopping times $S \le T$ we have nearly*

$$\mathbb{E}\big[M_T - M_S|\mathcal{F}_S\big] = 0 \ \ (\ge 0, \ \le 0) \,.$$

Proof. We show this for submartingales. Let $A \in \mathcal{F}_S$ and consider the reduced stopping times S_A, T_A. From equation (2.1.3) and proposition 2.5.10

$$0 \le \mathbb{E}\Big[\int (\!(S_A, T_A]\!]\, dM\Big] = \mathbb{E}\Big[M_{T_A} - M_{S_A}\Big] = \mathbb{E}\Big[\big(M_T - M_S\big)\cdot 1_A\Big]$$
$$= \mathbb{E}\Big[\Big(\mathbb{E}\big[M_T|\mathcal{F}_S\big] - M_S\Big)\cdot 1_A\Big] \,.$$

As $A \in \mathcal{F}_S$ was arbitrary, this shows that $\mathbb{E}\big[M_T|\mathcal{F}_S\big] \ge M_S$, except in a negligible set of $\mathcal{F}_T \subset \mathcal{F}_{\max T}$. ∎

Exercise 2.5.12 (i) An adapted integrable process M is a martingale (submartingale, supermartingale) if and only if $\mathbb{E}\,[M_T - M_S] = 0$ $(\ge 0,\ \le 0)$ for any two elementary stopping times $S \le T$.

(ii) The infimum of two supermartingales is a supermartingale.

Regularity of the Paths: Right-Continuity and Left Limits

Consider the estimate (2.3.3) of the number of upcrossings of the interval $[a, b]$ that the path of our martingale M performs on the finite subset

$$S = \{s_0 < s_1 < \ldots < s_N\} \ \text{ of } \ \mathbb{Q}_+^{u-} \stackrel{\text{def}}{=} \mathbb{Q}_+ \cap [0, u) :$$

$$\Big[U_S^{[a,b]} \ge n\Big] \le \frac{1}{n(b-a)}\left(\int \sum_{k=0}^{\infty} (\!(T_{2k}, T_{2k+1}]\!]\, dM \ + \ \big|M_u - a\big|\right) .$$

Applying the expectation and proposition 2.5.10 we obtain an estimate of the probability that this number exceeds $n \in \mathbb{N}$:

$$\mathbb{P}\left[U_S^{[a,b]} \ge n\right] \le \frac{1}{n(b-a)} \cdot \mathbb{E}\big[\,|M_u| + |a|\,\big] \,.$$

Taking the supremum over all finite subsets $S \subset \mathbb{Q}_+^{u-}$ and then over all $n \in \mathbb{N}$ and all pairs of rationals $a < b$ shows as on page 60 that the set

$$Osc \stackrel{\text{def}}{=} \bigcup_n \ \bigcup_{a,b\in\mathbb{Q}, a<b} \left[U_{\mathbb{Q}_+^{n-}}^{[a,b]} = \infty\right]$$

belongs to $\mathcal{A}_{\infty\sigma}$ and is $\mathbb{P}$-nearly empty. Let Ω_0 be its complement and define

$$M'_t(\omega) = \begin{cases} \lim\limits_{\mathbb{Q}\ni q\downarrow t} M_q(\omega) & \text{for } \omega \in \Omega_0, \\ 0 & \text{for } \omega \in Osc. \end{cases}$$

The limit is understood in the extended reals $\overline{\mathbb{R}}$ as nothing said so far prevents it from being $\pm\infty$. The process M' is right-continuous and has left limits at any finite instant. If M is right-continuous in probability, then clearly $M_t = M'_t$ nearly for all t. The same is true when the filtration is right-continuous. To see this notice that $M_q \to M'_t$ in $\| \|_1$-mean as $\mathbb{Q} \ni q \downarrow t$, since the collection $\{M_q : q \in \mathbb{Q}, t < q < t+1\}$ is uniformly integrable (example 2.5.2 and theorem A.8.6); both M_t and M'_t are measurable on $\mathcal{F}_t = \bigcap_{\mathbb{Q}\ni q>t} \mathcal{F}_q$ and have the same integral over every set $A \in \mathcal{F}_t$:

$$\int_A M_t \, d\mathbb{P} = \int_A M_q \, d\mathbb{P} \xrightarrow[t<q\to t]{} \int_A M'_t \, d\mathbb{P} .$$

That is to say, M' is a modification of M.

Consider the case that M is L^1-bounded. Then we can take ∞ for the time u of the proof and see that the paths of M that have an oscillatory discontinuity anywhere, including at ∞, are negligible. In other words, then

$$M'_\infty \stackrel{\text{def}}{=} \lim_{t\uparrow\infty} M'_t$$

exists almost surely.

A ***local martingale*** is, of course, a process that is locally a martingale. Localizing with a sequence T_n of stopping times that reduce M to uniformly integrable martingales, we arrive at the following conclusion:

Proposition 2.5.13 *Let M be a local $\mathbb{P}$-martingale on the filtration $\mathcal{F}_.$. If M is right-continuous in probability or $\mathcal{F}_.$ is right-continuous, then M has a modification adapted to the $\mathbb{P}$-regularization $\mathcal{F}^{\mathbb{P}}_.$, one all of whose paths are right-continuous and have left limits. If M is L^1-bounded, then this modification has almost surely a limit $M_\infty \in \overline{\mathbb{R}}$ at infinity.*

Exercise 2.5.14 M also has a modification that is adapted to $\mathcal{F}_.$ and whose paths are nearly right-continuous. If M is uniformly integrable, then it is L^1-bounded and is a modification of the martingale M^g of example 2.5.2, where $g = M_\infty$.

Example 2.5.15 The positive martingale M_t of example 1.3.45 on page 41 converges at every point, the limit being $+\infty$ at zero and zero elsewhere. It is L^1-bounded but not uniformly integrable, and its value at time t is not $\mathbb{E}[M_\infty|\mathcal{F}_t]$.

Exercise 2.5.16 Doob considered martingales first in discrete time: let $\{\mathcal{F}_n : n \in \mathbb{N}\}$ be an increasing collection of sub-σ-algebras of $\mathcal{F}$. The random variables M_n, $n \in \mathbb{N}$ form a ***martingale*** on $\{\mathcal{F}_n\}$ if $\mathbb{E}[F_{n+1}|\mathcal{F}_n] = F_n$ almost surely for all $n \ge 1$. He developed the upcrossing argument to show that an L^1-bounded martingale converges almost surely to a limit in the extended reals $\overline{\mathbb{R}}$ as $n \to \infty$, in $\mathbb{R}$ if $\{M_n\}$ is uniformly integrable.

Exercise 2.5.17 (A Strong Law of Large Numbers) The previous exercise allows a simple proof of an uncommonly strong version of the strong law of large numbers. Namely, let $F_1, F_2, \ldots$ be a sequence of square integrable random variables that all have expectation p and whose variances all are bounded by some σ^2. Assume that the conditional expectation of F_{n+1} given $F_1, F_2, \ldots, F_n$ equals p as well, for $n = 1, 2, 3, \ldots$. [To paraphrase: knowledge of previous executions of the experiment may influence the law of its current replica only to the extent that the expectation does not change and the variance does not increase overly much.] Then

$$\lim_{n\to\infty} \frac{1}{n}\sum_{\nu=1}^{n} F_\nu = p$$

almost surely. See exercise 4.2.14 for a generalization to the case that the F_n merely have bounded moments of order q for some $q > 1$ and [80] for the case that the random variables F_n are merely orthogonal in L^2.

Boundedness of the Paths

Lemma 2.5.18 (Doob's Maximal Lemma) *Let M be a right-continuous martingale. Then at any instant t and for any $\lambda > 0$*

$$\mathbb{P}\Big[M_t^\star > \lambda\Big] \le \frac{1}{\lambda}\int_{[M_t^\star > \lambda]} |M_t| \, d\mathbb{P} \le \frac{1}{\lambda} \cdot \mathbb{E}[|M_t|] \, .$$

Proof. Let $S = \{s_0 < s_1 < \ldots < s_N\}$ be a finite set of rationals contained in the interval $[0,t]$, let $u > t$, and set

$$M^S = \sup_{s\in S} |M_s| \quad \text{and} \quad U = \inf\big\{s \in S : |M_s| > \lambda\big\} \wedge u \, .$$

Clearly U is an elementary stopping time and $|M_U| = |M_{U\wedge t}| > \lambda$ on

$$[U < u] = [U \le t] = \Big[M^S > \lambda\Big] \subset \big[M_t^\star > \lambda\big] \, .$$

Therefore
$$1_{\big[M^S > \lambda\big]} \le \frac{|M_U|}{\lambda} \cdot 1_{\big[U\le t\big]} \in \mathcal{F}_t \, .$$

We apply the expectation; since $|M|$ is a submartingale,

$$\mathbb{P}\Big[M^S > \lambda\Big] \le \lambda^{-1} \cdot \int_{[U\le t]} |M_U| \, d\mathbb{P} = \lambda^{-1} \cdot \int_{[U\wedge t\le t]} |M_U| \, d\mathbb{P}$$

by corollary 2.5.11:
$$\le \lambda^{-1} \cdot \int_{[U\wedge t \le t]} \mathbb{E}\big[|M_t| \bigm| \mathcal{F}_{U\wedge t}\big] \, d\mathbb{P}$$

$$= \lambda^{-1} \cdot \int_{[M^S > \lambda]} |M_t| \, d\mathbb{P} \le \lambda^{-1} \cdot \int_{[M_t^\star > \lambda]} |M_t| \, d\mathbb{P} \, .$$

We take the supremum over all finite subsets S of $\{t\} \cup (\mathbb{Q} \cap [0,t])$ and use the right-continuity of M: Doob's inequality follows. ∎

Theorem 2.5.19 (Doob's Maximal Theorem) *Let M be a right-continuous martingale on $(\Omega, \mathcal{F}_., \mathbb{P})$ and p, p' conjugate exponents, that is to say, $1 \le p, p' \le \infty$ and $1/p + 1/p' = 1$. Then*

$$\| M_\infty^\star \|_{L^p(\mathbb{P})} \le p' \cdot \sup_t \| M_t \|_{L^p(\mathbb{P})} \, .$$

Proof. If $p = 1$, then $p' = \infty$ and the inequality is trivial; if $p = \infty$, then it is obvious. In the other cases consider an instant $t \in (0, \infty)$ and resurrect the finite set $S \subset [0, t]$ and the random variable M^S from the previous proof. From equation (A.3.9) and lemma 2.5.18,

$$\begin{aligned}\int (M^S)^p \, d\mathbb{P} &= p \int_0^\infty \lambda^{p-1} \mathbb{P}[M^S > \lambda] \, d\lambda \\ &\leq p \int_0^\infty \int \lambda^{p-2} |M_t| \cdot [M^S > \lambda] \, d\mathbb{P} \, d\lambda \\ &= \frac{p}{p-1} \int |M_t| (M^S)^{p-1} \, d\mathbb{P} \,.\end{aligned}$$

by A.8.4: $$\int (M^S)^p \, d\mathbb{P} \leq \frac{p}{p-1} \cdot \Big(\int |M_t|^p \, d\mathbb{P} \Big)^{\frac{1}{p}} \cdot \Big(\int (M^S)^p \, d\mathbb{P} \Big)^{\frac{p-1}{p}} \,.$$

Now $\int (M^S)^p \, d\mathbb{P}$ is finite if $\int |M_t|^p \, d\mathbb{P}$ is, and we may divide by the second factor on the right to obtain

$$\Big\| M^S \Big\|_{L^p(\mathbb{P})} \leq p' \cdot \big\| M_t \big\|_{L^p(\mathbb{P})} \,.$$

Taking the supremum over all finite subsets S of $\{t\} \cup (\mathbb{Q} \cap [0, t])$ and using the right-continuity of M produces $\|M_t^\star\|_{L^p(\mathbb{P})} \leq p' \cdot \|M_t\|_{L^p(\mathbb{P})}$. Now let $t \to \infty$. ∎

Exercise 2.5.20 For a vector $\boldsymbol{M} = (M^1, \ldots, M^d)$ of right-continuous martingales set

$$|\boldsymbol{M}_t|_\infty = \|\boldsymbol{M}_t\|_{\ell^\infty} \stackrel{\text{def}}{=} \sup_\eta |M_t^\eta| \quad \text{and} \quad \boldsymbol{M}_t^\star \stackrel{\text{def}}{=} \sup_{s \leq t} |\boldsymbol{M}_s|_\infty \,.$$

Using exercise 2.5.8 and the observation that the proofs above use only the property of $|M|$ of being a positive submartingale, show that

$$\big\| \boldsymbol{M}_\infty^\star \big\|_{L^p(\mathbb{P})} \leq p' \cdot \sup_{t < \infty} \big\| |\boldsymbol{M}_t|_\infty \big\|_{L^p(\mathbb{P})} \,.$$

Exercise 2.5.21 (i) For a standard Wiener process W and $\alpha, \beta \in \mathbb{R}_+$,

$$\mathbb{P}\big[\sup_t (W_t - \alpha t/2) > \beta\big] \leq e^{-\alpha\beta} \,.$$

(ii) $\lim_{t \to \infty} W_t / t = 0$.

Doob's Optional Stopping Theorem

In support of the vague principle "what holds at instants t holds at stopping times T," which the reader might be intuiting by now, we offer this generalization of the martingale property:

Theorem 2.5.22 (Doob) *Let M be a right-continuous uniformly integrable martingale. Then $\mathbb{E}\left[M_\infty | \mathcal{F}_T\right] = M_T$ almost surely at any stopping time T.*

Proof. We know from exercise 2.5.14 that $M_t = \mathbb{E}\Big[M_\infty\Big|\mathcal{F}_t\Big]$ for all t. To start with, assume that T takes only countably many values $0 \le t_0 \le t_1 \le \ldots$, among them possibly the value $t_\infty = \infty$. Then for any $A \in \mathcal{F}_T$

$$\int_A M_\infty \, d\mathbb{P} = \sum_{0 \le k \le \infty} \int_{A \cap [T=t_k]} M_\infty \, d\mathbb{P} = \sum_{0 \le k \le \infty} \int_{A \cap [T=t_k]} M_{t_k} \, d\mathbb{P}$$
$$= \sum_{0 \le k \le \infty} \int_{A \cap [T=t_k]} M_T \, d\mathbb{P} = \int_A M_T \, d\mathbb{P} .$$

The claim is thus true for such T. Given an arbitrary stopping time T we apply this to the discrete-valued stopping times $T^{(n)}$ of exercise 1.3.20. The right-continuity of M implies that

$$M_T = \lim_n \mathbb{E}[M_\infty | \mathcal{F}_{T^{(n)}}] .$$

This limit exists in mean, and the integral of it over any set $A \in \mathcal{F}_T$ is the same as the integral of M_∞ over A (exercise A.3.27). ∎

Exercise 2.5.23 (i) Let M be a right-continuous uniformly integrable martingale and $S \le T$ any two stopping times. Then $M_S = \mathbb{E}[M_T|\mathcal{F}_S]$ almost surely. (ii) If M is a right-continuous martingale and T a stopping time, then the stopped process M^T is a martingale; if T is bounded, then M^T is uniformly integrable. (iii) A local martingale is locally uniformly integrable. (iv) A positive local martingale M is a supermartingale; if $\mathbb{E}[M_t]$ is constant, M is a martingale. In any case, if $\mathbb{E}[M_S] = \mathbb{E}[M_T] = \mathbb{E}[M_0]$, then $\mathbb{E}[M_{S \vee T}] = \mathbb{E}[M_0]$.

Martingales Are Integrators

A simple but pivotal result is this:

Theorem 2.5.24 *A right-continuous square integrable martingale M is an L^2-integrator whose size at any instant t is given by*

$$\lceil\!\lceil M^t \rceil\!\rceil_{\mathcal{I}^2} = \| M_t \|_2 . \tag{2.5.1}$$

Proof. Let X be an elementary integrand as in (2.1.1):

$$X = f_0 \cdot [\![0]\!] + \sum_{n=1}^{N} f_n \cdot (\!(t_n, t_{n+1}]\!] , \qquad 0 = t_1 < \ldots, \; f_n \in \mathcal{F}_{t_n} ,$$

that vanishes past time t. Then

$$\Big(\int X \, dM\Big)^2 = \Big(f_0 M_0 + \sum_{n=1}^{N} f_n \cdot (M_{t_{n+1}} - M_{t_n})\Big)^2$$
$$= f_0^2 M_0^2 + 2 f_0 M_0 \cdot \sum_{n=1}^{N} f_n \cdot (M_{t_{n+1}} - M_{t_n})$$
$$+ \sum_{m,n=1}^{N} f_m (M_{t_{m+1}} - M_{t_m}) \cdot f_n (M_{t_{n+1}} - M_{t_n}) . \tag{$*$}$$

If $m \neq n$, say $m < n$, then $f_m(M_{t_{m+1}} - M_{t_m}) \cdot f_n$ is measurable on $\mathcal{F}_{t_n}$. Upon taking the expectation in $(*)$, terms with $m \neq n$ will vanish. At this point our particular choice of the elementary integrands pays off: had we allowed the steps to be measurable on a σ-algebra larger than the one attached to the left endpoint of the interval of constancy, then $f_n = X_{t_n}$ would not be measurable on $\mathcal{F}_{t_n}$, and the cancellation of terms would not occur. As it is we get

$$\mathbb{E}\Big[\Big(\int X\,dM\Big)^2\Big] = \mathbb{E}\Big[f_0^2 M_0^2 + \sum_{n=1}^{N} f_n^2 \cdot (M_{t_{n+1}} - M_{t_n})^2\Big]$$

$$\leq \mathbb{E}\Big[M_0^2 + \textstyle\sum_n (M_{t_{n+1}} - M_{t_n})^2\Big]$$

$$= \mathbb{E}\Big[M_0^2 + \textstyle\sum_n (M_{t_{n+1}}^2 - 2M_{t_{n+1}} M_{t_n} + M_{t_n}^2)\Big]$$

by exercise 2.5.3:
$$= \mathbb{E}\Big[M_0^2 + \textstyle\sum_{n=1}^{N} (M_{t_{n+1}}^2 - M_{t_n}^2)\Big] = \mathbb{E}\big[M_{t_{N+1}}^2\big] \leq \mathbb{E}[M_t^2]\,.$$

Taking the square root and the supremum over elementary integrands X that do not exceed[2] $[\![0,t]\!]$ results in equation (2.5.1). ∎

Exercise 2.5.25 If W is a standard Wiener process on the filtration $\mathcal{F}_.$, then it is an L^2-integrator on $\mathcal{F}_.$ and on its natural enlargement, and for every elementary integrand X

$$\lceil\!\lceil X \rceil\!\rceil_{W\text{-}2} = \|X\|_{W\text{-}2} = \Big(\int\!\!\int X_s^2\,ds\,d\mathbb{P}\Big)^{1/2}.$$

In particular $\qquad \lvert W^t \rvert_{\mathcal{I}^2} = \sqrt{t}\,. \qquad$ (For more see 4.2.20.)

Example 2.5.26 Let $X_1, X_2, \ldots$ be independent identically distributed Bernoulli random variables with $\mathbb{P}[X_k = \pm 1] = 1/2$. Fix a (large) natural number n and set

$$Z_t = \frac{1}{\sqrt{n}} \sum_{k \leq tn} X_k \qquad 0 \leq t \leq 1\,.$$

This process is right-continuous and constant on the intervals $[k/n, (k+1)/n)$, as is its basic filtration. Z is a process of finite variation. In fact, its variation process clearly is

$$\lvert Z \rvert_t = \frac{1}{\sqrt{n}} \cdot \lfloor tn \rfloor \;\approx t \cdot \sqrt{n}\,.$$

Here $\lfloor r \rfloor$ denotes the largest integer less than or equal to r. Thus if we estimate the size of Z as an L^2-integrator through its variation, using proposition 2.4.1 on page 68, we get the following estimate:

$$\lvert Z^t \rvert_{\mathcal{I}^2} \leq t\sqrt{n}\,. \tag{v}$$

Z is also evidently a martingale. Also, the L^2-mean of Z_t is easily seen to be $\sqrt{\lfloor tn \rfloor / n} \leq \sqrt{t}$. Theorem 2.5.24 yields the much superior estimate

$$\lvert Z^t \rvert_{\mathcal{I}^2} \leq \sqrt{t}\,, \tag{m}$$

which is, in particular, independent of n.

Let us use this example to continue the discussion of remark 1.2.9 on page 18 concerning the driver of Brownian motion. Consider a point mass on the line that receives at the instants k/n a kick of momentum $p_0 X_k$, i.e., either to the right or to the left with probability $1/2$ each. Let us scale the units so that the total energy transfer up to time 1 equals 1. An easy calculation shows that then $p_0 = 1/\sqrt{n}$. Assume that the point mass moves through a viscous medium. Then we are led to the stochastic differential equation

$$\begin{pmatrix} dx_t \\ dp_t \end{pmatrix} = \begin{pmatrix} p_t/m\,dt \\ -\alpha\, p_t\, dt + dZ_t \end{pmatrix}, \tag{2.5.2}$$

just as in equation (1.2.1). If we are interested in the solution at time 1, then the pertinent probability space is finite. It has 2^n elements. So the problem is to solve finitely many ordinary differential equations and to assemble their statistics. Imagine that n is on the order of 10^{23}, the number of molecules per mole. Then 2^n far exceeds the number of elementary particles in the universe! This makes it impossible to do the computations, and the estimates toward any procedure to solve the equation become useless if inequality (v) is used. Inequality (m) offers much better prospects in this regard but necessitates the development of stochastic integration theory.

An aside: if dt is large as compared with $1/n$, then $dZ_t = Z_{t+dt} - Z_t$ is the superposition of a large number of independent Bernoulli random variables and thus is distributed approximately $N(0, dt)$. It can be shown that Z tends to a Wiener process in law as $n \to \infty$ (theorem A.4.9) and that the solution of equation (2.5.2) accordingly tends in law to the solution of our idealized equation (1.2.1) for physical Brownian motion (see exercise A.4.14).

Martingales in L^p

The question arises whether perhaps a p-integrable martingale M is an L^p-integrator for exponents p other than 2. This is true in the range $1 < p < \infty$ (theorem 2.5.30) but not in general at $p = 1$, where M can only be shown to be a *local* L^1-integrator. For the proof of these claims some estimates are needed:

Lemma 2.5.27 *(i) Let Z be a bounded adapted process and set*

$$\lambda = \sup |Z| \quad \textit{and} \quad \mu = \sup \Big\{ \mathbb{E}\Big[\int X\, dZ \Big] : X \in \mathcal{E}_1 \Big\}.$$

Then for all X in the unit ball $\mathcal{E}_1$ of $\mathcal{E}$

$$\mathbb{E}\Big[\Big| \int X\, dZ \Big| \Big] \le \sqrt{2} \cdot (\lambda + \mu). \tag{2.5.3}$$

*In other words, Z has global L^1-integrator size $\lceil Z \rceil_{\mathcal{I}^1} \le \sqrt{2} \cdot (\lambda + \mu)$. Inequality (2.5.3) holds if $\mathbb{P}$ is merely a **subprobability**: $0 \le \mathbb{P}[\Omega] \le 1$.*

(ii) Suppose Z is a positive bounded supermartingale. Then for all $X \in \mathcal{E}_1$

$$\mathbb{E}\Big[\Big|\int X\, dZ\Big|^2\Big] \le 8 \cdot \sup Z \cdot \mathbb{E}[Z_0]\,. \tag{2.5.4}$$

That is to say, Z has global L^2-integrator size $\lceil Z \rceil_{\mathcal{I}^2} \le 2\sqrt{2 \sup Z \cdot \mathbb{E}[Z_0]}$.

Proof. It is easiest to argue if the elementary integrand $X \in \mathcal{E}_1$ of the claims (2.5.3) and (2.5.4) is written in the form (2.1.1) on page 46:

$$X = f_0 \cdot [\![0]\!] + \sum_{n=1}^{N} f_n \cdot (\!(t_n, t_{n+1}]\!]\,, \qquad 0 = t_1 < \ldots < t_{N+1},\ f_n \in \mathcal{F}_{t_n}\,.$$

Since X is in the unit ball $\mathcal{E}_1 \overset{\text{def}}{=} \{X \in \mathcal{E} : |X| \le 1\}$ of $\mathcal{E}$, the f_n all have absolute value less than 1. For $n = 1, \ldots, N$ let

$$\zeta_n \overset{\text{def}}{=} Z_{t_{n+1}} - Z_{t_n} \quad \text{and} \quad Z'_n \overset{\text{def}}{=} \mathbb{E}\Big[Z_{t_{n+1}} \big| \mathcal{F}_{t_n}\Big]\,;$$

$$\widehat{\zeta}_n \overset{\text{def}}{=} \mathbb{E}\Big[\zeta_n \big| \mathcal{F}_{t_n}\Big] = Z'_n - Z_{t_n} \quad \text{and} \quad \widetilde{\zeta}_n \overset{\text{def}}{=} \zeta_n - \widehat{\zeta}_n = Z_{t_{n+1}} - Z'_n\,.$$

Then $$\int X\, dZ = f_0 \cdot Z_0 + \sum_{n=1}^{N} f_n \cdot \big(Z_{t_{n+1}} - Z_{t_n}\big) = f_0 \cdot Z_0 + \sum_{n=1}^{N} f_n \cdot \zeta_n$$

$$= \Big(f_0 \cdot Z_0 + \sum_{n=1}^{N} f_n \cdot \widetilde{\zeta}_n\Big) + \Big(\sum_{n=1}^{N} f_n \cdot \widehat{\zeta}_n\Big)$$

$$= \qquad M \qquad + \qquad V\,.$$

The L^1-means of the two terms can be estimated separately. We start on M. Note that $\mathbb{E}\big[f_m \widetilde{\zeta}_m \cdot f_n \widetilde{\zeta}_n\big] = 0$ if $m \ne n$ and compute

$$\begin{aligned}
\mathbb{E}[M^2] &= \mathbb{E}\Big[f_0^2 \cdot Z_0^2 + \sum_{n=1}^{N} f_n^2 \cdot \widetilde{\zeta}_n^2\Big] \le \mathbb{E}\Big[Z_0^2 + \sum_{n=1}^{N} \widetilde{\zeta}_n^2\Big] \\
&= \mathbb{E}\Big[Z_{t_1}^2 + \textstyle\sum_n \big(Z_{t_{n+1}} - Z'_n\big)^2\Big] \\
&= \mathbb{E}\Big[Z_{t_1}^2 + \textstyle\sum_n \Big(Z_{t_{n+1}}^2 - 2Z_{t_{n+1}} Z'_n + Z'^2_n\Big)\Big] \\
&= \mathbb{E}\Big[Z_{t_1}^2 + \textstyle\sum_n \Big(Z_{t_{n+1}}^2 - Z'^2_n\Big)\Big] \\
&= \mathbb{E}\Big[Z_{t_1}^2 + \textstyle\sum_n \Big(Z_{t_{n+1}}^2 - Z_{t_n}^2\Big) + \textstyle\sum_n \big(Z_{t_n}^2 - Z'^2_n\big)\Big] \\
&= \mathbb{E}\Big[Z_{t_{N+1}}^2 + \textstyle\sum_n \big(Z_{t_n} + Z'_n\big) \cdot \big(Z_{t_n} - Z'_n\big)\Big] \\
&= \mathbb{E}\Big[Z_{t_{N+1}}^2 + \textstyle\sum_n \big(Z_{t_n} + Z'_n\big) \cdot \big(Z_{t_n} - Z_{t_{n+1}}\big)\Big] \\
&= \mathbb{E}\Big[Z_{t_{N+1}}^2\Big] - \mathbb{E}\Big[\int \Big(\textstyle\sum_{n=1}^{N} \big(Z_{t_n} + Z'_n\big) \cdot (\!(t_n, t_{n+1}]\!]\Big)\, dZ\Big] \\
&\le \lambda^2 + 2\lambda\mu\,.
\end{aligned} \tag{*}$$

After this preparation let us prove (i). Since $\mathbb{P}$ has mass less than 1, (*) results in

$$\mathbb{E}[|M|] \leq \left(\mathbb{E}[M^2]\right)^{1/2} \leq \sqrt{\lambda^2 + 2\lambda\mu}\,.$$

We add the estimate of the expectation of

$$|V| \leq \sum_n |f_n \widehat{\zeta}_n| = \sum_n |f_n| \operatorname{sgn}(\widehat{\zeta}_n) \cdot \widehat{\zeta}_n :$$

$$\begin{aligned}\mathbb{E}|V| &\leq \mathbb{E}\Big[\sum_{n=1}^{N} |f_n| \operatorname{sgn}(\widehat{\zeta}_n) \cdot \widehat{\zeta}_n\Big] = \mathbb{E}\Big[\sum_{n=1}^{N} |f_n| \operatorname{sgn}(\widehat{\zeta}_n) \cdot \zeta_n\Big] \\ &= \mathbb{E}\Big[\int \sum_{n=1}^{N} |f_n| \operatorname{sgn}(\widehat{\zeta}_n) \cdot (\!(t_n, t_{n+1}]\!]\, dZ\Big] \leq \mu\end{aligned}$$

to get $\mathbb{E}\Big[\Big|\int X\, dZ\Big|\Big] \leq \sqrt{\lambda^2 + 2\lambda\mu} + \mu \leq \sqrt{2} \cdot (\lambda + \mu)\,.$

We turn to claim (ii). Pick a $u > t_{N+1}$ and replace Z by $Z \cdot [\![0, u)\!)$. This is still a positive bounded supermartingale, and the left-hand side of inequality (2.5.4) has not changed. Since $X = 0$ on $(\!(t_{N+1}, u]\!]$, renaming the t_n so that $t_{N+1} = u$ does not change it either, so we may for convenience assume that $Z_{t_{N+1}} = 0$. Continuing (*) we find, using proposition 2.5.10, that

$$\begin{aligned}\mathbb{E}[M^2] &\leq -\mathbb{E}\Big[\int_{(\!(0, t_{N+1}]\!]} 2\lambda\, dZ\Big] \\ &= 2\lambda \cdot \mathbb{E}\Big[Z_0 - Z_{t_{N+1}}\Big] = 2 \sup Z \cdot \mathbb{E}\Big[Z_0\Big]\,. \qquad (**)\end{aligned}$$

To estimate $\mathbb{E}[V^2]$ note that the $\widehat{\zeta}_n$ are all negative: $\Big|\sum_n f_n \cdot \widehat{\zeta}_n\Big|$ is largest when all the f_n have the same sign. Thus, since $-1 \leq f_n \leq 1$,

$$\begin{aligned}\mathbb{E}[V^2] &= \mathbb{E}\Big[\Big(\sum_{n=1}^{N} f_n \cdot \widehat{\zeta}_n\Big)^2\Big] \leq \mathbb{E}\Big[\Big(\sum_{n=1}^{N} \widehat{\zeta}_n\Big)^2\Big] \\ &\leq 2 \sum_{1 \leq m \leq n \leq N} \mathbb{E}\big[\widehat{\zeta}_m \cdot \widehat{\zeta}_n\big] = 2 \sum_{1 \leq m \leq n \leq N} \mathbb{E}\big[\widehat{\zeta}_m \cdot \zeta_n\big] \\ &= 2 \sum_{1 \leq m \leq N} \mathbb{E}\Big[\widehat{\zeta}_m \cdot \big(Z_{t_{N+1}} - Z_{t_m}\big)\Big] = 2 \sum_{1 \leq m \leq N} \mathbb{E}\big[-\widehat{\zeta}_m \cdot Z_{t_m}\big] \\ &\leq 2 \sup Z \sum_{1 \leq m \leq N} \mathbb{E}\big[-\widehat{\zeta}_m\big] = -2 \sup Z \sum_{1 \leq m \leq N} \mathbb{E}\big[\zeta_m\big] \\ &= -2 \sup Z \cdot \big(Z_{t_{N+1}} - Z_0\big) = 2 \sup Z \cdot \mathbb{E}\big[Z_0\big]\,.\end{aligned}$$

Adding this to inequality (**) we find

$$\mathbb{E}\Big[\Big|\int X\, dZ\Big|^2\Big] \leq 2\mathbb{E}[M^2] + 2\mathbb{E}[V^2] \leq 8 \cdot \sup Z \cdot \mathbb{E}[Z_0]\,. \qquad \blacksquare$$

The following consequence of lemma 2.5.27 is the first step in showing that p-integrable martingales are L^p-integrators in the range $1 < p < \infty$ (theorem 2.5.30). It is a "weak-type" version of this result at $p = 1$:

Proposition 2.5.28 *An L^1-bounded right-continuous martingale M is a global L^0-integrator. In fact, for every elementary integrand X with $|X| \le 1$ and every $\lambda > 0$,*

$$\mathbb{P}\Big[\Big|\int X\,dM\Big| > \lambda\Big] \le \frac{2}{\lambda} \cdot \sup_t \|M_t\|_{L^1(\mathbb{P})} . \tag{2.5.5}$$

Proof. This inequality clearly implies that the linear map $X \mapsto \int X\,dM$ is bounded from $\mathcal{E}$ to L^0, in fact to the Lorentz space $L^{1,\infty}$. The argument is again easiest if X is written in the form (2.1.1):

$$X = f_0 \cdot [\![0]\!] + \sum_{n=1}^{N} f_n \cdot (\!(t_n, t_{n+1}]\!] , \qquad 0 = t_1 < \ldots,\ f_n \in \mathcal{F}_{t_n} .$$

Let U be a bounded stopping time strictly past t_{N+1}, and let us assume to start with that M is positive at and before time U. Set

$$T = \inf\{t_n : M_{t_n} \ge \lambda\} \wedge U .$$

This is an elementary stopping time (proposition 1.3.13). Let us estimate the probabilities of the disjoint events

$$B_1 = \Big[\Big|\int X\,dM\Big| > \lambda, T < U\Big] \text{ and } B_2 = \Big[\Big|\int X\,dM\Big| > \lambda, T = U\Big]$$

separately. B_1 is contained in the set $M_U^\star \ge \lambda$, and Doob's maximal lemma 2.5.18 gives the estimate

$$\mathbb{P}[B_1] \le \lambda^{-1} \cdot \mathbb{E}\big[|M_U|\big] . \tag{$*$}$$

To estimate the probability of B_2 consider the right-continuous process

$$Z = M \cdot [\![0, T)\!) .$$

This is a positive supermartingale bounded by λ; indeed, using A.1.5,

$$\begin{aligned}\mathbb{E}\big[Z_t|\mathcal{F}_s\big] &= \mathbb{E}\Big[M_t \cdot [T > t]\Big|\mathcal{F}_s\Big] \\ &\le \mathbb{E}\Big[M_t \cdot [T > s]\Big|\mathcal{F}_s\Big] = \mathbb{E}\Big[M_s \cdot [T > s]\Big|\mathcal{F}_s\Big] = Z_s .\end{aligned}$$

On B_2 the paths of M and Z coincide. Therefore $\int X\,dZ = \int X\,dM$ on B_2, and B_2 is contained in the set

$$\Big[\Big|\int X\,dZ\Big| > \lambda\Big] .$$

Due to Chebyschev's inequality and lemma 2.5.27, the probability of this set is less than

$$\lambda^{-2} \cdot \mathbb{E}\Big[\Big(\int X\, dZ\Big)^2\Big] \le \frac{8\lambda \cdot \mathbb{E}[Z_0]}{\lambda^2} = \frac{8\mathbb{E}[M_0]}{\lambda} \le \frac{8}{\lambda} \cdot \mathbb{E}[|M_U|] .$$

Together with (*) this produces

$$\mathbb{P}\Big[\Big|\int X\, dM\Big| > \lambda\Big] \le \frac{9}{\lambda} \cdot \mathbb{E}[M_U] .$$

In the general case we split M_U into its positive and negative parts $M_U^{\pm}$ and set $M_t^{\pm} = \mathbb{E}\big[M_U^{\pm}|\mathcal{F}_t\big]$, obtaining two positive martingales with difference M^U. We estimate

$$\begin{aligned}\mathbb{P}\Big[\Big|\int X\, dM\Big| \ge \lambda\Big] &\le \mathbb{P}\Big[\Big|\int X\, dM^+\Big| \ge \lambda/2\Big] + \mathbb{P}\Big[\Big|\int X\, dM^-\Big| \ge \lambda/2\Big] \\ &\le \frac{9}{\lambda/2} \cdot \Big(\mathbb{E}[M_U^+] + \mathbb{E}[M_U^-]\Big) = \frac{18}{\lambda} \cdot \mathbb{E}[|M_U|] \\ &\le \frac{18}{\lambda} \cdot \sup_t \mathbb{E}[|M_t|] .\end{aligned}$$

This is inequality (2.5.5), except for the factor of $1/\lambda$, which is 18 rather than 2, as claimed. We borrow the latter value from Burkholder [14], who showed that the following inequality holds and is best possible: for $|X| \le 1$

$$\mathbb{P}\Big[\sup_t \Big|\int_0^t X\, dM\Big| > \lambda\Big] \le \frac{2}{\lambda} \cdot \sup_t \|\, M_t \,\|_{L^1} . \qquad \blacksquare$$

The proof above can be used to get additional information about local martingales:

Corollary 2.5.29 *A right-continuous local martingale M is a local L^1-integrator. In fact, it can locally be written as the sum of an L^2-integrator and a process of integrable total variation.* (According to exercise 4.3.14, M can actually be written as the sum of a finite variation process and a locally square integrable ***martingale***.)

Proof. There is an arbitrarily large bounded stopping time U such that M^U is a uniformly integrable martingale and can be written as the difference of two positive martingales $M^{\pm}$. Both can be chosen right-continuous (proposition 2.5.13). The stopping time $T = \inf\{t : M_t^{\pm} \ge \lambda\} \wedge U$ can be made arbitrarily large by the choice of λ. Write

$$(M^{\pm})^T = M^{\pm} \cdot [\![0, T)\!) + M_T^{\pm} \cdot [\![T, \infty)\!) .$$

The first summand is a positive bounded supermartingale and thus is a global $L^2(\mathbb{P})$-integrator; the last summand evidently has integrable total variation $|M_T^{\mp}|$. Thus M^T is the sum of two global $L^2(\mathbb{P})$-integrators and two processes of integrable total variation. ∎

Theorem 2.5.30 *Let $1 < p < \infty$. A right-continuous L^p-integrable martingale M is an L^p-integrator. Moreover, there are universal constants A_p independent of M such that for all stopping times T*

$$\left\lceil M^T \right\rceil_{\mathcal{I}^p} \leq A_p \cdot \| M_T \|_p \,. \tag{2.5.6}$$

Proof. Let X be an elementary integrand with $|X| \leq 1$ and consider the following linear map U from $L^\infty(\mathcal{F}_\infty, \mathbb{P})$ to itself:

$$U(g) = \int X \, dM^g \,.$$

Here M^g is the right-continuous martingale $M_t^g = \mathbb{E}[g|\mathcal{F}_t]$ of example 2.5.2. We shall apply Marcinkiewicz interpolation to this map (see proposition A.8.24). By (2.5.5) U is of weak type 1–1:

$$\mathbb{P}\big[|U(g)| > \lambda\big] \leq \frac{2}{\lambda} \cdot \| g \|_1 \,.$$

By (2.5.1), U is also of strong type 2–2:

$$\| U(g) \|_2 \leq \| g \|_2 \,.$$

Also, U is self-adjoint: for $h \in L^\infty$ and $X \in \mathcal{E}_1$ written as in (2.1.1)

$$\begin{aligned}
\mathbb{E}[U(g) \cdot h] &= \mathbb{E}\left[\left(f_0 M_0^g + \sum_n f_n \left(M_{t_{n+1}}^g - M_{t_n}^g\right)\right) M_\infty^h\right] \\
&= \mathbb{E}\left[f_0 M_0^g M_0^h + \sum_n f_n \left(M_{t_{n+1}}^g M_{t_{n+1}}^h - M_{t_n}^g M_{t_n}^h\right)\right] \\
&= \mathbb{E}\left[\left(f_0 M_0^h + \sum_n f_n \left(M_{t_{n+1}}^h - M_{t_n}^h\right)\right) M_\infty^g\right] \\
&= \mathbb{E}[U(h) \cdot g] \,.
\end{aligned}$$

A little result from Marcinkiewicz interpolation, proved as corollary A.8.25, shows that U is of strong type p–p for all $p \in (1, \infty)$. That is to say, there are constants A_p with $\| \int X \, dM \|_p \leq A_p \cdot \| M_\infty \|_p$ for all elementary integrands X with $|X| \leq 1$. Now apply this to the stopped martingale M^T to obtain (2.5.6). ∎

Exercise 2.5.31 Provide an estimate for A_p from this proof.

Exercise 2.5.32 Let S_t be a positive bounded $\mathbb{P}$-supermartingale on the filtration $\mathcal{F}_.$ and assume that S is right-continuous in probability and almost surely strictly positive; that is to say, $\mathbb{P}[S_t = 0] = 0 \quad \forall\, t$. Then there exists a $\mathbb{P}$-nearly empty set N outside which the restriction of every path of S to the positive rationals is bounded away from zero on every bounded time-interval.

Repeated Footnotes: 47^2 56^5 65^7 66^{10} 70^{13}

3

Extension of the Integral

Recall our goal: if Z is an L^p-integrator, then there exists an extension of its associated elementary integral to a class of integrands on which the Dominated Convergence Theorem holds.

The reader with a firm grounding in Daniell's extension of the integral will be able to breeze through the next 40 pages, merely identifying the results presented with those he is familiar with; the presentation is fashioned so as to facilitate this transition from the ordinary to the stochastic integral. The reader not familiar with Daniell's extension can use them as a primer.

Daniell's Extension Procedure on the Line

As before we look for guidance at the half-line. Let z be a right-continuous distribution function of finite variation, let the integral be defined on the elementary functions by equation (2.2) on page 44, and let us review step 2 of the integration process, the extension theory. Daniell's idea was to apply Lebesgue's definition of an outer measure of sets to functions, thus obtaining an upper integral of functions. A short overview can be found on page 395 of appendix A. The upshot is this. Given a right-continuous distribution function z of finite variation on the half-line, Daniell first defines the associated elementary integral $\mathfrak{e} \to \mathbb{R}$ by equation (2.2) on page 44, and then defines a seminorm, the ***Daniell mean*** $\| \ \|_z^*$, on all functions $f : [0, \infty) \to \overline{\mathbb{R}}$ by

$$\| f \|_z^* = \inf_{|f| \le h \in \mathfrak{e}^\uparrow} \ \sup_{\phi \in \mathfrak{e}, |\phi| \le h} \left| \int \phi \, dz \right| . \tag{3.1}$$

Here $\mathfrak{e}^\uparrow$ is the collection of all those functions that are pointwise suprema of *countable* collections of elementary integrands. The integrable functions are simply the closure of $\mathfrak{e}$ under this seminorm, and the integral is the extension by continuity of the elementary integral. This is the Lebesgue–Stieltjes integral. The Dominated Convergence Theorem and the numerous beautiful features of the Lebesgue–Stieltjes integral are all due to only two properties of Daniell's mean $\| \ \|_z^*$; it is *countably subadditive:*

$$\left\| \sum_{n=1}^\infty f_n \right\|_z^* \le \sum_{n=1}^\infty \| f_n \|_z^* , \qquad f_n \ge 0 ,$$

and it is additive on $\mathfrak{e}_+$, as it agrees there with the variation measure $|dz|$.

Let us put this procedure in a general context. Much of modern analysis concerns linear maps on vector spaces. Given such, the analyst will most frequently start out by designing a seminorm on the given vector space, one with respect to which the given linear map is continuous, and then extend the linear map by continuity to the completion of the vector space under that seminorm. The analysis of the extended linear map is generally easier because of the completeness of its domain, which furnishes limit points to many arguments. Daniell's method is but an instance of this. The vector space is $\mathfrak{e}$, the linear map is $x \mapsto \int x\, dz$, and the Daniell mean $\| \ \|_z^*$ is a suitable, in fact superb, seminorm with respect to which the linear map is continuous. The completion of $\mathfrak{e}$ is the space $\mathcal{L}^1(dz)$ of integrable functions.

3.1 The Daniell Mean

We shall extend the elementary stochastic integral in literally the same way, by designing a seminorm under which it is continuous. In fact, we shall simply emulate Daniell's "up-and-down procedure" of equation (3.1) and thence follow our noses.

The first thing to do is to replace the absolute value, which measures the size of the real-valued integral in equation (3.1), by a suitable size measurement of the random variable-valued elementary stochastic integral that takes its place. Any of the means and gauges mentioned on pages 33–34 will suit. Now a right-continuous adapted process Z may be an $L^p(\mathbb{P})$-integrator for some pairs $(p, \mathbb{P})$ and not for others. We will pick a pair $(p, \mathbb{P})$ such that it is. The notation will generally reflect only the choice of p, and of course Z, but not of $\mathbb{P}$; so the size measurement in question is $\| \ \|_p$, $\lceil\!\lceil \ \rceil\!\rceil_p$, or $\| \ \|_{[\alpha]}$, depending on our predilection or need. The stochastic analog of definition (3.1) is

$$\lceil\!\lceil F \rceil\!\rceil_{Z-p}^* = \inf_{|F| \le H \in \mathcal{E}_+^\uparrow} \ \sup_{X \in \mathcal{E}, |X| \le H} \left\lceil\!\!\left\lceil \int X\, dZ \right\rceil\!\!\right\rceil_p , \text{ etc.} \tag{3.1.1}$$

Here $\mathcal{E}_+^\uparrow$ denotes the collection of positive processes that are pointwise suprema of a *sequence* of elementary integrands. Let us write separately the "up-part" and the "down-part" of (3.1.1): for $H \in \mathcal{E}_+^\uparrow$

$$\|H\|_{Z-p}^* = \sup\left\{ \left\| \int X\, dZ \right\|_p : X \in \mathcal{E}, |X| \le H \right\} \qquad (p \ge 1);$$

$$\lceil\!\lceil H \rceil\!\rceil_{Z-p}^* = \sup\left\{ \left\lceil\!\!\left\lceil \int X\, dZ \right\rceil\!\!\right\rceil_p : X \in \mathcal{E}, |X| \le H \right\} \qquad (p \ge 0);$$

$$\|H\|_{Z-[\alpha]}^* = \sup\left\{ \left\| \int X\, dZ \right\|_{[\alpha]} : X \in \mathcal{E}, |X| \le H \right\} \qquad (p = 0).$$

Then on an arbitrary numerical process F,

$$\|F\|^*_{Z\text{-}p} = \inf\Big\{\|H\|^*_{Z\text{-}p} : H \in \mathcal{E}^{\uparrow}_{+}\,,\ H \geq |F|\Big\} \qquad (p \geq 1)\,;$$

$$\lceil\!\lceil F\rceil\!\rceil^*_{Z\text{-}p} = \inf\Big\{\lceil\!\lceil H\rceil\!\rceil^*_{Z\text{-}p} : H \in \mathcal{E}^{\uparrow}_{+}\,,\ H \geq |F|\Big\} \qquad (p \geq 0)\,;$$

$$\|F\|^*_{Z\text{-}[\alpha]} = \inf\Big\{\|H\|^*_{Z\text{-}[\alpha]} : H \in \mathcal{E}^{\uparrow}_{+}\,,\ H \geq |F|\Big\} \qquad (p = 0)\,.$$

We shall refer to $\lceil\!\lceil\ \rceil\!\rceil^*_{Z\text{-}p}$ as ***THE Daniell mean.*** It goes with that semivariation which comes from the subadditive functional $\lceil\!\lceil\ \rceil\!\rceil_p$ – the subadditivity of $\lceil\!\lceil\ \rceil\!\rceil_p$ is the reason for singling it out. $\lceil\!\lceil\ \rceil\!\rceil^*_{Z\text{-}p}$, too, will turn out to be subadditive, even countably subadditive. This property makes it best suited for the extension of the integral. If the probability needs to be mentioned, we also write $\lceil\!\lceil\ \rceil\!\rceil^*_{Z\text{-}p;\mathbb{P}}$ etc.

As we would on the line we shall now establish the properties of the mean. Here as there, the Dominated Convergence Theorem and all of its beautiful corollaries are but consequences of these. The arguments are standard.

Exercise 3.1.1 $\lceil\!\lceil\ \rceil\!\rceil^*_{Z\text{-}p}$ agrees with the semivariation $\lceil\!\lceil\ \rceil\!\rceil_{Z\text{-}p}$ on $\mathcal{E}_+$. In fact, for $X \in \mathcal{E}$ we have $\lceil\!\lceil X\rceil\!\rceil^*_{Z\text{-}p} = \lceil\!\lceil |X|\rceil\!\rceil_{Z\text{-}p}$. The same holds for the means associated with the other gauges.

Exercise 3.1.2 The following comes in handy on several occasions: let S, T be stopping times and assume that the projection $[S < T]$ of the stochastic interval $(\!(S, T]\!]$ on Ω has measure less than ϵ. Then any process F that vanishes outside $(\!(S, T]\!]$ has $\lceil\!\lceil F\rceil\!\rceil^*_{Z\text{-}0} \leq \epsilon$.

Exercise 3.1.3 For a standard Wiener process W and arbitrary $F : \boldsymbol{B} \to \overline{\mathbb{R}}$,

$$\|F\|^*_{W\text{-}2} = \left(\int^* F^2(s,\omega)\, ds \times \mathbb{P}(d\omega)\right)^{1/2}.$$

$\|\ \|^*_{W\text{-}2}$ is simply the square mean for the measure $ds \times \mathbb{P}$ on $\mathcal{E}$. It is the mean originally employed by Itô and is still much in vogue (see definition (4.2.9)).

A Temporary Assumption

To start on the extension theory we have to place a temporary condition on the L^p-integrator Z, one that is at first sight rather more restrictive than the mere right-continuity in probability expressed in (IC-0); we have to require

Assumption 3.1.4 *The elementary integral is continuous in p-mean along increasing sequences. That is to say, for every increasing sequence $\big(X^{(n)}\big)$ of elementary integrands whose pointwise supremum X also happens to be an elementary integrand, we have*

$$\lim_n \int X^{(n)}\, dZ = \int X\, dZ \quad \textit{in p-mean.} \qquad \text{(IC-p)}$$

Exercise 3.1.5 This is equivalent with either of the following conditions:
(i) ***σ-continuity at*** 0: for every sequence $(X^{(n)})$ of elementary integrands that decreases pointwise to zero, $\lim_{n\to\infty}\int X^{(n)}\,dZ = 0$ in p-mean;
(ii) ***σ-additivity***: for every sequence $(X^{(n)})$ of positive elementary integrands whose sum is a priori an elementary integrand,

$$\sum_n \int X^{(n)}\,dZ = \int \sum_n X^{(n)}\,dZ \quad \text{in } p\text{-mean.}$$

Assumption 3.1.4 clearly implies (RC-0). In view of exercise 3.1.5 (ii), it is also reasonably called ***p-mean σ-additivity***. An L^p-integrator actually satisfies (IC-p) automatically; but when this fact is proved in proposition 3.3.2, the extension theory of the integral done under this assumption is needed. The reduction of (IC-p) to (RC-0) in section 3.3 will be made rather simple if the reader observes that

*In the extension theory of the elementary integral below, use is made only of the structure of the set $\mathcal{E}$ of elementary integrands – it is an algebra and vector lattice closed under chopping of bounded functions on some set, which is called the **base space** or **ambient set** – and of the properties* (B-p) *and* (IC-p) *of the vector measure*

$$\int \cdot\, dZ : \mathcal{E} \to L^p\,.$$

In particular, the structure of the ambient set is irrelevant to the extension procedure. The words "process" and "function" (on the base space) are used interchangeably.

Properties of the Daniell Mean

Theorem 3.1.6 *The Daniell mean $\lceil\!\lceil\ \rceil\!\rceil^*_{Z-p}$ has the following properties:*
(i) It is defined on all numerical functions on the base space and takes values in the positive extended reals $\overline{\mathbb{R}}_+$.
*(ii) It is **solid**: $|F| \le |G|$ implies $\lceil\!\lceil F \rceil\!\rceil^*_{Z-p} \le \lceil\!\lceil G \rceil\!\rceil^*_{Z-p}$.*
*(iii) It is **continuous along increasing sequences** $(H^{(n)})$ **of** $\mathcal{E}^{\uparrow}_+$:*

$$\left\lceil\!\!\left\lceil \sup_n H^{(n)} \right\rceil\!\!\right\rceil^*_{Z-p} = \sup_n \left\lceil\!\!\left\lceil H^{(n)} \right\rceil\!\!\right\rceil^*_{Z-p}\,.$$

*(iv) It is **countably subadditive**: for any sequence $(F^{(n)})$ of positive functions on the base space*

$$\left\lceil\!\!\left\lceil \sum_{n=1}^{\infty} F^{(n)} \right\rceil\!\!\right\rceil^*_{Z-p} \le \sum_{n=1}^{\infty} \left\lceil\!\!\left\lceil F^{(n)} \right\rceil\!\!\right\rceil^*_{Z-p}\,.$$

*(v) Elementary integrands are **finite for the mean**: $\lim_{r\to 0} \lceil\!\lceil rX \rceil\!\rceil^*_{Z-p} = 0$ for all $X \in \mathcal{E}$ – when $p > 0$ this simply reads $\lceil\!\lceil X \rceil\!\rceil^*_{Z-p} < \infty$.*

(vi) For any sequence $\left(X^{(n)}\right)$ of positive elementary integrands

$$\left(\lim_{r\to 0}\left\lceil\!\left\lceil r\cdot\sum_{n=1}^{\infty}X^{(n)}\right\rceil\!\right\rceil^{*}_{Z\text{-}p}=0\right)\quad\textit{implies}\quad\left(\left\lceil\!\left\lceil X^{(n)}\right\rceil\!\right\rceil^{*}_{Z\text{-}p}\xrightarrow[n\to\infty]{}0\right)\tag{M}$$

– when $p>0$ this simply reads:

$$\left(\left\lceil\!\left\lceil\sum_{n=1}^{\infty}X^{(n)}\right\rceil\!\right\rceil^{*}_{Z\text{-}p}<\infty\right)\quad\textit{implies}\quad\left(\left\lceil\!\left\lceil X^{(n)}\right\rceil\!\right\rceil^{*}_{Z\text{-}p}\xrightarrow[n\to\infty]{}0\right).$$

[It is this property which distinguishes the Daniell mean from an ordinary sup-norm and which is responsible for the Dominated Convergence Theorem and its beautiful consequences.]

(vii) The mean $\lceil\!\lceil\ \rceil\!\rceil^{*}_{Z\text{-}p}$ ***majorizes the elementary stochastic integral:***

$$\left\lceil\!\left\lceil\int X\,dZ\right\rceil\!\right\rceil_p\le\lceil\!\lceil X\rceil\!\rceil^{*}_{Z\text{-}p}\qquad\forall X\in\mathcal{E}.$$

Proof. The first property that is possibly not obvious is (iii). To prove it let $(H^{(n)})$ be an increasing sequence of $\mathcal{E}^{\uparrow}_{+}$. Its pointwise supremum H clearly belongs to $\mathcal{E}^{\uparrow}_{+}$ as well. From the solidity,

$$\lceil\!\lceil H\rceil\!\rceil^{*}_{Z\text{-}p}\ge\sup\left\lceil\!\left\lceil H^{(n)}\right\rceil\!\right\rceil^{*}_{Z\text{-}p}.$$

To show the reverse inequality assume that $\lceil\!\lceil H\rceil\!\rceil^{*}_{Z\text{-}p}>a$. There exists an $X\in\mathcal{E}$ with $|X|\le H$ and

$$\left\lceil\!\left\lceil\int X\,dZ\right\rceil\!\right\rceil_p>a.$$

Write X as the difference $X=X_+-X_-$ of its positive and negative parts. For every n there is a sequence $(X^{(n,k)})$ with pointwise supremum $H^{(n)}$. Set

$$X^{(N)}=\bigvee_{n,k\le N}X^{(n,k)}\quad\text{and}\quad X^{(N)}_{\pm}=X^{(N)}\wedge X_{\pm}.$$

Clearly $X^{(N)}_{\pm}\uparrow X_{\pm}$, and therefore, with $\overline{X}^{(N)}=X^{(N)}_{+}-X^{(N)}_{-}$,

$$\int\overline{X}^{(N)}\,dZ\to\int X\,dZ\qquad\text{in }p\text{-mean.}$$

It is here that assumption 3.1.4 is used. Thus $\lceil\!\lceil\int\overline{X}^{(N)}\,dZ\rceil\!\rceil_p>a$ for sufficiently large N. As $|\overline{X}^{(N)}|\le H^{(N)}$, $\lceil\!\lceil H^{(N)}\rceil\!\rceil^{*}_{Z\text{-}p}>a$ eventually. This argument applies to the Daniell extension of any other semivariation – associated with any other solid and continuous functional on L^p – as well and shows that $\|\ \|^{*}_{Z\text{-}p}$ and $\|\ \|^{*}_{Z\text{-}[\alpha]}$, too, are continuous along increasing sequences of $\mathcal{E}^{\uparrow}_{+}$.

(iv) We start by proving the subadditivity of $\lceil\!\lceil\ \rceil\!\rceil^*_{Z\text{-}p}$ on the class $\mathcal{E}^\uparrow_+$. Let $H^{(i)} \in \mathcal{E}^\uparrow_+$, $i = 1, 2$. There is a sequence $\left(X^{(i,n)}\right)_n$ in $\mathcal{E}_+$ whose pointwise supremum is $H^{(i)}$. Replacing $X^{(i,n)}$ by $\sup_{\nu\le n} X^{(i,\nu)}$, we may assume that $\left(X^{(i,n)}\right)$ is increasing. By (iii) and proposition 2.2.1,

$$\left\lceil\!\left\lceil H^{(1)} + H^{(2)} \right\rceil\!\right\rceil^*_{Z\text{-}p} = \lim_n \left\lceil\!\left\lceil X^{(1,n)} + X^{(2,n)} \right\rceil\!\right\rceil^*_{Z\text{-}p}$$

$$\le \lim_n \left(\left\lceil\!\left\lceil X^{(1,n)} \right\rceil\!\right\rceil^*_{Z\text{-}p} + \left\lceil\!\left\lceil X^{(2,n)} \right\rceil\!\right\rceil^*_{Z\text{-}p} \right) = \left\lceil\!\left\lceil H^{(1)} \right\rceil\!\right\rceil^*_{Z\text{-}p} + \left\lceil\!\left\lceil H^{(2)} \right\rceil\!\right\rceil^*_{Z\text{-}p} .$$

To prove the countable subadditivity in general let $\left(F^{(n)}\right)$ be a sequence of numerical functions on the base space with $\sum \lceil\!\lceil F^{(n)} \rceil\!\rceil^*_{Z\text{-}p} < a < \infty$ – if the sum is infinite, there is nothing to prove. There are $H^{(n)} \in \mathcal{E}^\uparrow_+$ with $F^{(n)} \le H^{(n)}$ and $\sum \lceil\!\lceil H^{(n)} \rceil\!\rceil^*_{Z\text{-}p} < a$. The process $H = \sum H^{(n)}$ belongs to $\mathcal{E}^\uparrow_+$ and exceeds F. Consequently

$$\lceil\!\lceil F \rceil\!\rceil^*_{Z\text{-}p} \le \lceil\!\lceil H \rceil\!\rceil^*_{Z\text{-}p} = \sup_N \left\lceil\!\left\lceil \sum_{n=1}^N H^{(n)} \right\rceil\!\right\rceil^*_{Z\text{-}p}$$

from first part of proof:
$$\le \sup_N \sum_{n=1}^N \left\lceil\!\left\lceil H^{(n)} \right\rceil\!\right\rceil^*_{Z\text{-}p} = \sum_{n=1}^\infty \left\lceil\!\left\lceil H^{(n)} \right\rceil\!\right\rceil^*_{Z\text{-}p} < a .$$

(v) follows from condition (B-p) on page 53, in view of exercise 3.1.1.

It remains to prove (M), which is the substitute for the additivity that holds in the scalar case. Note that it is a statement about the behavior of the mean $\lceil\!\lceil\ \rceil\!\rceil^*_{Z\text{-}p}$ on $\mathcal{E}$, where it equals the semivariation $\lceil\!\lceil\ \rceil\!\rceil_{Z\text{-}p}$ (see definition (2.2.1)).

We start with the case $p > 0$. Since $\lceil\!\lceil\ \rceil\!\rceil^*_{Z\text{-}p} = (\|\ \|^*_{Z\text{-}p})^{p\wedge 1}$, it suffices to show that

$$\left(\left\| \sum_{n=1}^\infty X^{(n)} \right\|^*_{Z\text{-}p} < \infty \right) \text{ implies } \left(\left\| X^{(n)} \right\|_{Z\text{-}p} \xrightarrow[n\to\infty]{} 0 \right) . \tag{$*$}$$

Now $\| X^{(n)} \|_{Z\text{-}p} \xrightarrow[n\to\infty]{} 0$ means that for any sequence $\left(X'^{(n)}\right)$ of elementary integrands with $|X'^{(n)}| \le X^{(n)}$

$$\left\| \int X'^{(n)}\, dZ \right\|_p \xrightarrow[n\to\infty]{} 0 . \tag{$**$}$$

For if $\| X^{(n)} \|_{Z\text{-}p} \not\to 0$, then the very definition of the semivariation would produce a sequence violating $(**)$.

Let $\epsilon_1(t), \epsilon_2(t), \ldots$ be independent identically distributed Bernoulli random variables, defined on a probability space $(D, \mathcal{D}, \tau)$, with $\tau([\epsilon_\nu = \pm 1]) = 1/2$. Then, with $f_n \stackrel{\text{def}}{=} \int X'^{(n)}\, dZ$,

$$\sum_{n\le N} \epsilon_n(t) f_n = \int \sum_{n \le N} \epsilon_n(t) X'^{(n)}\, dZ , \qquad t \in D .$$

The second of Khintchine's inequalities, proved as theorem A.8.26, provides a universal constant $K_p = K_p^{(A.8.5)}$ such that

$$\Big(\sum_{n\le N} f_n^2\Big)^{1/2} \le K_p \cdot \Big(\int \Big|\sum_{n\le N} \epsilon_n(t) f_n\Big|^p \,\tau(dt)\Big)^{1/p} .$$

Applying $\|\cdot\|_p$ and using Fubini's theorem A.3.18 on this results in

$$\begin{aligned}\Big\|\Big(\sum_{n\le N} f_n^2\Big)^{1/2}\Big\|_p &\le K_p \cdot \Big(\int\int\Big|\int \sum_{n\le N} \epsilon_n(t) X'^{(n)}\, dZ\Big|^p \,d\mathbb{P}\,\tau(dt)\Big)^{1/p}\\ &\le K_p \cdot \Big(\int \Big\|\sum_{n\le N} \epsilon_n(t) X'^{(n)}\Big\|_{Z-p}^p \,\tau(dt)\Big)^{1/p}\\ &\le K_p \cdot \sup_N \Big\|\sum_{n\le N} X^{(n)}\Big\|_{Z-p} \le K_p \cdot \Big\|\sum_{n=1}^{\infty} X^{(n)}\Big\|_{Z-p}^* < \infty .\end{aligned}$$

The function $h \overset{\text{def}}{=} \big(\sum_{n\in\mathbb{N}} f_n^2\big)^{1/2}$ therefore belongs to L^p. This implies that $f_n \to 0$ a.s. and dominatedly (by h); therefore $\|f_n\|_p \xrightarrow[n\to\infty]{} 0$.

If $p = 0$, we use inequality (A.8.6) instead:

$$\Big(\sum_{n\le N} f_n^2\Big)^{1/2} \le K_0 \cdot \Big\|\sum_{n\le N} \epsilon_n(t) f_n\Big\|_{[\kappa_0;\tau]} ;$$

and thus, applying $\| \ \|_{[\alpha;\mathbb{P}]}$ and exercise A.8.16,

$$\begin{aligned}\Big\|\Big(\sum_{n\le N} f_n^2\Big)^{\frac12}\Big\|_{[\alpha;\mathbb{P}]} &\le K_0 \cdot \Big\|\Big\|\sum_{n\le N} \epsilon_n(t) f_n\Big\|_{[\kappa_0;\tau]}\Big\|_{[\alpha;\mathbb{P}]}\\ &\le K_0 \cdot \Big\|\Big\|\int \sum_{n\le N} \epsilon_n X'^{(n)}\, dZ\Big\|_{[\gamma;\mathbb{P}]}\Big\|_{[\alpha\kappa_0-\gamma;\tau]}\\ &\le K_0 \cdot \Big\|\Big\|\sum_{n\le N} X^{(n)}\Big\|_{Z-[\gamma]}\Big\|_{[\alpha\kappa_0-\gamma;\tau]} \le K_0 \cdot \Big\|\sum_{n\le N} X^{(n)}\Big\|_{Z-[\gamma]} .\end{aligned}$$

This holds for all $\gamma < \alpha\kappa_0$, and therefore

$$\Big\|\Big(\sum_{n\le N} f_n^2\Big)^{1/2}\Big\|_{[\alpha]} \le K_0 \cdot \Big\|\sum_{n\le N} X^{(n)}\Big\|_{Z-[\alpha\kappa_0]} \le K_0 \cdot \Big\|\sum_{n=1}^{\infty} X^{(n)}\Big\|_{Z-[\alpha\kappa_0]}^* .$$

It is left to the reader to show that $(*)$ implies

$$\Big\|\sum_{n=1}^{\infty} X^{(n)}\Big\|_{Z-[\alpha\kappa_0]}^* < \infty \qquad \forall \alpha > 0 ,$$

which, in conjunction with the previous inequality, proves that

$$\Big(\sum_{n=1}^{\infty} f_n^2\Big)^{1/2} < \infty \text{ a.s.}$$

Thus clearly $f_n \to 0$ in L^0. ∎

It is worth keeping the quantitative information gathered above for an application to the square function on page 148.

Corollary 3.1.7 *Let Z be an adapted process and $X^{(1)}, X^{(2)}, \ldots \in \mathcal{E}$. Then*

$$\Big\| \Big(\sum_n \Big(\int X^{(n)}\, dZ\Big)^2\Big)^{1/2} \Big\|_p \le K_p \cdot \Big\| \sum_n |X^{(n)}| \Big\|^*_{Z\text{-}p}, \qquad p > 0;$$

and

$$\Big\| \Big(\sum_n \Big(\int X^{(n)}\, dZ\Big)^2\Big)^{1/2} \Big\|_{[\alpha]} \le K_0 \cdot \Big\| \sum_n |X^{(n)}| \Big\|^*_{Z\text{-}[\alpha\kappa_0]}, \qquad p = 0.$$

The constants K_p, κ_0 are the Khintchine constants of theorem A.8.26.

Exercise 3.1.8 The $\| \;\|^*_{Z\text{-}p}$ for $0 < p < 1$, and the $\| \;\|^*_{Z\text{-}[\alpha]}$ for $0 < \alpha$, too, have the properties listed in theorem 3.1.6, except countable subadditivity.

3.2 The Integration Theory of a Mean

Any functional $[\![\;]\!]^*$ satisfying (i)–(vi) of theorem 3.1.6 is called a mean on $\mathcal{E}$. This notion is so useful that a little repetition is justified:

Definition 3.2.1 *Let $\mathcal{E}$ be an algebra and vector lattice closed under chopping of bounded functions, all defined on some set $\boldsymbol{B}$. A* ***mean on*** *$\mathcal{E}$ is a positive $\overline{\mathbb{R}}$-valued functional $[\![\;]\!]^*$ that is defined on all numerical functions on $\boldsymbol{B}$ and has the following properties:*

(i) It is solid: $|F| \le |G|$ implies $[\![F]\!]^* \le [\![G]\!]^*$.

(ii) It is continuous along increasing sequences $(X^{(n)})$ of $\mathcal{E}_+$:

$$\Big[\!\!\Big[\sup_n X^{(n)}\Big]\!\!\Big]^* = \sup_n \Big[\!\!\Big[X^{(n)}\Big]\!\!\Big]^*.$$

(iii) It is countably subadditive: for any sequence $(F^{(n)})$ of positive functions on $\boldsymbol{B}$

$$\Big[\!\!\Big[\sum_{n=1}^{\infty} F^{(n)}\Big]\!\!\Big]^* \le \sum_{n=1}^{\infty} \Big[\!\!\Big[F^{(n)}\Big]\!\!\Big]^*. \qquad (CSA)$$

(iv) The functions of $\mathcal{E}$ are finite for the mean: for every $X \in \mathcal{E}$

$$\lim_{r \to 0} [\![rX]\!]^* = 0.$$

(v) For any sequence $(X^{(n)})$ of positive functions in $\mathcal{E}$

$$\left(\lim_{r\to 0}\left\lceil\!\left\lceil r\cdot\sum_{n=1}^{\infty} X^{(n)}\right\rceil\!\right\rceil^* = 0\right) \quad \textit{implies} \quad \left(\left\lceil\!\left\lceil X^{(n)}\right\rceil\!\right\rceil^* \xrightarrow[n\to\infty]{} 0\right) . \tag{M}$$

Let $\mathcal{V}$ be a topological vector space with a gauge $\lceil\!\lceil\ \rceil\!\rceil_{\mathcal{V}}$ defining its topology, and let $\mathcal{I} : \mathcal{E} \to \mathcal{V}$ be a linear map. The mean $\lceil\!\lceil\ \rceil\!\rceil^*$ is said to ***majorize*** *$\mathcal{I}$ if $\lceil\!\lceil \mathcal{I}(X)\rceil\!\rceil_{\mathcal{V}} \le \lceil\!\lceil X\rceil\!\rceil^*$ for all $X \in \mathcal{E}$. $\lceil\!\lceil\ \rceil\!\rceil^*$ is said to* ***control*** *$\mathcal{I}$ if there is a constant $C < \infty$ such that for all $X \in \mathcal{E}$*

$$\lceil\!\lceil \mathcal{I}(X)\rceil\!\rceil_{\mathcal{V}} \le C \cdot \lceil\!\lceil X\rceil\!\rceil^* . \tag{3.2.1}$$

The crossbars on top of the symbol $\lceil\!\lceil\ \rceil\!\rceil^*$ are a reminder that the mean is subadditive but possibly not homogeneous. THE Daniell mean was constructed so as to majorize the elementary stochastic integral.

This and the following two sections will use only the fact that the elementary integrands $\mathcal{E}$ form an algebra and vector lattice closed under chopping, of bounded functions, and that $\lceil\!\lceil\ \rceil\!\rceil^*$ is a mean on $\mathcal{E}$. The nature of the underlying set $\boldsymbol{B}$ in particular is immaterial, and so is the way in which the mean was constructed. In order to emphasize this point we shall develop in the next few sections the integration theory of a general mean $\lceil\!\lceil\ \rceil\!\rceil^*$. Later on we shall meet means other than Daniell's mean $\lceil\!\lceil\ \rceil\!\rceil^*_{Z\text{-}p}$, so that we may then use the results established here for them as well. In fact, Daniell's mean is unsuitable as a controlling device for Picard's scheme, which so far was the motivation for all of our proceedings. Other "pathwise" means controlling the elementary integral have to be found (see definition (4.2.9) and exercise 4.5.18).

Exercise 3.2.2 A mean is automatically continuous along increasing sequences $(H^{(n)})$ of $\mathcal{E}_+^{\uparrow}$:

$$\left\lceil\!\left\lceil \sup_n H^{(n)}\right\rceil\!\right\rceil^* = \sup_n \left\lceil\!\left\lceil H^{(n)}\right\rceil\!\right\rceil^* . \tag{$\uparrow$}$$

(Every mean that we shall encounter in this book, including Daniell's, is actually continuous along arbitrary increasing sequences. That is to say, ($\uparrow$) holds for any increasing sequence of positive numerical functions. See proposition 3.6.5.)

Negligible Functions and Sets

In this subsection only the solidity and countable subadditivity of the mean $\lceil\!\lceil\ \rceil\!\rceil^*$ are exploited.

Definition 3.2.3 *A numerical function F on the base space (a process) is called $\lceil\!\lceil\ \rceil\!\rceil^*$-**negligible** or **negligible** for short if $\lceil\!\lceil F\rceil\!\rceil^* = 0$. A subset of the base space is negligible if its indicator function is negligible.*[1]

[1] In accordance with convention A.1.5 on page 364 we write variously A or 1_A for the indicator function of A; the $\lceil\!\lceil\ \rceil\!\rceil^*$-size of the set A, $\lceil\!\lceil 1_A\rceil\!\rceil^*$, is mostly written $\lceil\!\lceil A\rceil\!\rceil^*$ when A is a subset of the ambient space.

A property of the points of the underlying set is said to hold ***almost everywhere****, or* ***a.e.*** *for short, if the set of points where it fails to hold is negligible.*

If we want to stress the point that these definitions refer to $\lceil\!\lceil\ \rceil\!\rceil^*$, *we shall talk about* $\lceil\!\lceil\ \rceil\!\rceil^*$*-negligible processes and* $\lceil\!\lceil\ \rceil\!\rceil^*$***-a.e. convergence****, etc. If we want to stress the point that these definitions refer in particular to Daniell's mean* $\lceil\!\lceil\ \rceil\!\rceil^*_{Z\text{-}p}$, *we shall talk about* $\lceil\!\lceil\ \rceil\!\rceil^*_{Z\text{-}p}$*-negligible processes and* $\lceil\!\lceil\ \rceil\!\rceil^*_{Z\text{-}p}$*-a.e. convergence, or also about* ***Z–p-negligible*** *processes,* ***Z–p-a.e. convergence****, etc.*

These notions behave as one expects from ordinary integration:

Proposition 3.2.4 *(i) The union of countably many negligible sets is negligible. Any subset of a negligible set is negligible.*

(ii) A process F is negligible if and only if it vanishes almost everywhere, that is to say, if and only if the set $[F \neq 0]$ is negligible.

(iii) If the real-valued functions F and F' agree almost everywhere, then they have the same mean.

Proof. For ease of reading we use the same symbol for a set and its indicator function. For instance, $A_1 \cup A_2 = A_1 \vee A_2$ in the sense that the indicator function on the left[1] is the pointwise maximum of the two indicator functions on the right.

(i) If N_n, $n = 1, \ldots$, are negligible sets, then due to inequality (CSA)[1]

$$\lceil\!\lceil N_1 \cup N_2 \cup \ldots \rceil\!\rceil^* = \Big\lceil\!\!\Big\lceil \bigvee_{n=1}^{\infty} N_n \Big\rceil\!\!\Big\rceil^*$$
$$\leq \Big\lceil\!\!\Big\lceil \sum_{n=1}^{\infty} N_n \Big\rceil\!\!\Big\rceil^* \leq \sum_{n=1}^{\infty} \lceil\!\lceil N_n \rceil\!\rceil^* = 0\,,$$

due to the countable subadditivity of $\lceil\!\lceil\ \rceil\!\rceil^*$.

(ii) Obviously[1] $[F \neq 0] \leq \sum_{n=1}^{\infty} |F|$. Thus if $\lceil\!\lceil F \rceil\!\rceil^* = 0$, then

$$\lceil\!\lceil\, [F \neq 0]\, \rceil\!\rceil^* \leq \sum_{n=1}^{\infty} \lceil\!\lceil F \rceil\!\rceil^* = 0.$$

Conversely, $|F| \leq \sum_{n=1}^{\infty} [F \neq 0]$, so that $\lceil\!\lceil\, [F \neq 0]\, \rceil\!\rceil^* = 0$ implies

$$\lceil\!\lceil F \rceil\!\rceil^* \leq \sum_{n=1}^{\infty} \lceil\!\lceil\, [F \neq 0]\, \rceil\!\rceil^* = 0.$$

(iii) Since by the previous argument $\lceil\!\lceil F \cdot [F \neq F'] \rceil\!\rceil^* \leq \lceil\!\lceil \infty \cdot [F \neq F'] \rceil\!\rceil^* = 0$,

$$\lceil\!\lceil F \rceil\!\rceil^* \leq \Big\lceil\!\!\Big\lceil F \cdot [F = F'] \Big\rceil\!\!\Big\rceil^* + \Big\lceil\!\!\Big\lceil F \cdot [F \neq F'] \Big\rceil\!\!\Big\rceil^* = \Big\lceil\!\!\Big\lceil F \cdot [F = F'] \Big\rceil\!\!\Big\rceil^*$$
$$= \Big\lceil\!\!\Big\lceil F' \cdot [F = F'] \Big\rceil\!\!\Big\rceil^* \leq \Big\lceil\!\!\Big\lceil F' \Big\rceil\!\!\Big\rceil^* \text{ and vice versa.}$$ ∎

Exercise 3.2.5 The filtration being regular, an evanescent process is Z–p-negligible.

Processes Finite for the Mean and Defined Almost Everywhere

Definition 3.2.6 *A process F is* ***finite for the mean*** $\lceil\!\lceil\ \rceil\!\rceil^*$ *provided*

$$\lceil\!\lceil r \cdot F \rceil\!\rceil^* \xrightarrow[r\to 0]{} 0\,.$$

The collection of processes finite for the mean $\lceil\!\lceil\ \rceil\!\rceil^*$ *is denoted by* $\mathfrak{F}[\lceil\!\lceil\ \rceil\!\rceil^*]$, *or simply by* $\mathfrak{F}$ *if there is no need to specify the mean.*

If $\lceil\!\lceil\ \rceil\!\rceil^*$ is the Daniell mean $\lceil\!\lceil\ \rceil\!\rceil^*_{Z\text{-}p}$ for some $p > 0$, then F is finite for the mean if and only if simply $\lceil\!\lceil F \rceil\!\rceil^* < \infty$. If $p = 0$ and $\lceil\!\lceil\ \rceil\!\rceil^* = \lceil\!\lceil\ \rceil\!\rceil^*_{Z\text{-}0}$, though, then $\lceil\!\lceil F \rceil\!\rceil^* \le 1$ for all F, and the somewhat clumsy looking condition $\lceil\!\lceil rF \rceil\!\rceil^* \xrightarrow[r\to 0]{} 0$ properly expresses finiteness (see exercise A.8.18).

Proposition 3.2.7 *A process F finite for the mean* $\lceil\!\lceil\ \rceil\!\rceil^*$ *is finite* $\lceil\!\lceil\ \rceil\!\rceil^*$*-a.e.*

Proof.[1] $[|F| = \infty] \le |F|/n$ for all $n \in \mathbb{N}$, and the solidity gives

$$\lceil\!\lceil\, [|F| = \infty]\, \rceil\!\rceil^* \le \lceil\!\lceil F/n \rceil\!\rceil^* \qquad \forall\, n \in \mathbb{N}.$$

Let $n \to \infty$ and conclude that $\lceil\!\lceil\, [|F| = \infty]\, \rceil\!\rceil^* = 0$. ∎

The only processes of interest are, of course, those finite for the mean. We should like to argue that the sum of any two of them has finite mean again, in view of the subadditivity of $\lceil\!\lceil\ \rceil\!\rceil^*$. A technical difficulty appears: even if F and G have finite mean, there may be points ϖ in the base space where $F(\varpi) = +\infty$ and $G(\varpi) = -\infty$ or vice versa; then $F(\varpi) + G(\varpi)$ is not defined. The solution to this tiny quandary is to notice that such ambiguities may happen at most in a negligible set of ϖ's. We simply extend $\lceil\!\lceil\ \rceil\!\rceil^*$ to processes that are defined merely $\lceil\!\lceil\ \rceil\!\rceil^*$-almost everywhere:

Definition 3.2.8 (Extending the Mean) *Let F be a process defined almost everywhere, i.e., such that the complement of* $\mathrm{dom}(F)$ *is* $\lceil\!\lceil\ \rceil\!\rceil^*$*-negligible. We set* $\lceil\!\lceil F \rceil\!\rceil^* \overset{\text{def}}{=} \lceil\!\lceil F' \rceil\!\rceil^*$, *where F' is any process defined everywhere and coinciding with F almost everywhere in the points where F is defined.*

Part (iii) of proposition 3.2.4 shows that this definition is good: it does not matter which process F' we choose to agree $\lceil\!\lceil\ \rceil\!\rceil^*$-a.e. with F; any two will differ negligibly and thus have the same mean. Given two processes F and G finite for the mean that are merely almost everywhere defined, we define their sum $F + G$ to equal $F(\varpi) + G(\varpi)$ where both $F(\varpi)$ and $G(\varpi)$ are finite. This process is almost everywhere defined, as the set of points where F or G are infinite or not defined is negligible. It is clear how to define the scalar multiple $r{\cdot}F$ of a process F that is a.e. defined.

From now on, "***process***" will stand for "almost everywhere defined process" if the context permits it. It is nearly obvious that propositions 3.2.4 and 3.2.7 stay. We leave this to the reader.

Exercise 3.2.9 $|\,\lceil\!\lceil F \rceil\!\rceil^* - \lceil\!\lceil G \rceil\!\rceil^*\,| \le \lceil\!\lceil F - G \rceil\!\rceil^*$ for any two $F, G \in \mathfrak{F}[\lceil\!\lceil\ \rceil\!\rceil^*]$.

Theorem 3.2.10 *A process finite for the mean is finite almost everywhere. The collection $\mathfrak{F}[\lceil\!\lceil\ \rceil\!\rceil^*]$ of processes finite for $\lceil\!\lceil\ \rceil\!\rceil^*$ is closed under taking finite linear combinations, finite maxima and minima, and under chopping, and $\lceil\!\lceil\ \rceil\!\rceil^*$ is a solid and countably subadditive functional on $\mathfrak{F}[\lceil\!\lceil\ \rceil\!\rceil^*]$. The space $\mathfrak{F}[\lceil\!\lceil\ \rceil\!\rceil^*]$ is complete under the translation-invariant pseudometric*

$$dist(F, F') \stackrel{\text{def}}{=} \lceil\!\lceil F - F' \rceil\!\rceil^* .$$

Moreover, any mean-Cauchy sequence in $\mathfrak{F}[\lceil\!\lceil\ \rceil\!\rceil^*]$ has a subsequence that converges $\lceil\!\lceil\ \rceil\!\rceil^*$-almost everywhere to a $\lceil\!\lceil\ \rceil\!\rceil^*$-mean limit.

Proof. The first two statements are left as exercise 3.2.11. For the last two let (F_n) be a mean-Cauchy sequence in $\mathfrak{F}[\lceil\!\lceil\ \rceil\!\rceil^*]$; that is to say

$$\sup_{m,n \ge N} \lceil\!\lceil F_m - F_n \rceil\!\rceil^* \xrightarrow[N\to\infty]{} 0 .$$

For $n = 1, 2, \ldots$ let F'_n be a process that is everywhere defined and finite and agrees with F_n a.e. Let N_n denote the negligible set of points where F_n is not defined or does not agree with F'_n. There is an increasing sequence (n_k) of indices such that $\lceil\!\lceil F'_n - F'_{n_k} \rceil\!\rceil^* \le 2^{-k}$ for $n \ge n_k$. Using them set

$$G \stackrel{\text{def}}{=} \sum_{k=1}^{\infty} \left| F'_{n_{k+1}} - F'_{n_k} \right| .$$

G is finite for the mean. Indeed, for $|r| \le 1$,

$$\lceil\!\lceil rG \rceil\!\rceil^* \le \sum_{k=1}^{K} \left\lceil\!\!\left\lceil r \cdot \left(F'_{n_{k+1}} - F'_{n_k}\right) \right\rceil\!\!\right\rceil^* + \sum_{k=K+1}^{\infty} \left\lceil\!\!\left\lceil F'_{n_{k+1}} - F'_{n_k} \right\rceil\!\!\right\rceil^* .$$

Given $\epsilon > 0$ we first choose K so large that the second summand is less than $\epsilon/2$ and then r so small that the first summand is also less than $\epsilon/2$. This shows that $\lim_{r\to 0} \lceil\!\lceil rG \rceil\!\rceil^* = 0$.

$$N \stackrel{\text{def}}{=} \bigcup_{n=1}^{\infty} N_n \cup [G = \infty]$$

is therefore a negligible set. If $\varpi \notin N$, then

$$F(\varpi) = F'_{n_1}(\varpi) + \sum_{k=1}^{\infty} \left(F'_{n_{k+1}}(\varpi) - F'_{n_k}(\varpi)\right) = \lim_{k\to\infty} F'_{n_k}(\varpi)$$

exists, since the infinite sum converges absolutely. Also,[1]

$$\begin{aligned}
\lceil\!\lceil F - F_{n_K} \rceil\!\rceil^* &= \lceil\!\lceil F - F'_{n_K} \rceil\!\rceil^* \\
&\le \lceil\!\lceil N \cdot (F - F'_{n_K}) \rceil\!\rceil^* + \lceil\!\lceil N^c \cdot (F - F'_{n_K}) \rceil\!\rceil^* \\
&\le \left\lceil\!\!\left\lceil \sum_{k=K+1}^{\infty} \left| F'_{n_{k+1}} - F'_{n_k} \right| \right\rceil\!\!\right\rceil^* \le 2^{-K} \xrightarrow[K\to\infty]{} 0 .
\end{aligned}$$

Thus $(F'_{n_k})_{k=1}^{\infty}$ converges to F not only pointwise but also in mean. Given $\epsilon > 0$, let K be so large that both

$$\lceil\!\lceil F_m - F_n \rceil\!\rceil^* < \epsilon/2 \quad \text{for } m, n \geq n_K$$

and
$$\lceil\!\lceil F - F_{n_k} \rceil\!\rceil^* = \lceil\!\lceil F - F'_{n_k} \rceil\!\rceil^* < \epsilon/2 \quad \text{for } k \geq K .$$

For any $n \geq N \stackrel{\text{def}}{=} n_K$

$$\lceil\!\lceil F - F_n \rceil\!\rceil^* < \lceil\!\lceil F - F_{n_K} \rceil\!\rceil^* + \lceil\!\lceil F_{n_K} - F_n \rceil\!\rceil^* < \epsilon ,$$

showing that the original sequence (F_n) converges to F in mean. Its subsequence (F_{n_k}) clearly converges $\lceil\!\lceil\ \rceil\!\rceil^*$-almost everywhere to F. ∎

Henceforth we shall not be so excruciatingly punctilious. If we have to perform algebraic or limit arguments on a sequence of processes that are defined merely almost everywhere, we shall without mention replace every one of them with a process that is defined and finite everywhere, and perform the arguments on the resulting sequence; this affects neither the means of the processes nor their convergence in mean or almost everywhere.

Exercise 3.2.11 Define the linear combination, minimum, maximum, and product of two processes defined a.e., and prove the first two statements of theorem 3.2.10. Show that $\mathfrak{F}[\lceil\!\lceil\ \rceil\!\rceil^*]$ is not in general an algebra.

Exercise 3.2.12 (i) Let (F_n) be a mean-convergent sequence with limit F. Any process differing negligibly from F is also a mean limit of (F_n). Any two mean limits of (F_n) differ negligibly. (ii) Suppose that the processes F_n are finite for the mean $\lceil\!\lceil\ \rceil\!\rceil^*$ and $\sum_n \lceil\!\lceil F_n \rceil\!\rceil^*$ is finite. Then $\sum_n |F_n|$ is finite for the mean $\lceil\!\lceil\ \rceil\!\rceil^*$.

Integrable Processes and the Stochastic Integral

Definition 3.2.13 *An $\lceil\!\lceil\ \rceil\!\rceil^*$-almost everywhere defined process F is* $\lceil\!\lceil\ \rceil\!\rceil^*$***-integrable*** *if there exists a sequence (X_n) of elementary integrands converging in $\lceil\!\lceil\ \rceil\!\rceil^*$-mean to F: $\lceil\!\lceil F - X_n \rceil\!\rceil^* \xrightarrow[n\to\infty]{} 0$.*

The collection of $\lceil\!\lceil\ \rceil\!\rceil^*$-integrable processes is denoted by $\mathfrak{L}^1[\lceil\!\lceil\ \rceil\!\rceil^*]$ or simply by $\mathfrak{L}^1$. In other words, $\mathfrak{L}^1$ is the $\lceil\!\lceil\ \rceil\!\rceil^*$-closure of $\mathcal{E}$ in $\mathfrak{F}$ (see exercise 3.2.15). If the mean is Daniell's mean $\lceil\!\lceil\ \rceil\!\rceil^*_{Z\text{-}p}$ and we want to stress this point, then we shall also talk about ***Z-p-integrable*** *processes and write $\mathfrak{L}^1[\lceil\!\lceil\ \rceil\!\rceil^*_{Z\text{-}p}]$ or $\mathfrak{L}^1[Z\text{-}p]$. If the probability also must be exhibited, we write $\mathfrak{L}^1[Z\text{-}p;\mathbb{P}]$ or $\mathfrak{L}^1[\lceil\!\lceil\ \rceil\!\rceil^*_{Z\text{-}p;\mathbb{P}}]$.*

Definition 3.2.14 *Suppose that the mean $\lceil\!\lceil\ \rceil\!\rceil^*$ is Daniell's mean $\lceil\!\lceil\ \rceil\!\rceil^*_{Z\text{-}p}$ or at least controls the elementary integral (definition 3.2.1), and suppose that F is an $\lceil\!\lceil\ \rceil\!\rceil^*$-integrable process. Let (X_n) be a sequence of elementary integrands converging in $\lceil\!\lceil\ \rceil\!\rceil^*$-mean to F; the integral $\int F\, dZ$ is defined as the limit in p-mean of the sequence $(\int X_n\, dZ)$ in L^p. In other words, the extended integral is the extension by $\lceil\!\lceil\ \rceil\!\rceil^*$-continuity of the elementary integral. It is also called the* ***Itô stochastic integral****.*

This is unequivocal except perhaps for the definition of the integral. How do we know that the sequence $\left(\int X_n\, dZ\right)$ has a limit? Since $⌈\!⌈\ ⌉\!⌉^*$ controls the elementary integral, we have

$$\left⌈\!\left⌈ \int X_n\, dZ - \int X_m\, dZ \right⌉\!\right⌉_p$$

by equation (3.2.1):
$$\leq C \cdot ⌈\!⌈ X_n - X_m ⌉\!⌉^*$$
$$\leq C \cdot ⌈\!⌈ F - X_n ⌉\!⌉^* + C \cdot ⌈\!⌈ F - X_m ⌉\!⌉^* \xrightarrow[n,m\to\infty]{} 0 .$$

The sequence $\left(\int X_n\ dZ\right)$ is therefore Cauchy in L^p and has a limit in p-mean (exercise A.8.1). How do we know that this limit does not depend on the particular sequence (X_n) of elementary integrands chosen to approximate F in $⌈\!⌈\ ⌉\!⌉^*$-mean? If (X'_n) is a second such sequence, then clearly $⌈\!⌈ X_n - X'_n ⌉\!⌉^* \to 0$, and since the mean controls the elementary integral, $⌈\!⌈ \int X_n\, dZ - \int X'_n\, dZ ⌉\!⌉_p \to 0$: the limits are the same.

Let us be punctilious about this. The integrals $\int X_n\, dZ$ are by definition random variables. They form a Cauchy sequence in p-mean. There is not only one p-mean limit but many, all differing negligibly. The integral $\int F\, dZ$ above is by nature a ***class*** in $L^p(\mathbb{P})$! We won't be overly religious about this point; for instance, we won't hesitate to multiply a random variable f with the class $\int X\ dZ$ and understand $f \cdot \int X\ dZ$ to be the class $\dot{f} \cdot \int X\ dZ$. Yet there are some occasions where the distinction is important (see definition 3.7.6). Later on we shall pick from the class $\int F\, dZ$ a random variable in a nearly unique manner (see page 134).

Exercise 3.2.15 (i) A process F is $⌈\!⌈\ ⌉\!⌉^*$-integrable if and only if there exist integrable processes F_n with $F = \sum_n F_n$ and $\sum_n ⌈\!⌈ F_n ⌉\!⌉^* < \infty$. (ii) An integrable process F is finite for the mean. (iii) The mean satisfies the all-important property (M) of definition 3.2.1 on sequences (X_n) of positive integrable processes.

Exercise 3.2.16 (i) Assume that the mean controls the elementary integral $\int \cdot\, dZ : \mathcal{E} \to L^p$ (see definition 3.2.1 on page 94). Then the extended integral is a linear map $\int \cdot\, dZ : \mathfrak{L}^1[⌈\!⌈\ ⌉\!⌉^*] \to L^p$ again controlled by the mean:

$$\left⌈\!\left⌈ \int F\, dZ \right⌉\!\right⌉_p \leq C^{(3.2.1)} \cdot ⌈\!⌈ F ⌉\!⌉^* , \qquad F \in \mathfrak{L}^1[⌈\!⌈\ ⌉\!⌉^*] .$$

(ii) Let $⌈\!⌈\ ⌉\!⌉^* \leq ⌈\!⌈\ ⌉\!⌉'^*$ be two means on $\mathcal{E}$. Then a $⌈\!⌈\ ⌉\!⌉'^*$-integrable process is $⌈\!⌈\ ⌉\!⌉^*$-integrable. If both means control the elementary stochastic integral, then their integral extensions coincide on $\mathfrak{L}^1[⌈\!⌈\ ⌉\!⌉'^*] \subset \mathfrak{L}^1[⌈\!⌈\ ⌉\!⌉^*]$.

(iii) If Z is an L^q-integrator and $0 \leq p < q < \infty$, then Z is an L^p-integrator; a Z–q-integrable process X is Z–p-integrable, and the integrals in either sense coincide.

Exercise 3.2.17 If the martingale M is an L^1-integrator, then $\mathbb{E}[\int X\, dM] = 0$ for any M–1-integrable process X with $X_0 = 0$.

Exercise 3.2.18 If $\mathcal{F}_\infty$ is countably generated, then the pseudometric space $\mathfrak{L}^1[⌈\!⌈\ ⌉\!⌉^*]$ is separable.

Exercise 3.2.19 Suppose that we start with a measured filtration $(\mathcal{F}_., \mathbb{P})$ and an L^p-integrator Z in the sense of the original definition 2.1.7 on page 49. To obtain path regularity and simple truths like exercise 3.2.5, we replace $\mathcal{F}_.$ by its natural enlargement $\mathcal{F}^{\mathbb{P}}_{.+}$ and Z by a nice modification. $\mathfrak{L}^1$ is then the closure of $\mathcal{E}^{\mathbb{P}} = \mathcal{E}[\mathcal{F}^{\mathbb{P}}_{.+}]$ under $\lceil\!\lceil\ \rceil\!\rceil^*_{Z\text{-}p}$. Show that the original set $\mathcal{E}$ of elementary integrands is dense in $\mathfrak{L}^1$.

Permanence Properties of Integrable Functions

From now on we shall make use of all of the properties that make $\lceil\!\lceil\ \rceil\!\rceil^*$ a mean. We continue to write simply "integrable" and "negligible" instead of the more precise "$\lceil\!\lceil\ \rceil\!\rceil^*$-integrable" and "$\lceil\!\lceil\ \rceil\!\rceil^*$-negligible," etc. The next result is obvious:

Proposition 3.2.20 *Let (F_n) be a sequence of integrable processes converging in $\lceil\!\lceil\ \rceil\!\rceil^*$-mean to F. Then F is integrable. If $\lceil\!\lceil\ \rceil\!\rceil^*$ controls the elementary integral in $\lceil\!\lceil\ \rceil\!\rceil_p$-mean, i.e., as a linear map to L^p, then*

$$\int F\, dZ = \lim_{n\to\infty} \int F_n\, dZ \quad \textit{in } \lceil\!\lceil\ \rceil\!\rceil_p\textit{-mean.}$$

Permanence Under Algebraic and Order Operations

Theorem 3.2.21 *Let $0 \le p < \infty$ and Z an L^p-integrator. Let F and F' be $\lceil\!\lceil\ \rceil\!\rceil^*$-integrable processes and $r \in \mathbb{R}$. Then the combinations $F + F'$, rF, $F \vee F'$, $F \wedge F'$, and $F \wedge 1$ are $\lceil\!\lceil\ \rceil\!\rceil^*$-integrable. So is the product $F{\cdot}F'$, provided that at least one of F, F' is bounded.*

Proof. We start with the sum. For any two elementary integrands X, X'

we have $$|(F + F') - (X + X')| \le |F - X| + |F' - X'|\,,$$

and so $$\lceil\!\lceil(F + F') - (X + X')\rceil\!\rceil^* \le \lceil\!\lceil F - X\rceil\!\rceil^* + \lceil\!\lceil F' - X'\rceil\!\rceil^*\,.$$

Since the right-hand side can be made as small as one pleases by the choice of X, X', so can the left-hand side. This says that $F + F'$ is integrable, inasmuch as $X + X'$ is an elementary integrand. The same argument applies to the other combinations:

$$\begin{aligned}
|(rF) - (rX)| &\le ([\lfloor r\rfloor] + 1)\cdot|F - X|\,;\\
|(F \vee F') - (X \vee X')| &\le |F - X| + |F' - X'|\,;\\
|(F \wedge F') - (X \wedge X')| &\le |F - X| + |F' - X'|\,;\\
\big||F| - |X|\big| &\le |F - X|\,;\ |F \wedge 1 - X \wedge 1| \le |F - X|\,;\\
|(F \cdot F') - (X \cdot X')| &\le |F|\cdot|F' - X'| + |X'|\cdot|F - X|\\
&\le \|F\|_\infty \cdot |F' - X'| + \|X'\|_\infty \cdot |F - X|\,.
\end{aligned}$$

We apply $\lceil\!\lceil\ \rceil\!\rceil^*$ to these inequalities and obtain

$$\lceil\!\lceil (rF) - (rX) \rceil\!\rceil^* \le (\lceil r \rceil + 1) \lceil\!\lceil F - X \rceil\!\rceil^* ;$$
$$\lceil\!\lceil (F \vee F') - (X \vee X') \rceil\!\rceil^* \le \lceil\!\lceil F - X \rceil\!\rceil^* + \lceil\!\lceil F' - X' \rceil\!\rceil^* ;$$
$$\lceil\!\lceil (F \wedge F') - (X \wedge X') \rceil\!\rceil^* \le \lceil\!\lceil F - X \rceil\!\rceil^* + \lceil\!\lceil F' - X' \rceil\!\rceil^* ;$$
$$\lceil\!\lceil |F| - |X| \rceil\!\rceil^* \le \lceil\!\lceil F - X \rceil\!\rceil^* ; \ \lceil\!\lceil F \wedge 1 - X \wedge 1 \rceil\!\rceil^* \le \lceil\!\lceil F - X \rceil\!\rceil^* ;$$
$$\lceil\!\lceil (F \cdot F') - (X \cdot X') \rceil\!\rceil^* \le \|F\|_\infty \cdot \lceil\!\lceil F' - X' \rceil\!\rceil^* + \|X'\|_\infty \cdot \lceil\!\lceil F - X \rceil\!\rceil^* .$$

Given an $\epsilon > 0$, we may choose elementary integrands X, X' so that the right-hand sides are less than ϵ. This is possible because the processes F, F' are integrable and shows that the processes $rF, F \vee F' \ldots$ are integrable as well, inasmuch as the processes $rX, X \vee X' \ldots$ appearing on the left are elementary.

The last case, that of the product, is marginally more complicated than the others. Given $\epsilon > 0$, we first choose X' elementary so that

$$\lceil\!\lceil F' - X' \rceil\!\rceil^* \le \frac{\epsilon}{2(1 + \|F\|_\infty)} ,$$

using the fact that the process F is bounded. Then we choose X elementary so t

$$\lceil\!\lceil F - X \rceil\!\rceil^* \le \frac{\epsilon}{2(1 + \|X'\|_\infty)} .$$

Then again $\lceil\!\lceil F \cdot F' - X \cdot X' \rceil\!\rceil^* \le \epsilon$, showing that $F \cdot F'$ is integrable, inasmuch as the product $X \cdot X'$ is an elementary integrand. ∎

Permanence Under Pointwise Limits of Sequences

The algebraic and order permanence properties of $\mathfrak{L}^1[\lceil\!\lceil\ \rceil\!\rceil^*]$ are thus as good as one might hope for, to wit as good as in the case of the Lebesgue integral. Let us now turn to the permanence properties concerning limits. The first result is plain from theorem 3.2.10.

Theorem 3.2.22 *$\mathfrak{L}^1[\lceil\!\lceil\ \rceil\!\rceil^*]$ is complete in $\lceil\!\lceil\ \rceil\!\rceil^*$-mean. Every mean Cauchy sequence (F_n) has a subsequence that converges pointwise $\lceil\!\lceil\ \rceil\!\rceil^*$-a.e. to a mean limit of (F_n).*

The existence of an a.e. convergent subsequence of a mean-convergent sequence (F_n) is frequently very helpful in identifying the limit, as we shall presently see. We know from ordinary integration theory that there is, in general, no hope that the sequence (F_n) itself converges almost everywhere.

Theorem 3.2.23 (The Monotone Convergence Theorem) *Let (F_n) be a monotone sequence of integrable processes with $\lim_{r \to 0} \sup_n \lceil\!\lceil rF_n \rceil\!\rceil^* = 0$. (For $p > 0$ and $\lceil\!\lceil\ \rceil\!\rceil^* = \lceil\!\lceil\ \rceil\!\rceil^*_{Z\text{-}p}$ this reads simply $\sup_n \lceil\!\lceil F_n \rceil\!\rceil^*_{Z\text{-}p} < \infty$.) Then (F_n) converges to its pointwise limit in mean.*

Proof. As $\big(F_n(\varpi)\big)$ is monotone it has a limit $F(\varpi)$ at all points ϖ of the base space, possibly $\pm\infty$. Let us assume first that the sequence $\big(F_n\big)$ is increasing. We start by showing that $\big(F_n\big)$ is mean-Cauchy. Indeed, assume it were not. There would then exist an $\epsilon > 0$ and a subsequence $\big(F_{n_k}\big)$ with $\lceil\!\lceil F_{n_{k+1}} - F_{n_k} \rceil\!\rceil^* > \epsilon$. There would further exist positive elementary integrands X_n with

$$\lceil\!\lceil (F_{n_{k+1}} - F_{n_k}) - X_k \rceil\!\rceil^* < 2^{-k}. \tag{$*$}$$

Let $|r| \le 1$ and $K < L \in \mathbb{N}$. Then

$$\begin{aligned}\left\lceil\!\!\left\lceil r\sum_{k=1}^{L} X_k \right\rceil\!\!\right\rceil^* &\le \left\lceil\!\!\left\lceil r\sum_{k=1}^{L}\big((F_{n_{k+1}} - F_{n_k}) - X_k\big) \right\rceil\!\!\right\rceil^* + \left\lceil\!\!\left\lceil r\sum_{k=1}^{L}(F_{n_{k+1}} - F_{n_k}) \right\rceil\!\!\right\rceil^* \\ &\le \left\lceil\!\!\left\lceil r\sum_{k=1}^{K}\big((F_{n_{k+1}} - F_{n_k}) - X_k\big) \right\rceil\!\!\right\rceil^* + 2^{-K} + \left\lceil\!\!\left\lceil rF_{n_{L+1}} \right\rceil\!\!\right\rceil^* + \lceil\!\lceil rF_{n_1} \rceil\!\rceil^*.\end{aligned}$$

Given $\epsilon > 0$ we first fix $K \in \mathbb{N}$ so large that $2^{-K} < \epsilon/4$. Then we find r_ϵ so that the other three terms are smaller than $\epsilon/4$ each, for $|r| \le r_\epsilon$. By assumption, r_ϵ can be so chosen independently of L. That is to say,

$$\sup_L \left\lceil\!\!\left\lceil r\textstyle\sum_{k=1}^{L} X_k \right\rceil\!\!\right\rceil^* \xrightarrow[r\to 0]{} 0\,.$$

Property (M) of the mean (see page 91) now implies that $\lceil\!\lceil X_k \rceil\!\rceil^* \to 0$. Thanks to $(*)$, $\lceil\!\lceil F_{n_{k+1}} - F_{n_k} \rceil\!\rceil^* \xrightarrow[k\to\infty]{} 0$, which is the desired contradiction. Now that we know that $\big(F_n\big)$ is Cauchy we employ theorem 3.2.10: there is a mean-limit F' and a subsequence[2] $\big(F_{n_k}\big)$ so that $F_{n_k}(\varpi)$ converges to $F'(\varpi)$ as $k \to \infty$, for all ϖ outside some negligible set N. For all ϖ, though, $F_n(\varpi) \xrightarrow[n\to\infty]{} F(\varpi)$. Thus

$$F(\varpi) = \lim_{n\to\infty} F_n(\varpi) = \lim_{k\to\infty} F_{n_k}(\varpi) = F'(\varpi) \quad \text{for } \varpi \notin N\ :$$

F is equal almost surely to the mean-limit F' and thus is a mean-limit itself. If $\big(F_n\big)$ is decreasing rather than increasing, $\big(-F_n\big)$ increases pointwise – and by the above in mean – to $-F$: again $F_n \to F$ in mean. ∎

Theorem 3.2.24 (The Dominated Convergence Theorem or DCT) *Let $\big(F_n\big)$ be a sequence of integrable processes. Assume both*

(i) $\big(F_n\big)$ converges pointwise $\lceil\!\lceil\ \rceil\!\rceil^*$-almost everywhere to a process F; and

(ii) there exists a process $G \in \mathfrak{F}[\lceil\!\lceil\ \rceil\!\rceil^*]$ with $|F_n| \le G$ for all indices $n \in \mathbb{N}$.

Then $\big(F_n\big)$ converges to F in $\lceil\!\lceil\ \rceil\!\rceil^*$-mean, and consequently F is integrable.

The Dominated Convergence Theorem is central. Most other results in integration theory follow from it. It is false without some domination condition like (ii), as is well known from ordinary integration theory.

---

[2] Not the same as in the previous argument, which was, after all, shown not to exist.

Proof. As in the proof of the Monotone Convergence Theorem we begin by showing that the sequence (F_n) is Cauchy. To this end consider the positive process

$$G_N = \sup\{|F_n - F_m| : m, n \geq N\} = \lim_{K\to\infty} \bigvee_{m,n=N}^{K} |F_n - F_m| \leq 2G .$$

Thanks to theorem 3.2.21 and the MCT, G_N is integrable. Moreover, $(G_N(\varpi))$ converges decreasingly to zero at all points ϖ at which $(F_n(\varpi))$ converges, that is to say, almost everywhere. Hence $\lceil\!\lceil G_N \rceil\!\rceil^* \to 0$. Now $\lceil\!\lceil F_n - F_m \rceil\!\rceil^* \leq \lceil\!\lceil G_N \rceil\!\rceil^*$ for $m, n \geq N$, so (F_n) is Cauchy in mean. Due to theorem 3.2.22 the sequence has a mean limit F' and a subsequence (F_{n_k}) that converges pointwise a.e. to F'. Since (F_{n_k}) also converges to F a.e., we have $F = F'$ a.e. Thus $\lceil\!\lceil F_n - F \rceil\!\rceil^* = \lceil\!\lceil F_n - F' \rceil\!\rceil^* \xrightarrow[n\to\infty]{} 0$. Now apply proposition 3.2.20. ∎

Integrable Sets

Definition 3.2.25 *A set is* ***integrable*** *if its indicator function is integrable.*[1]

Proposition 3.2.26 *The union and relative complement of two integrable sets are integrable. The intersection of a countable family of integrable sets is integrable. The union of a countable family of integrable sets is integrable provided that it is contained in an integrable set C.*

Proof. For ease of reading we use the same symbol for a set and its indicator function.[1] For instance, $A_1 \cup A_2 = A_1 \vee A_2$ in the sense that the indicator function on the left is the pointwise maximum of the two indicator functions on the right.

Let $A_1, A_2, \ldots$ be a countable family of integrable sets. Then

$$A_1 \cup A_2 = A_1 \vee A_2 ,$$

$$A_1 \backslash A_2 = A_1 - (A_1 \wedge A_2) ,$$

$$\bigcap_{n=1}^{\infty} A_n = \bigwedge_{n=1}^{\infty} A_n = \lim_{N\to\infty} \bigwedge_{n=1}^{N} A_n ,$$

and

$$\bigcup_{n=1}^{\infty} A_n = C - \bigwedge_{n=1}^{\infty} (C - A_n) ,$$

in the sense that the set on the left has the indicator function on the right, which is integrable by theorem 3.2.24. ∎

A collection of subsets of a set that is closed under taking finite unions, relative differences, and countable intersections is called a ***δ*-ring**. Proposition 3.2.26 can thus be read as saying that the integrable sets form a δ-ring.

Proposition 3.2.27 *Let F be an integrable process. (i) The sets $[F > r]$, $[F \ge r]$, $[F < -r]$, and $[F \le -r]$ are integrable, whenever $r \in \mathbb{R}$ is strictly positive. (ii) F is the limit a.e. and in mean of a sequence (F_n) of integrable step processes with $|F_n| \le |F|$.*

Proof. For the first claim, note that the set[1]

$$[F > 1] = \lim_{n\to\infty} 1 \wedge \big(n(F - F \wedge 1)\big)$$

is integrable. Namely, the processes $F_n = 1 \wedge \big(n(F - F \wedge 1)\big)$ are integrable and are dominated by $|F|$; by the Dominated Convergence Theorem, their limit is integrable. This limit is 0 at any point ϖ of the base space where $F(\varpi) \le 1$ and 1 at any point ϖ where $F(\varpi) > 1$; in other words, it is the (indicator function of the) set $[F > 1]$, which is therefore integrable. Note that here we use for the first (and only) time the fact that $\mathcal{E}$ is closed under chopping. The set $[F > r]$ equals $[F/r > 1]$ and is therefore integrable as well. Next, $[F \ge r] = \bigcap_{n>1/r}[F > r - 1/n]$, $[F < -r] = [-F > r]$, and $[F \le -r] = [-F \ge r]$.

For the next claim, let F_n be the step process over integrable sets[1]

$$F_n = \sum_{k=1}^{2^{2n}} k2^{-n} \cdot \big[k2^{-n} < F \le (k+1)2^{-n}\big]$$
$$+ \sum_{k=1}^{2^{2n}} -k2^{-n} \cdot \big[-k2^{-n} > F \ge -(k+1)2^{-n}\big] .$$

By (i), the sets

$$\big[k2^{-n} < F \le (k+1)2^{-n}\big] = \big[k2^{-n} < F\big] \setminus \big[(k+1)2^{-n} < F\big]$$

are integrable if $k \ne 0$. Thus F_n, being a linear combination of integrable processes, is integrable. Now (F_n) converges pointwise to F and is dominated by $|F|$, and the claim follows. ∎

Notation 3.2.28 *The integral of an integrable set A is written $\int A\ dZ$ or $\int_A dZ$. Let $F \in \mathfrak{L}^1[Z\text{-}p]$. With an integrable set A being a bounded (idempotent) process, the product $A \cdot F = 1_A \cdot F$ is integrable; its integral is variously written*

$$\int_A F\, dZ \quad \text{or} \quad \int A \cdot F\, dZ .$$

Exercise 3.2.29 Let (F_n) be a sequence of bounded integrable processes, all vanishing off the same integrable set A and converging uniformly to F. Then F is integrable.

Exercise 3.2.30 In the stochastic case there exists a countable collection of $\lceil\!\lceil\ \rceil\!\rceil^*$-integrable sets that covers the whole space, for example, $\{[\![0, k]\!] : k \in \mathbb{N}\}$. We say that the mean is ***σ-finite***. In consequence, any collection $\mathcal{M}$ of mutually disjoint non-negligible $\lceil\!\lceil\ \rceil\!\rceil^*$-integrable sets is at most countable.

3.3 Countable Additivity in p-Mean

The development so far rested on the assumption (IC-p) on page 89 that our L^p-integrator be continuous in L^p-mean along increasing sequences of $\mathcal{E}$, or σ-additive in p-mean. This assumption is, on the face of it, rather stronger than mere right-continuity in probability, and was needed to establish properties (iv) and (v) of Daniell's mean in theorem 3.1.6 on page 90. We show in this section that continuity in L^p-mean along increasing sequences is actually equivalent with right-continuity in probability, in the presence of the boundedness condition (B-p). First the case $p = 0$:

Lemma 3.3.1 *An L^0-integrator is σ-additive in probability.*

Proof. It is to be shown that for any decreasing sequence $\left(X^{(k)}\right)_{k=1}^{\infty}$ of elementary integrands with pointwise infimum zero, $\lim_k \int X^{(k)}\, dZ = 0$ in probability, under the assumptions (B-0) that Z is a bounded linear map from $\mathcal{E}$ to L^0 and (RC-0) that it is right-continuous in measure (exercise 3.1.5). As so often before, the argument is very nearly the same as in standard integration theory. Let us fix representations[1,3]

$$X_s^{(k)}(\omega) = f_0^{(k)}(\omega) \cdot [0,0]_s + \sum_{n=1}^{N(k)} f_n^{(k)}(\omega) \cdot \left(t_n^{(k)}, t_{n+1}^{(k)}\right]_s$$

as in equation (2.1.1) on page 46. Clearly

$$\int f_0^{(k)} \cdot [\![0]\!]\, dZ = f_0^{(k)} \cdot Z_0 \xrightarrow[k\to\infty]{} 0 \ :$$

we may as well assume that $f_0^{(k)} = 0$. Scaling reduces the situation to the case that $\left|X^{(k)}\right| \le 1$ for all $k \in \mathbb{N}$. It eases the argument further to assume that the partitions $\{t_1^{(k)}, \ldots, t_{N(k)}^{(k)}\}$ become finer as k increases.

Let then $\epsilon > 0$ be given. Let U be an instant past which the $X^{(k)}$ all vanish. The continuity condition (B-0) provides a $\delta > 0$ such that[1]

$$\lceil\!\lceil \delta \cdot [\![0, U]\!] \rceil\!\rceil_{Z\text{-}0} < \epsilon/3 \,. \qquad (*)$$

Next let us define instants $u_n^{(k)} < v_n^{(k)}$ as follows: for $k = 1$ we set $u_n^{(1)} = t_n^{(1)}$ and choose $v_n^{(1)} \in (t_n^{(1)}, t_{n+1}^{(1)})$ so that

$$\lceil\!\lceil Z_u - Z_t \rceil\!\rceil_0 < 3^{-1-n-1}\epsilon \ \text{ for } \ u_n^{(1)} \le t < u \le v_n^{(1)} \,; \qquad 1 \le n \le N(1) \,.$$

The right-continuity of Z makes this possible. The intervals $[u_n^{(1)}, v_n^{(1)}]$ are clearly mutually disjoint. We continue by induction. Suppose that $u_n^{(j)}$ and $v_n^{(j)}$ have been found for $1 \le j < k$ and $1 \le n \le N(j)$, and let $t_n^{(k)}$ be one

[3] $[0,0]_s$ is the indicator function of $\{0\}$ evaluated at s, etc.

of the partition points for $X^{(k)}$. If $t_n^{(k)}$ lies in one of the intervals previously constructed or is a left endpoint of one of them, say $t_n^{(k)} \in [u_m^{(j)}, v_m^{(j)})$, then we set $u_n^{(k)} = u_m^{(j)}$ and $v_n^{(k)} = v_m^{(j)}$; in the opposite case we set $u_n^{(k)} = t_n^{(k)}$ and choose $v_n^{(k)} \in (t_n^{(k)}, t_{n+1}^{(k)})$ so that

$$\lceil\!\lceil Z_u - Z_t \rceil\!\rceil_0 < 3^{-k-n-1}\epsilon \ \text{ for } \ u_n^{(k)} \le t < u \le v_n^{(k)} ; \qquad 1 \le n \le N(k) .$$

The right-continuity in probability of Z makes this possible. This being done we set

$$\mathring{N}^{(k)} \stackrel{\text{def}}{=} \bigcup_{n=1}^{N(k)} (u_n^{(k)}, v_n^{(k)}) \subset N^{(k)} \stackrel{\text{def}}{=} \bigcup_{n=1}^{N(k)} (u_n^{(k)}, v_n^{(k)}] \qquad k = 1, 2, \ldots .$$

Both $\mathring{N}^{(k)}$ and $N^{(k)}$ are finite unions of mutually disjoint intervals and increase with k. Furthermore

$$\sum_{n=1}^{N(k)} \left\lceil\!\!\left\lceil Z_{v_n'^{(k)}} - Z_{u_n'^{(k)}} \right\rceil\!\!\right\rceil_0 < \epsilon/3 , \quad \text{for any } \ u_n^{(k)} \le u_n'^{(k)} \le v_n'^{(k)} \le v_n^{(k)} . \qquad (**)$$

We shall estimate separately the integrals of the elementary integrands in the sum

$$X^{(k)} = X^{(k)} \cdot (N^{(k)} \times \Omega) \; + \; X^{(k)} \cdot \Big(1 - (N^{(k)} \times \Omega)\Big) .$$

$$\begin{aligned} \int X^{(k)} \cdot N^{(k)}\, dZ &= \sum_{n=1}^{N(k)} f_n^{(k)} \cdot \int \bigcup_{m=1}^{N(k)} (\!(t_n^{(k)}, t_{n+1}^{(k)}]\!] \cap (\!(u_m^{(k)}, v_m^{(k)}]\!]\, dZ \\ &= \sum_{n=1}^{N(k)} f_n^{(k)} \cdot \int (\!(t_n^{(k)} \vee u_n^{(k)}, t_{n+1}^{(k)} \wedge v_n^{(k)}]\!]\, dZ \\ &= \sum_{n=1}^{N(k)} f_n^{(k)} \cdot \Big(Z_{t_{n+1}^{(k)} \wedge v_n^{(k)}} - Z_{t_n^{(k)} \vee u_n^{(k)}} \Big) . \end{aligned}$$

Since $\big|f_n^{(k)}\big| \le 1$, inequality $(**)$ yields

$$\left\lceil\!\!\left\lceil \int X^{(k)} \cdot N^{(k)}\, dZ \right\rceil\!\!\right\rceil_0 \le \epsilon/3 . \qquad (***)$$

Let us next estimate the remaining summand $X'^{(k)} = X^{(k)} \cdot \big((1 - N^{(k)}) \times \Omega\big)$. We start on this by estimating the process

$$X^{(k)} \cdot \big((1 - \mathring{N}^{(k)}) \times \Omega\big) ,$$

which evidently majorizes $X'^{(k)}$. Since every partition point of $X^{(k)}$ lies either inside one of the intervals $(u_n^{(k)}, v_n^{(k)})$ that make up $N^{(k)}$ or is a left endpoint of one of them, the paths of $X^{(k)} \cdot \big((1 - \mathring{N}^{(k)}) \times \Omega\big)$ are upper semicontinuous (see page 376). That is to say, for every $\omega \in \Omega$ and $\alpha > 0$, the set

$$C_\alpha(\omega) = \left\{ s \in \mathbb{R}_+ : X_s^{(k)}(\omega) \cdot \bigl(1 - \mathring{N}^{(k)}\bigr) \geq \alpha \right\}$$

is a finite union of closed intervals and is thus compact. These sets shrink as k increases and have void intersection. For every $\omega \in \Omega$ there is therefore an index $K(\omega)$ such that $C_\alpha(\omega) = \emptyset$ for all $k \geq K(\omega)$. We conclude that the maximal function

$$\bigl(X'^{(k)}\bigr)_U^\star = \sup_{0 \leq s \leq U} X_s'^{(k)} \leq \sup_{0 \leq s \leq U} X^{(k)} \cdot \bigl((1 - \mathring{N}^{(k)}) \times \Omega\bigr)$$

decreases pointwise to zero, *a fortiori* in measure. Let then K be so large that for $k \geq K$ the set

$$B \stackrel{\text{def}}{=} \left[\bigl(X'^{(k)}\bigr)_U^\star > \delta \right] \text{ has } \mathbb{P}[B] < \epsilon/3\,.$$

The dZ-integrals of $X'^{(k)}$ and $X'^{(k)} \wedge \delta$ agree pathwise outside B. Measured with $\lceil\!\lceil \ \rceil\!\rceil_0$ they differ thus by at most $\epsilon/3$. Since $X'^{(k)} \wedge \delta \leq \delta \cdot [\![0, U]\!]$, inequality $(*)$ yields

$$\left\lceil\!\!\left\lceil \int X'^{(k)} \wedge \delta \, dZ \right\rceil\!\!\right\rceil_0 \leq \epsilon/3\ , \text{ and thus } \left\lceil\!\!\left\lceil \int X'^{(k)} \, dZ \right\rceil\!\!\right\rceil_0 \leq 2\epsilon/3\,, \quad k \geq K\,.$$

In view of $(***)$ we get $\lceil\!\lceil \int X^{(k)} \, dZ \rceil\!\rceil_0 \leq \epsilon$ for $k \geq K$. ∎

Proposition 3.3.2 *An L^p-integrator is σ-additive in p-mean,* $0 \leq p < \infty$.

Proof. For $p = 0$ this was done in lemma 3.3.1 above, so we need to consider only the case $p > 0$. Part (ii) of the Stone–Weierstraß theorem A.2.2 provides a locally compact Hausdorff space $\widehat{\boldsymbol{B}}$ and a map $j : \boldsymbol{B} \to \widehat{\boldsymbol{B}}$ with dense image such that every $X \in \mathcal{E}$ is of the form $\widehat{X} \circ j$ for some unique continuous function $\widehat{X}$ on $\widehat{\boldsymbol{B}}$. $\widehat{X}$ is called the Gelfand transform of X. The map $X \mapsto \widehat{X}$ is an algebraic and order isomorphism of $\mathcal{E}$ onto an algebra and vector lattice $\widehat{\mathcal{E}}$ closed under chopping of continuous bounded functions of compact support on $\widehat{\boldsymbol{B}}$ ($\widehat{X}$ has support in $\widehat{[X{\neq}0]} \in \widehat{\mathcal{E}}$). The Gelfand transform $\widehat{\int}$, defined by

$$\widehat{\int} \widehat{X} \stackrel{\text{def}}{=} \int X \, dZ\,, \qquad X \in \mathcal{E}\,.$$

is plainly a vector measure on $\widehat{\mathcal{E}}$ with values in L^p that satisfies (B-p). (IC-p) is also satisfied, thanks to Dini's theorem A.2.1. For if the sequence $\bigl(\widehat{X}^{(n)}\bigr)$ in $\widehat{\mathcal{E}}$ increases pointwise to the continuous (!) function $\widehat{X} \in \widehat{\mathcal{E}}$, then the convergence is uniform and (B-p) implies that $\widehat{\int} \widehat{X}^{(n)} \to \widehat{\int} \widehat{X}$ in p-mean. Daniell's procedure of the preceding pages provides an integral extension of $\widehat{\int}$ for which the Dominated Convergence Theorem holds.

Let us now consider an increasing sequence $(X^{(n)})$ in $\mathcal{E}_+$ that increases pointwise on $\boldsymbol{B}$ to $X \in \mathcal{E}$. The extensions $\widehat{X}^{(n)}$ will increase on $\widehat{\boldsymbol{B}}$ to some function $\widehat{H}$. While $\widehat{H}$ does not necessarily equal the extension $\widehat{X}$ (!), it is clearly less than or equal to it. By the Dominated Convergence Theorem for the integral extension of $\int$, $\int X^{(n)}\,dZ = \widehat{\int} \widehat{X}^{(n)}$ converges *in p-mean* to an element f of L^p. Now Z is certainly an L^0-integrator and thus $\int X^{(n)}\,dZ \to \int X\,dZ$ in measure (lemma 3.3.1). Thus $f = \int X\,dZ$, and $\int X^{(n)}\,dZ \to \int X\,dZ$ in p-mean. This very argument is repeated in slightly more generality in corollary A.2.7 on page 370. ∎

Exercise 3.3.3 Assume that for every $t \geq 0$, $\mathcal{A}_t$ is an algebra or vector lattice closed under chopping of $\mathcal{F}_t$-adapted bounded random variables that contains the constants and generates $\mathcal{F}_t$. Let $\mathcal{E}^0$ denote the collection of all elementary integrands X that have $X_t \in \mathcal{A}_t$ for all $t \geq 0$. Assume further that the right-continuous adapted process Z satisfies

$$\lceil Z^t \rceil^0_{\mathcal{I}^p} \stackrel{\text{def}}{=} \sup\left\{ \left\| \int X\,dZ \right\|_p : X \in \mathcal{E}^0,\ |X| \leq 1 \right\} < \infty$$

for some $p > 0$ and all $t \geq 0$. Then Z is an L^p-integrator, and $\lceil Z^t \rceil^0_{\mathcal{I}^p} = \lceil Z^t \rceil_{\mathcal{I}^p}$ for all t.

Exercise 3.3.4 Let $0 < p < \infty$. An L^0-integrator Z is a local L^p-integrator iff there are arbitrarily large stopping times T such that $\{\int [\![0, T]\!] \cdot X\,dZ : |X| \leq 1\}$ is bounded in L^p.

The Integration Theory of Vectors of Integrators

We have mentioned before that often whole vectors $\boldsymbol{Z} = (Z^1, Z^2, \ldots, Z^d)$ of integrators drive a stochastic differential equation. It is time to consider their integration theory. An obvious way is to regard every component Z^η as an L^p-integrator, to declare $\boldsymbol{X} = (X_1, X_2, \ldots, X_d)$ $\boldsymbol{Z}$-integrable if X_η is Z^η–p-integrable for every $\eta \in \{1, \ldots, d\}$, and to define

$$\int \boldsymbol{X}\,d\boldsymbol{Z} \stackrel{\text{def}}{=} \int X_\eta\,dZ^\eta = \sum_{1 \leq \eta \leq d} \int X_\eta\,dZ^\eta\,, \tag{3.3.1}$$

simply extending the definition (2.2.2). Let us take another point of view, one that leads to better constants in estimates and provides a guide to the integration theory of random measures (section 3.10). Denote by $\boldsymbol{H}$ the discrete space $\{1, \ldots, d\}$ and by $\check{\boldsymbol{B}}$ the set $\boldsymbol{H} \times \boldsymbol{B}$ equipped with its elementary integrands $\check{\mathcal{E}} \stackrel{\text{def}}{=} C_{00}(\boldsymbol{H}) \otimes \mathcal{E}$. Now read a d-tuple $\boldsymbol{Z} = (Z^1, Z^2, \ldots, Z^d)$ of processes on $\boldsymbol{B}$ not as a vector-valued function on $\boldsymbol{B}$ but rather as a ***scalar*** function $(\eta, \varpi) \mapsto Z^\eta(\varpi)$ on the d-fold product $\check{\boldsymbol{B}}$. In this interpretation $\boldsymbol{X} \mapsto \int \boldsymbol{X}\,d\boldsymbol{Z}$ is a vector measure $\check{\mathcal{E}} \to L^p(\mathbb{P})$, and the extension theory of the previous sections applies. In particular, THE Daniell mean is defined as

$$\lceil\!\lceil \boldsymbol{F} \rceil\!\rceil^*_{\boldsymbol{Z}-p} \stackrel{\text{def}}{=} \inf_{\substack{\boldsymbol{H} \in \check{\mathcal{E}}^\uparrow,\\ \boldsymbol{H} \geq |\boldsymbol{F}|}} \ \sup_{\substack{\boldsymbol{X} \in \check{\mathcal{E}},\\ |\boldsymbol{X}| \leq \boldsymbol{H}}} \left\| \int \boldsymbol{X}\,d\boldsymbol{Z} \right\|_p \tag{3.3.2}$$

on functions $F : \check{B} \to \overline{\mathbb{R}}$. It is a fine exercise toward checking one's understanding of Daniell's procedure to show that $\int \cdot\, dZ$ satisfies (IC-p), that therefore $\| \ \|^*_{Z\text{-}p}$ is a mean satisfying

$$\left\lceil\!\left\lceil \int X\, dZ \right\rceil\!\right\rceil_p \le \| X \|^*_{Z\text{-}p}\,, \tag{3.3.3}$$

and that not only the integration theory developed so far but its continuation in the subsequent sections applies *mutatis perpauculis mutandis*. In particular, inequality (3.3.3) will imply that there is a unique extension

$$\int \cdot\, dZ : \mathfrak{L}^1[\lceil\!\lceil\ \rceil\!\rceil^*_{Z\text{-}p}] \to L^p$$

satisfying the same inequality. That extension is actually given by equation (3.3.1). For more along these lines see section 3.10.

3.4 Measurability

Measurability describes the local structure of the integrable processes. Lusin observed that Lebesgue integrable functions on the line are uniformly continuous on arbitrarily large sets. It is rather intuitive to use this behavior to *define* measurability. It turns out to be efficient as well.

As before, $\lceil\!\lceil\ \rceil\!\rceil^*$ is an arbitrary mean on the algebra and vector lattice closed under chopping $\mathcal{E}$ of bounded functions that live on the ambient set B. In order to be able to speak about the uniform continuity of a function on a set $A \subset B$, the ambient space B is equipped with the ***$\mathcal{E}$-uniformity***, the smallest uniformity with respect to which the functions of $\mathcal{E}$ are all uniformly continuous. The reader not yet conversant with uniformities may wish to read page 373 up to lemma A.2.16 on page 375 and to note the following: to say that a real-valued function on $A \subset B$ is $\mathcal{E}$-uniformly continuous is the same as saying that it agrees with the restriction to A of a function in the uniform closure of $\mathcal{E} \oplus \mathbb{R}$ or that it is, on A, the uniform limit of functions in $\mathcal{E} \oplus \mathbb{R}$. To say that a numerical function on $A \subset B$ is $\mathcal{E}$-uniformly continuous is the same as saying that it is, on A, the uniform limit of functions in $\mathcal{E} \oplus \mathbb{R}$, *with respect to the arctan metric.*

By way of motivation of definition 3.4.2 we make the following observation, whose proof is left to the reader:

Observation 3.4.1 *Let $F : B \to \mathbb{R}$ be $(\mathcal{E}, \lceil\!\lceil\ \rceil\!\rceil^*)$-integrable and $\epsilon > 0$. There exists a set $U \in \mathcal{E}_+^\uparrow$ with $\lceil\!\lceil U \rceil\!\rceil^* \le \epsilon$ on whose complement F is the uniform limit of elementary integrands and thus is uniformly continuous.*

Definition 3.4.2 *Let A be a $\lceil\!\lceil\ \rceil\!\rceil^*$-integrable set. A process[4] F almost everywhere defined on A is called $\lceil\!\lceil\ \rceil\!\rceil^*$**-measurable on** A if for every $\epsilon > 0$*

[4] "Process" shall mean any $\lceil\!\lceil\ \rceil\!\rceil^*$-a.e. defined function on the ambient space that has values in some uniform space.

there is a $\lceil\!\lceil\ \rceil\!\rceil^*$-integrable subset A_0 of A with[1] $\lceil\!\lceil A \setminus A_0 \rceil\!\rceil^* < \epsilon$ *on which F is $\mathcal{E}$-uniformly continuous. A process F is called $\lceil\!\lceil\ \rceil\!\rceil^*$-**measurable** if it is measurable on every integrable set.*

*Unless there is need to stress that this definition refers to the mean $\lceil\!\lceil\ \rceil\!\rceil^*$, we shall simply talk about measurability. If we want to make the point that $\lceil\!\lceil\ \rceil\!\rceil^*$ is Daniell's mean $\lceil\!\lceil\ \rceil\!\rceil^*_{Z\text{-}p}$, we shall talk about **Z-p-measurability** (this is actually independent of p – see corollary 3.6.11 on page 128).*

This definition is quite intuitive, describing as it does a considerable degree of smoothness. It says that F is measurable if it is on arbitrarily large sets as smooth as an elementary integrand, in other words, that it is "***largely as smooth as an elementary integrand.***" It is also quite workable in that it admits fast proofs of the permanence properties. We start with a tiny result that will however facilitate the arguments greatly.

Lemma 3.4.3 *Let A be an integrable set and (F_n) a sequence of processes that are measurable on A. For every $\epsilon > 0$ there exists an integrable subset A_0 of A with $\lceil\!\lceil A \setminus A_0 \rceil\!\rceil^* \le \epsilon$ such that every one of the F_n is uniformly continuous on A_0.*

Proof. Let $A_1 \subset A$ be integrable with $\lceil\!\lceil A \setminus A_1 \rceil\!\rceil^* < \epsilon \cdot 2^{-1}$ and so that, on A_1, F_1 is uniformly continuous. Next let $A_2 \subset A_1$ be integrable with $\lceil\!\lceil A_1 \setminus A_2 \rceil\!\rceil^* < \epsilon \cdot 2^{-2}$ and so that, on A_2, F_2 is uniformly continuous. Continue by induction, and set $A_0 = \bigcap_{n=1}^{\infty} A_n$. Then A_0 is integrable due to proposition 3.2.26,

$$\lceil\!\lceil A \setminus A_0 \rceil\!\rceil^* = \left\lceil\!\!\left\lceil (A \setminus A_1) \cup \bigcup_{n>1} (A_n \setminus A_{n-1}) \right\rceil\!\!\right\rceil^* \le \sum \epsilon \cdot 2^{-n} = \epsilon \,,$$

by the countable subadditivity of $\lceil\!\lceil\ \rceil\!\rceil^*$, and every F_n is uniformly continuous on A_0, inasmuch as it is so on the larger set A_n. ∎

Permanence Under Limits of Sequences

Theorem 3.4.4 (Egoroff's Theorem) *Let (F_n) be a sequence of $\lceil\!\lceil\ \rceil\!\rceil^*$-measurable processes with values in a metric space (S, ρ), and assume that (F_n) converges $\lceil\!\lceil\ \rceil\!\rceil^*$-almost everywhere to a process F. Then F is $\lceil\!\lceil\ \rceil\!\rceil^*$-measurable.*

*Moreover, for every integrable set A and $\epsilon > 0$ there is an integrable subset A_0 of A with $\lceil\!\lceil A \setminus A_0 \rceil\!\rceil^* < \epsilon$ on which (F_n) converges uniformly to F – we shall describe this behavior by saying "(F_n) **converges uniformly on arbitrarily large sets,**" or even simply by "(F_n) **converges largely uniformly.**"*

Proof. Let an integrable set A and an $\epsilon > 0$ be given. There is an integrable set $A_1 \subset A$ with $\lceil\!\lceil A \setminus A_1 \rceil\!\rceil^* < \epsilon/2$ on which every one of the F_n is uniformly continuous. Then $\rho(F_m, F_n)$ is uniformly continuous on A_1, and therefore is, on A_1, the uniform limit of a sequence in $\mathcal{E}$, and thus $A_1 \cdot \rho(F_m, F_n)$ is integrable for every $m, n \in \mathbb{N}$ (exercise 3.2.29). Therefore

$$A_1 \cap \left[\rho(F_m, F_n) > \frac{1}{r}\right]$$

is an integrable set for $r = 1, 2, \ldots$, and then so is the set (see proposition 3.2.26)

$$B_p^r \stackrel{\text{def}}{=} A_1 \cap \bigcup_{m,n \geq p} \left[\rho(F_m, F_n) > \frac{1}{r}\right].$$

As p increases, B_p^r decreases, and the intersection $\bigcap_p B_p^r$ is contained in the negligible set of points where (F_n) does not converge. Thus $\lim_{p\to\infty} \lceil\!\lceil B_p^r \rceil\!\rceil^* = 0$. There is a natural number $p(r)$ such that $\lceil\!\lceil B_{p(r)}^r \rceil\!\rceil^* < 2^{-r-1}\epsilon$. Set

$$B \stackrel{\text{def}}{=} \bigcup_r B_{p(r)}^r \quad \text{and} \quad A_0 \stackrel{\text{def}}{=} A_1 \setminus B .$$

It is evident that $\lceil\!\lceil A_1 \setminus A_0 \rceil\!\rceil^* = \lceil\!\lceil B \rceil\!\rceil^* < \epsilon/2$ and thus $\lceil\!\lceil A \setminus A_0 \rceil\!\rceil^* < \epsilon$. It is left to be shown that (F_n) converges uniformly on A_0. The limit F is then clearly also uniformly continuous there. To this end, let $\delta > 0$ be given. We let $N = p(r)$, where r is chosen so that $1/r < \delta$. Now if ϖ is any point in A_0 and $m, n \geq N$, then ϖ is not in the "bad set" $B_{p(r)}^r$; therefore $\rho\big(F_n(\varpi), F_m(\varpi)\big) \leq 1/r < \delta$, and thus $\rho\big(F(\varpi), F_n(\varpi)\big) \leq \delta$ for all $\varpi \in A_0$ and $n \geq N$. ∎

Corollary 3.4.5 *A numerical process*[4] F *is* $\lceil\!\lceil\ \rceil\!\rceil^*$*-measurable if and only if it is* $\lceil\!\lceil\ \rceil\!\rceil^*$*-almost everywhere the limit of a sequence of elementary integrands.*

Proof. The condition is sufficient by Egoroff's theorem. Toward its necessity we must assume that the mean is σ-finite, in the sense that there exists a countable collection of $\lceil\!\lceil\ \rceil\!\rceil^*$-integrable subsets B_n that exhaust the ambient set. The B_n can and will be chosen increasing with n. In the case of the stochastic integral take $B_n = [\![0, n]\!]$. Then find, for every integer n, a $\lceil\!\lceil\ \rceil\!\rceil^*$-integrable subset G_n of B_n with $\lceil\!\lceil B_n \setminus G_n \rceil\!\rceil^* < 2^{-n}$ and an elementary integrand X_n that differs from F uniformly by less than 2^{-n} on G_n. The sequence (X_n) converges to F in every point of $G = \bigcup_N \bigcap_{n \geq N} G_n$, a set of $\lceil\!\lceil\ \rceil\!\rceil^*$-negligible complement. ∎

Permanence Under Algebraic and Order Operations

Theorem 3.4.6 *(i) Suppose that* $F_1, \ldots, F_N$ *are* $\lceil\!\lceil\ \rceil\!\rceil^*$*-measurable processes*[4] *with values in complete uniform spaces* $(S_1, \mathfrak{u}_1), \cdots, (S_N, \mathfrak{u}_N)$*, and* ϕ *is a continuous map from the product* $S_1 \times \ldots \times S_N$ *to another uniform space* $(S, \mathfrak{u})$*. Then the composition* $\phi(F_1, \ldots, F_N)$ *is* $\lceil\!\lceil\ \rceil\!\rceil^*$*-measurable. (ii) Algebraic and order combinations of measurable processes are measurable.*

Exercise 3.4.7 The conclusion (i) stays if ϕ is a Baire function.

Proof. (i) Let an integrable set A and an $\epsilon > 0$ be given. There is an integrable subset A_0 of A with $\lceil\!\lceil A - A_0 \rceil\!\rceil^* < \epsilon$ on which every one of the F_n is uniformly continuous. By lemma A.2.16 (iv) the sets $F_n(A_0) \subset S_n$ are relatively compact, and by exercise A.2.15 ϕ is uniformly continuous on the compact product Π of their closures. Thus $\phi(F_1, \ldots, F_N) : A_0 \to \Pi \to S$ is uniformly continuous as the composition of uniformly continuous maps.

(ii) Let F_1, F_2 be measurable. Inasmuch as $+ : \mathbb{R}^2 \to \mathbb{R}$ is continuous, $F_1 + F_2$ is measurable. The same argument applies with $+$ replaced by $\cdot, \wedge, \vee$, etc. ∎

Exercise 3.4.8 (Localization Principle) The notion of measurability is local: (i) A process F $\lceil\!\lceil\ \rceil\!\rceil^*$-measurable on the $\lceil\!\lceil\ \rceil\!\rceil^*$-integrable set A is $\lceil\!\lceil\ \rceil\!\rceil^*$-measurable on every integrable subset of A. (ii) A process F $\lceil\!\lceil\ \rceil\!\rceil^*$-measurable on the $\lceil\!\lceil\ \rceil\!\rceil^*$-integrable sets A_1, A_2 is measurable on their union. (iii) If the process F is $\lceil\!\lceil\ \rceil\!\rceil^*$-measurable on the $\lceil\!\lceil\ \rceil\!\rceil^*$-integrable sets $A_1, A_2, \ldots$, then it is measurable on every $\lceil\!\lceil\ \rceil\!\rceil^*$-integrable subset of their union $\bigcup_n A_n$.

Exercise 3.4.9 (i) Let $\mathcal{D}$ be any collection of bounded functions whose linear span is dense in $\mathfrak{L}^1[\lceil\!\lceil\ \rceil\!\rceil^*]$. Replacing the $\mathcal{E}$-uniformity on $\boldsymbol{B}$ by the $\mathcal{D}$-uniformity does not change the notion of measurability. In the case of the stochastic integral, therefore, a real-valued process is measurable if and only if it equals on arbitrarily large sets a continuous adapted process (take $\mathcal{D} = \mathfrak{C}$).

(ii) The notion of measurability of $F : \boldsymbol{B} \to S$ also does not change if the uniformity on S is replaced with another one that has the same topology, provided both uniformities are complete (apply theorem 3.4.6 to the identity map $S \to S$).

In particular, a process that is measurable as a numerical function and happens to take only real values is measurable as a real-valued function.

The Integrability Criterion

Let us now show that the notion of measurability captures exactly the "local smoothness" of the integrable processes:

Theorem 3.4.10 *A numerical process F is $\lceil\!\lceil\ \rceil\!\rceil^*$-integrable if and only if it is $\lceil\!\lceil\ \rceil\!\rceil^*$-measurable and finite in $\lceil\!\lceil\ \rceil\!\rceil^*$-mean.*

Proof. An integrable process is finite for the mean (exercise 3.2.15) and, being the pointwise a.e. limit of a sequence of elementary integrands (theorem 3.2.22), is measurable (theorem 3.4.4). The two conditions are therefore necessary.

To establish the sufficiency let $\mathcal{C}$ be a maximal collection of mutually disjoint non-negligible integrable sets on which F is uniformly continuous. Due to the stipulated σ-finiteness of our mean there exists a countable collection $\{B_k\}$ of integrable sets that cover the base space, and $\mathcal{C}$ is countable: $\mathcal{C} = \{A_1, A_2, \ldots\}$ (see exercise 3.2.30). Now the complement of $C \stackrel{\text{def}}{=} \bigcup_n A_n$ is negligible; if it were not, then one of the integrable sets $B_k \setminus C$ would not be negligible and would contain a non-negligible integrable subset on which F is uniformly continuous – this would contradict the maximality of $\mathcal{C}$. The processes $F_n = F\cdot(\bigcup_{k\le n} A_k)$ are integrable, converge a.e. to F, and are dominated by $|F| \in \mathfrak{F}[\lceil\!\lceil\ \rceil\!\rceil^*]$. Thanks to the Dominated Convergence Theorem, F is integrable. ∎

Measurable Sets

Definition 3.4.11 ***A set is measurable*** *if its indicator function is measurable*[1] *– we write "measurable" instead of "$\lceil\!\lceil\ \rceil\!\rceil^*$-measurable," etc.*

Since sets are but idempotent functions, it is easy to see how their measurability interacts with that of arbitrary functions:

Theorem 3.4.12 *(i) A set M is measurable if and only if its intersection with every integrable set A is integrable. The measurable sets form a σ-algebra.*

(ii) If F is a measurable process, then the sets $[F > r]$, $[F \ge r]$, $[F < r]$, and $[F \le r]$ are measurable for any number r, and F is almost everywhere the pointwise limit of a sequence (F_n) of step processes with measurable steps.

(iii) A numerical process F is measurable if and only if the sets $[F > d]$ are measurable for every dyadic rational d.

Proof. These are standard arguments. (i) if $M \cap A$ is integrable, then it is measurable on A. The condition is thus sufficient. Conversely, if M is measurable and A integrable, then $M \cap A$ is measurable and has finite mean; so it is integrable (3.4.10). For the second claim let $A_1, A_2, \ldots$ be a countable family of measurable sets.[1]

Then
$$(A_1)^c = 1 - A_1\ ,$$

$$\bigcap_{n=1}^{\infty} A_n = \bigwedge_{n=1}^{\infty} A_n = \lim_{N\to\infty} \bigwedge_{n=1}^{N} A_n,$$

and
$$\bigcup_{n=1}^{\infty} A_n = \bigvee_{n=1}^{\infty} A_n = \lim_{N\to\infty} \bigvee_{n=1}^{N} A_n,$$

in the sense that the set on the left has the indicator function on the right, which is measurable.

(ii) For the first claim, note that the process

$$\lim_{n\to\infty} 1 \wedge \big(n(F - F\wedge 1)\big)$$

is measurable, in view of the permanence properties. It vanishes at any point ϖ where $F(\varpi) \le 1$ and equals 1 at any point ϖ where $F(\varpi) > 1$; in other words, this limit is the (indicator function of the) set $[F > 1]$, which is therefore measurable. The set $[F > r]$ equals $[F/r > 1]$ when $r > 0$ and is thus measurable as well. $[F > 0] = \bigcup_{n=1}^{\infty}[F > 1/n]$ is measurable. Next, $[F \ge r] = \bigcap_{n>1/r}[F > r - 1/n]$, $[F < -r] = [-F > r]$, and $[F \le -r] = [-F \ge r]$. Finally, when $r \le 0$, then $[F > r]{=}[-F \ge -r]^c$, etc.

For the next claim, let F_n be the step process over measurable sets[1]

$$F_n = \sum_{k=-2^{2n}}^{2^{2n}} k2^{-n} \cdot \left[k2^{-n} < F \leq (k+1)2^{-n}\right] . \qquad (*)$$

The sets $[k2^{-n} < F \leq (k+1)2^{-n}] = [k2^{-n} < F] \cap [(k+1)2^{-n} < F]^c$ are measurable, and the claim follows by inspection.

(iii) The necessity follows from the previous result. So does the sufficiency: The sets appearing in $(*)$ are then measurable, and F is as well, being the limit of linear combinations of measurable processes. ∎

3.5 Predictable and Previsible Processes

The Borel functions on the line are measurable for every measure. They form the *smallest* class that contains the elementary functions and has the usual permanence properties for measurability: closure under algebraic and order combinations, and under pointwise limits of sequences – and therein lies their virtue. Namely, they lend themselves to this argument: a property of functions that holds for the elementary ones and persists under limits of sequences, etc., holds for Borel functions. For instance, if two measures μ, ν satisfy $\mu(\phi) \leq \nu(\phi)$ for step functions ϕ, then the same inequality is satisfied on Borel functions ϕ – observe that it makes no sense in general to state this inequality for integrable functions, inasmuch as a μ-integrable function may not even be ν-measurable. But the Borel functions also form a *large* class in the sense that every function measurable for some measure μ is μ-a.e. equal to a Borel function, and that takes the sting out of the previous observation: on that Borel function μ and ν can be compared.

It is the purpose of this section to identify and analyze the stochastic analog of the Borel functions.

Predictable Processes

The Borel functions on the line are the sequential closure[5] of the step functions or elementary integrands $\mathfrak{e}$. The analogous notion for processes is this:

Definition 3.5.1 *The sequential closure (in* $\overline{\mathbb{R}}^{\boldsymbol{B}}$ *!) of the elementary integrands* $\mathcal{E}$ *is the collection of* ***predictable processes*** *and is denoted by* $\mathcal{P}$. *The* σ*-algebra of sets in* $\mathcal{P}$ *is also denoted by* $\mathcal{P}$. *If there is need to indicate the filtration, we write* $\mathcal{P}[\mathcal{F}_.]$.

An elementary integrand X is prototypically predictable in the sense that its value X_t at any time t is measurable on some strictly earlier σ-algebra $\mathcal{F}_s$: at time s the value X_t can be foretold. This explains the choice of the word "predictable."

[5] See pages 391–393.

$\mathcal{P}$ is of course also the name of the σ-algebra generated by the idempotents (sets[6]) in $\mathcal{E}$. These are the finite unions of elementary stochastic intervals of the form $(\!(S,T]\!]$. This again is the difference of $[\![0,T]\!]$ and $[\![0,S]\!]$. Thus $\mathcal{P}$ also agrees with the σ-algebra spanned by the family of stochastic intervals $\{[\![0,T]\!] : \text{T an elementary stopping time}\}$.

Egoroff's theorem 3.4.4 implies that a predictable process is measurable for any mean $\lceil\!\lceil\ \rceil\!\rceil^*$. Conversely, any $\lceil\!\lceil\ \rceil\!\rceil^*$-measurable process F coincides $\lceil\!\lceil\ \rceil\!\rceil^*$-almost everywhere with some predictable process. Indeed, there is a sequence $(X^{(n)})$ of elementary integrands that converges $\lceil\!\lceil\ \rceil\!\rceil^*$-a.e. to F (see corollary 3.4.5); the predictable process $\liminf X^{(n)}$ qualifies.

The next proposition provides a stock-in-trade of predictable processes.

Proposition 3.5.2 *(i) Any left-open right-closed stochastic interval $(\!(S,T]\!]$, $S \le T$, is predictable. In fact, whenever f is a random variable measurable on $\mathcal{F}_S$, then $f \cdot (\!(S,T]\!]$ is predictable[6]; if it is Z–p-integrable, then its integral is as expected – see exercise 2.1.14:*

$$f \cdot (Z_T - Z_S) \in \int f \cdot (\!(S,T]\!]\, dZ\,. \tag{3.5.1}$$

(ii) A left-continuous adapted process X is predictable. The continuous adapted processes generate $\mathcal{P}$.

Proof. (i) Let $T^{(n)}$ be the stopping times of exercise 1.3.20:

$$T^{(n)} \stackrel{\text{def}}{=} \sum_{k=0}^{\infty} \frac{k+1}{n} \cdot \left[\frac{k}{n} < T \le \frac{k+1}{n}\right] \; + \; \infty \cdot [T = \infty]\,, \qquad n \in \mathbb{N}.$$

Recall that $(T^{(n)})$ decreases to T. Next let $f^{(m)}$ be $\mathcal{F}_S$-measurable simple functions with $|f^{(m)}| \le |f|$ that converge pointwise to f. Set[6]

$$\begin{aligned} X^{(m,n)} &\stackrel{\text{def}}{=} f^{(m)} \cdot (\!(S^{(n)} \wedge m, T^{(n)} \wedge m]\!] \\ &= f^{(m)} \cdot [S^{(n)} < m] \cdot (\!(S^{(n)} \wedge m, T^{(n)} \wedge m]\!]\,. \end{aligned}$$

Since $f^{(m)} \in \mathcal{F}_S \subset \mathcal{F}_{S^{(n)}}$ (exercise 1.3.16),

$$f^{(m)} \cdot [S^{(n)} \le m] \cdot [S^{(n)} \wedge m \le t] = \begin{cases} f^{(m)} \cdot [S^{(n)} \le m] \in \mathcal{F}_m \subset \mathcal{F}_t\,, & t \ge m, \\ f^{(m)} \cdot [S^{(n)} \le t] \in \mathcal{F}_t\,, & t < m, \end{cases}$$

and $f^{(m)} \cdot [S^{(n)} \le m]$ is measurable on $\mathcal{F}_{S^{(n)} \wedge m}$. By exercise 2.1.5, $X^{(m,n)}$ is an elementary integrand. Therefore

$$f \cdot (\!(S,T]\!] = \lim_{m \to \infty} \lim_{n \to \infty} X^{(m,n)}$$

---

[6] In accordance with convention A.1.5 on page 364, sets are identified with their (idempotent) indicator functions. A stochastic interval $(\!(S,T]\!]$, for instance, has at the instant s the value $(\!(S,T]\!]_s = [S < s \le T] = \begin{cases} 1 & \text{if } S(\omega) < s \le T(\omega), \\ 0 & \text{elsewhere.} \end{cases}$

is predictable. If Z is an L^p-integrator, then the integral of $X^{(m,n)}$ is $f^{(m)}\cdot\big(Z_{T^{(n)}\wedge m}-Z_{S^{(n)}\wedge m}\big)$ (ibidem). Now $|X^{(m,n)}|\le\|f^{(m)}\|_\infty\cdot[\![0,m]\!]$, so the Dominated Convergence Theorem in conjunction with the right-continuity of Z gives[6]

$$f^{(m)}\cdot\big(Z_{T\wedge m}-Z_{S\wedge m}\big)\in\int f^{(m)}\cdot(\!(S\wedge m,T\wedge m]\!]\,dZ$$

as $n\to\infty$. The integrands are dominated by $|f|\cdot(\!(S,T]\!]$; so if this process is Z–p-integrable, then a second application of the Dominated Convergence Theorem produces (3.5.1) as $m\to\infty$.

(ii) To start with assume that X is continuous. Thanks to corollary 1.3.12 the random times $T_0=0$,

$$T^n_{k+1}=\inf\{t:\ \big|X-X^{T^n_k}\big|^\star_t\ge 2^{-n}\}$$

are all stopping times. The process[6]

$$X_0\cdot[\![0]\!]+\sum_{k=0}^{\infty}X_{T^n_k}\cdot(\!(T^n_k,T^n_{k+1}]\!]$$

is predictable and uniformly as close as 2^{-n} to X. Hence X is predictable. Now to the case that X is merely left-continuous and adapted. Replacing X by $-k\vee X\wedge k$, $k\in\mathbb{N}$, and taking the limit we may assume that X is bounded. Let $\phi^{(n)}$ be a positive continuous function with support in $[0,1/n]$ and Lebesgue integral 1. Since the path of X is Lebesgue measurable and bounded, the convolution

$$X^{(n)}_t\stackrel{\text{def}}{=}\int_{-\infty}^{+\infty}X_s\cdot\phi^{(n)}(t-s)\,ds=\int_{t-1/n}^{t}X_s\cdot\phi^{(n)}(t-s)\,ds$$

exists (set $X_s=0$ for $s<0$). Every $X^{(n)}$ is an adapted process with continuous paths and is therefore predictable. The sequence $\big(X^{(n)}\big)$ converges pointwise to X, by the left-continuity of this process. Hence X is predictable.

Since in particular every elementary integrand can be approximated in this way by continuous adapted processes, the latter generate $\mathcal{P}$. ∎

Exercise 3.5.3 $\mathcal{F}_.$ and its right-continuous version $\mathcal{F}_{.+}$ have the same predictables.

Exercise 3.5.4 If Z is predictable and T a stopping time, then the stopped process Z^T is predictable. The variation process of a right-continuous predictable process of finite variation is predictable.

Exercise 3.5.5 Suppose G is Z–0-integrable; let S,T be two stopping times; and let $f\in L^0(\mathcal{F}_S)$. Then $f\cdot(\!(S,T]\!]\cdot G$ is Z–0-integrable and

$$\int f\cdot(\!(S,T]\!]\cdot G\,dZ\doteq \dot f\cdot\int(\!(S,T]\!]\cdot G\,dZ\,. \tag{3.5.2}$$

Previsible Processes

For the remainder of the chapter we resurrect our hitherto unused standing assumption 2.2.12 *that the measured filtration* $(\Omega, \mathcal{F}_., \mathbb{P})$ *satisfies the natural conditions.*

One should respect a process that tries to be predictable and *nearly* succeeds:

Definition 3.5.6 *A process X is called* **previsible** *with $\mathbb{P}$ if there exists a predictable process $X_{\mathbb{P}}$ that cannot be distinguished from X with $\mathbb{P}$. The collection of previsible processes is denoted by $\mathcal{P}^{\mathbb{P}}$, and so is the collection of sets in $\mathcal{P}^{\mathbb{P}}$.*

A previsible process is measurable for any of the Daniell means associated with an integrator. This follows from exercise 3.2.5 and uses the regularity of the filtration.

Exercise 3.5.7 (i) $\mathcal{P}^{\mathbb{P}}$ is sequentially closed and the idempotent functions (sets) in $\mathcal{P}^{\mathbb{P}}$ form a σ-algebra. (ii) In the presence of the natural conditions a measurable previsible process is predictable.

Exercise 3.5.8 Redo exercise 3.5.4 for previsible processes.

Predictable Stopping Times

On the half-line any singleton $\{t\}$ is integrable for any measure dz, since it is a compact Borel set. Its dz-measure is $\Delta z_t = z_t - z_{t-}$. The stochastic analog of a singleton is the graph of a random time, and the stochastic analog of the Borels are the predictable sets. It is natural to ask when the graph of a random time T is a predictable set and, if so, what the dZ-integral of its graph $[\![T]\!]$ is. The answer is given in theorem 3.5.13 in terms of the predictability of T:

Definition 3.5.9 *A random time T is called* **predictable** *if there exists a sequence of stopping times $T_n \le T$ that are strictly less than T on $[T > 0]$ and increase to T everywhere; the sequence (T_n) is said to* **predict** *or to* **announce** *T.*

A predictable time is a stopping time (exercise 1.3.15). Before showing that it is precisely the predictable stopping times that answer our question it is expedient to develop their properties.

Exercise 3.5.10 (i) Instants are predictable. If T is any stopping time, then $T+\epsilon$ is predictable, as long as $\epsilon > 0$. The infimum of a finite number of predictable times and the supremum of a countable number of predictable times are predictable.

(ii) For any $A \in \mathcal{F}_0$ the reduced time 0_A is predictable; if S is predictable, then so is its reduction S_A, in particular $S_{[S>0]}$. If S, T are stopping times, S predictable, then the reduction $S_{[S\le T]}$ is predictable.

Exercise 3.5.11 Let S, T be predictable stopping times. Then all stochastic intervals that have $S, T, 0$, or ∞ as endpoints are predictable sets. In particular, $[\![0, T)\!)$, the graph $[\![T]\!]$, and $[\![T, \infty)\!)$ are predictable.

Lemma 3.5.12 *(i) A random time T nearly equal to a predictable stopping time S is itself a predictable stopping time; the σ-algebras $\mathcal{F}_S$ and $\mathcal{F}_T$ agree.*
(ii) Let T be a stopping time, and assume that there exists a sequence (T_n) of stopping times that are almost surely less than T, almost surely strictly so on $[T > 0]$, and that increase almost surely to T. Then T is predictable.
(iii) The limit S of a decreasing sequence (S_n) of predictable stopping times is a predictable stopping time provided (S_n) is almost surely ultimately constant.

Proof. We employ – for the first time in this context – the natural conditions.

(i) Suppose that S is announced by (S_n) and that $[S \neq T]$ is nearly empty. Then, due to the regularity of the filtration, the random variables[6]

$$T_n \stackrel{\text{def}}{=} S_n \cdot [S = T] + \big(0 \vee (T - 1/n) \wedge n\big) \cdot [S \neq T]$$

are stopping times. The T_n evidently increase to T, strictly so on $[T > 0]$. If $A \in \mathcal{F}_T$, then $A \cap [S \le t]$ nearly equals $A \cap [T \le t] \in \mathcal{F}_t$, so $A \cap [S \le t] \in \mathcal{F}_t$ by regularity. This says that A belongs to $\mathcal{F}_S$.

(ii) Replacing T_n by $\bigvee_{m \le n} T_m$ we may assume that (T_n) increases everywhere. $T_\infty \stackrel{\text{def}}{=} \sup T_n$ is a stopping time (exercise 1.3.15) nearly equal to T (exercise 1.3.27). It suffices therefore to show that T_∞ is predictable. In other words, we may assume that (T_n) increases everywhere to T, almost surely strictly so on $[T > 0]$. The set $N \stackrel{\text{def}}{=} [T > 0] \cap \bigcup_n [T = T_n]$ is nearly empty, and the chopped reductions $T_{nN^c} \wedge n$ increase strictly to T_{N^c} on $[T_{N^c} > 0]$: T_{N^c} is predictable. T, being nearly equal to T_{N^c}, is predictable as well.

(iii) To say that $(S_n(\omega))$ is ***ultimately constant*** means of course that for every $\omega \in \Omega$ there is an $N(\omega)$ such that $S(\omega) = S_n(\omega)$ for all $n \ge N(\omega)$. To start with assume that S_1 is bounded, say $S_1 \le k$. For every n let S'_n be a stopping time less than or equal to S_n, strictly less than S_n where $S_n > 0$, and having

$$\mathbb{P}\big[S'_n < S_n - 2^{-n}\big] < 2^{-n-1} .$$

Such exist as S_n is predictable. Since $\mathcal{F}_.$ is right-continuous, the random variables $S''_n \stackrel{\text{def}}{=} \inf_{\nu \ge n} S'_\nu$ are stopping times (exercise 1.3.30). Clearly $S''_n < S$ almost surely on $[S > 0]$, namely, at all points $\omega \in [S > 0]$ where $S_n(\omega)$ is ultimately constant. Since $\mathbb{P}\big[S''_n < S\big] \le 2^{-n}$, (S''_n) increases almost surely to S. By (ii) S is predictable. In the general case we know now that $S \wedge k = \inf_n S_n \wedge k$ is predictable. Then so is the pointwise supremum $S = \bigvee_k S \wedge k$ (exercise 1.3.15). ∎

Theorem 3.5.13 *(i) Let $B \subset \boldsymbol{B}$ be previsible and $\epsilon > 0$. There is a predictable stopping time T whose graph is contained in B and such that $\mathbb{P}\big[\pi_\Omega[B]\big] < \mathbb{P}[T < \infty] + \epsilon$.*

(ii) A random time is predictable if and only if its graph is previsible.

Proof. (i) Let $B_{\mathbb{P}}$ be a predictable set that cannot be distinguished from B. Theorem A.5.14 on page 438 provides a predictable stopping time S whose graph lies inside $B_{\mathbb{P}}$ and satisfies $\mathbb{P}\big[\pi_\Omega[B_{\mathbb{P}}]\big] < \mathbb{P}[S < \infty] + \epsilon$ (see figure A.17 on page 436). The projection of $[\![S]\!] \setminus B$ is nearly empty and by regularity belongs to $\mathcal{F}_0$. The reduction of S to its complement is a predictable stopping time that meets the description.

(ii) The necessity of the condition was shown in exercise 3.5.11. Assume then that the graph $[\![T]\!]$ of the random time T is a previsible set. There are predictable stopping times S_k whose graphs are contained in that of T and so that $\mathbb{P}[T \ne S_k] \le 1/k$. Replacing S_k by $\inf_{\kappa \le k} S_\kappa$ we may assume the S_k to be decreasing. They are clearly ultimately constant. Thanks to lemma 3.5.12 their infimum S is predictable. The set $[S \ne T]$ is evidently nearly empty; so in view of lemma 3.5.12 (ii) T is predictable. ∎

The Strict Past of a Stopping Time The question at the beginning of the section is half resolved: the stochastic analog of a singleton $\{t\}$ *qua* integrand has been identified as the graph of a predictable time T. We have no analog yet of the fact that the measure of $\{t\}$ is $\Delta z_t = z_t - z_{t-}$. Of course in the stochastic case the right question to ask is this: for which random variables f is the process $f \cdot [\![T]\!]$ previsible, and what is its integral? Theorem 3.5.14 gives the answer in terms of the ***strict past*** of T. This is simply the σ-algebra $\mathcal{F}_{T-}$ generated by $\mathcal{F}_0$ and the collection

$$\big\{A \cap [t < T] : t \in \mathbb{R}_+, A \in \mathcal{F}_t\big\} .$$

A generator is "an event that occurs and is observable at some instant t strictly prior to T." A stopping time is evidently measurable on its strict past.

Theorem 3.5.14 *Let T be a stopping time, f a real-valued random variable, and Z an L^0-integrator. Then $f \cdot [\![T]\!]$ is a previsible process*[6] *if and only if both $f \cdot [T < \infty]$ is measurable on the strict past of T and the reduction $T_{[f \ne 0]}$ is predictable; and in this case*

$$f \cdot \Delta Z_T \in \int f \cdot [\![T]\!]\, dZ . \tag{3.5.3}$$

Before proving this theorem it is expedient to investigate the strict past of stopping times.

Lemma 3.5.15 *(i) If $S \le T$, then $\mathcal{F}_{S-} \subset \mathcal{F}_{T-} \subset \mathcal{F}_T$; and if in addition $S < T$ on $[T > 0]$, then $\mathcal{F}_S \subset \mathcal{F}_{T-}$.*

(ii) Let T_n be stopping times increasing to T. Then $\mathcal{F}_{T-} = \bigvee \mathcal{F}_{T_n-}$. If the T_n announce T, then $\mathcal{F}_{T-} = \bigvee \mathcal{F}_{T_n}$.

(iii) If X is a previsible process and T any stopping time, then X_T is measurable on $\mathcal{F}_{T-}$.

(iv) If T is a predictable stopping time and $A \in \mathcal{F}_{T-}$, then the reduction T_A is predictable.

Proof. (i) A generator $A \cap [t < S]$ of $\mathcal{F}_{S-}$ can be written as the intersection of $(A \cap [t < S])$ with $[t < T]$ and belongs to $\mathcal{F}_{T-}$ inasmuch as $[t < S] \in \mathcal{F}_t$. A generator $A \cap [t < T]$ of $\mathcal{F}_{T-}$ belongs to $\mathcal{F}_T$ since

$$(A \cap [t < T]) \cap [T \le u] = \begin{cases} \emptyset \in \mathcal{F}_u & \text{for } u \le t \\ A \cap [T \le u] \cap [T \le t]^c \in \mathcal{F}_u & \text{for } u > t. \end{cases}$$

Assume that $S < T$ on $[T > 0]$, and let $A \in \mathcal{F}_S$. Then $A \cap [T > 0] = A \cap \bigcup_{q \in \mathbb{Q}_+} [S < q] \cap [q < T]$ belongs to $\mathcal{F}_{T-}$, and so does $A \cap [T = 0] \in \mathcal{F}_0$. This proves the second claim of (i).

(ii) A generator $A \cap [t < T] = \bigcup_n A \cap [t < T_n]$ clearly lies in $\bigvee_n \mathcal{F}_{T_n -}$. If the T_n announce T, then by (i) $\mathcal{F}_{T-} \subset \bigvee_n \mathcal{F}_{T_n -} \subset \bigvee_n \mathcal{F}_{T_n} \subset \mathcal{F}_{T-}$.

(iii) Assume first that X is of the form $X = A \times (s, t]$ with $A \in \mathcal{F}_s$. Then $X_T = A \cap [s < T \le t] = (A \cap [s < T]) \setminus (\Omega \cap [t < T]) \in \mathcal{F}_{T-}$. By linearity, $X_T \in \mathcal{F}_{T-}$ for all $X \in \mathcal{E}$. The processes X with $X_T \in \mathcal{F}_{T-}$ evidently form a sequentially closed family, so every predictable process has this property. An evanescent process clearly has it as well, so every previsible process has it.

(iv) Let (T^n) be a sequence announcing T. Since $A \in \bigvee \mathcal{F}_{T^n}$, there are sets $A^n \in \bigcup \mathcal{F}_{T^n}$ with $\mathbb{P}[|A - A^n|] < 2^{-n-1}$. Taking a subsequence, we may assume that $A^n \in \mathcal{F}_{T^n}$. Then $A_N \stackrel{\text{def}}{=} \bigcap_{n > N} A^n \in \mathcal{F}_T$, and $T^n_{A^n} \wedge n$ announces T_{A_N}. This sequence of predictable stopping times is ultimately constant and decreases almost surely to T_A, so T_A is predictable. ∎

Proof of Theorem 3.5.14. If $X \stackrel{\text{def}}{=} f \cdot [\![T]\!]$ is previsible,[6] then $X_T = f \cdot [T < \infty]$ is measurable on $\mathcal{F}_{T-}$ (lemma 3.5.15 (iii)), and $T_{[f \ne 0]}$ is predictable since it has previsible graph $[X \ne 0]$ (theorem 3.5.13). The conditions listed are thus necessary.

To show their sufficiency we replace first of all f by $f \cdot [T < \infty]$, which does not change X. We may thus assume that f is measurable on $\mathcal{F}_{T-}$, and that $T = T_{[f \ne 0]}$ is predictable. If f is a set in $\mathcal{F}_{T-}$, then X is the graph of a predictable stopping time (ibidem) and thus is predictable (exercise 3.5.11). If f is a step function over $\mathcal{F}_{T-}$, a linear combination of sets, then X is predictable as a linear combination of predictable processes. The usual sequential closure argument shows that X is predictable in general.

It is left to be shown that equation (3.5.3) holds. We fix a sequence (T^n) announcing T and an L^0-integrator Z. Since f is measurable on the span of the $\mathcal{F}_{T^n}$, there are $\mathcal{F}_{T^n}$-measurable step functions f^n that converge in probability to f. Taking a subsequence we can arrange things so that $f^n \to f$ almost surely. The processes $X^n \stackrel{\text{def}}{=} f^n \cdot (\!(T^n, T]\!]$, previsible by proposition 3.5.2, converge to $X = f \cdot [\![T]\!]$ except possibly on the evanescent set $\mathbb{R}_+ \times [f_n \not\to f]$, so the limit is previsible. To establish equation (3.5.3) we note that $f^m \cdot (\!(T^n, T]\!]$ is Z–0-integrable for $m \le n$ (exercise 3.5.5) with

$$f^m \cdot (Z_T - Z_{T^n}) \in \int f^m \cdot (\!(T^n, T]\!] \, dZ .$$

We take $n \to \infty$ and get $f^m \cdot \Delta Z_T \in \int f^m \cdot [\![T]\!]\, dZ$. Now, as $m \to \infty$, the left-hand side converges almost surely to $f \cdot \Delta Z_T$. If the $|f^m|$ are uniformly bounded, say by M, then $f^m \cdot [\![T]\!]$ converges to Z–0-a.e., being dominated by $M \cdot [\![T]\!]$. Then $f^m \cdot [\![T]\!]$ converges to $f \cdot [\![T]\!]$ in Z–0-mean, thanks to the Dominated Convergence Theorem, and (3.5.3) holds. We leave to the reader the task of extending this argument to the case that f is almost surely finite (replace M by $\sup |f^m|$ and use corollary 3.6.10 to show that $\sup |f^m| \cdot [\![T]\!]$ is finite in Z–0-mean). ∎

Corollary 3.5.16 *A right-continuous previsible process X with finite maximal process is locally bounded.*

Proof. Let $t < \infty$ and $\epsilon > 0$ be given. By the choice of $\lambda > 0$ we can arrange things so that $T^\lambda = \inf\{t : |X_t| \geq \lambda\}$ has $\mathbb{P}[T^\lambda < t] < \epsilon/2$. The graph of T^λ is the intersection of the previsible sets $[|X| \geq \lambda]$ and $[\![0, T^\lambda]\!]$. Due to theorem 3.5.13, T^λ is predictable: there is a stopping time $S < T^\lambda$ with $\mathbb{P}[S < T^\lambda \wedge t] < \epsilon/2$. Then $\mathbb{P}[S < t] < \epsilon$ and $|X^S|$ is bounded by λ. ∎

Accessible Stopping Times

For an application in section 4.4 let us introduce stopping times that are "partly predictable" and those that are "nowhere predictable:"

Definition 3.5.17 *A stopping time T is* ***accessible on a set*** $A \in \mathcal{F}_T$ *of strictly positive measure if there exists a predictable stopping time S that agrees with T on A – clearly T is then accessible on the larger set $[S = T]$ in $\mathcal{F}_T \cap \mathcal{F}_S$. If there is a countable cover of Ω by sets on which T is accessible, then T is simply called* ***accessible****. On the other hand, if T agrees with no predictable stopping time on any set of positive probability, then T is called* ***totally inaccessible****.*

For example, in a realistic model for atomic decay, the first time T a Geiger counter detects a decay should be totally inaccessible: there is no circumstance in which the decay is foreseeable.

Given a stopping time T, let $\mathfrak{A}$ be a maximal collection of mutually disjoint sets on which T is accessible. Since the sets in $\mathfrak{A}$ have strictly positive measure, there are at most countably many of them, say $\mathfrak{A} = \{A_1, A_2, \ldots\}$. Set $A \stackrel{\text{def}}{=} \bigcup A_n$ and $I \stackrel{\text{def}}{=} A^c$. Then clearly the reduction T_A is accessible and T_I is totally inaccessible:

Proposition 3.5.18 *Any stopping time T is the infimum of two stopping times T_A, T_I having disjoint graphs, with T_A accessible – wherefore $[\![T_A]\!]$ is contained in the union of countably many previsible graphs – and T_I totally inaccessible.*

Exercise 3.5.19 Let $V \in \mathfrak{D}$ be previsible, and let $\lambda \geq 0$. (i) Then

$$T_V^\lambda = \inf\{t : |V_t| \geq \lambda\} \quad \text{and} \quad T_{\Delta V}^\lambda = \inf\{t : \Delta V_t \geq \lambda\}$$

are predictable stopping times. (ii) There exists a sequence $\{T_n\}$ of predictable stopping times with disjoint graphs such that $[\Delta V \neq 0] \subset \bigcup_n [\![T_n]\!]$.

Exercise 3.5.20 If M is a uniformly integrable martingale and T a predictable stopping time, then $M_{T-} \in \mathbb{E}[M_\infty|\mathcal{F}_{T-}]$ and thus $\mathbb{E}[\Delta M_T|\mathcal{F}_{T-}] \doteq 0$.

Exercise 3.5.21 For deterministic instants t, $\mathcal{F}_{t-}$ is the σ-algebra generated by $\{\mathcal{F}_s : s < t\}$. The σ-algebras $\mathcal{F}_{t-}$ make up the ***left-continuous version*** $\mathcal{F}_{.-}$ of $\mathcal{F}_.$. Its predictables and previsibles coincide with those of $\mathcal{F}_.$.

3.6 Special Properties of Daniell's Mean

In this section a probability $\mathbb{P}$, an exponent $p \geq 0$, and an $L^p(\mathbb{P})$-integrator Z are fixed. The mean is Daniell's mean $\lceil\!\lceil\ \rceil\!\rceil^*_{Z-p}$, computed with respect to $\mathbb{P}$. As usual, mention of $\mathbb{P}$ is suppressed in the notation. Recall that we often use the words *Z–p-integrable*, *Z–p-a.e.*, *Z–p-measurable*, etc., instead of $\lceil\!\lceil\ \rceil\!\rceil^*_{Z-p}$-integrable, $\lceil\!\lceil\ \rceil\!\rceil^*_{Z-p}$-a.e., $\lceil\!\lceil\ \rceil\!\rceil^*_{Z-p}$-measurable, etc.

Maximality

Proposition 3.6.1 $\lceil\!\lceil\ \rceil\!\rceil^*_{Z-p}$ *is **maximal**. That is to say, if* $\lceil\!\lceil\ \rceil\!\rceil^*$ *is any mean less than or equal to* $\lceil\!\lceil\ \rceil\!\rceil^*_{Z-p}$ *on positive elementary integrands, then the inequality* $\lceil\!\lceil F \rceil\!\rceil^* \leq \lceil\!\lceil F \rceil\!\rceil^*_{Z-p}$ *holds for all processes* F.

Proof. Suppose that $\lceil\!\lceil F \rceil\!\rceil^*_{Z-p} < a$. There exists an $H \in \mathcal{E}^\uparrow_+$, limit of an increasing sequence of positive elementary integrands $X^{(n)}$, with $|F| \leq H$ and $\lceil\!\lceil H \rceil\!\rceil^*_{Z-p} < a$. Then

$$\lceil\!\lceil F \rceil\!\rceil^* \leq \lceil\!\lceil H \rceil\!\rceil^* = \sup_n \left\lceil\!\!\left\lceil X^{(n)} \right\rceil\!\!\right\rceil^* \leq \sup_n \left\lceil\!\!\left\lceil X^{(n)} \right\rceil\!\!\right\rceil^*_{Z-p} = \lceil\!\lceil H \rceil\!\rceil^*_{Z-p} < a\,. \qquad \blacksquare$$

Exercise 3.6.2 $\|\ \|^*_{Z-p}$ and $\|\ \|^*_{Z-[\alpha]}$ are maximal as well.

Exercise 3.6.3 Suppose that Z is an L^p-integrator, $p \geq 1$, and $X \mapsto \int X\,dZ$ has been extended in some way to a vector lattice $\mathfrak{L}$ of processes such that the Dominated Convergence Theorem holds. Then there exists a mean $\lceil\!\lceil\ \rceil\!\rceil^*$ such that the integral is the extension by $\lceil\!\lceil\ \rceil\!\rceil^*$-continuity of the elementary integral, at least on the $\lceil\!\lceil\ \rceil\!\rceil^*$-closure of $\mathcal{E}$ in $\mathfrak{L}$.

Exercise 3.6.4 If $\lceil\!\lceil\ \rceil\!\rceil^*$ is any mean, then

$$\lceil\!\lceil F \rceil\!\rceil^{**} \stackrel{\text{def}}{=} \sup\{\lceil\!\lceil F \rceil\!\rceil' : \lceil\!\lceil\ \rceil\!\rceil' \text{ a mean with } \lceil\!\lceil\ \rceil\!\rceil' \leq \lceil\!\lceil\ \rceil\!\rceil^* \text{ on } \mathcal{E}_+\ \}$$

defines a mean $\lceil\!\lceil\ \rceil\!\rceil^{**}$, evidently a maximal one. It is given by Daniell's up-and-down procedure:

$$\lceil\!\lceil F \rceil\!\rceil^{**} = \begin{cases} \sup\{\lceil\!\lceil X \rceil\!\rceil^* : X \in \mathcal{E}_+\} & \text{if } F \in \mathcal{E}^\uparrow_+ \\ \inf\{\lceil\!\lceil H \rceil\!\rceil^{**} : |F| \leq H \in \mathcal{E}^\uparrow_+\} & \text{for arbitrary } F. \end{cases} \tag{3.6.1}$$

Exercise 3.6.3 says that an integral extension featuring the Dominated Convergence Theorem can be had essentially only by using a mean that controls the elementary integral. Other examples can be found in definition (4.2.9) and exercise 4.5.18: Daniell's procedure is not so ad hoc as it may seem at first. Exercise 3.6.4 implies that we might have also defined Daniell's mean as the maximal mean that agrees with the semivariation on $\mathcal{E}_+$. That would have left us, of course, with the onus to show that there exists at least one such mean.

It seems at this point, though, that Daniell's mean is the worst one to employ, whichever way it is constructed. Namely, the larger the mean, the smaller evidently the collection of integrable functions. In order to integrate as large a collection as possible of processes we should try to find as small a mean as possible that still controls the elementary integral. This can be done in various non-canonical and uninteresting ways. We prefer to develop some nice and useful properties that are direct consequences of the maximality of Daniell's mean.

Continuity Along Increasing Sequences

It is well known that the outer measure μ^* associated with a measure μ satisfies $0 \le A_n \uparrow A \implies \mu^*(A_n) \uparrow \mu^*(A)$, making it a capacity. The Daniell mean has the same property:

Proposition 3.6.5 *Let* $\lceil\!\lceil\ \rceil\!\rceil^*$ *be a maximal mean on* $\mathcal{E}$. *For any increasing sequence* $\big(F^{(n)}\big)$ *of positive numerical processes,*

$$\left\lceil\!\!\left\lceil \sup F^{(n)} \right\rceil\!\!\right\rceil^* = \sup_n \left\lceil\!\!\left\lceil F^{(n)} \right\rceil\!\!\right\rceil^* .$$

Proof. We start with an observation, which might be called *upper regularity: for every positive integrable process* F *and every* $\epsilon > 0$ *there exists a process* $H \in \mathcal{E}_+^{\uparrow}$ *with* $H > F$ *and* $\lceil\!\lceil H - F \rceil\!\rceil^* \le \epsilon$. Indeed, there exists an $X \in \mathcal{E}_+$ with $\lceil\!\lceil F - X \rceil\!\rceil^* < \epsilon/2$; equation (3.6.1) provides an $H^\epsilon \in \mathcal{E}_+^{\uparrow}$ with $|F - X| \le H^\epsilon$ and $\lceil\!\lceil H^\epsilon \rceil\!\rceil^* < \epsilon/2$; and evidently $H \stackrel{\text{def}}{=} X + H^\epsilon$ meets the description.

Now to the proof proper. Only the inequality

$$\left\lceil\!\!\left\lceil \sup F^{(n)} \right\rceil\!\!\right\rceil^* \le \sup_n \left\lceil\!\!\left\lceil F^{(n)} \right\rceil\!\!\right\rceil^* \tag{?}$$

needs to be shown, the reverse inequality being obvious from the solidity of $\lceil\!\lceil\ \rceil\!\rceil^*$. To start with, assume that the $F^{(n)}$ are $\lceil\!\lceil\ \rceil\!\rceil^*$-integrable. Let $\epsilon > 0$. Using the upper regularity choose for every n an $H^{(n)} \in \mathcal{E}_+^{\uparrow}$ with $F^{(n)} \le H^{(n)}$ and $\lceil\!\lceil H^{(n)} - F^{(n)} \rceil\!\rceil^* < \epsilon/2^n$, and set $F = \sup F^{(n)}$ and $\overline{H}^{(N)} = \sup_{n \le N} H^{(n)}$. Then $F \le H \stackrel{\text{def}}{=} \sup_N \overline{H}^{(N)} \in \mathcal{E}_+^{\uparrow}$.

$$\text{Now} \qquad \begin{aligned} \overline{H}^{(N)} &= \sup\nolimits_{n \le N} \big(F^{(n)} + (H^{(n)} - F^{(n)})\big) \\ &\le F^{(N)} + \textstyle\sum_{n \le N} H^{(n)} - F^{(n)} \end{aligned}$$

$$\text{and so} \qquad \left\lceil\!\!\left\lceil \overline{H}^{(N)} \right\rceil\!\!\right\rceil^* \le \sup\nolimits_N \left\lceil\!\!\left\lceil F^{(N)} \right\rceil\!\!\right\rceil^* + \epsilon .$$

Now $\lceil\!\lceil\ \rceil\!\rceil^*$ is continuous along the increasing sequence $\big(\overline{H}^{(N)}\big)$ of $\mathcal{E}_+^\uparrow$, so

$$\lceil\!\lceil F \rceil\!\rceil^* \le \lceil\!\lceil H \rceil\!\rceil^* = \sup_N \left\lceil\!\!\left\lceil \overline{H}^{(N)} \right\rceil\!\!\right\rceil^* \le \sup_N \left\lceil\!\!\left\lceil F^{(N)} \right\rceil\!\!\right\rceil^* + \epsilon ,$$

which in view of the arbitraryness of ϵ implies that $\lceil\!\lceil F \rceil\!\rceil^* \le \sup_n \lceil\!\lceil F^{(n)} \rceil\!\rceil^*$. Next assume that the $F^{(n)}$ are merely $\lceil\!\lceil\ \rceil\!\rceil^*$-measurable. Then

$$\underline{F}^{(n)} \stackrel{\text{def}}{=} \big(F^{(n)} \wedge n\big)\cdot[\![0, n]\!]$$

is $\lceil\!\lceil\ \rceil\!\rceil^*$-integrable (theorem 3.4.10). Since $\sup \underline{F}^{(n)} = \sup F^{(n)}$, the first part of the proof gives $\lceil\!\lceil \sup F^{(n)} \rceil\!\rceil^* = \lceil\!\lceil \sup \underline{F}^{(n)} \rceil\!\rceil^* \le \sup \lceil\!\lceil F^{(n)} \rceil\!\rceil^*$.

Now if the $F^{(n)}$ are arbitrary positive $\overline{\mathbb{R}}$-valued processes, choose for every $n, k \in \mathbb{N}$ a process $H^{(n,k)} \in \mathcal{E}_+^\uparrow$ with $F^{(n)} \le H^{(n,k)}$ and

$$\left\lceil\!\!\left\lceil H^{(n,k)} \right\rceil\!\!\right\rceil^* \le \left\lceil\!\!\left\lceil F^{(n)} \right\rceil\!\!\right\rceil^* + 1/k .$$

This is possible by the very definition of the Daniell mean; if $\lceil\!\lceil F^{(n)} \rceil\!\rceil^* = \infty$, then $H^{(n,k)} \stackrel{\text{def}}{=} \infty$ qualifies. Set

$$\overline{F}^{(N)} = \inf_{n \ge N} \inf_k H^{(n,k)} , \qquad N \in \mathbb{N} .$$

The $\overline{F}^{(n)}$ are $\lceil\!\lceil\ \rceil\!\rceil^*$-measurable, satisfy $\lceil\!\lceil F^{(n)} \rceil\!\rceil^* = \lceil\!\lceil \overline{F}^{(n)} \rceil\!\rceil^*$, and increase with n, whence the desired inequality (?):

$$\left\lceil\!\!\left\lceil \sup F^{(n)} \right\rceil\!\!\right\rceil^* \le \left\lceil\!\!\left\lceil \sup \overline{F}^{(n)} \right\rceil\!\!\right\rceil^* = \sup \left\lceil\!\!\left\lceil \overline{F}^{(n)} \right\rceil\!\!\right\rceil^* = \sup \left\lceil\!\!\left\lceil F^{(n)} \right\rceil\!\!\right\rceil^* . \qquad \blacksquare$$

Predictable Envelopes

A subset A of the line is contained in a Borel set $\widetilde{A}$ whose measure equals the outer measure of A. A similar statement holds for the Daniell mean:

Proposition 3.6.6 *Let $\lceil\!\lceil\ \rceil\!\rceil^*$ be a maximal mean on $\mathcal{E}$.*

(i) If F is a $\lceil\!\lceil\ \rceil\!\rceil^*$-negligible process, then there is a predictable process $\widetilde{F} \ge |F|$ that is also $\lceil\!\lceil\ \rceil\!\rceil^*$-negligible.

(ii) If F is a $\lceil\!\lceil\ \rceil\!\rceil^*$-measurable process, then there exist predictable processes $\underset{\sim}{F}$ and $\widetilde{F}$ that differ $\lceil\!\lceil\ \rceil\!\rceil^*$-negligibly and sandwich F: $\underset{\sim}{F} \le F \le \widetilde{F}$.

(iii) Let F be a non-negative process. There exists a predictable process $\widetilde{F} \ge F$ such that $\lceil\!\lceil r\widetilde{F} \rceil\!\rceil^* = \lceil\!\lceil rF \rceil\!\rceil^*$ for all $r \in \mathbb{R}$ and such that every $\lceil\!\lceil\ \rceil\!\rceil^*$-measurable process bigger than or equal to F is $\lceil\!\lceil\ \rceil\!\rceil^*$-a.e. bigger than or equal to $\widetilde{F}$. If F is a set,[7] then $\widetilde{F}$ can be chosen to be a set as well. If F is finite for $\lceil\!\lceil\ \rceil\!\rceil^*$, then $\widetilde{F}$ is $\lceil\!\lceil\ \rceil\!\rceil^*$-integrable. $\widetilde{F}$ is called a ***predictable $\lceil\!\lceil\ \rceil\!\rceil^*$-envelope*** *of F.*

[7] In accordance with convention A.1.5 on page 364, sets are identified with their (idempotent) indicator functions.

Proof. (i) For every $n \in \mathbb{N}$ there is an $H^{(n)} \in \mathcal{E}_+^\uparrow$ satisfying both $|F| \le H^{(n)}$ and $\lceil\!\lceil H^{(n)} \rceil\!\rceil^* \le 1/n$ (see equation (3.6.1)). $\widetilde{F} \stackrel{\text{def}}{=} \inf_n H^{(n)}$ meets the description.

(ii) To start with, assume that F is $\lceil\!\lceil\ \rceil\!\rceil^*$-integrable. Let $(X^{(n)})$ be a sequence of elementary integrands converging $\lceil\!\lceil\ \rceil\!\rceil^*$-almost everywhere to F. The process $Y = (\liminf X^{(n)} - F) \vee 0$ is $\lceil\!\lceil\ \rceil\!\rceil^*$-negligible. $\underset{\sim}{F} \stackrel{\text{def}}{=} \liminf X^{(n)} - \widetilde{Y}$ is less than or equal to F and differs $\lceil\!\lceil\ \rceil\!\rceil^*$-negligibly from F. $\widetilde{F}$ is constructed similarly. Next assume that F is positive and let $F^{(n)} = (n \wedge F)\cdot[\![0,n]\!]$. Then $\underset{\sim}{F} \stackrel{\text{def}}{=} \limsup \underset{\sim}{F^{(n)}}$ and $\widetilde{F} \stackrel{\text{def}}{=} \liminf \widetilde{F^{(n)}}$ qualify. Finally, if F is arbitrary, write it as the difference of two positive measurable processes: $F = F^+ - F^-$, and set $\underset{\sim}{F} = \underset{\sim}{F^-} - \widetilde{F^+}$ and $\widetilde{F} = \widetilde{F^+} - \underset{\sim}{F^-}$.

(iii) To start with, assume that F is finite for $\lceil\!\lceil\ \rceil\!\rceil^*$. For every $q \in \mathbb{Q}_+$ and $k \in \mathbb{N}$ there is an $H^{(q,k)} \in \mathcal{E}_+^\uparrow$ with $H^{(q,k)} \ge F$ and

$$\left\lceil\!\left\lceil q \cdot H^{(q,k)} \right\rceil\!\right\rceil^* \le \lceil\!\lceil q \cdot F \rceil\!\rceil^* + 2^{-k} .$$

(If $\lceil\!\lceil q \cdot F \rceil\!\rceil^* = \infty$, then $H^{(q,k)} = \infty$ clearly qualifies.) The predictable process $\widehat{F} \stackrel{\text{def}}{=} \bigwedge_{q,k} H^{(q,k)}$ is greater than or equal to F and has $\lceil\!\lceil q\widehat{F} \rceil\!\rceil^* = \lceil\!\lceil qF \rceil\!\rceil^*$ for all positive rationals q; since $\widehat{F}$ is evidently finite for $\lceil\!\lceil\ \rceil\!\rceil^*$, it is $\lceil\!\lceil\ \rceil\!\rceil^*$-integrable, and the previous equality extends by continuity to all positive reals. Next let $\{X_\alpha\}$ be a maximal collection of non-negative predictable and $\lceil\!\lceil\ \rceil\!\rceil^*$-non-negligible processes with the property that

$$F + \textstyle\sum_\alpha X_\alpha \le \widehat{F} .$$

Such a collection is necessarily countable (theorem 3.2.23). It is easy to see that

$$\widetilde{F} \stackrel{\text{def}}{=} \widehat{F} - \textstyle\sum_\alpha X_\alpha$$

meets the description. For if $H \ge F$ is a $\lceil\!\lceil\ \rceil\!\rceil^*$-measurable process, then $H \wedge \widetilde{F}$ is integrable; the envelope $\widetilde{H \wedge F}$ of part (ii) can be chosen to be smaller than $\widetilde{F}$; the positive process $\widetilde{F} - \widetilde{H \wedge F}$ is both predictable and $\lceil\!\lceil\ \rceil\!\rceil^*$-integrable; if it were not $\lceil\!\lceil\ \rceil\!\rceil^*$-negligible, it could be adjoined to $\{X_\alpha\}$, which would contradict the maximality of this family; thus $\widetilde{F} - \widetilde{H \wedge F}$ and $\widetilde{F} - H \wedge \widetilde{F}$ are $\lceil\!\lceil\ \rceil\!\rceil^*$-negligible, or, in other words, $H \ge \widetilde{F}$ $\lceil\!\lceil\ \rceil\!\rceil^*$-almost everywhere.

If F is not finite for $\lceil\!\lceil\ \rceil\!\rceil^*$, then let $\widetilde{F}^{(n)}$ be an envelope for $F \wedge (n \cdot [\![0,n]\!])$. This can evidently be arranged so that $\widetilde{F}^{(n)}$ increases with n. Set $\widetilde{F} = \sup_n \widetilde{F}^{(n)}$. If $H \ge F$ is $\lceil\!\lceil\ \rceil\!\rceil^*$-measurable, then $H \ge \widetilde{F}^{(n)}$ $\lceil\!\lceil\ \rceil\!\rceil^*$-a.e., and consequently $H \ge \widetilde{F}$ $\lceil\!\lceil\ \rceil\!\rceil^*$-a.e. It follows from equation (3.6.1) that $\lceil\!\lceil F \rceil\!\rceil^* = \lceil\!\lceil \widetilde{F} \rceil\!\rceil^*$.

To see the homogeneity let $r > 0$ and let $\widetilde{rF}$ be an envelope for rF. Since $r^{-1} \cdot \widetilde{rF} \ge F$, we have $\widetilde{rF} \ge r\widetilde{F}$ $\lceil\!\lceil \ \rceil\!\rceil^*$-a.e. and

$$\lceil\!\lceil rF \rceil\!\rceil^* = \left\lceil\!\!\left\lceil \widetilde{rF} \right\rceil\!\!\right\rceil^* \ge \left\lceil\!\!\left\lceil r\widetilde{F} \right\rceil\!\!\right\rceil^* \ge \lceil\!\lceil rF \rceil\!\rceil^* ,$$

whence equality throughout. Finally, if F is a set with envelope $\widetilde{F}$, then $[\widetilde{F} \ge 1]$ is a smaller envelope and a set. ∎

We apply this now in particular to the Daniell mean of an L^0-integrator Z.

Corollary 3.6.7 *Let A be a subset of Ω, not necessarily measurable. If $B \stackrel{\text{def}}{=} [0, \infty) \times A$ is Z–0-negligible, then the whole path of Z nearly vanishes on A.*

Proof. Let $\widetilde{B}$ be a predictable Z–0-envelope of B and C its complement. Since the natural conditions are in force, the debut T of C is a stopping time (corollary A.5.12). Replace $\widetilde{B}$ by $\widetilde{B} \setminus (\!(T, \infty)\!)$. This does not disturb T, but has the effect that now the graph of T separates $\widetilde{B}$ from its complement C. Now fix an instant $t < \infty$ and an $\epsilon > 0$ and set $T^\epsilon = \inf\{s : |Z_s| \ge \epsilon\}$.

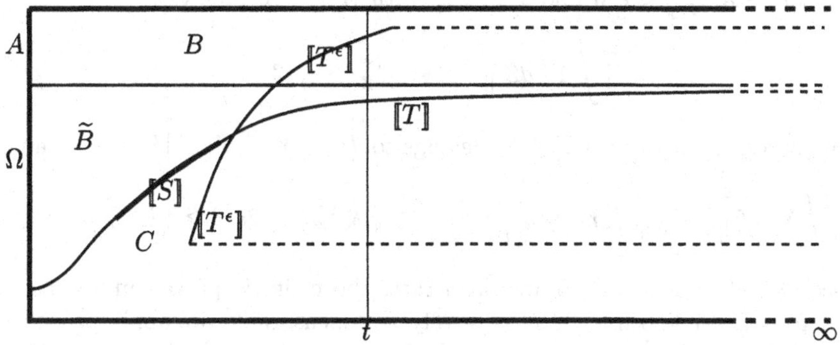

Figure 3.10

The stochastic interval $[\![0, T \wedge T^\epsilon \wedge t]\!]$ intersects C in a predictable subset of the graph of T, which is therefore the graph of a predictable stopping time S (theorem 3.5.13). The rest, $[\![0, T \wedge T^\epsilon \wedge t]\!] \setminus C \subset \widetilde{B}$, is Z–0-negligible. The random variable $Z_{T \wedge T^\epsilon \wedge t}$ is a member of the class $\int [\![0, T \wedge T^\epsilon \wedge t]\!] \, dZ = \int [\![S]\!] \, dZ$ (exercise 3.5.5), which also contains ΔZ_S (theorem 3.5.14). Now $\Delta Z_S = 0$ on A, so we conclude that, on A, $Z_{T^\epsilon \wedge t} = Z_{T \wedge T^\epsilon \wedge t} \doteq 0$. Since $|Z_{T^\epsilon}| \ge \epsilon$ on $[T^\epsilon \le t]$ (proposition 1.3.11), we must conclude that $A \cap [T^\epsilon \le t]$ is negligible. This holds for all $\epsilon > 0$ and $t < \infty$, so $A \cap [Z^\star_\infty > 0]$ is negligible. As $[Z^\star_\infty > 0] = \bigcup_n [Z^\star_n > 0] \in \mathcal{A}_{\infty\sigma}$, $A \cap [Z^\star_\infty > 0]$ is actually nearly empty. ∎

Exercise 3.6.8 Let $X \ge 0$ be predictable. Then $\widetilde{XF} = X\widetilde{F}$ Z–p-a.e.

Exercise 3.6.9 Let $\widetilde{B}$ be a predictable Z–0-envelope of $B \subset \boldsymbol{B}$. Any two Z–0-measurable processes X, X' that agree Z–0-almost everywhere on B agree Z–0-almost everywhere on $\widetilde{B}$.

Regularity

Here is an analog of the well-known fact that the measure of a Lebesgue integrable set is the supremum of the Lebesgue measures of the compact sets contained in it (exercise A.3.14). The role of the compact sets is taken by the collection $\mathcal{P}_{00}$ of predictable processes that are bounded and vanish after some instant.

Corollary 3.6.10 *For any Z–p-measurable process F,*

$$\lceil\!\lceil F \rceil\!\rceil^*_{Z\text{-}p} = \sup\left\{ \left\lceil\!\left\lceil \int Y\, dZ \right\rceil\!\right\rceil_p : Y \in \mathcal{P}_{00}, |Y| \le |F| \right\} .$$

Proof. Since $\lceil\!\lceil \int Y\, dZ \rceil\!\rceil_p \le \lceil\!\lceil Y \rceil\!\rceil^*_{Z\text{-}p} \le \lceil\!\lceil F \rceil\!\rceil^*_{Z\text{-}p}$, one inequality is obvious. For the other, the solidity of $\lceil\!\lceil\ \rceil\!\rceil^*_{Z\text{-}p}$ and proposition 3.6.6 allow us to assume that F is positive and predictable: if necessary, we replace F by $\widetilde{|F|}$. To start with, assume that F is Z–p-integrable and let $\epsilon > 0$. There are an $X \in \mathcal{E}_+$ with $\lceil\!\lceil F - X \rceil\!\rceil^*_{Z\text{-}p} < \epsilon/3$ and a $Y' \in \mathcal{E}$ with $|Y'| \le X$ such that

$$\left\lceil\!\left\lceil \int Y'\, dZ \right\rceil\!\right\rceil_p > \lceil\!\lceil X \rceil\!\rceil^*_{Z\text{-}p} - \epsilon/3 .$$

The process $Y \stackrel{\text{def}}{=} (-F) \vee Y' \wedge F$ belongs to $\mathcal{P}_{00}$, $|Y' - Y| \le |F - X|$, and

$$\left\lceil\!\left\lceil \int Y\, dZ \right\rceil\!\right\rceil_{L^p} \ge \left\lceil\!\left\lceil \int Y'\, dZ \right\rceil\!\right\rceil_{L^p} - \epsilon/3 \ge \lceil\!\lceil X \rceil\!\rceil^*_{Z\text{-}p} - 2\epsilon/3 \ge \lceil\!\lceil F \rceil\!\rceil^*_{Z\text{-}p} - \epsilon .$$

Since $|Y| \le F$ and $\epsilon > 0$ was arbitrary, the claim is proved in the case that F is Z–p-integrable. If F is merely Z–p-measurable, we apply proposition 3.6.5. The $F^{(n)} \stackrel{\text{def}}{=} (|F| \wedge n) \cdot [\![0, n]\!]$ increase to $|F|$. If $\lceil\!\lceil F \rceil\!\rceil^*_{Z\text{-}p} > a$, then $\lceil\!\lceil F^{(n)} \rceil\!\rceil^*_{Z\text{-}p} > a$ for large n, and the argument above produces a $Y \in \mathcal{P}_{00}$ with $|Y| \le F^{(n)} \le F$ and $\lceil\!\lceil \int Y\, dZ \rceil\!\rceil_p > a$. ∎

Corollary 3.6.11 *(i) A process F is Z–p-negligible if and only if it is Z–0-negligible, and is Z–p-measurable if and only if it is Z–0-measurable.*

(ii) Let $F \ge 0$ be any process and $\widetilde{F}$ predictable. Then $\widetilde{F}$ is a predictable Z–p-envelope of F if and only if it is a predictable Z–0-envelope.

Proof. (i) By the countable subadditivity of $\lceil\!\lceil\ \rceil\!\rceil^*_{Z\text{-}p}$ and $\lceil\!\lceil\ \rceil\!\rceil^*_{Z\text{-}0}$ it suffices to prove the first claim under the additional assumption that $|F|$ is majorized by an elementary integrand, say $|F| \le n \cdot [\![0, n]\!]$. The infimum of a predictable Z–p-envelope and a predictable Z–0-envelope is a predictable envelope in the sense both of $\lceil\!\lceil\ \rceil\!\rceil^*_{Z\text{-}p}$ and $\lceil\!\lceil\ \rceil\!\rceil^*_{Z\text{-}0}$ and is integrable in both senses, with the same integral. So if $\lceil\!\lceil F \rceil\!\rceil^*_{Z\text{-}0} = 0$, then $\lceil\!\lceil F \rceil\!\rceil^*_{Z\text{-}p} = 0$. In view of corollary 3.4.5, Z–p-measurability is determined entirely by the Z–p-negligible sets: the Z–p-measurable and Z–0-measurable real-valued processes are the same. We leave part (ii) to the reader. ∎

Definition 3.6.12 *In view of corollary 3.6.11 we shall talk henceforth about **Z-negligible** and **Z-measurable processes**, and about **predictable Z-envelopes**.*

Exercise 3.6.13 Let Z be an L^p-integrator, T a stopping time, and G a process.[6] Then

$$\lceil\!\lceil G \cdot [\![0,T]\!] \rceil\!\rceil^*_{Z\text{-}p} = \lceil\!\lceil G \rceil\!\rceil^*_{Z^T\text{-}p} .$$

Consequently, G is Z^T–p-integrable if and only if $G \cdot [\![0,T]\!]$ is Z–p-integrable, and in that case

$$\int G \, dZ^T = \int G \cdot [\![0,T]\!] \, dZ .$$

Exercise 3.6.14 Let Z, Z' be L^0-integrators. If F is both Z–0-integrable (Z–0-negligible, Z–0-measurable) and Z'–0-integrable (Z'–0-negligible, Z'–0-measurable), then it is $(Z{+}Z')$–0-integrable ($(Z{+}Z')$–0-negligible, $(Z{+}Z')$–0-measurable).

Exercise 3.6.15 Suppose Z is a local L^p-integrator. According to proposition 2.1.9, Z is an L^0-integrator, and the notions of negligibility and measurability for Z have been defined in section 3.2. On the other hand, given the definition of a local L^p-integrator one might want to define negligibility and measurability locally. No matter:

Let (T_n) be a sequence of stopping times that increase without bound and reduce Z to L^p-integrators. A process is Z-negligible or Z-measurable if and only if it is Z^{T_n}-negligible or Z^{T_n}-measurable, respectively, for every $n \in \mathbb{N}$.

Exercise 3.6.16 The Daniell mean is also ***minimal*** in this sense: if $\lceil\!\lceil \;\; \rceil\!\rceil^*$ is a mean such that $\lceil\!\lceil X \rceil\!\rceil^*_{Z\text{-}p} \le \lceil\!\lceil X \rceil\!\rceil^*$ for all elementary integrands X, then $\lceil\!\lceil F \rceil\!\rceil^*_{Z\text{-}p} \le \lceil\!\lceil F \rceil\!\rceil^*$ for all *predictable* F.

Exercise 3.6.17 A process X is Z–0-integrable if and only if for every $\epsilon > 0$ and α there is an $X' \in \mathcal{E}$ with $\|X - X'\|^*_{Z\text{-}[\alpha]} < \epsilon$.

Exercise 3.6.18 Let Z be an L^p-integrator, $0 \le p < \infty$. There exists a positive σ-additive measure μ on $\mathcal{P}$ that has the same negligible sets as $\lceil\!\lceil \;\; \rceil\!\rceil^*_{Z\text{-}p}$. If $p \ge 1$, then μ can be chosen so that $|\mu(X)| \le \lceil\!\lceil X \rceil\!\rceil^*_{Z\text{-}p}$. Such a measure is called a ***control measure*** for Z.

Exercise 3.6.19 Everything said so far in this chapter remains true *mutatis mutandis* if $L^p(\mathbb{P})$ is replaced by the closure $L^1(\|\;\;\|^*)$ of the step function over $\mathcal{F}_\infty$ under a mean $\|\;\;\|^*$ that has the same negligible sets as $\mathbb{P}$.

Stability Under Change of Measure

Let Z be an $L^0(\mathbb{P})$-integrator and $\mathbb{P}'$ a measure on $\mathcal{F}_\infty$ absolutely continuous with respect to $\mathbb{P}$. Since the injection of $L^0(\mathbb{P})$ into $L^0(\mathbb{P}')$ is bounded, Z is an $L^0(\mathbb{P}')$-integrator (proposition 2.1.9). How do the integrals compare?

Proposition 3.6.20 *A Z–0;$\mathbb{P}$-negligible (-measurable, -integrable) process is Z–0;$\mathbb{P}'$-negligible (-measurable, -integrable). The stochastic integral of a Z–0;$\mathbb{P}$-integrable process does not depend on the choice of the probability $\mathbb{P}$ within its equivalence class.*

Proof. For simplicity of reading let us write $\lceil\!\lceil \ \rceil\!\rceil$ for $\lceil\!\lceil \ \rceil\!\rceil_{Z\text{-}0;\mathbb{P}}$, $\lceil\!\lceil \ \rceil\!\rceil'$ for $\lceil\!\lceil \ \rceil\!\rceil_{Z\text{-}0;\mathbb{P}'}$, and $\lceil\!\lceil \ \rceil\!\rceil'^{*}_{Z\text{-}0}$ for the Daniell mean formed with $\lceil\!\lceil \ \rceil\!\rceil'$. Exercise A.8.12 on page 450 furnishes an increasing right-continuous function $\Phi : (0,1] \to (0,1]$ with $\Phi(r) \xrightarrow[r\to 0]{} 0$ such that

$$\lceil\!\lceil f \rceil\!\rceil' \le \Phi(\lceil\!\lceil f \rceil\!\rceil)\,, \qquad f \in L^0(\mathbb{P})\,.$$

The monotonicity of Φ causes the same inequality to hold on $\mathcal{E}^{\uparrow}_{+}$:

$$\begin{aligned}\lceil\!\lceil H \rceil\!\rceil'^{*}_{Z\text{-}0} &= \sup\Big\{\Big\lceil\!\!\Big\lceil \int X\,dZ \Big\rceil\!\!\Big\rceil' : X \in \mathcal{E}\,,\ |X| \le H\Big\}\\ &\le \sup\Big\{\Phi\Big(\Big\lceil\!\!\Big\lceil \int X\,dZ \Big\rceil\!\!\Big\rceil\Big) : X \in \mathcal{E}\,,\ |X| \le H\Big\} \le \Phi\Big(\lceil\!\lceil H \rceil\!\rceil^{*}_{Z\text{-}0}\Big)\end{aligned}$$

for $H \in \mathcal{E}^{\uparrow}_{+}$; the right-continuity of Φ allows its extension to all processes F:

$$\begin{aligned}\lceil\!\lceil F \rceil\!\rceil'^{*}_{Z\text{-}0} &= \inf\Big\{\lceil\!\lceil H \rceil\!\rceil'_{Z\text{-}0} : H \in \mathcal{E}^{\uparrow}_{+}\,,\ H \ge |F|\Big\}\\ &\le \inf\Big\{\Phi\Big(\lceil\!\lceil H \rceil\!\rceil^{*}_{Z\text{-}0}\Big) : H \in \mathcal{E}^{\uparrow}_{+}\,,\ H \ge |F|\Big\} = \Phi\Big(\lceil\!\lceil F \rceil\!\rceil^{*}_{Z\text{-}0}\Big)\,.\end{aligned}$$

Since $\Phi(r) \xrightarrow[r\to 0]{} 0$, a $\lceil\!\lceil \ \rceil\!\rceil^{*}_{Z\text{-}0}$-negligible process is $\lceil\!\lceil \ \rceil\!\rceil'^{*}_{Z\text{-}0}$-negligible, and a $\lceil\!\lceil \ \rceil\!\rceil^{*}_{Z\text{-}0}$-Cauchy sequence is $\lceil\!\lceil \ \rceil\!\rceil'^{*}_{Z\text{-}0}$-Cauchy. A process that is negligible, integrable, or measurable in the sense $\lceil\!\lceil \ \rceil\!\rceil^{*}_{Z\text{-}0}$ is thus negligible, integrable, or measurable, respectively, in the sense $\lceil\!\lceil \ \rceil\!\rceil'^{*}_{Z\text{-}0}$. ∎

Exercise 3.6.21 For the conclusion that Z is an $L^0(\mathbb{P}')$-integrator and that a Z-0;$\mathbb{P}$-negligible (-measurable) process is Z-0;$\mathbb{P}'$-negligible (-measurable) it suffices to know that $\mathbb{P}'$ is *locally* absolutely continuous with respect to $\mathbb{P}$.

Exercise 3.6.22 Modify the proof of proposition 3.6.20 to show in conjunction with exercise 3.2.16 that, whichever gauge on L^p is used to do Daniell's extension with – even if it is not subadditive – , the resulting stochastic integral will be the same.

3.7 The Indefinite Integral

Again a probability $\mathbb{P}$, an exponent $p \ge 0$, and an $L^p(\mathbb{P})$-integrator Z are fixed, and the filtration satisfies the natural conditions.

For motivation consider a measure dz on $\mathbb{R}_+$. The indefinite integral of a function g against dz is commonly defined as the function $t \mapsto \int_0^t g_s\,dz_s$. For this to make sense it suffices that g be locally integrable, i.e., dz-integrable on every bounded set. For instance, the exponential function is locally Lebesgue integrable but not integrable, and yet is of tremendous use. We seek the stochastic equivalent of the notions of local integrability and of the indefinite integral.

The stochastic analog of a bounded interval $[0,t] \subset \mathbb{R}_+$ is a finite stochastic interval $[\![0,T]\!]$. What should it mean to say "G is Z–p-integrable on the stochastic interval $[\![0,T]\!]$"? It is tempting to answer "the process $G \cdot [\![0,T]\!]$ is Z–p-integrable.[6]" This would not be adequate, though. Namely, if Z is not an L^p-integrator, merely a local one, then $[\![\ ]\!]^*_{Z-p}$ may fail to be finite on elementary integrands and so may be no mean; it may make no sense to talk about Z–p-integrable processes. Yet in some suitable sense, we feel, there ought to be many. We take our clue from the classical formula

$$\int_0^t g\,dz \stackrel{\text{def}}{=} \int g \cdot 1_{[0,t]}\,dz = \int g\,dz^t ,$$

where z^t is the stopped distribution function $s \mapsto z^t_s \stackrel{\text{def}}{=} z_{t\wedge s}$. This observation leads directly to the following definition:

Definition 3.7.1 *Let Z be a local L^p-integrator, $0 \le p < \infty$. The process G is* ***Z–p-integrable on the stochastic interval*** *$[\![0,T]\!]$ if T reduces Z to an L^p-integrator and G is Z^T–p-integrable. In this case we write*

$$\int_0^T G\,dZ = \int_{[\![0}^{T]\!]} G\,dZ \stackrel{\text{def}}{=} \int G\,dZ^T .$$

If S is another stopping time, then

$$\int_{S+}^T G\,dZ = \int_{(\!(S}^{T]\!]} G\,dZ \stackrel{\text{def}}{=} \int G \cdot (\!(S,\infty)\!)\,dZ^T . \tag{3.7.1}$$

The expressions in the middle are designed to indicate that the endpoint $[\![T]\!]$ is included in the interval of integration and $[\![S]\!]$ is not, just as it should be when one integrates on the line against a measure that charges points. We will however usually employ the notation on the left with the understanding that the endpoints are always included in the domain of integration, unless the contrary is explicitly indicated, as in (3.7.1). An exception is the point ∞, which is never included in the domain of integration, so that $\int_S^\infty$ and $\int_S^{\infty-}$ mean the same thing. Below we also consider cases where the left endpoint $[\![S]\!]$ is included in the domain of integration and the right endpoint $[\![T]\!]$ is not. *For (3.7.1) to make sense we must assume of course that Z^T is an L^p-integrator.*

Exercise 3.7.2 If G is Z–p-integrable on $(\!(S^{(i)},T^{(i)}]\!]$, $i = 1,2$, then it is Z–p-integrable on the union $(\!(S^{(1)} \wedge S^{(2)}, T^{(1)} \vee T^{(2)}]\!]$.

Definition 3.7.3 *Let Z be a local L^p-integrator, $0 \le p < \infty$. The process G is* ***locally Z–p-integrable*** *if it is Z–p-integrable on arbitrarily large stochastic intervals, that is to say, if for every $\epsilon > 0$ and $t < \infty$ there is a stopping time T with $\mathbb{P}[T < t] < \epsilon$ that reduces Z to an L^p-integrator such that G is Z^T–p-integrable.*

Here is yet another indication of the flexibility of L^0-integrators:

Proposition 3.7.4 *Let Z be a local L^0-integrator. A locally Z–0-integrable process is Z–0-integrable on every almost surely finite stochastic interval.*

Proof. The stochastic interval $[\![0, U]\!]$ is called almost surely finite, of course, if $\mathbb{P}[U = \infty] = 0$. We know from proposition 2.1.9 that Z is an L^0-integrator. Thanks to exercise 3.6.13 it suffices to show that $G' \stackrel{\text{def}}{=} G \cdot [\![0, U]\!]$ is Z–0-integrable. Let $\epsilon > 0$. There exists a stopping time T with $\mathbb{P}[T < U] < \epsilon$ so that G and then G' are Z^T–0-integrable.[6] Then $G'' \stackrel{\text{def}}{=} G' \cdot [\![0, T]\!] = G \cdot [\![0, U \wedge T]\!]$ is Z–0-integrable (ibidem). The difference $G''' = G' - G''$ is Z-measurable and vanishes off the stochastic interval $I \stackrel{\text{def}}{=} (\!(T, U]\!]$, whose projection on Ω has measure less than ϵ, and so $\lceil\!\lceil G''' \rceil\!\rceil^*_{Z\text{-}0} \le \epsilon$ (exercise 3.1.2). In other words, G' differs arbitrarily little (by less than ϵ) in $\lceil\!\lceil\ \rceil\!\rceil^*_{Z\text{-}0}$-mean from a Z–0-integrable process (G''). It is thus Z–0-integrable itself (proposition 3.2.20). ∎

The Indefinite Integral

Let Z be a local L^p-integrator, $0 \le p < \infty$, and G a locally Z–p-integrable process. Then Z is an L^0-integrator and G is Z–0-integrable on every finite deterministic interval $[\![0, t]\!]$ (proposition 3.7.4). It is tempting to define the indefinite integral as the function $t \mapsto \int G\, dZ^t$. This is for every t a *class* in L^0 (definition 3.2.14). We can be a little more precise: since $\int X\, dZ^t \in \mathcal{F}_t$ when $X \in \mathcal{E}$, the limit $\int G\, dZ^t$ of such elementary integrals can be viewed as an equivalence class of $\mathcal{F}_t$-measurable random variables. It is desirable to have for the indefinite integral a process rather than a mere slew of classes. This is possible of course by the simple expedient of selecting from every class $\int G\, dZ^t \subset L^0(\mathcal{F}_t, \mathbb{P})$ a random variable measurable on $\mathcal{F}_t$. Let us do that and *temporarily* call the process so obtained $G{*}Z$:

$$(G{*}Z)_t \in \int G\, dZ^t \quad \text{and} \quad (G{*}Z)_t \in \mathcal{F}_t \quad \forall t\,. \tag{3.7.2}$$

This is not really satisfactory, though, since two different people will in general come up with wildly differing modifications $G{*}Z$. Fortunately, this deficiency is easily repaired using the following observation:

Lemma 3.7.5 *Suppose that Z is an L^0-integrator and that G is locally Z–0-integrable. Then any process $G{*}Z$ satisfying (3.7.2) is an L^0-integrator and consequently has an adapted modification that is right-continuous with left limits. If $G{*}Z$ is such a version, then*

$$(G{*}Z)_T \in \int_0^T G\, dZ \tag{3.7.3}$$

for any stopping time T for which the integral on the right exists – in particular for all almost surely finite stopping times T.

Proof. It is immediate from the Dominated Convergence Theorem that $G{*}Z$ is right-continuous in probability. For if $t_n \downarrow t$, then $(G{*}Z)_{t_n} - (G{*}Z)_t \in \int G \cdot (\!(t, t_n]\!] \, dZ \to 0$ $\lceil\!\lceil \ \rceil\!\rceil_{L^0}$-mean. To see that the L^0-boundedness condition (B-0) of definition 2.1.7 is satisfied as well, take an elementary integrand X as in (2.1.1), to wit,[6]

$$X = f_0 \cdot [\![0]\!] + \sum_{n=1}^{N} f_n \cdot (\!(t_n, t_{n+1}]\!] \,, \qquad f_n \in \mathcal{F}_{t_n} \text{ simple.}$$

Then
$$\int X \, d(G{*}Z) = f_0 \cdot (G{*}Z)_0 + \sum_n f_n \cdot \big((G{*}Z)_{t_{n+1}} - (G{*}Z)_{t_n}\big)$$

$$\in \dot{f}_0 \cdot \dot{G}_0 \dot{Z}_0 + \sum_n \dot{f}_n \cdot \int_{t_n+}^{t_{n+1}} G \, dZ$$

by exercise 3.5.5:
$$= \dot{f}_0 \dot{G}_0 \cdot \dot{Z}_0 + \int \sum_n f_n \cdot (\!(t_n, t_{n+1}]\!] \cdot G \, dZ$$

$$= \int X \cdot G \, dZ \,. \tag{3.7.4}$$

Multiply with $\lambda > 0$ and measure both sides with $\lceil\!\lceil \ \rceil\!\rceil_{L^0}$ to obtain

$$\left\lceil\!\!\left\lceil \int \lambda X \, d(G{*}Z) \right\rceil\!\!\right\rceil_{L^0} = \left\lceil\!\!\left\lceil \int \lambda X \cdot G \, dZ \right\rceil\!\!\right\rceil_{L^0}$$
$$\le \lceil\!\lceil \lambda X \cdot G \rceil\!\rceil^*_{Z\text{-}0} \le \lceil\!\lceil \lambda \cdot G \rceil\!\rceil^*_{Z^t\text{-}0}$$

for all $X \in \mathcal{E}$ with $|X| \le 1$. The right-hand side tends to zero as $\lambda \to 0$: (B-0) is satisfied, and $G{*}Z$ indeed is an L^0-integrator.

Theorem 2.3.4 in conjunction with the natural conditions now furnishes the desired right-continuous modification with left limits. Henceforth $G{*}Z$ denotes such a version.

To prove equation (3.7.3) we start with the case that T is an elementary stopping time; it is then nothing but (3.7.4) applied to $X = [\![0, T]\!]$. For a general stopping time T we employ once again the stopping times $T^{(n)}$ of exercise 1.3.20. For any k they take only finitely many values less than k and decrease to T. In taking the limit as $n \to \infty$ in

$$(G{*}Z)_{T^{(n)} \wedge k} \in \int_0^{T^{(n)} \wedge k} G \, dZ \,,$$

the left-hand side converges to $(G{*}Z)_{T \wedge k}$ by right-continuity, the right-hand side to[6] $\int_0^{T \wedge k} G \, dZ = \int G \cdot [\![0, T \wedge k]\!] \, dZ$ by the Dominated Convergence Theorem. Now take $k \to \infty$ and use the domination $\big|G \cdot [\![0, T \wedge k]\!]\big| \le \big|G \cdot [\![0, T]\!]\big|$ to arrive at $(G{*}Z)_T \doteq \int G \cdot [\![0, T]\!] \, dZ$. In view of exercise 3.6.13, this is equation (3.7.3). ∎

Any two modifications produced by lemma 3.7.5 are of course indistinguishable. This observation leads directly to the following:

Definition 3.7.6 *Let Z be an L^0-integrator and G a locally Z–0-integrable process. The* ***indefinite integral*** *is a process $G{*}Z$ that is right-continuous with left limits and adapted to $\mathcal{F}_+^{\mathfrak{P}}$ and that satisfies*

$$(G{*}Z)_t \in \int_0^t G\,dZ \stackrel{\text{def}}{=} \int G\,dZ^t \quad \forall\, t \in [0,\infty)\,.$$

It is unique up to indistinguishability. If G is Z–0-integrable, it is understood that $G{*}Z$ is chosen so as to have almost surely a finite limit at infinity as well.

So far it was necessary to distinguish between random variables and their classes when talking about the stochastic integral, because the latter is by its very definition an equivalence class modulo negligible functions. Henceforth we shall do this: when we meet an L^0-integrator Z and a locally Z–0-integrable process G we shall pick once and for all an indefinite integral $G{*}Z$; then $\int_{S+}^T G\,dZ$ will denote the specific random variable $(G{*}Z)_T-(G{*}Z)_S$, etc. Two people doing this will not come up with precisely the same random variables $\int_{S+}^T G\,dZ$, but with nearly the same ones, since in fact the whole paths of their versions of $G{*}Z$ nearly agree. If G happens to be Z–0-integrable, then $\int G\,dZ$ is the almost surely defined random variable $(G{*}Z)_\infty$.

Vectors of integrators $\boldsymbol{Z} = (Z^1, Z^2, \ldots, Z^d)$ appear naturally as drivers of stochastic differentiable equations (pages 8 and 56). The gentle reader recalls from page 109 that the integral extension

$$\mathfrak{L}^1[\lceil\!\lceil\ \rceil\!\rceil^*_{\boldsymbol{Z}\text{–}p}] \ni \boldsymbol{X} \mapsto \int \boldsymbol{X}\,d\boldsymbol{Z}$$

of the elementary integral $\qquad \check{\mathcal{E}} \ni \boldsymbol{X} \mapsto \int \boldsymbol{X}\,d\boldsymbol{Z}$

is given by $\qquad (X_1, \ldots, X_d) = \boldsymbol{X} \mapsto \sum_{\eta=1}^d \int X_\eta\,dZ^\eta\,.$

Therefore $\qquad \boldsymbol{X}{*}\boldsymbol{Z} \stackrel{\text{def}}{=} \sum_{\eta=1}^d X_\eta{*}Z^\eta$

is reasonable notation for the indefinite integral of $\boldsymbol{X}$ against $d\boldsymbol{Z}$; the right-hand side is a càdlàg process unique up to indistinguishability and satisfies $(\boldsymbol{X}{*}\boldsymbol{Z})_T \in \int_0^T \boldsymbol{X}\,d\boldsymbol{Z} \quad \forall\, T \in \mathfrak{T}$. Henceforth $\int_0^T \boldsymbol{X}d\boldsymbol{Z}$ means the random variable $(\boldsymbol{X}{*}\boldsymbol{Z})_T$.

Exercise 3.7.7 Define $\lceil \boldsymbol{Z} \rceil_{I^p}$ and show that $\lceil \boldsymbol{Z} \rceil_{I^p} = \sup\{\lceil \boldsymbol{X}{*}\boldsymbol{Z} \rceil_{I^p} : \boldsymbol{X} \in \mathcal{E}_1^d\}$.

Exercise 3.7.8 Suppose we are faced with a whole collection $\mathfrak{P}$ of probabilities, the filtration $\mathcal{F}_.$ is right-continuous, and Z is an $L^0(\mathbb{P})$-integrator for every $\mathbb{P} \in \mathfrak{P}$. Let G be a predictable process that is locally Z–0;$\mathbb{P}$-integrable for every $\mathbb{P} \in \mathfrak{P}$. There is a right-continuous process $G{*}Z$ with left limits, adapted to the $\mathfrak{P}$***-regularization*** $\mathcal{F}^{\mathfrak{P}}_{.} \stackrel{\text{def}}{=} \bigcap_{\mathbb{P}\in\mathfrak{P}} \mathcal{F}^{\mathbb{P}}_{.}$, that is an indefinite integral in the sense of $L^0(\mathbb{P})$ for every $\mathbb{P} \in \mathfrak{P}$.

Exercise 3.7.9 If M is a right-continuous local martingale and G is locally M–1-integrable (see corollary 2.5.29), then $G{*}M$ is a local martingale.

Integration Theory of the Indefinite Integral

If a measure dy on $[0,\infty)$ has a density with respect to the measure dz, say $dy_t = g_t\, dz_t$, then a function f is dy-negligible (-integrable, -measurable) if and only if the product fg is dz-negligible (-integrable, -measurable). The corresponding statements are true in the stochastic case:

Theorem 3.7.10 *Let Z be an L^p-integrator, $p \in [0,\infty)$, and G a Z–p-integrable process. Then for all processes F*

$$\lceil\!\lceil F \rceil\!\rceil^*_{(G*Z)\text{–}p} = \lceil\!\lceil F\cdot G \rceil\!\rceil^*_{Z\text{–}p} \quad \textit{and} \quad \lceil G{*}Z \rceil_{\mathcal{I}^p} = \lceil\!\lceil G \rceil\!\rceil^*_{Z\text{–}p} . \tag{3.7.5}$$

Therefore a process F is $(G{*}Z)$–p-negligible (-integrable, -measurable) if and only if $F{\cdot}G$ is Z–p-negligible (-integrable, -measurable). If F is locally $(G{*}Z)$–p-integrable, then

$$F{*}(G{*}Z) = (FG){*}Z ,$$

in particular

$$\int F\, d(G{*}Z) \doteq \int F\cdot G\, dZ$$

when F is $(G{*}Z)$–p-integrable.

Proof. Let $Y = G{*}Z$ denote the indefinite integral. The family of bounded processes X with $\int X\, dY = \int XG\, dZ$ contains $\mathcal{E}$ (equation (3.7.4) on page 133) and is closed under pointwise limits of bounded sequences. It contains therefore the family $\mathcal{P}_b$ of all bounded predictable processes. The assignment $F \mapsto \lceil\!\lceil FG \rceil\!\rceil^*_{Z\text{–}p}$ is easily seen to be a mean: properties (i) and (iii) of definition 3.2.1 on page 94 are trivially satisfied, (ii) follows from proposition 3.6.5, (iv) from the Dominated Convergence Theorem, and (v) from exercise 3.2.15. If F is predictable, then, due to corollary 3.6.10,

$$\begin{aligned}
\lceil\!\lceil F \rceil\!\rceil^*_{Y\text{–}p} &= \sup\Big\{ \Big\lceil\!\Big\lceil \int X\, dY \Big\rceil\!\Big\rceil_p : X \in \mathcal{P}_b ,\, |X| \le |F| \Big\} \\
&= \sup\Big\{ \Big\lceil\!\Big\lceil \int XG\, dZ \Big\rceil\!\Big\rceil_p : X \in \mathcal{P}_b ,\, |X| \le |F| \Big\} \\
&= \sup\Big\{ \Big\lceil\!\Big\lceil \int X'\, dZ \Big\rceil\!\Big\rceil_p : X' \in \mathcal{P}_b ,\, |X'| \le |FG| \Big\} \\
&= \lceil\!\lceil FG \rceil\!\rceil^*_{Z\text{–}p} .
\end{aligned}$$

The maximality of Daniell's mean (proposition 3.6.1 on page 123) gives $\lceil\!\lceil FG \rceil\!\rceil^*_{Z\text{–}p} \le \lceil\!\lceil F \rceil\!\rceil^*_{Y\text{–}p}$ for all F. For the converse inequality let $\widetilde{F}$ be a predictable Y–p-envelope of $F \ge 0$ (proposition 3.6.6). Then

$$\lceil\!\lceil F \rceil\!\rceil^*_{Y\text{–}p} = \Big\lceil\!\Big\lceil \widetilde{F} \Big\rceil\!\Big\rceil^*_{Y\text{–}p} = \Big\lceil\!\Big\lceil \widetilde{F}G \Big\rceil\!\Big\rceil^*_{Z\text{–}p} \ge \lceil\!\lceil FG \rceil\!\rceil^*_{Z\text{–}p} .$$

This proves equation (3.7.5). The second claim is evident from this identity. The equality of the integrals in the last line holds for elementary integrands (equation (3.7.4)) and extends to $Y\text{–}p$-integrable processes by approximation in mean: if $\mathcal{E} \ni X^{(n)} \to F$ in $\lceil\!\lceil\ \rceil\!\rceil^*_{Y-p}$-mean, then $\mathcal{E} \ni X^{(n)} \cdot G \to F \cdot G$ in $\lceil\!\lceil\ \rceil\!\rceil^*_{Z-p}$-mean and so

$$\int F\, dY \doteq \lim \int X^{(n)}\, dY \doteq \lim \int X^{(n)} \cdot G\, dZ \doteq \int F \cdot G\, dZ$$

in the topology of L^p. We apply this to the processes[6] $F\cdot[\![0,t]\!]$ and find that $F{*}(G{*}Z)$ and $(FG){*}Z$ are modifications of each other. Being right-continuous and adapted, they are indistinguishable (exercise 1.3.28). ∎

Corollary 3.7.11 *Let $G^{(n)}$ and G be locally $Z\text{–}0$-integrable processes, T_k finite stopping times increasing to ∞, and assume that*

$$\left\lceil\!\!\left\lceil G - G^{(n)} \right\rceil\!\!\right\rceil^*_{Z^{T_k}\text{–}0} \xrightarrow[n\to\infty]{} 0$$

for every $k \in \mathbb{N}$. Then the paths of $G^{(n)}{*}Z$ converge to the paths of $G{*}Z$ uniformly on bounded intervals, in probability. There are a subsequence $\big(G^{(n_k)}\big)$ and a nearly empty set N outside which the path of $G^{(n_k)}{*}Z$ converges uniformly on bounded intervals to the path of $G{*}Z$.

Proof. By lemma 2.3.2 and equation (3.7.5),

$$\begin{aligned}\delta_k^{(n)}(\lambda) &\stackrel{\text{def}}{=} \mathbb{P}\Big[\big(G{*}Z - G^{(n)}{*}Z\big)^\star_{T_k} > \lambda\Big] = \mathbb{P}\Big[\big((G - G^{(n)}){*}Z\big)^\star_{T_k} > \lambda\Big] \\ &\le \left\lceil \lambda^{-1}\big((G - G^{(n)}){*}Z\big)^{T_k} \right\rceil_{\mathcal{I}^0} = \left\lceil\!\!\left\lceil \lambda^{-1}\big(G - G^{(n)}\big) \right\rceil\!\!\right\rceil^*_{Z^{T_k}\text{–}0} \xrightarrow[n\to\infty]{} 0 .\end{aligned}$$

We take a subsequence $\big(G^{(n_k)}\big)$ so that $\delta_k^{(n_k)}(2^{-k}) \le 2^{-k}$ and set

$$N \stackrel{\text{def}}{=} \limsup_k \Big[\big(G{*}Z - G^{(n_k)}{*}Z\big)^\star_{T_k} > 2^{-k}\Big] .$$

This set belongs to $\mathcal{A}_{\infty\sigma}$ and by the Borel–Cantelli lemma is negligible: it is nearly empty. If $\omega \notin N$, then the path $\big(G^{(n_k)}{*}Z\big)_.(\omega)$ converges evidently to the path $(G{*}Z)_.(\omega)$ uniformly on every one of the intervals $[0, T_k(\omega)]$. ∎

If $X \in \mathcal{E}$, then $X{*}Z$ jumps precisely where Z jumps. Therefore:

Corollary 3.7.12 *If Z has continuous paths, then every indefinite integral $G{*}Z$ has a modification with continuous paths – which will then of course be chosen.*

Corollary 3.7.13 *Let A be a subset of Ω, not necessarily measurable, and assume the paths of the locally $Z\text{–}0$-integrable process G vanish almost surely on A. Then the paths of $G{*}Z$ also vanish almost surely, in fact nearly, on A.*

Proof. The set $[0,\infty) \times A \subset \boldsymbol{B}$ is by equation (3.7.5) $(G{*}Z)$-negligible. Corollary 3.6.7 says that the paths of $G{*}Z$ nearly vanish on A.

Exercise 3.7.14 If G is Z–0-integrable, then $\lceil G{*}Z \rceil_{[\alpha]} = \| G \|^*_{Z\text{-}[\alpha]}$ for $\alpha > 0$.

Exercise 3.7.15 For any locally Z–0-integrable G and any almost surely finite stopping time T the processes $G{*}Z^T$ and $(G{*}Z)^T$ are indistinguishable.

Exercise 3.7.16 (P.–A. Meyer) Let Z, Z' be L^0-integrators and X, X' processes that are integrable for both. Let Ω_0 be a subset of Ω and $T : \Omega \to \mathbb{R}_+$ a time, neither of them necessarily measurable. If $X = X'$ and $Z = Z'$ up to and including (excluding) time T on Ω_0, then $X{*}Z = X'{*}Z'$ up to and including (excluding) time T on Ω_0, except possibly on an evanescent set.

A General Integrability Criterion

Theorem 3.7.17 *Let Z be an L^0-integrator, T an almost surely finite stopping time, and X a Z-measurable process. If $X^\star_T$ is almost surely finite, then X is Z–0-integrable on $[\![0,T]\!]$.*

This says – to put it plainly if a bit too strongly – that any reasonable process is Z–0-integrable. The assumptions concerning the integrand are often easy to check: X is usually given as a construct using algebraic and order combinations and limits of processes known to be Z-measurable, so the splendid permanence properties of measurability will make it obvious that X is Z-measurable; frequently it is also evident from inspection that the maximal process $X^\star$ is almost surely finite at any instant, and thus at any almost surely finite stopping time. In cases where the checks are that easy we shall not carry them out in detail but simply write down the integral without fear. That is the point of this theorem.

Proof. Let $\epsilon > 0$. Since $[X^\star_T \le K] \uparrow \Omega$ almost surely as $K \uparrow \infty$, and since outer measure $\mathbb{P}^*$ is continuous along increasing sequences, there is a number K with $\mathbb{P}^*[X^\star_T \le K] > 1 - \epsilon$. Write $X' = X \cdot [\![0,T]\!]$ and[1]

$$X' = X' \cdot \big[|X| \le K\big] + X' \cdot \big[|X| > K\big] = X^{(1)} + X^{(2)} .$$

Now Z^T is a global L^0-integrator, and so $X^{(1)}$ is Z^T–0-integrable, even Z–0-integrable (exercise 3.6.13). As to $X^{(2)}$, it is Z-measurable and its entire path vanishes on the set $[X^\star_T \le K]$. If Y is a process in $\mathcal{P}_{00}$ with $|Y| \le |X^{(2)}|$, then its entire path also vanishes on this set, and thanks to corollary 3.7.13 so does the path of $Y{*}Z$, at least almost surely. In particular, $\int Y\, dZ = 0$ almost surely on $[X^\star_T \le K]$. Thus $B \overset{\text{def}}{=} \big[\int Y\, dZ \ne 0\big]$ is a measurable set almost surely disjoint from $[X^\star_T \le K]$. Hence $\mathbb{P}[B] \le \epsilon$ and $\lceil\!\lceil \int Y\, dZ \rceil\!\rceil_0 \le \epsilon$. Corollary 3.6.10 shows that $\lceil\!\lceil X^{(2)} \rceil\!\rceil^*_{Z\text{-}0} \le \epsilon$. That is, X' differs from the Z–0-integrable process $X^{(1)}$ arbitrarily little in Z–0-mean and therefore is Z–0-integrable itself. That is to say, X is indeed Z–0-integrable on $[\![0,T]\!]$.

Exercise 3.7.18 Suppose that F is a process whose paths all vanish outside a set $\Omega_0 \subset \Omega$ with $\mathbb{P}^*(\Omega_0) < \epsilon$. Then $[\![F]\!]^*_{Z\text{-}0} < \epsilon$.

Exercise 3.7.19 If Z is previsible and $T \in \mathfrak{T}$, then the stopped process Z^T is previsible. If Z is a previsible integrator and X a Z-0-integrable process, then $X{*}Z$ is previsible.

Exercise 3.7.20 Let Z be an L^0-integrator and S, T two stopping times. (i) If G is a process Z-0-integrable on $(\!(S, T]\!]$ and $f \in L^0(\mathcal{F}_S, \mathbb{P})$, then the process $f \cdot G$ is Z-0-integrable on $(\!(S, T]\!]$

and $$\int_{S+}^{T} f \cdot G\, dZ = f \cdot \int_{S+}^{T} G\, dZ \in \mathcal{F}_T \text{ a.s.}$$

Also, for $f \in L^0(\mathcal{F}_0, \mathbb{P})$, $\displaystyle\int_0^0 f\, dZ = \int_{[\![0}^{0]\!]} f\, dZ = f \cdot Z_0$.

(ii) If G is Z-0-integrable on $[\![0, T]\!]$, S is predictable, and f is measurable on the strict past of S and almost surely finite, then $f \cdot G$ is Z-0-integrable on $[\![S, T]\!]$, and

$$\int_S^T f \cdot G\, dZ = f \cdot \int_S^T G\, dZ . \tag{3.7.6}$$

(iii) Let (S_k) be a sequence of finite stopping times that increases to ∞ and f_k almost surely finite random variables measurable on $\mathcal{F}_{S_k}$. Then $G \stackrel{\text{def}}{=} \sum_k f_k \cdot (\!(S_k, S_{k+1}]\!]$ is locally Z-0-integrable, and its indefinite integral is given by

$$(G{*}Z)_t \doteq \textstyle\sum_k f_k \cdot \left(Z^t_{S_{k+1}} - Z^t_{S_k}\right) = \sum_k f_k \cdot \left(Z_t^{S_{k+1}} - Z_t^{S_k}\right) .$$

Exercise 3.7.21 Suppose Z is an L^p-integrator for some $p \in [0, \infty)$, and $X, X^{(n)}$ are previsible processes Z-p-integrable on $[\![0, T]\!]$ and such that $|X^{(n)} - X|^\star_T \xrightarrow[n\to\infty]{} 0$ in probability. Then X is Z-p-integrable on $[\![0, T]\!]$, and $|X^{(n)}{*}Z - X{*}Z|^\star_T \xrightarrow[n\to\infty]{} 0$ in L^p-mean (cf. [91]).

Approximation of the Integral via Partitions

The Lebesgue integral of a càglàd integrand, being a Riemann integral, can be approximated via partitions. So can the stochastic integral:

Definition 3.7.22 *A **stochastic partition** or **random partition** of the stochastic interval $[\![0, \infty)\!)$ is a finite or countable collection*

$$\mathcal{S} = \{0 = S_0 \le S_1 \le S_2 \le \ldots \le S_\infty \le \infty\}$$

*of stopping times. $\mathcal{S}$ is assumed to contain the stopping time $S_\infty \stackrel{\text{def}}{=} \sup_k S_k$ – which is no assumption at all when $\mathcal{S}$ is finite or $S_\infty = \infty$. It simplifies the notation in some formulas to set $S_{\infty+1} \stackrel{\text{def}}{=} \infty$. We say that the random partition $\mathcal{T} = \{0 = T_0 \le T_1 \le T_2 \le \ldots \le T_\infty \le \infty\}$ **refines** $\mathcal{S}$ if*

$$\bigcup\{[\![S]\!] : S \in \mathcal{S}\} \subseteq \bigcup\{[\![T]\!] : T \in \mathcal{T}\} .$$

*The **mesh** of $\mathcal{S}$ is the (non-adapted) process* mesh$[\mathcal{S}]$ *that at $\varpi = (\omega, s) \in \boldsymbol{B}$ has the value*

$$\inf\left\{\overline{\rho}(S(\omega), S'(\omega)) : S \le S' \text{ in } \mathcal{S} ,\ S(\omega) < s \le S'(\omega)\right\} .$$

Here $\overline{\rho}$ is the arctan metric on $\overline{\mathbb{R}}_+$ (see item A.1.2 on page 363). With the random partition $\mathcal{S}$ and the process $Z \in \mathfrak{D}$ goes the ***$\mathcal{S}$-scalæfication***

$$Z^{\mathcal{S}} \stackrel{\text{def}}{=} \sum_{0\le k\le\infty} Z_{S_k}\cdot[\![S_k, S_{k+1})\!) \stackrel{\text{def}}{=} \sum_{0\le k<\infty} Z_{S_k}\cdot[\![S_k, S_{k+1})\!) + Z_{S_\infty}\cdot[\![S_\infty, \infty)\!) ,$$

defined so as to produce again a process $Z^{\mathcal{S}} \in \mathfrak{D}$.

The left-continuous version of the scalæfication of $X \in \mathfrak{D}$ evidently is

$$X_{\leftarrow}^{\mathcal{S}} = \sum\nolimits_{0\le k\le\infty} X_{S_k}\cdot(\!(S_k, S_{k+1}]\!]$$

with indefinite dZ-integral

$$\left(X_{\leftarrow}^{\mathcal{S}} * Z\right)_t = \sum\nolimits_{0\le k\le\infty} X_{S_k}\cdot\left(Z_t^{S_{k+1}} - Z_t^{S_k}\right) = \int_0^t X_{\leftarrow}^{\mathcal{S}}\, dZ . \tag{3.7.7}$$

Theorem 3.7.23 *Let $\mathcal{S}^n = \{0 = S_0^n \le S_1^n \le S_2^n \le \ldots \le \infty\}$ be a sequence of random partitions of $\boldsymbol{B}$ such that* mesh$[\mathcal{S}^n] \xrightarrow[n\to\infty]{} 0$ *except possibly on an evanescent set.*[8] *Assume that Z is an L^p-integrator for some $p \in [0,\infty)$, that $X \in \mathfrak{D}$, and that the maximal process of $X_{\leftarrow} \in \mathfrak{L}$ is Z–p-integrable on every interval $[\![0,u]\!]$ – recall from theorem 3.7.17 that this is automatic when $p = 0$. The indefinite integrals $X_{\leftarrow}^{\mathcal{S}^n} * Z$ of (3.7.7) then approximate the indefinite integral $X_{\leftarrow} * Z$ uniformly in p-mean, in the sense that for every instant $u < \infty$*

$$\left\|\!\left| \, |X_{\leftarrow} * Z - X_{\leftarrow}^{\mathcal{S}^n} * Z|_u^\star \right\|\!\right|_p \xrightarrow[n\to\infty]{} 0 . \tag{3.7.8}$$

Proof. Due to the left-continuity of $X_{\leftarrow}$, we evidently have

$$X_{\leftarrow}^{\mathcal{S}^n} \xrightarrow[n\to\infty]{} X_{\leftarrow} \quad \text{pointwise on } \boldsymbol{B}. \tag{3.7.9}$$

Also $\qquad |X_{\leftarrow}^{\mathcal{S}^n} - X_{\leftarrow}| \cdot [\![0,u]\!] \le 2|X_{\leftarrow}|_u^\star \cdot [\![0,u]\!] \in \mathfrak{L}^1[\lceil\!\lceil \ \rceil\!\rceil_{Z-p}^*] ,$

so the Dominated Convergence Theorem gives $\lceil\!\lceil |X_{\leftarrow}^{\mathcal{S}^n} - X_{\leftarrow}|\cdot[\![0,u]\!] \rceil\!\rceil_{Z-p}^* \xrightarrow[n\to\infty]{} 0$. An application of the maximal theorem 2.3.6 on page 63 leads to

$$\left\|\!\left| \, |X_{\leftarrow}^{\mathcal{S}^n} * Z - X_{\leftarrow} * Z|_u^\star \right\|\!\right|_p \le C_p^{\star(2.3.5)} \cdot \left\lceil (X_{\leftarrow}^{\mathcal{S}^n} - X_{\leftarrow}) * Z^u \right\rceil_{\mathcal{I}^p}$$

by theorem 3.7.10: $\qquad \le C_p^\star \cdot \left\|\!\left| \, |X_{\leftarrow}^{\mathcal{S}^n} - X_{\leftarrow}| \cdot [\![0,u]\!] \right\|\!\right|_{Z-p}^* \xrightarrow[n\to\infty]{} 0 .$

This proves the claim for $p > 0$. The case $p = 0$ is similar.

Note that equation (3.7.8) does not permit us to conclude that the approximants $X_{\leftarrow}^{\mathcal{S}^n} * Z$ converge to $X_{\leftarrow} * Z$ almost surely; for that, one has to choose the partitions $\mathcal{S}^n$ so that the convergence in equation (3.7.9) becomes uniform (theorem 3.7.26); even $\sup_{\varpi\in\boldsymbol{B}}$ mesh$[\mathcal{S}^n](\varpi) \xrightarrow[n\to\infty]{} 0$ does not guarantee that.

[8] A partition $\mathcal{S}$ is assumed to contain $S_\infty \stackrel{\text{def}}{=} \sup_{k<\infty} S_k$, and defining $S_{\infty+1} \stackrel{\text{def}}{=} \infty$ simplifies some formulas.

Exercise 3.7.24 For every $X \in \mathfrak{D}$ and $(\mathcal{S}^n)$ there exists a subsequence $(\mathcal{S}^{n_k})$ that depends on Z and is impossible to find in practice, so that $X_{.-}^{\mathcal{S}^{n_k}}*Z \to X_{.-}*Z$ almost surely uniformly on bounded intervals.

Exercise 3.7.25 Any two stochastic partitions have a common refinement.

Pathwise Computation of the Indefinite Integral

The discussion of chapter 1, in particular theorem 1.2.8 on page 17, seems to destroy any hope that $\int G\,dZ$ can be understood pathwise, not even when both the integrand G and the integrator Z are nice and continuous, say. On the other hand, there are indications that the path of the indefinite integral $G*Z$ is determined to a large extent by the paths of G and Z alone: this is certainly true if G is elementary, and corollary 3.7.11 seems to say that this is still so almost surely if G is any integrable process; and exercise 3.7.16 seems to say the same thing in a different way.

There is, in fact, an *algorithm* implementable on an (ideal) computer that takes the paths $s \mapsto X_s(\omega)$ and $s \mapsto Z_s(\omega)$ and computes from them an approximate path $s \mapsto Y_s^{(\delta)}(\omega)$ of the indefinite integral $X_{.-}*Z$. If the parameter $\delta > 0$ is taken through a sequence (δ_n) that converges to zero sufficiently fast, the approximate paths $Y_.^{(\delta_n)}(\omega)$ converge uniformly on every finite stochastic interval to the path $(X_{.-}*Z)_.(\omega)$ of the indefinite integral. Moreover, the rate of convergence can be estimated.

This applies only to certain integrands, so let us be precise about the data. The filtration is assumed right-continuous. $\mathbb{P}$ is a fixed probability on $\mathcal{F}_\infty$, and Z is a right-continuous $L^0(\mathbb{P})$-integrator. As to the integrand, it equals the left-continuous version $X_{.-}$ of some real-valued càdlàg adapted process X; its value at time 0 is 0. The integrand might be the left-continuous version of a continuous function of some integrator, for a typical example. Such a process is adapted and left-continuous, hence predictable (proposition 3.5.2). Since its maximal function is finite at any instant, it is locally Z–0-integrable (theorem 3.7.17). Here is the typical approximate $Y^{(\delta)}$ to the indefinite integral $Y = X_{.-}*Z$: fix a threshold $\delta > 0$. Set $S_0 \overset{\text{def}}{=} 0$ and $Y_0^{(\delta)} \overset{\text{def}}{=} 0$; then proceed recursively with

$$S_{k+1} \overset{\text{def}}{=} \inf\left\{t > S_k : \left|X_t - X_{S_k}\right| > \delta\right\} \tag{3.7.10}$$

and
$$Y_t^{(\delta)} \overset{\text{def}}{=} Y_{S_k} + X_{S_k}\cdot(Z_t - Z_{S_k}) \qquad \text{for } S_k < t \le S_{k+1}$$

by induction:
$$= \sum_{\kappa=1}^{k} X_{S_\kappa}\cdot\left(Z^t_{S_{\kappa+1}} - Z^t_{S_\kappa}\right). \tag{3.7.11}$$

In other words, the prescription is this: wait until the *change in the integrand* warrants a new computation, then do a linear approximation – the scheme above is an ***adaptive Riemann-sum scheme.***[9] Another way of looking at it is to note that (3.7.10) defines a stochastic partition $\mathcal{S} = \mathcal{S}^\delta$ and that by equation (3.7.7) the process $Y^{(\delta)}_.$ is but the indefinite dZ-integral of $X^{\mathcal{S}}_{.-}$.

The algorithm (3.7.10)–(3.7.11) converges pathwise provided δ is taken through a sequence (δ_n) that converges sufficiently quickly to zero:

Theorem 3.7.26 *Choose numbers $\delta_n > 0$ so that*

$$\sum_{n=1}^{\infty} \lceil n\delta_n \cdot Z^n \rceil_{\mathcal{I}^0} < \infty \tag{3.7.12}$$

(Z^n is Z stopped at the instant n). If X is any adapted càdlàg process, then $X_{.-}$ is locally Z–0-integrable; and for nearly all $\omega \in \Omega$ the approximates $Y^{(\delta_n)}_.(\omega)$ of (3.7.11) converge to the indefinite integral $(X_{.-}{*}Z)_.(\omega)$ uniformly on any finite interval, as $n \to \infty$.

Remarks 3.7.27 (i) The sequence (δ_n) depends only on the integrator sizes of the stopped processes Z^n. For instance, if $\lceil Z \rceil_{\mathcal{I}^p} < \infty$ for some $p > 0$, then the choice $\delta_n \overset{\text{def}}{=} n^{-q}$ will do as long as $q > 1 + 1 \vee 1/p$.

The algorithm (3.7.10)–(3.7.11) can be viewed as a black box – not hard to write as a program on a computer once the numbers δ_n are fixed – that takes two inputs and yields one output. One of the inputs is a path $Z_.(\omega)$ of any integrator Z satisfying inequality (3.7.12); the other is a path $X_.(\omega)$ of any $X \in \mathfrak{D}$. Its output is the path $(X_{.-}{*}Z)_.(\omega)$ of the indefinite integral – where the algorithm does not converge have the box produce the zero path.

(ii) Suppose we are not sure which probability $\mathbb{P}$ is pertinent and are faced with a whole collection $\mathfrak{P}$ of them. If the size of the integrator Z is bounded independently of $\mathbb{P} \in \mathfrak{P}$ in the sense that

$$f(\lambda) \overset{\text{def}}{=} \sup_{\mathbb{P} \in \mathfrak{P}} \lceil \lambda{\cdot}Z \rceil_{\mathcal{I}^p[\mathbb{P}]} \xrightarrow[\lambda \to 0]{} 0\,, \tag{3.7.13}$$

then we choose δ_n so that $\sum_n f(n\delta_n) < \infty$. The proof of theorem 3.7.26 shows that the set where the algorithm (3.7.11) does not converge belongs to $\mathcal{A}_{\infty\sigma}$ and is negligible for all $\mathbb{P} \in \mathfrak{P}$ simultaneously, and that the limit is $X_{.-}{*}Z$, understood as an indefinite integral in the sense $L^0(\mathbb{P})$ for all $\mathbb{P} \in \mathfrak{P}$.

(iii) Assume (3.7.13). By representing (X, Z) on canonical path space $\mathscr{D}^2$ (item 2.3.11), we can produce a ***universal integral.*** This is a bilinear map $\mathscr{D} \times \mathscr{D} \to \mathscr{D}$, adapted to the canonical filtrations and written as a binary operation $._- \circledast$, such that $X_{.-}{*}Z$ is but the composition $X_{.-} \circledast Z$ of (X, Z) with this operation. We leave the details as an exercise.

---

[9] This is of course what one should do when computing the Riemann integral $\int_\eta^\theta f(x)\,dx$ for a continuous integrand f that for lack of smoothness does not lend itself to a simplex method or any other method whose error control involves derivatives: chopping the x-axis into lots of little pieces as one is ordinarily taught in calculus merely incurs round-off errors when f is constant or varies slowly over long stretches.

(iv) The theorem also shows that the problem about the "meaning in mean" of the stochastic differential equation (1.1.5) raised on page 6 is really no problem at all: the stochastic integral appearing in (1.1.5) can be read as a ***pathwise***[10] integral, provided we do not insist on understanding it as a Lebesgue–Stieltjes integral but rather as defined by the limit of the algorithm (3.7.11) – which surely meets everyone's intuitive needs for an integral[11] – and provided the integrand $b(X)$ belongs to $\mathfrak{L}$.

(v) Another way of putting this point is this. Suppose we are, as is the case in the context of stochastic differential equations, only interested in stochastic integrals of integrands in $\mathfrak{L}$; such are on $(\!(0,\infty)\!)$ the left-continuous versions $X_{\leftarrow}$ of càdlàg processes $X \in \mathfrak{D}$. Then the limit of the algorithm (3.7.11) serves as a perfectly intuitive *definition* of the integral. From this point of view one might say that the definition 2.1.7 on page 49 of an integrator serves merely to identify the conditions[12] under which this limit exists and defines an integral with decent limit properties. It would be interesting to have a proof of this that does not invoke the whole machinery developed so far.

Proof of Theorem 3.7.26. Since the filtration is right-continuous, one sees recursively that the S_k are stopping times (exercise 1.3.30). They increase strictly with k and their limit is ∞. For on the set $\left[\sup_k S_k < \infty\right]$ X must have an oscillatory discontinuity or be unbounded, which is ruled out by the assumption that $X \in \mathfrak{D}$: this set is void. The key to all further arguments is the observation that $Y^{(\delta)}$ is nothing but the indefinite integral of[6]

$$X_{\leftarrow}^{\mathcal{S}} = \sum_{k=0}^{\infty} X_{S_k} \cdot (\!(S_k, S_{k+1}]\!] ,$$

with $\mathcal{S}$ denoting the partition $\{0 = S_0 \le S_1 \le \cdots\}$. This is a predictable process (see proposition 3.5.2), and in view of exercise 3.7.20

$$Y^{(\delta)} = X_{\leftarrow}^{\mathcal{S}} {*} Z .$$

The very construction of the stopping times S_k is such that $X_{\leftarrow}$ and $X_{\leftarrow}^{\mathcal{S}}$ differ uniformly by less than δ. The indefinite integral $X_{\leftarrow}{*}Z$ may not exist in the sense Z–p if $p > 0$, but it does exist in the sense Z–0, since the maximal process of an $X_{\leftarrow} \in \mathfrak{L}$ is finite at any finite instant t (use theorem 3.7.17). There is an immediate estimate of the difference $X_{\leftarrow}{*}Z - Y^{(\delta)}$. Namely, let U be any finite stopping time. If $X_{\leftarrow} \cdot [\![0, U]\!]$ is Z–p-integrable for some $p \in [0, \infty)$, then the maximal difference of the indefinite integral $X_{\leftarrow}{*}Z$ from $Y^{(\delta)}$ can be estimated as follows:

$$\mathbb{P}\left[\left|X_{\leftarrow}{*}Z - Y^{(\delta)}\right|_U^\star > \lambda\right] = \mathbb{P}\left[\left|(X_{\leftarrow} - X_{\leftarrow}^{\mathcal{S}}){*}Z\right|_U^\star > \lambda\right]$$

---

[10] That is, computed separately for every single path $t \mapsto (X_t(\omega), Z_t(\omega))$, $\omega \in \Omega$.
[11] See, however, pages 168–171 and 310 for further discussion of this point.
[12] They are (RC-0) and (B-p), ibidem.

by lemma 2.3.2:
$$\leq \left\lceil \frac{1}{\lambda} \cdot (X_{.-} - X_{.-}^{S}) * Z^U \right\rceil_{\mathcal{I}^p}$$

using (3.7.5) twice:
$$\leq \left\lceil \frac{\delta}{\lambda} \cdot Z^U \right\rceil_{\mathcal{I}^p} . \tag{3.7.14}$$

At $p = 0$ inequality (3.7.14) has the consequence

$$\mathbb{P}\Big[\big|X_{.-}*Z - Y^{(\delta_n)}\big|_u^\star > 1/n\Big] \leq \lceil n\delta_n \cdot Z^n \rceil_{\mathcal{I}^0} , \qquad n \geq u .$$

Since the right-hand side is summable over n by virtue of the choice (3.7.12) of the δ_n, the Borel-Cantelli lemma yields at any instant u

$$\mathbb{P}\Big[\limsup_{n\to\infty} \big|X_{.-}*Z - Y^{(\delta_n)}\big|_u^\star > 0\Big] = 0 . \qquad \blacksquare$$

Remark 3.7.28 The proof shows that the particular definition of the S_k in equation (3.7.10) is not important. What is needed is that $X_{.-}$ differ from X_{S_k} by less than δ on $(\!(S_k, S_{k+1}]\!]$ and of course that $\lim_k S_k = \infty$; and (3.7.10) is one way to obtain such S_k. We might, for instance, be confronted with several L^0-integrators $Z^1, \ldots, Z^d$ and left-continuous integrands $X_{1.-}, \ldots, X_{d.-}$. In that case we set $S_0 = 0$ and continue recursively by

$$S_{k+1} = \inf\Big\{t > S_k : \sup_{1\leq\eta\leq d} \big|X_{\eta t} - X_{\eta S_k}\big| > \delta\Big\}$$

and choose the δ_n so that $\sup_\eta \sum_{n=1}^\infty \lceil n\delta_n \cdot (Z^\eta)^n \rceil_{\mathcal{I}^0} < \infty$. Equation (3.7.10) then defines a black box that computes the integrals $X_{.-}^\eta * Z^\eta$ pathwise simultaneously for all $\eta \in \{1, \ldots, d\}$, and thus computes $\boldsymbol{X}_{.-}*\boldsymbol{Z} = \sum_\eta X_{\eta.-}*Z^\eta$.

Exercise 3.7.29 Suppose Z is a global L^0-integrator and the δ_n are chosen so that $\sum_n \lceil n\delta_n \cdot Z \rceil_{\mathcal{I}^0}$ is finite. If $X_{.-} \in \mathfrak{L}$ is Z–0-integrable, then the approximate path $Y_.^{(\delta_n)}(\omega)$ of (3.7.11) converges to the path of the indefinite integral $(X_{.-}*Z)_.(\omega)$ uniformly on $[0, \infty)$, for almost all $\omega \in \Omega$.

Exercise 3.7.30 We know from theorem 2.3.6 that $\| Z_\infty^\star \|_p \leq C_p^{\star(2.3.5)} \lceil Z \rceil_{\mathcal{I}^p}$. This can be used to establish the following strong version of the weak-type inequality (3.7.14), which is useful only when Z is $\mathcal{I}^p$-bounded on $[0, U]$:

$$\Big\| \big|X_{.-}*Z - Y^{(\delta)}\big|_U^\star \Big\|_p \leq \delta C_p^\star \cdot \lceil Z^U \rceil_{\mathcal{I}^p} , \qquad 0 < p < \infty.$$

Exercise 3.7.31 The rate at which the algorithm (3.7.11) converges as $\delta \to 0$ does not depend on the integrand $X_{.-}$ and depends on the integrator Z only through the function $\lambda \mapsto \lceil \lambda Z \rceil_{\mathcal{I}^p}$. Suppose Z is an L^p-integrator for some $p > 0$, let U be a stopping time, and suppose $X_{.-}$ is a priori known to be Z–p-integrable on $[0, U]$.
(i) With δ as in (3.7.11) derive the confidence estimate

$$\mathbb{P}\Big[\sup_{0\leq s\leq U} |X_{.-}*Z - Y^{(\delta)}|_s > \lambda\Big] \leq \Big[\frac{\delta}{\lambda}\Big]^{p\wedge 1} \cdot \lceil Z^U \rceil_{\mathcal{I}^p} .$$

How must δ be chosen, if with probability 0.9 the error is to be less than 0.05 units?

(ii) If the integrand $X_{.-}$ varies furiously, then the loop (3.7.10)–(3.7.11) inside our black box is run through very often, even when δ is moderately large, and round-off errors accumulate. It may even occur that the stopping times S_k follow each other so fast that the physical implementation of the loop cannot keep up. It is desirable to have an estimate of the number $N(U)$ of calculations needed before a given ultimate time U of interest is reached. Now, rather frequently the integrand $X_{.-}$ comes as follows: there are an L^q-integrator X' and a Lipschitz function[13] Φ such that $X_{.-}$ is the left-continuous version of $\Phi(X')$. In that case there is a simple estimate for $N(U)$: with $c_q = 1$ for $q \geq 2$ and $c_q \leq 2.00075$ for $0 < q < 2$,

$$\mathbb{P}[N(U) > K] \leq c_q \left(\frac{L \lceil X'^U \rceil_{\mathcal{I}^q}}{\delta\sqrt{K}} \right)^q .$$

Exercise 3.7.32 Let Z be an L^0-integrator. (i) If Z has continuous paths and X is Z–0-integrable, then $X*Z$ has continuous paths. (ii) If X is the uniform limit of elementary integrands, then $\Delta(X*Z) = X \cdot \Delta Z$. (iii) If $X \in \mathfrak{L}$, then $\Delta(X*Z) = X \cdot \Delta Z$. (See proposition 3.8.21 for a more general statement.)

Integrators of Finite Variation

Suppose our L^0-integrator Z is a process V of finite variation. Surely our faith in the merit of the stochastic integral would increase if in this case it were the same as the ordinary Lebesgue–Stieltjes integral computed path-by-path. In other words, we hope that for all instants t

$$(X*V)_t(\omega) = LS\text{–}\int_0^t X_s(\omega)\, dV_s(\omega) , \tag{3.7.15}$$

at least almost surely. Since both sides of the equation are right-continuous and adapted, $X*Z$ would then in fact be indistinguishable from the indefinite Lebesgue–Stieltjes integral.

There is of course no hope that equation (3.7.15) will be true for all integrands X. The left-hand side is only defined if X is locally V–0-integrable and thus "somewhat non-anticipating." And it also may happen that the left-hand side is defined but the right-hand side is not. The obstacle is that for the Lebesgue–Stieltjes integral on the right to exist in the usual sense it is necessary that the upper integral[6] $\int^* |X_s(\omega)| \cdot [0,t]_s |dV_s(\omega)|$ be finite; and for the equality itself the random variable $\omega \mapsto LS\text{–}\int_0^t X_s(\omega)\, dV_s(\omega)$ must be measurable on $\mathcal{F}_t$. The best we can hope for is that the class of integrands X for which equation (3.7.15) holds be rather large. Indeed it is:

Proposition 3.7.33 *Both sides of equation (3.7.15) are defined and agree almost surely in the following cases:*

(i) X is previsible and the right-hand side exists a.s.

(ii) V is increasing and X is locally V–0-integrable.

---

[13] $|\Phi(x) - \Phi(x')| \leq L|x - x'|$ for $x, x' \in \mathbb{R}$. The smallest such L is the Lipschitz constant of Φ.

Proof. (i) Equation (3.7.15) is true by definition if X is an elementary integrand. The class of processes X such that $LS\text{-}\int_0^t X_s\,dV_s$ belongs to the class of the stochastic integral $\int_0^t X\,dV$ and is thus almost surely equal to $(X{*}V)_t$ is evidently a vector space closed under limits of bounded monotone sequences. So, thanks to the monotone class theorem A.3.4, equation (3.7.15) holds for all bounded predictable X, and then evidently for all bounded previsible X. To say that $LS\text{-}\int_0^t X_s(\omega)\,dV_s(\omega)$ exists almost surely implies that $LS\text{-}\int_0^t |X|_s(\omega)\,|dV_s|(\omega)$ is finite almost surely. Then evidently $|X|$ is finite for the mean $\lceil\!\lceil\ \rceil\!\rceil^*_{V\text{-}0}$, so by the Dominated Convergence Theorem $-n\vee X\wedge n$ converges in $\lceil\!\lceil\ \rceil\!\rceil^*_{V\text{-}0}$-mean to X, and the dV-integrals of this sequence converge to both the right-hand side and the left-hand side of equation (3.7.15), which thus agree.

(ii) We split X into its positive and negative parts and prove the claim for them separately. In other words, we may assume $X \geq 0$. We sandwich X between two predictable processes $\underset{\sim}{X} \leq X \leq \widetilde{X}$ with $\lceil\!\lceil \widetilde{X} - \underset{\sim}{X} \rceil\!\rceil^*_{V\text{-}0} = 0$, as in proposition 3.6.6. Part (i) implies that $\int_{[0}^{\infty} (\widetilde{X}_s(\omega) - \underset{\sim}{X}_s(\omega)) \wedge n\,dV_s(\omega) = 0$ $\forall n$ and then $\int_{[0}^{\infty} \widetilde{X}_s(\omega) - \underset{\sim}{X}_s(\omega)\,dV_s(\omega) = 0$ for almost all $\omega \in \Omega$. Neither $(X{*}V)_t$ nor $LS\text{-}\int_0^t X_s\,dV_s$ change but negligibly if X is replaced by $\widetilde{X}$: we may assume that $X \geq 0$ is predictable. Equation (3.7.15) holds then for $X \wedge n$, and by the Monotone Convergence Theorem for X. ∎

Exercise 3.7.34 The conclusion continues to hold if X is $(\mathcal{E}, \lceil\!\lceil\ \rceil\!\rceil^*)$-integrable for the mean

$$F \mapsto \lceil\!\lceil F \rceil\!\rceil^* \stackrel{\text{def}}{=} \left\lceil\!\!\left\lceil \int^* |F_s| \cdot [0,t]\,|dV_s| \right\rceil\!\!\right\rceil^*_{L^0(\mathbb{P})} .$$

Exercise 3.7.35 Let V be an adapted process of integrable variation $|V|$. Let μ denote the σ-additive measure $X \mapsto \mathbb{E}[\int X\,d|V|]$ on $\mathcal{E}$. Its usual Daniell upper integral (page 396)

$$F \mapsto \int^* F\,d\mu = \inf\left\{\sum \mu(X^{(n)}) : X^{(n)} \in \mathcal{E},\ \sum X^{(n)} \geq F\right\}$$

gives rise to the usual Daniell mean $F \mapsto \|F\|^*_\mu \stackrel{\text{def}}{=} \int^* |F|\,d\mu$, which majorizes $\lceil\!\lceil\ \rceil\!\rceil^*_{V\text{-}1}$ and so gives rise to fewer integrable processes.

If X is integrable for the mean $\|\ \|^*_\mu$, then X is V–1-integrable (but not necessarily vice versa); its path $t \mapsto X_t(\omega)$ is a.s. integrable for the scalar measure $dV(\omega)$ on the line; the pathwise integral LS–$\int X\,dV$ is integrable and is a member of the class $\int X\,dV$.

3.8 Functions of Integrators

Consider the classical formula
$$f(t) - f(s) = \int_s^t f'(\sigma)\,d\sigma . \tag{3.8.1}$$

The equation
$$\Phi(Z_T) - \Phi(Z_S) = \int_{S+}^T \Phi'(Z)\,dZ$$

suggests itself as an appealing analog when Z is a stochastic integrator and Φ a differentiable function. Alas, life is not that easy. Equation (3.8.1) remains true if $d\sigma$ is replaced by an arbitrary measure μ on the line provided that provisions for jumps are made; yet the assumption that the distribution function of μ have finite variation is crucial to the usual argument. This is not at our disposal in the stochastic case, as the example of theorem 1.2.8 shows. What can be said? We take our clue from the following consideration: if we want a representation of $\Phi(Z_t)$ in a "Generalized Fundamental Theorem of Stochastic Calculus" similar to equation (3.8.1), then $\Phi(Z)$ must be an integrator (cf. lemma 3.7.5). So we ask for which Φ this is the case. It turns out that $\Phi(Z)$ is rather easily seen to be an L^0-integrator if Φ is *convex*. We show this next.

For the applications later results in higher dimension are needed. Accordingly, let D be a convex open subset of $\mathbb{R}^d$ and let

$$\boldsymbol{Z} = (Z^1, \ldots, Z^d)$$

be a vector of L^0-integrators. We follow the custom of denoting partial derivatives by subscripts that follow a semicolon:

$$\Phi_{;\eta} \stackrel{\text{def}}{=} \frac{\partial \Phi}{\partial x^\eta}\,, \quad \Phi_{;\eta\theta} \stackrel{\text{def}}{=} \frac{\partial^2 \Phi}{\partial x^\eta \partial x^\theta}\,, \text{ etc.,}$$

and use the ***Einstein convention***: if an index appears twice in a formula, once as a subscript and once as a superscript, then summation over this index is implied. For instance, $\Phi_{;\eta}G^\eta$ stands for the sum $\sum_\eta \Phi_{;\eta}G^\eta$. Recall the convention that $X_{0-} = 0$ for $X \in \mathfrak{D}$.

Theorem 3.8.1 *Assume that $\Phi : D \to \mathbb{R}$ is continuously differentiable and convex, and that the paths both of the L^0-integrator $\boldsymbol{Z}$. and of its left-continuous version $\boldsymbol{Z}_{.-}$ stay in D at all times. Then $\Phi(\boldsymbol{Z})$ is an L^0-integrator. There exists an adapted right-continuous increasing process $A = A[\Phi; \boldsymbol{Z}]$ with $A_0 = 0$ such that nearly*

$$\Phi(\boldsymbol{Z}) = \Phi(\boldsymbol{Z}_0) + \Phi_{;\eta}(\boldsymbol{Z})_{.-}*Z^\eta + A[\Phi; \boldsymbol{Z}]\,, \tag{3.8.2}$$

i.e.,
$$\Phi(\boldsymbol{Z}_t) = \Phi(\boldsymbol{Z}_0) + \sum_{1 \le \eta \le d} \int_{0+}^t \Phi_{;\eta}(\boldsymbol{Z}_{.-})\, dZ^\eta + A_t \qquad \forall\, t \ge 0\,.$$

Like every increasing process, A is the sum of a continuous increasing process $C = C[\Phi; \boldsymbol{Z}]$ that vanishes at $t = 0$ and an increasing pure jump process $J = J[\Phi; \boldsymbol{Z}]$, both adapted (see theorem 2.4.4). J is given at $t \ge 0$ by

$$J_t = \sum_{0<s\le t} \Big(\Phi(\boldsymbol{Z}_s) - \Phi(\boldsymbol{Z}_{s-}) - \Phi_{;\eta}(\boldsymbol{Z}_{s-}) \cdot \Delta Z^\eta_s \Big)\,, \tag{3.8.3}$$

the sum on the right being a sum of positive terms.

Proof. In view of theorem 3.7.17, the processes $\Phi_{;\eta}(Z)_{.-}$ are Z–0-integrable. Since Φ is convex,

$$\Phi(z_2) - \Phi(z_1) - \Phi_{;\eta}(z_1)(z_2^\eta - z_1^\eta)$$

is non-negative for any two points $z_1, z_2 \in D$. Consider now a stochastic partition[8] $\mathcal{S} = \{0 = S_0 \le S_1 \le S_2 \le \ldots \le \infty\}$ and let $0 \le t < \infty$. Set

$$A_t^{\mathcal{S}} \stackrel{\text{def}}{=} \Phi(Z_t) - \Phi(Z_0) - \sum_{0\le k\le\infty} \Phi_{;\eta}(Z_{S_k\wedge t}) \cdot (Z^\eta_{S_{k+1}\wedge t} - Z^\eta_{S_k\wedge t}) \tag{3.8.4}$$

$$= \sum_{0\le k\le\infty} \Big(\Phi(Z_{S_{k+1}\wedge t}) - \Phi(Z_{S_k\wedge t}) - \Phi_{;\eta}(Z_{S_k\wedge t}) \cdot (Z^\eta_{S_{k+1}\wedge t} - Z^\eta_{S_k\wedge t})\Big) .$$

On the interval of integration, which does not contain the point $t = 0$, $\Phi_{;\eta}(Z) = \Phi_{;\eta}(Z_{.-})_{.+}$. Thus the sum on the top line of this equation converges to the stochastic integral $\int_{0+}^t \Phi_{;\eta}(Z)_{.-}\, dZ^\eta$ as the partition $\mathcal{S}$ is taken through a sequence whose mesh goes to zero (theorem 3.7.23), and so $A_t^{\mathcal{S}}$ converges to

$$A_t \stackrel{\text{def}}{=} \Phi(Z_t) - \Phi(Z_0) - \int_{0+}^t \Phi_{;\eta}(Z_{.-})\, dZ^\eta .$$

In fact, the convergence is uniform in t on bounded intervals, in measure. The second line of (3.8.4) shows that A_t increases with t. We can and shall choose a modification of A that has right-continuous and increasing paths. Exercise 3.7.32 on page 144 identifies the jump part of A as $\Delta A_s = \Delta\Phi(Z)_s - \Phi_{;\eta}(Z)_{s-} \cdot \Delta Z_s^\eta$, $s > 0$. The terms on the right are positive and have a finite sum over $s \le t$ since $A_t < \infty$; this observation identifies the jump part of A as stated. ∎

Remarks 3.8.2 (i) If Φ is, instead, the difference of two convex functions of class C^1, then the theorem remains valid, except that the processes A, C, J are now of finite variation, with the expression for J converging absolutely.

(ii) It would be incorrect to write

$$J_t = \sum_{0<s\le t} \Big(\Phi(Z_s) - \Phi(Z_{s-})\Big) - \sum_{0<s\le t} \Phi_{;\eta}(Z_{s-}) \cdot \Delta Z_s^\eta ,$$

on the grounds that neither sum on the right converges in general by itself.

Exercise 3.8.3 Suppose Z is continuous. Set $\nabla_\eta|z| \stackrel{\text{def}}{=} \partial|z|/\partial z^\eta = z^\eta/|z|$ for $z \ne 0$, with $\nabla|z| \stackrel{\text{def}}{=} 0$ at $z = 0$. There exists an increasing process L so that

$$|Z|_t = |Z_0| + \int_0^t \nabla_\eta|Z_s|\, dZ^\eta + L_t .$$

If $d = 1$, then L is known as the ***local time of Z at zero***.

Square Bracket and Square Function of an Integrator

A most important process arises by taking $d = 1$ and $\Phi(z) = z^2$. In this case the process $\Phi(Z_0) + A[\Phi; Z]$ of equation (3.8.2) is denoted by $[Z, Z]$ and is called the ***square bracket*** or ***square variation*** of Z. It is thus defined by

$$Z^2 = 2Z_{.-}{*}Z + [Z, Z] \quad \text{or} \quad Z_t^2 = 2\int_{0+}^{t} Z_{.-}\, dZ + [Z, Z]_t\,, \qquad t \ge 0\,.$$

Note that the jump Z_0^2 is subsumed in $[Z, Z]$. For the simple function $\Phi(z) = z^2$, equation (3.8.4) reduces to $A_t^{\mathcal{S}} = \sum_{0 \le k \le \infty} \left(Z_{S_{k+1}}^t - Z_{S_k}^t\right)^2$, so that $[Z, Z]_t$ is the limit in measure of

$$Z_0^2 + \sum\nolimits_{0 \le k \le \infty} \left(Z_{S_{k+1}}^t - Z_{S_k}^t\right)^2\,,$$

taken as the random partition $\mathcal{S} = \{0 = S_0 \le S_1 \le S_2 \le \ldots \le \infty\}$ of $[\![0, \infty)\!)$ runs through a sequence whose mesh tends to zero. By equation (3.8.3), the jump part of $[Z, Z]$ is simply

$${}^j[Z, Z]_t = Z_0^2 + \sum\nolimits_{0 < s \le t} (\Delta Z_s)^2 = \sum\nolimits_{0 \le s \le t} (\Delta Z_s)^2\,.$$

Its continuous part is ${}^c[Z, Z]$, the ***continuous bracket***. Note that we make the convention $[Z, Z]_0 = {}^j[Z, Z]_0 = Z_0^2$ and ${}^c[Z, Z]_0 = 0$. Homogeneity mandates considering the square roots of these quantities. We set

$$S[Z] = \sqrt{[Z, Z]} \qquad \text{and} \qquad \sigma[Z] = \sqrt{{}^c[Z, Z]}$$

and call these processes the ***square function*** and the ***continuous square function*** of Z, respectively. Evidently

$$\sigma[Z] \le S[Z] \quad \text{and} \quad \sqrt{{}^j[Z, Z]} \le S[Z]\,.$$

The proof of theorem 3.8.1 exhibits $S_T[Z]$ as the limit in measure of the square roots

$$\sqrt{Z_0^2 + A_T^{\mathcal{S}}} = \left(Z_0^2 + \sum\nolimits_{0 \le k \le \infty} (Z_{S_{k+1}}^T - Z_{S_k}^T)^2\right)^{1/2}\,, \tag{3.8.5}$$

taken as the random partition $\mathcal{S} = \{0 = S_0 \le S_1 \le S_2 \le \ldots \le \infty\}$ runs through a sequence whose mesh tends to zero. For an estimate of the speed of the convergence see exercise 3.8.14.

Theorem 3.8.4 (The Size of the Square Function) *At all stopping times T*

for all exponents $p > 0$ $\qquad \|S_T[Z]\|_{L^p} \le K_p \cdot \lceil\!\lceil Z^T \rceil\!\rceil_{\mathcal{I}^p}\,.$ (3.8.6)

Also, for $p = 0$, $\qquad \|S_T[Z]\|_{[\alpha]} \le K_0 \cdot \lceil\!\lceil Z^T \rceil\!\rceil_{[\alpha\kappa_0]}\,.$ (3.8.7)

The universal constants K_p, κ_0 are bounded by the Khintchine constants of theorem A.8.26 on page 455: $K_p^{(3.8.6)} \le K_p^{(A.8.9)}$ when p is strictly positive; and for $p=0$, $K_0^{(3.8.7)} \le K_0^{(A.8.9)}$ and $\kappa_0^{(3.8.7)} \ge \kappa_0^{(A.8.9)}$.

Proof. In equation (3.8.5) set $X^{(0)} \overset{\text{def}}{=} [\![0]\!] = [\![S_0]\!]$ and $X^{(k+1)} \overset{\text{def}}{=} (\!(S_k, S_{k+1}]\!]$ for $k = 0, 1, \dots \infty$. Then

$$Z_0^2 + A_T^{\mathcal{S}} = \sum_{0 \le k \le \infty} \Big(\int X^{(k)}\, dZ^T \Big)^2 ;$$

and since $\sum_k |X^{(k)}| \le 1$, corollary 3.1.7 on page 94 results in

$$\Big\| \sqrt{Z_0^2 + A_T^{\mathcal{S}}} \Big\|_{L^p} \le K_p^{(A.8.5)} \cdot \lceil Z^T \rceil_{\mathcal{I}^p} , \qquad p > 0,$$

and

$$\Big\| \sqrt{Z_0^2 + A_T^{\mathcal{S}}} \Big\|_{[\alpha]} \le K_0^{(A.8.6)} \cdot \lceil Z^T \rceil_{[\alpha\kappa_0]} , \qquad p = 0.$$

As the partition is taken through a sequence whose mesh tends to zero, Fatou's lemma produces the inequalities of the statement and exercise A.8.29 the estimates of the constants. True, corollary 3.1.7 was proved for *elementary* integrands $X^{(k)}$ only – for lack of others at the time – but the reader will have no problem seeing that it holds for integrable $X^{(k)}$ as well. ▮

Remark 3.8.5 Consider a standard Wiener process W. The variation

$$\lim_{\mathcal{S}} \sum_k \big| W_{S_{k+1}} - W_{S_k} \big| ,$$

taken over partitions $\mathcal{S} = \{S_k\}$ of any interval, however small, is infinite (theorem 1.2.8). The proof above shows that the limit exists and is finite, provided the absolute value of the differences is replaced with their squares. This explains the name square variation. Proposition 3.8.16 on page 153 makes the identification $[W,W]_t = \lim_{\mathcal{S}} \sum_k |W_{S_{k+1}} - W_{S_k}|^2 = t$.

Exercise 3.8.6 The square function of a previsible integrator is previsible.

Exercise 3.8.7 For $p > 0$ and Z–p-integrable X,

$$\lceil\!\lceil \sigma_\infty[X{*}Z] \rceil\!\rceil_{L^p} \le \lceil\!\lceil S_\infty[X{*}Z] \rceil\!\rceil_{L^p} \le K_p^{p\wedge 1} \cdot \lceil\!\lceil X \rceil\!\rceil^*_{Z\text{–}p}$$

and

$$\Big\lceil\!\Big\lceil \sqrt{{}^j[Z,Z]} \Big\rceil\!\Big\rceil \le K_p^{p\wedge 1} \cdot \lceil\!\lceil X \rceil\!\rceil^*_{Z\text{–}p} .$$

Exercise 3.8.8 Let $\boldsymbol{Z} = (Z^1, \dots, Z^d)$ be L^0-integrators and $T \in \mathfrak{T}$. Then

for $p \in (0,\infty)$
$$\Big\| \Big(\sum_{\eta=1}^d [Z^\eta, Z^\eta]_T \Big)^{1/2} \Big\|_{L^p} \le K_p^{(3.8.6)} \cdot \lceil \boldsymbol{Z}^T \rceil_{\mathcal{I}^p} ;$$

and for $p = 0$
$$\Big\| \Big(\sum_{\eta=1}^d [Z^\eta, Z^\eta]_T \Big)^{1/2} \Big\|_{[\alpha]} \le K_0^{(3.8.7)} \cdot \lceil \boldsymbol{Z}^T \rceil_{[\alpha\kappa_0^{(3.8.7)}]} .$$

The Square Bracket of Two Integrators

This process associated with two integrators Y, Z is obtained by taking in theorem 3.8.1 the function $\Phi(y,z) = y \cdot z$ of two variables, which is the difference of two convex smooth functions:

$$y \cdot z = \frac{1}{2}\Big((y+z)^2 - (y^2+z^2)\Big) ,$$

and thus remark 3.8.2 (i) applies. The process $Y_0 Z_0 + A[\Phi; (Y,Z)]$ of finite variation that arises in this case is denoted by $[Y,Z]$ and is called the ***square bracket*** of Y and Z. It is thus defined by

$$YZ = Y_{.-}{*}Z + Z_{.-}{*}Y + [Y,Z] \tag{3.8.8}$$

or, equivalently, by

$$Y_t{\cdot}Z_t = \int_{0+}^{t} Y_{.-}\, dZ + \int_{0+}^{t} Z_{.-}\, dY + [Y,Z]_t , \qquad t \ge 0 .$$

For an algorithm computing $[Y,Z]$ see exercise 3.8.14. By equation (3.8.3) the jump part of $[Y,Z]$ is simply

$$^{j}[Y,Z]_t = Y_0 Z_0 + \sum_{0<s\le t} (\Delta Y_s \cdot \Delta Z_s) = \sum_{0\le s\le t} (\Delta Y_s \cdot \Delta Z_s) .$$

Its continuous part is denoted by $^{c}[Y,Z]$ and vanishes at $t = 0$. Both $[Y,Z]$ and $^{c}[Y,Z]$ are evidently linear in either argument, and so is their difference $^{j}[Y,Z]$. All three brackets have the structure of positive semidefinite inner products, so the usual Cauchy–Schwarz inequality holds. In fact, there is a slight generalization:

Theorem 3.8.9 (Kunita–Watanabe) *For any two L^0-integrators Y, Z there exists a nearly empty set N such that for all $\omega \in \Omega_0 \stackrel{\text{def}}{=} \Omega \setminus N$ and any two processes U, V with Borel measurable paths*

$$\int_0^\infty |UV|\, d\big|[Y,Z]\big| \le \Big(\int_0^\infty U^2\, d[Y,Y]\Big)^{1/2} \cdot \Big(\int_0^\infty V^2\, d[Z,Z]\Big)^{1/2} ,$$

$$\int_0^\infty |UV|\, d\big|{}^{c}[Y,Z]\big| \le \Big(\int_0^\infty U^2\, d^{c}[Y,Y]\Big)^{1/2} \cdot \Big(\int_0^\infty V^2\, d^{c}[Z,Z]\Big)^{1/2} ,$$

$$\int_0^\infty |UV|\, d\big|{}^{j}[Y,Z]\big| \le \Big(\int_0^\infty U^2\, d^{j}[Y,Y]\Big)^{1/2} \cdot \Big(\int_0^\infty V^2\, d^{j}[Z,Z]\Big)^{1/2} .$$

Proof. Consider the polynomial

$$\begin{aligned} p(\lambda) &\stackrel{\text{def}}{=} A + 2B\lambda + C\lambda^2 \\ &\stackrel{\text{def}}{=} ([Y,Y]_t - [Y,Y]_s) + 2([Y,Z]_t - [Y,Z]_s)\lambda + ([Z,Z]_t - [Z,Z]_s)\lambda^2 \\ &= [Y+\lambda Z, Y+\lambda Z]_t - [Y+\lambda Z, Y+\lambda Z]_s . \end{aligned} \tag{$*$}$$

There is a set $\Omega_0 \in \mathcal{F}_\infty$ of full measure $\mathbb{P}[\Omega_0] = 1$ on which for all rational λ and all rational pairs $s \le t$ the equality $(*)$ obtains and thus $p(\lambda)$ is positive. The description shows that its complement N belongs to $\mathcal{A}_{\infty\sigma}$. By right-continuity, etc., $(*)$ is true and $p(\lambda) \ge 0$ for all real λ, all pairs $s \le t$, and all $\omega \in \Omega_0$.

Henceforth an $\omega \in \Omega_0$ is fixed. The positivity of $p(\lambda)$ gives $B^2 \le AC$, which implies that

$$|[Y,Z]_{t_i} - [Y,Z]_{t_{i-1}}| \le ([Y,Y]_{t_i} - [Y,Y]_{t_{i-1}})^{1/2} \cdot ([Z,Z]_{t_i} - [Z,Z]_{t_{i-1}})^{1/2}$$

for any partition $\{\ldots t_{i-1} \le t_i \ldots\}$. For any $r_i, s_i \in \mathbb{R}$ we get therefore

$$\sum_i r_i s_i \cdot ([Y,Z]_{t_i} - [Y,Z]_{t_{i-1}}) \le \sum_i |r_i s_i| |([Y,Z]_{t_i} - [Y,Z]_{t_{i-1}})|$$

$$\le \sum_i |r_i| ([Y,Y]_{t_i} - [Y,Y]_{t_{i-1}})^{1/2} \cdot |s_i| ([Z,Z]_{t_i} - [Z,Z]_{t_{i-1}})^{1/2} .$$

Schwarz's inequality applied to the right-hand side yields

$$\int_0^\infty uv \, d[Y,Z] \le \Big(\int_0^\infty u^2 \, d[Y,Y]\Big)^{1/2} \cdot \Big(\int_0^\infty v^2 \, d[Z,Z]\Big)^{1/2} \qquad (**)$$

at ω for the special functions $u = \sum r_i \cdot (t_{i-1}, t_i]$ and $v = \sum s_i \cdot (t_{i-1}, t_i]$ on the half-line. The collection of processes for which the inequality $(**)$ holds is evidently closed under taking limits of convergent sequences, so it holds for all Borel functions (theorem A.3.4). Replacing u by $|u|$ and v by $|v|h$, where h is a Borel version of the Radon–Nikodym derivative $d[Y,Z]/d\,\vert[Y,Z]\vert$, produces

$$\int_0^\infty |uv| \, d\,\vert[Y,Z]\vert \le \Big(\int_0^\infty u^2 \, d[Y,Y]\Big)^{1/2} \cdot \Big(\int_0^\infty v^2 \, d[Z,Z]\Big)^{1/2} .$$

The proof of the other two inequalities is identical to this one. ∎

Exercise 3.8.10 (Kunita–Watanabe) (i) Except possibly on an evanescent set

$$S[Y+Z] \le S[Y] + S[Z] \quad \text{and} \quad |S[Y] - S[Z]| \le S[Y-Z] ;$$

$$\sigma[Y+Z] \le \sigma[Y] + \sigma[Z] \quad \text{and} \quad |\sigma[Y] - \sigma[Z]| \le \sigma[Y-Z] .$$

Consequently, $\big\| |\sigma[Y] - \sigma[Z]|_T^\star \big\|_{L^p} \le \big\| |S[Y] - S[Z]|_T^\star \big\|_{L^p} \le K_p^{(3.8.6)} \lceil (Y-Z)^T \rceil_{\mathcal{I}^p}$ for all $p > 0$ and all stopping times T.

(ii) For any stopping time T and $1/r = 1/p + 1/q > 0$,

$$\Big\| \vert[Y,Z]\vert_T \Big\|_{L^r} \le \|S_T[Y]\|_{L^p} \cdot \|S_T[Z]\|_{L^q} ,$$

$$\Big\| \vert{}^c[Y,Z]\vert_T \Big\|_{L^r} \le \|\sigma_T[Y]\|_{L^p} \cdot \|\sigma_T[Z]\|_{L^q} ,$$

$$\Big\| \vert{}^j[Y,Z]\vert_T \Big\|_{L^r} \le \Big\| \sqrt{{}^j[Y,Y]_T} \Big\|_{L^p} \cdot \Big\| \sqrt{{}^j[Z,Z]_T} \Big\|_{L^q} .$$

Exercise 3.8.11 Let M, N be local martingales. There are arbitrarily large stopping times T such that

$$\mathbb{E}[M_T \cdot N_T] = \mathbb{E}[[M,N]_T] \text{ and } \mathbb{E}[M_T^{\star 2}] \le 2\mathbb{E}[[M,M]_T] .$$

Exercise 3.8.12 Let V be an L^0-integrator whose paths have finite variation, and let $V = {}^cV + {}^jV$ be its decomposition into a continuous and a pure jump process (theorem 2.4.4). Then $\sigma[V] = [{}^cV, {}^cV] = 0$ and, since $\Delta V_0 = V_0$,

$$[V,V]_t = [{}^jV, {}^jV]_t = \sum_{0 \le s \le t} (\Delta V_s)^2 .$$

Also, ${}^c[Z,V] = 0, \quad [Z,V]_t = {}^j[Z,V]_t = \sum_{0 \le s < t} \Delta Z_s \Delta V_s$

and
$$Z_T \cdot V_T - Z_S \cdot V_S = \int_{S+}^{T} Z \, dV + \int_{S+}^{T} V_{.-} \, dZ$$

for any other L^0-integrator Z and any two stopping times $S \le T$.

Exercise 3.8.13 A continuous local martingale of finite variation is nearly constant.

Exercise 3.8.14 Let Y, Z be L^p-integrators, $p > 0$, and $\mathcal{S} = \{0 = S_0 \le S_1 \le \cdots\}$ a stochastic partition[8] with $S_\infty = \infty$. Then for any stopping time T

$$\Big\| \sup_{s \le T} \Big| [Y,Z]_s - \Big(Y_0 Z_0 + \sum_{0<k<\infty} (Y_s^{S_{k+1}} - Y_s^{S_k})(Z_s^{S_{k+1}} - Z_s^{S_k}) \Big) \Big| \Big\|_{L^p}$$

$$\le C_p^{\star(2.3.5)} \Big(\big\| (Y - Y^{\mathcal{S}})_{.-} \cdot ((0,T]] \big\|^*_{Z-p} + \big\| (Z - Z^{\mathcal{S}})_{.-} \cdot ((0,T]] \big\|^*_{Y-p} \Big) .$$

In particular, if $\mathcal{S}$ is chosen so that on its intervals both Y and Z vary by less than δ, then

$$\Big\| \sup_{s \le T} \Big| [Y,Z]_s - \Big(Y_0 Z_0 + \sum_{0<k<\infty} (Y_s^{S_{k+1}} - Y_s^{S_k})(Z_s^{S_{k+1}} - Z_s^{S_k}) \Big) \Big| \Big\|_{L^p}$$

$$\le C_p^\star \Big(\lceil \delta Z^T \rceil_{\mathcal{I}^p} + \lceil \delta Y^T \rceil_{\mathcal{I}^p} \Big) .$$

If δ runs through a sequence (δ_n) with $\sum_n \lceil \delta_n Z^n \rceil_{\mathcal{I}^p} + \lceil \delta_n Y^n \rceil_{\mathcal{I}^p} < \infty$, then the sums of equation (3.8.5) nearly converge to the square bracket uniformly on bounded time-intervals.

Exercise 3.8.15 A complex-valued process $Z = X + iY$ is an L^p-integrator if its real and imaginary parts are, and the size of Z shall be the size of the vector (X, Y). We define the square function $S[Z]$ as $[Z, \overline{Z}]^{1/2} = ([X,X] + [Y,Y])^{1/2}$. Using exercise 3.8.8 reprove the square function estimates of theorem 3.8.4:

$$\|S_T[Z]\|_p \le K_p \cdot \lceil Z^T \rceil_{\mathcal{I}^p} \tag{3.8.9}$$

and
$$\|S_T[Z]\|_{[\alpha]} \le K_0 \cdot \lceil Z^T \rceil_{[\alpha\kappa_0]}$$

and show that for two complex integrators Z, Z'

$$Z_t \cdot Z'_t = \int_{0+}^{t} Z_{.-} \, dZ' + \int_{0+}^{t} Z'_{.-} \, dZ + [Z, Z']_t , \tag{3.8.10}$$

with the stochastic integral and bracket being defined so as to be complex-linear in either of their two arguments.

Proposition 3.8.16 *(i) A standard Wiener process W has bracket $[W,W]_t = t$. (ii) A standard d-dimensional Wiener process $\boldsymbol{W} = (W^1, \ldots, W^d)$ has bracket*

$$[W^\eta, W^\theta]_t = \delta^{\eta\theta} \cdot t\,.$$

Proof. (i) By exercise 2.5.4, $W_t^2 - t$ is a continuous martingale on the natural filtration of W. So is $W_t^2 - [W,W]_t \;= 2\int_0^t W\,dW$, because for $T_n = n \wedge \inf\{t : |W_t| \ge n\} \xrightarrow[n\to\infty]{} \infty$ the stopped process $(W{*}W)^{T_n}$ is the indefinite integral in the sense L^2 of the bounded integrand[6] $W \cdot [\![0,T_n]\!])$ against the L^2-integrator W. Then their difference $[W,W]_t - t$ is a local martingale as well. Since this continuous process has finite variation, it equals the value 0 that it has at time 0 (exercise 3.8.13).

(ii) If $\eta \ne \theta$, then W^η and W^θ are independent martingales on the natural filtrations $\mathcal{F}_.[W^\eta]$ and $\mathcal{F}_.[W^\theta]$, respectively. It is trivial to check that then $W^\eta \cdot W^\theta$ is a martingale on the filtration $t \mapsto \mathcal{F}_t[W^\eta] \vee \mathcal{F}_t[W^\theta]$. Since $[W^\eta, W^\theta] = W^\eta \cdot W^\theta - W^\eta{*}W^\theta - W^\theta{*}W^\eta$ is a continuous local martingale of finite variation, it must vanish. ∎

Definition 3.8.17 *We shall say that two paths $X_.(\omega)$ and $X_.(\omega')$ of the continuous integrator X describe the same arc in $\mathbb{R}^n$ if there is an increasing invertible continuous function $t \mapsto t'$ from $[0,\infty)$ onto itself so that $X_t(\omega) = X_{t'}(\omega')$ $\forall t$. We shall also say that $X_.(\omega)$ and $X_.(\omega')$* ***describe the same arc via*** *$t \mapsto t'$.*

Exercise 3.8.18 (i) Suppose F is a $d \times n$-matrix of uniformly continuous functions on $\mathbb{R}^n$. There exist a version of $F(X){*}X$ and a nearly empty set after the removal of which the following holds: whenever $X_.(\omega)$ and $X_.(\omega')$ describe the same arc in $\mathbb{R}^n$ via $t \mapsto t'$ then $(F(X){*}X)_.(\omega)$ and $(F(X){*}X)_.(\omega')$ describe the same arc in $\mathbb{R}^d$, also via $t \mapsto t'$.

(ii) Let $\boldsymbol{W}$ be a standard d-dimensional Wiener process. There is a nearly empty set after removal of which any two paths $\boldsymbol{W}_.(\omega) = \boldsymbol{W}_.(\omega')$ describing the same arc actually agree: $\boldsymbol{W}_t(\omega) = \boldsymbol{W}_t(\omega')$ for all t.

The Square Bracket of an Indefinite Integral

Proposition 3.8.19 *Let Y, Z be L^0-integrators, T a stopping time, and X a locally Z–0-integrable process. Then*

(i) $$[Y,Z]^T = [Y^T, Z^T] = [Y, Z^T] \qquad \textit{almost surely, and}$$

(ii) $$[Y, X{*}Z] = X{*}[Y,Z] \qquad \textit{up to indistinguishability.}$$

Here $X{*}[Y,Z]$ is understood as an indefinite Lebesgue–Stieltjes integral.

Proof. (i) For the first equality write

$$[Y,Z]^T = (YZ)^T - (Y_{-}{*}Z)^T - (Z_{-}{*}Y)^T$$

by exercise 3.7.15: $$= Y^T Z^T - Y_{-}{*}Z^T - Z_{-}{*}Y^T$$

by exercise 3.7.16: $$= Y^T Z^T - Y_{-}^T{*}Z^T - Z_{-}^T{*}Y^T = [Y^T, Z^T]\,.$$

The second equality of (i) follows from the computation

$$[Y - Y^T, Z^T] = (Y - Y^T) \cdot Z^T - (Y_{.-} - Y^T_{.-}){*}Z^T - Z^T_{.-}{*}(Y - Y^T) = 0 .$$

(ii) Equality (i) applied to the stopping times $T \wedge t$ yields $[Y, [\![0,T]\!]{*}Z] = [\![0,T]\!]{*}[Y,Z]$. Taking differences gives $[Y, (\!(S,T]\!]{*}Z] = (\!(S,T]\!]{*}[Y,Z]$ for stopping times $S \le T$. Taking linear combinations shows that (ii) holds for elementary integrands X. Let $\mathfrak{L}^Z$ denote the class of locally Z–0-integrable predictable processes X that meet the following description: for any L^0-integrator Y there is a nearly empty set outside which the indefinite Lebesgue–Stieltjes integral $X{*}[Y,Z]$ exists and agrees with $[Y, X{*}Z]$. This is a vector space containing $\mathcal{E}$. Let $X^{(n)}$ be an increasing sequence in $\mathfrak{L}^Z_+$, and assume that its pointwise limit is locally Z–0-integrable. From the inequality of Kunita–Watanabe

$$\int_0^t X^{(n)}\, d\,\rvert[Y,Z]\lvert \le S_t[Y] \cdot \left(\int_0^t X^{(n)2}\, d[Z,Z]\right)^{1/2} .$$

Since $X^{(n)} \in \mathfrak{L}^Z$, the random variable on the far right equals $S_t\big[X^{(n)}{*}Z\big]$. Exercise 3.7.14 allows this estimate of its size: for all $\alpha > 0$

$$\left\| S_t\big[X^{(n)}{*}Z\big] \right\|_{[\alpha]} \le K_0 \cdot \left\| X^{(n)} \right\|^*_{Z^t\text{–}[\alpha\kappa_0]} \le K_0 \cdot \|X\|^*_{Z^t\text{–}[\alpha\kappa_0]} < \infty .$$

Thus $$\int_0^t X^{(n)}\, d\,\rvert[Y,Z]\lvert \le K_0 \cdot S_t[Y] \cdot \|X\|^*_{Z^t\text{–}[\alpha\kappa_0]} < \infty . \qquad (*)$$

Hence $\sup_n \int_0^t X^{(n)}\, d\,\rvert[Y,Z]\lvert < \infty$ almost surely for all t, and the indefinite Lebesgue–Stieltjes integral $X{*}[Y,Z]$ exists except possibly on a nearly empty set. Moreover, $X{*}[Y,Z] = \lim X^{(n)}{*}[Y,Z]$ up to indistinguishability.

Exercise 3.7.14 can be put to further use:

$$\big|[Y, X{*}Z] - [Y, X^{(n)}{*}Z]\big|^\star_t = \big|[Y, (X - X^{(n)}){*}Z]\big|^\star_t$$

by 3.8.9: $$\le S_t[Y] \cdot S_t\big[(X - X^{(n)}){*}Z\big]$$

and $$\left\| S_t\big[(X - X^{(n)}){*}Z\big] \right\|_{[\alpha]} \le K_0 \cdot \rvert (X - X^{(n)}){*}Z^t \lvert_{[\alpha\kappa_0]}$$

$$= K_0 \cdot \left\| X - X^{(n)} \right\|^*_{Z^t\text{–}[\alpha\kappa_0]} .$$

We replace $X^{(n)}$ by a subsequence such that $\|\, X - X^{(n)} \,\|^*_{Z^n\text{–}[2^{-n}]} < 2^{-n}$; the Borel–Cantelli lemma then allows us to conclude that $S_n\big[(X - X^{(n)}){*}Z\big] \to 0$ almost surely, so that

$$X{*}[Y,Z] = \lim X^{(n)}{*}[Y,Z] = \lim_{n\to\infty} [Y, X^{(n)}{*}Z] = [Y, X{*}Z]$$

uniformly on bounded time-intervals, except on a nearly empty set. Applying this to $X^{(n)} - X^{(1)}$ or $X^{(1)} - X^{(n)}$ shows that $\mathfrak{L}^Z$ is closed under pointwise convergent monotone sequences $X^{(n)}$ whose limits are locally Z–0-integrable. The Monotone Class Theorem A.3.5 then implies that $\mathfrak{L}^Z$ contains all bounded predictable processes, and the usual truncation argument shows that it contains in fact all predictable locally Z–0-integrable processes X.

If X is also Z–0-negligible, then thanks to $(*)$ $\int X \, d\left\vdots[Y,Z]\right\vdots$ is evanescent. If X is Z–0-negligible but not predictable, we apply this remark to a predictable envelope of $|X|$ and conclude again that $\int X \, d\left\vdots[Y,Z]\right\vdots$ is evanescent. The general case is done by sandwiching X between predictable lower and upper envelopes (page 125). ∎

Exercise 3.8.20 Let Y, Z be L^0-integrators. For any stopping time T and any locally Z–0-integrable process X

$$^c[Y,Z]^T = {}^c[Y^T, Z^T] = {}^c[Y, Z^T] \quad \text{and} \quad [Y,Z]^{jT} = {}^j[Y^T, Z^T] = {}^j[Y, Z^T] \; ;$$

$$\text{also} \quad {}^c[Y, X{*}Z] = X{*}{}^c[Y,Z] \qquad \text{and} \quad {}^j[Y, X{*}Z] = X{*}{}^j[Y,Z] \; .$$

Application: The Jump of an Indefinite Integral

Proposition 3.8.21 *Let Z be an L^0-integrator and X a Z–0-integrable process. There exists a nearly empty set $N \subset \Omega$ such that for all $\omega \notin N$*

$$\Delta(X{*}Z)_t(\omega) = X_t(\omega) \cdot \Delta Z_t(\omega) \, , \qquad 0 \le t < \infty \, .$$

In other words, $\Delta(X{*}Z)$ is indistinguishable from $X \cdot \Delta Z$.

Proof. If X is elementary, then the claim is obvious by inspection. In the general case we find a sequence $\big(X^{(n)}\big)$ of elementary integrands converging in Z–0-mean to X and so that their indefinite integrals nearly converge uniformly on bounded intervals to $X{*}Z$ (corollary 3.7.11). The path of $\Delta(X{*}Z)$ is thus nearly given by

$$\Delta(X{*}Z) = \lim_{n \to \infty} X^{(n)} \cdot \Delta Z \, .$$

It is left to be shown that the path on the right is nearly equal to the path of $X \cdot \Delta Z$:

$$\lim_{n \to \infty} X^{(n)} \cdot \Delta Z = X \cdot \Delta Z \, . \tag{?}$$

$$\begin{aligned} \text{Since} \quad \sup_{0 \le t < u} \left| X_t \cdot \Delta Z_t - X_t^{(n)} \cdot \Delta Z_t \right| &\le \Big(\sum_{t<u} |X - X^{(n)}|_t^2 \cdot (\Delta Z_t)^2 \Big)^{1/2} \\ &= \Big(\int_0^{u-} |X - X^{(n)}|^2 \, d^j[Z,Z] \Big)^{1/2} \\ &\le \Big(\int_0^{u-} |X - X^{(n)}|^2 \, d[Z,Z] \Big)^{1/2} \end{aligned}$$

by proposition 3.8.19: $= S_{u-}[(X - X^{(n)})*Z]$,

$$\Big\| \sup_{0\le t<u} \big| X_t \cdot \Delta Z_t - X_t^{(n)} \cdot \Delta Z_t \big| \Big\|_{[\alpha]} \le \Big\| S_{u-}[(X - X^{(n)})*Z] \Big\|_{[\alpha]}$$

by theorem 3.8.4: $\le K_0 \cdot \lceil\!\lceil (X - X^{(n)})*Z \rceil\!\rceil_{[\alpha\kappa_0]}$

by lemma 3.7.5: $\le K_0 \cdot \big\| X - X^{(n)} \big\|^*_{Z-[\alpha\kappa_0]}$.

If we insist that $\| X - X^{(n)} \|^*_{Z-[2^{-n}\kappa_0]} < 2^{-n}/K_0$, as we may by taking a subsequence, then the previous inequality turns into

$$\mathbb{P}\Big[\sup_{0\le t<u} \big| X_t \cdot \Delta Z_t - X_t^{(n)} \cdot \Delta Z_t \big| > 2^{-n} \Big] < 2^{-n} ,$$

and the Borel–Cantelli lemma implies that

$$\Big[\limsup_{n\to\infty} \sup_{0\le t<u} \big| X_t \cdot \Delta Z_t - X_t^{(n)} \cdot \Delta Z_t \big| > 0 \Big] \in \mathcal{F}_u$$

is negligible. Equation (?) thus holds nearly. ∎

Proposition 3.8.22 *Let Y, Z be L^0-integrators, with Y previsible. At any finite stopping time T,*

$$\int_0^T \Delta Y \, dZ = \sum_{0\le s\le T} \Delta Y_s \cdot \Delta Z_s , \tag{3.8.11}$$

the sum nearly converging absolutely. Consequently

$$Y_T \cdot Z_T = Y_0 \cdot Z_0 + \int_{0+}^T Y \, dZ + \int_{0+}^T Z_- \, dY + [Y, Z]_T$$
$$= \int_0^T Y \, dZ + \int_0^T Z_- \, dY + [Y, Z]_T .$$

Proof. The integral on the left exists since ΔY is previsible and has finite maximal function $\Delta Y_T^\star \le 2Y_T^\star$ (lemma 2.3.2 and theorem 3.7.17). Let $\epsilon > 0$ and define $S'_0 = 0$, $S'_{n+1} = \inf\{t > S'_n : |\Delta Y_t| \ge \epsilon\}$. Next fix an instant t and let S_n be the reduction of S'_n to $[S'_n \le t] \in \mathcal{F}_{S'_n-}$. Since the graph of S'_{n+1} is the intersection of the previsible sets $(\!(S'_n, S'_{n+1}]\!]$ and $[|\Delta Y| \ge \epsilon]$, the S'_n are predictable stopping times (theorem 3.5.13). Thanks to lemma 3.5.15 (iv), so are the S_n. Also, the S_n have disjoint graphs. Thanks to theorem 3.5.14,[6]

$$\int_0^t \bigcup_n [\![S_n]\!] \cdot \Delta Y \, dZ = \sum_n \int_0^t [\![S_n]\!] \cdot \Delta Y \, dZ = \sum_n \Delta Y_{S_n} \cdot \Delta Z_{S_n} .$$

We take the limit as $\epsilon \to 0$ and arrive at the claim, at least at the instant t. We apply this to the stopped process Z^T and let $t \to \infty$: equation (3.8.11) holds in general. The absolute convergence of the sum follows from the inequality of Kunita–Watanabe.

The second claim follows from the definition (3.8.8) of the continuous and pure jump square brackets by

$$Y_T \cdot Z_T = \int_0^T Y_{.-}\, dZ + \int_0^T Z_{.-}\, dY + {}^c[Y,Z]_T + \sum_{0 \le s \le T} \Delta Y_s \cdot \Delta Z_s \,. \qquad ❑$$

Corollary 3.8.23 *Let V be a right-continuous previsible process of integrable total variation. For any bounded martingale M and stopping times $S \le T$*

$$\mathbb{E}[M_T V_T - M_S V_S] = \mathbb{E}\Big[\int_{S+}^T M_{.-}\, dV\Big] \,. \tag{3.8.12}$$

Proof. Let $T' \le T$ be a bounded stopping time such that V is bounded on $[\![0, T']\!]$ (corollary 3.5.16). Since by exercise 3.8.12 ${}^c[M, V] = 0$, proposition 3.8.22 gives

$$M_{T'} V_{T'} = \int_{0+}^{T'} M_{.-}\, dV + \int_S^{T'} V\, dM \,.$$

The term on the far right has expectation $\mathbb{E}[V_S M_S]$. Now take $T' \uparrow T$. ❑

It is shown in proposition 4.3.2 on page 222 that equation (3.8.12) actually characterizes the predictable processes among the right-continuous processes of finite variation.

Exercise 3.8.24 (i) A previsible local martingale of finite variation is constant.

(ii) The bracket $[V, M]$ of a previsible process V of finite variation and a local martingale M is a local martingale of finite variation.

Exercise 3.8.25 (i) Let Z be an L^0-integrator, T a random time, and f a random variable. If $f \cdot [\![T]\!]$ is Z–0-integrable, then $\int f \cdot [\![T]\!]\, dZ = f \cdot \Delta Z_T$.

(ii) Let Y, Z be L^0-integrators and assume that Y is Z-measurable. Then for any almost surely finite stopping time T

$$Y_T \cdot Z_T = Y_0 \cdot Z_0 + \int_{0+}^T Y\, dZ + \int_{0+}^T Z_{.-}\, dY + {}^c[Y, Z]_T \,.$$

3.9 Itô's Formula

Itô's formula is the stochastic analog of the Fundamental Theorem of Calculus. It identifies the increasing process $A[\Phi; \boldsymbol{Z}]$ of theorem 3.8.1 in terms of the second derivatives of Φ and the square brackets of $\boldsymbol{Z}$. Namely,

Theorem 3.9.1 (Itô) *Let $D \subset \mathbb{R}^d$ be open and let $\Phi : D \to \mathbb{R}$ be a twice continuously differentiable function. Let $\boldsymbol{Z} = (Z^\eta)^{\eta=1\ldots d}$ be a d-vector of L^0-integrators such that the paths of both $\boldsymbol{Z}$ and its left-continuous ver-*

sion $Z_{.-}$ stay in D at all times. Then $\Phi(Z)$ is an L^0-integrator, and for any nearly finite stopping time T

$$\Phi(Z_T) = \Phi(Z_0) + \int_{0+}^{T} \Phi_{;\eta}(Z_{.-})\, dZ^\eta + \int_0^1 (1-\lambda) \int_{0+}^{T} \Phi_{;\eta\theta}\big(Z_{.-} + \lambda\Delta Z\big)\, d[Z^\eta, Z^\theta]\, d\lambda \tag{3.9.1}$$

$$= \Phi(Z_0) + \int_{0+}^{T} \Phi_{;\eta}(Z_{.-})\, dZ^\eta + \frac{1}{2}\int_{0+}^{T} \Phi_{;\eta\theta}(Z_{.-})\, d^c[Z^\eta, Z^\theta] + \sum_{0<s\le T}\Big(\Phi(Z_s) - \Phi(Z_{s-}) - \Phi_{;\eta}(Z_{s-})\cdot\Delta Z_s^\eta\Big)\,, \tag{3.9.2}$$

the last sum nearly converging absolutely.[14] It is often convenient to write equation (3.9.2) in its ***differential form***: in obvious notation

$$d\Phi(Z) = \Phi_{;\eta}(Z_{.-})dZ^\eta + \frac{1}{2}\Phi_{;\eta\theta}(Z_{.-})d^c[Z^\eta, Z^\theta] + \big(\Delta\Phi(Z) - \Phi_{;\eta}(Z_{.-})\Delta Z^\eta\big)\,.$$

Proof. That $\Phi \circ Z$ is an L^0-integrator is evident from (3.9.2) in conjunction with lemma 3.7.5. The $d[Z^\eta, Z^\theta]$-integral in (3.9.1) has to be read as a pathwise Lebesgue–Stieltjes integral, of course, since its integrand is not in general previsible. Note also that the two expressions (3.9.1) and (3.9.2) for $\Phi(Z_T)$ agree. Namely, since the continuous part $d^c[Z^\eta, Z^\theta]$ of the square bracket does not charge the instants t – at most countable in number – where $\Delta Z_t \ne 0$,

$$\int_0^1 (1-\lambda) \int_{0+}^{T} \Phi_{;\eta\theta}\big(Z_{.-} + \lambda\Delta Z\big)\, d^c[Z^\eta, Z^\theta]\, d\lambda$$
$$= \int_0^1 (1-\lambda) \int_{0+}^{T} \Phi_{;\eta\theta}\big(Z_{.-}\big)\, d^c[Z^\eta, Z^\theta]\, d\lambda = \frac{1}{2}\int_{0+}^{T} \Phi_{;\eta\theta}\big(Z_{.-}\big) d^c[Z^\eta, Z^\theta]\,.$$

This leaves
$$\int_0^1 (1-\lambda) \int_{0+}^{T} \Phi_{;\eta\theta}\big(Z_{.-} + \lambda\Delta Z\big)\, d^j[Z^\eta, Z^\theta]\, d\lambda$$
$$= \sum_{0<s\le T} \int_0^1 (1-\lambda)\Phi_{;\eta\theta}\big(Z_{s-} + \lambda\Delta Z_s\big)\Delta Z_s^\eta \Delta Z_s^\theta\, d\lambda\,,$$

the jump part, which by Taylor's formula of order two (A.2.42) equals the sum in (3.9.2).

To start on the proof proper observe that any linear combination of two functions in $C^2(D)$ (see proposition A.2.11 on page 372) that satisfy equation (3.9.2) again satisfies this equation: such functions form a vector space $\mathcal{I}$. We leave it as an exercise in bookkeeping to show that $\mathcal{I}$ is also closed

[14] Subscripts after semicolons denote partial derivatives, e.g., $\Phi_{;\eta} \stackrel{\text{def}}{=} \frac{\partial \Phi}{\partial x^\eta}$, $\Phi_{;\eta\theta} \stackrel{\text{def}}{=} \frac{\partial^2 \Phi}{\partial x^\eta \partial x^\theta}$.

under multiplication, so that it is actually an algebra. Since every coordinate function $z \mapsto z^\eta$ is evidently a member of $\mathcal{I}$, every polynomial belongs to $\mathcal{I}$.

By proposition A.2.11 on page 372 there exists a sequence of polynomials P_k that converges to Φ uniformly on compact subsets of D, and such that every first and second partial of P_k also converges to the corresponding first or second partial of Φ, uniformly on every compact subset of D. Now observe this: the image of the path $\boldsymbol{Z}_{.}(\omega)$ on $[0,t]$ has compact closure $\Gamma(\omega)$ in D, for every $\omega \in \Omega$ and $t > 0$ – this is immediate from the fact that the path is càdlàg and the assumption that it and its left-continuous version stay in D at all times t. Since $(P_k)_{;\eta} \to \Phi_{;\eta}$ uniformly on $\Gamma(\omega)$, the maximal process G_η of the process $\sup_k |(P_k)_{;\eta}(\boldsymbol{Z})|$ is finite at (t,ω), and this holds for all $t < \infty$, all $\omega \in \Omega$, and $\eta = 1, \ldots, d$ (see convention 2.3.5). By theorem 3.7.17 on page 137, the previsible processes $G_{\eta.-}$ are therefore Z^η-integrable in the sense L^0 and can serve as "the dominators" in the DCT: according to the latter $\Phi_{;\eta}(\boldsymbol{Z}_{.-}) = \lim (P_k)_{;\eta}(\boldsymbol{Z}_{.-})$ in $[\![\ ]\!]^*_{Z^\eta\text{-}0}$-mean

and $$\int_0^t (P_k)_{;\eta}(\boldsymbol{Z}_{.-})\, dZ^\eta \xrightarrow[k\to\infty]{} \int_0^t \Phi_{;\eta}(\boldsymbol{Z}_{.-})\, dZ^\eta \ \text{ in measure.} \tag{$*$}$$

Clearly $$(P_k)_{;\eta\theta}\big(\boldsymbol{Z}_{.-} + \lambda\Delta\boldsymbol{Z}\big) \xrightarrow[k\to\infty]{} \Phi_{;\eta\theta}\big(\boldsymbol{Z}_{.-} + \lambda\Delta\boldsymbol{Z}\big) \tag{$**$}$$

uniformly up to time t. Now equation (3.9.1) is known to hold with P_k replacing Φ. The facts $(*)$ and $(**)$ allow us to take the limit as $k \to \infty$ and to conclude that (3.9.1) persists on Φ. Finally, càdlàg processes that nearly agree at any instant t nearly agree at any nearly finite stopping time T: (3.9.1) is established. ∎

The Doléans–Dade Exponential

Here is a computation showing how Itô's theorem can be used to solve a stochastic differential equation.

Proposition 3.9.2 *Let Z be an L^0-integrator. There exists a unique right-continuous process $\mathscr{E} = \mathscr{E}[Z]$ with $\mathscr{E}_0 = 1$ satisfying $d\mathscr{E} = \mathscr{E}_{.-}\, dZ$ on $[\![0,\infty)\!)$,*

that is to say $$\mathscr{E}_t = 1 + \int_0^t \mathscr{E}_{.-}\, dZ \quad \textit{for } t \ge 0$$

or, equivalently, $$\mathscr{E} = 1 + \mathscr{E}_{.-} * Z\ . \tag{3.9.3}$$

It is given by $$\mathscr{E}_t[Z] = e^{Z_t - Z_0 - [Z,Z]_t/2} \cdot \prod_{0<s\le t} (1 + \Delta Z_s) e^{-\Delta Z_s} \tag{3.9.4}$$

and is called the ***Doléans–Dade****, or* ***stochastic****,* ***exponential*** *of Z.*

Proof. There is no loss of generality in assuming that $Z_0 = 0$; neither $\mathscr{E}$ nor the right-hand side $'\mathscr{E}$ of equation (3.9.4) change if Z is replaced by $Z - Z_0$.

Set $$T_1 \stackrel{\text{def}}{=} \inf\{t : {'\mathscr{E}_t} \le 0\} = \inf\{t : \Delta Z_t \le -1\}\ ,$$

$${}^1Z \stackrel{\text{def}}{=} Z^{T_1-} = (Z - \Delta Z_{T_1})^{T_1}\ , \text{ "the process } Z \textbf{\textit{ stopped just before }} T_1\text{,"}$$

and $${}^1L_t \stackrel{\text{def}}{=} {}^1Z_t - [{}^1Z, {}^1Z]_t/2 + \textstyle\sum_{0<s\le t}\big(\ln(1+\Delta {}^1Z_s) - \Delta {}^1Z_s\big)\ .$$

Since $\big|\ln(1+u)-u\big| \le 2u^2$ for $|u|<1/2$ and $\sum_{0<s\le t}\big|\Delta Z_s\big|^2 < \infty$, the sum on the right converges absolutely, exhibiting ${}^1L$ as an L^0-integrator. Therefore so is ${}^1\mathscr{E} \stackrel{\text{def}}{=} e^{{}^1L}$. A straightforward application of Itô's formula shows that ${}^1\mathscr{E} = {}'\mathscr{E}^{T_1-}$ satisfies ${}^1\mathscr{E} = 1 + {}^1\mathscr{E}_{.-}*{}^1Z$. A simple calculation of jumps invoking proposition 3.8.21 then reveals that ${}'\mathscr{E}$ satisfies ${}'\mathscr{E}^{T_1} = 1 + {}'\mathscr{E}^{T_1}_{.-}*Z^{T_1}$.

Next set $T_2 \stackrel{\text{def}}{=} \inf\{t > T_1 : \Delta Z_t \le -1\}$,

$${}^2Z \stackrel{\text{def}}{=} Z^{T_2-} - Z^{T_1}$$

and
$${}^2L_t \stackrel{\text{def}}{=} {}^2Z_t - {}^c[{}^2Z, {}^2Z]_t/2 + \sum_{0<s\le t}\big(\ln(1+\Delta {}^2Z_s) - \Delta {}^2Z_s\big).$$

The same argument as above shows that ${}^2\mathscr{E} \stackrel{\text{def}}{=} e^{{}^2L} = 1 + {}^2\mathscr{E}_{.-}*{}^2Z$. Now clearly ${}'\mathscr{E} = {}^1\mathscr{E}_{T_1}\cdot{}^2\mathscr{E}$ on $[\![T_1, T_2)\!)$, from which we conclude that ${}'\mathscr{E}$ satisfies (3.9.3) on $[\![0, T_2)\!)$, and by proposition 3.8.21 even on $[\![0, T_2]\!]$. We continue by induction and see that the right-hand side ${}'\mathscr{E}$ of (3.9.4) solves (3.9.3).

Exercise 3.9.3 Finish the proof by establishing the uniqueness and use the latter to show that $\mathscr{E}[Z]\cdot\mathscr{E}[Z'] = \mathscr{E}[Z + Z' + [Z, Z']]$ for any two L^0-integrators Z, Z'.

Exercise 3.9.4 The solution of $dX = X_{.-}\,dZ$, $X_0 = x$, is $X = x\cdot\mathscr{E}[Z]$.

Corollary 3.9.5 (Lévy's Characterization of a Standard Wiener Process) *Assume that $\boldsymbol{M} = (M^1,\ldots,M^d)$ is a d-vector of local martingales on some measured filtration $(\mathcal{F}_., \mathbb{P})$ and that $\boldsymbol{M}$ has the same bracket as a standard d-dimensional Wiener process: $M_0^\eta = 0$ and $[M^\eta, M^\theta]_t = \delta^{\eta\theta}\cdot t$ for $\eta, \theta = 1\ldots d$. Then $\boldsymbol{M}$ is a standard d-dimensional Wiener process.*

In fact, for any finite stopping time T, $\boldsymbol{N}_. \stackrel{\text{def}}{=} \boldsymbol{M}_{T+.} - \boldsymbol{M}_T$ is a standard Wiener process on $\mathcal{G}_. \stackrel{\text{def}}{=} \mathcal{F}_{T+.}$ and is independent of $\mathcal{F}_T$.

Proof. We do the case $d = 1$. The generalization to $d > 1$ is trivial (see example A.3.31 and proposition 3.8.16). Note first that $(N_0)^2 = (\Delta N_0)^2 = [N, N]_0 = 0$, so that $N_0 = 0$. Since $[N, N]_t = [M, M]_{T+t} - [M, M]_T = t$ and therefore ${}^j[N, N] = 0$, N is continuous. Thanks to Doob's optional stopping theorem 2.5.22, it is locally a bounded martingale on $\mathcal{G}_.$. Now let Γ denote the vector space of all functions $\gamma : [0, \infty) \to \mathbb{R}$ of compact support that have a continuous derivative $\dot\gamma$. We view $\gamma \in \Gamma$ as the cumulative distribution function of the measure $d\gamma_t = \dot\gamma_t dt$. Since γ has finite variation, $[N, \gamma] = 0$; since $N_0\gamma_0 = N_u\gamma_u = 0$ when u lies to the right of the support of $\dot\gamma$, $\int_0^\infty N\,d\gamma = -\int_0^\infty \gamma\,dN$. Proposition 3.9.2 exhibits

$$e^{i\int_0^\infty N\,d\gamma + \int_0^\infty \gamma_s^2\,ds/2} = e^{-(i\gamma*N)_\infty - [i\gamma*N, i\gamma*N]_\infty/2}$$

as the value at ∞ of a bounded $\mathcal{G}$-martingale with expectation 1.

Thus
$$\mathbb{E}\Big[e^{i\int_0^\infty M\,d\gamma}\cdot A\Big] = e^{-\int_0^\infty \gamma_s^2\,ds/2}\cdot\mathbb{E}[A] \tag{$*$}$$

for any $A \in \mathcal{G}_0 = \mathcal{F}_T$. Now the measures $d\gamma$, $\gamma \in \Gamma$, form the dual of the space $\mathscr{C}$: equation (*) simply says that that $N.$ is independent of $\mathcal{F}_T$ and that the characteristic function of the law of N is

$$\gamma \mapsto e^{-\int_0^t \gamma_s^2\, ds/2} .$$

The same calculation shows that this function is also the characteristic function of Wiener measure $\mathbb{W}$ (proposition 3.8.16). The law of N is thus $\mathbb{W}$ (exercise A.3.35). ∎

Additional Exercises

Exercise 3.9.6 (Wiener Processes with Covariance) (i) Let $\boldsymbol{W}$ be a standard n-dimensional Wiener process as in 3.9.5 and U a constant $d{\times}n$-matrix. Then[15] $\boldsymbol{W}' \stackrel{\text{def}}{=} U\boldsymbol{W} = (U^\eta_\nu W^\nu)$ is a d-vector of Wiener processes with ***covariance matrix*** $B = (B^{\eta\theta})$ defined by

$$\mathbb{E}[W_t'^\eta \cdot W_t'^\theta] = \mathbb{E}[W'^\eta, W'^\theta]_t = tB^{\eta\theta} \stackrel{\text{def}}{=} t(UU^T)^{\eta\theta} = t\textstyle\sum_\nu U^\eta_\nu U^\theta_\nu .$$

(ii) Conversely, suppose that $\boldsymbol{W}'$ is a d-vector of continuous local martingales that vanishes at 0 and has square function $[\boldsymbol{W}', \boldsymbol{W}']_t = tB$, B constant and (necessarily) symmetric and positive semidefinite. There exist an $n \in \mathbb{N}$ and a $d{\times}n$-matrix U such that $B^{\eta\theta} = \sum_{\nu=1}^n U^\eta_\nu U^\theta_\nu$. Then there exists a standard n-dimensional Wiener process $\boldsymbol{W}$ so that $\boldsymbol{W}' = U\boldsymbol{W}$.

(iii) The integrator size of $\boldsymbol{W}'$ can be estimated by $\lceil \boldsymbol{W}'^t \rceil_{\mathcal{I}^p} \approx \sqrt{t}\,\|B\|$ for $p > 0$, where $\|B\|$ is the operator size of $B : \ell^\infty \to \ell^1$, $\|B\| \stackrel{\text{def}}{=} \sup\{\zeta_\eta\zeta_\theta B^{\eta\theta} : |\boldsymbol{\zeta}|_{\ell^\infty} \le 1\}$.

Exercise 3.9.7 A standard Wiener process $\boldsymbol{W}$ starts over at any finite stopping time T; in fact, the process $t \mapsto \boldsymbol{W}_{T+t} - \boldsymbol{W}_T$ is again a standard Wiener process and is independent of $\mathcal{F}_T[\boldsymbol{W}]$.

Exercise 3.9.8 Let M be a continuous martingale on the right-continuous filtration $\mathcal{F}.$ and assume that $M_0 = 0$ and $[M, M]_t \xrightarrow[t\to\infty]{} \infty$ almost surely. Set

$$T^\lambda = \inf\{t : [M,M]_t \ge \lambda\} \qquad \text{and} \qquad T^{\lambda+} = \inf\{t : [M,M]_t > \lambda\} .$$

Then $W_\lambda \stackrel{\text{def}}{=} M_{T^{\lambda+}}$ is a continuous martingale on the filtration $\mathcal{G}_\lambda \stackrel{\text{def}}{=} \mathcal{F}_{T^{\lambda+}}$ with $W_0 = 0$ and $[W, W]_\lambda = \lambda$ and consequently is a standard Wiener process on $\mathcal{G}.$. Furthermore, if X is $\mathcal{G}$-predictable, then $X_{[M,M]}$ is $\mathcal{F}$-predictable; if X is also W–p-integrable, $0 \le p < \infty$, then $X_{[M,M]}$ is M–p-integrable and

$$\int X_\lambda\, dW_\lambda = \int X_{[M,M]_t}\, dM_t .$$

Conversely, if X is predictable on $\mathcal{F}.$, then $X_{T^\cdot}$ is predictable on $\mathcal{G}.$; and if X is M–p-integrable, then $X_{T^\cdot}$ is W–p-integrable and

$$\int X_t\, dM_t = \int X_{T^\lambda}\, dW_\lambda .$$

Definition 3.9.9 *Let X be an n-dimensional continuous integrator on the measured filtration $(\Omega, \mathcal{F}., \mathbb{P})$, T a stopping time, and $H = \{x : \langle\xi|x\rangle = a\}$ a hyperplane*

---

[15] Einstein's convention, adopted, implies summation over the same indices in opposite positions.

in $\mathbb{R}^n$ *with equation* $\langle\xi|x\rangle = a$. *We call* H ***transparent*** *for the path* $X_.(\omega)$ *at the time* T *if the following holds: if* $X_T(\omega) \in H$, *then* $T^\pm \stackrel{\text{def}}{=} \inf\{t > T : \langle\xi|X\rangle \gtrless a\} = T$ *at* ω. *This expresses the idea that immediately after* T *the path* $X_.(\omega)$ *can be found strictly on both sides of* H, *"or oscillates between the two strict sides of* H." *In the opposite case, when the path* $X_.(\omega)$ *stays on one side of* H *for a strictly positive length of time, we call* H ***opaque*** *for the path.*

Exercise 3.9.10 Suppose that X is a continuous martingale under $\mathbb{P} \in \mathfrak{P}$ such that $\xi_\mu\xi_\nu[X^\mu, X^\nu]_t$ increases strictly to ∞ as $t \to \infty$, for every non-zero $\xi \in \mathbb{R}^n$.

(i) For any hyperplane H and finite stopping time T with $X_T \in H$, the paths for which H is opaque form a $\mathbb{P}$-nearly empty set.

(ii) Let S be a finite stopping time such that X_S depends continuously on the path (in the topology of uniform convergence on compacta), and H a hyperplane with equation $\langle\xi|x\rangle = a$ and not containing X_S. Then the stopping times

$$T \stackrel{\text{def}}{=} \inf\{t > S : X_t \in H\} \quad \text{and} \quad T^\pm \stackrel{\text{def}}{=} \inf\{t > T : \langle\xi|X_t\rangle \gtrless a\}$$

$\mathbb{P}$-nearly are finite, agree, and are continuous.

Girsanov Theorems

Girsanov theorems are results to the effect that the sum of a standard Wiener process and a suitably smooth and small process of finite variation, a "slightly shifted Wiener process," is again a standard Wiener process, provided the original probability $\mathbb{P}$ is replaced with a properly chosen locally equivalent probability $\mathbb{P}'$.

We approach this subject by investigating how much a martingale under $\mathbb{P}' \approx_. \mathbb{P}$ deviates from being a $\mathbb{P}$-martingale. We assume that the filtration satisfies the natural conditions under either of $\mathbb{P}, \mathbb{P}'$ and then under both (exercise 1.3.42). The restrictions $\mathbb{P}_t, \mathbb{P}'_t$ of $\mathbb{P}, \mathbb{P}'$ to $\mathcal{F}_t$ being by definition mutually absolutely continuous at finite times t, there are Radon–Nikodym derivatives (theorem A.3.22): $\mathbb{P}'_t = G'_t\mathbb{P}_t$ and $\mathbb{P}_t = G_t\mathbb{P}'_t$. Then G' is a $\mathbb{P}$-martingale, and G is a $\mathbb{P}'$-martingale. G, G' can be chosen right-continuous (proposition 2.5.13), strictly positive, and so that $G \cdot G' \equiv 1$. They have expectations $\mathbb{E}[G'_t] = \mathbb{E}'[G_t] = 1$, $0 \le t < \infty$. Here $\mathbb{E}'$ denotes the expectation with respect to $\mathbb{P}'$, of course. $\mathbb{P}'$ is absolutely continuous with respect to $\mathbb{P}$ on $\mathcal{F}_\infty$ if and only if G' is uniformly $\mathbb{P}$-integrable (see exercises 2.5.2 and 2.5.14).

Lemma 3.9.11 (Girsanov–Meyer) *Suppose* M' *is a local* $\mathbb{P}'$*-martingale. Then* $M'G'$ *is a local* $\mathbb{P}$*-martingale, and*

$$M' = \Big(M'_0 + G_{.-}{*}[M', G']\Big) + \Big(G_{.-}{*}(M'G') - (M'G)_{.-}{*}G'\Big). \tag{3.9.5}$$

Reversing the roles of $\mathbb{P}, \mathbb{P}'$ *gives this information: if* M *is a local* $\mathbb{P}$*-martingale, then*

$$\begin{aligned} M + G_{.-}{*}[M, G'] &= M - G'_{.-}{*}[M, G] \\ &= M_0 + G'_{.-}{*}(MG) - (MG')_{.-}{*}G, \end{aligned} \tag{3.9.6}$$

every one of the processes in (3.9.6) being a local $\mathbb{P}'$*-martingale.*

The point, which will be used below and again in the proof of proposition 4.4.1, is that the first summand in (3.9.5) is a process of finite variation and the second a local $\mathbb{P}$-martingale, being as it is the difference of indefinite integrals against two local $\mathbb{P}$-martingales.

Proof. Two easy manipulations show that G, G' are martingales with respect to $\mathbb{P}', \mathbb{P}$, respectively, and that a process N' is a $\mathbb{P}'$-martingale if and only if the product $N'G'$ is a $\mathbb{P}$-martingale. Localization exhibits $M'G'$ as a local $\mathbb{P}$-martingale.

Now $$M'G' = G'_{.-}*M' + M'_{.-}*G' + [G', M']$$

gives $$G_{.-}*(M'G') = (\!(0, \infty)\!)*M' + (GM')_{.-}*G' + G_{.-}*[G', M'] ,$$

and exercise 3.7.9 produces the claim after sorting terms. The second equality in (3.9.6) is the same as equation (3.9.5) with the roles of $\mathbb{P}$, $\mathbb{P}'$ reversed and the finite variation process shifted to the other side. Inasmuch as $GG' = 1$, we have $0 = G_{.-}*G' + G'_{.-}*G + [G, G']$, whence $0 = G_{.-}*[G', M] + G'_{.-}*[G, M]$ for continuous M, which gives the first equality. ∎

Now to approach the classical Girsanov results concerning Wiener process, consider a standard d-dimensional Wiener process $\boldsymbol{W} = (W^1, \ldots, W^d)$ on the measured filtration $(\mathcal{F}_., \mathbb{P})$ and let $\boldsymbol{h} = (h^1, \ldots, h^d)$ be a locally bounded $\mathcal{F}_.$-previsible process. Then clearly the indefinite integral

$$M \stackrel{\text{def}}{=} \boldsymbol{h}*\boldsymbol{W} \stackrel{\text{def}}{=} \sum_{\eta=1}^{d} h^\eta * W^\eta$$

is a continuous locally bounded local martingale and so is its Doléans–Dade exponential (see proposition 3.9.2)

$$G'_t \stackrel{\text{def}}{=} \exp\Big(M_t - 1/2 \int_0^t |\boldsymbol{h}|_s^2 \, ds\Big) = 1 + \int_0^t G'_s \, dM_s \, .$$

G' is a strictly positive supermartingale and is a martingale if and only if $\mathbb{E}[G'_t] = 1$ for all $t > 0$ (exercise 2.5.23 (iv)). Its reciprocal $G \stackrel{\text{def}}{=} 1/G'$ is an L^0-integrator (exercise 2.5.32 and theorem 3.9.1).

Exercise 3.9.12 (i) If there is a locally Lebesgue square integrable function $\eta : [0, \infty) \to \mathbb{R}$ so that $|\boldsymbol{h}|_t \le \eta_t \quad \forall t$, then G' is a square integrable martingale; in fact, then clearly $\mathbb{E}[G_t'^2] \le \exp(\int_0^t \eta_s^2 \, ds)$. (ii) If it can merely be ascertained that the quantity

$$\mathbb{E}\Big[\exp\Big(\frac{1}{2}\int_0^t |\boldsymbol{h}|_s^2 \, ds\Big)\Big] = \mathbb{E}\big[\exp\big([M, M]_t/2\big)\big] \tag{3.9.7}$$

is finite at all instants t, then G' is still a martingale. Equation (3.9.7) is known as ***Novikov's condition***. The condition $\mathbb{E}[\exp([M, M]_t/b)] < \infty$ for some $b > 2$ and all $t \ge 0$ will not do in general.

After these preliminaries consider the "shifted Wiener process"

$$W' \overset{\text{def}}{=} W + H \quad , \text{ where } \; H_. \overset{\text{def}}{=} \int_0^. h_s \, ds = [M, W]_. \, .$$

Assume for the moment that G' is a uniformly integrable martingale, so that there is a limit G'_∞ in mean and almost surely (2.5.14). Then $\mathbb{P}' \overset{\text{def}}{=} G'_\infty \mathbb{P}$ defines a probability absolutely continuous with respect to $\mathbb{P}$ and locally equivalent to $\mathbb{P}$. Now H equals $G{*}[G', W]$ and thus W' is a vector of local $\mathbb{P}'$-martingales – see equation (3.9.6) in the Girsanov–Meyer lemma 3.9.11. Clearly W' vanishes at time 0 and has the same bracket as a standard Wiener process. Due to Lévy's characterization 3.9.5, W' is itself a standard Wiener process under $\mathbb{P}'$. The requirement of uniform integrability will be satisfied for instance when G' is $L^2(\mathbb{P})$-bounded, which in turn is guaranteed by part (i) of exercise 3.9.12 when the function η is Lebesgue square integrable. To summarize:

Proposition 3.9.13 (Girsanov — the Basic Result) *Assume that G' is uniformly integrable. Then $\mathbb{P}' \overset{\text{def}}{=} G'_\infty \mathbb{P}$ is absolutely continuous with respect to $\mathbb{P}$ on $\mathcal{F}_\infty$ and W' is a standard Wiener process under $\mathbb{P}'$.*

In particular, if there is a Lebesgue square integrable function η on $[0, \infty)$ such that $|h_t(\omega)| \le \eta_t$ for all t and all $\omega \in \Omega$, then G' is uniformly integrable and moreover $\mathbb{P}$ and $\mathbb{P}'$ are mutually absolutely continuous on $\mathcal{F}_\infty$.

Example 3.9.14 The assumption of uniform integrability in proposition 3.9.13 is rather restrictive. The simple shift $W'_t = W_t + t$ is not covered. Let us work out this simple one-dimensional example in order to see what might and might not be expected under less severe restrictions. Since here $h \equiv 1$, we have $G'_t = \exp(W_t - t/2)$, which is a square integrable – but not square bounded, not even uniformly integrable – martingale. Nevertheless there is, for every instant t, a probability $\mathbb{P}'_t$ on $\mathcal{F}_t$ equivalent with the restriction $\mathbb{P}_t$ of $\mathbb{P}$ to $\mathcal{F}_t$, to wit, $\mathbb{P}'_t \overset{\text{def}}{=} G'_t \mathbb{P}_t$. The pairs $(\mathcal{F}_t, \mathbb{P}'_t)$ form a ***consistent family of probabilities*** in the sense that for $s < t$ the restriction of $\mathbb{P}'_t$ to $\mathcal{F}_s$ equals $\mathbb{P}'_s$. There is therefore a unique measure $\mathbb{P}'$ on the algebra $\mathcal{A}_\infty \overset{\text{def}}{=} \bigcup_t \mathcal{F}_t$ of sets, the ***projective limit***, defined unequivocally by

$$\begin{aligned} \mathbb{P}'[A] &\overset{\text{def}}{=} \mathbb{P}'_s[A] \quad \text{if } \; A \in \mathcal{A}_\infty \; \text{ belongs to } \mathcal{F}_s \, . \\ &= \mathbb{P}'_t[A] \quad \text{if } \; A \quad \text{also belongs to } \mathcal{F}_t \, . \end{aligned}$$

Things are looking up. Here is a damper,[16] though: $\mathbb{P}'$ cannot be absolutely continuous with respect to $\mathbb{P}$. Namely, since $\lim_{t\to\infty} W_t/t = 0$ $\mathbb{P}$-almost surely, the set $[\lim_{t\to\infty} W_t/t = -1]$ is $\mathbb{P}$-negligible; yet this set has $\mathbb{P}'$-measure 1, since it coincides with the set $[\lim_t W'_t/t = 0]$. *Mutatis*

[16] This point is occasionally overlooked in the literature.

mutandis we see that $\mathbb{P}$ is not absolutely continuous with respect to $\mathbb{P}'$ either. In fact, these two measures are disjoint.

The situation is actually even worse. Namely, in the previous argument the σ-additivity of $\mathbb{P}'$ was used, but this is by no means assured.[16] Roughly, Σ-additivity requires that the ambient space be "not too sparse," a feature Ω may miss. Assume for example that the underlying set Ω is the path space $\mathscr{C}$ with the $\mathbb{P}$-negligible Borel set $\{\omega : \limsup |\omega_t/t| > 0\}$ removed, W_t is of course evaluation: $W_t(\omega_.) = \omega_t$, and $\mathbb{P}$ is Wiener measure $\mathbb{W}$ restricted to Ω. The set function $\mathbb{P}'$ is additive on $\mathcal{A}_\infty$ but cannot be σ-additive. If it were, it would have a unique extension to the σ-algebra generated by $\mathcal{A}_\infty$, which is the Borel σ-algebra on Ω; $t \mapsto \omega_t + t$ would be a standard Wiener process under $\mathbb{P}'$ with $\mathbb{P}'\big[\{\omega : \lim(\omega_t{+}t)/t = 0\}\big] = 1$, yet Ω does not contain a single path ω with $\lim(\omega_t + t)/t = 0$!

On the positive side, the discussion suggests that if Ω is the full path space $\mathscr{C}$, then $\mathbb{P}'$ might in fact be σ-additive as the projective limit of tight probabilities (see theorem A.7.1 (v)). As long as we are content with having $\mathbb{P}'$ absolutely continuous with respect to $\mathbb{P}$ merely locally, there ought to be some "non-sparseness" or "fullness" condition on Ω that permits a satisfactory conclusion even for somewhat large $\boldsymbol{h}$. ∎

Let us approach the Girsanov problem again, with example 3.9.14 in mind. Now the collection $\mathfrak{T}'$ of stopping times T with $\mathbb{E}[G'_T] = 1$ is increasingly directed (exercise 2.5.23 (iv)), and therefore $\mathcal{A} \stackrel{\text{def}}{=} \bigcup_{T\in\mathfrak{T}'} \mathcal{F}_T$ is an algebra of sets. On it we define unequivocally the additive measure $\mathbb{P}'$ by $\mathbb{P}'[A] \stackrel{\text{def}}{=} \mathbb{E}[G'_S A]$ if $A \in \mathcal{A}$ belongs to $\mathcal{F}_S$, $S \in \mathfrak{T}$. Due to the optional stopping theorem 2.5.22, this definition is consistent. It looks more general than it is, however:

Exercise 3.9.15 In the presence of the natural conditions $\mathcal{A}$ generates $\mathcal{F}_\infty$, and for $\mathbb{P}'$ to be σ-additive G' must be a martingale.

Now one might be willing to forgo the σ-additivity of $\mathbb{P}'$ on $\mathcal{F}_\infty$ given as it is that it holds on "arbitrarily large" σ-subalgebras $\mathcal{F}_T$, $T \in \mathfrak{T}$. But probabilists like to think in terms of σ-additive measures, and without the σ-additivity some of the cherished facts about a Wiener process W', such as $\lim_{t\to\infty} W'_t/t = 0$ a.s., for example, are lost. We shall therefore have to assume that G' is a martingale, for instance by requiring the Novikov condition (3.9.7) on $\boldsymbol{h}$.

Let us now go after the "non-sparseness" or "fullness" of $(\Omega, \mathcal{F}_.)$ mentioned above. One can formulate a technical condition essentially to the effect that each of the $\mathcal{F}_t$ contain lots of compact sets; we will go a different route and give a definition[17] that merely spells out the properties we need, and then provide a plethora of permanence properties ensuring that this definition is usually met.

[17] As far as I know first used in Ikeda–Watanabe [39, page 176].

Definition 3.9.16 *(i) The filtration $(\Omega, \mathcal{F}_.)$ is **full** if whenever $(\mathcal{F}_t, \mathbb{P}_t)$ is a consistent family of probabilities (see page 164) on $\mathcal{F}_.$, then there exists a σ-additive probability $\mathbb{P}$ on $\mathcal{F}_\infty$ whose restriction to $\mathcal{F}_t$ is $\mathbb{P}_t$, $t \geq 0$.*

*(ii) The measured filtration $(\Omega, \mathcal{F}_., \mathbb{P})$ is **full** if whenever $(\mathcal{F}_t, \overline{\mathbb{P}}_t)$ is a consistent family of probabilities with $\overline{\mathbb{P}}_t \ll \mathbb{P}$ on $\mathcal{F}_t$,[18] $t < \infty$, then there exists a σ-additive probability $\overline{\mathbb{P}}$ on $\mathcal{F}_\infty$ whose restriction to $\mathcal{F}_t$ is $\overline{\mathbb{P}}_t$, $t \geq 0$. The measured filtration $(\Omega, \mathcal{F}_., \mathfrak{P})$ is full if every one of the measured filtrations $(\Omega, \mathcal{F}_., \mathbb{P})$, $\mathbb{P} \in \mathfrak{P}$, is full.*

Proposition 3.9.17 (The Prime Examples) *Fix a polish space (P, ρ). The cartesian product $P^{[0,\infty)}$ equipped with its basic filtration is full. The path spaces $\mathscr{D}_P$ and $\mathscr{C}_P$ equipped with their basic filtrations are full.* ∎

When making a stochastic model for some physical phenomenon, financial phenomenon, etc., one usually has to begin by producing a filtered measured space that carries a model for the drivers of the stochastic behavior – in this book this happens for instance when Wiener process is constructed to drive Brownian motion (page 11), or when Lévy processes are constructed (page 267), or when a Markov process is associated with a semigroup (page 351). In these instances the naturally appearing ambient space Ω is a path space $\mathscr{D}_P$ or $\mathscr{C}^d$ equipped with its basic full filtration. Thereafter though, in order to facilitate the stochastic analysis, one wishes to discard inconsequential sets from Ω and to go to the natural enlargement. At this point one hopes that fullness has permanence properties good enough to survive these operations. Indeed it has:

Proposition 3.9.18 *(i) Suppose that $(\Omega, \mathcal{F}_.)$ is full, and let $N \in \mathcal{A}_{\infty\sigma}$. Set $\Omega' \stackrel{\text{def}}{=} \Omega \setminus N$, and let $\mathcal{F}'_.$ denote the filtration induced on Ω', that is to say, $\mathcal{F}'_t \stackrel{\text{def}}{=} \{A \cap \Omega' : A \in \mathcal{F}_t\}$. Then $(\Omega', \mathcal{F}'_.)$ is full. Similarly, if the measured filtration $(\Omega, \mathcal{F}_., \mathfrak{P})$ is full and a $\mathfrak{P}$-nearly empty set N is removed from Ω, then the measured filtration induced on $\Omega' \stackrel{\text{def}}{=} \Omega \setminus N$ is full. (ii) If the measured filtration $(\Omega, \mathcal{F}_., \mathfrak{P})$ is full, then so is its natural enlargement. In particular, the natural filtration on canonical path space is full.*

Proof. (i) Let $(\mathcal{F}'_t, \mathbb{P}'_t)$ be a consistent family of σ-additive probabilities, with additive projective limit $\mathbb{P}'$ on the algebra $\mathcal{A}'_\infty \stackrel{\text{def}}{=} \bigcup_t \mathcal{F}'_t$. For $t \geq 0$ and $A \in \mathcal{F}_t$ set $\mathbb{P}_t[A] \stackrel{\text{def}}{=} \mathbb{P}'_t[A \cap \Omega']$. Then $(\mathcal{F}_t, \mathbb{P}_t)$ is easily seen to be a consistent family of σ-additive probabilities. Since $\mathcal{F}_.$ is full there is a σ-additive probability $\mathbb{P}$ that coincides with $\mathbb{P}_t$ on $\mathcal{F}_t$, $t \geq 0$. Now let $\mathcal{A}'_\infty \ni A'_n \downarrow \emptyset$. It is to be shown that $\mathbb{P}'[A'_n] \to 0$; any of the usual extension procedures will then provide the required σ-additive $\mathbb{P}'$ on $\mathcal{F}'_.$ that agrees with $\mathbb{P}'_t$ on $\mathcal{F}'_t$, $t \geq 0$. Now there are $A_n \in \mathcal{A}_\infty$ such that $A'_n = A_n \cap \Omega'$; they can be chosen to decrease as n increases, by replacing A_n with $\bigcap_{\nu \leq n} A_\nu$ if necessary. Then there are $N_n \in \mathcal{A}_\infty$ with union N; they can be chosen to increase with n.

[18] I.e., a $\mathbb{P}$-negligible set belonging to $\mathcal{F}_t$ (!) is $\overline{\mathbb{P}}_t$-negligible.

There is an increasing sequence of instants t^n so that both $N_n \in \mathcal{F}_{t^n}$ and $A_n \in \mathcal{F}_{t^n}$, $n \in \mathbb{N}$. Now, since $\bigcap_n A_n \subseteq N$,

$$\begin{aligned} \lim \mathbb{P}'[A'_n] &= \lim \mathbb{P}'_{t^n}[A'_n] = \lim \mathbb{P}'_{t^n}[A_n \cap \Omega'] = \lim \mathbb{P}_{t^n}[A_n] \\ &= \lim \mathbb{P}[A_n] = \mathbb{P}[\bigcap A_n] \le \mathbb{P}[N] = \lim \mathbb{P}[N_n] \qquad (3.9.8) \\ &= \lim \mathbb{P}_{t^n}[N_n] = \lim \mathbb{P}'_{t^n}[N_n \cap \Omega'] = \lim \mathbb{P}'_{t^n}[\emptyset] = 0 . \end{aligned}$$

The proof of the second statement of (i) is left as an exercise.

(ii) It is easy to see that $(\Omega, \mathcal{F}_{.+})$ is full when $(\Omega, \mathcal{F}_.)$ is, so we may assume that $\mathcal{F}_.$ is right-continuous and only need to worry about the regularization. Let then $(\Omega, \mathcal{F}_., \mathfrak{P})$ be a full measured filtration and $(\mathcal{F}_t^{\mathfrak{P}}, \overline{\mathbb{P}}_t)$ a consistent family of σ-additive probabilities on $\mathcal{F}_.^{\mathfrak{P}}$, with additive projective limit $\overline{\mathbb{P}}$ on $\mathcal{A}_\infty^{\mathfrak{P}} \stackrel{\text{def}}{=} \bigcup_t \mathcal{F}_t^{\mathfrak{P}}$ and $\overline{\mathbb{P}}_t \ll \mathbb{P}$ on $\mathcal{F}_t^{\mathfrak{P}}$, $\mathbb{P} \in \mathfrak{P}, t \ge 0$. The restrictions $\overline{\mathbb{P}}_t^0$ of $\overline{\mathbb{P}}_t$ to $\mathcal{F}_t$ have a σ-additive extension $\overline{\mathbb{P}}^0$ to $\mathcal{F}_\infty$ that vanishes on $\mathbb{P}$-nearly empty sets, $\mathbb{P} \in \mathfrak{P}$, and thus is defined and σ-additive on $\mathcal{F}_\infty^{\mathbb{P}}$. On $\mathcal{A}_\infty^{\mathfrak{P}}$, $\overline{\mathbb{P}}^0$ coincides with $\overline{\mathbb{P}}$, which is therefore σ-additive. ∎

Imagine, for example, that we started off by representing a number of processes, among them perhaps a standard Wiener process $\boldsymbol{W}$ and a few Poisson point processes, canonically on the Skorohod path space: $\Omega = \mathscr{D}^n$. Having proved that the $\omega \in \Omega$ where the path $\boldsymbol{W}_.(\omega)$ is anywhere differentiable form a nearly empty set, we may simply throw them away; the remainder is still full. Similarly we may then toss out the ω where the Wiener paths violate the law of the iterated logarithm, the paths where the approximation scheme 3.7.26 for some stochastic integral fails to converge, etc. What we cannot throw away without risking complications are sets like $[\boldsymbol{W}_t(\cdot)/t \xrightarrow[t\to\infty]{} 0]$ that depend on the tail–σ-algebra of W; they may be negligible but may well not be nearly empty. With a modicum of precaution we have the Girsanov theorem in its most frequently stated form:

Theorem 3.9.19 (Girsanov's Theorem) *Assume that $\boldsymbol{W} = (W^1, \ldots, W^d)$ is a standard Wiener process on the full measured filtration $(\Omega, \mathcal{F}_., \mathbb{P})$, and let $\boldsymbol{h} = (h^1, \ldots, h^d)$ be a locally bounded previsible process. If the Doléans–Dade exponential G' of the local martingale $M \stackrel{\text{def}}{=} \boldsymbol{h}{*}\boldsymbol{W}$ is a martingale, then there is a unique σ-additive probability $\mathbb{P}'$ on $\mathcal{F}_\infty$ so that $\mathbb{P}' = G'_t\mathbb{P}$ on $\mathcal{F}_t$ at all finite instants t, and*

$$\boldsymbol{W}' \stackrel{\text{def}}{=} \boldsymbol{W} + [M, \boldsymbol{W}] = \boldsymbol{W} + \int_0^{\cdot} \boldsymbol{h}_s \, ds$$

is a standard Wiener process under $\mathbb{P}'$.

Warning 3.9.20 In order to ensure a plentiful supply of stopping times (see exercise 1.3.30 and items A.5.10–A.5.21) and the existence of modifications with regular paths (section 2.3) and of cross sections (pages 436–440), most every author requires right off the bat that the underlying filtration $\mathcal{F}_.$ satisfy the so-called ***usual conditions***, which say that $\mathcal{F}_.$ is right-continuous and that every $\mathcal{F}_t$ contains every negligible set of $\mathcal{F}_\infty$ (!). This is achieved by making the basic filtration right-continuous and by throwing into $\mathcal{F}_0$ all subsets of negligible sets in $\mathcal{F}_\infty$. If the enlargement is effected this way, then theorem 3.9.19 fails, even when Ω is the full path space $\mathscr{C}$ and the shift is as simple as $h \equiv 1$, i.e., $W'_t = W_t + t$, as witness example 3.9.14. In other words, the usual enlargement of a full measured filtration may well not be full. If the enlargement is effected by adding into $\mathcal{F}_0$ only the nearly empty sets,[19] then all of the benefits mentioned persist and theorem 3.9.19 turns true.

We hope the reader will at this point forgive the painstaking (unusual but natural) way we chose to regularize a measured filtration.

The Stratonovich Integral

Let us revisit the algorithm (3.7.11) on page 140 for the pathwise approximation of the integral $\int_0^T X_{.-}\, dZ$. Given a threshold δ we would define stopping times S_k, $k = 0, 1, \ldots$, partitioning $(\!(0, T]\!]$ such that on $\mathcal{I}_k \stackrel{\text{def}}{=} (\!(S_k, S_{k+1}]\!]$ the integrand $X_{.-}$ did not change by more than δ. On each of the intervals $\mathcal{I}_k$ we would approximate the integral by the value of the right-continuous process X at the left endpoint S_k multiplied with the change $Z^T_{S_{k+1}} - Z^T_{S_k}$ of Z^T over $\mathcal{I}_k$. Then we would approximate the integral over $(\!(0, T]\!]$ by the sum over k of these local approximations. We said in remarks 3.7.27 (iii)–(iv) that the limit of these approximations as $\delta \to 0$ would serve as a perfectly intuitive *definition* of the integral, if integrands in $\mathfrak{L}$ were all we had to contend with – definition 2.1.7 identifies the condition under which the limit exists.

Now the practical reader who remembers the trapezoidal rule from calculus might at this point offer the following suggestion. Since from the definition (3.7.10) of S_{k+1} we know the value of X at that time already, a better local approximation to X than its value at the left endpoint might be the average

$$1/2\big(X_{S_k} + X_{S_{k+1}}\big) = X_{S_k} + 1/2\big(X_{S_{k+1}} - X_{S_k}\big)$$

of its values at the two endpoints. He would accordingly propose to define $\int_{0+}^T X\, dZ$ as

$$\lim_{\delta \to 0} \sum_{0 \le k \le \infty} \frac{X_{S_k} + X_{S_{k+1}}}{2} \cdot \big(Z^T_{S_{k+1}} - Z^T_{S_k}\big)$$

$$= \lim_{\delta \to 0} \sum_{0 \le k \le \infty} X_{S_k}\big(Z^T_{S_{k+1}} - Z^T_{S_k}\big) + \frac{1}{2} \lim_{\delta \to 0} \sum_{0 \le k \le \infty} \big(X_{S_{k+1}} - X_{S_k}\big)\big(Z^T_{S_{k+1}} - Z^T_{S_k}\big) .$$

The merit of writing it as in the second line above is that the two limits are actually known: the first one equals the Itô integral $\int_{0+}^T X\, dZ$, thanks to

[19] Think of them as the sets whose negligibility can be detected before the expiration of time.

theorem 3.7.26, and the second limit is $[X, Z^T]_T - [X, Z^T]_0$ – at least when X is an L^0-integrator (page 150). Our practical reader would be lead to the following notion:

Definition 3.9.21 *Let X, Z be two L^0-integrators and T a finite stopping time. The* ***Stratonovich integral*** *is defined by*

$$\int_0^T X\,\delta Z \stackrel{\text{def}}{=} X_0 Z_0 + \frac{1}{2} \lim \sum_{0 \le k \le \infty} \big(X_{S_k} + X_{S_{k+1}}\big)\big(Z^T_{S_{k+1}} - Z^T_{S_k}\big)\,, \tag{3.9.9}$$

the limit being taken as the partition[8] $\mathcal{S} = \{0{=}S_0 \le S_1 \le S_2 \le \ldots \le S_\infty{=}\infty\}$ *runs through a sequence whose mesh goes to zero. It can be computed in terms of the Itô integral as*

$$\int_0^T X\,\delta Z = X_0 Z_0 + \int_0^T X_{\text{-}}\,dZ + \frac{1}{2}\big([X,Z]_T - [X,Z]_0\big)\,. \tag{3.9.10}$$

$X{\circ}Z$ *denotes the corresponding indefinite integral* $t \mapsto \int_0^t X\,\delta Z$ *:*

$$X{\circ}Z = X_0 Z_0 + X_{\text{-}}{*}Z + 1/2\,\big([X,Z] - [X,Z]_0\big)\,.$$

Remarks 3.9.22 (i) The Itô and Stratonovich integrals not only apply to different classes of integrands, they also give different results when they happen to apply to the same integrand. For instance, when both X and Z are continuous L^0-integrators, then $X{\circ}Z = X{*}Z + 1/2[X,Z]$. In particular, $(W{\circ}W)_t = (W{*}W)_t + t/2$ (proposition 3.8.16).

(ii) Which of the two integrals to use? The answer depends entirely on the purpose. Engineers and other applied scientists generally prefer the Stratonovich integral when the driver Z is continuous. This is partly due to the appeal of the "trapezoidal" definition (3.9.9) and partly to the simplicity of the formula governing coordinate transformations (theorem 3.9.24 below). The ultimate criterion is, of course, which integral better models the physical situation. It is claimed that the Stratonovich integral generally does. Even in pure mathematics – if there is such a thing – the Stratonovich integral is indispensable when it comes to coordinate-free constructions of Brownian motion on Riemannian manifolds, say.

(iii) So why not stick to Stratonovich's integral and forget Itô's? Well, the Dominated Convergence Theorem does not hold for Stratonovich's integral, so there are hardly any limit results that one can prove without resorting to equation (3.9.10), which connects it with Itô's. In fact, when it comes to a computation of a Stratonovich integral, it is generally turned into an Itô integral via (3.9.10), which is then evaluated.

(iv) An algorithm for the pathwise computation of the Stratonovich integral $X{\circ}Z$ is available just as for the Itô integral. We describe it in case both X and Z are L^p-integrators for some $p > 0$, leaving the case $p = 0$ to the reader. Fix a threshold $\delta > 0$. There is a partition $\mathcal{S} = \{0 = S_0 \le S_1 \le \cdots\}$ with $S_\infty \stackrel{\text{def}}{=} \sup_{k<\infty} S_k = \infty$ on whose intervals $[\![S_k, S_{k+1})\!)$ both X and Z vary by less than δ. For example, the recursive definition $S_{k+1} \stackrel{\text{def}}{=} \inf\{t > S_k : |X_t - X_{S_k}| \vee |Z_t - Z_{S_k}| > \delta\}$ produces one. The approximate

$$Y_t^{(\delta)} \stackrel{\text{def}}{=} \sum \frac{X_t^{S_k} + X_t^{S_{k+1}}}{2} \cdot \left(Z_t^{S_{k+1}} - Z_t^{S_k}\right)$$

$$= \sum X_{S_k}(Z_t^{S_{k+1}} - Z_t^{S_k}) + \frac{1}{2}\Big(X_0 Z_0 + \sum_{0<k<\infty} (X_t^{S_{k+1}} - X_t^{S_k})(Z_t^{S_{k+1}} - Z_t^{S_k})\Big)$$

has, by exercise 3.8.14,

$$\left\|\left| |(X{\circ}Z) - Y^{(\delta)}|_t^\star \right|\right\|_{L^p} \le 2C_p^{\star(2.3.5)} \left(\lceil \delta Z^t \rceil_{\mathcal{I}^p} + \lceil \delta X^t \rceil_{\mathcal{I}^p} \right) .$$

Thus, if δ runs through a sequence (δ_n) with

$$\textstyle\sum_n \lceil \delta_n Z^n \rceil_{\mathcal{I}^p} + \lceil \delta_n X^n \rceil_{\mathcal{I}^p} < \infty ,$$

then $Y^{(\delta)} \xrightarrow[n\to\infty]{} X{\circ}Z$ nearly, uniformly on bounded intervals. ∎

Practically, Stratonovich's integral is useful only when *the integrator is a continuous L^0-integrator*, so we shall assume this for the remainder of this subsection.

Given a partition $\mathcal{S}$ and a continuous L^0-integrator Z, let $\overline{Z}^{\mathcal{S}}$ denote the continuous process that agrees in the points S_k of $\mathcal{S}$ with Z and is linear in between. It is clearly not adapted in general: knowledge of $Z_{S_{k+1}}$ is contained in the definition of Z_t for $t \in (\!(S_k, S_{k+1}]\!]$. Nevertheless, the piecewise linear process $\overline{Z}^{\mathcal{S}}$ of finite variation is easy to visualize, and the approximate $Y_t^{(\delta)}$ above is nothing but the Lebesgue–Stieltjes integral $\int_0^t \overline{X}^{\mathcal{S}}\, d\overline{Z}^{\mathcal{S}}$, at least at the points $t = S_k$, $k = 1, 2, \ldots$. In other words, the approximation scheme above can be seen as an approximation of the Stratonovich integral by Lebesgue–Stieltjes integrals that are measurably parametrized by $\omega \in \Omega$.

Exercise 3.9.23 $X \cdot Z = X{\circ}Z + Z{\circ}X$, and so $\delta(XZ) = X\delta Z + Z\delta X$. Also, $X \circ (Y \circ Z) = (XY) \circ Z$.

Consider a differentiable curve $t \mapsto \zeta_t = (\zeta_t^1, \ldots, \zeta_t^d)$ in $\mathbb{R}^d$ and a smooth function $\Phi : \mathbb{R}^d \to \mathbb{R}$. The Fundamental Theorem of Calculus says that

$$\Phi(\zeta_t) = \Phi(\zeta_0) + \int_0^t \Phi_{;\eta}(\zeta_s)\, d\zeta_s^\eta .$$

It is perhaps the main virtue of the Stratonovich integral that a similarly simple formula holds for it:

Theorem 3.9.24 *Let $D \subset \mathbb{R}^d$ be open and convex, and let $\Phi : D \to \mathbb{R}$ be a twice continuously differentiable function. Let $\boldsymbol{Z} = (Z^\eta)_{\eta=1}^d$ be a d-vector* *of continuous L^0-integrators and assume that the path of $\boldsymbol{Z}$ stays in D at all times. Then $\Phi(\boldsymbol{Z})$ is an L^0-integrator, and for any almost surely finite stopping time T* [14]

$$\Phi(\boldsymbol{Z}_T) = \Phi(\boldsymbol{Z}_0) + \int_{0+}^{T} \Phi_{;\eta}(\boldsymbol{Z})\, \delta Z^\eta \,. \tag{3.9.11}$$

Proof. Recall our convention that $X_{0-} = 0$ for $X \in \mathfrak{D}$. Itô's formula gives

$$\Phi_{;\eta}(\boldsymbol{Z}) = \Phi_{;\eta}(\boldsymbol{Z}_0) + \Phi_{;\eta\theta}(\boldsymbol{Z})_{.-}{*}Z^\theta + 1/2\, \Phi_{;\eta\theta\iota}(\boldsymbol{Z}){*}{}^c[Z^\theta, Z^\iota] \,,$$

so by exercise 3.8.12 and proposition 3.8.19

$${}^c[\Phi_{;\eta}(\boldsymbol{Z}), Z^\eta] = {}^c[\Phi_{;\eta\theta}(\boldsymbol{Z}){*}Z^\theta, Z^\eta] = \Phi_{;\eta\theta}(\boldsymbol{Z}){*}{}^c[Z^\eta, Z^\theta] \,.$$

Equation (3.9.10) produces

$$\int_{0+}^{T} \Phi_{;\eta}(\boldsymbol{Z})\delta Z^\eta = \int_{0+}^{T} \Phi_{;\eta}(\boldsymbol{Z})\, dZ^\eta + 1/2 \int_{0+}^{T} \Phi_{;\eta\theta}(\boldsymbol{Z})\, d{}^c[Z^\eta, Z^\theta] \,,$$

thus $$\Phi(\boldsymbol{Z}_T) = \Phi(\boldsymbol{Z}_0) + \int_{0+}^{T} \Phi_{;\eta}(\boldsymbol{Z})\, dZ^\eta + 1/2 \int_{0+}^{T} \Phi_{;\eta\theta}(\boldsymbol{Z})\, d{}^c[Z^\eta, Z^\theta] \,,$$

i.e., $$\Phi(\boldsymbol{Z}_T) = \Phi(\boldsymbol{Z}_0) + \int_{0+}^{T} \Phi_{;\eta}(\boldsymbol{Z})\delta Z^\eta \,.$$

For this argument to work Φ must be thrice continuously differentiable. In the general case we find a sequence of smooth functions Φ^n that converge to Φ uniformly on compacta together with their first and second derivatives, use the penultimate equation above, and apply the Dominated Convergence Theorem. ∎

3.10 Random Measures

Very loosely speaking, a random measure is what one gets if the index η in a vector $\boldsymbol{Z} = (Z^\eta)$ of integrators is allowed to vary over a continuous set, the "auxiliary space," instead of the finite index set $\{1, \ldots, d\}$. Visualize for instance a drum pelted randomly by grains of sand (see [107]). At any surface element $d\eta$ and during any interval ds there is some random noise $\beta(d\eta, ds)$ acting on the surface; a suitable model for the noise β together with the appropriate differential equation should describe the effect of the action. We won't go into such a model here but only provide the mathematics to do so, since the integration theory of random measures is such a straightforward extension of the stochastic analysis above. The reader may look at this section as an overview or summary of the material offered so far, done through a slight generalization.

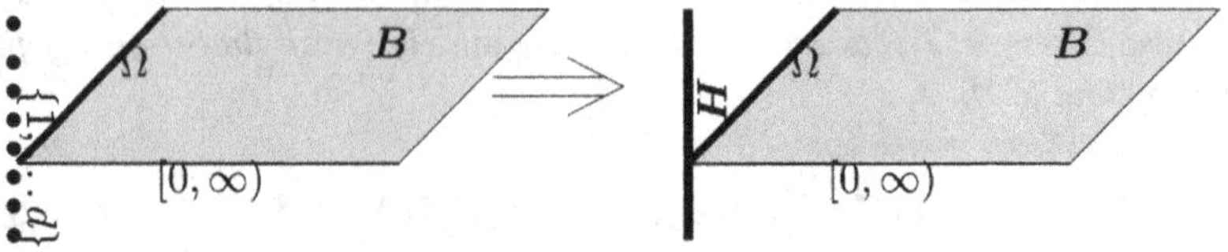

Figure 3.11 Going from a discrete to a continuous auxiliary space

Let us then fix an ***auxiliary space*** $\boldsymbol{H}$; this is to be a separable metrizable locally compact space[20] equipped with its natural class $\mathcal{E}[\boldsymbol{H}] \stackrel{\text{def}}{=} C_{00}(\boldsymbol{H})$ of elementary integrands and its Borel σ-algebra $\mathcal{B}^\bullet(\boldsymbol{H})$. This and the ambient measured filtration $(\Omega, \mathcal{F}_., \mathfrak{P})$ with accompanying base space $\boldsymbol{B}$ give rise to the ***auxiliary base space***

$$\check{\boldsymbol{B}} \stackrel{\text{def}}{=} \boldsymbol{H} \times \boldsymbol{B} \quad \text{with typical point} \quad \check{\eta} = (\eta, \varpi) = (\eta, s, \omega) ,$$

which is naturally equipped with the algebra of elementary functions

$$\check{\mathcal{E}} \stackrel{\text{def}}{=} \mathcal{E}[\boldsymbol{H}] \otimes \mathcal{E}[\mathcal{F}_.] = \Big\{ (\eta, \varpi) \mapsto \sum_i h^i(\eta) X^i(\varpi) : h^i \in \mathcal{E}[\boldsymbol{H}] \ , \ X^i \in \mathcal{E} \Big\} .$$

$\check{\mathcal{E}}$ is evidently self-confined, and its natural topology is the topology of confined uniform convergence (see item A.2.5 on page 370). Its confined uniform closure is a self-confined algebra and vector lattice closed under chopping (see exercise A.2.6). The sequential closure of $\check{\mathcal{E}}$ is denoted by $\check{\mathcal{P}}$ and consists of the ***predictable random functions***. One further bit of notation: for any stopping time T,

$$[\![0, T]\!] \stackrel{\text{def}}{=} \boldsymbol{H} \times [\![0, T]\!] = \{(\eta, s, \omega) : 0 \le s \le T(\omega)\} .$$

The essence of a function F of finite variation is the measure dF that goes with it. The essence of an integrator $\boldsymbol{Z} = (Z^1, Z^2, \ldots, Z^d)$ is the vector measure $d\boldsymbol{Z}$, which maps an elementary integrand $\boldsymbol{X} = \check{X} = (X^\eta)^{\eta = 1 \ldots d}$ to the random variable $\int \boldsymbol{X}\, d\boldsymbol{Z}$; the vector $\boldsymbol{Z}$ of cumulative distribution functions is really but a tool in the investigation of $d\boldsymbol{Z}$. Viewing matters this way leads to a straightforward generalization of the notion of an integrator: a random measure with auxiliary space $\boldsymbol{H}$ should be a non-anticipating continuous linear and σ-additive map ζ from $\check{\mathcal{E}}$ to L^p. Such a map will have an ***elementary indefinite integral***[6]

$$(\check{X} * \zeta)_t \stackrel{\text{def}}{=} \zeta\big([\![0, t]\!] \cdot \check{X}\big) , \qquad \check{X} \in \check{\mathcal{E}} \ , \ t \ge 0 .$$

It is convenient to replace the requirement of σ-additivity by a weaker condition, which is easier to check yet implies it, just as was done for integrators in definition 2.1.7, (RC-0), and proposition 3.3.2, and to employ this definition:

[20] For most of the analysis below it would suffice to have $\boldsymbol{H}$ Suslin, and to take for $\mathcal{E}[\boldsymbol{H}]$ a self-confined algebra or vector lattice closed under chopping of bounded Borel functions that generates the Borels – see [93].

Definition 3.10.1 *Let $0 \le p < \infty$. An L^p-random measure with auxiliary space $\boldsymbol{H}$ is a linear map $\zeta : \check{\mathcal{E}} \to L^p$ having the following properties:*

(i) ζ maps order-bounded sets of $\check{\mathcal{E}}$ to topologically bounded sets[21] in L^p.

(ii) The indefinite integral $\check{X}{*}\zeta$ of any $\check{X} \in \check{\mathcal{E}}$ is right-continuous in probability and adapted and satisfies

$$\left([\![0,T]\!] \cdot \check{X}\right){*}\zeta = \left(\check{X}{*}\zeta\right)^T, \qquad T \in \mathfrak{T}. \tag{3.10.1}$$

A few comments and amplifications are in order. If the probability $\mathbb{P}$ must be specified, we talk about an $L^p(\mathbb{P})$-random measure. If $p = 0$, we also speak simply of a ***random measure*** instead of an L^0-random measure.

The continuity condition (i) means that

for every order interval[6] $\quad [-\check{Y}, \check{Y}] \stackrel{\text{def}}{=} \{\check{X} \in \check{\mathcal{E}} : |\check{X}| \le \check{Y}\}$

its image $\quad \zeta([-\check{Y}, \check{Y}])$ is bounded[21] in L^p.

The integrator size of ζ is then naturally measured by the quantities[6]

$$\lceil \zeta^{h,t} \rceil_{\mathcal{I}^p} \stackrel{\text{def}}{=} \sup \left\{ \left\| \int \check{X}\, d\zeta \right\|_p : \check{X} \in \check{\mathcal{E}},\ |\check{X}(\eta, \varpi)| \le h(\eta) \cdot 1_{[\![0,t]\!]}(\varpi) \right\},$$

where $h \in \mathcal{E}_+[\boldsymbol{H}]$ and $t \ge 0$.

If $\quad \lceil \lambda \zeta^{h,\infty} \rceil_{\mathcal{I}^p} \xrightarrow[\lambda \to 0]{} 0 \quad$ for all $h \in \mathcal{E}_+[\boldsymbol{H}]$,

then ζ is reasonably called a ***global L^p-random measure.***

If $\quad \lceil \lambda \zeta^{1,t} \rceil_{\mathcal{I}^p} \xrightarrow[\lambda \to 0]{} 0 \quad$ at all t,

then ζ is ***spatially bounded.***

Equation (3.10.1) means that ζ is non-anticipating and generalizes with a little algebra to $(X{\cdot}\check{X}){*}\zeta = X{*}(\check{X}{*}\zeta)$ for $X \in \mathcal{E}$, $\check{X} \in \check{\mathcal{E}}$. This in conjunction with the continuity (i) shows that $\check{X}{*}\zeta$ is an L^p-integrator for every $\check{X} \in \check{\mathcal{E}}$ and is σ-additive in L^p-mean (proposition 3.3.2).

An L^0-random measure ζ is ***locally an L^p-random measure*** or is a ***local L^p-random measure*** if there are arbitrarily large stopping times T so that the ***stopped random measure***

$$\zeta^T \stackrel{\text{def}}{=} [\![0,T]\!]{\cdot}\zeta : \check{X} \mapsto \zeta([\![0,T]\!]\check{X})$$

has $\quad \lceil ([\![0,T]\!]\lambda\zeta)^{h,\infty} \rceil_{\mathcal{I}^p} \xrightarrow[\lambda \to 0]{} 0$

for all $h \in \mathcal{E}_+[\boldsymbol{H}]$. (This extends the notion for integrators – see exercise 3.3.4.) A random measure ζ ***vanishes at zero*** if $(\check{X}{*}\zeta)_0 = 0$ for every elementary integrand $\check{X} \in \check{\mathcal{E}}$.

---

[21] This amounts to saying that ζ is continuous from $\check{\mathcal{E}}$ to the target space, $\check{\mathcal{E}}$ being given the topology of confined uniform convergence (see item A.2.5 on page 370).

σ-Additivity

For the integration theory of a random measure its σ-additivity is indispensable, of course. It comes from the following result, which is the analog of proposition 3.3.2 and has a somewhat technical proof.

Lemma 3.10.2 *An L^p-random measure is σ-additive in p-mean,* $0 \le p < \infty$.

Proof. We shall use on two occasions the Gelfand transform $\widehat{\zeta}$ (see corollary A.2.7 on page 370). It is a linear map from a space $C_{00}(\widehat{\boldsymbol{B}})$, $\widehat{\boldsymbol{B}}$ locally compact, to L^p that maps order-bounded sets to bounded sets and therefore has the usual Daniell extension featuring the Dominated Convergence Theorem (pages 88–105).[22]

First the case $1 \le p < \infty$. Let g be any element of the dual $L^{p'}$ of L^p. Then $\theta(\check{X}) \stackrel{\text{def}}{=} \langle g|\zeta(\check{X})\rangle$ defines a scalar measure θ of finite variation on $\check{\mathcal{E}}$ that is marginally σ-additive on $\mathcal{E}$. Indeed, for every $H \in \mathcal{E}[\boldsymbol{H}]$ the functional $\mathcal{E} \ni X \mapsto \theta(H \otimes X) = \langle g| \int X \, d(H*\zeta)\rangle$ is σ-additive on the grounds that $H*\zeta$ is an L^p-integrator. By corollary A.2.8, $\theta = \langle g|\zeta\rangle$ is σ-additive on $\check{\mathcal{E}}$. As this holds for any g in the dual of L^p, ζ is σ-additive in the weak topology $\sigma(L^p, L^{p'})$. Due to corollary A.2.7, ζ is σ-additive in p-mean.

Now to the case $0 \le p < 1$. It is to be shown that $\zeta(\check{X}_n) \to 0$ in $L^p(\mathbb{P})$ whenever $\check{\mathcal{E}} \ni \check{X}_n \downarrow 0$ (exercise 3.1.5). There is a function $\check{H} \in \check{\mathcal{E}}$ that equals 1 on $[\check{X}_1 > 0]$. The random measure $\zeta' : \check{X} \to \zeta(\check{H} \cdot \check{X})$ has domain $\check{\mathcal{E}}' \stackrel{\text{def}}{=} \check{\mathcal{E}} + \mathbb{R}$, algebra of bounded functions containing the constants. On the $\check{X}_n$ both ζ and ζ' agree. According to exercise 4.1.8 on page 195 and proposition 4.1.12 on page 206, there is a probability $\mathbb{P}' \approx \mathbb{P}$ so that $\zeta' : \check{\mathcal{E}}' \to L^2(\mathbb{P}')$ is bounded. From proposition 3.3.2 and the first part of the proof we know now that ζ' and then ζ is σ-additive in the topology of $L^2(\mathbb{P}')$. Therefore $\zeta(\check{X}_n) \to 0$ in $L^0(\mathbb{P}) = L^0(\mathbb{P}')$: ζ is σ-additive in $\mathbb{P}$-probability. We invoke corollary A.2.7 again to produce the σ-additivity of ζ in $L^p(\mathbb{P})$. ∎

The extension theory of a random measure is entirely straightforward. We sketch here an overview, leaving most details to the reader – no novel argument is required, simply apply sections 3.1–3.6 *mutatis perpauculis mutandis.*

Suppose then that ζ is an L^p-random measure for some $p \in [0, \infty)$. In view of definition 3.10.1 and lemma 3.10.2, ζ is an L^p-valued linear map on a self-confined algebra of bounded functions, is continuous in the topology of confined uniform convergence, and is σ-additive. Thus there exists an extension of ζ that satisfies the Dominated Convergence Theorem. It is obtained by the utterly straightforward construction and application of THE Daniell mean

$$\check{F} \mapsto \left\lceil\!\left\lceil \check{F} \right\rceil\!\right\rceil^*_{\zeta-p} \stackrel{\text{def}}{=} \inf_{\substack{\check{H} \in \check{\mathcal{E}}^{\uparrow}, \\ \check{H} \ge |\check{F}|}} \sup_{\substack{\check{X} \in \check{\mathcal{E}}, \\ |\check{X}| \le \check{H}}} \left\lceil\!\left\lceil \int \check{X} \, d\zeta \right\rceil\!\right\rceil_{L^p}$$

[22] To be utterly precise, $\widehat{\zeta}$ is a linear map on $\widehat{\check{\mathcal{E}}} \subset C_{00}(\widehat{\boldsymbol{B}})$ that maps order-bounded sets to bounded sets, but it is easily seen to have an extension to C_{00} with the same property.

on $(\check{\boldsymbol{B}}, \check{\mathcal{E}})$ and is written in integral notation as

$$\check{F} \mapsto \int \check{F}\, d\zeta = \int_{\check{\boldsymbol{B}}} F(\eta, s)\, \zeta(d\eta, ds)\,, \qquad \check{F} \in \mathfrak{L}^1[\zeta\text{-}p]\,.$$

Here $\mathfrak{L}^1[\zeta\text{-}p] \stackrel{\text{def}}{=} \mathfrak{L}^1[\lceil\!\lceil\ \rceil\!\rceil^*_{\zeta\text{-}p}]$, the closure of $\check{\mathcal{E}}$ under $\lceil\!\lceil\ \rceil\!\rceil^*_{\zeta\text{-}p}$, is the collection of ***ζ-p-integrable random functions***. On it the DCT holds. For every $\check{F} \in \mathfrak{L}^1[\zeta\text{-}p]$ the process $\check{F}{*}\zeta$, whose value at t is variously written as

$$\check{F}{*}\zeta(t) = \int [\![0,t]\!]\check{F}\, d\zeta = \int_0^t \check{F}(\eta, s)\, \zeta(d\eta, ds)\,,$$

is an L^p-integrator (of classes) and thus has a nearly unique càdlàg representative, which is chosen for $\check{F}{*}\zeta$; and for all bounded predictable X

$$\int X\, d(\check{F}{*}\zeta) = \int X \cdot \check{F}\, d\zeta$$

or

$$X{*}(\check{F}{*}\zeta) = (X \cdot \check{F}){*}\zeta\,. \tag{3.10.2}$$

Hence

$$\lceil \check{F}{*}\zeta \rceil_{Z\text{-}p} = \lceil\!\lceil \check{F} \rceil\!\rceil^*_{\zeta\text{-}p} = \lceil\!\lceil \check{F} \rceil\!\rceil_{\mathfrak{L}^1[\zeta\text{-}p]}$$

and

$$\Big\| |\check{F}{*}\zeta|^\star_\infty \Big\|_{L^p} \le C_p^{\star(2.3.5)} \cdot \lceil\!\lceil \check{F} \rceil\!\rceil^*_{\zeta\text{-}p}\,. \tag{3.10.3}$$

A random function $\check{F}$ is of course defined to be ζ-p-measurable[23] if it is "largely as smooth as an elementary integrand from $\check{\mathcal{E}}$" (see definition 3.4.2). Egoroff's Theorem holds. A function $\check{F}$ in the algebra and vector lattice of measurable functions, which is sequentially closed and generated by its idempotents, is ζ-p-integrable if and only if $\lceil\!\lceil \lambda\check{F} \rceil\!\rceil^*_{\zeta\text{-}p} \xrightarrow[\lambda\to 0]{} 0$. The ***predictable random functions*** $\check{\mathcal{P}} \stackrel{\text{def}}{=} \check{\mathcal{E}}^\sigma$ provide upper and largest lower envelopes, and $\lceil\!\lceil\ \rceil\!\rceil^*_{\zeta\text{-}p}$ is regular in the sense of corollary 3.6.10 on page 128. And so on.

If ζ is merely a local L^p-random measure, then $\check{X}{*}\zeta$ is a local L^p-integrator for every bounded predictable $\check{X} \in \check{\mathcal{P}} \stackrel{\text{def}}{=} \check{\mathcal{E}}^\sigma$.

Law and Canonical Representation

For motivation consider first an integrator $\boldsymbol{Z} = (Z^1, \ldots, Z^d)$. Its essence is the random measure $d\boldsymbol{Z}$; yet its law was defined as the image of the pertinent probability $\mathbb{P}$ under the "vector of cumulative distribution functions" considered as a map Φ from Ω to the path space $\mathscr{D}^d$. In the case of a general random measure ζ there is no such thing as the "collection of cumulative distribution functions." But in the former case there is another way of looking at Φ. Let h^η denote the indicator function of the singleton set $\{\eta\} \subset \boldsymbol{H} = \{1, \ldots, d\}$. The collection $\mathcal{H} \stackrel{\text{def}}{=} \{h^\eta : \{\eta\} \subset \boldsymbol{H}\}$ has the property that its linear span is dense in (in fact is all of) $\mathcal{E}[\boldsymbol{H}]$, and we might as well interpret Φ as the map that sends $\omega \in \Omega$ to the vector $\big((h{*}\boldsymbol{Z}).(\omega) : h \in \mathcal{H}\big)$ of indefinite integrals.

[23] This notion does not depend on p; see corollary 3.6.11 on page 128.

This can be emulated in the case of a random measure. Namely, let $\mathcal{H} \subset \mathcal{E}[\boldsymbol{H}]$ be a collection of functions whose linear span is dense in $\mathcal{E}[\boldsymbol{H}]$, in the topology of confined uniform convergence. Such $\mathcal{H}$ can be chosen countable, due to the σ-compactness of $\boldsymbol{H}$, and most often will be. For every $h \in \mathcal{H}$ pick a càdlàg version of the indefinite integral $h{*}\zeta$ (theorem 2.3.4 on page 62). The map

$$\zeta^{\mathcal{H}} : \omega \mapsto \big((h{*}\zeta)_{\boldsymbol{.}}(\omega) : h \in \mathcal{H}\big)$$

sends Ω into the set $\mathscr{D}_{\mathbb{R}^{\mathcal{H}}}$ of càdlàg paths having values in $\mathbb{R}^{\mathcal{H}}$. In case $\mathcal{H}$ is countable $\mathbb{R}^{\mathcal{H}}$ equals ℓ^0, the Fréchet space of scalar sequences, and then $\mathscr{D}_{\mathbb{R}^{\mathcal{H}}} = \mathscr{D}_{\ell^0}$ is polish under the Skorohod topology (theorem A.7.1 on page 445); the map $\zeta^{\mathcal{H}}$ is measurable if $\mathscr{D}_{\mathbb{R}^{\mathcal{H}}}$ is equipped with the Borel σ-algebra for the Skorohod topology; and ***the law*** $\zeta^{\mathcal{H}}[\mathbb{P}]$ ***is a tight probability*** on $\mathscr{D}_{\mathbb{R}^{\mathcal{H}}}$.

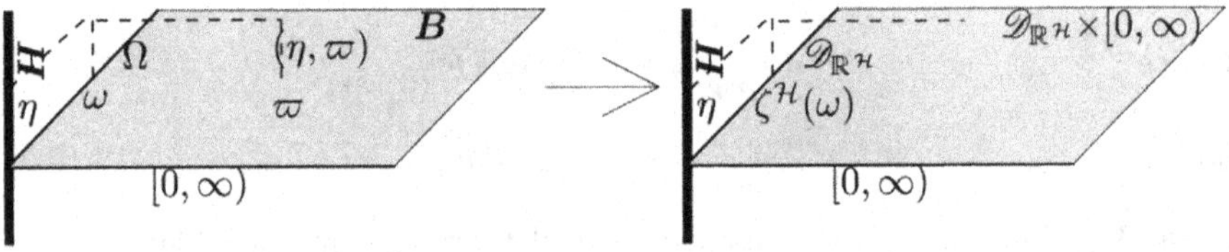

Figure 3.12 The canonical representation

Remark 3.10.3 Two different people will generally pick different càdlàg versions of the integrators $(h{*}\zeta)_{\boldsymbol{.}}$, $h \in \mathcal{H}$. This will affect the map $\zeta^{\mathcal{H}}$ only in a nearly empty set and thus will not change the law. More disturbing is this observation: two different people will generally pick different sets $\mathcal{H} \in \mathcal{E}[\boldsymbol{H}]$ with dense linear span, and that will affect both $\zeta^{\mathcal{H}}$ and the law. If $\boldsymbol{H}$ is discrete, there is a canonical choice of $\mathcal{H}$: take all singletons.[7] One might think of the choice $\mathcal{H} = \mathcal{E}[\boldsymbol{H}]$ in the general case, but that has the disadvantage that now path space $\mathscr{D}_{\mathbb{R}^{\mathcal{H}}}$ is not polish in general and the law cannot be ascertained to be tight. To see why it is desirable to have the path space polish, consider a Wiener random measure β (see definition 3.10.5). Then if β is realized canonically on a polish path space, whose basic filtration is full (theorem A.7.1), we have a chance at Girsanov's theorem (theorem 3.10.8).[24] We shall therefore do as above and simply state which choice of $\mathcal{H}$ enters the definition of the law.

For fixed $(\Omega, \mathcal{F}_{\boldsymbol{.}}, \mathbb{P})$, $\boldsymbol{H}$, $\mathcal{H}$ countable, and ζ, we have now an ad hoc map $\zeta^{\mathcal{H}}$ from Ω to $\mathscr{D}_{\mathbb{R}^{\mathcal{H}}}$ and have declared the law of ζ to be the image of $\mathbb{P}$ under this map. This is of course justified only if "the map $\zeta^{\mathcal{H}}$ of ζ is detailed enough" to capture the essence of ζ. Let us see that it does. To this end some notation. For $h \in \mathcal{H}$ let π^h denote the projection of $\ell^0 = \mathbb{R}^{\mathcal{H}}$ onto its hth component and $\pi^h_{\boldsymbol{.}}$ the projection of a path in $\mathscr{D}_{\mathbb{R}^{\mathcal{H}}}$ onto its hth component. Then $\zeta^h_{\boldsymbol{.}}(\omega) \stackrel{\text{def}}{=} \pi^h_{\boldsymbol{.}} \circ \zeta^{\mathcal{H}}(\omega)$ is the h-component of $\zeta^{\mathcal{H}}(\omega)$. This

[24] The measured filtration appearing in the construction of a Wiener random measure on page 177, is it by any chance full? I do not know.

is a scalar càdlàg path. The basic filtration on ${}^0\!\Omega \overset{\text{def}}{=} \mathscr{D}_{\mathbb{R}^{\mathcal{H}}}$ is denoted by ${}^0\mathcal{F}^0_.$, with elementary integrands ${}^0\mathcal{E} \overset{\text{def}}{=} \mathcal{E}[{}^0\mathcal{F}^0_.]$. Its counterpart on the given probability space is the ***basic filtration*** $\mathcal{F}^0_.[\zeta]$ of ζ, defined as the filtration generated by the càdlàg processes $h{*}\zeta$, $h \in \mathcal{H}$. A simple sequential closure argument shows that a function f on Ω is measurable on $\mathcal{F}^0_t[\zeta]$ if and only if it is of the form $f = \underline{f} \circ \zeta^{\mathcal{H}}$, $\underline{f} \in {}^0\mathcal{F}^0_t$. Indeed, the collection of functions f of this form is closed under pointwise limits of sequences and contains the functions $\omega \mapsto (h{*}\zeta)_s(\omega)$, $h \in \mathcal{H}$, $s \le t$, which generate $\mathcal{F}^0_t[\zeta]$. Also, if $f = \underline{f} \circ \zeta^{\mathcal{H}} = \underline{f}' \circ \zeta^{\mathcal{H}}$, then $[\underline{f} \ne \underline{f}']$ is negligible for the law $\zeta^{\mathcal{H}}[\mathbb{P}]$ of ζ. With $\zeta^{\mathcal{H}} : \Omega \to {}^0\!\Omega$ there go the maps

$$I{\times}\zeta^{\mathcal{H}} : \boldsymbol{B} \to {}^0\boldsymbol{B} \overset{\text{def}}{=} \mathbb{R}_+ \times {}^0\!\Omega ,$$

which does $$(s,\omega) \mapsto (s, \zeta^{\mathcal{H}}(\omega)) ,$$

and $$I{\times}I{\times}\zeta^{\mathcal{H}} : \check{\boldsymbol{B}} \to {}^0\check{\boldsymbol{B}} \overset{\text{def}}{=} \boldsymbol{H} \times [0,\infty) \times \Omega = \boldsymbol{H} \times {}^0\boldsymbol{B} ,$$

which does $$(\eta, s, \omega) \mapsto (\eta, s, \zeta^{\mathcal{H}}(\omega)) .$$

It is easily seen that a process X is $\mathcal{F}^0_.[\zeta]$-predictable if and only if it is of the form $X = \underline{X} \circ I{\times}\zeta^{\mathcal{H}}$, where $\underline{X}$ is ${}^0\mathcal{F}^0_.$-predictable, and that a random function $\check{F}$ is predictable if and only if it is of the form $\check{F} = \underline{\check{F}} \circ I{\times}I{\times}\zeta^{\mathcal{H}}$ with $\underline{\check{F}}$ ${}^0\mathcal{F}^0_.$-predictable. With these notations in place consider an elementary function $\underline{\check{X}} = \sum_i h^i \underline{X}^i \in \check{\mathcal{E}}[{}^0\mathcal{F}^0_.]$. Then $\check{X} \overset{\text{def}}{=} \underline{\check{X}} \circ I{\times}I{\times}\zeta^{\mathcal{H}} \in \check{\mathcal{E}}\big[\mathcal{F}^0_.[\zeta]\big]$ and

$$
\begin{aligned}
{}^0\zeta(\underline{\check{X}})(\zeta^{\mathcal{H}}(\omega)) &\overset{\text{def}}{=} \textstyle\sum_i (\underline{X}^i{*}\pi^{h_i}_.)(\zeta^{\mathcal{H}}(\omega)) \\
\text{with } X^i \overset{\text{def}}{=} \underline{X}^i \circ I{\times}\zeta^{\mathcal{H}}: \quad &= \textstyle\sum_i X^i{*}(h^i{*}\zeta)(\omega) = \check{X}{*}\zeta(\omega)
\end{aligned}
$$

for nearly every $\omega \in \Omega$. From this it is evident that

$${}^0\zeta(\underline{\check{X}}) \overset{\text{def}}{=} \textstyle\sum_i \underline{X}^i{*}\zeta^{h_i}$$

defines a random measure ${}^0\zeta$ on ${}^0\check{\mathcal{E}}$ that mirrors ζ in this sense:

$${}^0\zeta(\underline{\check{X}}) = \zeta(\underline{\check{X}} \circ I{\times}I{\times}\zeta^{\mathcal{H}}) , \qquad \underline{\check{X}} \in {}^0\check{\mathcal{E}} .$$

${}^0\zeta$ is defined on a *full filtration.* We call it the ***canonical representation of the random measure*** ζ – despite the arbitrariness in the choice of $\mathcal{H}$.

Exercise 3.10.4 The regularization of the right-continuous version of the basic filtration $\mathcal{F}^0_.[\zeta]$ is called the ***natural filtration*** of ζ and is denoted by $\mathcal{F}_.[\zeta]$. Every indefinite integral $\check{F}{*}\zeta$, $\check{F} \in \mathfrak{L}^1[\zeta{-}p]$, is adapted to it.

Example: Wiener Random Measure

Compare the following construction with exercise 1.2.16 on page 20. On the auxiliary space $\boldsymbol{H}$ let ν be a positive Radon measure, and on $\check{\boldsymbol{H}} \overset{\text{def}}{=} \boldsymbol{H} \times [0,\infty)$ let μ denote the product of ν and Lebesgue measure λ. Let $\{\phi_n\}$ be an orthonormal basis of $L^2(\mu)$ and ξ_n independent Gaussians distributed $N(0,1)$ and defined on some measured space $(\Omega, \mathcal{F}, \mathbb{P})$. The map $U : \phi_n \mapsto \xi_n$ extends to a linear isometry of $L^2(\mu)$ into $L^2(\mathbb{P})$ that takes orthogonal elements of $L^2(\mu)$ to independent random variables in $L^2(\mathbb{P})$; $U(f)$ has distribution $N(0, \|f\|^2_{L^2(\mu)})$ for $f \in L^2(\mu)$. The restriction of U to relatively compact Borel subsets[7] of $\check{\boldsymbol{H}}$ is an $L^2(\mathbb{P})$-valued set function, and aficionados of the Riesz representation theorem may wish to write the value of U at $f \in L^2(\mu)$ as

$$U(f) = \int f(\eta, s)\, U(d\eta, ds) \,.$$

We shall now equip Ω with a suitable filtration $\mathcal{F}^0_.$. To this end fix a countable subset $\mathcal{H} \subset \mathcal{E}[\boldsymbol{H}]$ whose linear span is dense in $\mathcal{E}[\boldsymbol{H}]$ in the topology of confined uniform convergence (item A.2.5). For $h \in \mathcal{H}$ set

$$U^h_t \stackrel{\text{def}}{=} \int h(\eta) 1_{[0,t]}(s)\, U(d\eta, ds) \,, \qquad t \geq 0 \,.$$

This produces a countable number of Wiener processes. By theorem 1.2.2 (ii) we can arrange things so that after removal of a negligible set every one of the $U^h_.$, $h \in \mathcal{H}$, has continuous paths. (If we want, we can employ Gram–Schmid and have the $U^h_.$ standard and independent.) We now let $\mathcal{F}^0_t$ be the σ-algebra generated by the random variables U^h_s, $h \in \mathcal{H}, s \leq t$, to obtain the sought-after filtration $\mathcal{F}^0_.$. It is left to the reader to check that, for every relatively compact Borel set $B \subset \boldsymbol{H} \times [0,t]$, $U(B)$ differs negligibly from some set in $\mathcal{F}^0_t$ [Hint: use the Hardy mean (3.10.4) below].

To construct the Wiener random measure β we take for the elementary functions $\mathcal{E}[\boldsymbol{H}]$ the step functions over the relatively compact Borels of $\boldsymbol{H}$ instead of $C_{00}[\boldsymbol{H}]$; this eases visualization a little. With that, an elementary integrand $\check{X} \in \mathcal{E}[\boldsymbol{H}] \otimes \mathcal{E}[\mathcal{F}_.]$ can be written as a finite sum

$$\check{X}(\eta, s; \omega) = \textstyle\sum_\iota h_\iota(\eta) \cdot X_\iota(s, \omega) \,, \qquad h_\iota \in \mathcal{E}[\boldsymbol{H}] \,,\ X_\iota \in \mathcal{E}[\mathcal{F}_.] \,,$$

or as
$$\check{X}(\eta, s, \omega) = \textstyle\sum_i f_i(\omega) \cdot R_i(\eta, s) \,,$$

where the R_i are mutually disjoint rectangles of $\check{\boldsymbol{H}} = \boldsymbol{H} \times [0, \infty)$ and f_i is a simple function measurable on the σ-algebra that goes with the left edge[25] of R_i. In fact, things can clearly be arranged so that the R_i all are rectangles of the form $B_i \times (t_{i-1}, t_i]$, where the B_i are from a fixed partition of $\boldsymbol{H}$ into disjoint relatively compact Borels and the t_i are from a fixed partition $0 \leq t_1 < \ldots < t_N < +\infty$ of $[0, \infty)$. For such $\check{X}$ define now

$$\beta(\check{X})(\omega) = \int \check{X}(\eta, s, \omega)\, U(d\eta, ds; \omega)$$
$$= \textstyle\sum_i f_i(\omega)\, U(R_i)(\omega) \,.$$

[25] Meaning the left endpoint of the projection of R_i on $[0, \infty)$.

The first line asks us to apply, for every fixed $\omega \in \Omega$, the $L^2(\mathbb{P})$-valued measure U to the integrand $(\eta, s) \mapsto \check{X}(\eta, s; \omega)$, showing that the definition does not depend on the particular representation of $\check{X}$, and implying the linearity[26] of β. The second line allows for an easy check that β is an L^2-random measure. Namely, consider

$$\mathbb{E}[(\beta(\check{X}))^2] = \sum_{i,j} \mathbb{E}[f_i f_j \cdot U(R_i) \cdot U(R_j)] . \tag{$*$}$$

Now if the left edge[25] of R_i is strictly less than the left edge of R_j, then even the right edge of R_i is less than the left edge of R_j, and the random variables $f_i f_j, U(R_i), U(R_j)$ are independent. Since the latter two have mean zero, the expectation $\mathbb{E}[f_i f_j \cdot U(R_i) \cdot U(R_j)]$ vanishes. It does so even if R_i and R_j have the same left edge[25] and $i \neq j$. Indeed, then R_i and R_j are disjoint, again $f_i f_j, U(R_i), U(R_j)$ are independent, and the previous argument applies. The cross-terms in $(*)$ thus vanish so that

$$\begin{aligned}\mathbb{E}\Big[(\beta(\check{X}))^2\Big] &= \sum_i \mathbb{E}\Big[f_i^2 (U(R_i))^2\Big] = \sum_i \mathbb{E}[f_i^2] \cdot \mathbb{E}\Big[(U(R_i))^2\Big] \\ &= \sum_i \mathbb{E}[f_i^2] \cdot \mu(R_i) = \mathbb{E}\Big[\int \check{X}^2(\eta, s; \cdot) \mu(d\eta, ds)\Big] .\end{aligned}$$

Therefore

$$F \mapsto \|F\|_{\beta\text{-}2}^{\mathcal{H}*} \overset{\text{def}}{=} \Big(\int^* F_t^2(\eta, s, \omega)\; \mu(d\eta, ds)\mathbb{P}(d\omega)\Big)^{1/2} \tag{3.10.4}$$

is a mean majorizing the linear map $\beta : \check{\mathcal{E}} \to L^2$, the ***Hardy mean***. From this it is obvious that β has an extension to all previsible $\check{X}$ with $\|\check{X}\|_{\beta\text{-}2}^{\mathcal{H}*} < \infty$, an extension reasonably denoted by $\check{X} \mapsto \int \check{X}\, d\beta$. β evidently meets the following description and shows that there are instances of it:

Definition 3.10.5 *A random measure β with auxiliary space $\boldsymbol{H}$ is a **Wiener random measure** if $h{*}\beta, h'{*}\beta$ are independent Wiener processes whenever the functions $h, h' \in \mathcal{E}[\boldsymbol{H}] \overset{\text{def}}{=} C_{00}[\boldsymbol{H}]$ have disjoint support.*

Here are a few features of Wiener random measure. Their proofs are left as exercises.

Theorem 3.10.6 (The Structure of Wiener Random Measures) *(i) The integral extension of β has the property that $h{*}\beta$, $h'{*}\beta$ are independent Wiener processes whenever h, h' are disjoint relatively compact Borel sets. Thus the set function $\nu : B \mapsto [B{*}\beta, B{*}\beta]_1$ is a positive σ-additive measure on $\mathcal{B}^\bullet[\boldsymbol{H}]$, called the **intensity rate of** β. For $h, h' \in L^2(\nu)$, $h{*}\beta$ and $h'{*}\beta$ are Wiener processes with $[h{*}\beta, h'{*}\beta]_t = t \cdot \int h(\eta)h'(\eta)\nu(d\eta)$; if h and h' are orthogonal in the Hilbert space $L^2(\nu)$, then $h{*}\beta, h'{*}\beta$ are independent.*

(ii) The Daniell mean $\|\ \|_{\beta\text{-}2}^*$ and the Hardy mean $\|\ \|_{\beta\text{-}2}^{\mathcal{H}*}$ agree on elementary integrands and therefore on $\check{\mathcal{P}}$. Consequently, $\mathfrak{L}^1[\beta\text{-}2]$ is the Hilbert space $L^2(\nu{\times}\lambda{\times}\mathbb{P})$, and the map $\check{X} \mapsto \int \check{X}\, d\beta$ is an isometry of $\mathfrak{L}^1[\beta\text{-}2]$ onto a closed

[26] As a map into *classes* modulo negligible functions.

subspace of $L^2[\mathbb{P}]$ *(its range is the subspace of all functions with expectation zero; see theorem 3.10.9). For* $\check{X} \in \mathfrak{L}^1[\beta\text{-}2]$, $\int \check{X}\, d\beta$ *is normal with mean zero and standard deviation* $\|\check{X}\|^{\mathcal{H}*}_{\beta\text{-}2}$.

(iii) β *is an* L^p*-random measure for all* $p < \infty$. *It is continuous in the sense that* $\check{X}*\beta$ *has a nearly continuous version, for all* $\check{X} \in \check{\mathcal{P}}_b$. *It is spatially bounded if and only if* $\nu(\boldsymbol{H})$ *is finite. For* $\boldsymbol{H} = \{1, \ldots, d\}$ *and* ν *counting measure,* β *is a standard d-dimensional Wiener process.*

Theorem 3.10.7 (Lévy Characterization of a Wiener Random Measure) *Suppose* β *is a local martingale random measure such that, for every* $h \in \mathcal{E}[\boldsymbol{H}]$, $[h*\beta, h*\beta]_t/t$ *is a constant, and* $[h*\beta, h'*\beta]_t = 0$ *if* $h, h' \in \mathcal{E}[\boldsymbol{H}]$ *have disjoint support. Then* β *is a Wiener random measure whose intensity rate is given by* $\nu(h) = \mathbb{E}[(h*\beta)_1^2] = [h*\beta, h*\beta]_1$.

Theorem 3.10.8 (The Girsanov Theorem for Wiener Random Measure) *Assume the measured filtration* $(\Omega, \mathcal{F}_., \mathbb{P})$ *is full, and* β *is a Wiener random measure on* $(\Omega, \mathcal{F}_., \mathbb{P})$ *with intensity rate* ν. *Suppose* $\check{H}$ *is a predictable random function such that the Doléans–Dade exponential* G' *of* $\check{H}*\beta$ *is a martingale. Then there is a unique* σ*-additive probability* $\mathbb{P}'$ *on* $\mathcal{F}_\infty$ *so that* $\mathbb{P}' = G'_t\mathbb{P}$ *on* $\mathcal{F}_t$ *at all finite instants* t, *and*

$$\beta'(d\eta, ds) \stackrel{\text{def}}{=} \beta(d\eta, ds) + \check{H}_s(\eta)\nu(d\eta)ds$$

is under $\mathbb{P}'$ *a Wiener random measure with intensity rate* ν.

Theorem 3.10.9 (Martingale Representation for Wiener Random Measure) *Every function* $f \in L^2(\mathcal{F}_\infty[\beta], \mathbb{P})$ *is the sum of the constant* $\mathbb{E}f$ *and a random variable of the form* $\int \check{X}\, d\beta$, $\check{X} \in \mathfrak{L}^1[\beta\text{-}2]$. *Thus every square integrable* $(\mathcal{F}_.[\beta], \mathbb{P})$*-martingale* $M_.$ *has the form*

$$M_t = M_0 + \int_0^t \check{X}(\eta, s)\, \beta(d\eta, ds)\,, \qquad \check{X}\cdot[\![0, t]\!] \in \mathfrak{L}^1[\beta\text{-}2]\,,\ t \geq 0\,.$$

For more on this interesting random measure see corollary 4.2.16 on page 219.

Example: The Jump Measure of an Integrator

Given a vector $\boldsymbol{Z} = (Z^1, Z^2, \ldots, Z^d)$ of L^0-integrators let us define, for any function that is measurable on $\mathcal{B}^\bullet(\mathbb{R}^d_*) \times \mathcal{P}$ [27] – a predictable random function – and any stopping time T, the number

$$\int_0^T H_s(\boldsymbol{y};\omega)\, \jmath_{\boldsymbol{Z}}(d\boldsymbol{y}, ds;\omega) \stackrel{\text{def}}{=} \sum_{0\leq s\leq T(\omega)} H_s(\Delta\boldsymbol{Z}_s(\omega);\omega)\,,$$

or, suppressing mention of ω as usual,

$$\int_0^T H_s(\boldsymbol{y})\, \jmath_{\boldsymbol{Z}}(d\boldsymbol{y}, ds) = \sum_{0\leq s\leq T} H_s(\Delta\boldsymbol{Z}_s)\,. \tag{3.10.5}$$

The sum will in general diverge. However, if H is the sure function

$$h_0(\boldsymbol{y}) \stackrel{\text{def}}{=} |\boldsymbol{y}|^2 \wedge 1\,,$$

[27] It is customary to denote by $\mathbb{R}^d_*$ the ***punctured d-space***: $\mathbb{R}^d_* \stackrel{\text{def}}{=} \mathbb{R}^d \setminus \{0\}$. We identify functions on $\mathbb{R}^d_*$ with functions on $\mathbb{R}^d$ that vanish at the origin. The generic point of the auxiliary space $\boldsymbol{H} = \mathbb{R}^d_*$ is denoted by $\boldsymbol{y} = (y^\eta)$ in this subsection.

then the sum will converge absolutely, since

$$|h_0(\Delta \boldsymbol{Z}_s)| \le \sum_{1\le\eta\le d} |\Delta Z^\eta_s|^2 = \sum_\eta \Delta[Z^\eta, Z^\eta]_s .$$

If H is a bounded predictable random function that is majorized in absolute value by a multiple of h_0, then the sum (3.10.5) will exist wherever T is finite. Let us call such a function a ***Hunt function***. Their collection clearly forms a vector lattice and algebra of predictable random functions. The integral notation in equation (3.10.5) is justified by the observation that the map

$$H \mapsto \sum_{0\le s\le t} H_s(\Delta \boldsymbol{Z}_s)$$

is, ω for ω, a positive σ-additive linear functional on the Hunt functions; in fact, it is a sum of point masses supported by the points $(\Delta \boldsymbol{Z}_s, s) \in \mathbb{R}^d_* \times [0,\infty)$ at the countable number of instants s at which the path $\boldsymbol{Z}_.(\omega)$ jumps:

$$\jmath_{\boldsymbol{Z}} = \sum_{\substack{0\le s\le\infty \\ \Delta \boldsymbol{Z}_s \ne 0}} \delta_{(\Delta \boldsymbol{Z}_s, s)} .$$

This ω-dependent measure $\jmath_{\boldsymbol{Z}}$ is called the ***jump measure*** of $\boldsymbol{Z}$. With $\boldsymbol{H} \overset{\text{def}}{=} \mathbb{R}^d_*$, the map $\check{X} \mapsto \int \check{X}(y,s)\, \jmath_{\boldsymbol{Z}}(dy,ds)$ is clearly a random measure. We identify it with $\jmath_{\boldsymbol{Z}}$. It integrates a priori more than the elementary integrands $\check{\mathcal{E}} \overset{\text{def}}{=} C_{00}[\mathbb{R}^d_*] \otimes \mathcal{E}$, to wit, any Hunt function whose carrier is bounded in time. For a Hunt function H the ***indefinite integral*** $H*\jmath_{\boldsymbol{Z}}$ is defined by

$$\left(H*\jmath_{\boldsymbol{Z}}\right)_t = \int_{[\![0,t]\!]} H_s(y)\, \jmath_{\boldsymbol{Z}}(dy,ds) , \quad \text{where } [\![0,t]\!] \overset{\text{def}}{=} \mathbb{R}^d_* \times [\![0,t]\!]$$

is the cartesian product of punctured d-space with the stochastic interval $[\![0,t]\!]$. Clearly $H*\jmath_{\boldsymbol{Z}}$ is an adapted[28] process of finite variation $|H|*\jmath_{\boldsymbol{Z}}$ and has bounded jumps. It is therefore a local L^p-integrator for all $p>0$ (exercise 4.3.8). Indeed, let $S_t = \sum_\eta [Z^\eta, Z^\eta]_t$; at the stopping times $T^n = \inf\{t : S_t \ge n\}$, which tend to infinity, $\left|H*\jmath_{\boldsymbol{Z}}\right|_{T^n} = \left(|H|*\jmath_{\boldsymbol{Z}}\right)_{T^n}$ is bounded by $n + \|H\|_\infty$. This fact explains the prominence of the Hunt functions. For any $q \ge 2$ and all $t<\infty$,

$$\int_{[\![0}^{t]\!]} |y|^q\, \jmath_{\boldsymbol{Z}}(dy,ds) \le \Big(\sum_{1\le\eta\le d} {}^j[Z^\eta, Z^\eta]_t\Big)^{q/2} \le \Big(\sum_{1\le\eta\le d} S_t[Z^\eta]\Big)^q$$

is nearly finite; if the components Z^η of $\boldsymbol{Z}$ are L^q-integrators, then this random variable is evidently integrable. The next result is left as an ecercise.

[28] Extend exercise 1.3.21 (iii) on page 31 slightly so as to cover *random* Hunt functions H.

Proposition 3.10.10 *Formula (3.9.2) can be rewritten in terms of* $\jmath_{\boldsymbol{Z}}$ *as*[15]

$$\Phi(\boldsymbol{Z}_T) = \Phi(\boldsymbol{Z}_0) + \int_{0+}^{T} \Phi_{;\eta}(\boldsymbol{Z}_{.-})\, dZ^\eta + \frac{1}{2}\int_{0+}^{T} \Phi_{;\eta\theta}(\boldsymbol{Z}_{.-})\, d{}^c[Z^\eta, Z^\theta] \tag{3.10.6}$$

$$+ \int_{0}^{T} \Big(\Phi(\boldsymbol{Z}_{s-} + \boldsymbol{y}) - \Phi(\boldsymbol{Z}_{s-}) - \Phi_{;\eta}(\boldsymbol{Z}_{s-}) \cdot y^\eta\Big)\, \jmath_{\boldsymbol{Z}}(d\boldsymbol{y}, ds)$$

$$= \Phi(\boldsymbol{Z}_0) + \int_{0+}^{T} \Phi_{;\eta}(\boldsymbol{Z}_{.-})\, dZ^\eta + \frac{1}{2}\int_{0+}^{T} \Phi_{;\eta\theta}(\boldsymbol{Z}_{.-})\, d[Z^\eta, Z^\theta]$$

$$+ \int_{0+}^{T} R^3_\Phi(\boldsymbol{Z}_{s-}, \boldsymbol{y})\, \jmath_{\boldsymbol{Z}}(d\boldsymbol{y}, ds)\,, \tag{3.10.7}$$

where

$$R^3_\Phi(\boldsymbol{z}, \boldsymbol{y}) = \Phi(\boldsymbol{z} + \boldsymbol{y}) - \Phi(\boldsymbol{z}) - \Phi_{;\eta}(\boldsymbol{z}) y^\eta - \frac{1}{2}\Phi_{;\eta\theta}(\boldsymbol{z}) y^\eta y^\theta$$

$$= \int_0^1 \frac{(1-\lambda)^2}{2} \Phi'_{;\eta\theta\iota}(\boldsymbol{z} + \lambda\boldsymbol{y}) y^\eta y^\theta y^\iota \, d\lambda\,,$$

or, when Φ *is n-times continuously differentiable:*

$$R^3_\Phi(\boldsymbol{z}, \boldsymbol{y}) = \sum_{\nu=3}^{n-1} \frac{1}{\nu!} \Phi_{;\eta_1 \ldots \eta_\nu}(\boldsymbol{z}) y^{\eta_1} \cdots y^{\eta_\nu}$$

$$+ \int_0^1 \frac{(1-\lambda)^{n-1}}{(n-1)!} \Phi_{\eta_1 \ldots \eta_n}(\boldsymbol{z}_1 + \lambda \boldsymbol{y}) y^{\eta_1} \cdots y^{\eta_n}\, d\lambda\,.$$

Exercise 3.10.11 $h_0' : \boldsymbol{y} \mapsto \int_{[|\zeta| \le 1]} \left| e^{i\langle\zeta|\boldsymbol{y}\rangle} - 1 \right|^2 d\boldsymbol{\zeta}$ defines another prototypical sure Hunt function in the sense that h_0'/h_0 is both bounded and bounded away from zero.

Exercise 3.10.12 Let H, H' be previsible Hunt functions and T a stopping time. Then (i)

$$[H{*}\jmath_{\boldsymbol{Z}}, H'{*}\jmath_{\boldsymbol{Z}}] = HH'{*}\jmath_{\boldsymbol{Z}}\,.$$

(ii) For any bounded predictable process X the product XH is a Hunt function and

$$\int_{[\![0,T]\!]} X_s H_s(\boldsymbol{y})\, \jmath_{\boldsymbol{Z}}(d\boldsymbol{y}, ds) = \int_{[\![0,T]\!]} X_s\, d(H{*}\jmath_{\boldsymbol{Z}})_s\,.$$

In fact, this equality holds whenever either side exists.

(iii)
$$\Delta(H{*}\jmath_{\boldsymbol{Z}})_t = H_t(\Delta \boldsymbol{Z}_t)\,, \qquad t \ge 0\,,$$

and
$$(H' * \jmath_{H*\jmath_{\boldsymbol{Z}}})_T = \int_{[\![0,T]\!]} H'_s(H_s(\boldsymbol{y}))\, \jmath_{\boldsymbol{Z}}(d\boldsymbol{y}, ds)$$

as long as merely $|H'_s(\boldsymbol{y})| \le const \cdot |\boldsymbol{y}|$. For any bounded predictable process X

(iv)
$$\int_{[\![0,T]\!]} H_s(\boldsymbol{y})\, \jmath_{X*\boldsymbol{Z}}(d\boldsymbol{y}, ds) = \int_{[\![0,T]\!]} H_s(X_s \cdot \boldsymbol{y})\, \jmath_{\boldsymbol{Z}}(d\boldsymbol{y}, ds)\,,$$

and if $X = X^2$ is a set, then[27] $\jmath_{X*\boldsymbol{Z}} = X \cdot \jmath_{\boldsymbol{Z}}\,.$

Strict Random Measures and Point Processes

The jump measure $\jmath_Z$ of an integrator Z actually is a strict random measure in this sense:

Definition 3.10.13 *Let* $\zeta : \Omega \to \mathfrak{M}^*[\check{H}]$ *be a family of* σ*-additive measures on* $\check{H} \stackrel{\text{def}}{=} H \times [0,\infty)$*, one for every* $\omega \in \Omega$*. If the ordinary integral*

$$\check{X} \mapsto \int_{\check{H}} \check{X}(\eta, s; \omega)\, \zeta(d\eta, ds; \omega)\,, \qquad \check{X} \in \check{\mathcal{E}}\,,$$

computed ω*-by-*ω*, is a random measure, then the linear map of the previous line is identified with* ζ *and is called a* ***strict random measure****.*

These are the random measures treated in [49] and [52]. The Wiener random measure of page 179 is in some sense as far from being strict as one can get. The definitions presented here follow [8]. Kurtz and Protter [60] call our random measures "standard semimartingale random measures" and investigate even more general objects.

Exercise 3.10.14 If ζ is a strict random measure, then $\check{F}*\zeta$ can be computed ω–by–ω when the random function $\check{F} \in \mathfrak{L}^1[\zeta\text{-}p]$ is predictable (meaning that $\check{F}$ belongs to the sequential closure $\check{\mathcal{P}} \stackrel{\text{def}}{=} \check{\mathcal{E}}^\sigma$ of $\check{\mathcal{E}}$, the collection of functions measurable on $\mathcal{B}^*(H) \otimes \mathcal{P}$). There is a nearly empty set outside which all the indefinite integrals (integrators) $\check{F}*\zeta$ can be chosen to be simultaneously càdlàg. Also the maps $\check{\mathcal{P}} \ni \check{X} \mapsto \check{X}*\zeta(\omega)$ are linear at every $\omega \in \Omega$ – not merely as maps from $\check{\mathcal{P}}$ to *classes* of measurable functions.

Exercise 3.10.15 An integrator is a random measure whose auxiliary space is a singleton, but it is a strict random measure only if it has finite variation.

Example 3.10.16 (Sure Random Measures) Let μ be a positive Radon measure on $\check{H} \stackrel{\text{def}}{=} H \times [0,\infty)$. The formula $\zeta(\check{X})(\omega) \stackrel{\text{def}}{=} \int_{\check{H}} \check{X}(\eta, s, \omega)\mu(d\eta, ds)$ defines a simple strict random measure ζ. In particular, when μ is the product of a Radon measure ν on H with Lebesgue measure ds then this reads

$$\zeta(\check{X})(\omega) = \int_0^\infty \int_H \check{X}(\eta, s, \omega)\, \nu(d\eta) ds\,. \tag{3.10.8}$$

Actually, the jump measure $\jmath_Z$ of an integrator Z is even more special. Namely, its value on a set $\check{A} \subset \check{B}$ is an integer, the number of jumps whose size lies in $\check{A}$. More specifically, $\jmath_Z(\cdot\,;\omega)$ is the sum of point masses on $\check{H}$:

Definition 3.10.17 *A positive strict random measure* ζ *is called a* ***point process*** *if* $\zeta(d\check{\eta};\omega)$ *is, for every* $\omega \in \Omega$*, the sum of point masses* $\delta_{\check{\eta}}$*. We call the point process* ζ ***simple*** *if almost surely* $\zeta\big(H \times \{t\}\big) \le 1$ *at all instants t – this means that* $\operatorname{supp}\zeta \cap \big(H \times \{t\}\big)$ *contains at most one point. A simple point process clearly is described entirely by the random point set* $\operatorname{supp}\zeta$*, whence the name.*

Exercise 3.10.18 For a simple point process ζ and $\check{F} \in \check{\mathcal{P}} \cap \mathfrak{L}^1[\zeta\text{-}p]$

$$\Delta(\check{F}*\zeta)_t = \int_{H \times \{t\}} \check{F}(\eta, s)\, \zeta(d\eta, ds) \quad \text{and} \quad [\check{F}*\zeta, \check{F}*\zeta] = \check{F}^2*\zeta\,.$$

Example: Poisson Point Processes

Suppose again that we are given on our separable metrizable locally compact auxiliary space $\boldsymbol{H}$ a positive Radon measure ν. Let $\check{B} \in \mathcal{B}^\bullet(\check{\boldsymbol{H}})$ with $\nu\times\lambda(\check{B}) < \infty$ and set $\mu \overset{\text{def}}{=} \check{B}\cdot(\nu\times\lambda)$. Next let N be a random variable distributed Poisson with mean $|\mu| \overset{\text{def}}{=} \mu(1) = \nu\times\lambda(\check{B})$ and let Y_i, $i = 0, 1, 2, \ldots$, be random variables with values in $\check{\boldsymbol{H}}$ that have distribution $\mu/|\mu|$. They and N are chosen to form an independent family and live on some probability space $(\Omega^\mu, \mathcal{F}^\mu, \mathbb{P}^\mu)$. We use these data to define a point process π^μ as follows: for $\check{F} : \check{\boldsymbol{H}} \to \mathbb{R}$ set

$$\pi^\mu(\check{F}) \overset{\text{def}}{=} \sum_{\nu=0}^{N} \check{F}(Y_\nu) = \sum_{\nu=0}^{N} \delta_{Y_\nu}(\check{F}) . \tag{3.10.9}$$

In other words, pick independently N points from $\check{\boldsymbol{H}}$ according to the distribution $\mu/|\mu|$, and let π^μ be the sum of the δ-masses at these points. To check the distribution of π^μ, let $\check{A}_k \subset \check{\boldsymbol{H}}$ be mutually disjoint and let us show that $\pi^\mu(\check{A}_1), \ldots, \pi^\mu(\check{A}_K)$ are independent and Poisson with means $\mu(\check{A}_1), \ldots, \mu(\check{A}_K)$, respectively. It is convenient to set $\check{A}_0 \overset{\text{def}}{=} \left(\bigcup_{k=1}^K \check{A}_k\right)^c$ and $p_k \overset{\text{def}}{=} \mathbb{P}^\mu[Y_0 \in \check{A}_k] = \mu(\check{A}_k)/|\mu|$. Fix natural numbers $n_0, \ldots, n_K$ and set $n = n_0 + \cdots + n_K$. The event

$$\left[\pi^\mu(\check{A}_0) = n_0, \ldots, \pi^\mu(\check{A}_K) = n_K\right]$$

occurs precisely when of the first n points Y_ν n_0 fall into $\check{A}_0$, n_1 fall into $\check{A}_1$, ..., and n_K fall into $\check{A}_K$, and has (multinomial) probability

$$\binom{n}{n_0 \cdots n_K} p_0^{n_0} \cdots p_K^{n_K}$$

of occurring given that $N = n$. Therefore

$$\begin{aligned}
&\mathbb{P}^\mu\left[\pi^\mu(\check{A}_0) = n_0, \ldots, \pi^\mu(\check{A}_K) = n_K\right] \\
&= \mathbb{P}^\mu\left[N = n, \pi^\mu(\check{A}_0) = n_0, \ldots, \pi^\mu(\check{A}_K) = n_K\right] \\
\text{by independence:}\quad &= e^{-|\mu|}\frac{|\mu|^n}{n!} \cdot \binom{n}{n_0 \cdots n_K} p_0^{n_0} \cdots p_K^{n_K} \\
&= e^{-\mu(\check{A}_0)}\frac{|\mu(\check{A}_0)|^{n_0}}{n_0!} \cdots e^{-\mu(\check{A}_K)}\frac{|\mu(\check{A}_K)|^{n_K}}{n_K!} \\
&= \prod_{k=0}^{K} e^{-\mu(\check{A}_k)}\frac{|\mu(\check{A}_k)|^{n_k}}{n_k!} .
\end{aligned}$$

Summing over n_0 produces

$$\mathbb{P}^\mu\left[\pi^\mu(\check{A}_1) = n_1, \ldots, \pi^\mu(\check{A}_K) = n_K\right] = \prod_{k=1}^{K} e^{-\mu(\check{A}_k)}\frac{|\mu(\check{A}_k)|^{n_k}}{n_k!} ,$$

showing that the random variables $\pi^\mu(\check{A}_1), \ldots, \pi^\mu(\check{A}_K)$ are independent Poisson random variables with means $\mu(\check{A}_1), \ldots, \mu(\check{A}_K)$, respectively.

To finish the construction we cover $\check{\boldsymbol{H}}$ by countably many mutually disjoint relatively compact Borel sets B^k, set $\mu^k \stackrel{\text{def}}{=} B^k(\nu\times\lambda)$, and denote by π^k the corresponding Poisson random measures just constructed, which live on probability spaces $(\Omega^k, \mathcal{F}^k, \mathbb{P}^k)$. Then we equip the cartesian product $\Omega \stackrel{\text{def}}{=} \prod_k \Omega^k$ with the product σ-algebra $\mathcal{F} \stackrel{\text{def}}{=} \bigotimes_k \mathcal{F}^k$, on which the natural probability is of course the product $\mathbb{P} \stackrel{\text{def}}{=} \prod_k \mathbb{P}^k$. It is left as an exercise in bookkeeping to show that $\pi \stackrel{\text{def}}{=} \sum_k \pi^k$ meets the following description:

Definition 3.10.19 *A point process π with auxiliary space $\boldsymbol{H}$ is called a **Poisson point process** if, for any two disjoint relatively compact Borel sets $B, B' \subset \boldsymbol{H}$, the processes $B{*}\pi, B'{*}\pi$ are independent and Poisson.*

Theorem 3.10.20 (Structure of Poisson Point Processes) $\nu : B \mapsto \mathbb{E}[(B{*}\pi)_1]$ *is a positive σ-additive measure on $\mathcal{B}^\bullet[\boldsymbol{H}]$, called the **intensity rate**. Whenever $h \in L^1(\nu)$, the indefinite integral $h{*}\pi$ is a process with independent stationary increments, is an L^p-integrator for all $p > 0$, and has square bracket $[h{*}\pi, h{*}\pi] = h^2{*}\pi$. If $h, h' \in L^1(\nu)$ have disjoint carriers, then $h{*}\pi, h'{*}\pi$ are independent. Furthermore,*

$$\widehat{\pi}(\check{X}) \stackrel{\text{def}}{=} \int \check{X}_s(\eta)\, \nu(d\eta)ds$$

*defines a strict random measure $\widehat{\pi}$, called the **compensator** of π. Also, $\widetilde{\pi} \stackrel{\text{def}}{=} \pi - \widehat{\pi}$ is a strict martingale random measure, called **compensated Poisson point process**. The $\pi, \widehat{\pi}, \widetilde{\pi}$ are L^p-random measures for all $p > 0$.*

The Girsanov Theorem for Poisson Point Processes

Let π be a Poisson point process with intensity rate ν on $\boldsymbol{H}$ and intensity $\widehat{\pi} = \nu \times \lambda$ on $\check{\boldsymbol{H}} \stackrel{\text{def}}{=} \boldsymbol{H} \times [0, \infty)$. A ***predictable transformation*** of $\boldsymbol{H}$ is a map $\Gamma : \check{\boldsymbol{B}} \to \check{\boldsymbol{B}}$ of the form

$$(\eta, s, \omega) \mapsto (\gamma(\eta, s; \omega), s, \omega) ,$$

where $\gamma : \check{\boldsymbol{B}} \to \boldsymbol{H}$ is predictable, i.e., $\check{\mathcal{P}}/\mathcal{B}^\bullet(\boldsymbol{H})$-measurable. Then Γ is clearly $\check{\mathcal{P}}/\check{\mathcal{P}}$-measurable. Let us fix such Γ, and assume the following:

(i) The given measured filtration $(\mathcal{F}_., \mathbb{P})$ is full (see definition 3.9.16).

(ii) Γ is invertible and Γ^{-1} is $\check{\mathcal{P}}/\check{\mathcal{P}}$-measurable.

(iii) $\gamma[\nu] \ll \nu$, with bounded Radon–Nikodym derivative $\check{D} \stackrel{\text{def}}{=} \dfrac{d\gamma[\nu]}{d\nu} \in \check{\mathcal{P}}$.

(iv) $\check{Y} \stackrel{\text{def}}{=} \check{D} - 1$ is a "Hunt function:" $\sup_{s,\omega} \int \check{Y}^2(\eta, s, \omega)\, \nu(d\eta) < \infty$.

Then $M \stackrel{\text{def}}{=} \check{Y}{*}\widetilde{\pi}$ is a martingale, and is a local L^p-integrator for all $p > 0$ on the grounds that its jumps are bounded (corollary 4.4.3). Consider the stochastic exponential $G' \stackrel{\text{def}}{=} 1 + G'_{\!-}{*}M = 1 + (G'_{\!-}\check{Y}){*}\widetilde{\pi}$ of M. Since $\Delta M \geq -1$, we have $G' \geq 0$.

$$\begin{aligned}\text{Now}\quad \mathbb{E}\big[[G', G']_t\big] &= \mathbb{E}\Big[1 + \big(G'^2_{\!-}{*}[M, M]\big)_t\Big]\\ &= \mathbb{E}\Big[1 + \big((G'_{\!-}\check{Y})^2{*}\pi\big)_t\Big] = \mathbb{E}\Big[1 + \big((G'_{\!-}\check{Y})^2{*}\widehat{\pi}\big)_t\Big]\\ &\leq \mathbb{E}\Big[1 + \int_0^t G'^2_{\!-}(s) \int_H \check{Y}^2(\eta, s)\, \nu(d\eta)\, ds\Big],\end{aligned}$$

$$\text{and so}\qquad \mathbb{E}\big[G'^{\star 2}_t\big] \leq const\Big(1 + const \int_0^t \mathbb{E}\big[G'^{\star 2}_s\big]\, ds\Big).$$

By Gronwall's lemma A.2.35, G' is a square integrable martingale, and the fullness provides a probability $\mathbb{P}'$ on $\mathcal{F}_\infty$ whose restriction to the $\mathcal{F}_t$ is $G'_t\mathbb{P}$.

Let us now compute the compensator $\widehat{\pi}'$ of π with respect to $\mathbb{P}'$. For $\check{H} \in \check{\mathcal{P}}_b$ vanishing after t we have

$$\begin{aligned}\mathbb{E}'\big[(\check{H}{*}\widehat{\pi}')_t\big] &= \mathbb{E}'\big[(\check{H}{*}\pi)_t\big] = \mathbb{E}\big[G'_t \cdot (\check{H}{*}\pi)_t\big]\\ &= \mathbb{E}\Big[\big((G'_{\!-}\check{H}){*}\pi\big)_t\Big] + \mathbb{E}\Big[\big((\check{H}{*}\pi)_{-}{*}G'\big)_t\Big] + \mathbb{E}\Big[[G', \check{H}{*}\pi]_t\Big]\\ &= \mathbb{E}\Big[\big((G'_{\!-}\check{H}){*}\widehat{\pi}\big)_t\Big] + 0 + \mathbb{E}\Big[G'_{\!-}{*}[\check{Y}{*}\widetilde{\pi}, \check{H}{*}\widetilde{\pi}]_t\Big]\\ \text{by 3.10.18:}\quad &= \mathbb{E}\Big[\big((G'_{\!-}\check{H}){*}\widehat{\pi}\big)_t\Big] + \mathbb{E}\Big[\big(G'_{\!-}\check{Y}\check{H}{*}\pi\big)_t\Big]\\ \text{as } 1+\check{Y} = \check{D}:\quad &= \mathbb{E}\Big[\big(G'_{\!-}\check{D}\check{H}{*}\widehat{\pi}\big)_t\Big] = \mathbb{E}\Big[\big(G'_{\!-}{*}(\check{D}\check{H}{*}\widehat{\pi})\big)_t\Big]\\ &= \mathbb{E}\Big[G'_t \cdot (\check{D}\check{H}{*}\widehat{\pi})_t\Big] = \mathbb{E}'\Big[(\check{D}\check{H}{*}\widehat{\pi})_t\Big];\end{aligned}$$

$$\text{so}\quad \mathbb{E}'\big[(\check{H}{*}\widehat{\pi}')_t\big] = \mathbb{E}'\Big[\big(\check{H}{*}(\check{D}\widehat{\pi})\big)_t\Big].$$

$$\text{Therefore}\qquad \widehat{\pi}' = \check{D}\widehat{\pi} = \Gamma[\widehat{\pi}]\ \ \text{, i.e.,}\ \ \widehat{\Gamma^{-1}[\pi]}' = \Gamma^{-1}[\widehat{\pi}'] = \widehat{\pi}.$$

In other words, replacing $\check{H}$ by $\check{H} \circ \Gamma^{-1} \in \check{\mathcal{P}}$ gives

$$\begin{aligned}\mathbb{E}'\big[(\check{H}{*}\Gamma^{-1}[\pi])_t\big] &= \mathbb{E}'\Big[\big((\check{H} \circ \Gamma^{-1}){*}(\check{D}\widehat{\pi})\big)_t\Big] = \mathbb{E}'\Big[\big(\check{H}{*}\Gamma^{-1}[\check{D}\widehat{\pi}]\big)_t\Big]\\ \text{as } \Gamma[\widehat{\pi}] = \check{D}\widehat{\pi}:\quad &= \mathbb{E}'\Big[\big(\check{H}{*}\widehat{\pi}\big)_t\Big]\ \ \text{, therefore}\end{aligned}$$

Theorem 3.10.21 *Under the assumptions above the "shifted Poisson point process" $\Gamma^{-1}[\pi]$ is a Poisson point process with respect to $\mathbb{P}'$, of the same intensity rate ν that π had under $\mathbb{P}$. Consequently, the law of $\Gamma^{-1}[\pi]$ under $\mathbb{P}'$ agrees with the law of π under $\mathbb{P}$.*

Repeated Footnotes: 95[1] 110[4] 116[6] 125[7] 139[8] 158[14] 161[15] 164[16] 173[21] 178[25] 180[27]

4

Control of Integral and Integrator

4.1 Change of Measure — Factorization

Let Z be a global $L^p(\mathbb{P})$-integrator and $0 \le p < q < \infty$. There is a probability $\mathbb{P}'$ equivalent with $\mathbb{P}$ such that Z is a global $L^q(\mathbb{P}')$-integrator; moreover, there is sufficient control over the change of measure from $\mathbb{P}$ to $\mathbb{P}'$ to turn estimates with respect to $\mathbb{P}'$ into estimates with respect to the original and presumably intrinsically relevant probability $\mathbb{P}$. In fact, all of this remains true for a whole vector $\boldsymbol{Z}$ of L^p-integrators. This is of great practical interest, since it is so much easier to compute and estimate in Hilbert space $L^2(\mathbb{P}')$, say, than in $L^p(\mathbb{P})$, which is not even locally convex when $0 \le p < 1$.

When $q \le 2$ or when $\boldsymbol{Z}$ is previsible, the universal constants that govern the change of measure are independent of the length of $\boldsymbol{Z}$, and that fact permits an easy extension of all of this to *random measures* (see corollary 4.1.14).

These facts are the goal of the present section.

A Simple Case

Here is a result that goes some way in this direction and is rather easily established (pages 188–190). It is due to Dellacherie [18] and the author [6], and, in conjunction with the Doob–Meyer decomposition of section 4.3 and the Girsanov–Meyer lemma 3.9.11, suffices to show that an L^0-integrator is a semimartingale (proposition 4.4.1).

Proposition 4.1.1 *Let Z be a global L^0-integrator on $(\Omega, \mathcal{F}_., \mathbb{P})$. There exists a probability $\mathbb{P}'$ equivalent with $\mathbb{P}$ on $\mathcal{F}_\infty$ such that Z is a global $L^1(\mathbb{P}')$-integrator. Moreover, let $\alpha_0 \in (0,1)$ have $\lceil Z \rceil_{[\alpha_0]} > 0$; the law $\mathbb{P}'$ can be chosen so that*

$$\lceil Z \rceil_{\mathcal{I}^1[\mathbb{P}']} \le \frac{3 \lceil Z \rceil_{[\alpha_0/2]}}{1-\alpha_0}$$

and so that the Radon–Nikodym derivative $g \stackrel{\text{def}}{=} d\mathbb{P}/d\mathbb{P}'$ satisfies

$$\|g\|_{[\alpha;\mathbb{P}]} \le \frac{16\,\alpha_0 \cdot \lceil Z \rceil_{[\alpha/8]}}{\alpha \cdot \lceil Z \rceil_{[\alpha_0/2]}} \quad \forall \alpha > 0\,, \tag{4.1.1}$$

$$\textit{which implies} \quad \|f\|_{[\alpha;\mathbb{P}]} \le \left(\frac{32\,\alpha_0 \cdot ⦀Z⦀_{[\alpha/16]}}{⦀Z⦀_{[\alpha_0/2]} \cdot \alpha^2} \right)^{1/r} \cdot \|f\|_{L^r(\mathbb{P}')} \tag{4.1.2}$$

for any $\alpha \in (0,1)$, $r \in (0,\infty)$, *and* $f \in \mathcal{F}_\infty$.

A remark about the utility of inequality (4.1.2) is in order. To fix ideas assume that f is a function computed from Z, for instance, the value at some time T of the solution of a stochastic differential equation driven by Z. First, it is rather easier to establish the existence and possibly uniqueness of the solution computing in the Banach space $L^1(\mathbb{P}')$ than in $L^0(\mathbb{P})$ – but generally still not as easy as in Hilbert space $L^2(\mathbb{P}')$. Second, it is generally very much easier to estimate the size of f in $L^r(\mathbb{P}')$ for $r > 1$, where Hölder and Minkowski inequalities are available, than in the non-locally convex space $L^0(\mathbb{P})$. Yet it is the original measure $\mathbb{P}$, which presumably models a physical or economical system and reflects the "true" probability of events, with respect to which one wants to obtain a relevant estimate of the size of f. Inequality (4.1.2) does that.

Apart from elevating the exponent from 0 to merely 1, there is another shortcoming of proposition 4.1.1. While it is quite easy to extend it to cover several integrators simultaneously, the constants of inequality (4.1.1) and (4.1.2) will increase linearly with their number. This prevents an application to a random measure, which can be viewed as an infinity of infinitesimal integrators (page 173). The comost general theorem, which overcomes these problems and is in some sense best possible, is theorem 4.1.2 below.

Proof of Proposition 4.1.1. This result follows from part (ii) of theorem 4.1.2, whose detailed proof takes 20 pages. The reader not daunted by the prospect of wading through them might still wish to read the following short proof of proposition 4.1.1, since it shares the strategy and major elements with the proof of theorem 4.1.2 and yields in its less general setup better constants.

The first step is the following claim: *For every* α *in* $(0,1)$ *there exists a function* k_α

$$\textit{with} \qquad 0 \le k_\alpha \le 1 \;\; \textit{and} \;\; \mathbb{E}\big[k_\alpha\big] \ge 1-\alpha$$

and such that the measure $\mu_\alpha = k_\alpha \cdot \mathbb{P}$ *satisfies*

$$\mathbb{E}^{\mu_\alpha}\Big[\Big|\int X\,dZ\Big|\Big] \le 3\,⦀Z⦀_{[\alpha/2]} \cdot \|X\|_{\mathcal{E}} \tag{4.1.3}$$

for $X \in \mathcal{E}$. Here $\mathbb{E}^{\mu_\alpha}[f] \stackrel{\text{def}}{=} \int f\,d\mu_\alpha$, of course. To see this fix an α in $(0,1)$ and set $T \stackrel{\text{def}}{=} \inf\{t : |Z_t| > ⦀Z⦀_{[\alpha/2]}\}$. Now $\mathbb{P}\Big[\Big|\int [\![0,T]\!]\,dZ\Big| \ge ⦀Z⦀_{[\alpha/2]}\Big] \le \alpha/2$ means that $\mathbb{P}\Big[\big|Z_T\big| \ge ⦀Z⦀_{[\alpha/2]}\Big] \le \alpha/2$ and produces

$$\mathbb{P}[T<\infty] \le \mathbb{P}\Big[\big|Z_T\big| \ge ⦀Z⦀_{[\alpha/2]}\Big] \le \alpha/2\,.$$

The complement $G \stackrel{\text{def}}{=} [T = \infty] \subset [Z^\star_\infty \le ⦚Z⦚_{[\alpha/2]}]$ has $\mathbb{P}[G] > 1 - \alpha/2$. Consider now the collection K of measurable functions k with $0 \le k \le G$ and $\mathbb{E}[k] \ge 1 - \alpha$. K is clearly a convex and weak*-compact subset of $L^\infty(\mathbb{P})$ (see A.2.32). As it contains G, it is not void. For every X in the unit ball $\mathcal{E}_1$ of $\mathcal{E}$ define a function h_X on K by

$$h_X(k) \stackrel{\text{def}}{=} ⦚Z⦚_{[\alpha/2]} - \mathbb{E}\Big[\int X\,dZ \cdot k\Big], \qquad k \in K .$$

Since, on G, $\int X\ dZ$ is a finite linear combination of bounded random variables, h_X is well-defined and real-valued. Every one of the functions h_X is evidently linear and continuous on K, and is non-negative at some point of K, to wit, at the set

$$k_X \stackrel{\text{def}}{=} G \cap \Big[\Big|\int X\,dZ\Big| \le ⦚Z⦚_{[\alpha/2]}\Big] .$$

Indeed, $\displaystyle h_X(k_X) = ⦚Z⦚_{[\alpha/2]} - \mathbb{E}\Big[\int X\,dZ \cdot G \cap \Big[\Big|\int X\,dZ\Big| \le ⦚Z⦚_{[\alpha/2]}\Big]\Big]$

$$\ge ⦚Z⦚_{[\alpha/2]} - \mathbb{E}\Big[\Big|\int X\,dZ\Big| \cdot \Big[\Big|\int X\,dZ\Big| \le ⦚Z⦚_{[\alpha/2]}\Big]\Big] \ge 0 ;$$

and, since $\displaystyle \mathbb{E}[k_X] = \mathbb{P}\Big[G \cap \Big[\Big|\int X\ dZ\Big| \le ⦚Z⦚_{[\alpha/2]}\Big]\Big]$

$$\ge 1 - \alpha/2 - \mathbb{P}\Big[\Big|\int X\ dZ\Big| > ⦚Z⦚_{[\alpha/2]}\Big] \ge 1 - \alpha ,$$

k_X belongs to K. The collection $\mathcal{H} \stackrel{\text{def}}{=} \{h_X : X \in \mathcal{E}_1\}$ is easily seen to be convex; indeed, $sh_X + (1-s)h_Y = h_{sX+(1-s)Y}$ for $0 \le s \le 1$. Thus Ky–Fan's minimax theorem A.2.34 applies and provides a common point $k_\alpha \in K$ at which every one of these functions is non-negative. This says that

$$\mathbb{E}^{\mu_\alpha}\Big[\int X\ dZ\Big] = \mathbb{E}\Big[k_\alpha \cdot \int X\ dZ\Big] \le ⦚Z⦚_{[\alpha/2]} \qquad \forall\, X \in \mathcal{E}_1 .$$

Note the lack of the absolute-sign under the expectation, which distinguishes this from (4.1.3). Since $|Z|$ is μ_α-a.s. bounded by $⦚Z⦚_{[\alpha/2]}$, though, part (i) of lemma 2.5.27 on page 80 applies and produces

$$\mathbb{E}^{\mu_\alpha}\Big[\Big|\int X\,dZ\Big|\Big] \le \sqrt{2}\Big(⦚Z⦚_{[\alpha/2]} + ⦚Z⦚_{[\alpha/2]}\Big) \cdot \|X\|_{\mathcal{E}} \le 3\,⦚Z⦚_{[\alpha/2]} \cdot \|X\|_{\mathcal{E}}$$

for all $X \in \mathcal{E}$, which is the desired inequality (4.1.3).

Now to the construction of $\mathbb{P}' = g'\mathbb{P}$. We pick an $\alpha_0 \in (0,1)$ with $\lceil Z \rceil_{[\alpha_0]} > 0$ and set $\alpha_n \stackrel{\text{def}}{=} \alpha_0/2^n$ and $\zeta_n \stackrel{\text{def}}{=} 3 \lceil Z \rceil_{[\alpha_n/2]}$ for $n \in \mathbb{N}$. Since $\mathbb{P}[k_\alpha = 0] \le \alpha$, the bounded function

$$g' \stackrel{\text{def}}{=} \gamma' \cdot \sum_{n=0}^{\infty} 2^{-n} . \frac{k_{\alpha_n}}{\zeta_n} \tag{4.1.4}$$

is $\mathbb{P}$-a.s. strictly positive and bounded, and with the proper choice of

$$\gamma' \in \left(\frac{\zeta_0}{2} , \frac{\zeta_0}{1-\alpha_0} \right)$$

it can be made to have $\mathbb{P}$-expectation one. The measure $\mathbb{P}' \stackrel{\text{def}}{=} g' \cdot \mathbb{P}$ is then a probability equivalent with $\mathbb{P}$. Let $\mathbb{E}'$ denote the expectation with respect to $\mathbb{P}'$. Inequality (4.1.3) implies that for every $X \in \mathcal{E}_1$

$$\mathbb{E}'\Big[\Big| \int X \, dZ \Big|\Big] \le \gamma' \le \frac{3 \lceil Z \rceil_{[\alpha_0/2]}}{1-\alpha_0} .$$

That is to say, Z is a global $L^1(\mathbb{P}')$-integrator of the size claimed. Towards the estimate (4.1.1) note that for $\alpha, \lambda \in (0,1)$

$$\mathbb{P}[k_\alpha \le \lambda] = \mathbb{P}[1 - k_\alpha \ge 1 - \lambda] \le \frac{\mathbb{E}[1-k_\alpha]}{1-\lambda} \le \frac{\alpha}{1-\lambda} ,$$

and thus $\quad \mathbb{P}[g \ge C] = \mathbb{P}[g' \le 1/C] \le \mathbb{P}\Big[\sum_{n=0}^{\infty} \frac{2^{-n} \, k_{\alpha_n}}{\zeta_n} \le \frac{1}{C\,\gamma'}\Big]$

for every single $n \in \mathbb{N}$: $\quad \le \mathbb{P}\Big[k_{\alpha_n} \le \frac{2^n \, \zeta_n}{C\,\gamma'}\Big] \le \mathbb{P}\Big[k_{\alpha_n} \le \frac{2^{n+1} \, \zeta_n}{C\,\zeta_0}\Big]$

$$\le \alpha_n \Big/ \left(1 - \frac{2^{n+1}\,\zeta_n}{C\,\zeta_0}\right) .$$

Given $\alpha \in (0,1)$, we choose n so that $\alpha/4 < \alpha_n \le \alpha/2$ and pick for C any number exceeding $16\alpha_0\zeta_n/\alpha\zeta_0$, for instance,

$$C \stackrel{\text{def}}{=} \frac{16\,\alpha_0 \cdot \lceil Z \rceil_{[\alpha/8]}}{\alpha \cdot \lceil Z \rceil_{[\alpha_0/2]}} .$$

Then $\quad \frac{2^{n+1}\,\zeta_n}{C\,\zeta_0} \le \frac{2^{n+1}\zeta_n\alpha\zeta_0}{16\alpha_0\zeta_n\zeta_0} = \frac{\alpha}{8\alpha_n} \le \frac{1}{2}$

and therefore $\quad \mathbb{P}[g \ge C] \le 2\alpha_n \le \alpha ,$

which is inequality (4.1.1). The last inequality (4.1.2) is a simple application of exercise A.8.17. ∎

The Main Factorization Theorem

Theorem 4.1.2 *(i) Let* $0 < p < q < \infty$ *and* Z *a d-tuple of global* $L^p(\mathbb{P})$*-integrators. There exists a probability* $\mathbb{P}'$ *equivalent with* $\mathbb{P}$ *on* $\mathcal{F}_\infty$ *with respect to which* Z *is a global* L^q*-integrator; furthermore* $d\mathbb{P}'/d\mathbb{P}$ *is bounded, and there exist universal constants* $D = D_{p,q,d}$ *and* $E = E_{p,q}$ *depending only on the subscripted quantities*

such that $$\lceil Z \rceil_{\mathcal{I}^q[\mathbb{P}']} \le D_{p,q,d} \cdot \lceil Z \rceil_{\mathcal{I}^p[\mathbb{P}]}, \tag{4.1.5}$$

and such that the Radon–Nikodym derivative $g \stackrel{\text{def}}{=} d\mathbb{P}/d\mathbb{P}'$ *satisfies*

$$\|g\|_{L^{p/(q-p)}(\mathbb{P})} \le E_{p,q} \tag{4.1.6}$$

– *this inequality has the consequence that for any* $r > 0$ *and* $f \in \mathcal{F}_\infty$

$$\|f\|_{L^r(\mathbb{P})} \le E_{p,q}^{p/qr} \cdot \|f\|_{L^{rq/p}(\mathbb{P}')} . \tag{4.1.7}$$

If $0 < p < q \le 2$ *or if* Z *is previsible, then* D *does not depend on* d.

(ii) Let $p = 0 < q < \infty$, *and let* Z *be a d-tuple of global* $L^0(\mathbb{P})$*-integrators with modulus of continuity*[1] $\lceil Z \rceil_{[.]}$. *There exists a probability* $\mathbb{P}' = \mathbb{P}/g$ *equivalent with* $\mathbb{P}$ *on* $\mathcal{F}_\infty$, *with respect to which* Z *is a global* L^q*-integrator; furthermore,* $d\mathbb{P}'/d\mathbb{P}$ *is bounded and there exist universal constants* $D = D_{q,d}[\lceil Z \rceil_{[.]}]$ *and* $E = E_{[\alpha],q}[\lceil Z \rceil_{[.]}]$, *depending only on* $q, d, \alpha \in (0,1)$ *and the modulus of continuity* $\lceil Z \rceil_{[.]}$,

such that $$\lceil Z \rceil_{\mathcal{I}^q[\mathbb{P}']} \le D_{q,d}[\lceil Z \rceil_{[.]}] , \tag{4.1.8}$$

and $$\|g\|_{[\alpha]} \le E_{[\alpha],q}[\lceil Z \rceil_{[.]}] \quad \forall \alpha \in (0,1) \tag{4.1.9}$$

– *this implies* $$\|f\|_{[\alpha+\beta;\mathbb{P}]} \le \Big(E_{[\alpha],q}[\lceil Z \rceil_{[.]}]/\beta\Big)^{1/r} \cdot \|f\|_{L^r(\mathbb{P}')} \tag{4.1.10}$$

for any $f \in \mathcal{F}_\infty$, $r > 0$, *and* $\alpha, \beta \in (0,1)$. *Again, in the range* $q \le 2$ *or when* Z *is previsible the constant* D *does not depend on* d.

Estimates independent of the length d of Z are used in the control of random measures – see corollary 4.1.14 and theorem 4.5.25.

The proof of theorem 4.1.2 varies with the range of p and of $q > p$, and will provide various estimates[2] for the constants D and E. The implication (4.1.6) $\Longrightarrow$ (4.1.7) results from a straightforward application of Hölder's inequality and is left to the reader:

Exercise 4.1.3 (i) Let μ be a positive σ-additive measure and $p < q < \infty$. The condition $1/g \le C$ has the effect that $\|f\|_{L^r(\mu/g)} \le C^{1/r}\|f\|_{L^r(\mu)}$ for all

---

[1] $\lceil Z \rceil_{[\alpha]} \stackrel{\text{def}}{=} \sup\{\|\int X\,dZ\|_{[\alpha;\mathbb{P}]} : X \in \mathcal{E}_1^d\}$ for $0 < \alpha < 1$; see page 56.

[2] See inequalities (4.1.6), (4.1.34), (4.1.35), (4.1.40), and (4.1.41).

measurable functions f and all $r > 0$. The condition $\|g\|_{L^{p/(q-p)}(\mu)} \le c$ has the effect that for all measurable functions f that vanish on $[g = 0]$ and all $r > 0$

$$\|f\|_{L^r(\mu)} \le c^{1/r} \cdot \|f\|_{L^{rq/p}(d\mu/g)} .$$

(ii) In the same vein prove that (4.1.9) implies (4.1.10).
(iii) $\mathfrak{L}^1[Z\text{–}q; \mathbb{P}'] \subset \mathfrak{L}^1[Z\text{–}p; \mathbb{P}]$, the injection being continuous.

The remainder of this section, which ends on page 209, is devoted to a detailed proof of this theorem. For both parts (i) and (ii) we shall employ several times the following

Criterion 4.1.4 (Rosenthal) *Let E be a normed linear space with norm $\|\ \|_E$, μ a positive σ-finite measure, $0 < p < q < \infty$, and $\mathcal{I} : E \to L^p(\mu)$ a linear map. For any constant $C > 0$ the following are equivalent:*

(i) There exists a measurable function $g \ge 0$ with $\|g\|_{L^{p/(q-p)}(\mu)} \le 1$ such that for all $x \in E$

$$\Big(\int |\mathcal{I}x|^q \, \frac{d\mu}{g}\Big)^{1/q} \le C \cdot \|x\|_E . \tag{4.1.11}$$

(ii) For any finite collection $\{x_1, \ldots, x_n\} \subset E$

$$\Big\| \Big(\sum_{\nu=1}^n |\mathcal{I}x_\nu|^q\Big)^{1/q} \Big\|_{L^p(\mu)} \le C \cdot \Big(\sum_{\nu=1}^n \|x_\nu\|_E^q\Big)^{1/q} . \tag{4.1.12}$$

(iii) For every measure space $(T, \mathcal{T}, \tau \ge 0)$ and q-integrable $f : T \to E$

$$\Big\| \|\mathcal{I}f\|_{L^q(\tau)} \Big\|_{L^p(\mu)} \le C \cdot \Big\| \|f\|_E \Big\|_{L^q(\tau)} . \tag{4.1.13}$$

The smallest constant C satisfying any and then all of (4.1.11), (4.1.12), and (4.1.13) is the ***p–q-factorization constant*** of $\mathcal{I}$ and will be denoted by

$$\eta_{p,q}(\mathcal{I}) .$$

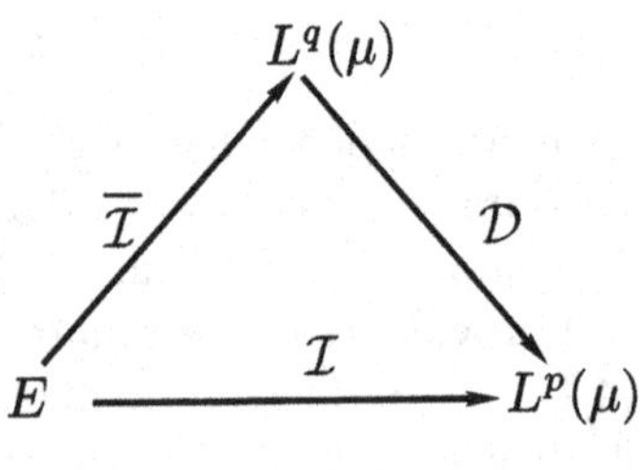

It may well be infinite, of course. Its name comes from the following way of looking at (i): the map $\mathcal{I}$ has been "factored as" $\mathcal{I} = \mathcal{D} \circ \overline{\mathcal{I}}$, where $\overline{\mathcal{I}} : E \to L^q(\mu)$ is defined by $\overline{\mathcal{I}}(x) = \mathcal{I}(x) \cdot g^{-1/q}$ and $\mathcal{D} : L^q(\mu) \to L^p(\mu)$ is the "diagonal map" $f \mapsto f \cdot g^{1/q}$. The number $\eta_{p,q}(\mathcal{I})$ is simply the operator (quasi)norm of $\overline{\mathcal{I}}$ – the operator (quasi)norm of $\mathcal{D}$ is $\|g\|_{L^{p/(q-p)}(\mu)}^{1/q} \le 1$. Thus, if $\eta_{p,q}(\mathcal{I})$ is finite, we also say that $\mathcal{I}$ ***factorizes through L^q***.

We are of course primarily interested in the case when $\mathcal{I}$ is the stochastic integral $\boldsymbol{X} \mapsto \int \boldsymbol{X}\, d\boldsymbol{Z}$, and the question arises whether μ/g is a probability when μ is. It won't be automatically but can be made into one:

Exercise 4.1.5 Assume in criterion 4.1.4 that μ is a *probability* $\mathbb{P}$ and $\eta_{p,q}(\mathcal{I}) < \infty$. Then there is a *probability* $\mathbb{P}'$ equivalent with $\mathbb{P}$ such that $\mathcal{I}$ is continuous as a map into $L^q(\mathbb{P}')$:

$$\|\mathcal{I}\|_{q;\mathbb{P}'} \le \eta_{p,q}(\mathcal{I}) ,$$

and such that $g' \stackrel{\text{def}}{=} d\mathbb{P}'/d\mathbb{P}$ is bounded and $g \stackrel{\text{def}}{=} d\mathbb{P}/d\mathbb{P}' = g'^{-1}$ satisfies

$$\|g\|_{L^{p/(q-p)}(\mathbb{P})} \le 2^{(p\vee(q-p))/p} , \tag{4.1.14}$$

$$\text{and therefore} \quad \|f\|_{L^r(\mathbb{P})} \le 2^{(p\vee(q-p))/rq}\|f\|_{L^{rq/p}(\mathbb{P}')} \le 2^{1/r}\|f\|_{L^{rq/p}(\mathbb{P}')} \tag{4.1.15}$$

for all measurable functions f and exponents $r > 0$.

Exercise 4.1.6 (i) $\eta_{p,q}(\mathcal{I})$ depends isotonically on q. (ii) For any two maps $\mathcal{I}, \mathcal{I}' : E \to L^p(\mu)$, we have $\eta_{p,q}(\mathcal{I}+\mathcal{I}') \le \eta_{p,q}(\mathcal{I}) + \eta_{p,q}(\mathcal{I}')$.

Proof of Criterion 4.1.4. If (i) holds, then $\int |\mathcal{I}x|^q \frac{d\mu}{g} \le C^q \cdot \|x\|_E^q$ for all $x \in E$, and consequently for any finite subcollection $\{x_1, \ldots, x_n\}$ of E,

$$\sum_{\nu=1}^{n} \int |\mathcal{I}x_\nu|^q \,\frac{d\mu}{g} \le C^q \cdot \sum_{\nu=1}^{n} \|x_\nu\|_E^q$$

$$\text{and} \qquad \Big\|\Big(\sum_{\nu=1}^{n} |\mathcal{I}x_\nu|^q\Big)^{1/q}\Big\|_{L^q(d\mu/g)} \le C \cdot \Big(\sum_{\nu=1}^{n} \|x_\nu\|_E^q\Big)^{1/q} .$$

Inequality (4.1.11) implies that $\mathcal{I}x$ vanishes μ-almost surely on $[g = 0]$, so exercise 4.1.3 applies with $r = p$ and $c = 1$, giving

$$\Big\|\Big(\sum_{\nu=1}^{n} |\mathcal{I}x_\nu|^q\Big)^{1/q}\Big\|_{L^p(\mu)} \le \Big\|\Big(\sum_{\nu=1}^{n} |\mathcal{I}x_\nu|^q\Big)^{1/q}\Big\|_{L^q(d\mu/g)} .$$

This together with the previous inequality results in (4.1.12).

The reverse implication (ii)⇒(i) is a bit more difficult to prove. To start with, consider the following collection of measurable functions:

$$K = \Big\{ k \ge 0 : \ \|k\|_{L^{q/(q-p)}(\mu)} \le 1 \Big\} .$$

Since $1 < q/(q-p) < \infty$, this convex set is weakly compact. Next let us define a host $\mathcal{H}$ of numerical functions on K, one for every finite collection $\{x_1, \ldots, x_n\} \subset E$, by

$$k \mapsto h_{x_1,\ldots,x_n}(k) \stackrel{\text{def}}{=} C^q \cdot \sum_{\nu=1}^{n} \|x_\nu\|_E^q - \int^* \sum_{\nu=1}^{n} |\mathcal{I}x_\nu|^q \cdot \frac{1}{k^{q/p}} \, d\mu . \tag{$*$}$$

The idea is to show that there is a point $k \in K$ at which every one of these functions is non-negative. Given that, we set $g = k^{q/p}$ and are done: $\|k\|_{L^{q/(q-p)}(\mu)} \le 1$ translates into $\|g\|_{L^{p/(q-p)}(\mu)} \le 1$, and $h_x(k) \ge 0$ is inequality (4.1.11). To prove the existence of the common point k of positivity we start with a few observations.

a) An $h = h_{x_1,\ldots,x_n} \in \mathcal{H}$ may take the value $-\infty$ on K, but never $+\infty$.
b) Every function $h \in \mathcal{H}$ is concave – simply observe the minus sign in front of the integral in $(*)$ and note that $k \mapsto 1/k^{q/p}$ is convex.
c) Every function $h = h_{x_1,\ldots,x_n} \in \mathcal{H}$ is upper semicontinuous (see page 376) in the weak topology $\sigma(L^{q/(q-p)}, L^{q/p})$. To see this note that the subset $[h_{x_1,\ldots,x_n} \ge r]$ of K is convex, so it is weakly closed if and only if it is norm-closed (theorem A.2.25). In other words, it suffices to show that $h_{x_1,\ldots,x_n}$ is upper semicontinuous in the norm topology of $L^{q/(q-p)}$ or, equivalently, that

$$k \mapsto \int \sum_{\nu=1}^{n} |\mathcal{I}x_\nu|^q \cdot \frac{1}{k^{q/p}}\, d\mu$$

is lower semicontinuous in the norm topology of $L^{q/(q-p)}$. Now

$$\int |\mathcal{I}x_\nu|^q \cdot k^{-q/p}\, d\mu = \sup_{\epsilon>0} \int \left(\epsilon^{-1} \wedge |\mathcal{I}x_\nu|^q\right) \cdot (\epsilon \vee |k|)^{-q/p}\, d\mu\,,$$

and the map that sends k to the integral on the right is norm-continuous on $L^{q/(q-p)}$, as a straightforward application of the Dominated Convergence Theorem shows. The characterization of semicontinuity in A.2.19 gives c).
d) For every one of the functions $h = h_{x_1,\ldots,x_n} \in \mathcal{H}$ there is a point $k_{x_1,\ldots,x_n} \in K$ (depending on h!) at which it is non-negative. Indeed,

$$k_{x_1,\ldots,x_n} = \left(\int \Big(\sum_{1\le\nu\le n} |\mathcal{I}x_\nu|^q\Big)^{p/q} d\mu\right)^{(p-q)/q} \cdot \Big(\sum_{1\le\nu\le n} |\mathcal{I}x_\nu|^q\Big)^{p(q-p)/4}$$

meets the description: raising this function to the power $q/(q-p)$ and integrating gives 1; hence $k_{x_1,\ldots,x_n}$ belongs to K. Next,

$$k_{x_1,\ldots,x_n}^{-q/p} = \left(\int \Big(\sum_{1\le\nu\le n} |\mathcal{I}x_\nu|^q\Big)^{p/q} d\mu\right)^{(q-p)/p} \cdot \Big(\sum_{1\le\nu\le n} |\mathcal{I}x_\nu|^q\Big)^{(p-q)/q};$$

thus

$$\sum_{1\le\nu\le n} |\mathcal{I}x_\nu|^q \cdot k_{x_1,\ldots,x_n}^{-q/p} = \left(\int \Big(\sum_{1\le\nu\le n} |\mathcal{I}x_\nu|^q\Big)^{p/q} d\mu\right)^{(q-p)/p} \cdot \Big(\sum_{1\le\nu\le n} |\mathcal{I}x_\nu|^q\Big)^{p/q},$$

and therefore

$$h_{x_1,\ldots,x_n}(k_{x_1,\ldots,x_n}) = C^q \cdot \sum_{1\le\nu\le n} \|x_\nu\|_E^q - \left(\int \Big(\sum_{1\le\nu\le n} |\mathcal{I}x_\nu|^q\Big)^{p/q} d\mu\right)^{q/p}.$$

Thanks to inequality (4.1.12), this number is non-negative.
e) Finally, observe that the collection $\mathcal{H}$ of concave upper semicontinuous functions defined in $(*)$ is convex. Indeed, for $\lambda, \lambda' \ge 0$ with sum $\lambda + \lambda' = 1$,

$$\lambda \cdot h_{x_1,\ldots,x_n} + \lambda' \cdot h_{x'_1,\ldots,x'_{n'}} = h_{\lambda^{1/q}x_1,\ldots,\lambda^{1/q}x_n,\lambda'^{1/q}x'_1,\ldots,\lambda'^{1/q}x'_{n'}}\,.$$

Ky–Fan's minimax theorem A.2.34 now guarantees the existence of the desired common point of positivity for all of the functions in $\mathcal{H}$.

The equivalence of (ii) with (iii) is left as an easy excercise. ▮

Proof for $p > 0$

Proof of Theorem 4.1.2 (i) for $0 < p < q \le 2$. We have to show that $\eta_{p,q}(\mathcal{I})$ is finite when $\mathcal{I} : \mathcal{E}^d \to L^p(\mathbb{P})$ is the stochastic integral $X \mapsto \int X\,dZ$, in fact, that $\eta_{p,q}(\mathcal{I}) \le D_{p,q,d} \cdot \lceil Z \rceil_{\mathcal{I}^p[\mathbb{P}]}$ with $D_{p,q,d}$ finite. Note that the domain $\mathcal{E}^d$ of the stochastic integral is the set of step functions over an algebra of sets. Therefore the following deep theorem from Banach space theory applies and provides in conjunction with exercises 4.1.6 and 4.1.5 for $0 < p < q \le 2$ the estimates

$$D^{(4.1.5)}_{p,q,d} < 3 \cdot 8^{1/p} \quad \text{and} \quad E_{p,q} \le 2^{(p \vee (q-p))/p} . \tag{4.1.16}$$

Theorem 4.1.7 *Let B be a set, $\mathcal{A}$ an algebra of subsets of B, and let $\mathcal{E}$ be the collection of step functions over $\mathcal{A}$. $\mathcal{E}$ is naturally equipped with the sup-norm $\|x\|_{\mathcal{E}} \overset{\text{def}}{=} \sup\{|x(\varpi)| : \varpi \in B\}$, $x \in \mathcal{E}$.*

Let μ be a σ-finite measure on some other space, let $0 < p < 2$, and let $\mathcal{I} : \mathcal{E} \to L^p(\mu)$ be a continuous linear map of size

$$\|\mathcal{I}\|_p \overset{\text{def}}{=} \sup\Big\{ \|\mathcal{I}x\|_{L^p(\mu)} : \|x\|_{\mathcal{E}} \le 1 \Big\} .$$

There exist a constant C_p and a measurable function $g \ge 0$ with

$$\|g\|_{L^{p/(2-p)}(\mu)} \le 1$$

such that

$$\left(\int |\mathcal{I}x|^2 \,\frac{d\mu}{g} \right)^{1/2} \le C_p \cdot \|\mathcal{I}\|_p \cdot \|x\|_{\mathcal{E}}$$

for all $x \in \mathcal{E}$. The universal constant C_p can be estimated in terms of the Khintchine constants of theorem A.8.26:

$$\begin{aligned} C_p &\le \Big(\big(2^{1/3} + 2^{-2/3}\big)\, 2^{0 \vee (1-p)/p}\, K_p^{(A.8.5)}\, K_1^{(A.8.5)} \Big)^{3/2} \\ &\le \big(2\sqrt{2}\big)^{1/p + 1 \vee 1/p} < 2^{3(2+p)/2p} < 3 \cdot 8^{1/p} . \end{aligned} \tag{4.1.17}$$

Exercise 4.1.8 The theorem persists if K is a compact space and $\mathcal{I}$ is a continuous linear map from $C(K)$ to $L^p(\mu)$, or if $\mathcal{E}$ is an algebra of bounded functions containing the constants and $\mathcal{I} : \mathcal{E} \to L^p(\mu)$ is a bounded linear map.

Theorem 4.1.7 was first proved by Rosenthal [95] in the range $1 \le p < q \le 2$ and was extended to the range $0 < p < q \le 2$ by Maurey [66] and Schwartz [67]. The remainder of this subsection is devoted to its proof. The next two lemmas, in fact the whole drift of the following argument, are from Pisier's paper [84]. We start by addressing the special case of theorem 4.1.7 in which $\mathcal{E}$

is $\ell^\infty(k)$, i.e., $\mathbb{R}^k$ equipped with the sup-norm. Note that $\ell^\infty(k)$ meets the description of $\mathcal{E}$ in theorem 4.1.7: with $\boldsymbol{B} = \{1, \ldots, k\}$, $\ell^\infty(k)$ consists exactly of the step functions over the algebra $\mathcal{A}$ of all subsets of $\boldsymbol{B}$. This case is prototypical; once theorem 4.1.7 is established for it, the general version is not far away (page 202).

If $\mathcal{I} : \ell^\infty(k) \to L^p(\mu)$ is continuous, then $\eta_{p,2}(\mathcal{I})$ is rather readily seen to be finite. In fact, a straightforward computation, whose details are left to the reader, shows that whenever the domain of $\mathcal{I} : E \to L^p(\mu)$ is k-dimensional $(k < \infty)$,

$$\left\| \Big(\sum_\nu |\mathcal{I}x_\nu|^2 \Big)^{1/2} \right\|_{L^p(\mu)} \le k^{1/p+1/2} \|\mathcal{I}\|_p \cdot \Big(\sum_\nu \|x_\nu\|^2_{\ell^\infty(k)} \Big)^{\frac{1}{2}} ,$$

which reads
$$\eta_{p,2}(\mathcal{I}) \le k^{1/p+1/2} \cdot \|\mathcal{I}\|_p < \infty . \tag{4.1.18}$$

Thus there is factorization in the sense of criterion 4.1.4 if $\mathcal{I} : \ell^\infty(k) \to L^p(\mu)$ is continuous. In order to parlay this result into theorem 4.1.7 in all its generality, an estimate of $\eta_{p,2}(\mathcal{I})$ better than (4.1.18) is needed, namely, one that is independent of the dimension k. The proof below of such an estimate uses the Bernoulli random variables ϵ_ν that were employed in the proof of Khintchine's inequality (A.8.4) and, in fact, uses this very inequality twice. Let us recall their definition:

We fix a natural number n – it is the number of vectors $x_\nu \in \ell^\infty(k)$ that appear in inequality (4.1.12) – and denote by T^n the n-fold product of two-point sets $\{1, -1\}$. Its elements are n-tuples $t = (t_1, t_2, \ldots, t_n)$ with $t_\nu = \pm 1$. $\epsilon_\nu : t \mapsto t_\nu$ is the νth coordinate function. The natural measure on T^n is the product τ of uniform measure on $\{1, -1\}$, so that $\tau(\{t\}) = 2^{-n}$ for $t \in T^n$. T^n is a compact abelian group and τ is its normalized Haar measure. There will be occasion to employ convolution on this group.

The $\epsilon_\nu, \nu = 1 \ldots n$, are independent and form an orthonormal set in $L^2(\tau)$, which is far from being a basis: since the σ-algebra $\mathcal{T}^n$ on T^n is generated by 2^n atoms, the dimension of $L^2(\tau)$ is 2^n. Here is a convenient extension to a basis for this Hilbert space: for any subset A of $\{1, \ldots, n\}$ set

$$w_A = \prod_{\nu \in A} \epsilon_\nu , \quad \text{with } w_\emptyset = 1 .$$

It is plain upon inspection that the w_A are characters[3] of the group T^n and form an orthonormal basis of $L^2(\tau)$, the ***Walsh basis***.

Consider now the Banach space $L^2(\tau, \ell^\infty)$ of $\ell^\infty(k)$-valued functions f on T^n having

$$\|f\|_{L^2(\tau,\ell^\infty)} \stackrel{\text{def}}{=} \Big(\int \|f(t)\|^2_{\ell^\infty(k)} \, \tau(dt) \Big)^{1/2} < \infty .$$

[3] A map χ from a group into the circle $\{z \in \mathbb{C} : |z| = 1\}$ is a ***character*** if it is multiplicative: $\chi(st) = \chi(s)\chi(t)$, with $\chi(1) = 1$.

Its dual can be identified isometrically with the Banach space $L^2(\tau,\ell^1)$ of $\ell^1(k)$-valued functions f^* on T^n for which

$$\|f^*\|_{L^2(\tau,\ell^1)} \stackrel{\text{def}}{=} \Big(\int \|f^*(t)\|_{\ell^1}^2 \,\tau(dt)\Big)^{1/2} < \infty\,,$$

under the pairing $\quad \langle f|f^*\rangle = \int \langle f(t)|f^*(t)\rangle\,\tau(dt)\,.$

Both spaces have finite dimension $k2^n$. $L^2(\tau,\ell^\infty)$ is the direct sum of the subspaces

$$E(\ell^\infty) \stackrel{\text{def}}{=} \Big\{\sum_{\nu=1}^n x_\nu\cdot\epsilon_\nu : x_\nu \in \ell^\infty(k)\Big\} \quad \text{and}$$

$$W(\ell^\infty) \stackrel{\text{def}}{=} \Big\{\sum x_A\cdot w_A : A\subset\{1,\ldots,n\}\,,\ |A|\neq 1\,,\ x_A\in\ell^\infty(k)\Big\}\,.$$

$|A|$ is, of course, the cardinality of the set $A\subset\{1,\ldots,n\}$, and $w_\emptyset = 1$. It is convenient to denote the corresponding projections of $L^2(\tau,\ell^\infty)$ onto $E(\ell^\infty)$ and $W(\ell^\infty)$ by E and W, respectively. Here is a little information about the geometry of these subspaces, used below to estimate the right-hand side of inequality (4.1.12) on page 192:

Lemma 4.1.9 *Let $x_1,\ldots,x_n \in \ell^\infty(k)$. There is a function $f\in L^2(\tau,\ell^\infty(k))$ of the form*

$$f = \sum_{\nu=1}^n x_\nu\,\epsilon_\nu + \sum_{|A|\neq 1} x_A\,w_A$$

such that

$$\|f\|_{L^2(\tau,\ell^\infty)} \le K_1^{(A.8.5)}\cdot\Big(\sum_{\nu=1}^n \|x_\nu\|_{\ell^\infty}^2\Big)^{1/2}. \tag{4.1.19}$$

Proof. Set $f^\epsilon = \sum_\nu x_\nu\epsilon_\nu$, and let a denote the norm of the class of f^ϵ in the quotient $L^2(\tau,\ell^\infty)/W(\ell^\infty)$:

$$a = \inf\Big\{\|f^\epsilon + w\|_{L^2(\tau,\ell^\infty)} \ :\ w\in W(\ell^\infty)\Big\}\,.$$

Since $L^2\big(\tau,\ell^\infty(k)\big)$ is finite-dimensional, there is a function of the form $f = f^\epsilon + w$, $w\in W(\ell^\infty)$, with $\|f\|_{L^2(\tau,\ell^\infty)} = a$. This is the function promised by the statement.

To prove inequality (4.1.19), let B_a denote the open ball of radius a about zero in $L^2(\tau,\ell^\infty)$. Since the open convex set

$$C \stackrel{\text{def}}{=} B_a - (f + W(\ell^\infty)) = \{g - (f+w) \ :\ g\in B_a\,,\ w\in W(\ell^\infty)\}$$

does not contain the origin, there is a linear functional f^* in the dual $L^2(\tau,\ell^1)$ of $L^2(\tau,\ell^\infty)$ that is negative on C, and without loss of generality f^* can be chosen to have norm 1:

$$\int \|f^*(t)\|_{\ell^1}^2\,\tau(dt) = 1\,. \tag{4.1.20}$$

Since $\quad \langle g|f^*\rangle \le \langle f+w|f^*\rangle \qquad \forall g\in B_a(0)\,,\ \forall w\in W(\ell^\infty)\,,$

it is evident that $\langle w|f^*\rangle = 0$ for all $w \in W(\ell^\infty)$, so that f^* is of the form

$$f^* = \sum_{\nu=1}^{n} x^*_\nu \,\epsilon_\nu\,, \qquad\qquad x^*_\nu \in \ell^1(k)\,.$$

Also, since

$$(1-\epsilon)a = \|(1-\epsilon)f\|_{L^2(\tau,\ell^\infty)} \le \langle f|f^*\rangle \le \|f\|_{L^2(\tau,\ell^\infty)} = a \qquad \forall\,\epsilon > 0\,,$$

we must have

$$\begin{aligned} a = \langle f|f^*\rangle &= \int \langle f(t)|f^*(t)\rangle\,\tau(dt) = \sum_{\nu=1}^{n}\langle x_\nu|x^*_\nu\rangle \\ &\le \sum_{\nu=1}^{n} \|x_\nu\|_{\ell^\infty}\|x^*_\nu\|_{\ell^1} \le \Big(\sum_{\nu=1}^{n}\|x_\nu\|^2_{\ell^\infty}\Big)^{1/2}\cdot\Big(\sum_{\nu=1}^{n}\|x^*_\nu\|^2_{\ell^1}\Big)^{1/2}. \end{aligned}$$

Now
$$\begin{aligned} \Big(\sum_{\nu=1}^{n}\|x^*_\nu\|^2_{\ell^1}\Big)^{1/2} &= \Big(\sum_{\nu=1}^{n}\Big(\sum_{\kappa=1}^{k}|x^{*\kappa}_\nu|\Big)^2\Big)^{1/2} \le \sum_{\kappa=1}^{k}\Big(\sum_{\nu=1}^{n}|x^{*\kappa}_\nu|^2\Big)^{1/2} \\ &\le K_1^{(A.8.5)}\cdot\int\sum_{\kappa=1}^{k}\Big|\sum_{\nu} x^{*\kappa}_\nu\epsilon_\nu(t)\Big|\,\tau(dt) \\ &= K_1\cdot\int\Big\|\sum_{\nu=1}^{n} x^*_\nu\epsilon_\nu(t)\Big\|_{\ell^1(k)}\tau(dt) \\ &\le K_1\cdot\Big(\int\Big\|\sum_{\nu} x^*_\nu\epsilon_\nu(t)\Big\|^2_{\ell^1(k)}\tau(dt)\Big)^{1/2} \end{aligned}$$

by equation (4.1.20):
$$= K_1\cdot\|f^*\|_{L^2(\tau,\ell^1)} = K_1\,,$$

and we get the desired inequality

$$\|f\|_{L^2(\tau,\ell^\infty)} = a \le K_1^{(A.8.5)}\cdot\Big(\sum_{\nu=1}^{n}\|x_\nu\|^2_{\ell^\infty}\Big)^{1/2}.$$ ∎

We return to the continuous map $\mathcal{I} : \ell^\infty(k) \to L^p(\mu)$. We want to estimate the smallest constant $\eta_{p,2}(\mathcal{I})$ such that for any n and $x_1, \ldots, x_n \in \ell^\infty(k)$

$$\Big\|\Big(\sum_{\nu=1}^{n}|\mathcal{I}x_\nu|^2\Big)^{1/2}\Big\|_{L^p(\mu)} \le \eta_{p,2}(\mathcal{I})\cdot\Big(\sum_{\nu=1}^{n}\|x_\nu\|^2_{\ell^\infty(k)}\Big)^{1/2}.$$

Lemma 4.1.10 *Let $x_1, \ldots, x_n \in \ell^\infty(k)$ be given, and let $f \in L^2(\tau, \ell^\infty)$ be any function with $Ef = \sum_{\nu=1}^{n} x_\nu\,\epsilon_\nu$, i.e.,*

$$f = \sum_{\nu=1}^{n} x_\nu\,\epsilon_\nu + \sum_{A\subset\{1,\ldots,n\}\,,|A|\ne 1} x_A\,w_A\,, \qquad\qquad x_A \in \ell^\infty(k)\,.$$

For any $\delta \in [0,1]$ *we have*

$$\Big\|\Big(\sum_{\nu=1}^{n}|\mathcal{I}x_\nu|^2\Big)^{1/2}\Big\|_{L^p(\mu)} \le 2^{0\vee(1-p)/p} K_p^{(A.8.5)} \tag{4.1.21}$$

$$\times\Big[\|\mathcal{I}\|_p/\sqrt{\delta} + \eta_{p,2}(\mathcal{I})\cdot\delta\Big]\cdot\|f\|_{L^2(\ell^\infty)}\,.$$

Proof. For $\theta \in [-1,1]$ set

$$\psi_\theta \overset{\text{def}}{=} \prod_{\nu=1}^{n}(1+\theta\epsilon_\nu) = \sum_{A\subset\{1,\ldots,n\}} \theta^{|A|} w_A\,.$$

Then $\psi_\theta \ge 0$, $\int|\psi_\theta(t)|\,\tau(dt) = \int\psi_\theta(t)\,\tau(dt) = 1$, and $\int w_A(t)\psi_\theta(t)\,\tau(dt) = \theta^{|A|}$. The function

$$\phi_\delta \overset{\text{def}}{=} \frac{1}{2\sqrt{\delta}}\Big(\psi_{\sqrt{\delta}} - \psi_{-\sqrt{\delta}}\Big)$$

is easily seen to have the following properties:

$$\int|\phi_\delta(t)|\,\tau(dt) \le 1/\sqrt{\delta}\,,$$

$$\int w_A(t)\phi_\delta(t)\,\tau(dt) = \begin{cases} 0 & \text{if } |A| \text{ is even,}\\ \sqrt{\delta}^{\,|A|-1} & \text{if } |A| \text{ is odd.}\end{cases} \tag{4.1.22}$$

For the proof proper of the lemma we analyze the convolution

$$f\star\phi_\delta(t) = \int f(st)\phi_\delta(s)\,\tau(ds)$$

of f with this function. For definiteness' sake write

$$f = \sum_{\nu=1}^{n} x_\nu\,\epsilon_\nu + \sum_{|A|\ne 1} x_A\,w_A = Ef + Wf\,.$$

As the w_A are characters,[3] including the $\epsilon_\nu = w_{\{\nu\}}$, (4.1.22) gives

$$w_A\star\phi_\delta(t) = \int w_A(st)\phi_\delta(s)\,\tau(ds) = w_A(t)\cdot\begin{cases} 0 & \text{if } |A| \text{ is even,}\\ \sqrt{\delta}^{\,|A|-1} & \text{if } |A| \text{ is odd;}\end{cases}$$

thus $\quad f\star\phi_\delta = \sum_{\nu=1}^{n} x_\nu\epsilon_\nu + \sum_{3\le|A|\ odd} x_A\,\sqrt{\delta}^{\,|A|-1} w_A = Ef + Wf\star\phi_\delta\,,$

whence $\quad Ef = f\star\phi_\delta - Wf\star\phi_\delta$ and $\mathcal{I}Ef = \mathcal{I}f\star\phi_\delta - \mathcal{I}Wf\star\phi_\delta\,.$

Here $\mathcal{I}f$ denotes the function $t \mapsto \mathcal{I}f(t)$ from T to $L^p(\mu)$, etc. Therefore

$$\Big\|\Big(\sum_{\nu=1}^{n}|\mathcal{I}x_\nu|^2\Big)^{1/2}\Big\|_{L^p(\mu)} = \Big\|\,\|\mathcal{I}Ef\|_{L^2(\tau)}\Big\|_{L^p(\mu)}\,.$$

Theorem A.8.26 permits the following estimate of the right-hand side:

$$\Big\| \|\mathcal{I}Ef\|_{L^2(\tau)} \Big\|_{L^p(\mu)} \le K_p \cdot \Big\| \|\mathcal{I}Ef\|_{L^p(\tau)} \Big\|_{L^p(\mu)}$$

$$\le 2^{0\vee(1-p)/p} K_p \cdot \Big[\Big\| \|\mathcal{I}f\star\phi_\delta\|_{L^p(\tau)} \Big\|_{L^p(\mu)} + \Big\| \|\mathcal{I}Wf\star\phi_\delta\|_{L^p(\tau)} \Big\|_{L^p(\mu)} \Big]$$

$$\le 2^{0\vee(1-p)/p} K_p \cdot \Big[\Big\| \|\mathcal{I}f\star\phi_\delta\|_{L^p(\mu)} \Big\|_{L^2(\tau)} + \Big\| \|\mathcal{I}Wf\star\phi_\delta\|_{L^2(\tau)} \Big\|_{L^p(\mu)} \Big]$$

$$\le 2^{0\vee(1-p)/p} K_p \cdot \Big[\|\mathcal{I}\|_p \cdot \|f\star\phi_\delta\|_{L^2(\tau,\ell^\infty)} + \Big\| \|\mathcal{I}Wf\star\phi_\delta\|_{L^2(\tau)} \Big\|_{L^p(\mu)} \Big]$$

$$= 2^{0\vee(1-p)/p} K_p \cdot [Q_1 + Q_s] \,. \tag{4.1.23}$$

The first term Q_1 can be bounded using Jensen's inequality for the measure $\phi_\delta(s)\tau(ds)$:

$$\int \|(f\star\phi_\delta)(t)\|^2_{\ell^\infty}\,\tau(dt) = \int \Big\| \int f(st)\phi_\delta(s)\,\tau(ds) \Big\|^2_{\ell^\infty}\,\tau(dt)$$

by A.3.28:

$$\le \int \Big(\int \|f(st)\|_{\ell^\infty} |\phi_\delta(s)|\,\tau(ds) \Big)^2\,\tau(dt)$$

$$\le \delta^{-1} \int \Big(\int \|f(st)\|_{\ell^\infty} |\phi_\delta(s)| \sqrt{\delta}\,\tau(ds) \Big)^2\,\tau(dt)$$

$$\le \delta^{-1} \int \int \|f(st)\|^2_{\ell^\infty} |\phi_\delta(s)| \sqrt{\delta}\,\tau(ds)\,\tau(dt)$$

$$= \delta^{-1} \int \|f(t)\|^2_{\ell^\infty}\,\tau(dt) \int |\phi_\delta(s)| \sqrt{\delta}\,\tau(ds)$$

$$\le \delta^{-1} \int \|f(t)\|^2_{\ell^\infty}\,\tau(dt)\,,$$

so that

$$\|f\star\phi_\delta\|_{L^2(\tau,\ell^\infty)} \le \frac{1}{\sqrt{\delta}} \cdot \|f\|_{L^2(\tau,\ell^\infty)} \,. \tag{4.1.24}$$

The function $Wf\star\phi_\delta$ in the second term Q_2 of inequality (4.1.23) has the form

$$(Wf\star\phi_\delta)(t) = \sum_{3\le|A|\ odd} x_A \sqrt{\delta}^{\,|A|-1} w_A(t)\,.$$

Thus

$$\|\mathcal{I}Wf\star\phi_\delta\|^2_{L^2(\tau)} = \int \Big(\sum_{3\le|A|\ odd} \mathcal{I}x_A \sqrt{\delta}^{\,|A|-1} w_A(t) \Big)^2\,\tau(dt)$$

$$= \sum_{3\le|A|\ odd} (\mathcal{I}x_A)^2 \cdot \delta^{|A|-1}$$

$$\le \delta^2 \cdot \sum_{A\subset\{1,\ldots,n\}} (\mathcal{I}x_A)^2 = \delta^2 \cdot \|\mathcal{I}f\|^2_{L^2(\tau)}\,,$$

and therefore, from inequality (4.1.13),

$$\Big\| \|\mathcal{I}W f \star \phi_\delta\|_{L^2(\tau)} \Big\|_{L^p(\mu)} \le \delta \cdot \eta_{p,2}(\mathcal{I}) \cdot \|f\|^2_{L^2(\tau)} \,.$$

Putting this and (4.1.24) into (4.1.23) yields inequality (4.1.21). ▮

If for the function f of lemma 4.1.10 we choose the one provided by lemma 4.1.9, then inequality (4.1.21) turns into

$$\Big\| \Big(\sum_{\nu=1}^{n} |\mathcal{I}x_\nu|^2 \Big)^{1/2} \Big\|_{L^p(\mu)} \le 2^{0 \vee (1-p)/p} K_p K_1$$

$$\times \Big[\frac{\|\mathcal{I}\|_p}{\sqrt{\delta}} + \eta_{p,2}(\mathcal{I}) \cdot \delta \Big] \cdot \Big(\sum_{\nu=1}^{n} \|x_\nu\|^2_{\ell^\infty} \Big)^{1/2} .$$

Since this inequality is true for all finite collections $\{x_1, \ldots, x_n\} \subset \mathcal{E}$, it implies

$$\eta_{p,2}(\mathcal{I}) \le 2^{0 \vee (1-p)/p} K_p K_1 \Big[\frac{\|\mathcal{I}\|_p}{\sqrt{\delta}} + \eta_{p,2}(\mathcal{I}) \cdot \delta \Big] \,. \qquad (*)$$

The function $\delta \mapsto \frac{\|\mathcal{I}\|_p}{\sqrt{\delta}} + \eta_{p,2}(\mathcal{I}) \cdot \delta$ takes its minimum at

$$\delta = \Big(\|\mathcal{I}\|_p / (2\eta_{p,2}) \Big)^{2/3} < 1 \,,$$

where its value is $(2^{1/3} + 2^{-2/3}) \|\mathcal{I}\|_p^{2/3} \eta_{p,2}^{1/3}$. Therefore $(*)$ gives

$$\eta_{p,2}(\mathcal{I}) \le (2^{1/3} + 2^{-2/3})\, 2^{0 \vee (1-p)/p}\, K_p K_1 \cdot \|\mathcal{I}\|_p^{2/3} \eta_{p,2}^{1/3} \,,$$

and so $\eta_{p,2}(\mathcal{I}) \le \Big((2^{1/3} + 2^{-2/3})\, 2^{0 \vee (1-p)/p}\, K_p K_1 \Big)^{3/2} \cdot \|\mathcal{I}\|_p$

by (A.8.9): $$\le \Big(2 \cdot 2^{0 \vee (1-p)/p}\, 2^{1/p - 1/2}\, 2^{1/2} \Big)^{3/2} \cdot \|\mathcal{I}\|_p$$

$$= \Big(2^{1 \vee 1/p}\, 2^{1/p} \Big)^{3/2} \cdot \|\mathcal{I}\|_p = \big(2\sqrt{2} \big)^{1/p + 1 \vee 1/p} \cdot \|\mathcal{I}\|_p \ :$$

Corollary 4.1.11 *A linear map $\mathcal{I} : \ell^\infty(k) \to L^p(\mu)$ is factorizable with*

$$\eta_{p,2}(\mathcal{I}) \le C_p \cdot \|\mathcal{I}\|_p \,, \qquad (4.1.25)$$

where $$C_p \le \Big((2^{1/3} + 2^{-2/3})\, 2^{0 \vee (1-p)/p}\, K_p K_1 \Big)^{3/2}$$

$$\le \big(2\sqrt{2} \big)^{1/p + 1 \vee 1/p} < 2^{3(2+p)/2p} \,.$$

Theorem 4.1.7, including the estimate (4.1.17) for C_p, is thus true when $\mathcal{E} = \ell^\infty(k)$.

Proof of Theorem 4.1.7. Given $\{x_1,\ldots,x_n\} \subset \mathcal{E}$, there is a finite subalgebra $\mathcal{A}'$ of $\mathcal{A}$, generated by finitely many atoms, say $\{A_1,\ldots,A_k\}$, such that every x_ν is a step function over $\mathcal{A}'$: $x_\nu = \sum_\kappa A_\kappa x_\nu^\kappa$. The linear map $\mathcal{I}' : \ell^\infty(k) \to L^p(\mu)$ that takes the κth standard basis vector of $\ell^\infty(k)$ to $\mathcal{I}A_\kappa$ has $\|\mathcal{I}'\|_p \le \|\mathcal{I}\|_p$ and takes $(x_\nu^\kappa)_{1\le\kappa\le k} \in \ell^\infty(k)$ to $\mathcal{I}x_\nu$. Since $\|(x_\nu^\kappa)_{1\le\kappa\le k}\|_{\ell^\infty(k)} = \|x_\nu\|_{\mathcal{E}}$, inequality (4.1.25) in corollary 4.1.11 gives

$$\left\|\Big(\sum_{\nu=1}^n |\mathcal{I}x_\nu|^2\Big)^{1/2}\right\|_{L^p(\mu)} \le C_p^{(4.1.25)} \cdot \|\mathcal{I}\|_p \cdot \Big(\sum_{\nu=1}^n \|x_\nu\|_{\mathcal{E}}^2\Big)^{1/2},$$

and another application of Rosenthal's criterion 4.1.4 yields the theorem. ∎

Proof of Theorem 4.1.2 for $0 < p \le 2 < q < \infty$. Theorem 4.1.7 does not extend to exponents $q > 2$ in general – it is due to the special nature of the stochastic integral, the "closeness of the arguments of $\mathcal{I}$ to its values" expressed for instance by exercise 2.1.14, that theorem 4.1.2 can be extended to $q > 2$. If $\boldsymbol{Z}$ is previsible, then "the values of $\mathcal{I}$ are *very close* to the arguments," and the factorization constant does not even depend on the length d of the vector $\boldsymbol{Z}$.

We start off by having what we know so far produce a probability $\mathbb{P}' = \mathbb{P}/g$ with $\|g\|_{L^{p/(2-p)}(\mathbb{P})} \le E_{p,2}$ for which $\boldsymbol{Z}$ is an L^2-integrator of size

$$\lceil\!\lceil \boldsymbol{Z} \rceil\!\rceil_{\mathcal{I}^2[\mathbb{P}']} \le D_{p,2}^{(4.1.5)} \cdot \lceil\!\lceil \boldsymbol{Z} \rceil\!\rceil_{\mathcal{I}^p[\mathbb{P}]}\,. \tag{4.1.26}$$

Then $\boldsymbol{Z}$ has a Doob–Meyer decomposition $\boldsymbol{Z} = \widehat{\boldsymbol{Z}} + \widetilde{\boldsymbol{Z}}$ with respect to $\mathbb{P}'$ (see theorem 4.3.1 on page 221) whose components have sizes

$$\lceil\!\lceil \widehat{\boldsymbol{Z}} \rceil\!\rceil_{\mathcal{I}^2[\mathbb{P}']} \le 2\,\lceil\!\lceil \boldsymbol{Z} \rceil\!\rceil_{\mathcal{I}^2[\mathbb{P}']} \quad\text{and}\quad \lceil\!\lceil \widetilde{\boldsymbol{Z}} \rceil\!\rceil_{\mathcal{I}^2[\mathbb{P}']} \le 2\,\lceil\!\lceil \boldsymbol{Z} \rceil\!\rceil_{\mathcal{I}^2[\mathbb{P}']}\,, \tag{4.1.27}$$

respectively. We shall estimate separately the factorization constants of the stochastic integrals driven by $\widehat{\boldsymbol{Z}}$ and by $\widetilde{\boldsymbol{Z}}$ and then apply exercise 4.1.6.

Our first claim is this: if $\mathcal{I} : \mathcal{E}^d \to L^p$ is the stochastic integral driven by a d-tuple $\boldsymbol{V}$ of finite variation processes, then, for $0 < p < q < \infty$,

$$\eta_{p,q}(\mathcal{I}) \le \Big\| \sum_{1\le\theta\le d} \lceil\!\lceil V^\theta \rceil\!\rceil_\infty \Big\|_{L^p}\,. \tag{4.1.28}$$

Since the right-hand side equals $\lceil\!\lceil \boldsymbol{V} \rceil\!\rceil_{\mathcal{I}^p}$ when $\boldsymbol{V}$ is previsible, this together with (4.1.27) and (4.1.26) will result in

$$\eta_{p,q}\Big(\int \cdot\, d\widehat{\boldsymbol{Z}}\Big) \le 2\,\lceil\!\lceil \boldsymbol{Z} \rceil\!\rceil_{\mathcal{I}^2[\mathbb{P}']} \le 2\,D_{p,2}^{(4.1.5)} \cdot \lceil\!\lceil \boldsymbol{Z} \rceil\!\rceil_{\mathcal{I}^p[\mathbb{P}]} \tag{4.1.29}$$

for $0 < p < q < \infty$. To prove equation (4.1.28), let $\boldsymbol{X}^1,\ldots,\boldsymbol{X}^n$ be a finite collection of elementary integrands in $\mathcal{E}^d$. Then

$$\sum_\nu \Big|\int \boldsymbol{X}^\nu\, d\boldsymbol{V}\Big|^q \le \sum_\nu \Big|\,\|\boldsymbol{X}^\nu\|_{\mathcal{E}^d} \sum_{1\le\theta\le d} \lceil\!\lceil V^\theta \rceil\!\rceil_\infty\Big|^q .$$

Applying the L^p-mean to the qth root gives

$$\left\| \Big(\sum_\nu \Big| \int \boldsymbol{X}^\nu \, d\boldsymbol{V} \Big|^q \Big)^{1/q} \right\|_{L^p} \le \left\| \sum_{1\le\theta\le d} |V^\theta|_\infty \right\|_{L^p} \cdot \Big(\sum_{\nu=1}^n \| \boldsymbol{X}_\nu \|^q_{\mathcal{E}^d} \Big)^{1/q} ,$$

which is equation (4.1.28).

Now to the martingale part. With $\boldsymbol{X}^\nu$ as above set

$$M^\nu \stackrel{\text{def}}{=} \boldsymbol{X}^\nu * \widetilde{\boldsymbol{Z}} , \qquad \nu = 1, \ldots, n ,$$

and for $\vec{m} = (m^1, \ldots, m^n)$ set

$$\Phi(\vec{m}) \stackrel{\text{def}}{=} Q^{2/q}(\vec{m}) \quad , \text{ where } Q(\vec{m}) \stackrel{\text{def}}{=} \sum_{\nu=1}^n |m^\nu|^q .$$

Then $\Phi'_\mu(\vec{m}) = 2Q^{\frac{2-q}{q}}(\vec{m}) \cdot |m^\mu|^{q-1} \operatorname{sgn}(m^\mu)$

and
$$\begin{aligned} \Phi''_{\mu\nu}(\vec{m}) &= 2(q-1)Q^{\frac{2-q}{q}}(\vec{m}) \cdot |m^\mu|^{q-2} \\ &\quad + 2(2-q)Q^{\frac{2-2q}{q}}(\vec{m}) \cdot |m^\mu|^{q-1} \operatorname{sgn}(m^\mu) |m^\nu|^{q-1} \operatorname{sgn}(m^\nu) \qquad (*) \\ &\le 2(q-1)Q^{\frac{2-q}{q}}(\vec{m}) \cdot |m^\mu|^{q-2} , \end{aligned}$$

on the grounds that the $\mu\nu$-matrix in $(*)$ is negative semidefinite for $q > 2$. Our interest in Φ derives from the fact that criterion 4.1.4 asks us to estimate the expectation of $\Phi(\vec{M}_\infty)$. With $M^\mu_\lambda \stackrel{\text{def}}{=} (1-\lambda)M^\mu_{\text{.}-} + \lambda M^\mu$ for short, Itô's formula results in the inequality

$$\begin{aligned} \Phi(\vec{M}_\infty) &\le \Phi(\vec{M}_0) + 2\int_{0+}^\infty Q^{\frac{2-q}{q}}(\vec{M}_{\text{.}-}) \cdot |M^\mu_{\text{.}-}|^{q-1} \operatorname{sgn}(M^\mu_{\text{.}-}) \, dM^\mu \\ &\quad + 2(q-1) \int_0^1 (1-\lambda)\, d\lambda \int_{0+}^\infty Q^{\frac{2-q}{q}}(\vec{M}_\lambda) \cdot \sum_\mu |M^\mu_\lambda|^{q-2} \, d[M^\mu, M^\mu] , \\ &\le 2\int_{0+}^\infty Q^{\frac{2-q}{q}}(\vec{M}_{\text{.}-}) \cdot |M^\mu_{\text{.}-}|^{q-1} \operatorname{sgn}(M^\mu_{\text{.}-}) \, dM^\mu \\ &\quad + 2(q-1) \int_0^1 (1-\lambda)\, d\lambda \int_0^\infty Q^{\frac{2-q}{q}}(\vec{M}_\lambda) \cdot \sum_\mu |M^\mu_\lambda|^{q-2} \, d[M^\mu, M^\mu] . \end{aligned}$$

Now
$$[M^\mu, M^\mu] = \sum_{1\le\eta,\theta\le d} X^\mu_\eta X^\mu_\theta * [\widetilde{Z}^\eta, \widetilde{Z}^\theta] , \tag{4.1.30}$$

whence[4]
$$\begin{aligned} \mathbb{E}'\Big[\Phi(\vec{M}_\infty)\Big] &\le 2(q-1) \int_0^1 (1-\lambda) \times \\ &\quad \times \mathbb{E}'\Big[\int_0^\infty Q^{\frac{2-q}{q}}(\vec{M}_\lambda) X^\mu_\eta X^\mu_\theta \, d[\widetilde{Z}^\eta, \widetilde{Z}^\theta]\Big] \, d\lambda . \end{aligned} \tag{4.1.31}$$

---

[4] Einstein's convention, adopted, implies summation over the same indices in opposite positions.

Consider first the case that $d = 1$, writing $\widetilde{Z}, X^\mu$ for the scalar processes $\widetilde{\boldsymbol{Z}}_1, \boldsymbol{X}^\mu_1$. Then $X^\mu_\eta X^\mu_\theta d[\widetilde{Z}^\eta, \widetilde{Z}^\theta] = (X^\mu)^2\, d[\widetilde{Z}, \widetilde{Z}] \le \|X^\mu\|^2_{\mathcal{E}}\, d[\widetilde{Z}, \widetilde{Z}]$, and using Hölder's inequality with conjugate exponents $q/2$ and $q/(q-2)$ in the sum over μ turns (4.1.31) into

$$\begin{aligned}\mathbb{E}'\Big[\Phi(\vec{M}_\infty)\Big] &\le 2(q-1)\int_0^1 (1-\lambda) \\ &\qquad \mathbb{E}'\bigg[\int_0^\infty Q^{\frac{2-q}{q}}(\vec{M}_\lambda)\Big(\sum_\mu |M^\mu_\lambda|^q\Big)^{\frac{q-2}{q}} \\ &\qquad\qquad \times \Big(\sum_\mu \|X^\mu\|^q_{\mathcal{E}}\Big)^{2/q} d[\widetilde{Z}, \widetilde{Z}]\bigg]\, d\lambda \\ &= 2(q-1)\int_0^1 (1-\lambda)\, d\lambda \Big(\sum_\mu \|X^\mu\|^q_{\mathcal{E}}\Big)^{2/q} \mathbb{E}'\Big[[\widetilde{Z}, \widetilde{Z}]_\infty\Big] \\ &= (q-1)\Big\|\widetilde{Z}_\infty\Big\|^2_{L^2(\mathbb{P}')} \Big(\sum_\mu \|X^\mu\|^q_{\mathcal{E}}\Big)^{2/q} \\ &= (q-1)\,\lceil\!\lceil \widetilde{Z} \rceil\!\rceil^2_{\mathcal{I}^2[\mathbb{P}']} \Big(\sum_\mu \|X^\mu\|^q_{\mathcal{E}}\Big)^{2/q} .\end{aligned} \tag{4.1.32}$$

Taking the square root results in

$$\eta_{2,q}\Big(\int \cdot\, d\widetilde{Z}\Big) \le \sqrt{q-1}\,\lceil\!\lceil \widetilde{Z} \rceil\!\rceil_{\mathcal{I}^2[\mathbb{P}']} \le 2\sqrt{q-1}\, D^{(4.1.5)}_{p,2} \cdot \lceil\!\lceil \boldsymbol{Z} \rceil\!\rceil_{\mathcal{I}^p[\mathbb{P}]} . \tag{4.1.33}$$

Now if $\boldsymbol{Z}$ and with it the $[\widetilde{Z}^\eta, \widetilde{Z}^\theta]$ are previsible, then the same inequality holds. To see this we pick an increasing previsible process V so that $d[\widetilde{Z}^\eta, \widetilde{Z}^\theta] \ll dV$, for instance the sum of the variations $\lceil\!\lceil [\widetilde{Z}^\eta, \widetilde{Z}^\theta] \rceil\!\rceil$, and let $G^{\eta\theta}$ be the previsible Radon–Nikodym derivative of $d[\widetilde{Z}^\eta, \widetilde{Z}^\theta]$ with respect to dV. According to lemma A.2.21 b), there is a Borel measurable function γ from the space $\mathcal{G}$ of $d \times d$-matrices to the unit box $\ell^\infty_1(d)$ such that

$$\sup\{x_\eta x_\theta G^{\eta\theta} : \boldsymbol{x} \in \ell^\infty_1\} = \gamma_\eta(G)\gamma_\theta(G) G^{\eta\theta} , \qquad G \in \mathcal{G} .$$

We compose this map γ with the $\mathcal{G}$-valued process $\big(G^{\eta\theta}\big)^d_{\eta,\theta=1}$ and obtain a previsible process $\boldsymbol{Y}$ with $|\boldsymbol{Y}| \le 1$. Let us write $M = \boldsymbol{Y} * \widetilde{\boldsymbol{Z}}$. Then

$$\mathbb{E}'[[M, M]_\infty] \le \lceil\!\lceil \widetilde{\boldsymbol{Z}} \rceil\!\rceil^2_{\mathcal{I}^2[\mathbb{P}']}$$

and in (4.1.30) $\qquad d[M^\mu, M^\mu] \le \|\boldsymbol{X}^\mu\|^2_{\mathcal{E}^d} \cdot d[M, M] .$

We can continue as in inequality (4.1.32), replacing $[\widetilde{Z}, \widetilde{Z}]$ by $[M, M]$, and arrive again at (4.1.33). Putting this inequality together with (4.1.29) into inequalities (4.1.26) and (4.1.27) gives

$$\eta_{p,q}\Big(\int \cdot \, d\boldsymbol{Z}\Big) \le 2(\sqrt{q-1}+1)D_{p,2} \, \lceil \boldsymbol{Z} \rceil_{\mathcal{I}^p[\mathbb{P}]}$$

or

$$D_{p,q,d} \le 2(1+\sqrt{q-1})D_{p,2} \le 3{\cdot}2^{1+3/p}{\cdot}(1+\sqrt{q-1}) \quad (4.1.34)$$

if $d = 1$ or if $\boldsymbol{Z}$ is previsible. We leave it to the reader to show that in general

$$D_{p,q,d} \le \sqrt{d} \cdot D_{p,2} \, . \quad (4.1.35)$$

Let $\mathbb{P}'' = \mathbb{P}/g'$ be the probability provided by exercise 4.1.5. It is not hard to see with Hölder's inequality that the estimates $\| g \|_{L^{p/(2-p)}(\mathbb{P})} < 2^{2/p}$ and $\| g \|_{L^{2/(q-2)}(\mathbb{P}')} < 2^{q/2}$ lead to

$$E_{p,q} \le 4^{q/p} \quad \text{for } 0 < p \le 2 < q \, .$$

Proof of Theorem 4.1.2 (i) Only the case $2 < p < q < \infty$ has not yet been covered. It is not too interesting in the first place and can easily be reduced to the previous case by considering $\boldsymbol{Z}$ an L^2-integrator. We leave this to the reader.

Proof for $p = 0$

If $\boldsymbol{Z}$ is a single L^0-integrator, then proposition 4.1.1 together with a suitable analog of exercise 4.1.6 provides a proof of theorem 4.1.2 when $p = 0$. This then can be extended rather simply to the case of finitely many integrators, except that the corresponding constants deteriorate with their number. This makes the argument inapplicable to random measures, which can be thought of as an infinity of infinitesimal integrators (page 173). So we go a different route.

Proof of Theorem 4.1.2 (ii) for $0 < q < 1$. Maurey [66] and Schwartz [67] have shown that the conclusion of theorem 4.1.2 (ii) holds for any continuous linear map $\mathcal{I} : E \to L^0(\mathbb{P})$, provided the exponent q is strictly less than one.[5] In this subsection we extract the pertinent arguments from their immense work. Later we can then prove the general case $0 < q < \infty$ by applying theorem 4.1.2 (i) to the stochastic integral regarded as an $L^q(\mathbb{P}')$-integrator for a probability $\mathbb{P}' \approx \mathbb{P}$ produced here. For the arguments we need the information on symmetric stable laws and on the stable type of a map or space that is provided in exercise A.8.31.

The role of theorem 4.1.7 in the proof of theorem 4.1.2 is played for $p = 0$ by the following result, reminiscent of proposition 4.1.1:

---

[5] Theorem 4.1.7 for $0 < p < 1$ is also first proved in their work.

Proposition 4.1.12 *Let E be a normed space, $(\Omega, \mathcal{F}, \mathbb{P})$ a probability space, $\mathcal{I} : E \to L^0(\mathbb{P})$ a continuous linear map, and let $0 < q < 1$.*

(i) For every $\alpha \in (0,1)$ there is a number $C_{[\alpha],q}[\|\mathcal{I}\|_{[.]}] < \infty$, depending only on the modulus of continuity $\|\mathcal{I}\|_{[.]}$, α, and q, such that for every finite subset $\{x_1, \ldots, x_n\}$ of E

$$\Big\| \Big(\sum_{\nu=1}^{n} |\mathcal{I}(x_\nu)|^q \Big)^{1/q} \Big\|_{[\alpha]} \le C_{[\alpha],q}[\|\mathcal{I}\|_{[.]}] \cdot \Big(\sum_{\nu=1}^{n} \| x_\nu \|_E^q \Big)^{1/q} . \tag{4.1.36}$$

(ii) There exists, for every $\alpha \in (0,1)$, a positive function $k_\alpha \le 1$ satisfying $\mathbb{E}[k_\alpha] \ge 1 - \alpha$ such that for all $x \in E$

$$\Big(\int |\mathcal{I}(x)|^q \, k_\alpha d\mathbb{P} \Big)^{1/q} \le C_{[\alpha],q}[\|\mathcal{I}\|_{[.]}] \cdot \| x \|_E .$$

(iii) There exists a probability $\mathbb{P}' = \mathbb{P}/g$ equivalent with $\mathbb{P}$ on $\mathcal{F}$ such that $\mathcal{I}$ is continuous as a map into $L^q(\mathbb{P}')$:

$$\forall x \in E \qquad \|\mathcal{I}(x)\|_{L^q(\mathbb{P}')} \le D_q[\|\mathcal{I}\|_{[.]}] \cdot \|x\|_E \tag{4.1.37}$$

$$\textit{and } \forall \alpha \in (0,1) \qquad \|g\|_{[\alpha;\mathbb{P}]} \le E_{[\alpha]}[\|\mathcal{I}\|_{[.]}] . \tag{4.1.38}$$

Proof. (i) Let $0 < p < q$ and let $(\gamma_\nu^{(q)})$ be a sequence of independent q-stable random variables, all defined on a probability space $(X, \mathcal{X}, dx)$ (see exercise A.8.31). In view of lemma A.8.33 (ii) we have, for every $\omega \in \Omega$,

$$\Big(\sum_\nu |\mathcal{I}(x_\nu)(\omega)|^q \Big)^{1/q} \le B_{[\beta],q}^{(A.8.12)} \cdot \Big\| \sum_\nu \mathcal{I}(x_\nu)(\omega) \gamma_\nu^{(q)} \Big\|_{[\beta;dx]} ,$$

and therefore

$$\Big\| \Big(\sum_\nu |\mathcal{I}(x_\nu)|^q \Big)^{1/q} \Big\|_{[\alpha;\mathbb{P}]} \le B_{[\beta],q} \cdot \Big\| \Big\| \sum_\nu \mathcal{I}(x_\nu) \gamma_\nu^{(q)} \Big\|_{[\beta;dx]} \Big\|_{[\alpha;\mathbb{P}]}$$

by A.8.16 with $0 < \delta < \alpha\beta$:

$$\le B_{[\beta],q} \cdot \Big\| \Big\| \mathcal{I}\Big(\sum_\nu x_\nu \gamma_\nu^{(q)} \Big) \Big\|_{[\delta;\mathbb{P}]} \Big\|_{[\alpha\beta - \delta;dx]}$$

$$\le B_{[\beta],q} \cdot \|\mathcal{I}\|_{[\delta]} \cdot \Big\| \Big\| \sum_\nu x_\nu \gamma_\nu^{(q)} \Big\|_E \Big\|_{[\alpha\beta - \delta;dx]}$$

by exercise A.8.15:

$$\le \frac{B_{[\beta],q} \cdot \|\mathcal{I}\|_{[\delta]}}{(\alpha\beta - \delta)^{1/p}} \cdot \Big\| \Big\| \sum_\nu x_\nu \gamma_\nu^{(q)} \Big\|_E \Big\|_{L^p(dx)}$$

by definition A.8.34:

$$\le \frac{B_{[\beta],q} \cdot \|\mathcal{I}\|_{[\delta]} \cdot T_{p,q}(E)}{(\alpha\beta - \delta)^{1/p}} \cdot \Big(\sum_\nu \|x_\nu\|_E^q \Big)^{1/q} .$$

Due to proposition A.8.39, the quantity $C_{[\alpha],q}[\|\mathcal{I}\|_{[.]}]^{(4.1.36)}$ is finite. Namely,

$$C_{[\alpha],q}[\|\mathcal{I}\|_{[.]}] \le \inf_{\substack{0<\beta<1 \\ 0<\delta<\alpha\beta \\ 0<p<q}} \frac{B_{[\beta],q}^{(A.8.12)} \cdot T_{p,q}(E) \cdot \|\mathcal{I}\|_{[\delta]}}{(\alpha\beta - \delta)^{1/p}}$$

with $\delta = \alpha/4$: $$\leq \frac{B_{[1/2],q} \cdot T_{q/2,q}(E) \|\mathcal{I}\|_{[\alpha/4]}}{(\alpha/4)^{2/q}} . \tag{4.1.39}$$

Exercise 4.1.13 $C_{[\alpha],.8,\|\mathcal{I}\|_{[.]}} \leq 218 \, \|\mathcal{I}\|_{[\alpha/8]} / \sqrt{\alpha}$.

(ii) Following the proof of theorem 4.1.7 we would now like to apply Rosenthal's criterion 4.1.4 to produce k_α. It does not apply as stated, but there is an easy extension of the proof, due to Nikisin and used once before in proposition 4.1.1, that yields the claim. Namely, inequality (4.1.36) can be read as follows: for every random variable of the form

$$\phi = \phi_{x_1,\ldots,x_n} \stackrel{\text{def}}{=} \sum_{\nu=1}^{n} |\mathcal{I}(x_\nu)|^q , \qquad n \in \mathbb{N} , \; x_\nu \in E ,$$

on Ω there is the set

$$k_{x_1,\ldots,x_n} \stackrel{\text{def}}{=} \left[|\phi_{x_1,\ldots,x_n}|^{1/q} \leq C_{[\alpha],q}[\|\mathcal{I}\|_{[.]}] \cdot \Big(\sum_{\nu=1}^{n} \|x_\nu\|_E \Big)^{1/q} \right]$$

of measure $\mathbb{P}\big[k_{x_1,\ldots,x_n}\big] \geq 1 - \alpha$ so that

$$\mathbb{E}[\phi_{x_1,\ldots,x_n} \cdot k_{x_1,\ldots,x_n}] \leq C_{[\alpha],q}[\|\mathcal{I}\|_{[.]}] \cdot \sum_{\nu=1}^{n} \|x_\nu\|_E^q .$$

Let K be the collection of positive random variables $k \leq 1$ on Ω satisfying $\mathbb{E}[k] \geq 1 - \alpha$. K is evidently convex and $\sigma(L^\infty, L^1)$-compact. The functions $k \mapsto \mathbb{E}[\phi_{x_1,\ldots,x_n} \cdot k]$ are lower semicontinuous as the suprema of integrals against chopped functions. Therefore the functions

$$h_{x_1,\ldots,x_n} : \; k \mapsto C_{[\alpha],q}[\|\mathcal{I}\|_{[.]}]^q \cdot \sum_{\nu=1}^{n} \|x_\nu\|_E^q - \mathbb{E}[\phi_{x_1,\ldots,x_n} \cdot k]$$

are upper semicontinuous on K, each one having a point of positivity $k_{x_1,\ldots,x_n} \in K$. Their collection clearly forms a convex cone $\mathcal{H}$. Ky–Fan's minimax theorem A.2.34 furnishes a common point $k_\alpha \in K$ of positivity, and this function evidently answers the second claim.

(iii) We pick an $\alpha_0 \in (0,1)$ with $C_{[\alpha_0],q}[\|\mathcal{I}\|_{[.]}] > 0$ and set $\alpha_n = \alpha_0/2^n$, $\zeta_n = C_{[\alpha_n],q}[\|\mathcal{I}\|_{[.]}]^q$, and $\mathbb{P}' = g'\mathbb{P}$ with

$$g' = \gamma' \sum_{n=0}^{\infty} 2^{-n} \frac{k_{\alpha_n}}{\zeta_n} , \text{ where } \gamma' \in \left(\frac{\zeta_0}{2} , \frac{\zeta_0}{1-\alpha_0} \right)$$

is chosen so as to make $\mathbb{P}'$ a probability. Then we proceed literally as in equation (4.1.4) on page 190 to estimate

$$\|\mathcal{I}(x)\|_{L^q(\mathbb{P}')} \leq \left(\frac{2}{1-\alpha_0} \right)^{1/q} C_{[\alpha_0],q}[\|\mathcal{I}\|_{[.]}] \cdot \|x\|_E$$

and $$\|g\|_{[\alpha;\mathbb{P}]} \leq \frac{16 \alpha_0 C^q_{[\alpha/4],q}[\|\mathcal{I}\|_{[.]}]}{\alpha C^q_{[\alpha_0],q}[\|\mathcal{I}\|_{[.]}]} .$$

This gives in the present range $0<q<1$ the estimates

$$D^{(4.1.8)}_{q,d}[\lceil Z\rceil_{[.]}] \le D^{(4.1.37)}_{q}[\|\mathcal{I}\|_{[.]}] \le \left(\frac{2}{1-\alpha_0}\right)^{1/q} C_{[\alpha_0],q}[\|\mathcal{I}\|_{[.]}]$$

and
$$E^{(4.1.9)}_{[\alpha],q} \le E^{(4.1.38)}_{[\alpha],q}[\|\mathcal{I}\|_{[.]}] \le \frac{16\alpha_0 C^q_{[\alpha/4],q}[\|\mathcal{I}\|_{[.]}]}{\alpha C^q_{[\alpha_0],q}[\|\mathcal{I}\|_{[.]}]} \tag{4.1.40}$$

by (4.1.39):
$$\le 16\frac{\alpha_0^2\cdot\|\mathcal{I}\|^q_{[\alpha/16]}}{\alpha^2\cdot\|\mathcal{I}\|^q_{[\alpha_0/4]}} .$$

Proof of Theorem 4.1.2 (ii) for $1<q<\infty$. Let $0<p<1$. We know now that there is a probability $\mathbb{P}'=\mathbb{P}/g$ with respect to which Z is a global L^p-integrator. Its size and that of g are controlled by the inequalities

$$\lceil Z\rceil_{\mathcal{I}^p[\mathbb{P}']} \le D^{(4.1.8)}_{p,d}[\lceil Z\rceil_{[.]}] \quad\text{and}\quad \|g\|_{[\alpha]} \le E^{(4.1.9)}_{[\alpha],p}[\lceil Z\rceil_{[.]}] .$$

By part (i) of theorem 4.1.2 there exists a probability $\mathbb{P}''=\mathbb{P}'/g'$ with respect to which Z is an L^q-integrator of size

$$\lceil Z\rceil_{\mathcal{I}^q[\mathbb{P}'']} \le D^{(4.1.5)}_{p,q,d}\cdot D_{p,d}[\lceil Z\rceil_{[.]}] .$$

The Radon–Nikodym derivative $d\mathbb{P}/d\mathbb{P}''=g'g$ satisfies by A.8.17 with $f=g'$ and $r=p/(q-p)$, and by inequalities (4.1.16) and (4.1.40)

$$E^{(4.1.9)}_{[\alpha],q} \le 2^{\frac{p\vee(q-p)}{p}}\cdot\|g\|^{\frac{q}{p}}_{[\alpha/2]}\cdot\left(\frac{2}{\alpha}\right)^{\frac{q-p}{p}} \tag{4.1.41}$$

$$\le 2^{\frac{p\vee(q-p)}{p}}\cdot\left(\frac{16\alpha_0 C^q_{[\alpha/4],q}[\|\mathcal{I}\|_{[.]}]}{\alpha C^q_{[\alpha_0],q}[\|\mathcal{I}\|_{[.]}]}\right)^{\frac{q}{p}}\cdot\left(\frac{2}{\alpha}\right)^{\frac{q-p}{p}} .$$

The following corollary makes a good exercise to test our understanding of the flow of arguments above; it is also needed in chapter 5.

Corollary 4.1.14 (Factorization for Random Measures) *(i) Let ζ be a spatially bounded global $L^p(\mathbb{P})$-random measure, where $0<p<2$. There exists a probability $\mathbb{P}'$ equivalent with $\mathbb{P}$ on $\mathcal{F}_\infty$ with respect to which ζ is a spatially bounded global L^2-random measure; furthermore, $d\mathbb{P}'/d\mathbb{P}$ is bounded and there exist universal constants D_p and E_p depending only on p such that*

$$\lceil\zeta\rceil_{\mathcal{I}^2[\mathbb{P}']} \le D_p\cdot\lceil\zeta\rceil_{\mathcal{I}^p[\mathbb{P}]} , \qquad D_p = D^{(4.1.5)}_{p,2,d} ,$$

and such that the Radon–Nikodym derivative $g \stackrel{\text{def}}{=} d\mathbb{P}/d\mathbb{P}'$ is bounded away from zero and satisfies

$$\|g\|_{L^{p/(2-p)}(\mathbb{P})} \le E_p , \qquad E_p = E^{(4.1.6)}_{p,2} ,$$

which has the consequence that for any $r>0$ and $f\in\mathcal{F}_\infty$

$$\|f\|_{L^r(\mathbb{P})} \le E_p^{p/2r}\cdot\|f\|_{L^{2r/p}[\mathbb{P}']} .$$

(ii) Let ζ be a spatially bounded global $L^0(\mathbb{P})$-random measure with modulus of continuity[6] $\lceil\zeta\rceil_{[.]}$. *There exists a probability $\mathbb{P}' = \mathbb{P}/g$ equivalent with $\mathbb{P}$ on $\mathcal{F}_\infty$ with respect to which ζ is a global L^2-integrator; furthermore, there exist universal constants $D[\lceil\zeta\rceil_{[.]}]$ and $E = E_{[\alpha]}[\lceil\zeta\rceil_{[.]}]$, depending only on $\alpha \in (0,1)$ and the modulus of continuity $\lceil\zeta\rceil_{[.]}$,*

such that $$\lceil\zeta\rceil_{\mathcal{I}^q[\mathbb{P}']} \le D[\lceil\zeta\rceil_{[.]}]\,, \qquad D = D^{(4.1.8)}_{2,d}\,,$$

and $$\|g\|_{[\alpha]} \le E_{[\alpha]}[\lceil\zeta\rceil_{[.]}] \quad \forall \alpha \in (0,1)\,,\ E = E^{(4.1.9)}_{[\alpha],2}[\lceil\zeta\rceil_{[.]}]\,,$$

– this implies $$\|f\|_{[\alpha+\beta;\mathbb{P}]} \le \Big(E_{[\alpha]}[\lceil\zeta\rceil_{[.]}]/\beta\Big)^{1/r} \|f\|_{L^r(\mathbb{P}')}$$

for any $f \in \mathcal{F}_\infty$, $r > 0$ and $\alpha, \beta \in (0,1)$.

(iii) When ζ is previsible, the exponent 2 can be replaced by any $q < \infty$.

4.2 Martingale Inequalities

Exercise 3.8.12 shows that the square bracket of an integrator of finite variation is just the sum of the squares of the jumps, a quantity of modest interest. For a martingale integrator M, though, the picture is entirely different: the size of the square function controls the size of the martingale, even of its integrator norm. In fact, in the range $1 \le p < \infty$ the quantities $\lceil M\rceil_{\mathcal{I}^p}$, $\|M^\star_\infty\|_{L^p}$, and $\|S_\infty[M]\|_{L^p}$ are all equivalent in the sense that there are universal constants C_p such that

$$\lceil M\rceil_{\mathcal{I}^p} \le C_p\cdot\|M^\star_\infty\|_{L^p}\,,\quad \|M^\star_\infty\|_{L^p} \le C_p\cdot\|S_\infty[M]\|_{L^p}\,,$$

and $$\|S_\infty[M]\|_{L^p} \le C_p\cdot\lceil M\rceil_{\mathcal{I}^p} \tag{4.2.1}$$

for all martingales M. These and related inequalities are proved in this section.

Fefferman's Inequality

The K^q-seminorms are auxiliary seminorms on L^q-integrators, defined for $2 \le q \le \infty$. They appear in Fefferman's famed inequality (4.2.2), and they simplify the proof of inequality (4.2.1) and other inequalities of interest. Towards their definition let us introduce, for every L^0-integrator Z, the class $\mathcal{K}[Z]$ of all $g \in L^2[\mathcal{F}_\infty]$ having the property that

$$\mathbb{E}\Big[[Z,Z]_\infty - [Z,Z]_{T-} \mid \mathcal{F}_T\Big] \le \mathbb{E}[g^2 \mid \mathcal{F}_T]$$

at all stopping times T. $\mathcal{K}[Z]$ is used to define the seminorm $\|Z\|_{\mathcal{K}^q}$ by

$$\|Z\|_{\mathcal{K}^q} = \inf\{\|g\|_{L^q} : g \in \mathcal{K}[Z]\}\,, \qquad 2 \le q \le \infty.$$

[6] $\lceil\zeta\rceil_{[\alpha]} \stackrel{\text{def}}{=} \sup\{\|\int \check X\,d\zeta\|_{[\alpha;\mathbb{P}]} : \check X \in \check{\mathcal{E}}\,,\ |\check X| \le 1\}$ for $0 < \alpha < 1$; see page 56.

As usual, this number is ∞ if $\mathcal{K}[Z]$ is void. When $q = \infty$ it is customary to write $\| Z \|_{\mathcal{K}^\infty} = \| Z \|_{BMO}$ and to say Z has ***bounded mean oscillation*** if this number is finite. We collect now a few properties of the seminorm $\| \ \|_{\mathcal{K}^q}$.

Exercise 4.2.1 $\| Z \|_{\mathcal{K}^q} \le \| S_\infty[Z] \|_{L^q}$. $d[Z,Z] \le d[Z',Z']$ implies $\| Z \|_{\mathcal{K}^q} \le \| Z' \|_{\mathcal{K}^q}$.

Lemma 4.2.2 *Let Z be an L^0-integrator and $I \ge 0$ an adapted increasing right-continuous process. Then, for $1 \le p < 2$,*

$$\mathbb{E}\Big[\int_0^\infty I \, d[Z,Z]\Big] \le \inf\Big\{ \mathbb{E}[I_\infty \cdot g^2] : g \in \mathcal{K}[Z] \Big\}$$
$$\le \Big(\mathbb{E}[I_\infty^{p/(2-p)}]\Big)^{(2-p)/p} \cdot \|Z\|^2_{\mathcal{K}^{p'}} .$$

Proof. With the usual understanding that $[Z,Z]_{0-} = 0 = I_{0-}$, integration by parts gives

$$\int_0^\infty I \, d[Z,Z] = \int_0^\infty \big([Z,Z]_\infty - [Z,Z]_{.-}\big) \, dI$$
$$= \int_0^\infty \big([Z,Z]_\infty - [Z,Z]_{T^\lambda -}\big) \cdot [T^\lambda < \infty] \, d\lambda ,$$

where $T^\lambda = \inf\{t : I_t \ge \lambda\}$ are the stopping times appearing in the change-of-variable theorem 2.4.7. Since $[T^\lambda < \infty] \in \mathcal{F}_{T^\lambda}$,

$$\mathbb{E}\Big[\int_0^\infty I \, d[Z,Z]\Big] = \mathbb{E}\Big[\int_0^\infty \mathbb{E}\big[\, [Z,Z]_\infty - [Z,Z]_{T^\lambda -} \mid \mathcal{F}_{T^\lambda}\big] \cdot [T^\lambda < \infty] \, d\lambda\Big]$$
$$\le \inf\Big\{\mathbb{E}\Big[\int_0^\infty g^2 \cdot [T^\lambda < \infty] \, d\lambda\Big] : g \in \mathcal{K}[Z]\Big\}$$
$$\le \inf\Big\{\mathbb{E}\Big[g^2 \cdot \int_0^\infty [I_\infty > \lambda] \, d\lambda\Big] : g \in \mathcal{K}[Z]\Big\}$$
$$= \inf\big\{\mathbb{E}\big[I_\infty \cdot g^2\big] : g \in \mathcal{K}[Z]\big\}$$
$$\le \Big(\mathbb{E}\big[I_\infty^{p/(2-p)}\big]\Big)^{(2-p)/p} \cdot \|Z\|^2_{\mathcal{K}^{p'}} .$$

The last inequality comes from an application of Hölder's inequality with conjugate exponents $p/(2-p)$ and $p'/2$. ∎

On an L^2-bounded martingale M the norm $\| M \|_{\mathcal{K}^q}$ can be rewritten in a useful way:

Lemma 4.2.3 *For any stopping time T*

$$\mathbb{E}\big[\, [M,M]_\infty - [M,M]_{T-} \mid \mathcal{F}_T \,\big] = \mathbb{E}\big[\, (M_\infty - M_{T-})^2 \mid \mathcal{F}_T \,\big] ,$$

and consequently $\mathcal{K}[M]$ equals the collection

$$\Big\{ g \in \mathcal{F}_\infty : \mathbb{E}\big[(M_\infty - M_{T-})^2 \mid \mathcal{F}_T\big] \le \mathbb{E}\big[g^2 \mid \mathcal{F}_T\big] \quad \forall \text{ stopping times } T \Big\} .$$

Proof. Let $U \ge T$ be a bounded stopping time such that $(M^U - M^T)_{.-}$ is bounded. By proposition 3.8.19

$$(M^U - M^T)^2 = 2(M^U - M^T)_{.-}{*}(M^U - M^T) + [M,M]^U - [M,M]^T ,$$

and so

$$[M,M]_U - [M,M]_{T-} = (M_U - M_T)^2 + (\Delta M_T)^2 \\ - 2\int_{T+}^{U} (M - M^T)_{.-}\, d(M - M^T) .$$

The integral is the value at U of a martingale that vanishes at T, so its conditional expectation on $\mathcal{F}_T$ vanishes (theorem 2.5.22). Consequently,

$$\begin{aligned} &\mathbb{E}\big[\, [M,M]_U - [M,M]_{T-} \,\big|\, \mathcal{F}_T \,\big] = \mathbb{E}\big[\, (M_U - M_T)^2 + (\Delta M_T)^2 \,\big|\, \mathcal{F}_T \,\big] \\ =&\mathbb{E}\big[\, (M_U - M_{T-} - \Delta M_T)^2 + (\Delta M_T)^2 \,\big|\, \mathcal{F}_T \,\big] \\ =&\mathbb{E}\big[\, (M_U - M_{T-})^2 - 2(M_U - M_{T-})\Delta M_T + 2(\Delta M_T)^2 \,\big|\, \mathcal{F}_T \,\big] \\ =&\mathbb{E}\big[\, (M_U - M_{T-})^2 \,\big|\, \mathcal{F}_T \,\big] . \end{aligned}$$

Now U can be chosen arbitrarily large. As $U \to \infty$, $[M,M]_U \to [M,M]_\infty$ and $M_U^2 \to M_\infty^2$ in L^1-mean, whence the claim. ∎

Since $|M_\infty - M_{T-}| \le 2M_\infty^\star$, the following consequence is immediate:

Corollary 4.2.4 *For any martingale M, $2M_\infty^\star \in \mathcal{K}^q[M]$ and consequently*

$$\|\, M \,\|_{\mathcal{K}^q} \le 2\, \|\, M_\infty^\star \,\|_{L^q} , \qquad 2 \le q < \infty .$$

Lemma 4.2.5 *Let I, D be positive bounded right-continuous adapted processes, with I increasing, D decreasing, and such that $I{\cdot}D$ is still increasing. Then for any bounded martingale N with $|N_\infty| \le D_\infty$ and any $q \in [2,\infty)$*

$$2I_\infty{\cdot}D_\infty \in \mathcal{K}[I_{.-}{*}N] \quad \textit{and thus} \quad \|\, I_{.-}{*}N \,\|_{\mathcal{K}^q} \le 2\, \|\, I_\infty \cdot D_\infty \,\|_{L^q} .$$

Proof. Let T be a stopping time. The stochastic integral in the quantity

$$Q \overset{\text{def}}{=} \mathbb{E}\Big[\, \big((I_{.-}{*}N)_\infty - (I_{.-}{*}N)_{T-}\big)^2 \,\Big|\, \mathcal{F}_T \Big]$$

that must be estimated is the limit of

$$I_{T-}{\cdot}\Delta N_T + \sum_{k=0}^{\infty} I_{S_k}{\cdot}(N^T_{S_{k+1}} - N^T_{S_k})$$

as the partition $\mathcal{S} = \{T = S_0 \le S_1 \le S_2 \le \ldots\}$ runs through a sequence $(\mathcal{S}^n)$ whose mesh tends to zero. (See theorem 3.7.23 and proposition 3.8.21.)

If we square this and take the expectation, the usual cancellation occurs, and

$$\begin{aligned}
Q &\le \lim_S \mathbb{E}\Big[(I_{T-}\cdot\Delta N_T)^2 + \sum_{0\le k} I^2_{S_k}\cdot(N^{T\,2}_{S_{k+1}} - N^{T\,2}_{S_k}) \mid \mathcal{F}_T\Big] \\
&\le \lim_S \mathbb{E}\Big[(I_{T-}\cdot\Delta N_T)^2 + \sum_{0\le k} I^2_{S_{k+1}}\cdot N^{T\,2}_{S_{k+1}} - \sum_{0\le k} I^2_{S_k}\cdot N^{T\,2}_{S_k} \mid \mathcal{F}_T\Big] \\
&= \mathbb{E}\Big[I^2_{T-}\cdot(N_T - N_{T-})^2 + I^2_\infty\cdot N^2_\infty - I^2_T\cdot N^2_T \mid \mathcal{F}_T\Big] \\
&\le \mathbb{E}\Big[I^2_{T-}\cdot(N^2_T - 2N_T N_{T-} + N^2_{T-}) + I^2_\infty\cdot N^2_\infty - I^2_{T-}\cdot N^2_T \mid \mathcal{F}_T\Big] \\
&\le \mathbb{E}\Big[I^2_{T-}\cdot(2|N_T||N_{T-}| + |N^2_{T-}|) + I^2_\infty\cdot N^2_\infty \mid \mathcal{F}_T\Big]
\end{aligned}$$

results. Now $|N_T| \le D_T \le D_{T-}$ and $|N_{T-}| \le D_{T-}$, so we continue

$$Q \le \mathbb{E}[3I^2_{T-}\cdot D^2_{T-} + I^2_\infty\cdot D^2_\infty \mid \mathcal{F}_T] \le 4\cdot\mathbb{E}[I^2_\infty\cdot D^2_\infty \mid \mathcal{F}_T]\,.$$

This says, in view of lemma 4.2.3, that $2I_\infty\cdot D_\infty \in \mathcal{K}[I_{.-}*N]$. ∎

Exercise 4.2.6 The conclusion persists for unbounded I and D.

Theorem 4.2.7 (Fefferman's Inequality) *For any two L^0-integrators Y, Z and $1 \le p \le 2$*

$$\mathbb{E}[\,|[Y,Z]|_\infty] \le \sqrt{2/p}\cdot\|S_\infty[Y]\|_{L^p}\cdot\|Z\|_{\mathcal{K}^{p'}}\,. \tag{4.2.2}$$

Proof. Let us abbreviate $S = S[Y]$. The mean value theorem produces

$$S^p_t - S^p_s = \left(S^2_t\right)^{p/2} - \left(S^2_s\right)^{p/2} = (p/2)\,\sigma^{p/2\,-1}\cdot\left(S^2_t - S^2_s\right),$$

where σ is a point between S^2_s and S^2_t. Since $p \le 2$, we have

$$p/2 - 1 \le 0$$

and $$(S^2_t)^{p/2\,-1} \le \sigma^{p/2\,-1}\,;$$

thus $$(p/2)\cdot S^{p-2}_t\cdot\left(S^2_t - S^2_s\right) \le S^p_t - S^p_s\,,$$

and by the same token $$(p/2)\cdot S^{p-2}_0 \le S^p_0\,.$$

We read this as a statement about the measures $d(S^p)$ and $d(S^2)$:

$$S^{p-2}\cdot d(S^2) \le (2/p)\,d(S^p)$$

(exercise 4.2.9). In conjunction with the theorem 3.8.9 of Kunita–Watanabe, this yields the estimate

$$\begin{aligned}
|[Y,Z]|_\infty &= \int_0^\infty S^{p/2\,-1}\cdot S^{1-\,p/2}\,d|[Y,Z]| \\
&\le \Big(\int_0^\infty S^{p-2}\cdot d(S^2)\Big)^{1/2}\cdot\Big(\int_0^\infty S^{2-p}\,d[Z,Z]\Big)^{1/2} \\
&\le \sqrt{2/p}\cdot\Big(\int_0^\infty d(S^p)\Big)^{1/2}\cdot\Big(\int_0^\infty S^{2-p}\,d[Z,Z]\Big)^{1/2}.
\end{aligned}$$

Upon taking the expectation and applying the Cauchy–Schwarz inequality,

$$\mathbb{E}[\,\rvert [Y,Z] \lvert_\infty] \le \sqrt{2/p}\cdot \left(\mathbb{E}[S^p_\infty]\right)^{1/2} \cdot \left(\mathbb{E}\Big[\int_0^\infty S^{2-p}\, d[Z,Z]\Big]\right)^{1/2}$$

follows. From lemma 4.2.2 with $I = S^{2-p}$,

$$\begin{aligned}\mathbb{E}[\,\rvert [Y,Z] \lvert_\infty] &\le \sqrt{2/p}\cdot \left(\mathbb{E}[S^p_\infty]\right)^{1/2} \cdot \left(\left(\mathbb{E}\big[S^p_\infty\big]\right)^{(2-p)/p} \cdot \|Z\|^2_{\mathcal{K}^{p'}}\right)^{1/2} \\ &= \sqrt{2/p}\cdot \|S_\infty\|_{L^p} \cdot \|Z\|_{\mathcal{K}^{p'}} . \end{aligned}$$

From corollary 4.2.4 and Doob's maximal theorem 2.5.19, the following consequence is immediate:

Corollary 4.2.8 *Let Z be an L^0-integrator and M a martingale. Then*

$$\begin{aligned}\mathbb{E}\big[\,\rvert [Z,M] \lvert_\infty\big] &\le 2\sqrt{2/p}\cdot \|S_\infty[Z]\|_{L^p} \cdot \|M^\star_\infty\|_{L^{p'}} \\ &\le 2\sqrt{2p}\cdot \|S_\infty[Z]\|_{L^p} \cdot \|M_\infty\|_{L^{p'}} . \end{aligned} \qquad 1 \le p \le 2$$

Exercise 4.2.9 Let $y, z, f : [0,\infty) \to [0,\infty)$ be right-continuous with left limits, y and z increasing. If $f_0 \cdot y_0 \le z_0$ and $f_t \cdot (y_t - y_s) \le z_t - z_s \quad \forall\, s < t$, then

$$f \cdot dy \le dz .$$

Exercise 4.2.10 Let M, N be two locally square integrable martingales. Then

$$\mathbb{E}[\ \rvert [M,N] \lvert_\infty\ ] \le \sqrt{2}\cdot \mathbb{E}[S_\infty[M]] \cdot \|N\|_{BMO} .$$

This was Fefferman's original result, enabling him to show that the martingales N with $\|N\|_{BMO} < \infty$ form the dual of the subspace of martingales in $\mathcal{I}^1$.

Exercise 4.2.11 For any local martingale M and $1 \le p \le 2$,

$$\rvert M \lvert_{\mathcal{I}^p} \le C_p \cdot \|S_\infty[M]\|_{L^p} \tag{4.2.3}$$

with $C_2 = 1$ and $C_p \le 2\sqrt{2p}$.

The Burkholder–Davis–Gundy Inequalities

Theorem 4.2.12 *Let $1 \le p < \infty$ and M a local martingale. Then*

$$\|M^\star_\infty\|_{L^p} \le C_p \cdot \|S_\infty[M]\|_{L^p} \tag{4.2.4}$$

and

$$\|S_\infty[M]\|_{L^p} \le C_p \cdot \|M^\star_\infty\|_{L^p} . \tag{4.2.5}$$

The arguments below provide the following bounds for the constants C_p:

$$C_p^{(4.2.4)} \le \begin{cases} \sqrt{10p}, & 1 \le p < 2, \\ 2, & p = 2, \\ \sqrt{e/2}\,p, & 2 < p < \infty \end{cases} ; \quad C_p^{(4.2.5)} \le \begin{cases} 6/\sqrt{p}, & 1 \le p \le 2, \\ 1, & p = 2, \\ \sqrt{2p}, & 2 \le p < \infty. \end{cases} \tag{4.2.6}$$

Proof of (4.2.4) for $2 \le p < \infty$. Let $K > 0$ and set $T = \inf\{t : |M_t| > K\}$. Itô's formula gives

$$|M_T|^p = |M_0|^p + p\int_{0+}^{T} |M_{.-}|^{p-1} \operatorname{sgn} M_{.-}\, dM$$
$$+ p(p-1)\int_0^1 (1-\lambda)\int_{0+}^{T} \big|(1-\lambda)M_{.-} + \lambda M\big|^{p-2}\, d[M,M]\, d\lambda$$
$$\le p\int_{0+}^{T} |M_{.-}|^{p-1} \operatorname{sgn} M_{.-}\, dM + \frac{p(p-1)}{2}\int_0^T |M|^{\star(p-2)}\, d[M,M] \,.$$

If $\| S_\infty[M] \|_{L^p} = \infty$, there is nothing to prove. And in the opposite case, $M_T^\star \le K + S_T[M]$ belongs to L^p, and M^T is a global L^p-integrator (theorem 2.5.30). The stochastic integral on the left in the last line above has a bounded integrand and is the value at T of a martingale vanishing at zero (exercise 3.7.9), and thus has expectation zero. Applying Doob's maximal theorem 2.5.19 and Hölder's inequality with conjugate exponents $p/(p-2)$ and $p/2$, we get

$$\mathbb{E}\big[|M_T|^{\star p}\big] \le \frac{p^p}{(p-1)^p} \cdot \mathbb{E}\big[|M_T|^p\big]$$
$$\le \frac{p^p \cdot p(p-1)}{(p-1)^p \cdot 2} \cdot \mathbb{E}\Big[\int_0^T |M|^{\star(p-2)}\, d[M,M]\Big]$$
$$\le \frac{p^2 \cdot p^{p-1}}{2 \cdot (p-1)^{p-1}} \cdot \mathbb{E}\Big[M_T^{\star(p-2)} \cdot \big(S_T[M]\big)^2\Big] \tag{4.2.7}$$
$$\le \frac{p^2}{2}\Big(1 + \frac{1}{p-1}\Big)^{p-1} \cdot \Big(\mathbb{E}[M_T^{\star p}]\Big)^{(p-2)/p} \Big(\mathbb{E}\big[(S_T[M])^p\big]\Big)^{2/p} < \infty \,.$$

Division by $\mathbb{E}\big[M_T^{\star p}\big]^{1-\,2/p}$ and taking the square root results in

$$\| M_T^\star \|_{L^p} \le p \cdot \sqrt{\Big(1 + \frac{1}{p-1}\Big)^{p-1}\Big/ 2} \cdot \| S_T[M] \|_{L^p} \le p\sqrt{e/2} \cdot \| S_T[M] \|_{L^p} \,.$$

Now we let K and with it T increase without bound. ∎

Exercise 4.2.13 For $2 \le q < \infty$, $\| M \|_{\mathcal{K}^q}/2 \le \big\| M_\infty^\star \big\|_{L^q} \le \sqrt{e/2}\, q \cdot \| M \|_{\mathcal{K}^q}$.

Proof of (4.2.4) for $1 \le p \le 2$. Doob's maximal theorem 2.5.19 and exercise 4.2.11 produce

$$\| M_\infty^\star \|_{L^p} \le p' \cdot \| M_\infty \|_{L^p} \le p' C_p^{(4.2.3)} \cdot \| S_\infty[M] \|_{L^p} \,.$$

This applies only for $p > 1$, though, and the estimate $p'\ 2\sqrt{2p}$ of the constant has a pole at $p = 1$. So we must argue differently for the general case. We use the maximal theorem for integrators 2.3.6: an application of exercise 4.2.11 gives

$$\| M^\star_\infty \|_{L^p} \le C_p^{\star(2.3.5)} C_p^{(4.2.3)} \cdot \| S_\infty[M] \|_{L^p} \le 6.7 \cdot 2^{1/p}\sqrt{p} \cdot \| S_\infty[M] \|_{L^p} .$$

The constant is a factor of 4 larger than the $\sqrt{10p}$ of the statement. We borrow the latter value from Garsia [35]. ∎

Proof of (4.2.5) for $1 \le p \le 2$. By homogeneity we may assume that $\|M^\star_\infty\|_{L^p}=1$. Then, using Hölder's inequality with conjugate exponents $2/p$ and $2/(2-p)$,

$$\begin{aligned}\mathbb{E}[(S_\infty[M])^p] &= \mathbb{E}\Big[\Big(M_\infty^{\star(p-2)} \cdot [M,M]_\infty\Big)^{p/2} \cdot M_\infty^{\star p(2-p)/2}\Big] \\ &\le \Big(\mathbb{E}\big[M_\infty^{\star(p-2)} \cdot [M,M]_\infty\big]\Big)^{p/2} ,\end{aligned}$$

i.e., $$\|S_\infty[M]\|^2_{L^p} \le \mathbb{E}\big[M_\infty^{\star(p-2)} \cdot [M,M]_\infty\big] .$$

With $$\begin{aligned}[M,M]_\infty &= M^2_\infty - 2\int_0^\infty M_{\!-}\, dM \\ &\le M_\infty^{\star 2} - 2\int_0^\infty M_{\!-}\, dM ,\end{aligned}$$

this turns into $$\|S_\infty[M]\|^2_{L^p} \le 1 + 2 \cdot \Big|\mathbb{E}\Big[M_\infty^{\star(p-2)} \cdot \int_0^\infty M_{\!-}\, dM\Big]\Big| .$$

Now let N be the martingale that has $N_\infty = M_\infty^{\star(p-2)}$. We employ lemma 4.2.5 with $I = M^\star$ and $D = M^{\star(p-2)}$. The previous inequality can be continued as

$$\begin{aligned}\|S_\infty[M]\|^2_{L^p} &\le 1 + 2 \cdot \Big|\mathbb{E}\big[N_\infty \cdot M_{\!-}{*}M_\infty\big]\Big| \\ &= 1 + 2 \cdot \Big|\mathbb{E}\big[[M, M_{\!-}{*}N]_\infty\big]\Big| \\ \text{by theorem 4.2.7:} \quad &\le 1 + 2\sqrt{2/p} \cdot \|S_\infty[M]\|_{L^p} \cdot \|M_{\!-}{*}N\|_{\mathcal{K}^{p'}} \\ \text{by exercise 4.2.1:} \quad &\le 1 + 2\sqrt{2/p} \cdot \|S_\infty[M]\|_{L^p} \cdot \big\|M^\star_{\!-}{*}N\big\|_{\mathcal{K}^{p'}} \\ \text{by lemma 4.2.5:} \quad &\le 1 + 4\sqrt{2/p} \cdot \|S_\infty[M]\|_{L^p} \cdot \Big\|M_\infty^{\star(p-1)}\Big\|_{L^{p'}} \\ &= 1 + 4\sqrt{2/p} \cdot \|S_\infty[M]\|_{L^p} .\end{aligned}$$

Completing the square we get

$$\|S_\infty[M]\|_{L^p} \le \sqrt{1+8/p} + \sqrt{8/p} < 6/\sqrt{p} .$$

If $M_\infty^{\star(p-2)}$ is not bounded, then taking the supremum over bounded martingales N with $N_\infty \le M_\infty^{\star(p-2)}$ achieves the same thing. ∎

Proof of (4.2.5) for $2 \le p < \infty$. Set $S = S[M]$. An easy argument as in the proof of theorem 4.2.7 gives

$$d(S^p) \le \frac{p}{2} \cdot S^{p-2}\, d(S^2) = \frac{p}{2} \cdot S^{p-2}\, d[M,M] \,,$$

and thus $$\mathbb{E}[S^p_\infty] \le \frac{p}{2} \cdot \mathbb{E}\Big[\int_0^\infty S^{p-2}\, d[M,M]\Big] \,.$$

Lemma 4.2.2 and corollary 4.2.4 allow us to continue with

$$\mathbb{E}[S^p_\infty] \le 2p \cdot \mathbb{E}\big[S^{p-2}_\infty \cdot M^{\star 2}_\infty\big] \le 2p \cdot \big(\mathbb{E}[S^p_\infty]\big)^{(p-2)/p} \cdot \big(\mathbb{E}[M^{\star p}_\infty]\big)^{2/p} \,.$$

We divide by $\mathbb{E}\big[S^p_\infty\big]^{1-2/p}$, take the square root, and arrive at

$$\| S_\infty[M] \|_{L^p} \le \sqrt{2p} \cdot \| M^\star_\infty \|_{L^p} \,.$$

Here is an interesting little application of the Burkholder–Davis–Gundy inequalities:

Exercise 4.2.14 (A Strong Law of Large Numbers) In a generalization of exercise 2.5.17 on page 75 prove the following: let $F_1, F_2, \ldots$ be a sequence of random variables that have bounded q^{th} moments for some fixed $q > 1$: $\| F_\nu \|_{L^q} \le \sigma_q$, all having the same expectation p. Assume that the conditional expectation of F_{n+1} given $F_1, F_2, \ldots, F_n$ equals p as well, for $n = 1, 2, 3, \ldots$. [To paraphrase: knowledge of previous executions of the experiment may influence the law of its current replica only to the extent that the expectation does not change and the qth moments do not increase overly much.] Then

$$\lim_{n\to\infty} \frac{1}{n} \sum_{\nu=1}^{n} F_\nu = p \quad \text{almost surely.}$$

The Hardy Mean

The following observation will throw some light on the merit of inequality (4.2.4). Proposition 3.8.19 gives $[X{*}M, X{*}M] = X^2{*}[M,M]$ for elementary integrands X. Inequality (4.2.4) applied to the local martingale $X{*}M$ therefore yields

$$\Big\| \int X\, dM \Big\|_{L^p} \le C_p \cdot \Big(\int \Big(\int_0^\infty X^2\, d[M,M] \Big)^{p/2} d\mathbb{P} \Big)^{1/p} \tag{4.2.8}$$

for $1 \le p < \infty$. The corresponding assignment

$$F \mapsto \|F\|^{\mathcal{H}*}_{M\text{–}p} \overset{\text{def}}{=} \Big(\int^* \Big(\int^* F_t^2(\omega)\, d[M,M]_t(\omega) \Big)^{p/2} \mathbb{P}(d\omega) \Big)^{1/p} \tag{4.2.9}$$

is a ***pathwise mean*** in the sense that it computes first for every single path $t \mapsto F_t(\omega)$ separately a quantity, $\big(\int^* F_t^2(\omega)\, d[M,M]_t(\omega)\big)^{1/2}$ in this case, and then applies a p-mean to the resulting random variable. It is called the ***Hardy mean.*** It controls the integral in the sense that

$$\Big\| \int X \, dM \Big\|_{L^p} \le C_p^{(4.2.4)} \cdot \| X \|_{M\text{-}p}^{\mathcal{H}*}$$

for elementary integrands X, and it can therefore be used to extend the elementary integral just as well as Daniell's mean can. It offers "pathwise" or "ω–by–ω" control of the integrand, and such is of paramount importance for the solution of stochastic differential equations – for more on this see sections 4.5 and 5.2.

The Hardy mean is the favorite mean of most authors who treat stochastic integration *for martingale integrators* and exponents $p \ge 1$.

How do the Hardy mean and Daniell's mean compare? The minimality of Daniell's mean on previsible processes (exercises 3.6.16 and 3.5.7(ii)) gives

$$\| F \|_{M\text{-}p}^{*} \le C_p^{(4.2.4)} \cdot \| F \|_{M\text{-}p}^{\mathcal{H}*} \tag{4.2.10}$$

for $1 \le p < \infty$ and *all previsible* F. In fact, if M is continuous, so that $S[M]$ agrees with the previsible square function $s[M]$, then inequality (4.2.10) extends to all $p > 0$ (exercise 4.3.20). On the other hand, proposition 3.8.19 and equation (3.7.5) produce for all elementary integrands X

$$\Big(\int \Big(\int_0^\infty X^2 \, d[M,M] \Big)^{p/2} d\mathbb{P} \Big)^{1/p} \le K_p^{(3.8.6)} \cdot \lceil X{*}M \rceil_{\mathcal{I}^p} = K_p \cdot \| X \|_{M\text{-}p}^{*} \,,$$

which due to proposition 3.6.1 results in the converse of inequality (4.2.10):

$$\| F \|_{M\text{-}p}^{\mathcal{H}*} \le K_p \cdot \| F \|_{M\text{-}p}^{*}$$

for $0 < p < \infty$ and for all functions F on the ambient space. In view of the integrability criterion 3.4.10, both $\| \; \|_{M\text{-}p}^{*}$ and $\| \; \|_{M\text{-}p}^{\mathcal{H}*}$ have the same *previsible* integrable processes:

$$\mathcal{P} \cap \mathfrak{L}^1[\lceil\!\lceil \; \rceil\!\rceil_{M\text{-}p}^{*}] = \mathcal{P} \cap \mathfrak{L}^1[\| \; \|_{M\text{-}p}^{\mathcal{H}*}] \text{ and } \lceil\!\lceil \; \rceil\!\rceil_{M\text{-}p}^{*} \approx \| \; \|_{M\text{-}p}^{\mathcal{H}*}$$

on this space for $1 \le p < \infty$, and even for $0 < p < \infty$ if M happens to be continuous. Here is an instance where $\| \; \|_{M\text{-}p}^{\mathcal{H}*}$ is nevertheless preferable: Suppose that M is *continuous*; then so is $[M,M]$, and $\| \; \|_{M\text{-}p}^{\mathcal{H}*}$ annihilates the graph of any random time – Daniell's mean $\| \; \|_{M\text{-}p}^{*}$ may well fail to do so. Now a well-measurable process differs from a predictable one only on the graphs of countably many stopping times (exercise A.5.18). Thus a well-measurable process is $\| \; \|_{M\text{-}p}^{\mathcal{H}*}$-measurable, and all well-measurable processes with finite mean are integrable if the mean $\| \; \|_{M\text{-}p}^{\mathcal{H}*}$ is employed. For another instance see the proof of theorem 4.2.15. ∎

Martingale Representation on Wiener Space

Consider an L^p-integrator $\boldsymbol{Z}$. The definite integral $\boldsymbol{X} \mapsto \int \boldsymbol{X}\, d\boldsymbol{Z}$ is a map from $\mathfrak{L}^1[\boldsymbol{Z}\text{–}p]$ to L^p. One might reasonably ask what its kernel and range are. Not much can be said when $\boldsymbol{Z}$ is arbitrary; but if it is a Wiener process and $p > 1$, then there is a complete answer (see also theorem 4.6.10 on page 261):

Theorem 4.2.15 *Assume that $\boldsymbol{W} = (W^1, \ldots, W^d)$ is a standard d-dimensional Wiener process on its natural filtration $\mathcal{F}_.[\boldsymbol{W}]$, and let $1 < p < \infty$. Then for every $f \in L^p(\mathcal{F}_\infty[\boldsymbol{W}])$ there is a unique $\boldsymbol{W}$–p-integrable vector $\boldsymbol{X} = (X^1, \ldots, X^d)$ of previsible processes so that*

$$f = \mathbb{E}[f] + \int_0^\infty \boldsymbol{X}\, d\boldsymbol{W} .$$

Put slightly differently, the martingale $M^f_. \overset{\text{def}}{=} \mathbb{E}[f|\mathcal{F}_.[\boldsymbol{W}]]$ has the representation

$$M^f_t = \mathbb{E}[f] + \int_0^t \boldsymbol{X}\, d\boldsymbol{W} .$$

Proof (Si–Jian Lin). Denote by $\boldsymbol{H}$ the discrete space $\{1, \ldots, d\}$ and by $\check{\boldsymbol{B}}$ the set $\boldsymbol{H} \times \boldsymbol{B}$ equipped with its elementary integrands $\check{\mathcal{E}} \overset{\text{def}}{=} C(\boldsymbol{H}) \otimes \mathcal{E}$. As on page 109, a d-vector of processes on $\boldsymbol{B}$ is identified with a function on $\check{\boldsymbol{B}}$. According to theorem 2.5.19 and exercise 4.2.18, the stochastic integral

$$\boldsymbol{X} \mapsto \int \boldsymbol{X}\, d\boldsymbol{W} = \sum_{\eta=1}^{d} \int_0^\infty X^\eta\, dW^\eta$$

is up to a constant an isometry of $\mathcal{P} \cap \mathfrak{L}^1[\| \ \|^*_{\boldsymbol{W}\text{–}p}]$ onto a subspace of $L^p(\mathcal{F}_\infty[\boldsymbol{W}])$. Namely, for $1 < p < \infty$ and with $M \overset{\text{def}}{=} \boldsymbol{X}{*}\boldsymbol{W}$

$$\left\| \int \boldsymbol{X}\, d\boldsymbol{W} \right\|_p = \|M_\infty\|_p$$

by theorems 2.5.19 and 4.2.12: $\quad \sim \|M^\star_\infty\|_p \sim \|S_\infty[M]\|_p = \|\boldsymbol{X}\|^{\mathcal{H}*}_{\boldsymbol{W}\text{–}p}$

by definition (4.2.9): $\quad \sim \|\boldsymbol{X}\|^*_{\boldsymbol{W}\text{–}p}$.

The image of $\mathfrak{L}^1[\| \ \|^*_{\boldsymbol{W}\text{–}p}]$ under the stochastic integral is thus a complete and therefore closed subspace $\mathcal{S} \subset \{f \in L^p(\mathcal{F}_\infty[\boldsymbol{W}]) : \mathbb{E}[f] = 0\}$. Since bounded pointwise convergence implies mean convergence in L^p, the subspace $\mathcal{S}_b(\mathbb{C})$ of bounded functions in the complexification of $\mathbb{R} \oplus \mathcal{S}$ forms a bounded monotone class. According to exercise A.3.5, it suffices to show that $\mathcal{S}_b(\mathbb{C})$ contains a complex multiplicative class $\mathcal{M}$ that generates $\mathcal{F}_\infty[\boldsymbol{W}]$. $\mathcal{S}_b(\mathbb{C})$ will then contain all bounded $\mathcal{F}_\infty[\boldsymbol{W}]$-measurable random variables and its closure all of $L^p_{\mathbb{C}}$. We take for $\mathcal{M}$ the multiples of random variables of the form

$$\exp\left(i\boldsymbol{\phi}{*}\boldsymbol{W}_\infty\right) = e^{i \int_0^\infty \sum_\eta \phi^\eta(s)\, dW^\eta_s} ,$$

the ϕ^η being bounded Borel and vanishing past some instant each. $\mathcal{M}$ is clearly closed under multiplication and complex conjugation and contains the constants. To see that it is contained in $\mathcal{S}_b(\mathbb{C})$, consider the Doléans–Dade exponential

$$\mathscr{E}_t = 1 + \int_0^t i\mathscr{E}_s \sum_\eta \phi^\eta(s)\, dW_s^\eta = 1 + \big(\mathscr{E}{*}i(\phi{*}\boldsymbol{W})\big)_t$$

by (3.9.4):
$$= \exp\Big(i\phi{*}\boldsymbol{W}_t + 1/2\int_0^t |\phi(s)|^2\, ds\Big)$$

of $i\phi{*}\boldsymbol{W}$. Clearly $\mathscr{E}_\infty = \exp\big(i\phi{*}\boldsymbol{W}_\infty + c\big)$ belongs to $\mathcal{S}_b(\mathbb{C})$, and so does the scalar multiple $\exp\big(i\phi{*}\boldsymbol{W}_\infty\big)$. To see that the σ-algebra $\mathcal{F}$ generated by $\mathcal{M}$ contains $\mathcal{F}_\infty[W]$, differentiate $\exp\big(i\tau\int\phi\, d\boldsymbol{W}\big)$ at $\tau = 0$ to conclude that $\int\phi\, d\boldsymbol{W} \in \mathcal{F}$. Then take $\phi^\eta = [0,t]$ for $\eta = 1,\ldots,d$ to see that $\boldsymbol{W}_t \in \mathcal{F}$ for all t. We leave to the reader the following straightforward generalization from finite auxiliary space to continuous auxiliary space: ∎

Corollary 4.2.16 (Generalization to Wiener Random Measure) *Let β be a Wiener random measure with intensity rate ν on the auxiliary space $\boldsymbol{H}$, as in definition 3.10.5. The filtration $\mathcal{F}_.$ is the one generated by β (ibidem).*

(i) For $0 < p < \infty$, THE *Daniell mean and the* ***Hardy mean***

$$\check{F} \mapsto \|F\|_{\beta\text{-}p}^{\mathcal{H}*} \stackrel{\text{def}}{=} \Big(\int^* \Big(\int^* \check{F}_s^2(\eta;\omega)\,\nu(d\eta)ds\Big)^{p/2}\,\mathbb{P}(d\omega)\Big)^{1/p}$$

agree on the previsibles $\check{F} \in \check{\mathcal{P}} \stackrel{\text{def}}{=} \mathcal{B}^\bullet(\boldsymbol{H}) \otimes \mathcal{P}$, up to a multiplicative constant.

(ii) For every $f \in L^p(\mathcal{F}_\infty)$, $1 < p < \infty$, there is a β–p-integrable predictable random function $\check{X}$, unique up to indistinguishability, so that

$$f = \mathbb{E}[f] + \int_{\check{\boldsymbol{B}}} \check{X}(\eta,s)\,\beta(d\eta,ds)\,.$$

Additional Exercises

Exercise 4.2.17 $\|M\|_{\mathcal{K}^q} \le \|S_\infty[M]\|_{L^q} \le \sqrt{q/2}\cdot\|M\|_{\mathcal{K}^q}$ for $2 \le q \le \infty$.

Exercise 4.2.18 Let $1 \le p < \infty$ and M a local martingale. Then

$$\big\|M_\infty^\star\big\|_{L^p} \le C_p \cdot \lceil M \rceil_{\mathcal{I}^p}\,, \tag{4.2.11}$$

$$\lceil M \rceil_{\mathcal{I}^p} \le C_p \cdot \|S_\infty[M]\|_{L^p}\,, \tag{4.2.12}$$

and
$$\big\|(X{*}M)_T^\star\big\|_{L^p} \le C_p^{(4.2.4)} \cdot \Big\|\Big(\int_0^T X^2\, d[M,M]\Big)^{1/2}\Big\|_{L^p}$$

for any previsible X and stopping time T, with

$$C_p^{(4.2.11)} \le \begin{cases} C_p^{(4.2.4)}\cdot K_p^{(3.8.6)} \le 2^{1/p}\sqrt{5p} \le 5 & \text{for } 1 \le p \le 1.3, \\ p' = p/(p-1) \le 5 & \text{for } 1.3 \le p \le 2, \\ p' = p/(p-1) \le 2 & \text{for } 2 \le p < \infty, \end{cases}$$

$$\text{and}\quad C_p^{(4.2.12)} \le C_p^{(4.2.3)} \wedge C_p^{(4.2.4)} \le \begin{cases} 2\sqrt{2p} & \text{for } 1 \le p < 2, \\ 1 & \text{for } p = 2, \\ \sqrt{e/2} \cdot p & \text{for } 2 < p < \infty. \end{cases}$$

Inequality (4.2.6) permits an estimate of the constant $A_p^{(2.5.6)}$:

$$A_p^{(2.5.6)} \le C_p^{(4.2.4)} \cdot C_p^{(4.2.5)} \cdot p' \le \begin{cases} 19p' & \text{for } 1 < p < 2, \\ 1 & \text{for } p = 2, \\ \sqrt{e}p^{3/2}p' \vee 19p & \text{for } 2 < p < \infty. \end{cases}$$

Exercise 4.2.19 Let $p, q, r > 1$ with $1/r = 1/q + 1/p$ and M an L^p-bounded martingale. If X is previsible and its maximal function is measurable and finite in L^q-mean, then X is M–p-integrable.

Exercise 4.2.20 A standard Wiener process W is an L^p-integrator for all $p < \infty$, of size $\lceil W^t \rceil_{\mathcal{I}^p} \le p\sqrt{et/2}$ for $p > 2$ and $\lceil W^t \rceil_{\mathcal{I}^p} \le \sqrt{t}$ for $0 < p \le 2$.

Exercise 4.2.21 Let $T^{c+} = \inf\{t : |W_t| > c\}$ and $T^c = \inf\{t : |W_t| \ge c\}$, where W is a standard Wiener process and $c \ge 0$. Then $\mathbb{E}[T^{c+}] = \mathbb{E}[T^c] = c^2$.

Exercise 4.2.22 (Martingale Representation in General) For $1 \le p < \infty$ let $\mathcal{H}_0^p$ denote the Banach space of $\mathbb{P}$-martingales M on $\mathcal{F}_.$ that have $M_0 = 0$ and that are global L^p-integrators. The ***Hardy space*** $\mathcal{H}_0^p$ carries the integrator norm $M \mapsto \lceil M \rceil_{\mathcal{I}^p} \sim \|S_\infty[M]\|_p$ (see inequality (4.2.1)). A closed linear subspace $\mathcal{S}$ of $\mathcal{H}_0^p$ is called ***stable*** if it is closed under stopping ($M \in \mathcal{S} \implies M^T \in \mathcal{S} \quad \forall T \in \mathfrak{T}$). The ***stable span*** $\mathcal{A}^{\|}$ of a set $\mathcal{A} \subset \mathcal{H}_0^p$ is defined as the smallest closed stable subspace containing $\mathcal{A}$. It contains with every finite collection $\boldsymbol{M} = \{M^1, \ldots, M^n\} \subset \mathcal{A}$, considered as a random measure having auxiliary space $\{1, \ldots, n\}$, and with every $\boldsymbol{X} = (X_i) \in \mathfrak{L}^1[\boldsymbol{M}\text{–}p]$, the indefinite integral $\boldsymbol{X} * \boldsymbol{M} = \sum_i X_i * M^i$; in fact, $\mathcal{A}^{\|}$ is the closure of the collection of all such indefinite integrals.

If $\mathcal{A}$ is finite, say $\mathcal{A} = \{M^1, \ldots, M^n\}$, and a) $[M^i, M^j] = 0$ for $i \ne j$ or b) $\boldsymbol{M}$ is previsible or b') the $[M^i, M^j]$ are previsible or c) $p = 2$ or d) $n = 1$, then the set $\{\boldsymbol{X} * \boldsymbol{M} : \boldsymbol{X} \in \mathfrak{L}^1[\boldsymbol{M}\text{–}p]\}$ of indefinite integrals is closed in $\mathcal{H}_0^p$ and therefore equals $\mathcal{A}^{\|}$; in other words, then every martingale in $\mathcal{A}^{\|}$ has a representation as an indefinite integral against the M^i.

Exercise 4.2.23 (Characterization of $\mathcal{A}^{\|}$) The dual $\mathcal{H}_0^{p*}$ of $\mathcal{H}_0^p$ equals $\mathcal{H}_0^{p'}$ when the conjugate exponent p' is finite and equals BMO_0 when $p = 1$ and then $p' = \infty$; the pairing is $(M, M') \mapsto \langle M | M' \rangle \stackrel{\text{def}}{=} \mathbb{E}[M_\infty \cdot M'_\infty]$ in both cases ($M \in \mathcal{H}_0^p, M' \in \mathcal{H}_0^{p*}$). A martingale M' in $\mathcal{H}_0^{p*}$ is called ***strongly perpendicular*** to $M \in \mathcal{H}_0^p$, denoted $M \perp\!\!\!\perp M'$, if $[M, M']$ is a (then automatically uniformly integrable) martingale. M' is strongly perpendicular to all $M \in \mathcal{A} \subset \mathcal{H}_0^p$ if and only if it is perpendicular to every martingale in $\mathcal{A}^{\|}$. The collection of all such martingales $M' \in \mathcal{H}_0^{p*}$ is denoted by $\mathcal{A}^{\perp\!\!\!\perp}$. It is a stable subspace of $\mathcal{H}_0^{p*}$, and $(\mathcal{A}^{\perp\!\!\!\perp})^{\perp\!\!\!\perp} = \mathcal{A}^{\|}$.

Exercise 4.2.24 (Continuation: Martingale Measures) Let $G' \stackrel{\text{def}}{=} 1 + M'$, with $\mathcal{A}^{\perp\!\!\!\perp} \ni M' > -1$. Then $\mathbb{P}' \stackrel{\text{def}}{=} G'\mathbb{P}$ is a probability, eqivalent with $\mathbb{P}$ and equal to $\mathbb{P}$ on $\mathcal{F}_0$, for which every element of $\mathcal{A}^{\|}$ is a martingale. For this reason such $\mathbb{P}'$ is called a ***martingale measure*** for $\mathcal{A}$. The set $\mathfrak{M}[\mathcal{A}]$ of martingale measures for $\mathcal{A}$ is evidently convex and contains $\mathbb{P}$. $\mathcal{A}^{\perp\!\!\!\perp}$ contains no bounded martingale other than zero if and only if $\mathbb{P}$ is an extremal point of $\mathfrak{M}[\mathcal{A}]$.

Assume now $\boldsymbol{M} = \{M^1, \ldots, M^n\} \subset \mathcal{H}_0^p$ has bounded jumps, and $M^i \perp\!\!\!\perp M^j$ for $i \ne j$. Then every martingale $M \in \mathcal{H}_0^p$ has a representation $M = \boldsymbol{X} * \boldsymbol{M}$ with $\boldsymbol{X} \in \mathfrak{L}^1[\boldsymbol{M}\text{–}p]$ if and only if $\mathbb{P}$ is an extremal point of $\mathfrak{M}[\boldsymbol{M}]$.

4.3 The Doob–Meyer Decomposition

Throughout the remainder of the chapter the probability $\mathbb{P}$ is fixed, and the filtration $(\mathcal{F}_., \mathbb{P})$ satisfies the natural conditions. As usual, mention of $\mathbb{P}$ is suppressed in the notation.

In this section we address the question of finding a canonical decomposition for an L^p-integrator Z. The classes in which the constituents of Z are sought are the finite variation processes and the local martingales. The next result is about as good as one might expect. Its estimates hold only in the range $1 \le p < \infty$.

Theorem 4.3.1 *An adapted process Z is a local L^1-integrator if and only if it is the sum of a right-continuous previsible process $\widehat{Z}$ of finite variation and a local martingale $\widetilde{Z}$ that vanishes at time zero. The decomposition*

$$Z = \widehat{Z} + \widetilde{Z}$$

is unique up to indistinguishability and is termed the ***Doob–Meyer decomposition*** *of Z. If Z has continuous paths, then so do $\widehat{Z}$ and $\widetilde{Z}$. For $1 \le p < \infty$ there are universal constants $\widehat{C}_p$ and $\widetilde{C}_p$ such that*

$$\lceil\!\lceil \widehat{Z} \rceil\!\rceil_{\mathcal{I}^p} \le \widehat{C}_p \cdot \lceil\!\lceil Z \rceil\!\rceil_{\mathcal{I}^p} \quad \textit{and} \quad \lceil\!\lceil \widetilde{Z} \rceil\!\rceil_{\mathcal{I}^p} \le \widetilde{C}_p \cdot \lceil\!\lceil Z \rceil\!\rceil_{\mathcal{I}^p} \,. \tag{4.3.1}$$

The size of the martingale part $\widetilde{Z}$ is actually controlled by the square function of Z alone:

$$\lceil\!\lceil \widetilde{Z} \rceil\!\rceil_{\mathcal{I}^p} \le \widetilde{C}'_p \cdot \| S_\infty[Z] \|_{L^p} \,. \tag{4.3.2}$$

The previsible finite variation part $\widehat{Z}$ is also called the ***compensator*** *or* ***dual previsible projection*** *of Z, and the local martingale part $\widetilde{Z}$ is called its* ***compensatrix*** *or "**Z compensated.**"*

The proof below (see page 227 ff.) furnishes the estimates

$$\widetilde{C}'^{(4.3.2)}_p \le \begin{cases} 2\sqrt{2p} < 4 & \text{for } 1 \le p < 2, \\ 2 & \text{for } p = 2, \\ C^{(4.2.5)}_{p'} \le 6/\sqrt{p'} & \text{for } 2 < p < \infty, \end{cases}$$

$$\widetilde{C}^{(4.3.1)}_p \le \begin{cases} 4 & \text{for } 1 \le p < 2, \\ 2 & \text{for } p = 2, \\ 6p & \text{for } 2 < p < \infty, \end{cases} \tag{4.3.3}$$

$$\widehat{C}^{(4.3.1)}_p \le \begin{cases} 1 & \text{for } p = 1, \\ 5 & \text{for } 1 < p < 2, \\ 2 & \text{for } p = 2, \\ 6p & \text{for } 2 < p < \infty. \end{cases}$$

In the range $0 \le p < 1$, a weaker statement is true: an L^p-integrator is the sum of a local martingale and a process of finite variation; but the decomposition is neither canonical nor unique, and the sizes of the summands cannot in general be estimated. These matters are taken up below (section 4.4). Processes that do have a Doob–Meyer decomposition are known in the literature also as ***processes of class D***.

Doléans–Dade Measures and Processes

The main idea in the construction of the Doob–Meyer decomposition 4.3.1 of a local L^1-integrator Z is to analyze its ***Doléans-Dade measure*** μ_Z. This is defined on all bounded previsible and locally Z–1-integrable processes X by

$$\mu_Z(X) = \mathbb{E}\Big[\int X\, dZ\Big]$$

and is evidently a σ-finite σ-additive measure on the previsibles $\mathcal{P}$ that vanishes on evanescent processes. Suppose it were known that every measure μ on $\mathcal{P}$ with these properties has a ***predictable representation*** in the form

$$\mu(X) = \mathbb{E}\Big[\int X\, dV^\mu\Big], \qquad X \in \mathcal{P}_b ,$$

where V^μ is a right-continuous predictable process of finite variation – such V^μ is known as a ***Doléans–Dade process for*** μ. Then we would simply set $\widehat{Z} \stackrel{\text{def}}{=} V^{\mu_Z}$ and $\widetilde{Z} \stackrel{\text{def}}{=} Z - \widehat{Z}$. Inasmuch as

$$\mathbb{E}\Big[\int X\, d\widetilde{Z}\Big] = \mathbb{E}\Big[\int X\, dZ\Big] - \mathbb{E}\Big[\int X\, dV^{\mu_Z}\Big] = 0$$

on (many) previsibles $X \in \mathcal{P}_b$, the difference $\widetilde{Z}$ would be a (local) martingale and $Z = \widehat{Z} + \widetilde{Z}$ would be a Doob–Meyer decomposition of Z: the battle plan is laid out.[7]

It is convenient to investigate first the case when μ is totally finite:

Proposition 4.3.2 *Let μ be a σ-additive measure of bounded variation on the σ-algebra $\mathcal{P}$ of predictable sets and assume that μ vanishes on evanescent sets in $\mathcal{P}$. There exists a right-continuous predictable process V^μ of integrable total variation $|V^\mu|_\infty$, unique up to indistinguishability, such that for all bounded previsible processes X*

$$\mu(X) = \mathbb{E}\Big[\int_0^\infty X\, dV^\mu\Big]. \tag{4.3.4}$$

Proof. Let us start with a little argument showing that if such a Doléans–Dade process V^μ exists, then it is unique. To this end fix t and $g \in L^\infty(\mathcal{F}_t)$, and let M^g be the bounded right-continuous martingale whose value at any instant s is $M^g_s = \mathbb{E}[g|\mathcal{F}_s]$ (example 2.5.2). Let $M^g_{\cdot-}$ be the left-continuous

[7] There are other ways to establish theorem 4.3.1. This particular construction, via the correspondence $Z \to \mu_Z$ and $\mu \to V^\mu$, is however used several times in section 4.5.

version of M^g and in (4.3.4) set $X = M_0^g \cdot [\![0]\!] + M_{\cdot-}^g \cdot (\!(0,t]\!]$. Then from corollary 3.8.23

$$\mu(X) = \mathbb{E}[M_0^g V_0^\mu] + \mathbb{E}\Big[\int_{0+}^t M_{\cdot-}^g \, dV^\mu\Big] = \mathbb{E}[gV_t^\mu] .$$

In other words, V_t^μ is a Radon–Nikodym derivative of the measure

$$\mu^t : g \mapsto \mu\big(M_0^g \cdot [\![0]\!] + M_{\cdot-}^g \cdot (\!(0,t]\!]\big) , \qquad g \in L^\infty(\mathcal{F}_t) ,$$

with respect to $\mathbb{P}$, both μ^t and $\mathbb{P}$ being regarded as measures on $\mathcal{F}_t$. This determines V_t^μ up to a modification. Since V^μ is also right-continuous, it is unique up to indistinguishability (exercise 1.3.28).

For the existence we reduce first of all the situation to the case that μ is positive, by splitting μ into its positive and negative parts. We want to show that then there exists an increasing right-continuous predictable process I with $\mathbb{E}[I_\infty] < \infty$ that satisfies (4.3.4) for all $X \in \mathcal{P}_b$. To do that we stand the uniqueness argument above on its head and *define* the random variable $I_t \in L^1_+(\mathcal{F}_t, \mathbb{P})$ as the Radon–Nikodym derivative of the measure μ^t on $\mathcal{F}_t$ with respect to $\mathbb{P}$. Such a derivative does exist: μ^t is clearly additive. And if (g_n) is a sequence in $L^\infty(\mathcal{F}_t)$ that decreases pointwise $\mathbb{P}$-a.s. to zero, then (M^{g_n}) decreases pointwise, and thanks to Doob's maximal lemma 2.5.18, $\inf_n (M_{\cdot-}^{g_n})$ is zero except on an evanescent set. Consequently,

$$\lim_{n\to\infty} \mu^t(g_n) = \lim_{n\to\infty} \mu\big(M_0^g \cdot [\![0]\!] + M_{\cdot-}^{g_n} \cdot (\!(0,t]\!]\big) = 0 .$$

This shows at the same time that μ^t is σ-additive and that it is absolutely continuous with respect to the restriction of $\mathbb{P}$ to $\mathcal{F}_t$. The Radon–Nikodym theorem A.3.22 provides a derivative $I_t = d\mu^t/d\mathbb{P} \in L^1_+(\mathcal{F}_t, \mathbb{P})$. In other words, I_t is defined by the equation

$$\mu\big(M_0^g \cdot [\![0]\!] + M_{\cdot-}^g \cdot (\!(0,t]\!]\big) = \mathbb{E}[M_t^g \cdot I_t] , \qquad g \in L^\infty(\mathcal{F}_\infty) .$$

Taking differences in this equation results in

$$\mu\big(M_{\cdot-}^g \cdot (\!(s,t]\!]\big) = \mathbb{E}\big[M_t^g I_t - M_s^g I_s\big] = \mathbb{E}\big[g \cdot (I_t - I_s)\big] \tag{4.3.5}$$
$$= \mathbb{E}\Big[\int g \cdot (\!(s,t]\!] \, dI\Big]$$

for $0 \le s < t \le \infty$. Taking $g = [I_s > I_t]$ we see that I is increasing. Namely, the left-hand side of equation (4.3.5) is then positive and the right-hand side negative, so that both must vanish. This says that $I_t \ge I_s$ a.s. Taking $t_n \downarrow s$ and $g = [\inf_n I_{t_n} > I_s]$ we see similarly that I is right-continuous in L^1-mean. I is thus a global L^1-integrator, and we may and shall replace it by its right-continuous modification (theorem 2.3.4). Another look at (4.3.5) reveals that μ equals the Doléans–Dade measure of I, at least on processes of the form $g \cdot (\!(s,t]\!]$, $g \in \mathcal{F}_s$. These processes generate the predictables, and so $\mu = \mu^I$ on all of $\mathcal{P}$. In particular,

$$\mathbb{E}[M_t I_t - M_0 I_0] = \mu(M_{\leftarrow} \cdot (\!(0,t]\!]) = \mathbb{E}\Big[\int_{0+}^{t} M_{\leftarrow}\, dI\Big]$$

for bounded martingales M. Taking differences turns this into

$$\mathbb{E}\Big[\int g \cdot (\!(t,\infty)\!)\, dI\Big] = \mathbb{E}\Big[\int M^g_{\leftarrow} \cdot (\!(t,\infty)\!)\, dI\Big]$$

for all bounded random variables g with attached right-continuous martingales $M^g_t = \mathbb{E}[g|\mathcal{F}_t]$. Now $M^g_{\leftarrow} \cdot (\!(t,\infty)\!)$ is the predictable projection of $(\!(t,\infty)\!) \cdot g$ (corollary A.5.15 on page 439), so the assumption on I can be read as

$$\mathbb{E}\Big[\int X\, dI\Big] = \mathbb{E}\Big[\int X^{\mathcal{P},\mathbb{P}}\, dI\Big], \tag{$*$}$$

at least for X of the form $(\!(t,\infty)\!) \cdot g$. Now such X generate the measurable σ-algebra on $\boldsymbol{B}$, and the bounded monotone class theorem implies that $(*)$ holds for all bounded measurable processes X (ibidem).

On the way to proving that I is predictable another observation is useful: *at a predictable time S the jump ΔI_S is measurable on $\mathcal{F}_{S-}$*:

$$\Delta I_S \in \mathcal{F}_{S-}\ . \tag{$**$}$$

To see this, let f be a bounded $\mathcal{F}_S$-measurable function and set $g \stackrel{\text{def}}{=} f - \mathbb{E}[f|\mathcal{F}_{S-}]$ and $M^g_t \stackrel{\text{def}}{=} \mathbb{E}[g|\mathcal{F}_t]$. Then M^g is a bounded martingale that vanishes at any time strictly prior to S and is constant after S. Thus $M^g \cdot [\![0,S]\!] = M^g \cdot [\![S]\!]$ has predictable projection $M^g_{\leftarrow}[\![S]\!] = 0$ and

$$\mathbb{E}\big[f \cdot (\Delta I_S - \mathbb{E}[\Delta I_S|\mathcal{F}_{S-}])\big] = \mathbb{E}[g\Delta I_S] = \mathbb{E}\Big[\int M^g[\![0,S]\!]\, dI\Big] = 0\ .$$

This is true for all $f \in \mathcal{F}_S$, so $\Delta I_S = \mathbb{E}[\Delta I_S|\mathcal{F}_{S-}]$.

Now let $a \geq 0$ and let P be a previsible subset of $[\Delta I > a]$, chosen so that $\mathbb{E}[\int P\, dI]$ is maximal. We want to show that $N \stackrel{\text{def}}{=} [\Delta I > a] \setminus P$ is evanescent. Suppose it were not. Then

$$0 < \mathbb{E}\Big[\int N\, dI\Big] = \mathbb{E}\Big[\int N^{\mathcal{P},\mathbb{P}}\, dI\Big]\ ,$$

so that $N^{\mathcal{P},\mathbb{P}}$ could not be evanescent. According to the predictable section theorem A.5.14, there would exist a predictable stopping time S with $[\![S]\!] \subset [N^{\mathcal{P},\mathbb{P}} > 0]$ and $\mathbb{P}[S < \infty] > 0$. Then

$$0 < \mathbb{E}\big[N^{\mathcal{P},\mathbb{P}}_S [S < \infty]\big] = \mathbb{E}\big[N_S [S < \infty]\big]\ .$$

Now either $N_S = 0$ or $\Delta I_S > a$. The predictable[8] reduction $S' \stackrel{\text{def}}{=} S_{[\Delta I_S > a]}$ still would have $\mathbb{E}\big[N_{S'}[S' < \infty]\big] > 0$, and consequently

$$\mathbb{E}\Big[\int N \cap [\![S']\!]\, dI\Big] > 0 .$$

Then $P_0 \stackrel{\text{def}}{=} [\![S']\!] \setminus P$ would be a previsible non-evanescent subset of N with $\mathbb{E}[\int P_0\, dI] > 0$, in contradiction to the maximality of P.

That is to say, $[\Delta I > a] = P$ is previsible, for all $a \ge 0$: ΔI is previsible. Then so is $I = I_- + \Delta I$; and since this process is right-continuous, it is even predictable. ∎

Exercise 4.3.3 A right-continuous increasing process $I \in \mathfrak{D}$ is previsible if and only if its jumps occur only at predictable stopping times and if, in addition, the jump ΔI_T at a stopping time T is measurable on the strict past $\mathcal{F}_{T-}$ of T.

Exercise 4.3.4 Let $V = {}^{c}V + {}^{j}V$ be the decomposition of the càdlàg predictable finite variation process V into continuous and jump parts (see exercise 2.4.6). Then the sparse set $[\Delta V \neq 0] = [\Delta\, {}^{j}V \neq 0]$ is previsible and is, in fact, the disjoint union of the graphs of countably many predictable stopping times [use theorem A.5.14].

Exercise 4.3.5 A supermartingale Z_t that is right-continuous in probability and has $\mathbb{E}[Z_t] \xrightarrow[t\to\infty]{} 0$ is called a ***potential***. A potential Z is of class D if and only if the random variables $\{Z_T : T \text{ an a.s. finite stopping time }\}$ are uniformly integrable.

Proof of Theorem 4.3.1: Necessity, Uniqueness, and Existence

Since a local martingale is a local L^1-integrator (corollary 2.5.29) and a predictable process of finite variation has locally bounded variation (exercise 3.5.4 and corollary 3.5.16) and is therefore a local L^p-integrator for every $p > 0$ (proposition 2.4.1), a process having a Doob–Meyer decomposition is necessarily a local L^1-integrator.

Next the uniqueness. Suppose that $Z = \widehat{Z} + \widetilde{Z} = \widehat{Z}' + \widetilde{Z}'$ are two Doob–Meyer decompositions of Z. Then $M \stackrel{\text{def}}{=} \widetilde{Z} - \widetilde{Z}' = \widehat{Z} - \widehat{Z}'$ is a predictable local martingale of finite variation that vanishes at zero. We know from exercise 3.8.24 (i) that M is evanescent.

Let us make here an observation to be used in the existence proof. Suppose that Z stops at the time T: $Z = Z^T$. Then $Z = \widehat{Z} + \widetilde{Z}$ and $Z = \big(\widehat{Z}\big)^T + \big(\widetilde{Z}\big)^T$ are both Doob–Meyer decompositions of Z, so they coincide. That is to say, if Z has a Doob–Meyer decomposition at all, then its predictable finite variation and martingale parts also stop at time T. Doing a little algebra one deduces from this that if Z vanishes strictly before time S, i.e., on $[\![0, S)\!)$, and is constant after time T, i.e., on $[\![T, \infty)\!)$, then the parts of its Doob–Meyer decomposition, should it have any, show the same behavior.

Now to the existence. Let (T_n) be a sequence of stopping times that reduce Z to global L^1-integrators and increase to infinity. If we can produce Doob–Meyer decompositions

$$Z^{T_{n+1}} - Z^{T_n} = V^n + M^n$$

[8] See (**) and lemma 3.5.15 (iv).

for the global L^1-integrators on the left, then $Z = \sum_n V^n + \sum_n M^n$ will be a Doob–Meyer decomposition for Z – note that at every point $\varpi \in \boldsymbol{B}$ this is a finite sum. In other words, *we may assume that Z is a global L^1-integrator.*

Consider then its Doléans–Dade measure μ:

$$\mu(X) = \mathbb{E}\Big[\int X\, dZ\Big], \qquad X \in \mathcal{P}_b ,$$

and let $\widehat{Z}$ be the predictable process V^μ of finite variation provided by proposition 4.3.2. From

$$\mathbb{E}\Big[\int X\, d(Z - \widehat{Z})\Big] = 0$$

it follows that $\widetilde{Z} \stackrel{\text{def}}{=} Z - \widehat{Z}$ is a martingale. $Z = \widehat{Z} + \widetilde{Z}$ is the sought-after Doob–Meyer decomposition. ∎

Exercise 4.3.6 Let $T > 0$ be a predictable stopping time and Z a global L^1-integrator with Doob–Meyer decomposition $Z = \widehat{Z} + \widetilde{Z}$. Then the jump $\Delta \widehat{Z}_T$ equals $\mathbb{E}\left[\Delta Z_T | \mathcal{F}_{T-}\right]$. The predictable finite variation and martingale parts of any continuous local L^1-integrator are again continuous. In the general case

$$\Big\| S_\infty[\widehat{Z}] \Big\|_{L^q} \le \sqrt{q/2} \cdot \| S_\infty[Z] \|_{L^q} , \qquad 2 \le q < \infty .$$

Exercise 4.3.7 If I and J are increasing processes with $\mu_I \le \mu_J$ – which we also write $dI \le dJ$, see page 406 – then $d\widehat{I} \le d\widehat{J}$ and $\widehat{I} \le \widehat{J}$.

Exercise 4.3.8 A local L^1-integrator with bounded jumps is a local L^q-integrator for any $q \in (0, \infty)$ (see corollary 4.4.3 on page 234 for much more).

Exercise 4.3.9 Let Z be a local L^1-integrator. There are arbitrarily large stopping times U such that Z agrees on the right-open interval $[\![0, U)\!)$ with a process that is a global L^q-integrator for all $q \in (0, \infty)$.

Exercise 4.3.10 Let X be a bounded previsible process. The Doob–Meyer decomposition of $X{*}Z$ is $X{*}Z = X{*}\widehat{Z} + X{*}\widetilde{Z}$.

Exercise 4.3.11 Let μ, V^μ be as in proposition 4.3.2 and let D be a bounded previsible process. The Doléans–Dade process of the measure $D \cdot \mu$ is $D{*}V^\mu$.

Exercise 4.3.12 Let V, V' be previsible positive increasing processes with associated Doléans–Dade measures $\mu_V, \mu_{V'}$ on $\mathcal{P}$. The following are equivalent: (i) for almost every $\omega \in \Omega$ the measure $dV_t(\omega)$ is absolutely continuous with respect to $dV'_t(\omega)$ on $\mathcal{B}^\bullet(\mathbb{R}_+)$; (ii) μ_V is absolutely continuous with respect to $\mu_{V'}$; and (iii) there exists a previsible process G such that $\mu_V = G{\cdot}\mu_{V'}$. In this case $dV_t(\omega) = G_t(\omega){\cdot}dV'_t(\omega)$ on $\mathbb{R}_+$, for almost all $\omega \in \Omega$.

Exercise 4.3.13 Let V be an adapted right-continuous process of integrable total variation $|V|$ and with $V_0 = 0$, and let $\mu = \mu_V$ be its Doléans–Dade measure. We know from proposition 2.4.1 that $|V|_{\mathcal{I}^p} \le \| |V|_\infty \|_{L^p}$, $0 < p < \infty$. If V is previsible, then the variation process $|V|$ is indistinguishable from the Doléans–Dade process of $|\mu|$, and equality obtains in this inequality: for $0 < p < \infty$

$$\|Y\|_{V-p} = \Big\| \int Y\, d|V| \Big\|_{L^p} \quad \text{and} \quad |V|_{\mathcal{I}^p} = \Big\| |V|_\infty \Big\|_{L^p} .$$

Exercise 4.3.14 (Fundamental Theorem of Local Martingales [75]) A local martingale M is the sum of a finite variation process and a locally square integrable local martingale (for more see corollary 4.4.3 and proposition 4.4.1).

Exercise 4.3.15 (Emery) For $0 < p < \infty$, $0 < q < \infty$, and $\frac{1}{r} = \frac{1}{p} + \frac{1}{q}$ there are universal constants $C_{p,q}$ such that for every global L^q-integrator and previsible integrand X with measurable maximal function $X^\star_\infty$

$$\lceil X{*}Z \rceil_{\mathcal{I}^r} \le C_{p,q} \cdot \| X^\star_\infty \|_{L^p} \cdot \lceil Z \rceil_{\mathcal{I}^q} .$$

Proof of Theorem 4.3.1: The Inequalities

Let $Z = \widehat{Z} + \widetilde{Z}$ be the Doob–Meyer decomposition of Z. We may assume that Z is a global L^p-integrator, else there is nothing to prove. If $p = 1$, then

$$\lceil \widehat{Z} \rceil_{\mathcal{I}^1} = \sup\Big\{ \mathbb{E}\Big[\int X\, dZ\Big] : X \in \mathcal{E}_1 \Big\} \le \lceil Z \rceil_{\mathcal{I}^1} ,$$

so inequality (4.3.1) holds with $\widehat{C}_1 = 1$.

For $p \ne 1$ we go after the martingale term instead. Since $\widetilde{Z}$ vanishes at time zero, it suffices to estimate the size of $(X{*}\widetilde{Z})_\infty$ for $X \in \mathcal{E}_1$ with $X_0 = 0$. Let then M be a martingale with $\| M_\infty \|_{p'} \le 1$. Let T be a stopping time such that $(X{*}\widetilde{Z}_{.-})^T$ and $M^T_{.-}$ are bounded and $[\widehat{Z}, M]^T$ is a martingale (exercise 3.8.24 (ii)). Then the first two terms on the right of

$$(X{*}\widetilde{Z}^T)M^T = X{*}\widetilde{Z}_{.-}{*}M^T + (XM_{.-}){*}\widetilde{Z}^T + X{*}[\widetilde{Z}, M]^T$$

are martingales and vanish at zero. Further, $X{*}[\widetilde{Z}, M]^T$ and $X{*}[Z, M]^T$ differ by the martingale $X{*}[\widehat{Z}, M]^T$. Therefore

$$\mathbb{E}\big[(X{*}\widetilde{Z})_T M_T\big] = \mathbb{E}\Big[\int_{0+}^T X\, d[Z, M]\Big] \le \mathbb{E}\Big[\lceil [Z, M] \rceil_\infty \Big] .$$

Now $X{*}\widetilde{Z}$ is constant after some instant. Taking the supremum over T and $X \in \mathcal{E}_1$ thus gives

$$\lceil \widetilde{Z} \rceil_{\mathcal{I}^p} \le \sup\Big\{ \mathbb{E}\Big[\lceil [Z, M] \rceil_\infty \Big] : \| M_\infty \|_{L^{p'}} \le 1 \Big\} . \tag{$*$}$$

If $1 < p < 2$, we continue this inequality using corollary 4.2.8:

$$\lceil \widetilde{Z} \rceil_{\mathcal{I}^p} \le 2\sqrt{2p}\, \| S_\infty[Z] \|_{L^p} \le 2\sqrt{2p}\, K_p^{(3.8.6)} \lceil Z \rceil_{\mathcal{I}^p} \le 4 \lceil Z \rceil_{\mathcal{I}^p} .$$

If $2 \le p$, we continue at $(*)$ instead with an application of exercise 3.8.10:

$$\lceil \widetilde{Z} \rceil_{\mathcal{I}^p} \le \| S_\infty[Z] \|_{L^p} \cdot \sup\{ \| S_\infty[M] \|_{L^{p'}} : \| M_\infty \|_{L^{p'}} \le 1 \} \tag{$**$}$$

$$\le \| S_\infty[Z] \|_{L^p} \cdot C_{p'}^{(4.2.5)} \cdot \sup\{ \| M^\star_\infty \|_{L^{p'}} : \| M_\infty \|_{L^{p'}} \le 1 \}$$

by 2.5.19: $$\le \| S_\infty[Z] \|_{L^p} \cdot C_{p'} \cdot p$$

by 3.8.4: $$\le p C_{p'}^{(4.2.5)} \lceil Z \rceil_{\mathcal{I}^p} \le 6p \lceil Z \rceil_{\mathcal{I}^p} .$$

For $p = 2$ use $\|S_\infty[M]\|_{L^{p'}} \le 1$ at (**) instead. This proves the stated inequalities for $\widetilde{Z}$; the ones for $\widehat{Z}$ follow by subtraction. Theorem 4.3.1 is proved in its entirety. ∎

Remark 4.3.16 The main ingredient in the proof of proposition 4.3.2 and thus of theorem 4.3.1 was the fact that an increasing process I that satisfies

$$\mathbb{E}\big[M_t I_t\big] = \mathbb{E}\Big[\int_0^t M_{.-}\, dI\Big] \tag{$*$}$$

for all bounded martingales M is previsible. It was Paul–André Meyer who called increasing processes with (∗) ***natural*** and then proceeded to show that they are previsible [70], [71]. At first sight there is actually something *unnatural* about all this [72, page 111]. Namely, while the interest in previsible processes as *integrands* is perfectly natural in view of our experience with Borel functions, of which they are the stochastic analogs, it may not be altogether obvious what good there is in having *integrators* previsible. In answer let us remark first that the previsibility of $\widehat{Z}$ enters essentially into the proof of the estimates (4.3.1)–(4.3.2). Furthermore, it will lead to *previsible pathwise control* of integrators, which permits a controlled analysis of stochastic differential equations driven by integrators with jumps (section 4.5).

The Previsible Square Function

If Z is an L^p-integrator with $p \ge 2$, then $[Z,Z]$ is an L^1-integrator and therefore has a Doob–Meyer decomposition

$$[Z,Z] = \widehat{[Z,Z]} + \widetilde{[Z,Z]}\,.$$

Its previsible finite variation part $\widehat{[Z,Z]}$ is called the ***previsible*** or ***oblique bracket*** or ***angle bracket*** and is denoted by $\langle Z,Z\rangle$. Note that $\langle Z,Z\rangle_0 = [Z,Z]_0 = Z_0^2$. The square root $s[Z] \stackrel{\text{def}}{=} \sqrt{\langle Z,Z\rangle}$ is called the ***previsible square function*** of Z. The processes $\langle Z,Z\rangle$ and $s[Z]$ evidently can be defined unequivocally also in case Z is merely a local L^2-integrator. If Z is continuous, then clearly $\langle Z,Z\rangle = [Z,Z]$ and $s[Z] = S[Z]$.

Let Y,Z be local L^2-integrators. According to the inequality of Kunita–Watanabe (theorem 3.8.9), $[Y,Z]$ is a local L^1-integrator and has a Doob–Meyer decomposition

$$[Y,Z] = \widehat{[Y,Z]} + \widetilde{[Y,Z]}$$

with

$$\widehat{[Y,Z]}_0 = [Y,Z]_0 = Y_0{\cdot}Z_0\,.$$

Its previsible finite variation part $\widehat{[Y,Z]}$ is called the ***previsible*** or ***oblique bracket*** or ***angle bracket*** and is denoted by $\langle Y,Z\rangle$. Clearly if either of Y,Z is continuous, then $\langle Y,Z\rangle = [Y,Z]$.

Exercise 4.3.17 The previsible bracket has the same general properties as $[\cdot,\cdot]$:

(i) The theorem of Kunita–Watanabe holds for it: for any two local L^2-integrators Y, Z there exists a set $\Omega_0 \in \mathcal{F}_\infty$ of full measure $\mathbb{P}[\Omega_0] = 1$ such that for all $\omega \in \Omega_0$ and any two $\mathcal{B}^\bullet(\mathbb{R}) \otimes \mathcal{F}_\infty$-measurable processes U, V

$$\int_0^\infty |UV|\, d\,\vdots\langle Y,Z\rangle\vdots \le \Big(\int_0^\infty U^2\, d\langle Y,Y\rangle\Big)^{1/2} \cdot \Big(\int_0^\infty V^2\, d\langle Z,Z\rangle\Big)^{1/2} .$$

(ii) $s[Y+Z] \le s[Y] + s[Z]$, except possibly on an evanescent set.

(iii) For any stopping time T and $p,q,r > 0$ with $1/r = 1/p + 1/q$

$$\Big\| \vdots\langle Y,Z\rangle\vdots_T \Big\|_{L^r} \le \| s_T[Z] \|_{L^p} \cdot \| s_T[Y] \|_{L^q}$$

(iv) Let Z^1, Z^2 be local L^2-integrators and X^1, X^2 processes integrable for both. Then

$$\langle X^1 {*} Z^1, X^2 {*} Z^2\rangle = (X^1 \cdot X^2) {*} \langle Z^1, Z^2\rangle .$$

Exercise 4.3.18 With respect to the previsible bracket the martingales M with $M_0 = 0$ and the previsible finite variation processes V are perpendicular: $\langle M, V\rangle = 0$. If Z is a local L^2-integrator with Doob–Meyer decomposition $Z = \widehat{Z} + \widetilde{Z}$, then

$$\langle Z, Z\rangle = \langle \widehat{Z}, \widehat{Z}\rangle + \langle \widetilde{Z}, \widetilde{Z}\rangle .$$

For $0 < p \le 2$ the previsible square function $s[M]$ can be used as a control for the integrator size of a local martingale M much as the square function $S[M]$ controls it in the range $1 \le p < \infty$ (theorem 4.2.12). Namely,

Proposition 4.3.19 *For a locally L^2-integrable martingale M and $0 < p \le 2$*

$$\|M^\star_\infty\|_{L^p} \le C_p \cdot \|s_\infty[M]\|_{L^p} , \tag{4.3.6}$$

with universal constants

$$C_2^{(4.3.6)} \le 2$$

and

$$C_p^{(4.3.6)} \le 4\sqrt{2/p}\,, \qquad p \ne 2 .$$

Exercise 4.3.20 For a *continuous* local martingale M the Burkholder–Davis–Gundy inequality (4.2.4) extends to all $p \in (0,\infty)$ and implies

$$\lceil M^t \rceil_{\mathcal{I}^p} \le C_p \cdot \Big\| s_\infty[M^t] \Big\|_{L^p} \tag{4.3.7}$$

for all t, with

$$C_p^{(4.3.7)} \le \begin{cases} C_p^{(4.3.6)} & \text{for } 0 < p \le 2 \\ C_p^{(4.2.4)} & \text{for } 1 \le p < \infty. \end{cases}$$

Proof of Proposition 4.3.19. First the case $p = 2$: thanks to Doob's maximal theorem 2.5.19 and exercise 3.8.11

$$\mathbb{E}\Big[M_T^{\star 2}\Big] \le 4\,\mathbb{E}\Big[M_T^2\Big] = 4\,\mathbb{E}\Big[[M,M]_T\Big] = 4\,\mathbb{E}\Big[\langle M,M\rangle_T\Big]$$

for arbitrarily large stopping times T. Upon letting $T \to \infty$ we get $\mathbb{E}\big[M_\infty^{\star 2}\big] \le 4\,\mathbb{E}\big[\langle M,M\rangle_\infty\big]$.

Now the case $p < 2$. By reduction to arbitrarily large stopping times we may assume that M is a global L^2-integrator. Let $s = s[M]$. Literally as in the proof of Fefferman's inequality 4.2.7 one shows that $s^{p-2} \cdot d(s^2) \le (2/p)\, d(s^p)$ and so

$$\int_0^T s^{p-2}\, d\langle M, M\rangle \le (2/p) \cdot s_t^p \,. \tag{$*$}$$

Next let $\epsilon > 0$ and define $\overline{s} = s[M] + \epsilon$ and $\overline{M} = \overline{s}^{(p-2)/2} * M$. From the first part of the proof

$$\mathbb{E}\Big[\overline{M}_t^{\star 2}\Big] \le 4 \cdot \mathbb{E}\Big[\overline{M}_t^2\Big] = 4 \cdot \mathbb{E}\Big[\langle \overline{M}, \overline{M}\rangle_t\Big] = 4 \cdot \mathbb{E}\Big[\int_0^T \overline{s}^{p-2}\, d\langle M, M\rangle\Big]$$

by $(*)$: $$\le 4 \cdot \mathbb{E}\Big[\int_0^T s^{p-2}\, d\langle M, M\rangle\Big] \le (8/p) \cdot \mathbb{E}\big[s_t^p\big] \,. \tag{$**$}$$

Next observe that for $t \ge 0$

$$\begin{aligned} M_t &= \int_0^T \overline{s}^{(2-p)/2}\, d\overline{M} = \overline{s}_t^{(2-p)/2} \cdot \overline{M}_t - \int_{0+}^t \overline{M}_{.-}\, d\overline{s}^{(2-p)/2} \\ &\le 2 \cdot \overline{s}_t^{(2-p)/2} \cdot \overline{M}_t^\star \,. \end{aligned}$$

The same inequality holds for $-M$, and since the process on the previous line increases with t,

$$M_t^\star \le 2 \cdot \overline{s}_t^{(2-p)/2} \cdot \overline{M}_t^\star \,.$$

From this, using Hölder's inequality with conjugate exponents $2/(2-p)$ and $2/p$ and inequality $(**)$,

$$\begin{aligned} \mathbb{E}\big[M_t^{\star p}\big] &\le 2^p \cdot \mathbb{E}\Big[\overline{s}_t^{p(2-p)/2} \cdot \overline{M}_t^{\star p}\Big] \le 2^p \cdot \Big(\mathbb{E}\big[\overline{s}_t^p\big]\Big)^{(2-p)/2} \cdot \Big(\mathbb{E}\big[\overline{M}_t^{\star 2}\big]\Big)^{p/2} \\ &\le 2^p (8/p)^{p/2} \cdot \Big(\mathbb{E}\big[\overline{s}_t^p\big]\Big)^{(2-p)/2} \cdot \Big(\mathbb{E}\big[s_t^p\big]\Big)^{p/2} \xrightarrow[\epsilon \to 0]{} \Big(4\sqrt{2/p}\Big)^p \cdot \mathbb{E}\Big[s_t^p\Big] \,. \end{aligned}$$

We take the pth root and get $\| M_t^\star \|_{L^p} \le 4\sqrt{2/p} \cdot \| s_t[M] \|_{L^p}$. ∎

Exercise 4.3.21 Suppose that Z is a global L^1-integrator with Doob–Meyer decomposition $Z = \widehat{Z} + \widetilde{Z}$. Here is an a priori L^p-mean estimate of the compensator $\widehat{Z}$ for $1 \le p < \infty$: let $\mathcal{P}_b$ denote the bounded previsible processes and set

$$\|Z\|_p^\wedge \stackrel{\text{def}}{=} \sup\Big\{\mathbb{E}\Big[\int X\, dZ\Big] : X \in \mathcal{P}_b\,,\ \|X_\infty^\star\|_{L^{p'}} \le 1\Big\}\,.$$

Then $$\|Z\|_p^\wedge \le \Big\| \,\rvert\widehat{Z}\lvert_\infty \Big\|_{L^p} = \rvert\widehat{Z}\lvert_{\mathcal{I}^p} \le p \cdot \|Z\|_p^\wedge \,.$$

Exercise 4.3.22 Let I be a positive increasing process with Doob–Meyer decomposition $I = \widehat{I} + \widetilde{I}$. In this case there is a better estimate of $\widehat{C}_p^{(4.3.1)}$ and $\widetilde{C}_p^{(4.3.1)}$ than inequality (4.3.3) provides. Namely, for $1 \le p < \infty$,

$$\rvert\widehat{I}\lvert_{\mathcal{I}^p} = \Big\|\widehat{I}_\infty\Big\|_{L^p} \le p \cdot \rvert I\lvert_{\mathcal{I}^p} \quad \text{and} \quad \rvert\widetilde{I}\lvert_{\mathcal{I}^p} \le (p+1) \cdot \rvert I\lvert_{\mathcal{I}^p} \,.$$

Exercise 4.3.23 Suppose that Z is a continuous L^p-integrator. Then $S[\widetilde{Z}] = S[Z]$ and inequality (4.3.3) can be improved to

$$\widetilde{C}_p \le \begin{cases} C_p^{(4.2.3)} K_p^{(3.8.6)} \le 2^{1+1/p} \sqrt{p} \le 4 & \text{for } 0 < p < 1, \\ C_p^{(4.2.4)} \le \sqrt{e/2}\, p & \text{for } 2 < p < \infty. \end{cases}$$

The Doob–Meyer Decomposition of a Random Measure

Let ζ be a random measure with auxiliary space $\boldsymbol{H}$ and elementary integrands $\check{\mathcal{E}}$ (see section 3.10). There is a straightforward generalization of theorem 4.3.1 to ζ.

Theorem 4.3.24 *Suppose ζ is a local L^1-random measure. There exist a unique previsible* ***strict*** *random measure $\widehat{\zeta}$ and a unique local martingale random measure $\widetilde{\zeta}$ that vanishes at zero, both local L^1-random measures, so that*

$$\zeta = \widehat{\zeta} + \widetilde{\zeta} .$$

In fact, there exist an increasing predictable process V and Radon measures $\nu_\varpi = \nu_{s,\omega}$ on $\boldsymbol{H}$, one for every $\varpi = (s, \omega) \in \boldsymbol{B}$ and usually written $\nu_s = \nu_s^\zeta$, so that $\widehat{\zeta}$ has the disintegration

$$\int_{\check{\boldsymbol{B}}} \check{H}_s(\eta)\, \widehat{\zeta}(d\eta, ds) = \int_0^\infty \int_{\boldsymbol{H}} \check{H}_s(\eta)\, \nu_s(d\eta)\, dV_s , \tag{4.3.8}$$

which is valid for every $\check{H} \in \check{\mathcal{P}}$. We call $\widehat{\zeta}$ the ***intensity*** *or* ***intensity measure*** *or* ***compensator*** *of ζ, and ν_s its* ***intensity rate****. $\widetilde{\zeta}$ is the* ***compensated random measure****. For $1 \le p < \infty$ and all $h \in \mathcal{E}_+[\boldsymbol{H}]$ and $t \ge 0$ we have the estimates (see definition 3.10.1)*

$$\left\lceil\!\left\lceil \widehat{\zeta}^{t,h} \right\rceil\!\right\rceil_{\mathcal{I}^p} \le \widehat{C}_p^{(4.3.1)} \left\lceil\!\left\lceil \zeta^{t,h} \right\rceil\!\right\rceil_{\mathcal{I}^p} \quad \textit{and} \quad \left\lceil\!\left\lceil \widetilde{\zeta}^{t,h} \right\rceil\!\right\rceil_{\mathcal{I}^p} \le \widetilde{C}_p^{(4.3.1)} \left\lceil\!\left\lceil \zeta^{t,h} \right\rceil\!\right\rceil_{\mathcal{I}^p} .$$

Proof. Regard the measure $\theta : \check{H} \mapsto \mathbb{E}[\int \check{H}\, d\zeta]$, $\check{H} \in \check{\mathcal{P}}$, as a σ-finite scalar measure on the product $\check{\boldsymbol{B}} \overset{\text{def}}{=} \boldsymbol{H} \times \boldsymbol{B}$ equipped with $C_{00}(\boldsymbol{H}) \otimes \mathcal{E}$. According to corollary A.3.42 on page 418 there is a disintegration $\theta = \int_{\boldsymbol{B}} \nu_\varpi\, \mu(d\varpi)$, where μ is a positive σ-additive measure on $\mathcal{E}$ and, for every $\varpi \in \boldsymbol{B}$, ν_ϖ a Radon measure on $\boldsymbol{H}$, so that

$$\int_{\boldsymbol{H} \times \boldsymbol{B}} \check{H}(\eta, \varpi)\, \theta(d\eta, d\varpi) = \int_{\boldsymbol{B}} \int_{\boldsymbol{H}} \check{H}(\eta, \varpi) \nu_\varpi(d\eta)\, \mu(d\varpi)$$

for all θ-integrable functions $\check{H} \in \check{\mathcal{P}}$. Since μ clearly annihilates evanescent sets, it has a Doléans–Dade process V^μ. We simply define $\widehat{\zeta}$ by

$$\int \check{H}_s(\eta, \omega)\, \widehat{\zeta}(d\eta, ds; \omega) = \int \int \check{H}_s(\eta, \omega)\, \nu_{(s,\omega)}(d\eta)\, dV_s^\mu(\omega) , \qquad \omega \in \Omega ,$$

set $\widetilde{\zeta} \overset{\text{def}}{=} \zeta - \widehat{\zeta}$, and leave verifying the properties claimed to the reader. ❚

If ζ is the jump measure $\jmath_Z$ of an integrator Z, then $\widehat{\zeta} = \widehat{\jmath_Z}$ is called the ***jump intensity*** of Z and $\nu_s = \nu_s^Z$ the ***jump intensity rate***. In this case both $\widehat{\zeta} = \widehat{\jmath_Z}$ and $\widetilde{\zeta} = \widetilde{\jmath_Z}$ are strict random measures. We say that Z has ***continuous jump intensity*** $\widehat{\jmath_Z}$ if $\widehat{Y}_. \overset{\text{def}}{=} \int_{[\![0,.]\!]} |y^2| \wedge 1 \; \widehat{\jmath_Z}(dy, ds)$ has continuous paths.

Proposition 4.3.25 *The following are equivalent: (i) Z has continuous jump intensity; (ii) the jumps of Z, if any, occur only at totally inaccessible stopping times; (iii) $H{*}\widehat{\jmath_Z}$ has continuous paths for every previsible Hunt function H.*

Definition 4.3.26 *A process with these properties, in other words, a process that has negligible jumps at any predictable stopping time is called* ***quasi-left-continuous****. A random measure ζ is quasi-left-continuous if and only if all of its indefinite integrals $\check{X}{*}\zeta$ are, $\check{X} \in \check{\mathcal{E}}$.*

Proof. *(i)* $\Longrightarrow$ *(ii)* Let S be a predictable stopping time. If ΔZ_S is non-negligible, then clearly neither is the jump $\Delta Y_S = |\Delta Z_S|^2 \wedge 1$ of the increasing process $Y_t \overset{\text{def}}{=} \int_{[\![0,t]\!]} |y|^2 \wedge 1 \; \jmath_Z(dy, ds)$. Since $\Delta Y_S \geq 0$, then $\Delta\widehat{Y}_S = \mathbb{E}\big[\Delta Y_S | \mathcal{F}_{S-}\big]$ is not negligible either (see exercise 4.3.6) and Z does not have continuous jump intensity. The other implications are even simpler to see. ∎

Exercise 4.3.27 If Z consists of local L^1-integrators, then $(\widehat{Z}, ⌉Z^\eta, Z^\theta], \widehat{\jmath_Z})$ is called the ***characteristic triple*** of Z. The expectation of any random variable of the form $\Phi(Z_t)$ can be expressed in terms of the characteristic triple.

4.4 Semimartingales

A process Z is called a ***semimartingale*** if it can be written as the sum of a process V of finite variation and a local martingale M. A semimartingale is clearly an L^0-integrator (proposition 2.4.1, corollary 2.5.29, and proposition 2.1.9). It is shown in proposition 4.4.1 below that the converse is also true: an L^0-integrator is a semimartingale. Stochastic integration in some generality was first developed for semimartingales $Z = V + M$. It was an amalgam of integration with respect to a finite variation process V, known forever, and of integration with respect to a square integrable martingale, known since Courrège [16] and Kunita–Watanabe [59] generalized Itô's procedure. A succinct account can be found in [74]. Here is a rough description: the dZ-integral of a process F is defined as $\int F\, dV + \int F\, dM$, the first summand being understood as a pathwise Lebesgue–Stieltjes integral, and the second as the extension of the elementary M-integral under the Hardy mean of definition (4.2.9). A problem with this approach is that the decomposition $Z = V + M$ is not unique, so that the results of any calculation have to be proven independent of it. There is a very simple example which shows that the class of processes F that can be so integrated depends on the decomposition (example 4.4.4 on page 234).

Integrators Are Semimartingales

Proposition 4.4.1 *An L^0-integrator Z is a semimartingale; in fact, there is a decomposition $Z = V + M$ with $|\Delta M| \le 1$.*

Proof. Recall that Z^n is Z stopped at n. ${}^nZ \stackrel{\text{def}}{=} Z^{n+1} - Z^n$ is a global $L^0(\mathbb{P})$-integrator that vanishes on $[\![0,n]\!]$, $n = 0,1,\ldots$. According to proposition 4.1.1 or theorem 4.1.2, there is a probability ${}^n\mathbb{P}$ equivalent with $\mathbb{P}$ on $\mathcal{F}_\infty$ such that ${}^nZ$ is a global $L^1({}^n\mathbb{P})$-integrator, which then has a Doob–Meyer decomposition ${}^nZ = \widehat{{}^nZ} + \widetilde{{}^nZ}$ with respect to ${}^n\mathbb{P}$. Due to lemma 3.9.11, $\widetilde{{}^nZ}$ is the sum of a finite variation process and a local $\mathbb{P}$-martingale. Clearly then so is ${}^nZ$, say ${}^nZ = {}^nV + {}^nM$. Both ${}^nV$ and ${}^nM$ vanish on $[\![0,n]\!]$ and are constant after time $n+1$. The (ultimately constant) sum $Z = \sum {}^nV + \sum {}^nM$ exhibits Z as a $\mathbb{P}$-semimartingale.

We prove the second claim "locally" and leave its "globalization" as an exercise. Let then an instant $t > 0$ and an $\epsilon > 0$ be given. There exists a stopping time T_1 with $\mathbb{P}[T_1 < t] < \epsilon/3$ such that Z^{T_1} is the sum of a finite variation process $V^{(1)}$ and a martingale $M^{(1)}$. Now corollary 2.5.29 provides a stopping time T_2 with $\mathbb{P}[T_2 < t] < \epsilon/3$ and such that the stopped martingale $M^{(1)T_2}$ is the sum of a process $V^{(2)}$ of finite variation and a global L^2-integrator $Z^{(2)}$. $Z^{(2)}$ has a Doob–Meyer decomposition $Z^{(2)} = \widehat{Z}^{(2)} + \widetilde{Z}^{(2)}$ whose constituents are global L^2-integrators. The following little lemma 4.4.2 furnishes a stopping time T_3 with $\mathbb{P}[T_3 < t] < \epsilon/3$ and such that $\widetilde{Z}^{(2)T_3} = V^{(3)} + M$, where V^3 is a process of finite variation and M a martingale whose jumps are uniformly bounded by 1. Then $T = T_1 \wedge T_2 \wedge T_3$ has $\mathbb{P}[T < t] < \epsilon$, and

$$Z^T = V + M \quad , \text{ where } V = \left(V^{(3)} + \widehat{Z}^{(2)T} + V^{(2)T} + V^{(1)T}\right)$$

is a process of finite variation: Z^T meets the description of the statement. ∎

Lemma 4.4.2 *Any L^2-bounded martingale M can be written as a sum $M = V + M'$, where V is a right-continuous process with integrable total variation $|V|_\infty$ and M' a locally square integrable globally $\mathcal{I}^1$-bounded martingale whose jumps are uniformly bounded by 1.*

Proof. Define the finite variation process V' by

$$V'_t = \sum\{\Delta M_s : s \le t,\ |\Delta M_s| \ge 1/2\}\,, \qquad t \le \infty\,.$$

This sum converges a.s. absolutely, since by theorem 3.8.4

$$\begin{aligned} |V'|_\infty &= \sum\{|\Delta M_s| : |\Delta M_s| > 1/2\} \\ &\le 2 \cdot \sum\nolimits_{s<\infty} (\Delta M_s)^2 \le 1/2 \cdot [M,M]_\infty \end{aligned}$$

is integrable. V' is thus a global L^1-integrator, and so is $Z = M - V'$, a process whose jumps are uniformly bounded by $1/2$. Z has a Doob–Meyer decomposition $Z = \widehat{Z} + \widetilde{Z}$. By exercise 4.3.6 the jump of $\widehat{Z}$ is uniformly bounded by $1/2$, and therefore by subtraction $|\Delta\widetilde{Z}| \le 1$. The desired decomposition is $M = (\widehat{Z} + V') + \widetilde{Z}$: $M' = \widetilde{Z}$ is reduced to a uniformly bounded martingale by the stopping times $\inf\{t : |M'|_t \ge K\}$, which can be made arbitrarily large by the choice of K (lemma 2.5.18). V' has integrable total variation as remarked above, and clearly so does $\widehat{Z}$ (inequality (4.3.1) and exercise 4.3.13).

Corollary 4.4.3 *Let $p > 0$. An L^0-integrator Z is a local L^p-integrator if and only if $|\Delta Z|_T^\star \in L^p$ at arbitrarily large stopping times T or, equivalently, if and only if its square function $S[Z]$ is a local L^p-integrator. In particular, an L^0-integrator with bounded jumps is a local L^p-integrator for all $p < \infty$.*

Proof. Note first that $|\Delta Z|_t^\star$ is in fact measurable on $\mathcal{F}_t$ (corollary A.5.13). Next write $Z = V + M$ with $|\Delta M| < 1$. By the choice of K we can make the time $T \overset{\text{def}}{=} \inf\{t : \lvert V \rvert_t \vee M_t^\star > K\} \wedge K$ arbitrarily large. Clearly $M_T^\star < K + 1$, so M^T is an L^p-integrator for all $p < \infty$ (theorem 2.5.30). Since $\Delta \lvert V \rvert \le 1 + |\Delta Z|$, we have $\lvert V \rvert_T \le K + 1 + |\Delta Z|_K^\star \in L^p$, so that V^T is an L^p-integrator as well (proposition 2.4.1).

Example 4.4.4 (S. J. Lin) Let N be a Poisson process that jumps at the times $T_1, T_2, \ldots$ by 1. It is an increasing process that at time T_n has the value n, so it is a local L^q-integrator for all $q < \infty$ and has a Doob–Meyer decomposition $N = \widehat{N} + \widetilde{N}$; in fact $\widehat{N}_t = t$. Considered as a semimartingale, there are two representations of the form $N = V + M$: $N = N + 0$ and $N = \widehat{N} + \widetilde{N}$.

Now let $H_t = ((0, T_1]]_t / t$. This predictable process is pathwise Lebesgue–Stieltjes–integrable against N, with integral $1/T_1$. So the disciple choosing the decomposition $N = N + 0$ has no problem with the definition of the integral $\int H\, dN$. A person viewing N as the semimartingale $\widehat{N} + \widetilde{N}$ – which is a very *natural* thing to do[9] – and attempting to integrate H with $d\widehat{N}_t$ and with $d\widetilde{N}_t$ and then to add the results will fail, however, since $\int H_t(\omega)\, d\widehat{N}_t(\omega) = \int H_t(\omega)\, dt = \infty$ for all $\omega \in \Omega$. In other words, the class of processes integrable for a semimartingale Z depends in general on its representation $Z = V + M$ if such an ad hoc integration scheme is used.

We leave to the reader the following mitigating fact: if there exists some representation $Z = V + M$ such that the previsible process F is pathwise dV-integrable and is dM-integrable in the sense of the Hardy mean of definition (4.2.9), then F is Z–0-integrable in the sense of chapter 3, and the integrals coincide.

Various Decompositions of an Integrator

While there is nothing unique about the finite variation and martingale parts in the decomposition $Z = V + M$ of an L^0-integrator, there are in fact some canonical parts and decompositions, all related to the location and size of its

9 See remark 4.3.16 on page 228.

jumps. Consider first the increasing L^0-integrator $Y \stackrel{\text{def}}{=} h_0 * \jmath_Z$, where h_0 is the prototypical sure Hunt function $y \mapsto |y|^2 \wedge 1$ (page 180). Clearly Y and Z jump at exactly the same times (by different amounts, of course). According to corollary 4.4.3, Y is a local L^2-integrator and therefore has a Doob–Meyer decomposition $Y = \widehat{Y} + \widetilde{Y}$, whose only use at present is to produce the sparse previsible set $P \stackrel{\text{def}}{=} [\Delta\widehat{Y} \neq 0]$ (see exercise 4.3.4). Let us set

$$ {}^pZ \stackrel{\text{def}}{=} P*Z \quad \text{and} \quad {}^qZ \stackrel{\text{def}}{=} Z - {}^pZ = (1-P)*Z \,. \tag{4.4.1} $$

By exercise 3.10.12 (iv) we have $\jmath_{{}^qZ} = (1-P)\cdot \jmath_Z$, and this random measure has continuous previsible part $\widehat{\jmath_{{}^qZ}} = (1-P)\cdot \widehat{\jmath_Z}$. In other words, ${}^qZ$ has continuous jump intensity. Thanks to proposition 4.3.25, $\Delta {}^qZ_S = 0$ at all predictable stopping times S:

Proposition 4.4.5 *Every L^0-integrator Z has a unique decomposition*

$$ Z = {}^pZ + {}^qZ $$

with the following properties: there exists a previsible set P, a union of the graphs of countably many predictable stopping times, such that ${}^pZ = P*{}^pZ$ [10]*; and ${}^qZ$ jumps at totally inaccessible stopping times only, which is to say that ${}^qZ$ is quasi-left-continuous. For $0 < p < \infty$, the maps $Z \mapsto {}^pZ$ and $Z \mapsto {}^qZ$ are linear contractive projections on $\mathcal{I}^p$.*

Proof. If $Z = {}^pZ + {}^qZ = Z = {}^pZ' + {}^qZ'$, then ${}^pZ - {}^pZ' = {}^qZ' - {}^qZ$ is supported by a sparse previsible set yet jumps at no predictable stopping time, so must vanish. This proves the uniqueness. The linearity and contractivity follow from this and the construction (4.4.1) of ${}^pZ$ and ${}^qZ$, which is therefore canonical. ∎

Exercise 4.4.6 Every random measure ζ has a unique decomposition $\zeta = {}^p\zeta + {}^q\zeta$ with the following properties: there exists a previsible set P, a union of the graphs of countably many predictable stopping times, such that ${}^p\zeta = (\boldsymbol{H}\times P)\cdot\zeta$ [10,11]; and ${}^q\zeta$ jumps at totally inaccessible stopping times only, in the sense that $\check{H}*{}^q\zeta$ does for all $\check{H} \in \check{\mathcal{P}}$. For $0 < p < \infty$, the maps $\zeta \mapsto {}^p\zeta$ and $\zeta \mapsto {}^q\zeta$ are linear contractive projections on the space of L^p-random measures.

Proposition 4.4.7 (The Continuous Martingale Part of an Integrator) *An L^0-integrator Z has a canonical decomposition*

$$ Z = {}^{\tilde{c}}Z + {}^rZ \,, $$

where ${}^{\tilde{c}}Z$ is a continuous local martingale with ${}^{\tilde{c}}Z_0 = 0$ and with continuous bracket $[{}^{\tilde{c}}Z, {}^{\tilde{c}}Z] = {}^c[Z,Z]$ and where the remainder ${}^rZ$ has continuous bracket ${}^c[{}^rZ, {}^rZ] = 0$. There are universal constants C_p such that at all instants t

$$ \lceil {}^{\tilde{c}}Z^t \rceil_{\mathcal{I}^p} \le C_p \lceil Z^t \rceil_{\mathcal{I}^p} \quad \textit{and} \quad \lceil {}^rZ^t \rceil_{\mathcal{I}^p} \le C_p \lceil Z^t \rceil_{\mathcal{I}^p} \,, \; 0 < p < \infty \,. \tag{4.4.2} $$

---

[10] We might paraphrase this by saying "$d{}^pZ$ is supported by a ***sparse previsible set***."
[11] This is to mean of course that $\check{H}*{}^p\zeta = (\check{H}\cdot(\boldsymbol{H}\times P))*\zeta$ for all $\check{H} \in \check{\mathcal{P}}$.

Exercise 4.4.8 Z and ${}^qZ$ have the same continuous martingale part. Taking the continuous martingale part is stable under stopping: ${}^{\tilde c}(Z^T) = ({}^{\tilde c}Z)^T$ at all $T \in \mathfrak{T}$. Consequently (4.4.2) persists if t is replaced by a stopping time T.

Exercise 4.4.9 Every random measure ζ has a canonical decomposition as $\zeta = {}^{\tilde c}\zeta + {}^r\zeta$, given by $\check H * {}^{\tilde c}\zeta = {}^{\tilde c}(\check H * \zeta)$ and $\check H * {}^r\zeta = {}^r(\check H * \zeta)$. The maps $\zeta \mapsto {}^{\tilde c}\zeta$ and $\zeta \mapsto {}^r\zeta$ are linear contractive projections on the space of L^p-random measures, for $0 < p < \infty$.

Proof of 4.4.7. First the uniqueness. If also $Z = {}^{\tilde c}Z' + {}^rZ'$, then ${}^{\tilde c}Z - {}^{\tilde c}Z'$ is a continuous martingale whose continuous bracket vanishes, since it is that of ${}^rZ' - {}^rZ$; thus ${}^{\tilde c}Z - {}^{\tilde c}Z'$ must be constant, in fact, since ${}^{\tilde c}Z_0 = {}^{\tilde c}Z'_0$, it must vanish.

Next the inequalities:

$$\lceil {}^{\tilde c}Z^t \rceil_{\mathcal{I}^p} \le C_p^{(4.3.7)} \left\| S_t[{}^{\tilde c}Z] \right\|_p = C_p \llbracket \sigma_t[Z] \rrbracket_p \le C_p \llbracket S_t[Z] \rrbracket_p$$

$$\le C_p K_p^{(3.8.6)} \lceil Z^t \rceil_{\mathcal{I}^p} .$$

Now to the existence. There is a sequence (T^i) of bounded stopping times with disjoint graphs so that every jump of Z occurs at one of them (exercise 1.3.21). Every T^i can be decomposed as the infimum of an accessible stopping time T^i_A and a totally inaccessible stopping time T^i_I (see page 122). For every i let $(S^{i,j})$ be a sequence of predictable stopping times so that $[\![T^i_A]\!] \subset \bigcup_j [\![S^{i,j}]\!]$, and let Δ denote the union of the graphs of the $S^{i,j}$. This is a previsible set, and $\Delta * Z$ is an integrator whose jumps occur only on Δ (proposition 3.8.21) and whose continuous square function vanishes. The jumps of $Z' \overset{\text{def}}{=} Z - \Delta * Z = (1-\Delta) * Z$ occur at the totally inaccessible times T^i_I.

Assume now for the moment that Z is an L^2-integrator; then clearly so is Z'. Fix an i and set $J^i \overset{\text{def}}{=} \Delta Z'_{T^i_I} \cdot [\![T^i, \infty))$. This is an L^2-integrator of total variation $|\Delta Z'_{T^i_I}| \in L^2$ and has a Doob–Meyer decomposition $J^i = \widehat{J^i} + \widetilde{J^i}$. The previsible part $\widehat{J^i}$ is continuous: if $[\Delta \widehat{J^i} > 0]$ were non-evanescent, it would contain the non-evanescent graph of a previsible stopping time (theorem A.5.14), at which the jump of Z' could not vanish (exercise 4.3.6), which is impossible due to the total inaccessibility of the jump times of Z'. Therefore $\widetilde{J^i}$ has exactly the same single jump as J^i, namely $\Delta Z'_{T^i_I}$ at T^i_I. Now at all instants t

$$\left\| \left| \sum_{I<i\le J} \widetilde{J^i} \right|_t^\star \right\|_{L^2} \le 2 \left\| S_t\Big[\sum_{I<i\le J} \widetilde{J^i} \Big] \right\|_{L^2} = 2 \left(\mathbb{E}\Big[\sum_{I<i\le J} |\Delta Z'_{T^i_I}|^2 \Big] \right)^{1/2}$$

$$\le 2 \left(\mathbb{E}\Big[\sum_{I<i\le J} |\Delta Z_{T^i}|^2 \Big] \right)^{1/2} \xrightarrow[I<J\,;\,I,J\to\infty]{} 0 .$$

The sum $M \overset{\text{def}}{=} \sum_i \widetilde{J^i}$ therefore converges uniformly almost surely and in L^2 and defines a martingale M that has exactly the same jumps as Z'. Then

$$Z'' \overset{\text{def}}{=} Z' - M = Z - (1-\Delta) * Z - M$$

is a continuous L^2-integrator whose square function is ${}^c[Z,Z]$. Its martingale part evidently meets the description of $\tilde{^c}Z$.

Now if Z is merely an $L^0(\mathbb{P})$-integrator, we fix a t and find a probability $\mathbb{P}' \approx \mathbb{P}$ on $\mathcal{F}_t$ with respect to which Z^t is an L^2-integrator. We then write Z^t as ${}^{\tilde{c}}Z'^t + {}^rZ'^t$, where ${}^{\tilde{c}}Z'^t$ is the canonical $\mathbb{P}'$-martingale part of Z^t. Thanks to the Girsanov–Meyer lemma 3.9.11, whose notations we employ here again,

$$ {}^{\tilde{c}}Z'^t = \left({}^{\tilde{c}}Z_0'^t + G_{.-}{*}[{}^{\tilde{c}}Z'^t, G']\right) + \left(G_{.-}{*}({}^{\tilde{c}}Z'^t G') - ({}^{\tilde{c}}Z'^t G)_{.-}{*}G'\right) $$

is the sum of two continuous processes, of which the first has finite variation and the second is a local $\mathbb{P}$-martingale, which we call ${}^{\tilde{c}}Z^t$. Clearly ${}^{\tilde{c}}Z^t$ is the canonical local $\mathbb{P}$-martingale part of the stopped process Z^t. We can do this for arbitrarily large times t. From the uniqueness, established first, we see that the sequence $({}^{\tilde{c}}Z^n)$ is ultimately constant, except possibly on an evanescent set. Clearly ${}^{\tilde{c}}Z \stackrel{\text{def}}{=} \lim_{n\to\infty} {}^{\tilde{c}}Z^n$ is the desired canonical continuous martingale part of Z. Now ${}^rZ$ is of course defined as the difference $Z - {}^{\tilde{c}}Z$.

∎

In view of exercise 4.4.8 we now have a canonical decomposition

$$ \boldsymbol{Z} = {}^p\boldsymbol{Z} + {}^{\tilde{c}}\boldsymbol{Z} + {}^r\boldsymbol{Z} $$

of an L^0-integrator $\boldsymbol{Z}$, with the linear projections

$$ Z \mapsto {}^p\boldsymbol{Z}\,,\ Z \mapsto {}^{\tilde{c}}\boldsymbol{Z}\ ,\ \text{and}\ \ Z \mapsto {}^r\boldsymbol{Z} $$

continuous from $\mathcal{I}^0$ to $\mathcal{I}^0$ and contractive from $\mathcal{I}^p$ to $\mathcal{I}^p$ for all $p > 0$.

Exercise 4.4.10 ${}^r\boldsymbol{Z}$ can be decomposed further. Set

$$ {}^{\tilde{s}}\boldsymbol{Z}_t \stackrel{\text{def}}{=} \int_{[\![0,t]\!]} y \cdot [|y| \le 1]\, d\widetilde{\jmath_{{}^r\boldsymbol{Z}}} = \int_{[\![0,t]\!]} y \cdot [|y| \le 1]\, d\widetilde{\jmath_{\boldsymbol{Z}}}\,, $$

$$ {}^l\boldsymbol{Z}_t \stackrel{\text{def}}{=} \int_{[\![0,t]\!]} y \cdot [|y| > 1]\, d\jmath_{{}^r\boldsymbol{Z}} = \int_{[\![0,t]\!]} y \cdot [|y| > 1]\, d\widetilde{\jmath_{\boldsymbol{Z}}}\,, $$

and $$ {}^{\hat{v}}\boldsymbol{Z} \stackrel{\text{def}}{=} {}^r\boldsymbol{Z} - {}^{\tilde{s}}\boldsymbol{Z} - {}^l\boldsymbol{Z}\,. $$

Then ${}^{\tilde{s}}\boldsymbol{Z}$ is a martingale with zero continuous part but jumps bounded by 1, the ***small jump martingale part***; ${}^l\boldsymbol{Z}$ is a finite variation process without a continuous part and constant between its jumps, which are of size at least 1 and occur at discrete times, the ***large jump part***; and ${}^{\hat{v}}\boldsymbol{Z}$ is a continuous finite variation process. The projections ${}^r\boldsymbol{Z} \mapsto {}^{\tilde{s}}\boldsymbol{Z}$, ${}^r\boldsymbol{Z} \mapsto {}^l\boldsymbol{Z}$, and ${}^r\boldsymbol{Z} \mapsto {}^{\hat{v}}\boldsymbol{Z}$ are not even linear, much less continuous in $\mathcal{I}^p$. From this obtain a decomposition

$$ \boldsymbol{Z} = ({}^{lp}\boldsymbol{Z} + {}^{\tilde{s}p}\boldsymbol{Z}) + ({}^{\tilde{c}}\boldsymbol{Z}) + ({}^{\hat{v}}\boldsymbol{Z} + {}^{\tilde{s}}\boldsymbol{Z} + {}^l\boldsymbol{Z}) \tag{4.4.3} $$

and describe its ingredients.

4.5 Previsible Control of Integrators

A general L^p-integrator's integral is controlled by Daniell's mean; if the integrator happens to be a martingale M and $p \geq 1$, then its integral can be controlled pathwise by the finite variation process $[M,M]$ (see definition (4.2.9) on page 216). For the solution of stochastic differential equations it is desirable to have such pathwise control of the integral not only for general integrators instead of merely for martingales but also *by a previsible* increasing process instead of $[M,M]$, and *for a whole vector of integrators simultaneously.* Here is what we can accomplish in this regard – see also theorem 4.5.25 concerning random measures:

Theorem 4.5.1 *Let* $\boldsymbol{Z}$ *be a d-tuple of local* L^q*-integrators, where* $2 \leq q < \infty$. *Fix a (small)* $\alpha > 0$. *There exists a strictly increasing previsible process* $\Lambda{=}\Lambda^{\langle q\rangle}[\boldsymbol{Z}]$ *such that for every* $p{\in}[2,q]$, *every stopping time* T, *and every d-tuple* $\boldsymbol{X} = (X_1, \ldots, X_d)$ *of previsible processes*

$$\left\|\,|\boldsymbol{X}{*}\boldsymbol{Z}|_T^\star\right\|_{L^p} \leq C_p^\diamond \cdot \max_{\rho=1^\diamond,p^\diamond} \left\|\left(\int_0^T |\boldsymbol{X}|_s^\rho \, d\Lambda_s\right)^{1/\rho}\right\|_{L^p}, \tag{4.5.1}$$

with $|\boldsymbol{X}| \stackrel{\text{def}}{=} |\boldsymbol{X}|_\infty = \max_{1\leq\eta\leq d} |X_\eta|$ *and universal constant* $C_p^\diamond \leq 9.5p$.

$$\textit{Here} \quad 1^\diamond = 1^\diamond[\boldsymbol{Z}] \stackrel{\text{def}}{=} \begin{cases} 2 & \textit{if } \boldsymbol{Z} \textit{ is a martingale} \\ 1 & \textit{otherwise,} \end{cases}$$

$$\textit{and} \quad p^\diamond = p^\diamond[\boldsymbol{Z}] \stackrel{\text{def}}{=} \begin{cases} 1 & \textit{if } \boldsymbol{Z} \textit{ is continuous and has finite variation} \\ 2 & \textit{if } \boldsymbol{Z} \textit{ is continuous and } \widetilde{\boldsymbol{Z}} \neq 0 \\ p & \textit{if } \boldsymbol{Z} \textit{ has jumps.} \end{cases}$$

Furthermore, the previsible controller Λ *can be estimated by*

$$\mathbb{E}\Big[\Lambda_T^{\langle q\rangle}[\boldsymbol{Z}]\Big] \leq \alpha \mathbb{E}[T] + 3\Big(\,\lceil \boldsymbol{Z}^T \rceil_{\mathcal{I}^q} \vee \lceil \boldsymbol{Z}^T \rceil_{\mathcal{I}^q}^q\Big) \tag{4.5.2}$$

$$\textit{and} \qquad \Lambda_T^{\langle q\rangle}[\boldsymbol{Z}] \geq \alpha T \quad \textit{at all stopping times } T\,.$$

Remark 4.5.2 The controller Λ constructed in the proof below satisfies the four inequalities

$$\alpha \cdot dt \leq d\Lambda_t\,; \tag{4.5.3}$$

and

$$|X_\eta|_s \, d\lceil \widehat{Z}^\eta \rceil_s \leq |\boldsymbol{X}|_s \, d\Lambda_s\,,$$

$$d\big|X_\eta X_\theta {*} \langle Z^\eta, Z^\theta\rangle\big|_s^\star \leq |\boldsymbol{X}|_s^2 \, d\Lambda_s\,,$$

$$\int_{\mathbb{R}_*^d} |\langle \boldsymbol{X}|\boldsymbol{y}\rangle|^q \, \widehat{\jmath_{\boldsymbol{Z}}}(d\boldsymbol{y}, ds) \leq |\boldsymbol{X}|_s^q \, d\Lambda_s\,,$$

for all previsible $\boldsymbol{X}$, and is in fact the smallest increasing process doing so. This makes it somewhat canonical (when α is fixed) and justifies naming it ***THE previsible controller of*** $\boldsymbol{Z}$.

The imposition of inequality (4.5.3) above is somewhat artificial. Its purpose is to ensure that Λ is strictly increasing and that the predictable (see exercise 3.5.19) stopping times

$$T^\lambda \stackrel{\text{def}}{=} \inf\{t : \Lambda_t \ge \lambda\} \quad \text{and} \quad T^{\lambda+} \stackrel{\text{def}}{=} \inf\{t : \Lambda_t > \lambda\} \tag{4.5.4}$$

agree and are bounded (by λ/α); in most naturally occurring situations one of the drivers Z^η is time, and then this inequality is automatically satisfied. The collection $T^\cdot = \{T^\lambda : \lambda > 0\}$ will henceforth simply be called ***THE time transformation for Z*** (or for Λ). The process $\Lambda^{\langle q \rangle}[\boldsymbol{Z}]$ or the parameter λ, its value, is occasionally referred to as the ***intrinsic time***. It and the time transformation $T^\cdot$ are the main tools in the existence, uniqueness, and stability proofs for stochastic differential equations driven by $\boldsymbol{Z}$ in chapter 5.

The proof of theorem 4.5.1 will show that if $\boldsymbol{Z}$ is quasi-left-continuous, then Λ is continuous. This happens in particular when $\boldsymbol{Z}$ is a Lévy process (see section 4.6 below) or is the solution of a differential equation driven by a Lévy process (exercise 5.2.17 and page 349). If Λ is continuous, then the time transformation $\lambda \mapsto T^\lambda$ is evidently strictly increasing without bound.

Exercise 4.5.3 Suppose that inequality (4.5.1) holds whenever $\boldsymbol{X} \in \mathcal{P}^d$ is bounded and T reduces $\boldsymbol{Z}$ to a global L^p-integrator. Then it holds in the generality stated.

Controlling a Single Integrator

The remainder of the section up to page 249 is devoted to the proof of theorem 4.5.1. We start with the case $d = 1$; in other words, $\boldsymbol{Z}$ is a single integrator Z.

The main tools are the ***higher order brackets*** $Z^{[\rho]}$ defined for all t by[12]

$$Z_t^{[1]} \stackrel{\text{def}}{=} Z_t \ , \quad Z_t^{[2]} \stackrel{\text{def}}{=} [Z,Z]_t = {}^c[Z,Z]_t + \int_{[\![0,t]\!]} y^2 \, \jmath_Z(dy,ds) \ ,$$

$$Z_t^{[\rho]} \stackrel{\text{def}}{=} \sum_{0 \le s \le t} (\Delta Z_s)^\rho = \int_{[\![0,t]\!]} y^\rho \, \jmath_Z(dy,ds) \ , \qquad \text{for } \rho = 3,4,\ldots,$$

and

$$\left|Z^{[\rho]}\right|_t \stackrel{\text{def}}{=} \sum_{0 \le s \le t} |\Delta Z_s|^\rho = \int_{[\![0,t]\!]} |y|^\rho \, \jmath_Z(dy,ds) \ ,$$

defined for any real $\rho > 2$ and satisfying

$$\left|Z^{[\rho]}\right|_t^{1/\rho} \le \Big(\sum_{0 \le s \le t} |\Delta Z_s|^2 \Big)^{1/2} \le S_t[Z] \, . \tag{4.5.5}$$

For integer ρ, $\left|Z^{[\rho]}\right|$ is the variation process of $Z^{[\rho]}$. Observe now that equation (3.10.7) can be rewritten in terms of the $Z^{[\rho]}$ as follows:

---

[12] $[\![0,t]\!]$ is the product $\mathbb{R}_*^d \times [0,t]$ of ***auxiliary space*** $\mathbb{R}_*^d$ with the stochastic interval $[\![0,t]\!]$.

Lemma 4.5.4 *For an n-times continuously differentiable function Φ on $\mathbb{R}$ and any stopping time T*

$$\Phi(Z_T) = \Phi(Z_0) + \sum_{\nu=1}^{n-1} \frac{1}{\nu!} \int_{0+}^{T} \Phi^{(\nu)}(Z_{.-})\, dZ^{[\nu]}$$

$$+ \int_0^1 \frac{(1-\lambda)^{n-1}}{(n-1)!} \int_{0+}^{T} \Phi^{(n)}(Z_{.-} + \lambda \Delta Z)\, dZ^{[n]}\, d\lambda\,.$$

Let us apply lemma 4.5.4 to the function $\Phi(z) = |z|^p$, with $1 < p < \infty$. If n is a natural number strictly less than p, then Φ is n-times continuously differentiable. With $\epsilon = p - n$ we find, using item A.2.43 on page 388,

$$|Z_t|^p = |Z_0|^p + \sum_{\nu=1}^{n-1} \binom{p}{\nu} \int_{0+}^{t} |Z|_{.-}^{p-\nu} \cdot (\operatorname{sgn} Z_{.-})^\nu\, dZ^{[\nu]}$$

$$+ \int_0^1 n(1-\lambda)^{n-1} \int_{0+}^{t} \binom{p}{n} |Z_{.-} + \lambda \Delta Z|^\epsilon \big(\operatorname{sgn}(Z_{.-} + \lambda\Delta Z)\big)^n\, dZ^{[n]}\, d\lambda\,.$$

Writing $|Z_0|^p$ as $\int_{[\![0]\!]} d\,¦Z^{[p]}¦$ produces this useful estimate:

Corollary 4.5.5 *For every L^0-integrator Z, stopping time T, and $p > 1$ let $n = \lfloor p \rfloor$ be the largest integer less than or equal to p and set $\epsilon \stackrel{\text{def}}{=} p - n < 1$.*

$$\textit{Then}\quad |Z|_T^p \le p \int_0^T |Z|_{.-}^{p-1} \cdot \operatorname{sgn} Z_{.-}\, dZ + \sum_{\nu=2}^{n-1} \binom{p}{\nu} \int_0^T |Z|_{.-}^{p-\nu}\, d\,¦Z^{[\nu]}¦$$

$$+ \int_0^1 n(1-\lambda)^{n-1} \int_{0+}^{T} \binom{p}{n} \big(|Z|_{.-} + \lambda |\Delta Z|\big)^\epsilon\, d\,¦Z^{[n]}¦\, d\lambda\,. \quad (4.5.6)$$

Proof. This is clear when $p > \lfloor p \rfloor$. In the case that p is an integer, apply this to a sequence of $p_n > p$ that decrease to p and take the limit.

Now thanks to inequality (4.5.5) and theorem 3.8.4, $¦Z^{[\rho]}¦$ is locally integrable and therefore has a Doob–Meyer decomposition when ρ is any real number between 2 and q. We use this observation to define positive increasing previsible processes $Z^{\langle\rho\rangle}$ as follows: $Z^{\langle 1\rangle} = ¦\widehat{Z}¦$; and for $\rho \in [2, q]$, $Z^{\langle\rho\rangle}$ is the previsible part in the Doob–Meyer decomposition of $¦Z^{[\rho]}¦$. For instance, $Z^{\langle 2\rangle} = \langle Z, Z\rangle$. In summary

$$Z^{\langle 1\rangle} \stackrel{\text{def}}{=} ¦\widehat{Z}¦\,,\; Z^{\langle 2\rangle} \stackrel{\text{def}}{=} \langle Z, Z\rangle\,, \text{ and } Z^{\langle\rho\rangle} \stackrel{\text{def}}{=} \widehat{¦Z^{[\rho]}¦} \qquad \text{for } 2 \le \rho \le q\,.$$

Exercise 4.5.6 $(X{*}Z)^{\langle\rho\rangle} = |X|^\rho{*}Z^{\langle\rho\rangle}$ for $X \in \mathcal{P}_b$ and $\rho \in \{1\} \cup [2, q]$.

In the following keep in mind that $Z^{\langle\rho\rangle} = 0$ for $\rho > 2$ if Z is continuous, and $Z^{\langle\rho\rangle} = 0$ for $\rho > 1$ if in addition Z has no martingale component, i.e., if Z is a continuous finite variation process. The desired previsible controller $\Lambda^{\langle q\rangle}[Z]$

will be constructed from the processes $Z^{\langle\rho\rangle}$, which we call the ***previsible higher order brackets***. On the way to the construction and estimate three auxiliary results are needed:

Lemma 4.5.7 *For $2 \le \rho < \sigma < \tau \le q$, we have both $Z^{\langle\sigma\rangle} \le Z^{\langle\rho\rangle} \vee Z^{\langle\tau\rangle}$ and $\big(Z^{\langle\sigma\rangle}\big)^{1/\sigma} \le \big(Z^{\langle\rho\rangle}\big)^{1/\rho} \vee \big(Z^{\langle\tau\rangle}\big)^{1/\tau}$, except possibly on an evanescent set. Also,*

$$\left\| \big(Z_T^{\langle\sigma\rangle}\big)^{1/\sigma} \right\|_{L^p} \le \left\| \big(Z_T^{\langle\rho\rangle}\big)^{1/\rho} \right\|_{L^p} \vee \left\| \big(Z_T^{\langle\tau\rangle}\big)^{1/\tau} \right\|_{L^p} \tag{4.5.7}$$

for any stopping time T and $p \in (0,\infty)$ – the right-hand side is finite for sure if Z^T is $\mathcal{I}^q$-bounded and $p \le q$.

Proof. A little exercise in calculus furnishes the equality

$$\inf_{\lambda>0} A\lambda^{\rho-\sigma} + B\lambda^{\tau-\sigma} = C \cdot A^{\frac{\tau-\sigma}{\tau-\rho}} B^{\frac{\sigma-\rho}{\tau-\rho}} , \tag{4.5.8}$$

with
$$C = \Big(\frac{\sigma-\rho}{\tau-\sigma}\Big)^{\frac{\rho-\sigma}{\tau-\rho}} + \Big(\frac{\sigma-\rho}{\tau-\sigma}\Big)^{\frac{\tau-\sigma}{\tau-\rho}} .$$

The choice $A = B = 1$ and $\lambda = |\Delta Z_s|$ gives

$$C \cdot |\Delta Z_s|^\sigma \le |\Delta Z_s|^\rho + |\Delta Z_s|^\tau , \qquad 0 \le s < \infty ,$$

which says $\quad C \cdot d\,\dot{|}Z^{[\sigma]}\dot{|} \le d\,\dot{|}Z^{[\rho]}\dot{|} + d\,\dot{|}Z^{[\tau]}\dot{|}$

and implies
$$C \cdot dZ^{\langle\sigma\rangle} \le dZ^{\langle\rho\rangle} + dZ^{\langle\tau\rangle} \tag{4.5.9}$$

and
$$C \cdot Z^{\langle\sigma\rangle} \le Z^{\langle\rho\rangle} + Z^{\langle\tau\rangle} ,$$

except possibly on an evanescent set. Homogeneity produces

$$C \cdot \lambda^\sigma Z^{\langle\sigma\rangle} \le \lambda^\rho Z^{\langle\rho\rangle} + \lambda^\tau Z^{\langle\tau\rangle} , \qquad \lambda > 0 .$$

By changing $Z^{\langle\sigma\rangle}$ on an evanescent set we can arrange things so that this inequality holds at all points of the base space $\boldsymbol{B}$ and for all $\lambda > 0$. Equation (4.5.8) implies

$$C \cdot Z^{\langle\sigma\rangle} \le C \cdot \big(Z^{\langle\rho\rangle}\big)^{\frac{\tau-\sigma}{\tau-\rho}} \cdot \big(Z^{\langle\tau\rangle}\big)^{\frac{\sigma-\rho}{\tau-\rho}} ,$$

i.e.,
$$Z^{\langle\sigma\rangle} \le \big(Z^{\langle\rho\rangle}\big)^{\frac{\tau-\sigma}{\tau-\rho}} \cdot \big(Z^{\langle\tau\rangle}\big)^{\frac{\sigma-\rho}{\tau-\rho}}$$

and
$$\big(Z^{\langle\sigma\rangle}\big)^{1/\sigma} \le \big(Z^{\langle\rho\rangle 1/\rho}\big)^{\frac{\rho(\tau-\sigma)}{\sigma(\tau-\rho)}} \cdot \big(Z^{\langle\tau\rangle 1/\tau}\big)^{\frac{\tau(\sigma-\rho)}{\sigma(\tau-\rho)}} . \tag{$*$}$$

The two exponents e_ρ and e_τ on the right-hand side sum to 1, in either of the previous two inequalities, and this produces the first two inequalities of lemma 4.5.7; the third one follows from Hölder's inequality applied to the pth power of $(*)$ with conjugate exponents $1/(pe_\rho)$ and $1/(pe_\tau)$. ∎

Exercise 4.5.8 Let $\mu^{\langle\rho\rangle}$ denote the Doléans–Dade measure of $Z^{\langle\rho\rangle}$. Then $\mu^{\langle\sigma\rangle} \le \mu^{\langle\rho\rangle} \vee \mu^{\langle\tau\rangle}$ whenever $2 \le \rho < \sigma < \tau \le q$.

Lemma 4.5.9 *At any stopping time T and for all $X \in \mathcal{P}_b$ and all $p \in [2, q]$*

$$\left\| |X{*}Z|^\star_T \right\|_{L^p} \le C_p^\diamond \cdot \max_{\rho=1^\diamond,2,p^\diamond} \left\| \left(\int_0^T |X|^\rho \, dZ^{\langle\rho\rangle} \right)^{1/\rho} \right\|_{L^p}, \tag{4.5.10}$$

$$\le C_p^\diamond \cdot \max_{\rho=1^\diamond,2,q^\diamond} \left\| \left(\int_0^T |X|^\rho \, dZ^{\langle\rho\rangle} \right)^{1/\rho} \right\|_{L^p}, \tag{4.5.11}$$

with universal constant $C_p^\diamond \le 9.5p$.

Proof. Let $n = \lfloor p \rfloor$ be the largest integer less than or equal to p, set $\epsilon = p - n$

and
$$\zeta = \zeta[Z] \stackrel{\text{def}}{=} \max_{\rho=1,\ldots,n,p} \left\| \left(Z_\infty^{\langle\rho\rangle} \right)^{1/\rho} \right\|_{L^p} \tag{4.5.12}$$

by inequality (4.5.7):
$$= \max_{\rho=1,2,p} \left\| \left(Z_\infty^{\langle\rho\rangle} \right)^{1/\rho} \right\|_{L^p} = \max_{\rho=1^\diamond,2,p^\diamond} \left\| \left(Z_\infty^{\langle\rho\rangle} \right)^{1/\rho} \right\|_{L^p}$$

$$\le \max_{\rho=1^\diamond,2,q^\diamond} \left\| \left(Z_\infty^{\langle\rho\rangle} \right)^{1/\rho} \right\|_{L^p}. \tag{4.5.13}$$

The last equality follows from the fact that $Z^{\langle\rho\rangle} = 0$ for $\rho > 2$ if Z is continuous and for $\rho = 1$ if it is a martingale, and the previous inequality follows again from (4.5.7). Applying the expectation to inequality (4.5.6) on page 240 produces

$$\mathbb{E}[|Z|_\infty^p] \le p\mathbb{E}\Big[\int_0^\infty |Z|_{\leftarrow}^{p-1} \cdot \operatorname{sgn} Z_{\leftarrow} \, dZ\Big] + \sum_{\nu=2}^{n-1} \binom{p}{\nu} \mathbb{E}\Big[\int_0^\infty |Z|_{\leftarrow}^{p-\nu} \, d{\rm ¦}Z^{[\nu]}{\rm ¦}\Big]$$

$$+ \int_0^1 n(1-\lambda)^{n-1} \binom{p}{n} \mathbb{E}\Big[\int_{0+}^\infty \big(|Z|_{\leftarrow} + \lambda|\Delta Z|\big)^\epsilon \, d{\rm ¦}Z^{[n]}{\rm ¦}\Big] \, d\lambda\,.$$

$$\le \sum_{\nu=1}^{n-1} \binom{p}{\nu} \mathbb{E}\Big[\int_0^\infty |Z|_{\leftarrow}^{p-\nu} \, dZ^{\langle\nu\rangle}\Big]$$

$$+ \int_0^1 n(1-\lambda)^{n-1} \binom{p}{n} \mathbb{E}\Big[\int_{0+}^\infty \big(|Z|_{\leftarrow} + \lambda|\Delta Z|\big)^\epsilon \, d{\rm ¦}Z^{[n]}{\rm ¦}\Big] \, d\lambda$$

$$= Q_1 + Q_2\,. \tag{4.5.14}$$

Let us estimate the expectations in the first quantity Q_1:

$$\mathbb{E}\Big[\int_0^\infty |Z|_{\leftarrow}^{p-\nu} \, dZ^{\langle\nu\rangle}\Big] \le \mathbb{E}\Big[|Z_\infty^\star|^{p-\nu} \cdot Z_\infty^{\langle\nu\rangle}\Big]$$

using Hölder's inequality:
$$\le \|Z_\infty^\star\|_{L^p}^{p-\nu} \cdot \left\| \left(Z_\infty^{\langle\nu\rangle} \right)^{1/\nu} \right\|_{L^p}^\nu$$

by definition (4.5.12):
$$\le \|Z_\infty^\star\|_{L^p}^{p-\nu} \cdot \zeta^\nu\,.$$

Therefore
$$Q_1 \le \sum_{\nu=1}^{n-1} \binom{p}{\nu} \|Z_\infty^\star\|_{L^p}^{p-\nu} \cdot \zeta^\nu\,. \tag{4.5.15}$$

To treat Q_2 we assume to start with that $\zeta = 1$. This implies that the measure $X \mapsto \mathbb{E}[\int X\, d\vert Z^{[p]}\vert]$ on measurable processes X has total mass

$$\mathbb{E}\Big[\int 1\, d\vert Z^{[p]}\vert\Big] = \mathbb{E}\Big[Z^{\langle p\rangle}_\infty\Big] = \Big\| \big(Z^{\langle p\rangle}_\infty\big)^{1/p} \Big\|^p_{L^p} \le \zeta^p = 1$$

and makes Jensen's inequality A.3.25 applicable to the concave function $\mathbb{R}_+ \ni z \mapsto z^\epsilon$ in $(*)$ below:

$$\begin{aligned}
&\mathbb{E}\Big[\int_{0+}^\infty \big(\vert Z\vert_- + \lambda\vert\Delta Z\vert\big)^\epsilon\, d\vert Z^{[n]}\vert\Big] \\
&= \mathbb{E}\Big[\int_{0+}^\infty \big(\vert Z\vert_- \vert\Delta Z\vert^{-1} + \lambda\big)^\epsilon\, d\vert Z^{[p]}\vert\Big] \qquad (*)\\
&\le \Big(\mathbb{E}\Big[\int_{0+}^\infty \vert Z\vert_- \vert\Delta Z^{-1}\, d\vert Z^{[p]}\vert\Big] + \lambda\mathbb{E}\Big[\int_{0+}^\infty d\vert Z^{[p]}\vert\Big]\Big)^\epsilon \\
&= \Big(\mathbb{E}\Big[\int_{0+}^\infty \vert Z\vert_-\, d\vert Z^{[p-1]}\vert\Big] + \lambda\mathbb{E}\Big[Z^{\langle p\rangle}_\infty\Big]\Big)^\epsilon \\
&\le \Big(\mathbb{E}\Big[Z^\star_\infty Z^{\langle p-1\rangle}_\infty\Big] + \lambda\Big)^\epsilon \\
\text{by Hölder:}\quad &\le \Big(\|Z^\star_\infty\|_{L^p} \cdot \Big\| \big\vert Z^{\langle p-1\rangle}_\infty\big\vert^{1/(p-1)} \Big\|^{p-1}_{L^p} + \lambda\Big)^\epsilon \\
\text{as } \zeta = 1:\quad &\le \big(\|Z^\star_\infty\|_{L^p} \cdot \zeta^{p-1} + \lambda\big)^\epsilon = \big(\|Z^\star_\infty\|_{L^p} + \lambda\zeta\big)^\epsilon \cdot \zeta^n .
\end{aligned}$$

We now put this and inequality (4.5.15) into inequality (4.5.14) and obtain

$$\begin{aligned}
\mathbb{E}[\vert Z\vert^p_\infty] &\le \sum_{\nu=1}^{n-1} \binom{p}{\nu} \|Z^\star_\infty\|^{p-\nu}_{L^p} \cdot \zeta^\nu \\
&\quad + \int_0^1 n(1-\lambda)^{n-1} \binom{p}{n} \big(\|Z^\star_\infty\|_{L^p} + \lambda\zeta\big)^\epsilon \cdot \zeta^n\, d\lambda \\
\text{by A.2.43:}\quad &= \big(\|Z^\star_\infty\|_{L^p} + \zeta\big)^p - \|Z^\star_\infty\|^p_{L^p} ,
\end{aligned}$$

which we rewrite as

$$\| Z_\infty \|^p_{L^p} + \| Z^\star_\infty \|^p_{L^p} \le \big(\| Z^\star_\infty \|_{L^p} + \zeta[Z]\big)^p . \tag{4.5.16}$$

If $\zeta[Z] \ne 1$, we obtain $\zeta[\rho Z] = 1$ and with it inequality (4.5.16) for a suitable multiple ρZ of Z; division by ρ^p produces that inequality for the given Z.

We leave it to the reader to convince herself with the aid of theorem 2.3.6 that (4.5.10) and (4.5.11) hold, with $C^\diamond_p \le 4^p$. Since this constant increases exponentially with p rather than linearly, we go a different, more labor-intensive route:

If Z is a positive increasing process I, then $I = I^\star$ and (4.5.16) gives

$$2^{1/p} \cdot \|I^\star_\infty\|_{L^p} \le \|I^\star_\infty\|_{L^p} + \zeta[I] ,$$

i.e.,
$$\|I^\star_\infty\|_{L^p} \le \left(2^{1/p} - 1\right)^{-1} \cdot \zeta[I] . \tag{4.5.17}$$

It is easy to see that $\left(2^{1/p} - 1\right)^{-1} \le p/\ln 2 \le 3p/2$. If I instead is predictable and has finite variation, then inequality (4.5.17) still obtains. Namely, there is a previsible process D of absolute value 1 such that $D{*}I$ is increasing. We can choose for D the Radon–Nikodym derivative of the Doléans–Dade measure of I with respect to that of $\vert I \vert$. Since $I^{\langle\rho\rangle} = \vert I \vert^{\langle\rho\rangle}$ for all ρ and therefore $\zeta[I] = \zeta[\vert I \vert]$, we arrive again at inequality (4.5.17):

$$\| I^\star_\infty \|_{L^p} \le \left\| \vert I \vert^\star_\infty \right\|_{L^p} \le (3p/2)\, \zeta[\vert I \vert] = (3p/2)\, \zeta[I] . \tag{4.5.18}$$

Next consider the case that Z is a q-integrable martingale M. Doob's maximal theorem 2.5.19, rewritten as

$$((1/p')^p + 1) \cdot \|M^\star_\infty\|^p_{L^p} \le \|M_\infty\|^p_{L^p} + \|M^\star_\infty\|^p_{L^p} ,$$

turns (4.5.16) into

$$((1/p')^p + 1)^{1/p} \cdot \|M^\star_\infty\|_{L^p} \le \|M^\star_\infty\|_{L^p} + \zeta[M] ,$$

which reads
$$\|M^\star_\infty\|_{L^p} \le \left(((1/p')^p + 1)^{1/p} - 1\right)^{-1} \cdot \zeta[M] . \tag{4.5.19}$$

We leave as an exercise the estimate $\left(((1/p')^p + 1)^{1/p} - 1\right)^{-1} \le 5p$ for $p > 2$.

Let us return to the general L^p-integrator Z. Let $Z = \widehat{Z} + \widetilde{Z}$ be its Doob–Meyer decomposition. Due to lemma 4.5.10 below, $\zeta[\widehat{Z}] \le \zeta[Z]$ and $\zeta[\widetilde{Z}] \le 2\zeta[Z]$. Consequently,

$$\|Z^\star_\infty\|_{L^p} \le \left\|\widehat{Z}^\star_\infty\right\|_{L^p} + \left\|\widetilde{Z}^\star_\infty\right\|_{L^p}$$

by (4.5.18) and (4.5.19):
$$\le (3p/2 + 2 \cdot 5p)\, \zeta[Z] \le 12p \cdot \zeta[Z] .$$

The number 12 can be replaced by 9.5 if slightly more fastidious estimates are used. In view of the definition (4.5.12) of ζ and the bound (4.5.13) for it we have arrived at

$$\| Z^\star_\infty \|_{L^p} \le 9.5p \max_{\rho=1,2,p^\diamond} \left\| \left(Z^{\langle\rho\rangle}_\infty\right)^{1/\rho} \right\|_{L^p} \le 9.5p \max_{\rho=1,2,q^\diamond} \left\| \left(Z^{\langle\rho\rangle}_\infty\right)^{1/\rho} \right\|_{L^p} .$$

Inequality (4.5.10) follows from an application of this and exercise 4.5.6 to $X{*}Z^{T\wedge T_n}$, for which the quantities ζ, etc., are finite if T_n reduces Z to a global L^q-integrator, and letting $T_n \uparrow \infty$. This establishes (4.5.10) and (4.5.11). ∎

Lemma 4.5.10 *Let $Z = \widehat{Z} + \widetilde{Z}$ be the Doob–Meyer decomposition of Z. Then*

$$\widehat{Z}^{\langle\rho\rangle} \le Z^{\langle\rho\rangle} \quad \text{and} \quad \widetilde{Z}^{\langle\rho\rangle} \le 2^\rho Z^{\langle\rho\rangle} , \qquad \rho \in \{1\} \cup [2, q] .$$

Proof. First the case $\rho = 1$. Since $\widehat{Z}^{[1]} = \widehat{Z}$ is previsible, $\widehat{Z}^{\langle 1\rangle} = |\widehat{Z}| = Z^{\langle 1\rangle}$.

Next let $2 \le \rho \le q$. Then $|\widehat{Z}^{[\rho]}|_t = \sum_{s\le t} |\Delta\widehat{Z}_s|^\rho$ is increasing and predictable on the grounds (cf. 4.3.3) that it jumps only at predictable stopping times S and there has the jump (see exercise 4.3.6)

$$\Delta |\widehat{Z}^{[\rho]}|_S = |\Delta\widehat{Z}_S|^\rho = |\mathbb{E}[\Delta Z_S | \mathcal{F}_{S-}]|^\rho ,$$

which is measurable on the strict past $\mathcal{F}_{S-}$. Now this jump is by Jensen's inequality (A.3.10) less than

$$\mathbb{E}[|\Delta Z_S|^\rho | \mathcal{F}_{S-}] = \mathbb{E}[\Delta |Z^{[\rho]}|_S | \mathcal{F}_{S-}] = \Delta Z_S^{\langle\rho\rangle} .$$

That is to say, the predictable increasing process $\widehat{Z}^{\langle\rho\rangle} = |\widehat{Z}^{[\rho]}|$, which has no continuous part, jumps only at predictable times S and there by less than the predictable increasing process $Z^{\langle\rho\rangle}$. Consequently (see exercise 4.3.7) $d\widehat{Z}^{\langle\rho\rangle} \le dZ^{\langle\rho\rangle}$ and the first inequality is established.

Now to the martingale part. Clearly $\widetilde{Z}^{\langle 1\rangle} = 0$. At $\rho = 2$ we observe that $[Z, Z]$ and $[\widetilde{Z}, \widetilde{Z}] + [\widehat{Z}, \widehat{Z}]$ differ by the local martingale $2[\widehat{Z}, \widetilde{Z}]$ – see exercise 3.8.24(iii) – and therefore

$$\widetilde{Z}^{\langle 2\rangle} = \widehat{[\widetilde{Z}, \widetilde{Z}]} \le \widehat{[Z, Z]} = \langle Z, Z\rangle = Z^{\langle 2\rangle} .$$

If $\rho > 2$, then $\quad |\Delta\widetilde{Z}_s|^\rho \le 2^{\rho-1}\big(|\Delta Z_s|^\rho + |\Delta\widehat{Z}_s|^\rho\big)$,

which reads $\quad d|\widetilde{Z}^{[\rho]}| \le 2^{\rho-1}\big(d|Z^{[\rho]}| + d|\widehat{Z}^{[\rho]}|\big)$

by part 1: $\quad \le 2^{\rho-1}\big(d|Z^{[\rho]}| + dZ^{\langle\rho\rangle}\big)$.

The predictable parts of the Doob–Meyer decomposition are thus related by

$$\widetilde{Z}^{\langle\rho\rangle} \le 2^{\rho-1}\big(Z^{\langle\rho\rangle} + Z^{\langle\rho\rangle}\big) = 2^\rho Z^{\langle\rho\rangle} .$$ ∎

Proof of Theorem 4.5.1 for a Single Integrator. While lemma 4.5.9 affords pathwise and solid control of the indefinite integral by previsible processes of finite variation, it is still bothersome to have to contend with two or three different previsible processes $Z^{\langle\rho\rangle}$. Fortunately it is possible to reduce their number to only one. Namely, for each of the $Z^{\langle\rho\rangle}$, $\rho = 1^\diamond, 2, q^\diamond$, let $\mu^{\langle\rho\rangle}$ denote its Doléans–Dade measure. To this collection add (artificially) the measure $\mu^{\langle 0\rangle} \stackrel{\text{def}}{=} \alpha \cdot dt \times \mathbb{P}$. Since the measures on $\mathcal{P}$ form a vector lattice (page 406), there is a least upper bound $\nu \stackrel{\text{def}}{=} \mu^{\langle 0\rangle} \vee \mu^{\langle 1\rangle} \vee \mu^{\langle 2\rangle} \vee \mu^{\langle q^\diamond\rangle}$. If Z is a martingale, then $\mu^{\langle 1\rangle} = 0$, so that $\nu = \mu^{\langle 0\rangle} \vee \mu^{\langle 1^\diamond\rangle} \vee \mu^{\langle 2\rangle} \vee \mu^{\langle q^\diamond\rangle}$, always. Let $\Lambda^{\langle q\rangle}[Z]$ denote the Doléans–Dade process of ν. It provides the pathwise and solid control of the indefinite integral $X{*}Z$ promised in theorem 4.5.1. Indeed, since by exercise 4.5.8

$$dZ^{\langle\rho\rangle} \le d\Lambda^{\langle q\rangle}[Z] \,, \qquad \rho \in \{1^\diamond, 2, p^\diamond, q^\diamond\} \,,$$

each of the $Z^{\langle\rho\rangle}$ in (4.5.10) and (4.5.11) can be replaced by $\Lambda^{\langle q\rangle}[Z]$ without disturbing the inequalities; with exercise A.8.5, inequality (4.5.1) is then immediate from (4.5.10).

Except for inequality (4.5.2), which we save for later, the proof of theorem 4.5.1 is complete in the case $d = 1$.

Exercise 4.5.11 Assume that Z is a continuous integrator. Then Z is a local L^q-integrator for any $q > 0$. Then $\Lambda = \Lambda^{\langle q\rangle} = \Lambda^{\langle 2\rangle}$ is a controller for $[Z, Z]$. Next let f be a function with two continuous derivatives, both bounded by L. Then Λ also controls $f(Z)$. In fact for all $T \in \mathfrak{T}$, $X \in \mathcal{P}$, and $p \in [2, q]$

$$\big\| \, |X{*}f(Z)|_T^\star \big\|_{L^p} \le (C_p^\diamond + 1)L \cdot \max_{\rho=1,2} \Big\| \Big(\int_0^T |X|_s^\rho \, d\Lambda_s \Big)^{1/\rho} \Big\|_{L^p} .$$

Previsible Control of Vectors of Integrators

A stochastic differential equation frequently is driven by not one or two but by a whole slew $\boldsymbol{Z} = (Z^1, Z^2, \ldots, Z^d)$ of integrators – see equation (1.1.9) on page 8 or equation (5.1.3) on page 271, and page 56. Its solution requires a single previsible control for all the Z^η simultaneously. This can of course simply be had by adding the $\Lambda^{\langle q\rangle}[Z^\eta]$; but that introduces their number d into the estimates, sacrificing sharpness of estimates and rendering them inapplicable to random measures. So we shall go a different if slightly more labor-intensive route.

We are after control of $\boldsymbol{Z}$ as expressed in inequality (4.5.1); the problem is to find and estimate a suitable previsible controller $\Lambda = \Lambda^{\langle q\rangle}[\boldsymbol{Z}]$ as in the scalar case. The idea is simple. Write $\boldsymbol{X} = |\boldsymbol{X}| \cdot \boldsymbol{X}'$, where $\boldsymbol{X}'$ is a vector field of previsible processes with $|\boldsymbol{X}'_s|(\omega) \stackrel{\mathrm{def}}{=} \sup_\eta |X'_{\eta\, s}(\omega)| \le 1$ for all $(s, \omega) \in \boldsymbol{B}$. Then $\boldsymbol{X}{*}\boldsymbol{Z} = |\boldsymbol{X}|{*}(\boldsymbol{X}'{*}\boldsymbol{Z})$, and so in view of inequality (4.5.11)

$$\| \boldsymbol{X}{*}\boldsymbol{Z}_T^\star \|_{L^p} \le C_p^\diamond \cdot \max_{\rho=1^\diamond,2,q^\diamond} \Big\| \Big(\int_0^T |\boldsymbol{X}|^\rho \, d(\boldsymbol{X}'{*}\boldsymbol{Z})^{\langle\rho\rangle} \Big)^{1/\rho} \Big\|_{L^p}$$

whenever $2 \le p \le q$. It turns out that there are increasing previsible processes $\boldsymbol{Z}^{\langle\rho\rangle}$, $\rho = 1^\diamond, 2, q^\diamond$, that satisfy

$$d(\boldsymbol{X}'{*}\boldsymbol{Z})^{\langle\rho\rangle} \le d\boldsymbol{Z}^{\langle\rho\rangle}$$

simultaneously for all predictable $\boldsymbol{X}' = (X'_1, \ldots, X'_d)$ with $|\boldsymbol{X}'| \le 1$. Then

$$\big\| \, |\boldsymbol{X}{*}\boldsymbol{Z}|_T^\star \big\|_{L^p} \le C_p^\diamond \cdot \max_{\rho=1^\diamond,2,q^\diamond} \Big\| \Big(\int_0^T |\boldsymbol{X}|^\rho \, d\boldsymbol{Z}^{\langle\rho\rangle} \Big)^{1/\rho} \Big\|_{L^p} . \tag{4.5.20}$$

The $Z^{\langle\rho\rangle}$ can take the role of the $Z^{\langle\rho\rangle}$ in lemma 4.5.9. They can in fact be chosen to be of the form $\boldsymbol{Z}^{\langle\rho\rangle} = {}^\rho\boldsymbol{X}{*}\boldsymbol{Z}^{\langle\rho\rangle}$ with ${}^\rho\boldsymbol{X}$ predictable and having $|{}^\rho\boldsymbol{X}| \le 1$; this latter fact will lead to the estimate (4.5.2).

1) To find $\boldsymbol{Z}^{\langle 1\rangle}$ we look at the Doob–Meyer decomposition $\boldsymbol{Z} = \widehat{\boldsymbol{Z}} + \widetilde{\boldsymbol{Z}}$, in obvious notation. Clearly

$$d(\boldsymbol{X}'{*}\boldsymbol{Z})^{\langle 1\rangle} = \sum\nolimits_\eta X'_\eta \, d\widehat{Z^\eta} \le \sum\nolimits_\eta |X'_\eta| \, d\,\rvert\widehat{Z^\eta}\rvert \le d\boldsymbol{Z}^{\langle 1\rangle} \tag{4.5.21}$$

for all $\boldsymbol{X}' \in \mathcal{P}^d$ having $|\boldsymbol{X}'| \le 1$, provided that we define

$$\boldsymbol{Z}^{\langle 1\rangle} \stackrel{\text{def}}{=} \sum_{1\le\eta\le d} \rvert\widehat{Z^\eta}\rvert \, .$$

To estimate the size of this controller let G^η be a previsible Radon–Nikodym derivative of the Doléans–Dade measure of $\widehat{Z}^\eta$ with respect to that of $\rvert\widehat{Z}^\eta\rvert$. These are previsible processes of absolute value 1, which we assemble into a d-tuple to make up the vector field ${}^1\boldsymbol{X}$. Then

$$\mathbb{E}\Big[\boldsymbol{Z}^{\langle 1\rangle}_\infty\Big] = \mathbb{E}\Big[\int {}^1\boldsymbol{X}\, d\boldsymbol{Z}\Big] \le \rvert{}^1\boldsymbol{X}{*}\widehat{\boldsymbol{Z}}\rvert_{\mathcal{I}^1}$$

by inequality (4.3.1):

$$\le \rvert{}^1\boldsymbol{X}{*}\boldsymbol{Z}\rvert_{\mathcal{I}^1} \le \rvert\boldsymbol{Z}\rvert_{\mathcal{I}^1} \, . \tag{4.5.22}$$

Exercise 4.5.12 Assume $\boldsymbol{Z}$ is a global L^q-integrator. Then the Doléans–Dade measure of $\boldsymbol{Z}^{\langle 1\rangle}$ is the maximum in the vector lattice $\mathfrak{M}^*[\mathcal{P}]$ (see page 406) of the Doléans–Dade measures of the processes $\{\boldsymbol{X}'{*}\boldsymbol{Z} : \boldsymbol{X}' \in \mathcal{E}^d, |\boldsymbol{X}'| \le 1\}$.

2) To find next the previsible controller $\boldsymbol{Z}^{\langle 2\rangle}$, consider the equality

$$d(\boldsymbol{X}'{*}\boldsymbol{Z})^{\langle 2\rangle} = \sum_{1\le\eta,\theta\le d} X'_\eta X'_\theta \, d\langle Z^\eta, Z^\theta\rangle \, .$$

Let $\mu^{\eta,\theta}$ be the Doléans–Dade measure of the previsible bracket $\langle Z^\eta, Z^\theta\rangle$. There exists a positive σ-additive measure μ on the previsibles with respect to which every one of the $\mu^{\eta,\theta}$ is absolutely continuous, for instance, the sum of their variations. Let $G^{\eta,\theta}$ be a previsible Radon–Nikodym derivative of $\mu^{\eta,\theta}$ with respect to μ, and V the Doléans–Dade process of μ. Then

$$\langle Z^\eta, Z^\theta\rangle = G^{\eta,\theta}{*}V \, .$$

On the product of the vector space $\mathcal{G}$ of $d \times d$-matrices g with the unit ball (box) of $\ell^\infty(d)$ define the function Φ by $\Phi(g,y) \stackrel{\text{def}}{=} \sum_{\eta,\theta} y_\eta y_\theta \, g^{\eta,\theta}$ and the function σ by $\sigma(g) \stackrel{\text{def}}{=} \sup\{\Phi(g,y) : y \in \ell^\infty_1\}$. This is a continuous function of $g \in \mathcal{G}$, so the process $\sigma(G)$ is previsible. The previous equality gives

$$d\boldsymbol{X}'{*}\boldsymbol{Z}^{\langle 2\rangle} = \sum_{\eta,\theta} X'_\eta X'_\theta \, G^{\eta,\theta} \, dV \le \sigma(G)\, dV = d\boldsymbol{Z}^{\langle 2\rangle} \, ,$$

for all $\boldsymbol{X}' \in \mathcal{P}^d$ with $|\boldsymbol{X}'| \le 1$, provided we define

$$\boldsymbol{Z}^{\langle 2\rangle} \stackrel{\text{def}}{=} \sigma(G){*}V \, .$$

To estimate the size of $Z^{\langle 2\rangle}$, we use the Borel function $\gamma : \mathcal{G} \to \ell_1^\infty$ with $\sigma(g) = \Phi\big(g, \gamma(g)\big)$ that is provided by lemma A.2.21 (b). Since ${}^2\boldsymbol{X} \stackrel{\text{def}}{=} \gamma \circ G$ is a previsible vector field with $|{}^2\boldsymbol{X}| \le 1$,

$$\begin{aligned}
\mathbb{E}\Big[Z_\infty^{\langle 2\rangle}\Big] &= \mathbb{E}\Big[\int_0^\infty \sigma(G)\, dV\Big] = \int \sigma(G)\, d\mu \\
&= \int \sum_{\eta,\theta} {}^2X_\eta {}^2X_\theta\, G^{\eta,\theta}\, d\mu = \mathbb{E}\Big[\int_0^\infty \sum_{\eta,\theta} {}^2X_\eta {}^2X_\theta\, d\langle Z^\eta, Z^\theta\rangle\Big] \\
&= \mathbb{E}[\langle {}^2\boldsymbol{X}{*}\boldsymbol{Z}, {}^2\boldsymbol{X}{*}\boldsymbol{Z}\rangle_\infty] = \mathbb{E}\big[[{}^2\boldsymbol{X}{*}\boldsymbol{Z}, {}^2\boldsymbol{X}{*}\boldsymbol{Z}]_\infty\big] && (4.5.23) \\
&= \mathbb{E}\Big[\big(S_\infty[{}^2\boldsymbol{X}{*}\boldsymbol{Z}]\big)^2\Big] \le \lceil {}^2\boldsymbol{X}{*}\boldsymbol{Z} \rceil_{\mathcal{I}^2}^2 \le \lceil \boldsymbol{Z} \rceil_{\mathcal{I}^2}^2 . && (4.5.24)
\end{aligned}$$

Exercise 4.5.13 Assume $\boldsymbol{Z}$ is a global L^q-integrator, $q \ge 2$. Then the Doléans–Dade measure of $\boldsymbol{Z}^{\langle 2\rangle}$ is the maximum in the vector lattice $\mathfrak{M}^*[\mathcal{P}]$ of the Doléans–Dade measures of the brackets $\big\{[\boldsymbol{X}'{*}\boldsymbol{Z}, \boldsymbol{X}'{*}\boldsymbol{Z}] : \boldsymbol{X}' \in \mathcal{E}^d, |\boldsymbol{X}'| \le 1\big\}$.

q) To find a useful previsible controller $\boldsymbol{Z}^{\langle q\rangle}$ now that $\boldsymbol{Z}^{\langle 1\rangle}$ and $\boldsymbol{Z}^{\langle 2\rangle}$ have been identified, we employ the Doob–Meyer decomposition of the jump measure $\jmath_{\boldsymbol{Z}}$ from page 232. According to it,

$$\begin{aligned}
\mathbb{E}\Big[\int_0^\infty |X|^q\, d\lceil \boldsymbol{X}'{*}\boldsymbol{Z}^{[q]} \rceil\Big] &= \mathbb{E}\Big[\int_{\mathbb{R}_*^d \times [0,\infty)} |X_s|^q |\langle \boldsymbol{X}_s' | \boldsymbol{y}\rangle|^q\, \jmath_{\boldsymbol{Z}}(d\boldsymbol{y}, ds)\Big] \\
&= \mathbb{E}\Big[\int_{[0,\infty)} |X_s|^q \int_{\mathbb{R}_*^d} |\langle \boldsymbol{X}_s' | \boldsymbol{y}\rangle|^q\, \nu_s(d\boldsymbol{y})\, dV_s\Big] .
\end{aligned}$$

Now the process $\sigma^{\langle q\rangle}$ defined by

$$\sigma_s^{\langle q\rangle} = \sup_{|\boldsymbol{x}'| \le 1} \int |\langle \boldsymbol{x}' | \boldsymbol{y}\rangle|^q\, \nu_s(d\boldsymbol{y}) = \sup_{|\boldsymbol{x}'| \le 1} \|\boldsymbol{x}'\|_{L^q(\nu_s)}^q$$

is previsible, inasmuch as it suffices to extend the supremum over $\boldsymbol{x}' \in \ell_1^\infty(d)$ with rational components. Therefore

$$\begin{aligned}
\mathbb{E}\Big[\int_0^\infty |X|^q\, d(\boldsymbol{X}'{*}\boldsymbol{Z})^{\langle q\rangle}\Big] &= \mathbb{E}\Big[\int_0^\infty |X|_s^q \int_{\mathbb{R}_*^d} |\langle \boldsymbol{X}_s' | \boldsymbol{y}\rangle|^q\, \nu_s(d\boldsymbol{y})\, dV_s\Big] \\
&\le \mathbb{E}\Big[\int_0^\infty |X|^q\, \sigma_s^{\langle q\rangle}\, dV_s\Big] .
\end{aligned}$$

From this inequality we read off the fact that $d(\boldsymbol{X}'{*}\boldsymbol{Z})^{\langle q\rangle} \le d\boldsymbol{Z}^{\langle q\rangle}$ for all $\boldsymbol{X}' \in \mathcal{P}^d$ with $|\boldsymbol{X}'| \le 1$, provided we define

$$\boldsymbol{Z}^{\langle q\rangle} \stackrel{\text{def}}{=} \sigma^{\langle q\rangle}{*}V .$$

To estimate $\boldsymbol{Z}^{\langle q\rangle}$ observe that the supremum in the definition of $\sigma^{\langle q\rangle}$ is assumed in one of the 2^d extreme points (corners) of $\ell_1^\infty(d)$, on the grounds that the function

$$\boldsymbol{x} \mapsto \phi_\varpi(\boldsymbol{x}) \stackrel{\text{def}}{=} \int |\langle \boldsymbol{x} | \boldsymbol{y}\rangle|^q\, \nu_\varpi(d\boldsymbol{y})$$

is convex.[13] Enumerate the corners: $\boldsymbol{c}_1, \boldsymbol{c}_2, \ldots$ and consider the previsible sets $P_k \stackrel{\mathrm{def}}{=} \{\varpi : \phi_\varpi(\boldsymbol{c}_k) = \sigma(\varpi)\}$ and $P'_k \stackrel{\mathrm{def}}{=} P_k \setminus \bigcup_{i<k} P_i$, $k = 1, \ldots, 2^d$. The vector field ${}^q\boldsymbol{X}$ which on P'_k has the value $\boldsymbol{c}_k$ clearly satisfies

$$\boldsymbol{Z}^{\langle q \rangle} = {}^q\boldsymbol{X}*\boldsymbol{Z}^{\langle q \rangle} .$$

Thanks to inequality (4.5.5) and theorem 3.8.4,

$$\mathbb{E}\Big[\boldsymbol{Z}^{\langle q \rangle}_\infty\Big] = \mathbb{E}\Big[({}^q\boldsymbol{X}*\boldsymbol{Z})^{\langle q \rangle}_\infty\Big] = \mathbb{E}\Big[\big|{}^q\boldsymbol{X}*\boldsymbol{Z}\big|^{[q]}_\infty\Big]$$

by 3.8.21:
$$= \mathbb{E}\Big[\sum_{s<\infty} |\langle {}^q\boldsymbol{X}_s | \Delta \boldsymbol{Z}_s \rangle|^q\Big] = \mathbb{E}\Big[\int_{\check{\boldsymbol{B}}} |\langle {}^q\boldsymbol{X}_s | \boldsymbol{y}\rangle|^q \, \jmath_{\boldsymbol{Z}}(d\boldsymbol{y}, ds)\Big] \qquad (4.5.25)$$

$$\le \mathbb{E}\big[S^q_\infty[{}^q\boldsymbol{X}*\boldsymbol{Z}]\big] \le \big\lceil {}^q\boldsymbol{X}*\boldsymbol{Z} \big\rceil^q_{\mathcal{I}^q} \le \big\lceil \boldsymbol{Z} \big\rceil^q_{\mathcal{I}^q} . \qquad (4.5.26)$$

Proof of Theorem 4.5.1. We now define the desired previsible controller $\Lambda^{\langle q \rangle}[\boldsymbol{Z}]$ as before to be the Doléans–Dade process of the supremum of $\mu^{\langle 0 \rangle}$ and the Doléans–Dade measures of $\boldsymbol{Z}^{\langle 1 \rangle}$, $\boldsymbol{Z}^{\langle 2 \rangle}$, and $\boldsymbol{Z}^{\langle q \rangle}$ and continue as in the proof of theorem 4.5.1 on page 245, replacing $Z^{\langle \rho \rangle}$ by $\boldsymbol{Z}^{\langle \rho \rangle}$ for $\rho = 1, 2, q$.

To establish the estimate (4.5.2) of $\Lambda^{\langle q \rangle}[\boldsymbol{Z}]$ observe that

$$\Lambda^{\langle q \rangle}_T[\boldsymbol{Z}] \le \alpha \cdot T + \boldsymbol{Z}^{\langle 1 \rangle}_T + \boldsymbol{Z}^{\langle 2 \rangle}_T + \boldsymbol{Z}^{\langle q \rangle}_T ,$$

so that
$$\mathbb{E}\Big[\Lambda^{\langle q \rangle}_T[\boldsymbol{Z}]\Big] \le \alpha \cdot \mathbb{E}[T] + \mathbb{E}\Big[\boldsymbol{Z}^{\langle 1 \rangle}_T\Big] + \mathbb{E}\Big[\boldsymbol{Z}^{\langle 2 \rangle}_T\Big] + \mathbb{E}\Big[\boldsymbol{Z}^{\langle q \rangle}_T\Big]$$

by (4.5.22), (4.5.24), and (4.5.26):
$$\le \alpha \cdot \mathbb{E}[T] + \big\lceil \boldsymbol{Z}^T \big\rceil_{\mathcal{I}^1} + \big\lceil \boldsymbol{Z} \big\rceil^2_{\mathcal{I}^2} + \big\lceil \boldsymbol{Z}^T \big\rceil^q_{\mathcal{I}^q}$$

$$\le \alpha \cdot \mathbb{E}[T] + \big\lceil \boldsymbol{Z}^T \big\rceil_{\mathcal{I}^q} + \big\lceil \boldsymbol{Z}^T \big\rceil^2_{\mathcal{I}^q} + \big\lceil \boldsymbol{Z}^T \big\rceil^q_{\mathcal{I}^q}$$

$$\le \alpha \cdot \mathbb{E}[T] + 3\Big(\big\lceil \boldsymbol{Z} \big\rceil_{\mathcal{I}^q} \vee \big\lceil \boldsymbol{Z} \big\rceil^q_{\mathcal{I}^q}\Big) . \qquad ■$$

Exercise 4.5.14 Assume $\boldsymbol{Z}$ is a global L^q-integrator, $q > 2$. Then the Doléans–Dade measure of $\boldsymbol{Z}^{\langle q \rangle}$ is the maximum in the vector lattice $\mathfrak{M}^*[\mathcal{P}]$ of the Doléans–Dade measures of the processes $\big\{|\check{H}|^q * \jmath_{\boldsymbol{Z}} : \check{H}(\boldsymbol{y}, s) \stackrel{\mathrm{def}}{=} \langle \boldsymbol{X}'_s | \boldsymbol{y} \rangle ,\ \boldsymbol{X}' \in \mathcal{E}^d, |\boldsymbol{X}'| \le 1\big\}$.

Exercise 4.5.15 Repeat exercise 4.5.11 for a slew $\boldsymbol{Z}$ of continuous integrators: let Λ be its previsible controller as constructed above. Then Λ is a controller for the slew $[Z^\eta, Z^\theta]$, $\eta, \theta = 1, \ldots, d$. Next let f be a function with continuous first and second partial derivatives:[14]

$$|f_{;\eta}(x) u^\eta| \le L \cdot |u|_p \ \text{ and } \ |f_{;\eta\theta} u^\eta u^\theta| \le L \cdot |u|^2_p , \qquad u \in \mathbb{R}^d .$$

Then Λ also controls $f(\boldsymbol{Z})$. In fact, for all $T \in \mathfrak{T}$ and $X \in \mathcal{P}$

$$\big\| |X*f(\boldsymbol{Z})|^\star_T \big\|_{L^p} \le (C^\diamond_p + 1) L \cdot \max_{\rho=1,2} \Big\| \Big(\int_0^T |X|^\rho_s \, d\Lambda_s \Big)^{1/\rho} \Big\|_{L^p} .$$

---

[13] Recall that $\varpi = (s, \omega)$. Thus ν_ϖ is ν_s with ω in evidence.

[14] Subscripts after semicolons denote partial derivatives, e.g., $\Phi_{;\eta} \stackrel{\mathrm{def}}{=} \frac{\partial \Phi}{\partial x^\eta}$, $\Phi_{;\eta\theta} \stackrel{\mathrm{def}}{=} \frac{\partial^2 \Phi}{\partial x^\eta \partial x^\theta}$.

Exercise 4.5.16 If $\boldsymbol{Z}$ is quasi-left-continuous (${}^{p}\boldsymbol{Z}=0$), then $\Lambda^{\langle q\rangle}[\boldsymbol{Z}]$ is continuous. Conversely, if $\Lambda^{\langle q\rangle}[\boldsymbol{Z}]$ is continuous, then the jump of $\boldsymbol{X}*\boldsymbol{Z}$ at any predictable stopping time is negligible, for every $\boldsymbol{X}\in\mathcal{P}^d$.

Exercise 4.5.17 Inequality (4.5.1) extends to $0<p\le 2$ with

$$C_p^{\diamond}\le 2^{0\vee(1-p)/p}\big(1+C_p^{(4.3.6)}\big)\,.$$

Exercise 4.5.18 In the case $d=1$ and when Z is an L^p-integrator (not merely a local one) then the functional $\|\ \|_{p\text{-}Z}^{\diamond}$ on processes $F:\boldsymbol{B}\to\overline{\mathbb{R}}$ defined by

$$\|F\|_{p\text{-}Z}^{\diamond}\stackrel{\text{def}}{=}C_p^{\diamond}\cdot\max_{\rho=1^{\diamond},p^{\diamond}}\Big\|\Big(\int^{*}|F|^{\rho}\,d\Lambda\Big)^{1/\rho}\Big\|_{L^p}^{*}$$

is a mean on $\mathcal{E}$ that majorizes the elementary dZ-integral in the sense of L^p.

If Z is a continuous local martingale, then $1^{\diamond}=p^{\diamond}=2$, the constant $C_p^{\diamond}$ can be estimated by $7p/6$, and $\|\ \|_{p\text{-}Z}^{\diamond}$ is an extension to $p>2$ of the Hardy mean of definition (4.2.9).

Exercise 4.5.19 For a d-dimensional Wiener process $\boldsymbol{W}$ and $q\ge 2$,

$$\Lambda_t^{\langle q\rangle}[\boldsymbol{W}]=\big\bracevert \boldsymbol{W}^t\big\bracevert_{\ell^2}^2=d\cdot t\,.$$

Exercise 4.5.20 Suppose $\boldsymbol{Z}$ is continuous and use the previsible controller Λ of remark 4.5.2 on page 238 to define THE time transformation (4.5.4). Let $0\le g_\eta\in\mathcal{F}_{T^\kappa}$, and $1\le\eta,\theta,\iota\le d$. (i) For $\ell=0,1,\ldots$ there are constants $C_\ell\le \ell!(C_p^{\diamond})^\ell$ such that,

$$\text{for } \kappa<\mu<\kappa{+}1,\quad \Big\|g_\eta\cdot\big|Z^\eta-Z^{\eta T^\kappa}\big|_{T^\mu}^{\star\ell}\Big\|_{L^p}\le C_\ell\,(\mu{-}\kappa)^{\ell/2}\cdot\|g\|_{L^p}\tag{4.5.27}$$

$$\text{and}\quad \Big\|\int_{T^\kappa}^{T^\mu}|g_\iota|\cdot\big|Z^\iota{-}Z^\iota_{T^\kappa}\big|_s^{\star\ell}\,d\big\bracevert[Z^\eta,Z^\theta]\big\bracevert_s\Big\|_{L^p}\le\frac{C_\ell}{\ell/2+1}\,(\mu{-}\kappa)^{\ell/2+1}\cdot\|g\|_{L^p}\,.\tag{4.5.28}$$

(ii) For $\ell=0,1,\ldots$ there are polynomials P_ℓ such that for any $\mu>\kappa$

$$\Big\|g_\eta\cdot\big|Z^\eta-Z^{\eta T^\kappa}\big|_{T^\mu}^{\star\ell}\Big\|_{L^p}\le P_\ell(\sqrt{\mu{-}\kappa})\,.\tag{4.5.29}$$

Exercise 4.5.21 (Emery) Let Y,A be positive, adapted, increasing, and right-continuous processes, with A also previsible. If $\mathbb{E}[Y_T]\le\mathbb{E}[A_T]$ for all finite stopping times T, then for all $y,a\ge 0$

$$\mathbb{P}[Y_\infty\ge y,A_\infty\le a]\le\mathbb{E}[A_\infty\wedge a]/y\,;\tag{4.5.30}$$

$$\text{in particular,}\qquad [Y_\infty=\infty]\subseteq[A_\infty=\infty]\ \ \mathbb{P}\text{-almost surely.}\tag{4.5.31}$$

Exercise 4.5.22 (Yor) Let Y,A be positive random variables satisfying the inequality $\mathbb{P}[Y\ge y,A\le a]\le\mathbb{E}[A\wedge a]/y$ for $y,a>0$. Next let $\phi,\psi:\mathbb{R}_+\to\mathbb{R}_+$ be càglàd increasing functions, and set

$$\Phi(x)\stackrel{\text{def}}{=}\phi(x)+x\int_{(x}^{\infty}\frac{d\phi(x)}{x}\,.$$

$$\text{Then}\quad \mathbb{E}[\phi(Y)\cdot\psi(\tfrac{1}{A})]\le\mathbb{E}\Big[\big(\Phi(A){+}\phi(A)\big)\cdot\psi(\tfrac{1}{A})+\int_{\frac{1}{A}}^{\infty}\Phi(\tfrac{1}{y})\,d\psi(y)\Big]\,.\tag{4.5.32}$$

In particular, for $0\le\alpha<\beta<1$,

$$\mathbb{E}[Y^\beta/A^\alpha]\le\Big(\frac{2}{(1-\beta)(\beta-\alpha)}\Big)\cdot\mathbb{E}[A^{\beta-\alpha}]\tag{4.5.33}$$

$$\text{and}\qquad \mathbb{E}[Y^\beta]\le\frac{2-\beta}{1-\beta}\cdot\mathbb{E}[A^\beta]\,.\tag{4.5.34}$$

Exercise 4.5.23 From THE previsible controller Λ of the L^q-integrator $\boldsymbol{Z}$ define

$$A \stackrel{\text{def}}{=} C_q^\diamond \cdot (\Lambda^{1/1^\diamond} \vee \Lambda^{1/q^\diamond})$$

and show that $\quad \mathbb{E}\Big[|\boldsymbol{Z}|_T^{\star p\beta} / A_T^{p\alpha}\Big] \le \dfrac{2}{(1-\beta)(\beta-\alpha)} \cdot \mathbb{E}\Big[A_T^{p(\beta-\alpha)}\Big]$

for all finite stopping times T, all $p \le q$, and all $0 \le \alpha < \beta < 1$. Use this to estimate from below Λ at the first time $\boldsymbol{Z}$ leaves the ball of a given radius r about its starting point $\boldsymbol{Z}_0$. Deduce that the first time a Wiener process leaves an open ball about the origin has moments of all (positive and negative) orders.

Previsible Control of Random Measures

Our definition 3.10.1 of a random measure was a straightforward generalization of a d-tuple of integrators; we would simply replace the auxiliary space $\{1,\dots,d\}$ of indices by a locally compact space $\boldsymbol{H}$, and regard a random measure as an $\boldsymbol{H}$-tuple of (infinitesimal) integrators. This view has already paid off in a simple proof of the Doob–Meyer decomposition 4.3.24 for random measures. It does so again in a straightforward generalization of the control theorem 4.5.1 to random measures. On the way we need a small technical result, from which the desired control follows with but a little soft analysis:

Exercise 4.5.24 Let $\boldsymbol{Z}$ be a global L^q-integrator of length d, view it as a column vector, let $C : \ell^1(d) \to \ell^1('d)$ be a contractive linear map, and set $'\boldsymbol{Z} \stackrel{\text{def}}{=} C\boldsymbol{Z}$. Then $\jmath_{'\boldsymbol{Z}} = C[\jmath_{\boldsymbol{Z}}]$. Next, let μ be the Doléans–Dade measure for any one of the previsible controllers $\boldsymbol{Z}^{\langle 1\rangle}, \boldsymbol{Z}^{\langle 2\rangle}, \boldsymbol{Z}^{\langle q\rangle}, \Lambda^{\langle q\rangle}[\boldsymbol{Z}]$ and $'\mu$ the Doléans–Dade measure for the corresponding controller of $'\boldsymbol{Z}$. Then $'\mu \le \mu$.

Theorem 4.5.25 (Revesz [93]) *Let $q \ge 2$ and suppose ζ is a spatially bounded L^q-random measure with auxiliary space $\boldsymbol{H}$. There exist a previsible increasing process $\Lambda = \Lambda^{\langle q\rangle}[\zeta]$ and a universal constant $C_p^\diamond$ that control ζ in the following sense: for every $\check{X} \in \check{\mathcal{P}}$, every stopping time T, and every $p \in [2,q]$*

$$\Big\| (\check{X}*\zeta)_T^\star \Big\|_{L^p} \le C_p^\diamond \cdot \max_{\rho=1^\diamond,p^\diamond} \Big\| \Big(\int^* [\![0,T]\!] \cdot |\check{X}_s|_\infty^\rho \, d\Lambda_s \Big)^{1/\rho} \Big\|_{L^p}^* . \tag{4.5.35}$$

Here $|\check{X}_s|_\infty \stackrel{\text{def}}{=} \sup_{\eta\in\boldsymbol{H}} |\check{X}(\eta,s)|$ is $\mathcal{P}$-analytic and hence universally $\mathcal{P}$-measurable. The meaning of $1^\diamond, p^\diamond$ is mutatis mutandis *as in theorem 4.5.1 on page 238. Part of the claim is that the left-hand side makes sense, i.e., that $[\![0,T]\!]\cdot\check{X}$ is ζ–p-integrable, whenever the right-hand side is finite.*

Proof. Denote by $\mathcal{K}$ the paving of $\boldsymbol{H}$ by compacta. Let (K_ν) be a sequence in $\mathcal{K}$ whose interiors cover $\boldsymbol{H}$, cover each K_ν by a finite collection B_ν of balls of radius less than $1/\nu$, and let P_n denote the collection of atoms of the algebra of sets generated by $B_1\cup\ldots\cup B_n$. This yields a sequence (P^n) of partitions of $\boldsymbol{H}$ into mutually disjoint Borel subsets such that P^n refines P^{n-1} and such that $\mathcal{B}^\bullet(\boldsymbol{H})$ is generated by $\bigcup_n P^n$. Suppose P^n has d_n members $B_1^n,\dots,B_{d_n}^n$. Then $Z_i^n \stackrel{\text{def}}{=} B_i^n*\zeta$ defines a vector $\boldsymbol{Z}^n$ of L^q-integrators of length d_n, of integrator size less than $\lceil\zeta\rceil_{\mathcal{I}^q}$, and controllable by a previsible increasing process $\Lambda^n \stackrel{\text{def}}{=} \Lambda^{\langle q\rangle}[\boldsymbol{Z}^n]$. Collapsing P^n to P^{n-1} gives rise to a contractive linear map $C^n_{n-1} : \ell^1(d_n) \to \ell^1(d_{n-1})$ in an obvious way. By exercise 4.5.24, the Doléans–Dade measures of the Λ^n increase with n. They have a least upper bound μ in the order-complete vector lattice $\mathfrak{M}^*[\mathcal{P}]$, and the Doléans–Dade process Λ of μ will satisfy the description.

To see that it does, let $\check{\mathcal{E}}'$ denote the collection of functions of the form $\sum h_i \otimes X_i$, with h_i step functions over the algebra generated by $\bigcup P^n$ and $X_i \in \mathcal{P}_{00}$. It is evident that (4.5.35) is satisfied when $\check{X} \in \check{\mathcal{E}}'$, with $C_p^\diamond = C_p^{\diamond(4.5.1)}$. Since both sides depend continuously on their arguments in the topology of confined uniform convergence, the inequality will stay even for $\check{X}$ in the confined uniform closure of $\check{\mathcal{E}}' = \check{\mathcal{E}}'_{00}$, which contains $C_{00}[\boldsymbol{H}]\otimes\mathcal{P}$ and is both an algebra and a vector lattice (theorem A.2.2). Since the right-hand side is not a mean in its argument $\check{X}$, in view of the appearance of the sup-norm $|\ |_\infty$, it is not possible to extend to $\check{X} \in \check{\mathcal{P}}$ by the usual sequential closure argument and we have to go a more circuitous route.

To begin with, let us show that $|\check{X}|_\infty$ is measurable on the universal completion $\mathcal{P}^*$ whenever $\check{X} \in \check{\mathcal{P}}$. Indeed, for any $a \in \mathbb{R}$, $[|\check{X}|_\infty > a]$ is the projection on $\boldsymbol{B}$ of the $\mathcal{K}\times\mathcal{P}$-analytic (see A.5.3) set $[|\check{X}| > a]$ of $\mathcal{B}^\bullet[\boldsymbol{H}]\otimes\mathcal{P}$, and is therefore $\mathcal{P}$-analytic (proposition A.5.4) and $\mathcal{P}^*$-measurable (apply theorem A.5.9 to every outer measure on $\mathcal{P}$). Hence $|\check{X}|_\infty \in \mathcal{P}^*$. In fact, this argument shows that $|\check{X}|_\infty$ is measurable for any mean on $\mathcal{P}$ that is continuous along arbitrary increasing sequences, since such a mean is a $\mathcal{P}$-capacity. In particular (see proposition 3.6.5 or equation (A.3.2)), $|\check{X}|_\infty$ is measurable for the mean $\|\ \|^\diamond$ that is defined on $F : \boldsymbol{B} \to \overline{\mathbb{R}}$ by

$$\|F\|^\diamond \stackrel{\text{def}}{=} C_p^\diamond \cdot \max_{\rho=1^\diamond,p^\diamond} \left\| \left(\int^* |F|^\rho \, d\Lambda\right)^{1/\rho} \right\|^*_{L^p} .$$

Next let $g \in L^{p'}$ be an element of the unit ball of the Banach-space dual $L^{p'}$ of L^p, and define θ on $\check{\mathcal{P}}_{00}$ by $\theta(\check{X}) \stackrel{\text{def}}{=} \mathbb{E}[g\cdot\zeta(\check{X})]$. This is a σ-additive measure of finite variation: $\lceil\theta\rceil(|\check{X}|) \le \|\check{X}\|_{\zeta-p}$. There is a $\check{\mathcal{P}}$-measurable Radon–Nikodym derivative $\check{G} = d\theta/d\lceil\theta\rceil$ with $|\check{G}| = 1$. Also, $\lceil\theta\rceil$ has a disintegration (see corollary A.3.42), so that

$$\theta(\check{X}) = \int_{\boldsymbol{B}}\int_{\boldsymbol{H}} \check{X}(\eta,\varpi)\check{G}(\eta,\varpi)\nu_\varpi(d\eta)\ \mu(d\varpi) \le \big\||\check{X}|_\infty\big\|^\diamond . \tag{4.5.36}$$

The equality in (4.5.36) holds for all $\check{X} \in \check{\mathcal{P}} \cap \mathcal{L}^1(\theta)$ (ibidem), while the inequality so far is known only for $\check{X} \in \overline{\check{\mathcal{E}}'}_{00}$. Now there exists a sequence $(\check{G}_n)$ in $\check{\mathcal{E}}'$ that converges in $\|\ \|^*_\theta$-mean to $\check{G} = 1/\check{G}$ and has $|\check{G}_n| \le 1$. Replacing $\check{X}$ by $\check{X}\cdot\check{G}_n$ in (4.5.36) and taking the limit produces

$$\lceil\theta\rceil(\check{X}) = \int_{\boldsymbol{B}}\int_{\boldsymbol{H}} \check{X}(\eta,\varpi)\nu_\varpi(d\eta)\ \mu(d\varpi) \le \big\||\check{X}|_\infty\big\|^\diamond , \qquad \check{X} \in \overline{\check{\mathcal{E}}'}_{00} .$$

In particular, when $\check{X}$ does not depend on $\eta \in \boldsymbol{H}$, $\check{X} = 1_{\boldsymbol{H}} \otimes X$ with $X \in \mathcal{P}_{00}$, say, then this inequality, in view of $\nu_\varpi(\boldsymbol{H}) = 1$, results in

$$\int_{\boldsymbol{B}} X(\varpi)\, \mu(d\varpi) \le \big\| |X|_\infty \big\|^\diamond = \|X\|^\diamond ,$$

and by exercise 3.6.16 in: $$\int^* |X(\varpi)|\, \mu(d\varpi) \le \|X\|^\diamond \qquad \forall X \in \mathcal{P}^* .$$

Thus for $X' \in \mathcal{P}$ with $|X'| \le 1$ and $\check{X} \in \check{\mathcal{P}} \cap \mathcal{L}^1[\zeta]$ we have

$$\mathbb{E}\Big[g \cdot \int X'\, d(\check{X}*\zeta)\Big] = \theta(X' \cdot \check{X}) \le \left\lceil\theta\right\rceil(|X' \cdot \check{X}|)$$

$$= \int_{\boldsymbol{B}} \int_{\boldsymbol{H}} |X'(\eta,\varpi)|\, |\check{X}(\eta,\varpi)| \nu_\varpi(d\eta)\, \mu(d\varpi)$$

as $|\check{X}'| \le 1$: $$\le \int_{\boldsymbol{B}} \int_{\boldsymbol{H}} |\check{X}(\eta,\varpi)| \nu_\varpi(d\eta)\, \mu(d\varpi)$$

as $\nu_\varpi(\boldsymbol{H}) = 1$: $$\le \int_{\boldsymbol{B}}^* |\check{X}|_\infty(\varpi)\, \mu(d\varpi) \le \big\| |\check{X}|_\infty \big\|^\diamond .$$

Taking the supremum over $g \in L_1^{p'}$ and $X' \in \mathcal{E}_1'$ gives

$$\left\lceil \check{X}*\zeta \right\rceil_{\mathcal{I}^p} \le \big\| |\check{X}|_\infty \big\|^\diamond$$

and, finally, $$\Big\| (\check{X}*\zeta)^\star_\infty \Big\|_{L^p} \le C_p^{\star(2.3.5)} \cdot \big\| |\check{X}|_\infty \big\|^\diamond .$$

This inequality was established under the assumption that $\check{X} \in \check{\mathcal{P}}$ is ζ–p-integrable. It is left to be shown that this is the case whenever the right-hand side is finite. Now by corollary 3.6.10, $\|\check{X}\|^*_{\zeta-p}$ is the supremum of $\|\int \check{Y}\, d\zeta\|_{L^p}$, taken over $\check{Y} \in \mathcal{L}^1[\zeta\text{–}p]$ with $|\check{Y}| \le |cX|$. Such $\check{Y}$ have $\|\check{Y}\|^\diamond \le \|\check{X}\|^\diamond$, whence $\|\check{X}\|^*_{\zeta-p} \le \|\check{X}\|^\diamond < \infty$.

Now replace $\check{X}$ by $[\![0,T]\!] \cdot \check{X}$ to obtain inequality (4.5.35). ∎

Project 4.5.26 *Make a theory of time transformations.*

4.6 Lévy Processes

Let $(\Omega, \mathcal{F}_., \mathbb{P})$ be a measured filtration and $\boldsymbol{Z}_.$ an adapted $\mathbb{R}^d$-valued process that is right-continuous in probability. $\boldsymbol{Z}_.$ is a ***Lévy process*** on $\mathcal{F}_.$ if it has independent identically distributed and stationary increments and $\boldsymbol{Z}_0 = 0$. To say that the increments of $\boldsymbol{Z}$ are independent means that for any $0 \le s < t$ the increment $\boldsymbol{Z}_t - \boldsymbol{Z}_s$ is independent of $\mathcal{F}_s$; to say that the increments of $\boldsymbol{Z}$ are stationary and identically distributed means that the increments $\boldsymbol{Z}_t - \boldsymbol{Z}_s$ and $\boldsymbol{Z}_{t'} - \boldsymbol{Z}_{s'}$ have the same law whenever the elapsed times $t-s$ and $t'-s'$ are the same. If the filtration is not specified, a Lévy process $\boldsymbol{Z}$ is understood to be a Lévy process on its own basic filtration $\mathcal{F}^0_.[Z]$. Here are a few simple observations:

Exercise 4.6.1 A Lévy process Z on $\mathcal{F}_.$ is a Lévy process both on its basic filtration $\mathcal{F}^0_.[Z]$ and on the natural enlargement of $\mathcal{F}_.$. At any instant s, $\mathcal{F}_s$ and $\mathcal{F}^0_\infty[Z - Z^s]$ are independent. At any $\mathcal{F}_.$-stopping time T, $Z'_. \stackrel{\text{def}}{=} Z_{T+.} - Z_T$ is independent of $\mathcal{F}_T$; in fact $Z'_.$ is a Lévy process.

Exercise 4.6.2 If $Z_.$ and $Z'_.$ are $\mathbb{R}^d$-valued Lévy processes on the measured filtration $(\Omega, \mathcal{F}_., \mathbb{P})$, then so is any linear combination $\alpha Z + \beta Z'$ with constant coefficients. If $Z^{(n)}$ are Lévy processes on $(\Omega, \mathcal{F}_., \mathbb{P})$ and $|Z^{(n)} - Z|^\star_t \xrightarrow[n\to\infty]{} 0$ in probability at all instants t, then Z is a Lévy process.

Exercise 4.6.3 If the Lévy process Z is an L^1-integrator, then the previsible part $\widehat{Z}_.$ of its Doob–Meyer decomposition has $\widehat{Z}_t = A \cdot t$, with $A = \mathbb{E}[Z_1]$; thus then both $\widehat{Z}$ and $\widetilde{Z}$ are Lévy processes.

We now have sufficiently many tools at our disposal to analyze this important class of processes. The idea is to look at them as integrators. The stochastic calculus developed so far eases their analysis considerably, and, on the other hand, Lévy processes provide fine examples of various applications and serve to illuminate some of the previously developed notions.

In view of exercise 4.6.1 we may and do assume the natural conditions.

Let us denote the inner product on $\mathbb{R}^d$ variously by juxtaposition or by $\langle\ |\ \rangle$: for $\zeta \in \mathbb{R}^d$,

$$\zeta Z_t = \langle \zeta | Z_t \rangle = \sum_{\eta=1}^{d} \zeta_\eta Z^\eta_t .$$

It is convenient to start by analyzing the characteristic functions of the distributions μ_t of the Z_t. For $\zeta \in \mathbb{R}^d$ and $s, t \geq 0$

$$\begin{aligned}
\widehat{\mu_{s+t}}(\zeta) &\stackrel{\text{def}}{=} \mathbb{E}\Big[e^{i\langle\zeta|Z_{s+t}\rangle}\Big] = \mathbb{E}\Big[e^{i\langle\zeta|Z_{s+t}-Z_s\rangle} e^{i\langle\zeta|Z_s\rangle}\Big] \\
&= \mathbb{E}\Big[\mathbb{E}\Big[e^{i\langle\zeta|Z_{s+t}-Z_s\rangle}\big|\mathcal{F}^0_s[Z]\Big] \cdot e^{i\langle\zeta|Z_s\rangle}\Big] \\
\text{by independence:} \quad &= \mathbb{E}\Big[e^{i\langle\zeta|Z_{s+t}-Z_s\rangle}\Big]\mathbb{E}\Big[e^{i\langle\zeta|Z_s\rangle}\Big] \\
\text{by stationarity:} \quad &= \mathbb{E}\Big[e^{i\langle\zeta|Z_t\rangle}\Big]\mathbb{E}\Big[e^{i\langle\zeta|Z_s\rangle}\Big] = \widehat{\mu_s}(\zeta) \cdot \widehat{\mu_t}(\zeta) .
\end{aligned} \tag{4.6.1}$$

From (A.3.16), $\mu_{s+t} = \mu_s \star \mu_t$.

That is to say, $\{\mu_t : t \geq 0\}$ is a convolution semigroup. Equation (4.6.1) says that $t \mapsto \widehat{\mu_t}(\zeta)$ is multiplicative. As this function is evidently right-continuous in t, it is of the form $\widehat{\mu_t}(\zeta) = e^{t\cdot\psi(\zeta)}$ for some number $\psi(\zeta) \in \mathbb{C}$:

$$\widehat{\mu_t}(\zeta) = \mathbb{E}\Big[e^{i\langle\zeta|Z_t\rangle}\Big] = e^{t\cdot\psi(\zeta)} , \qquad 0 \leq t < \infty .$$

But then $t \mapsto \widehat{\mu_t}(\zeta)$ is even continuous, so by the Continuity Theorem A.4.3 the convolution semigroup $\{\mu_t : 0 \leq t\}$ is weakly continuous.

Since $\widehat{\mu_t}(\zeta)$ depends continuously on ζ, so does ψ; and since $\widehat{\mu_t}$ is bounded, the real part of $\psi(\zeta)$ is negative. Also evidently $\psi(0) = 0$. Equation (4.6.1) generalizes immediately to

$$\mathbb{E}\Big[e^{i\langle\zeta|\boldsymbol{Z}_t-\boldsymbol{Z}_s\rangle}\big|\mathcal{F}_s\Big] = e^{(t-s)\cdot\psi(\zeta)} ,$$

or, equivalently, $$\mathbb{E}\Big[e^{i\langle\zeta|\boldsymbol{Z}_t\rangle}A\Big] = e^{(t-s)\cdot\psi(\zeta)}\mathbb{E}\Big[e^{i\langle\zeta|\boldsymbol{Z}_s\rangle}A\Big] \tag{4.6.2}$$

for $0 \le s \le t$ and $A \in \mathcal{F}_s$. Consider now the process M^ζ that is defined by

$$e^{i\langle\zeta|\boldsymbol{Z}_t\rangle} = M_t^\zeta + \psi(\zeta)\int_0^t e^{i\langle\zeta|\boldsymbol{Z}_s\rangle}\,ds \tag{4.6.3}$$

– read the integral as the Bochner integral of a right-continuous $L^1(\mathbb{P})$-valued curve. M^ζ is a martingale. Indeed, for $0 \le s < t$ and $A \in \mathcal{F}_s$, repeated applications of (4.6.2) and Fubini's theorem A.3.18 give

$$\mathbb{E}\Big[\big(M_t^\zeta - M_s^\zeta\big)\cdot A\Big] = \mathbb{E}\Big[e^{i\langle\zeta|\boldsymbol{Z}_s\rangle}A\Big]\Big(e^{(t-s)\cdot\psi(\zeta)} - 1 - \psi(\zeta)\int_s^t e^{(\tau-s)\cdot\psi(\zeta)}\,d\tau\Big) = 0 .$$

Since $|M_t^\zeta| \le 1 + t|\psi(\zeta)|$, the real and imaginary parts of these martingales are L^p-integrators of (stopped) sizes less than $C_p^{(2.5.6)}(1+t|\psi(\zeta)|)$, for $p > 1$, $t < \infty$, and $\zeta \in \mathbb{R}^d$, and the processes $e^{i\langle\zeta|\boldsymbol{Z}\rangle}$ are L^p-integrators of (stopped) sizes

$$\big\lceil e^{i\langle\zeta|\boldsymbol{Z}^t\rangle}\big\rceil_{\mathcal{I}^p} \le 2C_p^{(2.5.6)}\big(1 + 2t|\psi(\zeta)|\big) . \tag{4.6.4}$$

Suppose $X \in \mathcal{E}_1$ vanishes after time t and let γ_I be the Gaussian density of mean 0 and variance I on $\mathbb{R}^d$ (see definition A.3.50). Integrating the inequality

$$\Big\|\gamma_I(\zeta)\int X\,d\big(e^{i\langle\zeta|\boldsymbol{Z}\rangle}\big)\Big\|_p \le C_t^{(4.6.4)}\gamma_I(\zeta)$$

over $|\zeta| \le 1$ gives, with the help of Jensen's inequality A.3.28 and exercise A.3.51,

$$\Big\|\int X\,d\big(e^{-|\boldsymbol{Z}|^2/2}\big)\Big\|_p \le C_t^{(4.6.4)}\int_{[|\zeta|\le 1]}\gamma_I(\zeta)\,d\zeta \le C_t$$

and shows that $e^{-|\boldsymbol{Z}|^2/2}$ is an L^p-integrator. Due to Ito's theorem, $|\boldsymbol{Z}|^2$ is an L^0-integrator. Now $|\boldsymbol{Z}|^2$ and the $e^{i\langle\zeta|\boldsymbol{Z}\rangle}$, $\zeta \in \mathbb{R}^d$, have càdlàg modifications that are bounded on bounded intervals and adapted to the natural enlargement $\mathcal{F}_.[\boldsymbol{Z}]$ (theorem 2.3.4) and that nearly agree with $|\boldsymbol{Z}|^2$ and $e^{i\langle\zeta|\boldsymbol{Z}\rangle}$, respectively, at all rational instants and for all $\zeta \in \mathbb{Q}^d$. Now inasmuch as on any bounded subset of $\mathbb{R}^d$ the usual topology agrees with the topology generated by the functions $\boldsymbol{z} \mapsto e^{i\langle\zeta|\boldsymbol{z}\rangle}$, $\zeta \in \mathbb{Q}^d$, the limit $\boldsymbol{Z}'_t \stackrel{\text{def}}{=} \lim_{\mathbb{Q}\ni q\downarrow t}\boldsymbol{Z}_q$ exists at all instants t and defines a càdlàg version $\boldsymbol{Z}'$ of $\boldsymbol{Z}$. We rename this version $\boldsymbol{Z}$ and arrive at the following situation: we may, and therefore shall, henceforth assume that *a Lévy process is càdlàg.*

Lemma 4.6.4 *(i) Z is an L^0-integrator. (ii) For any bounded continuous function $F : [0,\infty) \to \mathbb{R}^d$ whose components have finite variation and compact support*

$$\mathbb{E}\Big[e^{i\int_0^\infty \langle Z|dF\rangle}\Big] = e^{\int_0^\infty \psi(-F_s)\,ds} . \tag{4.6.5}$$

*(iii) The **logcharacteristic function** $\psi_Z \stackrel{\text{def}}{=} \psi : \zeta \mapsto t^{-1}\ln\big(\widehat{\mu_t}(\zeta)\big)$ therefore determines the law of Z.*

Proof. (i) The stopping times $T^n \stackrel{\text{def}}{=} \inf\{t : |Z|_t \ge n\}$ increase without bound, since $Z \in \mathfrak{D}$. For $|\zeta| < 1/n$, the process $e^{i\langle\zeta|Z\rangle}\cdot[\![0,T^n)\!) + [\![T_n,\infty)\!)$ is an L^0-integrator whose values lie in a disk of radius 1 about $1 \in \mathbb{C}$. Applying the main branch of the logarithm produces $i\langle\zeta|Z\rangle\cdot[\![0,T^n)\!)$. By Itô's theorem 3.9.1 this is an L^0-integrator for all such ζ. Then so is $Z\cdot[\![0,T^n)\!)$. Since $T^n \uparrow \infty$, Z is a local L^0-integrator. The claim follows from proposition 2.1.9 on page 52.

(ii) Assume for the moment that F is a left-continuous step function with steps at $0 = s_0 < s_1 < \cdots < s_K$, and let $t > 0$. Then, with $\sigma_k = s_k \wedge t$,

$$\begin{aligned}
\mathbb{E}\Big[e^{i\int_0^t \langle F|dZ\rangle}\Big] &= \mathbb{E}\Big[e^{i\sum_{1\le k\le K}\langle F_{s_{k-1}}|Z_{\sigma_k}-Z_{\sigma_{k-1}}\rangle}\Big] \\
&= \prod_{k=1}^{K}\mathbb{E}\Big[e^{i\langle F_{s_{k-1}}|Z_{\sigma_k}-Z_{\sigma_{k-1}}\rangle}\Big] \\
&= \prod_{k=1}^{K} e^{\psi(F_{s_{k-1}})(\sigma_k-\sigma_{k-1})} = e^{\int_0^t \psi(F_s)\,ds} .
\end{aligned}$$

Now the class of bounded functions $F : [0,\infty) \to \mathbb{R}^d$ for which the equality

$$\mathbb{E}\Big[e^{i\int_0^t \langle F|dZ\rangle}\Big] = e^{\int_0^t \psi(F_s)\,ds} \qquad (t \ge 0)$$

holds is closed under pointwise limits of dominated sequences. This follows from the Dominated Convergence Theorem, applied to the stochastic integral with respect to dZ and to the ordinary Lebesgue integral with respect to ds. So this class contains all bounded $\mathbb{R}^d$-valued Borel functions on the half-line. We apply the previous equality to a continuous function F of finite variation and compact support. Then $\int_0^\infty \langle Z|dF\rangle = \int_0^\infty \langle -F|dZ\rangle$ and (4.6.5) follows.

(iii) The functions $\mathscr{D}^d \ni \zeta. \mapsto e^{i\int_0^\infty \langle\zeta|dF\rangle}$, where $F : [0,\infty) \to \mathbb{R}^d$ is continuously differentiable and has compact support, say, form a multiplicative class $\mathcal{M}$ that generates a σ-algebra $\mathcal{F}$ on path space $\mathscr{D}^d$. Any two measures that agree on $\mathcal{M}$ agree on $\mathcal{F}$. ∎

Exercise 4.6.5 (The Zero-One Law) The regularization of the basic filtration $\mathcal{F}^0_.[Z]$ is right-continuous and thus equals $\mathcal{F}_.[Z]$.

The Lévy–Khintchine Formula

This formula – equation (4.6.12) on page 259 – is a description of the logcharacteristic function ψ. We approach it by analyzing the jump measure $\jmath_Z$ of Z (see page 180). The finite variation process

$$V^\zeta \stackrel{\text{def}}{=} \left[e^{i\langle\zeta|Z\rangle}, e^{-i\langle\zeta|Z\rangle}\right]$$

has continuous part
$$^cV^\zeta = {}^c\!\left[e^{i\langle\zeta|Z\rangle}, e^{-i\langle\zeta|Z\rangle}\right] = {}^c[\langle\zeta|Z\rangle, \langle\zeta|Z\rangle] \tag{4.6.6}$$

and jump part
$$^jV_t^\zeta = \sum_{s\le t}\left|\Delta e_s^{i\langle\zeta|Z\rangle}\right|^2 = \sum_{s\le t}\left|e^{i\langle\zeta|\Delta Z_s\rangle} - 1\right|^2$$
$$= \int_{[\![0,t]\!]}\left|e^{i\langle\zeta|y\rangle} - 1\right|^2 \jmath_Z(dy, ds)\,. \tag{4.6.7}$$

Taking the L^p-norm, $1 < p < \infty$, in (4.6.7) results in

$$\left\|{}^jV_t^\zeta\right\|_{L^p} \le 2K_p^{(3.8.9)}C_p^{(4.6.4)}\left(1 + 2t|\psi(\zeta)|\right).$$

By A.3.29, $\displaystyle \left\|\int_{[|\zeta|\le 1]} {}^jV_t^\zeta\, d\zeta\right\|_{L^p} \le \int_{[|\zeta|\le 1]}\left\|{}^jV_t^\zeta\right\|_{L^p} d\zeta < \infty\,.$

Setting
$$h_0'(y) \stackrel{\text{def}}{=} \int_{[|\zeta|\le 1]}\left|e^{i\langle\zeta|y\rangle} - 1\right|^2 d\zeta\,,$$

we obtain
$$\left\lceil\!\left\lceil (h_0' {*} \jmath_Z)_t \right\rceil\!\right\rceil_{\mathcal{I}^p} < \infty\,, \qquad \forall t < \infty, \quad \forall p < \infty\,.$$

That is to say, $h_0' {*} \jmath_Z$ is an L^p-integrator for all p. Now h_0' is a prototypical Hunt function (see exercise 3.10.11). From this we read off the next result.

Lemma 4.6.6 *The indefinite integral $H {*} \jmath_Z$ is an L^p-integrator, for every previsible Hunt function H and every $p \in (0, \infty)$.*

Now fix a *sure and time-independent* Hunt function $h : \mathbb{R}^d \to \mathbb{R}$. Then the indefinite integral $h {*} \jmath_Z$ has increment

$$(h{*}\jmath_Z)_t - (h{*}\jmath_Z)_s = \sum_{s<\sigma\le t} h(\Delta Z_\sigma) = \sum_{\sigma\le t} h(\Delta[Z - Z_s]_\sigma)\,, \qquad s < t\,,$$

which in view of exercises 1.3.21 and 4.6.1 is $\mathcal{F}_t$-measurable and independent of $\mathcal{F}_s$; and $h{*}\jmath_Z$ is clearly stationary. Now exercise 4.6.3 says that $\widehat{h{*}\jmath_Z} = h{*}\widehat{\jmath_Z}$ has at the instant t the value

$$\left(\widehat{h{*}\jmath_Z}\right)_t = \nu(h)\cdot t\,, \quad \text{with } \nu(h) \stackrel{\text{def}}{=} \mathbb{E}[J_1^h]\,.$$

Clearly $\nu(h)$ depends linearly and increasingly on h: ν is a Radon measure on punctured d-space $\mathbb{R}_*^d \stackrel{\text{def}}{=} \mathbb{R}^d \setminus \{0\}$ and integrates the Hunt function $h_0 : y \mapsto |y|^2 \wedge 1$. An easy consequence of this is the following:

Lemma 4.6.7 *For any sure and time-independent Hunt function h on $\mathbb{R}^d$, $h{*}\jmath_Z$ is a Lévy process. The previsible part $\widehat{\jmath_Z}$ of $\jmath_Z$ has the form*

$$\widehat{\jmath_Z}(dy, ds) = \nu(dy) \times ds \;, \tag{4.6.8}$$

*where ν, the **Lévy measure** of Z, is a measure on punctured d-space that integrates h_0. Consequently, the jumps of $\boldsymbol{Z}$, if any, occur only at totally inaccessible stopping times; that is to say, $\boldsymbol{Z}$ is quasi-left-continuous (see proposition 4.3.25).*

In the terms of page 231, equation (4.6.8) says that the jump intensity measure $\widehat{\jmath_Z}$ has the disintegration[15]

$$\int_{[\![0,\infty)\!)} h_s(dy)\, \widehat{\jmath_Z}(dy, ds) = \int_0^\infty \int_{\mathbb{R}^d_*} h_s(y)\, \nu(dy)\; ds \;,$$

with the jump intensity rate ν independent of $\varpi \in \boldsymbol{B}$. Therefore

$$(H{*}\widehat{\jmath_Z})_t = \int_0^t \int_{\mathbb{R}^d_*} H_s(y)\, \nu(y)\; ds$$

for any random Hunt function H. Next, since $e^{i\langle\zeta|Z\rangle} \cdot e^{-i\langle\zeta|Z\rangle} = 1$, equation (3.8.10) gives

$$-dV^\zeta = e_-^{i\langle\zeta|Z\rangle} de^{-i\langle\zeta|Z\rangle} + e_-^{-i\langle\zeta|Z\rangle} de^{i\langle\zeta|Z\rangle}$$

by (4.6.3):

$$= e_-^{i\langle\zeta|Z\rangle}\; dM^{-\zeta} + e_-^{-i\langle\zeta|Z\rangle}\; dM^\zeta$$
$$+\ \psi(-\zeta)\, dt \ +\ \psi(\zeta)\, dt \;,$$

whence $$\widehat{V_t^\zeta} = -t \cdot \big(\psi(-\zeta) + \psi(\zeta)\big) \;.$$

From (4.6.7) $$\widehat{{}^{j}V_t^\zeta} = t \cdot \int \left| e^{i\zeta y} - 1 \right|^2 \nu(dy) \;.$$

By (4.6.6) $\;{}^{c}\!\big[\langle\zeta|Z\rangle, \langle\zeta|Z\rangle\big] = \widehat{{}^{c}V^\zeta} = \widehat{V^\zeta} - \widehat{{}^{j}V^\zeta} = t \cdot g(\zeta)$,

where the constant $g(\zeta)$ must be of the form $\zeta_\eta \zeta_\theta B^{\eta\theta}$, in view of the bilinearity of $\zeta \mapsto {}^{c}\!\big[\langle\zeta|Z\rangle, \langle\zeta|Z\rangle\big]$. To summarize:

Lemma 4.6.8 *There is a constant symmetric positive semidefinite matrix B*

with $${}^{c}\!\big[\langle\zeta|Z\rangle, \langle\zeta|Z\rangle\big]_t = \sum_{1 \le \eta,\theta \le d} \zeta_\eta \zeta_\theta B^{\eta\theta} \cdot t \;,$$

which is to say, $${}^{c}[Z^\eta, Z^\theta] = B^{\eta\theta} \cdot t \;. \tag{4.6.9}$$

Consequently the continuous martingale part ${}^{\tilde c}\boldsymbol{Z}$ of $\boldsymbol{Z}$ (proposition 4.4.7) is a Wiener process with covariance matrix B (exercise 3.9.6).

[15] $[\![0,\infty)\!) \overset{\text{def}}{=} \mathbb{R}^d_* \times [0,\infty)$ and $[\![0,t]\!] \overset{\text{def}}{=} \mathbb{R}^d_* \times [0,t]$.

We are now in position to establish the renowned Lévy–Khintchine formula.

Since
$$e^{i\langle\zeta|Z_t\rangle}-1=\int_{0+}^{t} ie_{-s}^{i\langle\zeta|Z\rangle}\,d\langle\zeta|Z\rangle_s$$
$$-1/2\int_{0+}^{t} e_{-s}^{i\langle\zeta|Z\rangle}\,\zeta_\eta\zeta_\theta d{}^c[Z^\eta,Z^\theta]_s$$
$$+\sum_{0<s\le t} e_{-s}^{i\langle\zeta|Z\rangle}\Big(e^{i\langle\zeta|\Delta Z_s\rangle}-1-i\langle\zeta|\Delta Z_s\rangle\Big)\,,$$

$$\Big(e_{\cdot-}^{-i\langle\zeta|Z\rangle}{*}e^{i\langle\zeta|Z\rangle}\Big)_t=\Big(i\langle\zeta|Z\rangle_t-\int_0^t\!\!\int i\langle\zeta|y\rangle\cdot[\![|y|>1]\!]\,\jmath_Z(dy,ds)\Big)$$
$$-1/2\zeta_\eta\zeta_\theta{}^c[Z^\eta,Z^\theta]_t$$
$$+\int_0^t\!\!\int\Big(e^{i\langle\zeta|y\rangle}-1-i\langle\zeta|y\rangle\,[\![|y|\le 1]\!]\Big)\jmath_Z(dy,ds)\,,$$

the integrand of the previous line being a Hunt function. Now from (4.6.3)

$$Mart_t+t\,\psi(\zeta)=i\langle\zeta|{}^sZ_t\rangle-\frac{t}{2}\,\zeta_\eta\zeta_\theta B^{\eta\theta}$$
$$+t\int\Big(e^{i\langle\zeta|y\rangle}-1-i\langle\zeta|y\rangle\,[\![|y|\le 1]\!]\Big)\nu(dy)\,,$$

where
$$ {}^sZ_t\stackrel{\text{def}}{=}\Big(Z_t-\int_0^t\!\!\int y\cdot[\![|y|>1]\!]\,\jmath_Z(dy,ds)\Big)\,. \tag{4.6.10}$$

Taking previsible parts of this L^2-integrator results in

$$t\cdot\psi(\zeta)=i\langle\zeta|\widehat{{}^sZ}_t\rangle-\frac{t}{2}\cdot\zeta_\eta\zeta_\theta B^{\eta\theta}$$
$$+t\cdot\int\Big(e^{i\langle\zeta|y\rangle}-1-i\langle\zeta|y\rangle\,[\![|y|\le 1]\!]\Big)\,\nu(dy)\,.$$

Now three of the last four terms are of the form $t\cdot const$. Then so must be the fourth; that is to say,

$$\widehat{{}^sZ}_t=t\,A \tag{4.6.11}$$

for some constant vector A. We have finally arrived at the promised description of ψ:

Theorem 4.6.9 (The Lévy–Khintchine Formula) *There are a vector $A\in\mathbb{R}^d$, a constant positive semidefinite matrix B, and on punctured d-space $\mathbb{R}^d_*$ a positive measure ν that integrates $h_0: y\mapsto|y|^2\wedge 1$, such that*

$$\psi(\zeta)=i\langle\zeta|A\rangle-\frac{1}{2}\,\zeta_\eta\zeta_\theta B^{\eta\theta}+\int\Big(e^{i\langle\zeta|y\rangle}-1-i\langle\zeta|y\rangle\,[\![|y|\le 1]\!]\Big)\nu(dy)\,. \tag{4.6.12}$$

(A,B,ν) is called the ***characteristic triple*** of the Lévy process Z. According to lemma 4.6.4, it determines the law of Z.

We now embark on a little calculation that forms the basis for the next two results. Let $\boldsymbol{F} = (F_\eta)_{\eta=1\ldots d} : [0,\infty) \to \mathbb{R}^d$ be a right-continuous function whose components have finite variation and vanish after some fixed time, and let h be a Borel function of relatively compact carrier on $\mathbb{R}^d_* \times [0,\infty)$. Set

$$M = -\boldsymbol{F}{*}^{\tilde{c}}\boldsymbol{Z} \ , \quad \text{i.e.,} \quad M_t \stackrel{\text{def}}{=} -\int_0^t \langle \boldsymbol{F}_s | d^{\tilde{c}}\boldsymbol{Z}_s \rangle \ ,$$

which is a continuous martingale with square function

$$v_t \stackrel{\text{def}}{=} [M,M]_t = {}^c[M,M]_t = \int_0^t F_{\eta_s} F_{\theta_s} B^{\eta\theta} \, ds \ , \tag{4.6.13}$$

and set[15] $\quad V \stackrel{\text{def}}{=} h{*}\jmath_{\boldsymbol{Z}}$, i.e., $V_t = \int_{[\![0,t]\!]} h_s(\boldsymbol{y}) \, \jmath_{\boldsymbol{Z}}(d\boldsymbol{y}, ds)$.

Since the carrier $[h \neq 0]$ is relatively compact, there is an $\epsilon > 0$ such that $h_s(\boldsymbol{y}) = 0$ for $|\boldsymbol{y}| < \epsilon$. Therefore V is a finite variation process without continuous component and is constant between its jumps

$$\Delta V_s = h_s(\Delta \boldsymbol{Z}_s) \ , \tag{4.6.14}$$

which have size at least $\epsilon > 0$. We compute:

$$E_t \stackrel{\text{def}}{=} \exp\left(iM_t \ + \ v_t/2 \ + \ iV_t\right)$$

by Itô: $$= 1 + i\int_0^t E_{s-} \, dM_s + i^2/2 \int_0^t E_{s-} \, d{}^c[M,M]_s$$
$$+ 1/2 \int_0^t E_{s-} \, dv_s$$
$$+ i\int_{[\![0,T]\!]} E_{s-} \, dV_s + \sum_{0<s\le t} E_s - E_{s-} - iE_{s-}\Delta V_s$$

by (4.6.13): $$= 1 + i\int_0^t E_{s-} \, dM_s$$

and (4.6.14): $$+ i\sum_{0<s\le t} E_{s-}\Delta V_s + \sum_{0<s\le t} E_{s-}\left(e^{i\Delta V_s} - 1 - i\Delta V_s\right)$$
$$= 1 + i\int_0^t E_{s-} \, dM_s + \sum_{0<s\le t} E_{s-}\left(e^{i\Delta V_s} - 1\right) .$$

Thus $$E_t = 1 - i\int_0^t E_{s-} \, \langle \boldsymbol{F} | d^{\tilde{c}}\boldsymbol{Z} \rangle + \int_{[\![0,t]\!]} E_{s-} \cdot \left(e^{ih_s(\boldsymbol{y})} - 1\right) \jmath_{\boldsymbol{Z}}(d\boldsymbol{y}, ds) \ . \tag{4.6.15}$$

The Martingale Representation Theorem

The martingale representation theorem 4.2.15 for Wiener processes extends to Levy processes. It comes as a first application of equation (4.6.15): take $t = \infty$ and multiply (4.6.15) by the complex constant

$$\exp\Big(-v_\infty/2 \;-\; i\int_{[\![0,\infty)\!)} h_s(y)\,\widehat{\jmath_Z}(dy,ds)\Big)$$

to obtain

$$\exp\Big(i\Big(\int_0^\infty \langle {}^c\!Z|dF\rangle + \int_{[\![0,\infty)\!)} h_s(y)\,\widetilde{\jmath_Z}(dy,ds)\Big)\Big) \tag{4.6.16}$$

$$= c + \int_0^\infty \langle X_s|d^{\tilde c}Z_s\rangle + \int_{[\![0,\infty)\!)} H_s(y)\,\widetilde{\jmath_Z}(dy,ds)\;, \tag{4.6.17}$$

where c is a constant, X is some bounded previsible $\mathbb{R}^d$-valued process that vanishes after some instant, and $H : [\![0,\infty)\!) \to \mathbb{R}$ is some bounded previsible random function that vanishes on $[|y| < \epsilon]$ for some $\epsilon > 0$. Now observe that the exponentials in (4.6.16), with F and h as specified above, form a multiplicative class[16] $\mathcal{M}$ of bounded $\mathcal{F}^0_\infty[Z]$-measurable random variables on Ω. As F and h vary, the random variables

$$\int_0^\infty \langle {}^{\tilde c}Z|dF\rangle + \int_{[\![0,\infty)\!)} h_s(y)\,\widetilde{\jmath_Z}(dy,ds)$$

form a vector space Γ that generates precisely the same σ-algebra as $\mathcal{M} = e^{i\Gamma}$ (page 410); it is nearly evident that this σ-algebra is $\mathcal{F}^0_\infty[Z]$. The point of these observations is this: *$\mathcal{M}$ is a multiplicative class that generates $\mathcal{F}^0_\infty[Z]$ and consists of stochastic integrals of the form appearing in* (4.6.17).

Theorem 4.6.10 *Suppose Z is a Lévy process. Every random variable $F \in L^2(\mathcal{F}^0_\infty[Z])$ is the sum of the constant $c = \mathbb{E}[F]$ and a stochastic integral of the form*

$$\int_0^\infty \langle X_s|d^{\tilde c}Z_s\rangle + \int_{[\![0,\infty)\!)} H_s(y)\,\widetilde{\jmath_Z}(dy,ds)\;, \tag{4.6.18}$$

where $X = (X_\eta)_{\eta=1\ldots d}$ is a vector of predictable processes and H a predictable random function on $[\![0,\infty)\!) = \mathbb{R}^d_* \times [\![0,\infty)\!)$, with the pair (X,H) satisfying

$$\|X,H\|_2^* \stackrel{\text{def}}{=} \Big(\mathbb{E}\Big[\int^* X_{\eta s}\overline{X_{\theta s}}B^{\eta\theta}\,ds + \int^* |H|_s^2(y)\,\nu(dy)\,ds\Big]\Big)^{1/2} < \infty\,.$$

Proof. Let us denote by ${}^cM$ and ${}^jM$ the martingales whose limits at infinity appear as the second and third entries of equation (4.6.17), by M their sum,

---

[16] A multiplicative class is by (our) definition closed under complex conjugation. The complex-conjugate $\overline{X_\eta}$ equals X_η, of course, when X_η is real.

and let us compute $\mathbb{E}[|M|_\infty^2]$: since $[{}^cM, {}^jM] = 0$,

$$\mathbb{E}[|M|_\infty^2] = \mathbb{E}\Big[[M, \overline{M}]_\infty\Big] = \mathbb{E}\Big[[{}^cM, {}^c\overline{M}]_\infty + [{}^jM, {}^j\overline{M}]_\infty\Big]$$

by 4.6.7 (iii):
$$= \mathbb{E}\left[{}^c[{}^cM, {}^c\overline{M}]_\infty + \int_{[\![0,\infty))} |H|_s^2(\boldsymbol{y}) \jmath_{\boldsymbol{Z}}(d\boldsymbol{y}, ds)\right]$$

as $\widehat{\jmath_{\boldsymbol{Z}}} = \nu \times \lambda$:
$$= \mathbb{E}\left[\int X_{\eta_s} \overline{X}_{\theta_s} B^{\eta\theta}\, ds + \int |H|_s^2(d\boldsymbol{y})\ \nu(d\boldsymbol{y}) ds\right]$$

$$= \left(\|\boldsymbol{X}, H\|_2^*\right)^2 .$$

Now the vector space of previsible pairs $(\boldsymbol{X}, H)$ with $\|\boldsymbol{X}, H\|_2^* < \infty$ is evidently complete – it is simply the cartesian product of two $\mathcal{L}^2$-spaces – and the linear map $\mathcal{U}$ that associates with every pair $(\boldsymbol{X}, H)$ the stochastic integral (4.6.18) is an isometry of that set into $L^2_{\mathbb{C}}(\mathcal{F}^0_\infty[\boldsymbol{Z}])$; its image $\mathcal{I}$ is therefore a closed subspace of $L^2_{\mathbb{C}}(\mathcal{F}^0_\infty[\boldsymbol{Z}])$ and contains the multiplicative class $\mathcal{M}$, which generates $\mathcal{F}^0_\infty[\boldsymbol{Z}]$. We conclude with exercise A.3.5 on page 393 that $\mathcal{I}$ contains all bounded complex-valued $\mathcal{F}^0_\infty[\boldsymbol{Z}]$-measurable functions and, as it is closed, is all of $L^2_{\mathbb{C}}(\mathcal{F}^0_\infty[\boldsymbol{Z}])$. The restriction of $\mathcal{U}$ to real integrands will exhaust all of $L^2_{\mathbb{R}}(\mathcal{F}^0_\infty[\boldsymbol{Z}])$. ∎

Corollary 4.6.11 *For $H \in L^2(\nu \times \lambda)$, $H * \widetilde{\jmath_{\boldsymbol{Z}}}$ is a square integrable martingale.*

Project 4.6.12 [93] *Extend theorem 4.6.10 to exponents p other than 2.*

The Characteristic Function of the Jump Measure, in fact of the pair $({}^{\tilde c}\boldsymbol{Z}, \jmath_{\boldsymbol{Z}})$, can be computed from equation (4.6.15). We take the expectation:

$$e_t \stackrel{\text{def}}{=} \mathbb{E}[E_t] = 1 + \mathbb{E}\left[\int_{[\![0,t]\!]} E_{s-}\cdot\left(e^{ih_s(\boldsymbol{y})} - 1\right) \jmath_{\boldsymbol{Z}}(d\boldsymbol{y}, ds)\right]$$

as $\widehat{\jmath_{\boldsymbol{Z}}} = \nu \times \lambda$:
$$= 1 + \int_0^t e_s \int_{\mathbb{R}^d_*} \left(e^{ih_s(\boldsymbol{y})} - 1\right)\nu(d\boldsymbol{y})\ ds\ ,$$

whence
$$e_t' = e_t \cdot \phi_t \quad \text{with} \quad \phi_t = \int_{\mathbb{R}^d_*} \left(e^{ih_t(\boldsymbol{y})} - 1\right)\nu(d\boldsymbol{y})\ ,$$

and so
$$e_t = e^{\int_0^t \phi_s\, ds} = \exp\left(\int_0^t \int_{\mathbb{R}^d_*} \left(e^{ih_s(\boldsymbol{y})} - 1\right)\nu(d\boldsymbol{y})\ ds\right) .$$

Evaluating this at $t = \infty$ and multiplying with $\exp\left(-v_\infty/2\right)$ gives

$$\mathbb{E}\left[\exp\left(i \int {}^{\tilde c}\boldsymbol{Z}\, d\boldsymbol{F} + i \int h_s(\boldsymbol{y}) \jmath_{\boldsymbol{Z}}(d\boldsymbol{y}, ds)\right)\right]$$

$$= \exp\left(\frac{-1}{2} \int F_{\eta_s} F_{\theta_s} B^{\eta\theta}\, ds\right) \times \exp\left(\iint \left(e^{ih_s(\boldsymbol{y})} - 1\right)\nu(d\boldsymbol{y})\ ds\right) . \quad (4.6.19)$$

In order to illuminate this equation, let $\mathcal{V}$ denote the cartesian product of the path space $\mathscr{C}^d$ with the space $\mathfrak{M}^\cdot \stackrel{\text{def}}{=} \mathfrak{M}^\cdot(\mathbb{R}^d_* \times [0,\infty))$ of Radon measures on $\mathbb{R}^d_* \times [0,\infty)$, the former given the topology of uniform convergence on compacta and the latter the weak* topology $\sigma\big(\mathfrak{M}^\cdot, C_{00}(\mathbb{R}^d_* \times [0,\infty))\big)$ (see A.2.32). We equip $\mathcal{V}$ with the product of these two locally convex topologies, which is metrizable and locally convex and, in fact, Fréchet. For every pair $(\boldsymbol{F}, h)$, where $\boldsymbol{F}$ is a vector of distribution functions on $[0,\infty)$ that vanish ultimately and $h : \mathbb{R}^d_* \times [0,\infty) \to \mathbb{R}$ is continuous and of compact support, consider the function

$$\gamma^{\boldsymbol{F},h} : (\boldsymbol{z}, \mu) \mapsto \int_0^\infty \langle \boldsymbol{z} | d\boldsymbol{F} \rangle + \int_{\mathbb{R}^d_* \times [0,\infty)} h_s(\boldsymbol{y})\, \mu(d\boldsymbol{y}, ds) .$$

The $\gamma^{\boldsymbol{F},h}$ are evidently continuous linear functionals on $\mathcal{V}$, and their collection forms a vector space[17] Γ that generates the Borels on $\mathcal{V}$. If we consider the pair $({}^{\tilde{c}}\boldsymbol{Z}, \jmath_{\boldsymbol{Z}})$ as a $\mathcal{V}$-valued random variable, then equation (4.6.19) simply says that the law $\mathbb{L}$ of this random variable has the characteristic function $\widehat{\mathbb{L}}^\Gamma(\gamma^{\boldsymbol{F},h})$ given by the right-hand side of equation (4.6.19). Now this is the product of the characteristic function of the law of ${}^{\tilde{c}}\boldsymbol{Z}$ with that of $\jmath_{\boldsymbol{Z}}$. We conclude that *the Wiener process* ${}^{\tilde{c}}\boldsymbol{Z}$ *and the random measure* $\jmath_{\boldsymbol{Z}}$ *are independent.* The same argument gives

Proposition 4.6.13 *Suppose $\boldsymbol{Z}^{(1)}, \ldots, \boldsymbol{Z}^{(K)}$ are $\mathbb{R}^d$-valued Lévy processes on $(\Omega, \mathcal{F}_., \mathbb{P})$. If the brackets $[Z^{(k)\eta}, Z^{(l)\theta}]$ are evanescent whenever $1 \le k \ne l \le K$ and $1 \le \eta, \theta \le d$, then the $\boldsymbol{Z}^{(1)}, \ldots, \boldsymbol{Z}^{(K)}$ are independent.*

Proof. Denote by $(\boldsymbol{A}^{(k)}, B^{(k)}, \nu^{(k)})$ the characteristic triple of $\boldsymbol{Z}^{(k)}$ and by $\mathbb{L}^{(k)}$ its law on $\mathcal{V}$, and define $\boldsymbol{Z} = \sum_k \boldsymbol{Z}^{(k)}$. This is a Lévy process, whose characteristic triple shall be denoted by $(\boldsymbol{A}, B, \nu)$. Since by assumption no two of the $\boldsymbol{Z}^{(k)}$ ever jump at the same time, we have

$$\sum_{s \le t} H_s(\Delta \boldsymbol{Z}_s) = \sum_k \sum_{s \le t} H_s(\Delta \boldsymbol{Z}^{(k)}_s)$$

for any Hunt function H, which signifies that $\jmath_{\boldsymbol{Z}} = \sum_k \jmath_{\boldsymbol{Z}^{(k)}}$

and implies that $$\nu = \textstyle\sum_k \nu^{(k)} .$$

From (4.6.9) $$B = \textstyle\sum_k B^{(k)} ,$$

since ${}^c[\boldsymbol{Z}^{(k)}, \boldsymbol{Z}^{(l)}] = 0$. Let $\boldsymbol{F}^{(k)} : [0,\infty) \to \mathbb{R}^d$, $1 \le k, l \le K$, be right-continuous functions whose components have finite variation and vanish after some fixed time, and let $h^{(k)}$ be Borel functions of relatively compact carrier on $\mathbb{R}^d_* \times [0,\infty)$. Set

$$M = - \sum_{1 \le k \le K} \boldsymbol{F}^{(k)} * {}^{\tilde{c}}\boldsymbol{Z}^{(k)} ,$$

[17] In fact, Γ is the whole dual $\mathcal{V}^*$ of $\mathcal{V}$ (see exercise A.2.33).

which is a continuous martingale with square function

$$v_t \overset{\text{def}}{=} [M,M]_t = \lceil M,M\rceil_t = \sum_k \int_0^t F^{(k)}_{\eta s} F^{(k)}_{\theta s} B^{(k)\eta\theta}\, ds\,,$$

and set $\quad V = \sum_k h^{(k)} * \jmath_{Z^{(k)}}$, i.e., $V_t = \sum_k \int_{[\![0,t]\!]} h^{(k)}_s(y)\, \jmath_{Z^{(k)}}(dy,ds)$.

A straightforward repetition of the computation leading to equation (4.6.15) on page 260 produces

$$\begin{aligned} E_t &\overset{\text{def}}{=} \exp\left(iM_t \;+\; v_t/2 \;+\; iV_t\right) \\ &= 1 - i\int_0^t E_{s-}\, dM_s + \sum_k \int_{[\![0,t]\!]} E_{s-}\cdot\left(e^{ih^{(k)}_s(y)}-1\right) \jmath_{Z^{(k)}}(dy,ds) \end{aligned}$$

and
$$\begin{aligned} e_t &\overset{\text{def}}{=} \mathbb{E}[E_t] = 1 + \mathbb{E}\Big[\sum_k \int_{[\![0,t]\!]} E_{s-}\cdot\left(e^{ih^{(k)}_s(y)}-1\right) \jmath_{Z^{(k)}}(dy,ds)\Big] \\ &= 1 + \int_0^t e_s\phi_s\, ds \;\text{ with }\; \phi_s = \sum_k \int_{\mathbb{R}^d_*} \left(e^{ih^{(k)}_s(y)}-1\right)\nu^{(k)}(dy)\,, \end{aligned}$$

whence
$$e_t = e^{\int_0^t \phi_s\, ds} = \exp\Big(\sum_k \int_0^t \int_{\mathbb{R}^d_*} \left(e^{ih^{(k)}_s(y)}-1\right)\nu^{(k)}(dy)\, ds\Big)\,.$$

Evaluating this at $t = \infty$ and dividing by $\exp(v_\infty/2)$ gives

$$\begin{aligned} &\mathbb{E}\Big[\exp\Big(i\sum_k\Big(\int {}^{\tilde{c}}\!Z^{(k)}\, dF^{(k)} + \int h^{(k)}_s(y) \jmath_{Z^{(k)}}(dy,ds)\Big)\Big)\Big] \\ &= \prod_k \exp\Big(\frac{-1}{2}\int F^{(k)}_{\eta s} F^{(k)}_{\theta s} B^{(k)\eta\theta}\, ds\Big) \times \exp\Big(\int\!\!\int \left(e^{ih^{(k)}_s(y)}-1\right)\nu^{(k)}(dy)\, ds\Big) \\ &= \prod_k \widehat{\mathbb{L}}^{(k)}\big(\gamma^{F^{(k)},h^{(k)}}\big) \;: \end{aligned} \tag{4.6.20}$$

the characteristic function of the k-tuple $\big(({}^{\tilde{c}}\!Z^{(k)}, \jmath_{Z^{(k)}})\big)_{1\le k\le K}$ is the product of the characteristic functions of the components. Now apply A.3.36. ❚

Exercise 4.6.14 (i) Suppose $A^{(k)}$ are mutually disjoint relatively compact Borel subsets of $\mathbb{R}^d_* \times [0,\infty)$. Applying equation (4.6.20) with $h^{(k)} \overset{\text{def}}{=} \alpha^{(k)} A^{(k)}$, show that the random variables $\jmath_Z(A^{(k)})$ are independent Poisson with means $(\nu{\times}\lambda)(A^{(k)})$. In other words, the jump measure of our Lévy process is a Poisson point process on $\mathbb{R}^d_*$ with intensity rate ν, in the sense of definition 3.10.19 on page 185.

(ii) Suppose next that $h^{(k)} : \mathbb{R}^d_* \to \mathbb{R}$ are time-independent sure Hunt functions with disjoint carriers. Then the indefinite integrals $Z^{(k)} \overset{\text{def}}{=} h^{(k)} * \widetilde{\jmath_Z}$ are independent Lévy processes with characteristic triples $(0, 0, h^{(k)}\nu)$.

Exercise 4.6.15 If ζ is a Poisson point process with intensity rate ν and $f \in \mathcal{L}^1(\nu)$, then $f{*}\zeta$ is a Lévy process with values in $\mathbb{R}$ and with characteristic triple $(\int f[|f| \le 1]d\nu, 0, f[\nu])$.

Canonical Components of a Lévy Process

Since our Lévy process $\boldsymbol{Z}$ with characteristic triple $(\boldsymbol{A}, B, \nu)$ is quasi-left-continuous, the sparsely and previsibly supported part ${}^p Z$ of proposition 4.4.5 vanishes and the decomposition (4.4.3) of exercise 4.4.10 boils down to

$$\boldsymbol{Z} = {}^{\hat v}\boldsymbol{Z} + {}^{\tilde c}\boldsymbol{Z} + {}^{\tilde s}\boldsymbol{Z} + {}^{l}\boldsymbol{Z} . \tag{4.6.21}$$

The following table shows the features of the various parts.

Part	Given by	Char. Triple	Prev. Control
${}^{\hat v}\boldsymbol{Z}_t$	$t\boldsymbol{A} = \widehat{{}^s\boldsymbol{Z}_t}$, see (4.6.10)	$(\boldsymbol{A},0,0)$	via (4.6.22)
${}^{\tilde c}\boldsymbol{Z}_t$	$\sigma \boldsymbol{W}$, see lemma 4.6.8	$(0,B,0)$	via (4.6.23)
${}^{\tilde s}\boldsymbol{Z}_t$	$\int \boldsymbol{y}\cdot [\lvert \boldsymbol{y}\rvert \le 1]\times [\![0,t]\!]\, \widetilde{\jmath_{\boldsymbol{Z}}}(d\boldsymbol{y},ds)$	$(0,0,{}^s\nu)$	via (4.6.24)
${}^{l}\boldsymbol{Z}_t$	$\int \boldsymbol{y}\cdot [\lvert \boldsymbol{y}\rvert > 1]\times [\![0,t]\!]\, \jmath_{\boldsymbol{Z}}(d\boldsymbol{y},ds)$	$(0,0,{}^l\nu)$	via (4.6.29)

To discuss the items in this table it is convenient to introduce some notation. We write $\lvert\ \rvert$ for the sup-norm $\lvert\ \rvert_\infty$ on vectors or sequences and $\lvert\ \rvert_1$ for the ℓ^1-norm: $\lvert \boldsymbol{x}\rvert_1 = \sum_\eta \lvert x_\eta \rvert$. Furthermore we write

$${}^s\nu(d\boldsymbol{y}) \stackrel{\text{def}}{=} [\lvert y\rvert \le 1]\, \nu(d\boldsymbol{y}) \qquad \text{for the small-jump intensity rate,}$$

$${}^l\nu(d\boldsymbol{y}) \stackrel{\text{def}}{=} [\lvert y\rvert > 1]\, \nu(d\boldsymbol{y}) \qquad \text{for the large-jump intensity rate,}$$

and

$$\lvert \mu \rvert_\rho \stackrel{\text{def}}{=} \sup_{\lvert \boldsymbol{x}'\rvert \le 1} \left(\int \lvert \langle \boldsymbol{x}' | \boldsymbol{y}\rangle \rvert^\rho\, \mu(d\boldsymbol{y}) \right)^{1/\rho}, \qquad 0<\rho<\infty ,$$

for any positive measure μ on $\mathbb{R}^d_*$. Now the previsible controllers $\boldsymbol{Z}^{\langle\rho\rangle}$ of inequality (4.5.20) on page 246 are given in terms of the characteristic triple $(\boldsymbol{A}, B, \nu)$ by

$$\boldsymbol{Z}^{\langle\rho\rangle}_t = t\cdot \begin{cases} \sup\limits_{\lvert \boldsymbol{x}'\rvert \le 1} \left(\langle \boldsymbol{x}' | \boldsymbol{A}\rangle + \int \langle \boldsymbol{x}' | \boldsymbol{y}\rangle \cdot [\lvert \boldsymbol{y}\rvert > 1]\, \nu(d\boldsymbol{y}) \right) \le \lvert \boldsymbol{A}\rvert_1 + \lvert {}^l\nu\rvert_1^1 , & \rho = 1, \\ \sup\limits_{\lvert \boldsymbol{x}'\rvert \le 1} \left(x'_\eta x'_\theta B^{\eta\theta} + \int \lvert \langle \boldsymbol{x}' | \boldsymbol{y}\rangle \rvert^2\, \nu(d\boldsymbol{y}) \right) \le \lvert B\rvert + \lvert \nu\rvert_2^2 , & \rho = 2, \\ \sup\limits_{\lvert \boldsymbol{x}'\rvert \le 1} \left(\int \lvert \langle \boldsymbol{x}' | \boldsymbol{y}\rangle \rvert^\rho\, \nu(d\boldsymbol{y}) \right) = \lvert \nu\rvert_\rho^\rho , & \rho > 2, \end{cases}$$

provided of course that $\boldsymbol{Z}$ is a local L^q-integrator for some $q \ge 2$ and $\rho \le q$. Here $\lvert B\rvert \stackrel{\text{def}}{=} \sup\{x'_\eta x'_\theta B^{\eta\theta} : \lvert \boldsymbol{x}'\rvert_\infty \le 1\}$. Now the first three Lévy processes ${}^{\hat v}\boldsymbol{Z}$, ${}^{\tilde c}\boldsymbol{Z}$, and ${}^{\tilde s}\boldsymbol{Z}$ all have bounded jumps and thus are L^q-integrators for all $q<\infty$. Inequality (4.5.20) and the second column of the table above therefore result in the following inequalities: for any previsible $\boldsymbol{X}$, $2 \le p \le q < \infty$, and instant t

$$\left\|X{*}^{\hat{v}}Z_t^\star\right\|_{L^p} \le \left(|A|_1 + |{}^{l}\!\nu|_1\right)\left\|\int_0^t |X|_s\,ds\right\|_{L^p}, \tag{4.6.22}$$

$$\left\|X{*}^{\tilde{c}}Z_t^\star\right\|_{L^p} \le C_p^{\diamond(4.5.11)}\cdot|B|\cdot\left\|\left(\int_0^t |X|_s^2\,ds\right)^{1/2}\right\|_{L^p}, \tag{4.6.23}$$

and
$$\left\|X{*}^{\tilde{s}}Z_t^\star\right\|_{L^p} \le C_p^\diamond\cdot\max_{\rho=2,q}|{}^{s}\!\nu|_\rho\cdot\left\|\left(\int_0^t |X|_s^\rho\,ds\right)^{1/\rho}\right\|_{L^p}. \tag{4.6.24}$$

Lastly, let us estimate the large-jump part ${}^{l}Z$, first for $0 < p \le 1$. To this end let X be previsible and set $Y \stackrel{\text{def}}{=} X{*}{}^{l}Z$. Then

$$|Y_t^\star|^p \le \sum_{s\le t}\left(|Y_{s-}| + |\Delta Y_s|\right)^p - |Y_{s-}|^p$$

as $p \le 1$:
$$\le \sum_{s\le t}|\Delta Y_s|^p = \sum_{s\le t}\left|\langle X_s|\Delta^{l}Z_s\rangle\right|^p$$

$$= \int_{[\![0,t]\!]} |\langle X_s|y\rangle|^p\cdot[|y|>1]\;\jmath_Z(dy,ds)\,.$$

Hence
$$\left\|X{*}^{l}Z_t^\star\right\|_{L^p} \le |{}^{l}\!\nu|_p\cdot\left(\int\int_0^t |X|_s^p\,ds\,d\mathbb{P}\right)^{1/p} \quad\text{for } 0<p\le 1\,. \tag{4.6.25}$$

This inequality of course says nothing at all unless every linear functional x' on $\mathbb{R}^d$ has p^{th} moments with respect to ${}^{l}\!\nu$, so that $|{}^{l}\!\nu|_p$ is finite. Let us next address the case that $|{}^{l}\!\nu|_p < \infty$ for some $p \in [1,2]$. Then ${}^{l}Z$ is an L^p-integrator with Doob–Meyer decomposition

$${}^{l}Z = t\cdot\int y\,{}^{l}\!\nu(dy) + \int_{[\![0,t]\!]} y\cdot[|y|>1]\;\widetilde{\jmath_Z}(dy,ds)\,.$$

With $Y \stackrel{\text{def}}{=} X{*}{}^{l}Z$, a little computation produces

$$\left\|\widehat{Y}_t^\star\right\|_{L^p} \le |{}^{l}\!\nu|_p\cdot\left(\int\int |X|_s^p\,ds\,d\mathbb{P}\right)^{1/p}, \tag{4.6.26}$$

and theorem 4.2.12 gives

$$\left\|\widetilde{Y}_t^\star\right\|_{L^p} \le C_p^{(4.2.4)}\cdot\left\|S_t[\widetilde{Y}]\right\|_{L^p} = C_p\cdot\left\|\left(\sum_{s\le t}\left|\langle X_s|\Delta^{l}\widetilde{Z}_s\rangle\right|^2\right)^{1/2}\right\|_{L^p}$$

$$\le C_p\cdot\left\|\left(\sum_{s\le t}\left|\langle X_s|\Delta^{l}\widetilde{Z}_s\rangle\right|^p\right)^{1/p}\right\|_{L^p}$$

$$= C_p\cdot\left(\mathbb{E}\left[\int_{[\![0,t]\!]} |\langle X_s|y\rangle|^p\cdot[|y|>1]\;\jmath_Z(dy,ds)\right]\right)^{1/p}$$

$$\le C_p\,|{}^{l}\!\nu|_p\cdot\left(\int\int_0^t |X|_s^p\,ds\,d\mathbb{P}\right)^{1/p}. \tag{4.6.27}$$

Putting (4.6.26) and (4.6.27) together yields, for $1 \le p \le 2$,

$$\left\| X{*}{}^{l}Z^{\star}_{t} \right\|_{L^p} \le 2^{p-1} C_p \,|{}^{l}\nu|_p \cdot \Big(\int \int_0^t |X|^p_s \, ds \, d\mathbb{P} \Big)^{1/p} . \tag{4.6.28}$$

If $p \ge 2$ and $|{}^{l}\nu|_p < \infty$, then we use inequality (4.6.26) to estimate the predictable finite variation part ${}^{l}\widehat{Z}$, and inequality (4.5.20) for the martingale part ${}^{l}\widetilde{Z}$. Writing down the resulting inequality together with (4.6.25) and (4.6.28) gives

$$\left\| X{*}{}^{l}Z^{\star}_{t} \right\|_{L^p} \le \begin{cases} C_p \cdot |{}^{l}\nu|_p \cdot \Big\| \Big(\int_0^t |X|^p_s \, ds \Big)^{1/p} \Big\|_{L^p} & \text{for } 0 < p \le 2, \\ C_p \cdot \max\limits_{\rho = 2, q} |{}^{l}\nu|_\rho \cdot \Big\| \Big(\int_0^t |X|^\rho_s \, ds \Big)^{1/\rho} \Big\|_{L^p} & \text{for } 2 \le p \le q, \end{cases} \tag{4.6.29}$$

where $$C_p = \begin{cases} 1 & \text{for } 0 < p \le 1, \\ C_p^{(4.2.4)} + 1 & \text{for } 1 < p \le 2, \\ C^{\diamond(4.5.11)} & \text{for } 2 \le p. \end{cases}$$

We leave to the reader the proof of the necessity part and the estimation of the universal constants $C_p^{(\rho)}(\mathfrak{t})$ in the following proposition – the sufficiency has been established above.

Proposition 4.6.16 *Let Z be a Lévy process with characteristic triple $\mathfrak{t} = (A, B, \nu)$ and let $0 < q < \infty$. Then Z is an L^q-integrator if and only if its Lévy measure ν has q^{th} moments away from zero:*

$$|{}^{l}\nu|_q \stackrel{\text{def}}{=} \sup_{|x'| \le 1} \Big(\int |\langle x' | y \rangle|^q \cdot [|y| > 1] \, \nu(dy) \Big)^{1/q} < \infty .$$

If so, then there is the following estimate for the stochastic integral of a previsible integrand X with respect to Z: for any stopping time T and any $p \in (0, q)$

$$\| |X{*}Z|^{\star}_{T} \|_{L^p} \le \max_{\rho = 1^{\diamond}, 2, p^{\diamond}} C_p^{(\rho)}(\mathfrak{t}) \cdot \Big\| \Big(\int_0^T |X_s|^\rho \, ds \Big)^{1/\rho} \Big\|_{L^p} . \tag{4.6.30}$$

Construction of Lévy Processes

Do Lévy processes exist? For all we know so far we could have been investigating the void situation in the previous 11 pages.

Theorem 4.6.17 *Let (A, B, ν) be a triple having the properties spelled out in theorem 4.6.9 on page 259. There exist a probability space $(\Omega, \mathcal{F}, \mathbb{P})$ and on it a Lévy process Z with characteristic triple (A, B, ν); any two such processes have the same law.*

Proof. The uniqueness of the law has been shown in lemma 4.6.4. We prove the existence piecemeal.

First the continuous martingale part $^{\tilde{c}}Z$. By exercise 3.9.6 on page 161, there is a probability space $(^{\tilde{c}}\Omega, ^{\tilde{c}}\mathcal{F}, ^{\tilde{c}}\mathbb{P})$ on which there lives a d-dimensional Wiener process with covariance matrix B. This Lévy process is clearly a good candidate for the continuous martingale part $^{\tilde{c}}Z$ of the prospective Lévy process Z.

Next, the idea leading to the "jump part" $^{j}Z \stackrel{\text{def}}{=} {}^{\tilde{s}}Z + {}^{l}Z$ is this: we construct the jump measure $\jmath_Z$ and, roughly (!), write $^{j}Z_t$ as $\int_{[\![0,t]\!]} y \, \jmath_Z(dy, ds)$. Exercise 4.6.14 (i) shows that $\jmath_Z$ should be a Poisson point process with intensity rate ν on $\mathbb{R}^d_*$. The construction following definition (3.10.9) on page 184 provides such a Poisson point process π. The proof of the theorem will be finished once the following facts have been established – they are left as an exercise in bookkeeping:

Exercise 4.6.18 π has intensity $\nu\times\lambda$ and is independent of $^{\tilde{c}}Z$.

$$ ^{k}Z_t \stackrel{\text{def}}{=} \int_{[0,t]} y \cdot [2^k < |y| \le 2^{k+1}] \, \pi(dy, ds) \,, \qquad k \in \mathbb{Z} \,, $$

has independent stationary increments and is continuous in q-mean for all $q < \infty$; it is an L^q-integrator, càdlàg after suitable modification. The sum

$$ ^{\tilde{s}}Z \stackrel{\text{def}}{=} \sum_{k<0} \widetilde{^{k}Z} $$

converges in the $\mathcal{I}^q$-norm and is a martingale and a Lévy process with characteristic triple $(0, 0, {}^{s}\nu)$, càdlàg after modification. Since the set $[1 < |y|]$ is ν-integrable, $\pi([1 < |y|] \times [0,t]) < \infty$ a.s. Now as π is a sum of point masses, this implies that

$$ ^{l}Z_t \stackrel{\text{def}}{=} \int_{[0,t]} y \cdot [1 < |y|] \, \pi(dy, ds) = \sum_{0 \le k} {}^{k}Z_t $$

is almost surely a finite sum of points in $\mathbb{R}^d_*$ and defines a Lévy process ^{l}Z with characteristic triple $(0, 0, {}^{l}\nu)$. Finally, setting $^{\hat{v}}Z_t \stackrel{\text{def}}{=} A \cdot t$,

$$ Z_t \stackrel{\text{def}}{=} {}^{\hat{v}}Z_t + {}^{\tilde{c}}Z_t + {}^{\tilde{s}}Z_t + {}^{l}Z_t $$

is a Lévy process with the given characteristic triple (A, B, ν), written in its canonical decomposition (4.6.21). ∎

Exercise 4.6.19 For $d = 1$ and $\nu = \delta_1$ the point mass at $1 \in \mathbb{R}$, Z_t is a Poisson process and has Doob–Meyer decomposition $Z_t = t + \widetilde{Z}_t$.

Feller Semigroup and Generator

We continue to consider the Lévy process Z in $\mathbb{R}^d$ with convolution semigroup $\mu_.$ of distributions. We employ them to define bounded linear operators[18]

$$ \phi \mapsto T_t\phi\,(y) \stackrel{\text{def}}{=} \mathbb{E}\big[\phi(z{+}Z_t)\big] = \int_{\mathbb{R}^n} \phi(y{+}z) \, \mu_t(dz) = \big(\overset{*}{\mu}_t \star \phi\big)(y) \,. \qquad (4.6.31) $$

[18] $\overset{*}{\phi}(x) \stackrel{\text{def}}{=} \phi(-x)$ and $\overset{*}{\mu}(\phi) \stackrel{\text{def}}{=} \mu(\overset{*}{\phi})$ define the ***reflections through the origin*** $\overset{*}{\phi}$ and $\overset{*}{\mu}$.

They are easily seen to form a conservative Feller semigroup on $C_0(\mathbb{R}^d)$ (see exercise A.9.4). Here are a few straightforward observations. $T_.$ is ***translation-invariant*** or ***commutes with translation***. This means the following: for any $\boldsymbol{a} \in \mathbb{R}^d$ and $\phi \in C(\mathbb{R}^d)$ define the ***translate*** $\phi_{\boldsymbol{a}}$ by $\phi_{\boldsymbol{a}}(\boldsymbol{x}) = \phi(\boldsymbol{x} - \boldsymbol{a})$; then $T_t(\phi_{\boldsymbol{a}}) = (T_t\phi)_{\boldsymbol{a}}$.

The dissipative generator $\mathcal{A} \stackrel{\text{def}}{=} dT_t/dt_{|t=0}$ of $T_.$ obeys the positive maximum principle (see item A.9.8) and is again translation-invariant: for $\phi \in \text{dom}(\mathcal{A})$ we have $\phi_a \in \text{dom}(\mathcal{A})$ and $\mathcal{A}\phi_a = (\mathcal{A}\phi)_a$. Let us compute $\mathcal{A}$. For instance, on a function ϕ in ***Schwartz space***[19] $\mathcal{S}$, Itô's formula (3.10.6) gives[4]

$$\phi(\boldsymbol{Z}_t^{\boldsymbol{z}}) = \phi(\boldsymbol{z}) + \int_{0+}^t \phi_{;\eta}(\boldsymbol{Z}_{s-}^{\boldsymbol{z}})\, d^s Z_s^\eta + \frac{1}{2}\int_{0+}^t \phi_{;\eta\theta}(\boldsymbol{Z}_{.-}^{\boldsymbol{z}})\, d^c[Z^\eta, Z^\theta]_s$$

$$+ \int_0^t \Big(\phi(\boldsymbol{Z}_{s-}^{\boldsymbol{z}} + \boldsymbol{y}) - \phi(\boldsymbol{Z}_{s-}^{\boldsymbol{z}}) - \phi_{;\eta}(\boldsymbol{Z}_{s-}^{\boldsymbol{z}}) \cdot y^\eta [|\boldsymbol{y}| \le 1]\Big)\, \jmath_{\boldsymbol{Z}}(d\boldsymbol{y}, ds)$$

$$= Mart_t + \int_0^t \boldsymbol{A}^\eta \phi_{;\eta}(\boldsymbol{Z}_s^{\boldsymbol{z}})\, ds + \frac{1}{2}\int_0^t B^{\eta\theta}\phi_{;\eta\theta}(\boldsymbol{Z}_s^{\boldsymbol{z}})\, ds$$

$$+ \int_0^t \int_{\mathbb{R}^d_*} \Big(\phi(\boldsymbol{Z}_s^{\boldsymbol{z}} + \boldsymbol{y}) - \phi(\boldsymbol{Z}_s^{\boldsymbol{z}}) - \phi_{;\eta}(\boldsymbol{Z}_s^{\boldsymbol{z}}) \cdot y^\eta [|\boldsymbol{y}| \le 1]\Big)\, \nu(d\boldsymbol{y})\, ds\,.$$

Here $\boldsymbol{Z}_.^{\boldsymbol{z}} \stackrel{\text{def}}{=} \boldsymbol{z} + \boldsymbol{Z}_.$, and part of the jump-measure integral was shifted into the first term using definition (4.6.10). The second equality comes from the identifications (4.6.11), (4.6.8), and (4.6.9). Since ϕ is bounded, the martingale part $Mart_.$ is an integrable martingale; so taking the expectation is permissible, and differentiating the result in t at $t = 0$ yields[4]

$$\mathcal{A}\phi(\boldsymbol{z}) = \boldsymbol{A}^\eta \phi_{;\eta}(\boldsymbol{z}) + \frac{1}{2} B^{\eta\theta}\phi_{;\eta\theta}(\boldsymbol{z}) \tag{4.6.32}$$

$$+ \int \Big(\phi(\boldsymbol{z}+\boldsymbol{y}) - \phi(\boldsymbol{z}) - \phi_{;\eta}(\boldsymbol{z})\, y^\eta [|\boldsymbol{y}| \le 1]\Big)\, \nu(d\boldsymbol{y})\,.$$

Example 4.6.20 Suppose the Lévy measure ν has finite mass, $|\nu| \stackrel{\text{def}}{=} \nu(1) < \infty$,

and, with $$\underline{\boldsymbol{A}} \stackrel{\text{def}}{=} \boldsymbol{A} - \int \boldsymbol{y}\, [|\boldsymbol{y}| \le 1]\, \nu(d\boldsymbol{y})\,,$$

set[4] $$\mathcal{D}\phi(\boldsymbol{z}) \stackrel{\text{def}}{=} \underline{A}^\eta \frac{\partial\phi(\boldsymbol{z})}{\partial z^\eta} + \frac{B^{\eta\theta}}{2}\frac{\partial^2\phi(\boldsymbol{z})}{\partial z^\eta \partial z^\theta}$$

and $$\mathcal{J}\phi(\boldsymbol{z}) \stackrel{\text{def}}{=} \int \Big(\phi(\boldsymbol{z}+\boldsymbol{y}) - \phi(\boldsymbol{z})\Big)\, \nu(d\boldsymbol{y}) = \overset{*}{\nu} \star \phi\,(\boldsymbol{z}) - |\nu|\phi(\boldsymbol{z})\,.$$

Then evidently $$\mathcal{A} = \mathcal{D} + \mathcal{J}\,. \tag{4.6.33}$$

---

[19] $\mathcal{S} = \{\phi \in C_b^\infty(\mathbb{R}^n) : \sup_x |x|^k \cdot |\phi(x)| < \infty \quad \forall k \in \mathbb{N}\}$.

In the case that ν is atomic, $\mathcal{J}$ is but a linear combination of difference operators, and in the general case we view it as a "continuous superposition" of difference operators. Now $\mathcal{J}$ is also a bounded linear operator on $C_0(\mathbb{R}^n)$, and so

$$T_t^{\mathcal{J}} = e^{t\mathcal{J}} \stackrel{\text{def}}{=} \sum (t\mathcal{J})^k / k!$$

is a contractive, in fact Feller, semigroup. With $\nu^0 \stackrel{\text{def}}{=} \delta_0$ and $\nu^k \stackrel{\text{def}}{=} \nu^{k-1} \star \nu$ denoting the k-fold convolution of ν with itself, it can be written

$$T_t^{\mathcal{J}} \phi = e^{-t|\nu|} \sum_{k=0}^{\infty} \frac{t^k \overset{*}{\nu}{}^k \star \phi}{k!} .$$

If $\mathcal{D} = 0$, then the characteristic triple is $(0, 0, \nu)$, with $\nu(1) < \infty$. A Lévy process with such a special characteristic triple is a ***compound Poisson process***. In the even more special case that $\mathcal{D} = 0$ and ν is the Dirac measure $\delta_{\boldsymbol{a}}$ at $\boldsymbol{a}$, the Lévy process is the ***Poisson process*** with jump $\boldsymbol{a}$. Then equation (4.6.33) reads $\mathcal{A}\phi(\boldsymbol{z}) = \phi(\boldsymbol{z} + \boldsymbol{a}) - \phi(\boldsymbol{z})$ and integrates to

$$T_t \phi(\boldsymbol{z}) = e^{-t} \sum_{k \geq 0} \frac{t^k \phi(\boldsymbol{z} + k\boldsymbol{a})}{k!} .$$

$\mathcal{D}$ can be exponentiated explicitly as well: with γ_{tB} denoting the Gaussian with covariance matrix B from definition A.3.50 on page 420, we have

$$T_t^{\mathcal{D}} \phi(\boldsymbol{z}) = \int \phi(\boldsymbol{z} + \boldsymbol{A}t + \boldsymbol{y}) \gamma_{tB}(d\boldsymbol{y}) = \overset{*}{\gamma}_{tB} \star \phi_{-\boldsymbol{A}t}(\boldsymbol{z}) .$$

In general, the operators $\mathcal{D}$ and $\mathcal{J}$ commute on Schwartz space, which is a core for either operator, and then so do the corresponding semigroups. Hence

$$T_t^{\mathcal{A}}[\phi] = T_t^{\mathcal{J}}\left[T_t^{\mathcal{D}}[\phi]\right] = T_t^{\mathcal{D}}\left[T_t^{\mathcal{J}}[\phi]\right] , \qquad \phi \in C_0(\mathbb{R}^d) .$$

Exercise 4.6.21 (A Special Case of the Hille–Yosida Theorem) Let $(\boldsymbol{A}, B, \nu)$ be a triple having the properties spelled out in theorem 4.6.9 on page 259. Then the operator $\mathcal{A}$ on $\mathcal{S}$ defined in equation (4.6.32) from this triple is conservative and dissipative on $C_0(\mathbb{R}^d)$. $\mathcal{A}$ is closable and $T_.$ is the unique Feller semigroup on $C_0(\mathbb{R}^d)$ with generator $\overline{\mathcal{A}}$, which has $\mathcal{S}$ for a core.

Repeated Footnotes: 196[3] 203[4] 235[10] 239[12] 249[14] 258[15] 261[16] 268[18]

5

Stochastic Differential Equations

We shall now solve the stochastic differential equation of section 1.1, which served as the motivation for the stochastic integration theory developed so far.

5.1 Introduction

The stochastic differential equation (1.1.9) reads[1]

$$X_t = C_t + \int_0^t F_\eta[X]\, dZ^\eta \,. \tag{5.1.1}$$

The previous chapters were devoted to giving a meaning to the dZ^η-integrals and to providing the tools to handle them. As it stands, though, the equation above still does not make sense; for the solution, if any, will be a right-continuous adapted process – but then so will the $F_\eta[X]$, and we cannot in general integrate right-continuous integrands. What does make sense in general is the equation

$$X_t = C_t + \int_0^t F_\eta[X]_{s-}\, dZ^\eta_s \tag{5.1.2}$$

or, equivalently, $$X = C + F_\eta[X]_{.-}{*}Z^\eta = C + \boldsymbol{F}[X]_{.-}{*}\boldsymbol{Z} \tag{5.1.3}$$

in the notation of definition 3.7.6, and with $\boldsymbol{F}[X]_{.-}$ denoting the left-continuous version of the matrix

$$\boldsymbol{F}[X] = \big(F_\eta[X]\big)_{\eta=1\ldots d} = \big(F^\nu_\eta[X]\big)^{\nu=1\ldots n}_{\eta=1\ldots d} \,.$$

In (5.1.3) now the integrands are left-continuous, and therefore previsible, and thus are integrable if not too big (theorem 3.7.17).

The intuition behind equation (5.1.2) is that $X_t = (X^\nu_t)^{\nu=1\ldots n}$ is a vector in $\mathbb{R}^n$ representing the state at time t of some system whose evolution is

[1] Einstein's convention is adopted throughout; it implies summation over the same indices in opposite positions, for instance, the η in (5.1.1).

driven by a collection $\boldsymbol{Z} = \{Z^\eta : 1 \le \eta \le d\}$ of scalar integrators. The F_η are the coupling coefficients, which describe the effect of the background noises Z^η on the change of X. $C = (C^\nu)^{\nu=1\ldots n}$ is the initial condition. Z^1 is typically time, so that $F_1[X]_{t-}\, dZ^1_t = F_1[X]_{t-}\, dt$ represents the systematic drift of the system.

First Assumptions on the Data and Definition of Solution

The ingredients X, C, F_η, Z^η of equation (5.1.2) are for now as general as possible, just so the equation makes sense. It is well to state their nature precisely:

(i) $Z^1, \ldots, Z^d$ are L^0-integrators. This is the very least one must assume to give meaning to the integrals in (5.1.2). It is no restriction to assume that $Z^\eta_0 = 0$, and we shall in fact do so. Namely, by convention $F_\eta[X]_{0-} = 0$, so that Z^η and $Z^\eta - Z^\eta_0$ has the same effect in driving X.

(ii) The ***coupling coefficients*** *F_η are random vector fields.* This means that each of them associates with every n-vector of processes[2] $X \in \mathfrak{D}^n$ another n-vector $F_\eta[X] \in \mathfrak{D}^n$. Most frequently they are ***markovian***; that is to say, there are ordinary ***vector fields*** $f_\eta : \mathbb{R}^n \to \mathbb{R}^n$ such that F_η is simply the composition with f_η: $F_\eta[X]_t = f_\eta(X_t) = f_\eta \circ X_t$. It is, however, not necessary, and indeed would be insufficient for the stability theory, to assume that the correspondences $X \mapsto F_\eta[X]$ are always markovian. Other coefficients arising naturally are of the form $X \mapsto A_\eta \cdot X$, where A_η is a bounded matrix-valued process[2] in $\mathfrak{D}^{n\times n}$. Genuinely non-markovian coefficients also arise in context with approximation schemes (see equation (5.4.29) on page 323.)

Below, Lipschitz conditions will be placed on $\boldsymbol{F}$ that will ensure automatically that the F_η are ***non-anticipating*** in the sense that at all stopping times T

$$F_\eta[X]_T = F_\eta[X^T]_T \ . \tag{5.1.4}$$

To paraphrase: "at any time T the value of the coupling coefficient is determined by the history of its argument up to that time." It is sometimes convenient to require that $F_\eta[0] = 0$ for $1 \le \eta \le d$. This is no loss of generality. Namely, X solves equation (5.1.3):

$$X = C + F_\eta[X]_{.-} {*} Z^\eta = C + \boldsymbol{F}[X]_{.-} {*} \boldsymbol{Z} \tag{5.1.5}$$

iff it solves

$$X = {}^0\!C + {}^0\!F_\eta[X]_{.-} {*} Z^\eta = {}^0\!C + {}^0\!\boldsymbol{F}[X]_{.-} {*} \boldsymbol{Z} \ , \tag{5.1.6}$$

where ${}^0\!C \overset{\text{def}}{=} C + \boldsymbol{F}[0]_{.-} {*} \boldsymbol{Z}$ is the adjusted initial condition and where ${}^0\!\boldsymbol{F}[X] \overset{\text{def}}{=} \boldsymbol{F}[X]_{.-} - \boldsymbol{F}[0]_{.-}$ is the adjusted coupling coefficient, which vanishes at $X = 0$. Note that ${}^0\!C$ will not, in general, be constant in time, even if C is.

---

[2] Recall that $\mathfrak{D}$ is the vector space of all càdlàg adapted processes and $\mathfrak{D}^n$ is its n-fold product, identified with the càdlàg $\mathbb{R}^n$-valued processes.

(iii) $C \in \mathfrak{D}^n$. In other words, C is an n-vector of adapted right-continuous processes with left limits. We shall refer to C as the ***initial condition***, despite the fact that it need not be constant in time. The reason for this generality is that in the stability theory of the system (5.1.2) time-dependent additive random inputs C appear automatically – see equation (5.2.32) on page 293. Also, in the form (5.1.6) the random input ${}^0\!C$ is time-dependent even if the C in the original equation is not.

(iv) ***X solves the stochastic differential equation on the stochastic interval*** $[\![0,T]\!]$ if the stopped process X^T belongs to $\mathfrak{D}^n$ and[3]

$$X^T = C^T + \boldsymbol{F}[X]_{\cdot-} * \boldsymbol{Z}^T$$

or, equivalently, if $\qquad X^T = {}^0\!C^T + {}^0\boldsymbol{F}[X]_{\cdot-} * \boldsymbol{Z}^T$;

we also say that X ***solves the equation up to time*** T, or that X is a ***strong solution*** on $[\![0,T]\!]$. In view of theorem 3.7.17 on page 137, there is no question that the indefinite integrals on the right-hand side exist, at least in the sense L^0. Clearly, if X solves our equation on both $[\![0,T_1]\!]$ and $[\![0,T_2]\!]$, then it also solves it on the union $[\![0,T_1 \vee T_2]\!]$. The supremum of the (classes of the) stopping times T so that X solves our equation on $[\![0,T]\!]$ is called the ***lifetime*** of the solution X and is denoted by $\zeta[X]$. If $\zeta[X] = \infty$, then X is a ***(strong) global solution.***

As announced in chapter 1, stochastic differential equations will be solved using Picard's iterative scheme. There is an analog of Euler's method that rests on compactness and works when the coupling coefficients are merely continuous. But the solution exists only in a weak sense (section 5.5), and to extract uniqueness in the presence of some positivity is a formidable task (ibidem, [104]).

We next work out an elementary example. Both the results and most of the arguments explained here will be used in the stochastic case later on.

Example: The Ordinary Differential Equation (ODE)

Let us recall how Picard's scheme, which was sketched on pages 1 and 5, works in the deterministic case, when there is only one driving term, time. The stochastic differential equation (5.1.2) is handled along the same lines; in fact, we shall refer below, in the stochastic case, to the arguments described here without carrying them out again. A few slightly advanced classical results are developed here in detail, for use in the approximation of Stratonovich equations (see page 321 ff.).

The ordinary differential equation corresponding to (5.1.2) is

$$x_t = c_t + \int_0^t f(x_s)\, ds \,. \tag{5.1.7}$$

[3] If X is only defined on $[\![0,T]\!]$, then X^T shall mean that extension of X which is constant on $[\![T,\infty)\!)$.

To paraphrase: "time drives the state in the direction of the vector field f." To solve it one defines for every path $x. : [0,\infty) \to \mathbb{R}^n$ the path $\mathfrak{u}[x.].$ by

$$\mathfrak{u}[x.]_t = c_t + \int_0^t f(x_s)\, ds\,, \qquad t \geq 0\,.$$

Equation (5.1.7) asks for a fixed point of the map $\mathfrak{u}$. Picard's method of finding it is to design a complete norm $\| \ \|$ on the space of paths with respect to which $\mathfrak{u}$ is ***strictly contractive***. This means that there is a $\gamma < 1$ such that for any two paths $x'_., x.$

$$\big\|\mathfrak{u}[x'_.] - \mathfrak{u}[x.]\big\| \leq \gamma \cdot \big\|x'_. - x.\big\|\,.$$

Such a norm is called a ***Picard norm*** for $\mathfrak{u}$ or for (5.1.7), and the least γ satisfying the previous inequality is the ***contractivity modulus*** of $\mathfrak{u}$ for that norm. The map $\mathfrak{u}$ will then have a fixed point, solution of equation (5.1.7) – see page 275 for how this comes about.

The strict contractivity of $\mathfrak{u}$ usually derives from the assumption that the vector field f on $\mathbb{R}^n$ is ***Lipschitz***; that is to say, that there is a constant $L < \infty$, the ***Lipschitz constant*** of f, such that[4]

$$\big|f(x'_t) - f(x_t)\big| \leq L \cdot \big|x'_t - x_t\big| \quad \forall\, t \geq 0\,. \tag{5.1.8}$$

For a path $x.$ define as usual its maximal path $|x|^\star_.$ by $|x|^\star_t = \sup_{s \leq t} |x_s|$, $t \geq 0$. Then inequality (5.1.8) has the consequence that for all t

$$\big|f(x'_.) - f(x.)\big|^\star_t \leq L \cdot \big|x'_. - x.\big|^\star_t\,. \tag{5.1.9}$$

If this is satisfied, then

$$\|x.\| = \|x.\|_M \overset{\text{def}}{=} \sup_{t>0} e^{-Mt} \cdot |x|^\star_t = \sup_{t>0} e^{-Mt} \cdot |x|_t \tag{5.1.10}$$

is a clever choice of the Picard norm sought, just as long as M is chosen strictly greater than L. Namely, inequality (5.1.9) implies

$$\big|f(x') - f(x)\big|^\star_t \leq Le^{Mt} \cdot e^{-Mt}\big|x' - x\big|^\star_t \leq Le^{Mt} \cdot \big\|x'_. - x.\big\|_M\,. \tag{5.1.11}$$

Therefore, multiplying the ensuing inequality

$$\begin{aligned}\Big|\mathfrak{u}[x'_.]_t - \mathfrak{u}[x.]_t\Big| &\leq \Big|\int_0^t f(x'_s) - f(x_s)\, ds\Big| \leq \int_0^t \big|f(x') - f(x)\big|^\star_s\, ds \\ &\leq L\cdot\big\|x'_. - x.\big\|_M \cdot \int_0^t e^{Ms}\, ds \leq \frac{L}{M} \cdot \big\|x'_. - x.\big\|_M \cdot e^{Mt}\end{aligned}$$

by e^{-Mt} and taking the supremum over t results in

$$\big\|\mathfrak{u}[x'_.] - \mathfrak{u}[x.]\big\|_M \leq \gamma \cdot \big\|x'_. - x.\big\|_M\,, \tag{5.1.12}$$

[4] $|\ |$ denotes the absolute value on $\mathbb{R}$ and also any of the usual and equivalent ℓ^p-norms $|\ |_p$ on $\mathbb{R}^k$, $\mathbb{R}^n$, $\mathbb{R}^{k\times n}$, etc., whenever p does not matter.

with $\gamma \overset{\text{def}}{=} L/M < 1$. Thus $\mathfrak{u}$ is indeed strictly contractive for $\| \ \|_M$.

The strict contractivity implies that $\mathfrak{u}$ has a unique fixed point. Let us review how this comes about. One picks an arbitrary starting path $x^{(0)}_.$, for instance $x^{(0)}_. \equiv 0$, and defines the Picard iterates by $x^{(n+1)}_. \overset{\text{def}}{=} \mathfrak{u}[x^{(n)}_.]_.$, $n = 0, 1, \ldots$. A simple induction on inequality (5.1.12) yields

$$\left\| x^{(n+1)}_. - x^{(n)}_. \right\|_M \le \gamma^n \cdot \left\| \mathfrak{u}[x^{(0)}_.] - x^{(0)}_. \right\|_M$$

and
$$\sum_{n=1}^{\infty} \left\| x^{(n+1)}_. - x^{(n)}_. \right\|_M \le \frac{\gamma}{1-\gamma} \cdot \left\| \mathfrak{u}[x^{(0)}_.] - x^{(0)}_. \right\|_M . \tag{5.1.13}$$

Provided that
$$\left\| \mathfrak{u}[x^{(0)}_.] - x^{(0)}_. \right\|_M < \infty , \tag{5.1.14}$$

the collapsing sum
$$x_. \overset{\text{def}}{=} x^{(1)}_. + \sum_{n=1}^{\infty} \left(x^{(n+1)}_. - x^{(n)}_. \right) = \lim_{n\to\infty} x^{(n)}_. \tag{5.1.15}$$

converges in the Banach space $\mathfrak{s}_M$ of paths $y_. : [0,\infty) \to \mathbb{R}^n$ that have $\|y_.\|_M < \infty$. Since $\mathfrak{u}[x_.] = \lim_n \mathfrak{u}[x^{(n)}] = \lim_n x^{(n+1)} = x$, the limit $x_.$ is a fixed point of $\mathfrak{u}$. If $x'_.$ is any other fixed point in $\mathfrak{s}_M$, then $\|x'_. - x_.\| \le \|\mathfrak{u}[x'_.] - \mathfrak{u}[x_.]\| \le \gamma \cdot \|x'_. - x_.\|$ implies that $\|x'_. - x_.\| = 0$: inside $\mathfrak{s}_M$, $x_.$ is the only solution of equation (5.1.7). There is a priori the possibility that there exists another solution in some space larger than $\mathfrak{s}_M$; generally some other but related reasoning is needed to rule this out.

Exercise 5.1.1 The norm $\| \ \|$ used above is by no means the only one that does the job. Setting instead $\|x_.\| \overset{\text{def}}{=} \int_0^\infty |x|^\star_t \cdot Me^{-Mt}\,dt$, with $M > L$, defines a complete norm on continuous paths, and $\mathfrak{u}$ is strictly contractive for it.

Let us discuss six consequences of the argument above — they concern only the action of $\mathfrak{u}$ on the Banach space $\mathfrak{s}_M$ and can be used literally or minutely modified in the general stochastic case later on.

5.1.2 General Coupling Coefficients To show the strict contractivity of $\mathfrak{u}$, only the consequence (5.1.9) of the Lipschitz condition (5.1.8) was used. Suppose that f is a map that associates with every path $x_.$ another one, but not necessarily by the simple expedient of evaluating a vector field at the values x_t. For instance, $f(x_.)_.$ could be the path $t \mapsto \phi(t, x_t)$, where ϕ is a measurable function on $\mathbb{R}_+ \times \mathbb{R}^n$ with values in $\mathbb{R}^n$, or it could be convolution with a fixed function, or even the composition of such maps. As long as inequality (5.1.9) is satisfied and $\mathfrak{u}[0]$ belongs to $\mathfrak{s}_M$, our arguments all apply and produce a unique solution in $\mathfrak{s}_M$.

5.1.3 The Range of $\mathfrak{u}$ Inequality (5.1.14) states that $\mathfrak{u}$ maps at least one – and then every – element of $\mathfrak{s}_M$ into $\mathfrak{s}_M$. This is another requirement on the system equation (5.1.7). In the present simple sure case it means that $\|c_. + \int_0^. f(0)\,ds\|_M < \infty$ and is satisfied if $c_. \in \mathfrak{s}_M$ and $f(0)_. \in \mathfrak{s}_M$, since $\|\int_0^. f(0)_s\,ds\|_M \le \|f(0)_.\|_M / M$.

5.1.4 Growth Control The arguments in (5.1.12)–(5.1.15) produce an a priori estimate on the growth of the solution $x_.$ in terms of the initial condition and u[0]. Namely, if the choice $x^{(0)}_. = 0$ is made, then equation (5.1.15) in conjunction with inequality (5.1.13) gives

$$\|x_.\|_M \le \left\|x^{(1)}_.\right\|_M + \frac{\gamma}{1-\gamma}\cdot\left\|x^{(1)}_.\right\|_M = \frac{1}{1-\gamma}\cdot\left\|c_. + \int_0^. f(0)_s\,ds\right\|_M . \tag{5.1.16}$$

The very structure of the norm $\| \;\|_M$ shows that $|x_t|$ grows at most exponentially with time t.

5.1.5 Speed of Convergence The choice $x^{(0)}_. \equiv 0$ for the zeroth iterate is popular but not always the most cunning. Namely, equation (5.1.15) in conjunction with inequality (5.1.13) also gives

$$\left\|x_. - x^{(1)}_.\right\|_M \le \frac{\gamma}{1-\gamma}\cdot\left\|x^{(1)}_. - x^{(0)}_.\right\|_M$$

and

$$\left\|x_. - x^{(0)}_.\right\|_M \le \frac{1}{1-\gamma}\cdot\left\|x^{(1)}_. - x^{(0)}_.\right\|_M .$$

We learn from this that if $x^{(0)}_.$ and the first iterate $x^{(1)}_.$ do not differ much, then both are already good approximations of the solution $x_.$. This innocent remark can be parlayed into various schemes for the pathwise solution of a stochastic differential equation (section 5.4). For the choice $x^{(0)}_. = c_.$ the second line produces an estimate of the deviation of the solution from the initial condition:

$$\|x_. - c_.\|_M \le \frac{1}{1-\gamma}\cdot\left\|\int_0^. f(c)_s\,ds\right\|_M \le \frac{1}{M(1-\gamma)}\cdot\|f(c)_.\|_M . \tag{5.1.17}$$

5.1.6 Stability Suppose f' is a second vector field on $\mathbb{R}^n$ that has the same Lipschitz constant L as f, and $c'_.$ is a second initial condition. If the corresponding map u$'$ maps 0 to $\mathfrak{s}_M$, then the differential equation $x'_t = c'_t + \int_0^t f'(x'_s)\,ds$ has a unique solution $x'_.$ in $\mathfrak{s}_M$. The difference $\delta_. \stackrel{\text{def}}{=} x'_. - x_.$ is easily seen to satisfy the differential equation

$$\delta_t = (c'-c)_t + \int_0^t g(\delta_s)\,ds ,$$

where $g : \delta \mapsto f'(\delta + x) - f(x)$ has Lipschitz constant L. Inequality (5.1.16) results in the estimates

$$\|x'_. - x_.\|_M \le \frac{1}{1-\gamma}\left\|(c'-c)_. + \int_0^. f'(x_s) - f(x_s)\,ds\right\|_M , \tag{5.1.18}$$

and

$$\|x_. - x'_.\|_M \le \frac{1}{1-\gamma}\left\|(c-c')_. + \int_0^. f(x'_s) - f'(x'_s)\,ds\right\|_M ,$$

reversing roles. Both exhibit neatly the dependence of the solution x on the ingredients c, f of the differential equation. It depends, in particular, Lipschitz-continuously on the initial condition c.

5.1.7 Differentiability in Parameters If initial condition and coupling coefficient of equation (5.1.7) depend differentiably on a parameter u that ranges over an open subset $U \subset \mathbb{R}^k$, then so does the solution. We sketch a proof of this, using the notation and terminology of definition A.2.45 on page 388. The arguments carry over to the stochastic case (section 5.3), and some of the results developed here will be used there.

Formally differentiating the equation $x[u]_t = c[u]_t + \int_0^t f(u, x[u]_s)\, ds$ gives

$$Dx[u]_t = \Big(Dc[u]_t + \int_0^t D_1 f(u, x[u]_s) ds\Big) + \int_0^t D_2 f(u, x[u]_s) \cdot Dx[u]_s ds \ . \quad (5.1.19)$$

This is a linear differential equation for an $n \times k$-matrix-valued path $Dx[u]_.$. It is a matter of a smidgen of bookkeeping to see that the remainder

$$Rx[v; u]_s \stackrel{\text{def}}{=} x[v]_s - x[u]_s - Dx[u]_s \cdot (v-u)$$

satisfies the linear differential equation

$$Rx[v; u]_t = \Big(Rc[v; u]_t + \int_0^t Rf\big(u, x[u]_s; v, x[v]_s\big)\, ds\Big) \quad (5.1.20)$$

$$+ \int_0^t D_2 f\big(u, x[u]_s\big) \cdot Rx[v; u]_s\, ds \ .$$

At this point we should show that[5] $\big|Rx[v; u]_t\big| = o(|v-u|)$ as $v \to u$; but the much stronger conclusion

$$\|Rx[v; u]_.\|_M = o(|v-u|) \quad (5.1.21)$$

seems to be in reach. Namely, if we can show that both

$$\|Rc[v; u]_.\| = o(|v-u|) \quad (5.1.22)$$

and

$$\|Rf(u, x[u]_.; v, x[v]_.)\| = o(|v-u|) \ ,$$

then (5.1.21) will follow immediately upon applying (5.1.16) to (5.1.20). Now (5.1.22) will hold if we simply require that $v \mapsto c[v]_.$, considered as a map from U to $\mathfrak{s}_M$, be uniformly differentiable; and this will for instance be the case if *the family* $\{c[\cdot]_t : t \geq 0\}$ *is uniformly equidifferentiable.*

Let us then require that $f : U \times \mathbb{R}^n \to \mathbb{R}^n$ be continuously differentiable with bounded derivative $Df = (D_1 f, D_2 f)$. Then the common coupling coefficient $D_2 f(u, x)$ of (5.1.19) and (5.1.20) is bounded by $L \stackrel{\text{def}}{=} \sup_{u,x} \big\|D_2 f(u, x)\big\|_{\mathbb{R}^{k \times n} \to \mathbb{R}^n}$ (see exercise A.2.46 (iii)), and for every $M > L$ the solutions $x[u]_.$, $u \in U$, lie in a common ball of $\mathfrak{s}_M$. One hopes of course that the coupling coefficient

$$F : U \times \mathfrak{s}_M \to \mathfrak{s}_M \quad , \quad F : (u, x_.) \mapsto f(u, x_.)$$

[5] For $o(\cdot)$ and $O(\cdot)$ see definition A.2.44 on page 388.

is differentiable. Alas, it is not, in general. We leave it to the reader (i) to fashion a counterexample and (ii) to establish that F *is weakly differentiable from* $U \times \mathfrak{s}_M$ *to* $\mathfrak{s}_M$, *uniformly on every ball* [Hint: see example A.2.48 on page 389]. When this is done a first application of inequality (5.1.18) shows that $u \to x[u].$ is Lipschitz from U to $\mathfrak{s}_M$, and a second one that $Rx[v;u]_t = o(|v-u|)$ on any ball of U: the solution $Dx[u].$ of (5.1.19) really is the derivative of $v \mapsto x[v].$ at u. This argument generalizes without too much ado to the stochastic case (see section 5.3 on page 298).

ODE: Flows and Actions

In the discussion of higher order approximation schemes for stochastic differential equations on page 321 ff. we need a few classical results concerning flows on $\mathbb{R}^n$ that are driven by different vector fields. They appear here in the form of propositions whose proofs are mostly left as exercises. We assume that the vector fields $f, g : \mathbb{R}^n \to \mathbb{R}^n$ appearing below are at least once differentiable with bounded and Lipschitz-continuous partial derivatives.

For every $x \in \mathbb{R}^n$ let $\xi. = \xi^f_.(x) = \xi[x, \cdot\,; f]$ denote the unique solution of $dX_t = f(X_t)\,dt$, $X_0 = x$, and extend to negative times t via $\xi^f_t(x) \overset{\text{def}}{=} \xi^{-f}_{-t}(x)$. Then

$$\frac{d\,\xi^f_t(x)}{dt} = f\big(\xi^f_t(x)\big) \quad \forall\, t \in (-\infty, +\infty) \ , \text{ with } \ \xi^f_0(x) = x\,. \tag{5.1.23}$$

This is the ***flow generated*** by f on $\mathbb{R}^n$. Namely,

Proposition 5.1.8 *(i) For every $t \in \mathbb{R}$, $\xi^f_t : x \mapsto \xi^f_t(x)$ is a Lipschitz-continuous map from $\mathbb{R}^n$ to $\mathbb{R}^n$, and $t \mapsto \xi^f_t$ is a group under composition; i.e., for all $s, t \in \mathbb{R}$*

$$\xi^f_{t+s} = \xi^f_t \circ \xi^f_s\,.$$

(ii) In fact, every one of the maps $\xi^f_t : \mathbb{R}^n \to \mathbb{R}^n$ is differentiable, and the $n\times n$-matrix $\big(D\xi^f_t[x]\big)^\mu_\nu \overset{\text{def}}{=} \partial\xi^{f\mu}_t(x)/\partial x^\nu$ of partial derivatives satisfies the following linear differential equation, obtained by formal differentiation of (5.1.23) in x:[6]

$$\frac{d\,D\xi^f_t[x]}{dt} = Df\big(\xi^f_t(x)\big)\cdot D\xi^f_t[x]\ ,\ D\xi^f_0[x] = I_n\,. \tag{5.1.24}$$

Consider now two vector fields f and g. Their ***Lie bracket*** is the vector field[7]

$$[f,g](x) \overset{\text{def}}{=} Df(x)\cdot g(x) - Dg(x)\cdot f(x)\,,$$

or[1]

$$[f,g]^\mu \overset{\text{def}}{=} f^\mu_{;\nu}g^\nu - g^\mu_{;\nu}f^\nu\,, \qquad \mu = 1,\ldots,n\,.$$

[6] I_n is the identity matrix on $\mathbb{R}^n$.

[7] Subscripts after semicolons denote partial derivatives, e.g., $f_{;\nu} \overset{\text{def}}{=} \frac{\partial f}{\partial x^\nu}$, $f_{;\mu\nu} \overset{\text{def}}{=} \frac{\partial^2 f}{\partial x^\nu \partial x^\mu}$. Einstein's convention is in force: summation over repeated indices in opposite positions is implied.

The fields f, g are said to ***commute*** if $[f,g] = 0$. ***Their flows*** ξ^f, ξ^g are said to ***commute*** if

$$\xi^g_t \circ \xi^f_s = \xi^f_s \circ \xi^g_t \,, \qquad s,t \in \mathbb{R} \,.$$

Proposition 5.1.9 *The flows generated by* f, g *commute if and only if* f, g *do.*

Proof. We shall prove only the harder implication, the sufficiency, which is needed in theorem 5.4.23 on page 326. Assume then that $[f,g] = 0$. The $\mathbb{R}^n$-valued path

$$\Delta_t \overset{\text{def}}{=} D\xi^f_t(x) \cdot g(x) - g\big(\xi^f_t(x)\big) \,, \qquad t \geq 0 \,,$$

satisfies
$$\frac{d\,\Delta_t}{dt} = Df\big(\xi^f_t(x)\big) \cdot D\xi^f_t(x) \cdot g(x) - Dg\big(\xi^f_t(x)\big) \cdot f\big(\xi^f_t(x)\big)$$

as $[f,g]=0$:
$$= Df\big(\xi^f_t(x)\big) \cdot D\xi^f_t(x) \cdot g(x) - Df\big(\xi^f_t(x)\big) \cdot g\big(\xi^f_t(x)\big)$$
$$= Df\big(\xi^f_t(x)\big) \cdot \Delta_t \,.$$

Since $\Delta_0 = 0$, the unique global solution of this linear equation is $\Delta_. \equiv 0$,

whence $D\xi^f_t(x) \cdot g(x) = g\big(\xi^f_t(x)\big) \quad \forall\, t \in \mathbb{R}\,.$ $\qquad (*)$

Fix a t and set
$$\Delta'_s \overset{\text{def}}{=} \xi^g_s\big(\xi^f_t(x)\big) - \xi^f_t\big(\xi^g_s(x)\big) \,, \qquad s \geq 0 \,.$$

Then
$$\frac{d\,\Delta'_s}{ds} = g\big(\xi^g_s\big(\xi^f_t(x)\big)\big) - D\xi^f_t\big(\xi^g_s(x)\big) \cdot g\big(\xi^g_s(x)\big)$$

by $(*)$:
$$= g\big(\xi^g_s\big(\xi^f_t(x)\big)\big) - g\big(\xi^f_t\big(\xi^g_s(x)\big)\big) \,,$$

and so
$$|\Delta'_s| \leq \int_0^s \Big| g\big(\xi^g_\sigma\big(\xi^f_t(x)\big)\big) - g\big(\xi^f_t\big(\xi^g_\sigma(x)\big)\big) \Big| \, d\sigma$$
$$\leq L \cdot \int_0^s |\Delta'_\sigma| \, d\sigma \,.$$

By lemma A.2.35 $\Delta'_. \equiv 0$: the flows ξ^f and ξ^g commute. ∎

Now let $f_1, \ldots, f_d$ be vector fields on $\mathbb{R}^n$ that have bounded and Lipschitz partial derivatives and that commute with each other; and let $\xi^{f_1}, \ldots, \xi^{f_d}$ be their associated flows. For any $\boldsymbol{z} = (z^1, \ldots, z^d) \in \mathbb{R}^d$ let

$$\Xi^{\boldsymbol{f}}[\,\cdot\,, \boldsymbol{z}] : \mathbb{R}^n \to \mathbb{R}^n$$

denote the composition in any order (see proposition 5.1.9) of $\xi^{f_1}_{z^1}, \ldots, \xi^{f_d}_{z^d}$.

Proposition 5.1.10 *(i)* $\Xi^{\boldsymbol{f}}$ *is a* ***differentiable action*** *of* $\mathbb{R}^d$ *on* $\mathbb{R}^n$ *in the sense that the maps* $\Xi^{\boldsymbol{f}}[\,\cdot\,, \boldsymbol{z}] : \mathbb{R}^n \to \mathbb{R}^n$ *are differentiable and*

$$\Xi^{\boldsymbol{f}}[\,\cdot\,, \boldsymbol{z} + \boldsymbol{z}'] = \Xi^{\boldsymbol{f}}[\,\cdot\,, \boldsymbol{z}] \circ \Xi^{\boldsymbol{f}}[\,\cdot\,, \boldsymbol{z}'] \,, \qquad \boldsymbol{z}, \boldsymbol{z}' \in \mathbb{R}^d \,.$$

$\Xi^{\boldsymbol{f}}$ *solves the initial value problem* $\Xi^{\boldsymbol{f}}[x, 0] = x$,

$$\frac{\partial\, \Xi^{\boldsymbol{f}}[x, \boldsymbol{z}]}{\partial z^\theta} = f_\theta\big(\Xi^{\boldsymbol{f}}[x, \boldsymbol{z}]\big) \,, \qquad \theta = 1, \ldots, d \,.$$

(ii) For a given $\boldsymbol{z} \in \mathbb{R}^d$ let $\boldsymbol{z}_. : [0, \tau] \to \mathbb{R}^d$ be any continuous and piecewise continuously differentiable curve that connects the origin with $\boldsymbol{z}$: $\boldsymbol{z}_0 = 0$ and $\boldsymbol{z}_\tau = \boldsymbol{z}$. Then $\Xi^{\boldsymbol{f}}[x, \boldsymbol{z}_.]$ is the unique (see item 5.1.2) solution of the initial value problem[1]

$$x_. = x + \int_0^. f_\eta(x_\sigma) \frac{dz_\sigma^\eta}{d\sigma}\, d\sigma \,, \tag{5.1.25}$$

and consequently $\Xi^{\boldsymbol{f}}[x, \boldsymbol{z}]$ equals the value x_τ at τ of that solution.

In particular, for fixed $\boldsymbol{z} \in \mathbb{R}^d$ set $\tau \stackrel{\text{def}}{=} |\boldsymbol{z}|$, $\boldsymbol{z}_\sigma \stackrel{\text{def}}{=} \sigma \boldsymbol{z}/\tau$, and $f \stackrel{\text{def}}{=} f_\eta\, z^\eta/\tau$. Then $\Xi^{\boldsymbol{f}}[x, \boldsymbol{z}]$ is the value x_τ at τ of the solution $x_.$ of the ordinary initial value problem $dx_\sigma = f(x_\sigma)\, d\sigma$, $x_0 = x$: $\Xi^{\boldsymbol{f}}[x, \boldsymbol{z}] = \xi[x, \tau; f]$.

ODE: Approximation

Picard's method constructs the solution $x_.$ of equation (5.1.7),

$$x_t = c + \int_0^t f(x_s)\, ds \,, \tag{5.1.26}$$

as an iterated limit. Namely, every Picard iterate $x_.^{(n)}$ is a limit, by virtue of being an integral, and $x_.$ is the limit of the $x_.^{(n)}$. As we have seen, this fact does not complicate questions of existence, uniqueness, and stability of the solution. It does, however, render nearly impossible the numerical computation of $x_.$.

There is of course a plethora of approximation schemes that overcome this conundrum, from Euler's method of little straight steps to complex multistep methods of high accuracy. We give here a common description of most single-step methods. This is meant to lay down a foundation for the generalization in section 5.4 to the stochastic case, and to provide the classical results needed there. We assume for simplicity's sake that the initial condition is a constant $c \in \mathbb{R}^n$.

A ***single-step method*** is a procedure that from a ***threshold*** or ***step size*** δ and from the coefficient[8] f produces both a partition $0 = t_0 < t_1 < \ldots$ of time and a function

$$(x, t) \mapsto \xi'[x, t] = \xi'[x, t; f]$$

that has the following purpose: when the approximate solution x'_t has been constructed for $0 \le t \le t_k$, then ξ' is used to extend it to $[0, t_{k+1}]$ via

$$x'_t \stackrel{\text{def}}{=} \xi'[x'_{t_k}, t - t_k] \quad \text{for} \quad t_k \le t \le t_{k+1} \,. \tag{5.1.27}$$

(ξ' is typically evaluated only once per step, to compute the next point $x'_{t_{k+1}}$.) If the approximation scheme at hand satisfies this description, then we talk about ***the method ξ'***. If the t_k are set in advance, usually by $t_k \stackrel{\text{def}}{=} \delta \cdot k$,

---

[8] If the coupling coefficient depends explicitely on time, apply the time rectification of example 5.2.6.

then it is a ***non-adaptive method***; if the next time t_{k+1} is determined from δ, the situation at time t_k, and its outlook $\xi'[x_{t_k}, t - t_k]$ at that time, then the method ξ' is ***adaptive***. For instance, Euler's method of little straight steps is defined by $\xi'[x, t] = x + f(x)t$; it can be made adaptive by defining the stop for the next iteration as

$$t_{k+1} \stackrel{\text{def}}{=} \inf\{t > t_k : |\xi'[x'_{t_k}, t - t_k] - x'_{t_k}| \geq \delta\} :$$

"proceed to the next calculation only when the increment is large enough to warrant a new computation."

For the remainder of this short discussion of numerical approximation a non-adaptive single-step method ξ' is fixed. We shall say that ξ' has ***local order*** r on the coupling coefficient f if there exists a constant $\underline{m}$ such that[4,9] for $t \geq 0$

$$\left|\xi'[c, t; f] - \xi[c, t; f]\right| \leq (|c|+1) \times (\underline{m}t)^r e^{\underline{m}t} . \tag{5.1.28}$$

The smallest such $\underline{m}$ will be denoted by $\underline{m}[f] = \underline{m}[f; \xi']$. If ξ' has local order r on all coefficients of class[10] C_b^∞, then it is simply said to have local order r. Inequality (5.1.28) will then usually hold on all coefficients of class C_b^k provided that k is sufficiently large.

We say that ξ' is of ***global order*** $r > 0$ on f if the difference of the exact solution $x_. = \xi[c, \cdot\,; f]$ of (5.1.26) from its approximate $x'_.$, made for the threshold δ via (5.1.27), satisfies an estimate $\|x'_. - x_.\|_{\overline{m}} = (|c|+1) \cdot O(\delta^r)$ for some constant $\overline{m} = \overline{m}[f; \xi']$. This amounts to saying that there exists a constant $\overline{b} = \overline{b}[f, \xi']$ such that

$$\left|x'_t - \xi[c, t; f]\right| \leq \overline{b} \cdot (|c|+1) \times \delta^r e^{\overline{m}t}$$

for all sufficiently small $\delta > 0$, all $t \geq 0$, and all $c \in \mathbb{R}^n$. Euler's method for example is locally of order 2 on $f \in C_b^2$, and therefore is globally of order 1 according to the following criterion, whose proof is left to the reader:

Criterion 5.1.11 *(i) Suppose that the growth of* ξ' *is limited by the inequality* $|\xi'[c, t]| \leq C' \cdot (|c|+1)\, e^{M't}$, *with* C', M' *constants. If* $|\xi'[c, t; f] - \xi[c, t; f]| = (|c|+1) \cdot O(t^r)$ *as* $t \to 0$, *then* ξ' *has local order* r. *The usual Runge–Kutta and Taylor methods meet this description.*

(ii) If the second-order mixed partials of ξ' *are bounded on* $\mathbb{R}^n \times [0, 1]$, *say by the constant* $L' < \infty$, *then, for* $\delta \leq 1$,

$$|\xi'[c', \cdot] - \xi'[c, \cdot]|_\delta^\star \leq e^{L'\delta} \cdot |c' - c| .$$

(iii) If ξ' *satisfies this inequality and has local order* r, *then it has global order* $r-1$.

[9] This definition is not entirely standard – see however criterion 5.1.11 (i). In the present formulation, though, the notion is best used in, and generalized to, stochastic equations.
[10] A function f is of class C^k if it has continuous partial derivatives of order $1, \ldots, k$. It is of class C_b^k if it and these partials are bounded. One also writes $f \in C_b^k$; f is of class C_b^∞ if it is of class C_b^k for all $k \in \mathbb{N}$.

Note 5.1.12 ***Scaling*** provides a cheap way to produce new single-step methods from old. Here is how. With the substitutions $s \stackrel{\text{def}}{=} \alpha\sigma$ and $y_\sigma \stackrel{\text{def}}{=} x_{\alpha\sigma}$, equation (5.1.26) turns into

$$y_\tau = c + \int_0^\tau \alpha f(y_\sigma)\, d\sigma .$$

Now $$|y_\tau - \xi'[c,\tau;\alpha f]| \le (|c|+1) \times (\underline{m}[\alpha f]\tau)^r \, e^{\underline{m}[\alpha f]\tau}$$

begets $$|x_t - \xi'[c,t/\alpha;\alpha f]| \le (|c|+1) \times \left(\frac{\underline{m}[\alpha f]}{\alpha}\cdot t\right)^r \times e^{\frac{\underline{m}[\alpha f]}{\alpha}\cdot t} .$$

That is to say, $\xi'_\alpha : (c,t;f) \mapsto \xi'[c,t/\alpha;\alpha f]$ is another single-step method of local order $r+1$ and constant $\underline{m}[\alpha f]/\alpha$. If this constant is strictly smaller than $\underline{m}[f]$, then the new method ξ'_α is clearly preferable to ξ'.

It is easily seen that Taylor methods and Runge–Kutta methods are in fact ***scale-invariant*** in the sense that $\xi'_\alpha = \xi'$ for all $\alpha > 0$. The constant $\underline{m}[f;\xi']$ due to its minimality then equals the infimum over α of the constants $\underline{m}[\alpha f]/\alpha$, and this evidently has the following effect whenever the method ξ' has local order r on f:

If ξ' is scale-invariant, then $\underline{m}[\alpha f;\xi'] = \underline{m}[f;\xi']/\alpha$ for all $\alpha > 0$.

5.2 Existence and Uniqueness of the Solution

We shall in principle repeat the arguments of pages 274–281 to solve and estimate the general stochastic differential equation (5.1.2), which we recall here: [1]

$$X = C + F_\eta[X]_{.-}{*}Z^\eta \quad \text{, or} \quad X = C + \boldsymbol{F}[X]_{.-}{*}\boldsymbol{Z} . \tag{5.2.1}$$

To solve this equation we consider of course the map $\mathfrak{U}$ from the vector space $\mathfrak{D}^n$ of càdlàg adapted $\mathbb{R}^n$-valued processes to itself that is given by

$$\mathfrak{U}[X] \stackrel{\text{def}}{=} C + \boldsymbol{F}[X]_{.-}{*}\boldsymbol{Z} .$$

The problem (5.2.1) amounts to asking for a fixed point of $\mathfrak{U}$. As in the example of the previous section, its solution lies in designing complete norms[11] with respect to which $\mathfrak{U}$ is strictly contractive, Picard norms.

Henceforth the minimal assumptions (i)–(iii) of page 272 are in effect. In addition we will require throughout this and the next three sections that $\boldsymbol{Z} = (Z^1,\dots,Z^d)$ is a local $L^q(\mathbb{P})$-integrator for some[12] $q \ge 2$ – except when this stipulation is explicitely rescinded on occasion.

[11] They are actually seminorms that vanish on evanescent processes, but we shall follow established custom and gloss over this point (see exercise A.2.31 on page 381).

[12] This requirement can of course always be satisfied provided we are willing to trade the given probability $\mathbb{P}$ for a suitable equivalent probability $\mathbb{P}'$ and to argue only up to some finite stopping time (see theorem 4.1.2). Estimates with respect to $\mathbb{P}'$ can be turned into estimates with respect to $\mathbb{P}$.

The Picard Norms

We will usually have selected a suitable exponent $p \in [2, q]$, and with it a norm $\| \ \|^*_{L^p}$ on random variables. To simplify the notation let us write

$$|F^\nu|_\infty \stackrel{\text{def}}{=} \sup_{1\le\eta\le d} |F^\nu_\eta| \, , \text{ and } |F|_{\infty p} \stackrel{\text{def}}{=} \Big(\sum_{1\le\nu\le n} |F^\nu|^p_\infty \Big)^{1/p} \tag{5.2.2}$$

for the size of a d-tuple $F = (F_1, \ldots, F_d)$ of n-vectors. Recall also that the maximal process of a vector $X \in \mathfrak{D}^n$ is the vector composed of the maximal functions of its components[13].

In the ordinary differential equation of page 274 both driver and controller were the same, to wit, time. In the presence of several drivers a common controller should be found and used to clock a common time transformation. THE strictly increasing previsible controller $\Lambda = \Lambda^{\langle q\rangle}[Z]$ of theorem 4.5.1 and THE associated continuous time transformation $T^\bullet$ by predictable stopping times

$$T^\lambda \stackrel{\text{def}}{=} \inf\{t : \Lambda_t \ge \lambda\} \tag{5.2.3}$$

of remark 4.5.2 come to mind.[14] Since $\Lambda_t \ge \alpha \cdot t$, the T^λ are bounded, so that a negligible set in $\mathcal{F}_{T^\lambda}$ is nearly empty. Since $\Lambda_t < \infty \quad \forall\, t$, the T^λ increase without bound as $\lambda \to \infty$.

We use THE time transformation and (5.2.2) to define, for any $p \in [2, q]$ and any $M \ge 1$, functionals $\| \ \|_{p,M}$ and $\| \ \|^\star_{p,M}$, the ***Picard norms***[11]

on vectors[2] $\qquad X = (X^1, \ldots, X^n) \in \mathfrak{D}^n$

by $\qquad \|X\|_{p,M} \stackrel{\text{def}}{=} \sup_{\lambda>0} e^{-M\lambda} \cdot \Big\| |X_{T^\lambda -}|_p \Big\|^*_{L^p(\mathbb{P})} ,$

which is less than

$$\|X\|^\star_{p,M} \stackrel{\text{def}}{=} \sup_{\lambda>0} e^{-M\lambda} \cdot \Big\| |X^\star_{T^\lambda -}|_p \Big\|^*_{L^p(\mathbb{P})} . \tag{5.2.4}$$

Then we set $\qquad \mathfrak{S}^n_{p,M} \stackrel{\text{def}}{=} \Big\{ X \in \mathfrak{D}^n : \|X\|_{p,M} < \infty \Big\} ,$

which clearly contains $\qquad \mathfrak{S}^{\star n}_{p,M} \stackrel{\text{def}}{=} \Big\{ X \in \mathfrak{D}^n : \|X\|^\star_{p,M} < \infty \Big\} .$

For the meaning of $\| \ \|^*_{L^p(\mathbb{P})}$ see item A.8.21 on page 452; it is used instead of $\| \ \|_{L^p(\mathbb{P})}$ to avoid worries about finiteness and measurability of its argument. Definition (5.2.4) is a straighforward generalization of (5.1.10).

The next lemma illuminates the role of the functionals (5.2.4) in the control of the driver Z:

---

13 $X^\star \stackrel{\text{def}}{=} (|X^1|^\star, \ldots |X^n|^\star)$ has size $|X|^\star_p \le |X^\star|_p \le n^{1/p} \cdot |X|^\star_p$.

14 This is disingenuous; Λ and $T^\bullet$ were of course specifically designed for the problem at hand.

Lemma 5.2.1 *(i)* $\|\ \|_{p,M}$ *and* $\|\ \|^\star_{p,M}$ *are seminorms on* $\mathfrak{S}^n_{p,M}$ *and* $\mathfrak{S}^{\star n}_{p,M}$*, respectively.* $\mathfrak{S}^{\star n}_{p,M}$ *is complete under* $\|\ \|^\star_{p,M}$. *A process* X *has Picard norm*[11] $\|X\|^\star_{p,M} = 0$ *if and only if it is evanescent.*

(ii) $\|X\|^\star_{p,M}$ *increases as* p *increases and as* M *decreases and*

$$\|X\|^\star_{p,M} = \sup\left\{\|X\|^\star_{p^\circ,M^\circ} : p^\circ < p, M^\circ > M\right\}, \qquad X \in \mathfrak{D}^n .$$

(iii) On any d-tuple $\boldsymbol{F} = (F_1, \ldots, F_d)$ *of adapted càdlàg* $\mathbb{R}^n$*-valued processes*

$$\|\boldsymbol{F}_{.-}{*}\boldsymbol{Z}\|^\star_{p,M} \le \frac{C_p^{\diamond(4.5.1)}}{M^{1/p^\diamond}} \tag{5.2.5}$$

Proof. Parts (i) and (ii) are quite straightforward and are left to the reader.

(iii) Pick a $\mu > 0$ and let S be a stopping time that is strictly prior to T^μ on $[T^\mu > 0]$. Fix a $\nu \in \{1, \ldots, n\}$. With that, theorem 4.5.1 gives

$$\left\|(\boldsymbol{F}^\nu_{.-}{*}\boldsymbol{Z})^\star_S\right\|^*_{L^p} \le C_p^{\diamond(4.5.1)} \cdot \max_{\rho=1^\diamond,p^\diamond} \left\|\left(\int_0^S |\boldsymbol{F}^\nu_{s-}|^\rho_\infty \, d\Lambda_s\right)^{1/\rho}\right\|^*_{L^p(\mathbb{P})}$$

by theorem 2.4.7: $$\le C_p^\diamond \cdot \max_\rho \left\|\left(\int_{[T^\lambda \le S]} |\boldsymbol{F}^\nu_{T^\lambda-}|^\rho_\infty \, d\lambda\right)^{1/\rho}\right\|^*_{L^p(\mathbb{P})}$$

(!) $$\le C_p^\diamond \cdot \max_\rho \left\|\left(\int_{[\lambda \le \mu]} |\boldsymbol{F}^\nu_{T^\lambda-}|^\rho_\infty \, d\lambda\right)^{1/\rho}\right\|^*_{L^p(\mathbb{P})} . \qquad (!)$$

The previsibility of the controller Λ is used in an essential way at (!). Namely, $T^\lambda \le T^\mu$ does not in general imply $\lambda \le \mu$ – in fact, λ could exceed μ by as much as the jump of Λ at T^μ. However, $T^\lambda < T^\mu$ does imply $\lambda < \mu$, and we produce this inequality by calculating only up to a stopping time S *strictly prior* to T^μ. That such exist arbitrarily close to T^μ is due to the previsibility of Λ, which implies the predictability of T^μ (exercise 3.5.19). We use this now: letting the S above run through a sequence announcing T^μ yields

$$\left\|(\boldsymbol{F}^\nu_{.-}{*}\boldsymbol{Z})^\star_{T^\mu-}\right\|^*_{L^p(\mathbb{P})} \le C_p^\diamond \cdot \max_{\rho=1^\diamond,p^\diamond} \left\|\left(\int_0^\mu |\boldsymbol{F}^\nu_{T^\lambda-}|^\rho_\infty \, d\lambda\right)^{1/\rho}\right\|^*_{L^p(\mathbb{P})} . \tag{5.2.6}$$

Applying the ℓ^p-norm $|\ |_p$ to these n-vectors and using Fubini produces

$$\left\|\left|(\boldsymbol{F}_{.-}{*}\boldsymbol{Z})^\star_{T^\mu-}\right|_p\right\|^*_{L^p(\mathbb{P})} \le C_p^\diamond \cdot \max_\rho \left\|\left|\left(\int_0^\mu |\boldsymbol{F}_{T^\lambda-}|^\rho_\infty \, d\lambda\right)^{1/\rho}\right|_p\right\|^*_{L^p(\mathbb{P})}$$

by exercise A.3.29: $$\le C_p^\diamond \cdot \max_\rho \left(\int_0^\mu \left\||\boldsymbol{F}_{T^\lambda-}|_{\infty p}\right\|^{*\rho}_{L^p(\mathbb{P})} d\lambda\right)^{1/\rho}$$

by definition (5.2.4): $$\le C_p^\diamond \cdot \max_\rho \left(\int_0^\mu \left\||\boldsymbol{F}|_\infty\right\|^\rho_{p,M} \cdot e^{M\lambda\rho} \, d\lambda\right)^{1/\rho}$$

$$= C_p^\diamond \cdot \left\||\boldsymbol{F}|_\infty\right\|_{p,M} \cdot \max_\rho \left(\int_0^\mu e^{M\lambda\rho} \, d\lambda\right)^{1/\rho}$$

with a little calculus: $$< C_p^\diamond \cdot \left\| |F|_\infty \right\|_{p,M} \cdot \max_{\rho=1^\diamond,p^\diamond} \frac{e^{M\mu}}{(M\rho)^{1/\rho}}$$

since $M \ge 1 \le \rho^{1/\rho}$: $$\le \frac{C_p^\diamond \, e^{M\mu}}{M^{1/p^\diamond}} \cdot \left\| |F|_\infty \right\|_{p,M} .$$

Multiplying this by $e^{-M\mu}$ and taking the supremum over $\mu > 0$ results in inequality (5.2.5).

Note here that the use of a sequence announcing T^μ provides information only about the left-continuous version of $(F_{\cdot-}{*}Z)^\star$ at T^μ; this explains why we chose to define $\|X\|_{p,M}$ and $\|X\|^\star_{p,M}$ using the left-continuous versions $X_{\cdot-}$ and $X^\star_{\cdot-}$ rather than $X_{\cdot}$ and $X^\star_{\cdot}$ itself. However, if Z is quasi-left-continuous and therefore Λ is continuous (exercise 4.5.16) and $T^{\cdot}$ strictly increasing,[15] then we can define $\| \ \|_{p,M}$ and $\| \ \|^\star_{p,M}$ using $X^\star_{\cdot}$ itself, with inequality (5.2.5) persisting. In fact, the computation leading to this inequality then simplifies a little, since we can take $S = T^\mu$ to start with. ∎

Here are further useful facts about the functionals $\| \ \|_{p,M}$ and $\| \ \|^\star_{p,M}$:

Exercise 5.2.2 Let $\Delta_\eta \in \mathfrak{L}$ with $|\Delta_\eta| \le \delta \in \mathbb{R}_+$. Then $\|\Delta_\eta{*}Z^\eta\|^\star_{p,M} \le \delta C_p^\diamond / eM$.

Exercise 5.2.3 Let $p \in [2, q]$ and $M > 0$. The seminorm

$$X \mapsto \|X\|^{\cdot}_{p,M} \stackrel{\text{def}}{=} \Big(\sum_{1\le\nu\le n} \int_0^\infty \|X^{\nu\star}_{T^\lambda-}\|^{*p}_{L^p} \, Me^{-M\lambda} d\lambda \Big)^{1/p}$$

is complete on $$\mathfrak{S}^{\cdot}_{p,M} \stackrel{\text{def}}{=} \Big\{ X \in \mathfrak{D}^n : \|X\|^{\cdot}_{p,M} < \infty \Big\}$$

and satisfies the following analog of inequality (5.2.5):

$$\|F_{\cdot-}{*}Z\|^{\cdot}_{p,M} \le C_p^{\diamond(4.5.1)} \Big(\frac{1}{M^{1/p}} \vee \frac{p}{M} \Big) \cdot \left\| |F|_\infty \right\|^{\cdot}_{p,M} .$$

Furthermore, $p \mapsto \|X\|^{\cdot}_{p,M}$ is increasing, $M \mapsto \|X\|^{\cdot}_{p,M}$ is decreasing, and

$$\|X\|^{\cdot}_{p,M'} \le \Big(\frac{M'}{M' - Mp} \Big)^{1/p} \|X\|^\star_{p,M} \qquad \text{for } M' > Mp ,$$

and $$\|X\|^\star_{p,M} \le \|X\|^{\cdot}_{p,M'} \qquad \text{for } M' \le Mp .$$

The seminorms $\| \ \|^{\cdot}_{p,M}$ are just as good as the $\| \ \|^\star_{p,M}$ for the development.

Lipschitz Conditions

As above in section 5.1 on ODEs, Lipschitz conditions on the coupling coefficient F are needed to solve (5.2.1) with Picard's scheme. A rather restrictive

[15] This happens for instance when Z is a Lévy process or the solution of a stochastic differential equation driven by a Lévy process (see exercise 5.2.17).

one is this ***strong Lipschitz condition***: there exists a constant $L < \infty$ such that for any two $X, Y \in \mathfrak{D}^n$

$$\left|(\boldsymbol{F}[Y] - \boldsymbol{F}[X])_{.-}\right|_{\infty p} \le L \cdot \left|Y - X\right|_p \tag{5.2.7}$$

up to evanescence. It clearly implies the slightly weaker Lipschitz condition

$$\left|(\boldsymbol{F}[Y] - \boldsymbol{F}[X])_{.-}\right|_{\infty p} \le L \cdot \left|(Y - X)^\star\right|_p , \tag{5.2.8}$$

which is to say that at any finite stopping time T almost surely

$$\Big(\sum_\nu \sup_\eta \left|(F^\nu_\eta[Y] - F^\nu_\eta[X])_{T-}\right|^p\Big)^{1/p} \le L \cdot \Big(\sum_\nu \sup_{s \le T} |Y^\nu - X^\nu|^p_s\Big)^{1/p} .$$

These conditions are independent of p in the sense that if one of them holds for one exponent $p \in (0, \infty]$, then it holds for any other, except that the Lipschitz constant may change with p. Inequality (5.2.8) implies the following rather much weaker ***p*-*mean Lipschitz condition:***

$$\Big\| \left|(\boldsymbol{F}[Y] - \boldsymbol{F}[X])_{T-}\right|_{\infty p} \Big\|_{L^p(\mathbb{P})} \le L \cdot \Big\| \left|(Y - X)^\star_t\right|_p \Big\|_{L^p(\mathbb{P})} , \tag{5.2.9}$$

which in turn implies that at any predictable stopping time T

$$\Big\| \left|(\boldsymbol{F}[Y] - \boldsymbol{F}[X])_{T-}\right|_{\infty p} \Big\|_{L^p(\mathbb{P})} \le L \cdot \Big\| \left|(Y - X)^\star_{T-}\right|_p \Big\|_{L^p(\mathbb{P})} , \tag{5.2.10}$$

in particular for the stopping times T^λ of THE time transformation (5.2.3)

$$\Big\| \left|(\boldsymbol{F}[Y] - \boldsymbol{F}[X])_{T^\lambda -}\right|_{\infty p} \Big\|_{L^p(\mathbb{P})} \le L \cdot \Big\| \left|(Y - X)^\star_{T^\lambda -}\right|_p \Big\|_{L^p(\mathbb{P})} , \tag{5.2.11}$$

whenever $X, Y \in \mathfrak{D}^n$. Inequality (5.2.10) can be had simply by applying (5.2.9) to a sequence that announces T and taking the limit. Finally, multiplying (5.2.11) with $e^{-M\lambda}$ and taking the supremum over $\lambda > 0$ results in

$$\Big\| \left|\boldsymbol{F}[Y] - \boldsymbol{F}[X]\right|_\infty \Big\|_{p,M} \le L \cdot \big\| X - Y \big\|^\star_{p,M} \tag{5.2.12}$$

for $X, Y \in \mathfrak{D}^n$. This is the form in which any Lipschitz condition will enter the existence, uniqueness, and stability proofs below. If it is satisfied, we say that $\boldsymbol{F}$ is ***Lipschitz in the Picard norm.***[11] See remark 5.2.20 on page 294 for an example of a coupling coefficient that is Lipschitz in the Picard norm without being Lipschitz in the sense of inequality (5.2.8).

The adjusted coupling coefficient ${}^0\boldsymbol{F}$ of equation (5.1.6) shares any or all of these Lipschitz conditions with $\boldsymbol{F}$, and any of them guarantees the non-anticipating nature (5.1.4) of $\boldsymbol{F}$ and of ${}^0\boldsymbol{F}$, at least at the stopping times entering their definition.

Here are a few examples of coupling coefficients that are strongly Lipschitz in the sense of inequality (5.2.7) and therefore also in the weaker senses of (5.2.8)–(5.2.12). The verifications are quite straightforward and are left to the reader.

Example 5.2.4 Suppose the F_η are ***markovian***, that is to say, are of the form $F_\eta[X] = f_\eta \circ X$, with $f_\eta : \mathbb{R}^n \to \mathbb{R}^n$ vector fields. If the f_η are ***Lipschitz***, meaning that[4]

$$\big| f_\eta(x) - f_\eta(y) \big| \le L_\eta \cdot |x - y| \tag{5.2.13}$$

for some constants L_η and all $x, y \in \mathbb{R}^n$, then $\boldsymbol{F}$ is Lipschitz in the sense of (5.2.7). This will be the case in particular when the partial derivatives[16] $f^\mu_{\eta;\nu}$ exist and are bounded for every η, ν, μ. Most Lipschitz coupling coefficients appearing at present in physical models, financial models, etc., are of this description. They are used when only the current state X_t of X influences its evolution, when information of how it got there is irrelevant. Markovian coefficients are also called ***autonomous***.

Example 5.2.5 In a slight generalization, we call F_η an ***instantaneous coupling coefficient*** if there exists a Borel vector field $f_\eta : [0, \infty) \times \mathbb{R}^n \to \mathbb{R}^n$ so that $F_\eta[X]_s = f_\eta(s, X_s)$ for $s \in [0, \infty)$ and $X \in \mathfrak{D}^n$. If f_η is ***equi-Lipschitz in its spacial arguments***, meaning that[4]

$$\sup_s \big| f_\eta(s, x) - f_\eta(s, y) \big| \le L_\eta \cdot |x - y|$$

for some constants L_η and all $x, y \in \mathbb{R}^n$, then $\boldsymbol{F}$ is strongly Lipschitz in the sense of (5.2.7). A markovian coupling coefficient clearly is instantaneous.

Example 5.2.6 (Time Rectification of Instantaneous Equations) The two previous examples are actually not too far apart. Suppose the instantaneous coefficients $(s, x) \mapsto f_\eta(s, x)$ happen to be Lipschitz in all of their arguments, which is to say

$$\big| f_\eta(s, x) - f_\eta(t, y) \big| \le L_\eta \cdot \big| (s, x) - (t, y) \big| \,. \tag{5.2.14}$$

Then we expand the driver by giving it the zeroth component $Z^0_t = t$ and set $\quad \boldsymbol{Z}^{\vdash} \overset{\text{def}}{=} (Z^0, Z^1, \ldots, Z^d)\,,$

expand the state space from $\mathbb{R}^n$ to $\mathbb{R}^{n+1} = (-\infty, \infty) \times \mathbb{R}^n$,

setting $\quad X^{\vdash} \overset{\text{def}}{=} (X^0, X^1, \ldots, X^n)\,,$

expand the initial state to

$$C^{\vdash} \overset{\text{def}}{=} (0, C^1, \ldots, C^n)\,,$$

and consider the expanded *and now markovian* differential equation

$$\begin{pmatrix} X^0 \\ X^1 \\ \vdots \\ X^n \end{pmatrix} = \begin{pmatrix} C^0 \\ C^1 \\ \vdots \\ C^n \end{pmatrix} + \begin{pmatrix} 1 & 0 & \cdots & 0 \\ 0 & f^1_1(X^{\vdash})_- & \cdots & f^1_d(X^{\vdash})_- \\ \vdots & \vdots & \ddots & \vdots \\ 0 & f^n_1(X^{\vdash})_- & \cdots & f^n_d(X^{\vdash})_- \end{pmatrix} * \begin{pmatrix} Z^0 \\ Z^1 \\ \vdots \\ Z^d \end{pmatrix}$$

or $$X^{\vdash} = C^{\vdash} + \boldsymbol{f}^{\vdash}(X^{\vdash})_- * \boldsymbol{Z}^{\vdash}\,,$$

[16] Subscripts after semicolons denote partial derivatives, e.g., $f_{\eta;\nu} \overset{\text{def}}{=} \frac{\partial f_\eta}{\partial x^\nu}$, $f_{\eta;\mu\nu} \overset{\text{def}}{=} \frac{\partial^2 f_\eta}{\partial x^\nu \partial x^\mu}$.

in obvious notation. The first line of this equation simply reads $X^0_t = t$; the others combine to the original equation $X_t = C_t + \sum_{\eta=1}^{d} \int_0^t f_\eta(s, X_s)\, dZ^\eta_s$. In this way it is possible to generalize very cheaply results about markovian stochastic differential equations to instantaneous stochastic differential equations, at least in the presence of inequality (5.2.14).

Example 5.2.7 We call F an ***autologous coupling coefficient*** if there exists an adapted map[17] $f : \mathscr{D}^n \to \mathscr{D}^n$ so that $F[X].(\omega) = f\big[X.(\omega)\big]$ for nearly all $\omega \in \Omega$. We say that such f is ***Lipschitz*** with constant L if[4] for any two paths $x., y. \in \mathscr{D}^n$ and all $t \geq 0$

$$\big| f\big[x.\big] - f\big[y.\big] \big|_{t-} \leq L \cdot |x. - y.|^\star_t \,. \tag{5.2.15}$$

In this case the coupling coefficient F is evidently strongly Lipschitz, and thus is Lipschitz in any of the Picard norms as well. Autologous[18] coupling coefficients might be used to model the influence of the whole past of a path of $X.$ on its future evolution. Instantaneous autologous coupling coefficients are evidently autonomous.

Example 5.2.8 A particular instance of an autologous Lipschitz coupling coefficient is this: let $v = (v^\nu_\mu)$ be an $n{\times}n$-matrix of deterministic càdlàg functions on the half-line that have uniformly bounded total variation, and let v act by convolution:

$$F^\nu[X]_t(\omega) \stackrel{\text{def}}{=} \int_{-\infty}^{\infty} \textstyle\sum_\mu X^\mu_{t-s}(\omega)\, dv^\nu_{\mu s} = \displaystyle\int_0^t \textstyle\sum_\mu X^\mu_{t-s}(\omega)\, dv^\nu_{\mu s} \,.$$

(As usual we think $X_s = v_s = 0$ for $s < 0$.) Such autologous Lipschitz coupling coefficients could model systematic influences of the past of X on its evolution that abate as time elapses. Technical stock analysts who believe in trends might use such coupling coefficients to model the evolution of stock prices.

Example 5.2.9 We call F a ***randomly autologous coupling coefficient*** if there exists a function $f : \Omega \times \mathscr{D}^n \to \mathscr{D}^n$, adapted to $\mathcal{F}. \otimes \mathcal{F}.[\mathscr{D}^n]$, such that $F[X].(\omega) = f\big[\omega, X.(\omega)\big]$ for nearly all $\omega \in \Omega$. We say that such f is ***Lipschitz*** with constant L if[4] for any two paths $x., y. \in \mathscr{D}^n$ and all $t \geq 0$

$$\big| f\big[\omega, x.\big] - f\big[\omega, y.\big] \big|_{t-} \leq L \cdot |x. - y.|^\star_t \tag{5.2.16}$$

at nearly every $\omega \in \Omega$. In this case the coupling coefficient F is evidently strongly Lipschitz, and thus is Lipschitz in any of the Picard norms as well. Here are several examples of randomly autologous coupling coefficients:

---

[17] See item 2.3.8. We may equip path space with its canonical or its natural filtration, *ad libitum*; consistency behooves, however.

[18] To the choice of word: if at the time of the operation the patient's own blood is used, usually collected previously on many occasions, then one talks of an autologous blood supply.

Example 5.2.10 Let $D = (D^\nu_\mu)$ be an $n \times n$-matrix of uniformly bounded adapted càdlàg processes. Then $X \mapsto \sum_\mu D^\nu_\mu X^\mu$ is Lipschitz in the sense of (5.2.7). Such coupling coefficients appear automatically in the stability theory of stochastic differential equations, even of those that start out with markovian coefficients (see section 5.3 on page 298). They would generally be used to model random influences that the past of X has on its future evolution.

Example 5.2.11 Let $V = (V^\nu_\mu)$ be a matrix of adapted càdlàg processes that have bounded total variation. Define F by

$$F^\nu[X]_t(\omega) \stackrel{\text{def}}{=} \int_0^t \textstyle\sum_\mu X^\mu_s(\omega)\, dV^\nu_{\mu s}(\omega) .$$

Such F is evidently randomly autologous and might again be used to model random influences that the past of X has on its future evolution.

Example 5.2.12 We call F an ***endogenous coupling coefficient*** if there exists an adapted function[17] $f : \mathscr{D}^d \times \mathscr{D}^n \to \mathscr{D}^n$ so that

$$F[X]_.(\omega) = f\big[\boldsymbol{Z}_.(\omega), X_.(\omega)\big]$$

for nearly all $\omega \in \Omega$. We say that such f is ***Lipschitz*** with constant L if[4] for any two paths $x_., y_. \in \mathscr{D}^n$ and all $\boldsymbol{z}_. \in \mathscr{D}^d$ and $t \ge 0$

$$\big| f\big[\boldsymbol{z}_., x_.\big] - f\big[\boldsymbol{z}_., y_.\big] \big|_{t-} \le L \cdot |x_. - y_.|^\star_t . \tag{5.2.17}$$

In this case the coupling coefficient F is evidently strongly Lipschitz, and thus is Lipschitz also in any of the Picard norms. Autologous coupling coefficients are evidently endogenous. Conversely, simply adding the equation $Z^\eta = \delta^\eta_\theta * Z^\theta$ to (5.2.1) turns that equation into an autologous equation for the vector $(X, \boldsymbol{Z}) \in \mathfrak{D}^{n+d}$. Equations with endogenous coefficients can be solved numerically by an algorithm (see theorem 5.4.5 on page 316).

Example 5.2.13 (Permanence Properties) If F, F' are strongly Lipschitz coupling coefficients with $d = 1$, then so is their composition. If the $F_1, F_2, \ldots$ each are strongly Lipschitz with $d = 1$, then the finite collection $\boldsymbol{F} \stackrel{\text{def}}{=} (F_1, \ldots, F_d)$ is strongly Lipschitz.

Existence and Uniqueness of the Solution

Let us now observe how our Picard norms[11] (5.2.4) and the Lipschitz condition (5.2.12) cooperate to produce a solution of equation (5.2.1), which we recall as

$$X = C + F_\eta[X]_{.-} * Z^\eta ; \tag{5.2.18}$$

i.e., how they furnish a fixed point of

$$\mathfrak{U} : X \mapsto C + \boldsymbol{F}[X]_{.-} * \boldsymbol{Z} .$$

We are of course after the contractivity of $\mathfrak{U}$.

To establish it consider two elements X, Y in $\mathfrak{S}^{\star n}_{p,M}$ and estimate:

$$\left\|\mathfrak{U}[Y] - \mathfrak{U}[X]\right\|^{\star}_{p,M} = \left\|(\boldsymbol{F}[Y] - \boldsymbol{F}[X])_{\cdot-}{*}\boldsymbol{Z}\right\|^{\star}_{p,M}$$

by inequality (5.2.5):
$$\leq \frac{C^{\diamond}_{p}}{M^{1/p^{\diamond}}} \cdot \left\|\left|\boldsymbol{F}[Y] - \boldsymbol{F}[X]\right|_{\infty}\right\|_{p,M}$$

by inequality (5.2.12):
$$\leq \frac{LC^{\diamond}_{p}}{M^{1/p^{\diamond}}} \cdot \left\|Y - X\right\|^{\star}_{p,M} . \tag{5.2.19}$$

Thus $\mathfrak{U}$ is strictly contractive provided M is sufficiently large, say

$$M > M^{\diamond}_{p,L} \stackrel{\text{def}}{=} \left(C^{\diamond}_{p}\, L\right)^{p^{\diamond}} . \tag{5.2.20}$$

$\mathfrak{U}$ then has modulus of contractivity

$$\gamma = \gamma_{p,M,L} \stackrel{\text{def}}{=} \left(M^{\diamond}_{p,L}/M\right)^{1/p^{\diamond}} \tag{5.2.21}$$

strictly less than 1. The arguments of items 5.1.3–5.1.5, adapted to the present situation, show that $\mathfrak{S}^{\star n}_{p,M}$ contains a unique fixed point X of $\mathfrak{U}$

provided
$$^{0}C \stackrel{\text{def}}{=} C + \boldsymbol{F}[0]_{\cdot-}{*}\boldsymbol{Z} \in \mathfrak{S}^{\star n}_{p,M} , \tag{5.2.22}$$

which is the same as saying that
$$\mathfrak{U}[0] \in \mathfrak{S}^{\star n}_{p,M} ;$$

and then they even furnish a priori estimates of the size of the solution X and its deviation from the initial condition, namely,

$$\left\|X\right\|^{\star}_{p,M} \leq \frac{1}{1-\gamma} \cdot \left\|{}^{0}C\right\|^{\star}_{p,M} , \tag{5.2.23}$$

and
$$\left\|X - C\right\|^{\star}_{p,M} \leq \frac{1}{1-\gamma} \cdot \left\|\boldsymbol{F}[C]_{\cdot-}{*}\boldsymbol{Z}\right\|^{\star}_{p,M} \tag{5.2.24}$$

by inequality (5.2.5):
$$\leq \frac{C^{\diamond}_{p}}{(1-\gamma)M^{1/p^{\diamond}}} \cdot \left\|\boldsymbol{F}[C]\right\|^{\star}_{p,M} . \tag{5.2.25}$$

Alternatively, by solving equation (5.2.21) for M we may specify a modulus of contractivity $\gamma \in (0,1)$ in advance:

if we set
$$M_{L:\gamma} \stackrel{\text{def}}{=} (10qL/\gamma)^{q} , \tag{5.2.26}$$

then
$$M_{L:\gamma} \geq M^{\diamond(5.2.20)}_{p,L} / \gamma^{p^{\diamond}} ,$$

and (5.2.5) turns into
$$\left\|\boldsymbol{F}_{\cdot-}{*}\boldsymbol{Z}\right\|^{\star}_{p,M} \leq \frac{\gamma}{L} \cdot \left\|\left|\boldsymbol{F}\right|_{\infty}\right\|_{p,M} \tag{5.2.27}$$
$$\leq \frac{\gamma}{L} \cdot \left\|\left|\boldsymbol{F}\right|_{\infty}\right\|^{\star}_{p,M} ,$$

and (5.2.19) into
$$\left\|\mathfrak{U}[Y] - \mathfrak{U}[X]\right\|^{\star}_{p,M} \leq \gamma \cdot \left\|Y - X\right\|^{\star}_{p,M}$$

for all $p \in [2, q]$ and all $M \geq M_{L:\gamma}$ simultaneously. To summarize:

Proposition 5.2.14 *In addition to the minimal assumptions (ii)–(iii) on page 272 assume that Z is a local L^q-integrator for some $q \ge 2$ and that F satisfies*[19] *the Picard norm Lipschitz condition (5.2.12) for some $p \in [2, q]$ and some $M > M^{\diamond(5.2.20)}_{p,L}$. If*

$$ {}^0C \overset{\text{def}}{=} C + F[0]_{.-}{*}Z \ \text{ belongs to } \ \mathfrak{S}^{\star n}_{p,M}, \tag{5.2.28} $$

then $\mathfrak{S}^{\star n}_{p,M}$ contains one and only one strong global solution X of the stochastic differential equation (5.2.18).

We are now in position to establish a rather general existence and uniqueness theorem for stochastic differential equations, without assuming more about Z than that it be an L^0-integrator:

Theorem 5.2.15 *Under the minimal assumptions (i)–(iii) on page 272 and the strong Lipschitz condition (5.2.8) there exists a strong global solution X of equation (5.2.18), and up to indistinguishability only one.*

Proof. Note first that (5.2.8) for some $p > 0$ implies (5.2.8) and then (5.2.10) for any probability equivalent with $\mathbb{P}$ and for any $p \in (0, \infty)$, in particular for $p = 2$, except that the Lipschitz L constant may change as p is altered. Let U be a finite stopping time. There is a probability $\mathbb{P}'$ equivalent to the given probability $\mathbb{P}$ such that the 2+d stopped processes $|C^{\star U}|_2$, $\left|(F[0]_{.-}{*}Z)^{\star U}\right|_2$, and $Z^{\eta U}$, $\eta = 1 \ldots d$, are global $L^2(\mathbb{P}')$-integrators. Namely, all of these processes are global L^0-integrators (proposition 2.4.1 and proposition 2.1.9), and theorem 4.1.2 provides $\mathbb{P}'$. According to proposition 3.6.20, if X satisfies

$$ X = C^U + F[X]_{.-}{*}Z^U \tag{5.2.29} $$

in the sense of the stochastic integral with respect to $\mathbb{P}$, it satisfies the same equation in the sense of the stochastic integral with respect to $\mathbb{P}'$, and vice versa: as long as we want to solve the stopped equation (5.2.29) we might as well assume that $|C^{\star U}|_2$, $\left|(F[0]_{.-}{*}Z)^{\star U}\right|_2$, and Z^U are global L^2-integrators. Then condition (5.2.22) is clearly satisfied, whatever $M > 0$ (lemma 5.2.1 (iii)). We apply inequality (5.2.19) with $p = q = 2$ and $M > M^{\diamond(5.2.20)}_{2,L}$ to make $\mathfrak{U}$ strictly contractive and see that there is a solution of (5.2.29). Suppose there are two solutions X, X'. Then we can choose $\mathbb{P}'$ so that in addition the difference $X - X'$, which stops after time U, is a global $L^2(\mathbb{P}')$-integrator as well and thus belongs to $\mathfrak{S}^{\star n}_{2,M}(\mathbb{P}')$. Since the strictly contractive map $\mathfrak{U}$ has at most one fixed point, we must have $\|X - X'\|^{\star}_{2,M} = 0$, which means that X and X' are indistinguishable. Let X^U denote the unique solution of equation (5.2.29). We let U run through a sequence (U_n) increasing to ∞ and set $X = \lim X^{U_n}$. This is clearly a global strong solution of equation (5.1.5), and up to indistinguishability the only one. ∎

[19] This is guaranteed by any of the inequalities (5.2.7)–(5.2.11). For (5.2.12) to make sense and to hold, F needs to be defined only on $X, Y \in \mathfrak{S}^{n}_{p,M}$. See also remark 5.2.20.

Exercise 5.2.16 Suppose $\boldsymbol{Z}$ is a quasi-left-continuous L^p-integrator for some $p \geq 2$. Then its previsible controller Λ is continuous and can be chosen strictly increasing; THE time transformation $T^{\cdot}$ is then continuous and strictly increasing as well. Then the Picard iterates $X^{(n)}$ for equation (5.2.18) converge to the solution X in the sense that for all $\lambda < \infty$

$$\left\| |X^{(n)} - X|^{\star}_{T^\lambda} \right\|_{L^p(\mathbb{P})} \xrightarrow[n\to\infty]{} 0 \,.$$

(i) Conclude from this that if both C and $\boldsymbol{Z}$ are local martingales, then so is X. (ii) Use factorization to extend the previous statement to $p \geq 1$. (iii) Suppose the T^λ are chosen bounded as they can be, and both C and $\boldsymbol{Z}$ are p-integrable martingales. Then X_{T^λ} is a p-integrable martingale on $\{\mathcal{F}_{T^\lambda} : \lambda \geq 0\}$.

Exercise 5.2.17 We say $\boldsymbol{Z}$ is ***surely controlled*** if there exists a right-continuous increasing sure (deterministic) function $\eta : \mathbb{R}_+ \to \mathbb{R}_+$ with $\eta_0 = 0$ so that $d\Lambda^{\langle q \rangle}[\boldsymbol{Z}] \leq d\eta$. In this case the stopping times T^λ of equation (5.2.3) are surely bounded from below by the instants

$$t^\lambda \stackrel{\text{def}}{=} \inf\{t : \eta_t \geq \lambda\} \xrightarrow[\lambda\to\infty]{} \infty \,,$$

which can be viewed as a deterministic time transform; and if ${}^0\!C$ is a surely controlled L^q-integrator and ${}^0\!\boldsymbol{F}$ is Lipschitz, then the unique solution of theorem 5.2.15 is a surely controlled L^q-integrator as well.

An example of a surely controlled integrator is a Lévy process, in particular a Wiener process. Its previsible controller is of the form $\Lambda_t = C \cdot t$, with $C = \sup_{\rho, p \leq q} C_p^{(\rho)(4.6.30)}$. Here $T^\lambda = \lambda / C$.

Exercise 5.2.18 (i) Let W be a standard scalar Wiener process. Exercises 5.2.16 and 5.2.17 together show that the Doléans–Dade exponential of any multiple of W is a martingale. Deduce the following estimates: for $m \in \mathbb{R}$, $p > 1$, and $t \geq 0$

$$\| \mathscr{E}_t[mW] \|_{L^p} = e^{m^2(p-1)t/2} \quad \text{and} \quad \left\| e^{|mW|^\star_t} \right\|_{L^p} \leq 2p' \cdot e^{m^2 pt/2} \,,$$

$p' \stackrel{\text{def}}{=} p/(p-1)$ being the conjugate of p. (ii) Next let $\boldsymbol{Z}_t = (t, W_t^2, \ldots, W_t^d)$, where $\boldsymbol{W}$ is a $d-1$-dimensional standard Wiener process. There are constants B', M' depending only on d, $m \in \mathbb{R}$, $p > 1$, $r > 1$ so that

$$\left\| |\boldsymbol{Z}^\star|^r_t \cdot e^{m|\boldsymbol{Z}^\star|_t} \right\|_{L^p} \leq B' \cdot t^{r/2} e^{M't} \,, \tag{5.2.30}$$

$$\| \mathscr{E}_t[mW] \|_{L^p} = e^{m^2(p-1)t/2} \,, \text{ and } \left\| e^{|mW|^\star_t} \right\|_{L^p} \leq 2p' \cdot e^{m^2 pt/2} \,.$$

Exercise 5.2.19 The autologous coefficients f_η of example 5.2.7 form a ***locally Lipschitz*** family if (5.2.17) is satisfied at least on bounded paths: for every $n \in \mathbb{N}$ there exists a constant L_n so that

$$| f_\eta[x_.] - f_\eta[y_.] |_{\_} \leq L_n \cdot |x_. - y_.|^\star$$

whenever the paths $x_., y_.$ satisfy[4] $|x|^\star_\infty \leq n$ and $|y|^\star_\infty \leq n$. For instance, markovian coupling coefficients that are continuously differentiable evidently form a locally Lipschitz family. Given such f_η, set ${}^n\!f_\eta[x] \stackrel{\text{def}}{=} f_\eta[{}^n\!x]$, where ${}^n\!x$ is the path x stopped just before the first time T^n its length exceeds n: ${}^n\!x \stackrel{\text{def}}{=} (x - \Delta_{T^n} x)^{T^n}$. Let ${}^n\!X$ denote the unique solution of the Lipschitz system

$$X = C + {}^n\!f_\eta[X]_{.-} {*} Z^\eta \quad \text{and set} \quad {}^n T \stackrel{\text{def}}{=} \inf\{t : {}^l\!X_t \geq n\}$$

for $l > n$. On $[\![0, {}^n T)\!)$, ${}^l\!X$ and ${}^n\!X$ agree. Set $\zeta \stackrel{\text{def}}{=} \sup {}^n T$. There is a unique limit X of ${}^n\!X$ on $[\![0, \zeta)\!)$. It solves $X = C + f_\eta[X]_{.-} {*} Z^\eta$ there, and ζ is its lifetime.

Stability

The solution of the stochastic differential equation (5.2.31) depends of course on the initial condition C, the coupling coefficient $\boldsymbol{F}$, and on the driver $\boldsymbol{Z}$. How?

We follow the lead provided by item 5.1.6 and consider the difference

$$\Delta \stackrel{\text{def}}{=} X' - X$$

of the solutions to the two equations

$$X = C + F_\eta[X]_{.-}{*}Z^\eta \tag{5.2.31}$$

and $$X' = C' + F'_\eta[X']_{.-}{*}Z'^\eta \,.$$

Δ satisfies itself a stochastic differential equation, namely,

$$\Delta = D + G_\eta[\Delta]_{.-}{*}Z'^\eta \,, \tag{5.2.32}$$

with initial condition

$$\begin{aligned} D &= (C' - C) + \big(F'_\eta[X]_{.-}{*}Z'^\eta - F_\eta[X]_{.-}{*}Z^\eta\big) \\ &= (C' - C) + (F'_\eta[X] - F_\eta[X])_{.-}{*}Z'^\eta + F_\eta[X]_{.-}{*}(Z'^\eta - Z^\eta) \end{aligned}$$

and coupling coefficients

$$\Delta \mapsto G_\eta[\Delta] \stackrel{\text{def}}{=} F'_\eta[\Delta + X] - F_\eta[X] \,.$$

To answer our question, how?, we study the size of the difference Δ in terms of the differences of the initial conditions, the coupling coefficients, and the drivers. This is rather easy in the following frequent situation: both $\boldsymbol{Z}$ and $\boldsymbol{Z}'$ are local L^q-integrators for some[12] $q \geq 2$, and the seminorms $\| \;\|^\star_{p,M}$ are defined via (5.2.3) and (5.2.4) from a previsible controller Λ common[20] to both $\boldsymbol{Z}$ and $\boldsymbol{Z}'$; and for some fixed $\gamma < 1$ and $M \geq M^{(5.2.26)}_{L:\gamma}$, $\boldsymbol{F}'$ and with it $\boldsymbol{G}$ satisfies the Lipschitz condition (5.2.12) on page 286, with constant L. In this situation inequality (5.2.23) on page 290 immediately gives the following estimate of Δ:

$$\begin{aligned} \|\Delta\|^\star_{p,M} &\leq \frac{1}{1-\gamma} \cdot \big\|(C'-C)+(\boldsymbol{F}'[X]_{.-}{*}\boldsymbol{Z}'-\boldsymbol{F}[X]_{.-}{*}\boldsymbol{Z})\big\|^\star_{p,M} \\ &= \frac{1}{1-\gamma} \cdot \big\|(C'-C)+(\boldsymbol{F}'[X]-\boldsymbol{F}[X])_{.-}{*}\boldsymbol{Z}'+\boldsymbol{F}[X]_{.-}{*}(\boldsymbol{Z}'-\boldsymbol{Z})\big\|^\star_{p,M}, \end{aligned}$$

which with (5.2.27) implies

$$\begin{aligned} \|X'-X\|^\star_{p,M} \leq \frac{1}{1-\gamma} \cdot \Big(\|(C'-C)\|^\star_{p,M} &+ \frac{\gamma}{L} \cdot \big\| |\boldsymbol{F}'[X]-\boldsymbol{F}[X]|_\infty \big\|_{p,M} \\ &+ \big\|\boldsymbol{F}[X]_{.-}{*}(\boldsymbol{Z}'-\boldsymbol{Z})\big\|^\star_{p,M}\Big) \,. \end{aligned} \tag{5.2.33}$$

[20] Take, for instance, for Λ the sum of the canonical controllers of the two integrators.

This inequality exhibits plainly how the solution X depends on the ingredients $C, \boldsymbol{F}, \boldsymbol{Z}$ of the stochastic differential equation (5.2.31). We shall make repeated use of it, to produce an algorithm for the pathwise solution of (5.2.31) in section 5.4, and to study the differentiability of the solution in parameters in section 5.3. Very frequently only the initial condition and coupling coefficient are perturbed, $\boldsymbol{Z} = \boldsymbol{Z}'$ staying unaltered. In this special case (5.2.33) becomes

$$\left\|X'-X\right\|^{\star}_{p,M} \le \frac{1}{1-\gamma} \cdot \left(\left\|(C'-C)\right\|^{\star}_{p,M} + \frac{\gamma}{L} \cdot \left\| |\boldsymbol{F}'[X]-\boldsymbol{F}[X]|_{\infty} \right\|_{p,M}\right) \tag{5.2.34}$$

or, with the roles of X, X' reversed,

$$\left\|X'-X\right\|^{\star}_{p,M} \le \frac{1}{1-\gamma} \cdot \left(\left\|(C'-C)\right\|^{\star}_{p,M} + \frac{\gamma}{L} \cdot \left\| |\boldsymbol{F}'[X']-\boldsymbol{F}[X']|_{\infty} \right\|_{p,M}\right). \tag{5.2.35}$$

Remark 5.2.20 In the case $\boldsymbol{F} = \boldsymbol{F}'$ and $\boldsymbol{Z} = \boldsymbol{Z}'$, (5.2.34) boils down to

$$\left\|X' - X\right\|^{\star}_{p,M} \le \frac{1}{1-\gamma} \cdot \left\|(C' - C)\right\|^{\star}_{p,M} . \tag{5.2.36}$$

Assume for example that $\boldsymbol{Z}, \underline{\boldsymbol{Z}}$ are two L^q-integrators and Λ a common controller.[20] Consider the equations $F = X + H^\eta[F]_{.-}{*}\underline{Z}^\eta$, $\eta = 1, \ldots d$, H^η Lipschitz. The map that associates with X the unique solution $F_\eta[X]$ is according to (5.2.36) a Lipschitz coupling coefficient in the weak sense of inequality (5.2.12) on page 286. To paraphrase: "the solution of a Lipschitz stochastic differential equation is a Lipschitz coupling coefficient in its initial value and as such can function in another stochastic differential equation." In fact, this Picard-norm Lipschitz coupling coefficient is even differentiable, provided the H^η are (exercise 5.3.7).

Exercise 5.2.21 If $\boldsymbol{F}[0] = 0$, then $\left\|C\right\|^{\star}_{p,M} \le 2 \cdot \left\|X\right\|^{\star}_{p,M}$.

Lipschitz and Pathwise Continuous Versions Consider the situation that the initial condition C and the coupling coefficient $\boldsymbol{F}$ depend on a parameter u that ranges over an open subset U of some seminormed space $(E, \|\ \|_E)$. Then the solution of equation (5.2.18) will depend on u as well: in obvious notation

$$X[u] = C[u] + \boldsymbol{F}\big[u, X[u]\big]_{.-}{*}\boldsymbol{Z} . \tag{5.2.37}$$

A cheap consequence of the stability results above is the following observation, which is used on several occasions in the sequel. Suppose that the initial condition and coupling coefficient in (5.2.37) are jointly Lipschitz, in the sense that there exists a constant L such that for all $u, v \in U$ and all $X \in \mathfrak{S}^{\star n}_{p,M}$ (where $2 \le p \le q$ and $M > M^{(5.2.20)}_{p,L}$)

$$\left\|C[v] - C[u]\right\|^{\star}_{p,M} \le L \cdot \|v - u\|_E \tag{5.2.38}$$

$$\text{and} \quad \left\| |\boldsymbol{F}[v,Y] - \boldsymbol{F}[u,X]|_{\infty} \right\|^{\star}_{p,M} \le L \cdot \left(\|v - u\|_E + \left\|Y - X\right\|^{\star}_{p,M}\right) ;$$

$$\text{hence} \quad \left\| |\boldsymbol{F}[v,X] - \boldsymbol{F}[u,X]|_{\infty} \right\|^{\star}_{p,M} \le L \cdot \|v - u\|_E . \tag{5.2.39}$$

Then inequality (5.2.34) implies the Lipschitz dependence of $X[u]$ on $u \in U$:

Proposition 5.2.22 *In the presence of (5.2.38) and (5.2.39) we have for all* $u, v \in U$

$$\|X[v] - X[u]\|^{\star}_{p,M} \le \frac{L+\gamma}{1-\gamma} \cdot \|v - u\|_E . \tag{5.2.40}$$

Corollary 5.2.23 *Assume that the parameter domain* U *is finite-dimensional,* Z *is a local* L^p*-integrator for some* p *strictly larger than* $\dim U$*, and for all* $u, v \in U$ [4]

$$\|X[v] - X[u]\|^{\star}_{p,M} \le const \cdot |v - u| . \tag{5.2.41}$$

Then the solution processes $X_{.}[u]$ *can be chosen in such a way that for every* $\omega \in \Omega$ *the map* $u \mapsto X_{.}[u](\omega)$ *from* U *to* $\mathscr{D}^n$ *is continuous.*[21]

Proof. For fixed t and λ the Lipschitz condition (5.2.41) gives

$$\Big\| \sup_{s \le T^\lambda} \|X[v] - X[u]\|_s \Big\|_{L^p} \le const \cdot e^{M\lambda} \cdot |v - u| ,$$

which implies $\mathbb{E}\Big[\sup_{s\le t} |X[v] - X[u]|_s^p \cdot [t < T^\lambda]\Big] \le const \cdot |v - u|^p$.

According to Kolmogorov's lemma A.2.37 on page 384, there is a negligible subset N_λ of $[t < T^\lambda] \in \mathcal{F}_t$ outside which the maps $u \mapsto X[u]^t_{.}(\omega)$ from U to paths in $\mathscr{D}^n$ stopped at t are, after modification, all continuous in the topology of uniform convergence. We let λ run through a sequence (λ_n) that increases without bound. We then either throw away the nearly empty set $\bigcup_n N_{\lambda(n)}$ or set $X[u]_{.} = 0$ there. ∎

In particular, when Z is continuous it is a local L^q-integrator for all $q < \infty$, and $u \mapsto X_{.}[u](\omega)$ can be had continuous for every $\omega \in \Omega$.

If Z is merely an L^0-integrator, then a change of measure as in the proof of theorem 5.2.15 allows the same conclusion, except that we need Lipschitz conditions that do not change with the measure:

Theorem 5.2.24 *In (5.2.37) assume that* C *and* F *are Lipschitz in the finite-dimensional parameter* u *in the sense that for all* $u, v \in U$ *and all* $X \in \mathfrak{D}^n$

both[4]
$$\big|C[v] - C[u]\big|^\star \le L \cdot |v - u|$$
and
$$\sup_\eta \big|F_\eta[v, X] - F_\eta[u, X]\big| \le L \cdot |v - u| ,$$

nearly. Then the solutions $X_{.}[u]$ *of (5.2.37) can be chosen in such a way that for every* $\omega \in \Omega$ *the map* $u \mapsto X_{.}[u](\omega)$ *from* U *to* $\mathscr{D}^n$ *is continuous.*[21] ∎

---

[21] The pertinent topolgy on path spaces is the topology of uniform convergence on compacta.

Differential Equations Driven by Random Measures

By definition 3.10.1 on page 173, a random measure ζ is but an "$\boldsymbol{H}$-tuple of integrators," $\boldsymbol{H}$ being the auxiliary space. Instead of $\int \sum_\eta F_{\eta s} dZ^\eta_s$ or $\int \int_\eta F_{\eta s} dZ^{d\eta}_s$ we write $\int F(\eta, s)\, \zeta(d\eta, ds)$, but that is in spirit the sum total of the difference. Looking at a random measure this way, as "a long vector of tiny integrators" as it were, has already paid nicely (see theorem 4.3.24 and theorem 4.5.25). We shall now see that stochastic differential equations driven by random measures can be handled just like the ones driven by slews of integrators that we have treated so far. In fact, the following is but a somewhat repetitious reprise of the arguments developed above. It simplifies things a little to ***assume that ζ is spatially bounded.***

From this point of view the stochastic differential equation driven by ζ is

$$X_t = C_t + \int_0^t F[\eta, X_.]_{s-}\, \zeta(d\eta, ds)$$

or, equivalently,
$$X = C + F[\,\cdot\,, X_.]_{.-} * \zeta\,, \tag{5.2.42}$$

with
$$F : \boldsymbol{H} \times \mathfrak{D}^n \to \mathfrak{D}^n \qquad \text{suitable.}$$

We expect to solve (5.2.42) under the strong Lipschitz condition (5.2.8) on page 286, which reads here as follows: at any stopping time T

$$\left|\big(F[\,\cdot\,,Y] - F[\,\cdot\,,X]\big)_T\right|_{\infty p} \le L \cdot \left|(Y-X)^\star_t\right|_p \quad \text{a.s.,}$$

where $\quad \left|F[\,\cdot\,,X]_s\right|_\infty$ is the n-vector $\left(\sup\left|F^\nu[\eta,X]_s\right| : \eta \in \boldsymbol{H}\right)_{\nu=1}^n$,

$\left|F[\,\cdot\,,X]^\star_.\right|_\infty$ its (vector-valued) maximal process,

and $\quad \left|F[\,\cdot\,,X]^\star_.\right|_{\infty p} \stackrel{\text{def}}{=} \left(\sum_\nu \left|F^\nu[\,\cdot\,,X]^\star_.\right|^p_\infty\right)^{1/p}$ the length of the latter.

As a matter of fact, if ζ is an L^q-integrator and $2 \le p \le q$, then the following rather much weaker "p-mean Lipschitz condition," analog of inequality (5.2.10), should suffice: for any $X, Y \in \mathfrak{D}^n$ and any predictable stopping time T

$$\left\| \left|F[\,\cdot\,,Y]_{T-} - F[\,\cdot\,,X]_{T-}\right|_{\infty p} \right\|_{L^p} \le L \cdot \left\| \left|(Y-X)^\star_{T-}\right|_p \right\|_{L^p} .$$

Assume this then and let $\Lambda = \Lambda^{\langle q \rangle}[\zeta]$ be the previsible controller provided by theorem 4.5.25 on page 251. With it goes THE time transformation (5.2.3), and with the help of the latter we define the seminorms $\|\ \|_{p,M}$ and $\|\ \|^\star_{p,M}$ as in definition (5.2.4) on page 283. It is a simple matter of shifting the η from a subscript on $\boldsymbol{F}$ to an argument in F, to see that lemma 5.2.1 persists. Using definition (5.2.26) on page 290, we have for any $\gamma \in (0,1)$

$$\left\|F_{.-} * \zeta\right\|^\star_{p,M} \le \frac{\gamma}{L} \cdot \left\| |F|_\infty \right\|^\star_{p,M}$$

and $$\left\|\mathfrak{U}[Y]-\mathfrak{U}[X]\right\|^{\star}_{p,M} \le \gamma\cdot\left\|Y-X\right\|^{\star}_{p,M} ,$$

where of course $$\mathfrak{U}[X]_t \stackrel{\mathrm{def}}{=} C_t + \int_0^t F[\eta,X]_{s-}\ \zeta(d\eta,ds)$$

and $$M > M_{L:\gamma} \stackrel{\mathrm{def}}{=} (10qL/\gamma)^q \vee 1 .$$

We see that *as long as* $\mathfrak{S}^{\star n}_{p,M}$ *contains* ${}^0C \stackrel{\mathrm{def}}{=} C+F[\,\cdot\,,0]_{-}{*}\zeta$ *it contains a unique solution of equation* (5.2.42). If ζ is merely an L^0-random measure, then we reduce as in theorem 5.2.15 the situation to the previous one by invoking a factorization, this time using corollary 4.1.14 on page 208:

Proposition 5.2.25 *Equation (5.2.42) has a unique global solution.*

The stability inequality (5.2.34) for the difference $\Delta \stackrel{\mathrm{def}}{=} X'-X$ of the solutions of two stochastic differential equations

$$X' = C' + F'[\,\cdot\,,X']_{-}{*}\zeta'$$

and $$X = C + F[\,\cdot\,,X]_{-}{*}\zeta ,$$

which satisfies $$\Delta = D + G[\,\cdot\,,\Delta]_{-}{*}\zeta'$$

with $$D = (C'-C) + (F'[\,\cdot\,,X] - F[\,\cdot\,,X])_{-}{*}\zeta' + F[\,\cdot\,,X]_{-}{*}(\zeta'-\zeta)$$

and $$G[\,\cdot\,,\Delta] = F'[\,\cdot\,,\Delta+X] - F'[\,\cdot\,,X] ,$$

persists *mutatis mutandis.* Assuming that both F and F' are Lipschitz with constant L, that $\|\ \|^{\star}_{p,M}$ is defined from a previsible controller Λ common to both ζ and ζ',[22] and that M has been chosen strictly larger[23] than $M^{\diamond(5.2.20)}_{p,L}$, the analog of inequality (5.2.23) results in these estimates of Δ:

$$\begin{aligned}\left\|\Delta\right\|^{\star}_{p,M} &\le \frac{1}{1-\gamma}\cdot\left\|(C'-C) + (F'[\,\cdot\,,X]_{-}{*}\zeta' - F[\,\cdot\,,X]_{-}{*}\zeta)\right\|^{\star}_{p,M} \\ &\le \frac{1}{1-\gamma}\cdot\Big(\left\|(C'-C)\right\|^{\star}_{p,M} + \frac{\gamma}{L}\cdot\left\|\left|F'[\,\cdot\,,X]-F[\,\cdot\,,X]\right|_{\infty}\right\|_{p,M} \\ &\qquad + \left\|F[\,\cdot\,,X]_{-}{*}(\zeta'-\zeta)\right\|^{\star}_{p,M}\Big) .\end{aligned}$$

This inequality shows neatly how the solution X depends on the ingredients C, F, ζ of the equation. Additional assumptions, such as that $\zeta' = \zeta$ or that both $\zeta = \zeta'$ and $C' = C$ simplify these inequalities in the manner of the inequalities (5.2.34)–(5.2.36) on page 294.

---

[22] Take, for instance, $\Lambda_t \stackrel{\mathrm{def}}{=} \Lambda^{\langle q\rangle}_t[\zeta] + \Lambda^{\langle q\rangle}_t[\zeta'] + \epsilon t$, $0<\epsilon\ll 1$.

[23] The point being that p and M must be chosen so that $\mathfrak{U}$ is strictly contractive on $\mathfrak{S}^{\star n}_{p,M}$.

The Classical SDE

The "classical" stochastic differential equation is the markovian equation

$$X = C + \boldsymbol{f}(X){*}\boldsymbol{Z} \quad \text{or} \quad X_t = C + \int_0^t f_\eta(X)_s \, dZ_s^\eta \,, \qquad t \geq 0 \,,$$

where the initial condition C is constant in time, $\boldsymbol{f} = (f_0, f_1, \ldots, f_d)$ are (at least measurable) vector fields on the state space $\mathbb{R}^n$ of X, and the driver $\boldsymbol{Z}$ is the $d{+}1$-tuple $\boldsymbol{Z}_t = (t, W_t^1, \ldots, W_t^d)$, $\boldsymbol{W}$ being a standard Wiener process on a filtration to which both $\boldsymbol{W}$ and the solution are adapted. The classical SDE thus takes the form

$$X_t = C + \int_0^t f_0(X_s)\, ds + \textstyle\sum_{\eta=1}^d \displaystyle\int_0^t f_\eta(X_s)\, dW_s^\eta \,. \tag{5.2.43}$$

In this case the controller is simply $\Lambda_t = d \cdot t$ (exercise 4.5.19) and thus THE time transformation is simply $T^\lambda = \lambda/d$. The Picard norms of a process X become simply

$$\|X\|_{p,M}^\star = \sup_{t>0} e^{-Mdt} \cdot \Big\| \left|X_t^\star\right|_p \Big\|_{L^p(\mathbb{P})}^* \,.$$

If the coupling coefficient $\boldsymbol{f}$ is Lipschitz, then the solution of (5.2.43) grows at most exponentially in the sense that $\| \left|X_t^\star\right|_p \|_{L^p(\mathbb{P})} \leq Const \cdot e^{Mdt}$. The stability estimates, etc., translate similarly.

5.3 Stability: Differentiability in Parameters

We consider here the situation that the initial condition C and the coupling coefficient $\boldsymbol{F}$ depend on a parameter u that ranges over an open subset U of some seminormed space $(E, \| \ \|_E)$. Then the solution of equation (5.2.18) will depend on u as well: in obvious notation

$$X[u] = C[u] + \boldsymbol{F}\big[u, X[u]\big]_{\leftarrow}{*}\boldsymbol{Z} \,. \tag{5.3.1}$$

We have seen in item 5.1.7 that in the case of an ordinary differential equation the solution depends differentiably on the initial condition and the coupling coefficient. This encourages the hope that our $X[u]$, too, will depend differentiably on u when both $C[u]$ and $\boldsymbol{F}[u, \cdot\,]$ do. This is true, and the goal of this section is to prove several versions of this fact.

Throughout the section ***the minimal assumptions (i)–(iii)*** of page 272 are in effect. In addition we will require that $\boldsymbol{Z} = (Z^1, \ldots, Z^d)$ is a local $L^q(\mathbb{P})$-integrator for some[12] ***q strictly larger than 2*** – except when this is explicitely rescinded on occasion. This requirement provides us with the previsible controller $\Lambda^{\langle q \rangle}[\boldsymbol{Z}]$, with THE time transformation (5.2.3), and the Picard norms[11] $\| \ \|_{p,M}^\star$ of (5.2.4). We also have settled on a modulus of contractivity $\gamma \in (0,1)$ to our liking. The coupling coefficients $\boldsymbol{F}[u,\cdot]$ are assumed to be Lipschitz in the sense of inequality (5.2.12) on page 286:

$$\Big\| \big| \boldsymbol{F}[u,Y] - \boldsymbol{F}[u,X] \big|_\infty \Big\|_{p,M} \le L \cdot \big\| X - Y \big\|^\star_{p,M} , \tag{5.3.2}$$

with ***Lipschitz constant*** L ***independent*** of the parameter u and of

$$p \in (2,q] , \quad \text{and} \quad M \ge M^{(5.2.26)}_{L:\gamma} . \tag{5.3.3}$$

Then any stochastic differential equation driven by $\boldsymbol{Z}$ and satisfying the Picard-norm Lipschitz condition (5.3.2) and equation (5.2.28) has its solution in $\mathfrak{S}^\star_{p,M}$, whatever such p and M. In particular, $X[u] \in \mathfrak{S}^{\star n}_{p,M}$ for all $u \in U$ and all (p,M), as in (5.3.3).

For the notation and terminology concerning differentiation refer to definitions A.2.45 on page 388 and A.2.49 on page 390.

Example 5.3.1 Consider first the case that U is an open convex subset of $\mathbb{R}^k$ and the coupling coefficient $\boldsymbol{F}$ of (5.3.1) is markovian (see example 5.2.4): $F_\eta[u,X] = f_\eta(u,X)$. Suppose the $f_\eta : U \times \mathbb{R}^n \to \mathbb{R}^n$ have a continuous bounded derivative $Df_\eta = (D_1 f_\eta, D_2 f_\eta)$. Then just as in example A.2.48 on page 389, F_η is not necessarily Fréchet differentiable as a map from $U \times \mathfrak{S}^{\star n}_{p,M}$ to $\mathfrak{S}^{\star n}_{p,M}$. It is, however, weakly uniformly differentiable, and the partial $D_2 F_\eta[u,X]$ is the continuous linear operator from $\mathfrak{S}^{\star n}_{p,M}$ to itself that operates on a $\xi \in \mathfrak{S}^{\star n}_{p,M}$ by applying the $n \times n$-matrix $D_2 f_\eta(u,X)$ to the vector $\xi \in \mathbb{R}^n$: for $\varpi \in \boldsymbol{B}$

$$\big(D_2 F_\eta[u, X(\varpi)] \cdot \xi\big)(\varpi) = D_2 f_\eta\big(u, X(\varpi)\big) \cdot \xi(\varpi) .$$

The operator norm of $D_2 F_\eta[u,X]$ is bounded by $\sup_{u,x} |Df_\eta(u,x)|_1$, independently of $u \in U$ and p, where $|D|_1 \stackrel{\text{def}}{=} \sum_{\nu,\kappa} |D^\nu_\kappa|$ on matrices D.

Example 5.3.2 The previous example has an extension to autologous coupling coefficients. Suppose the adapted map[17] $f : U \times \mathscr{D}^n \to \mathscr{D}^n$ has a continuous bounded Fréchet derivative. Let us make this precise: at any point $(u, x_.)$ in $U \times \mathscr{D}^n$ there exists a linear map $Df[u,x_.] : E \times \mathscr{D}^n \to \mathscr{D}^n$ such that for all t

$$\big| Df[u,x_.] \big|^\star_t \stackrel{\text{def}}{=} \sup \left\{ \left| Df[u,x_.] \cdot \begin{pmatrix} \xi \\ \Xi_. \end{pmatrix} \right|^\star_t : \|\xi\|_E + |\Xi|^\star_t \le 1 \right\} \le L < \infty$$

$$\text{and} \qquad \big| Df[v,y_.] - Df[u,x_.] \big|^\star_t \to 0 \quad \text{as} \quad \|v-u\|_E + |y_. - x_.|^\star_t \to 0 ,$$

with L independent of $(u, x_.)$, and such that

$$Rf[u,x_.;v,y_.] \stackrel{\text{def}}{=} f[v,y_.] - f[u,x_.] - Df[u,x_.] \cdot \begin{pmatrix} v-u \\ y_. - x_. \end{pmatrix} \quad \text{has}$$

$$\big| Rf[u,x_.;v,y_.] \big|^\star_t = o\big(\|v-u\|_E + |y_. - x_.|^\star_t \big) .$$

According to Taylor's formula of order one (see lemma A.2.42), we have[24]

$$Rf[u, x.; v, y.] = \int_0^1 \Big(Df[u^\sigma, x^\sigma_.] - Df[u, x.]\Big)\, d\sigma \cdot \begin{pmatrix} v-u \\ y.-x. \end{pmatrix},$$

where $u^\sigma \stackrel{\text{def}}{=} u + \sigma(v-u)$ and $x^\sigma_. \stackrel{\text{def}}{=} x. + \sigma(y.-x.)$.

Now the coupling coefficient F corresponding to f is defined at $X \in \mathfrak{D}^n$ by

$$F[u, X].(\omega) \stackrel{\text{def}}{=} f[u, X.(\omega)]$$

and has weak derivative $DF[u, X] : \begin{pmatrix} \xi \\ \Xi \end{pmatrix} \mapsto Df[u, X]\cdot \begin{pmatrix} \xi \\ \Xi. \end{pmatrix}$, $\Xi \in \mathfrak{S}^{\star n}_{p,M}$.

Indeed, for $1/p^\circ = 1/p + 1/r$ and $M^\circ = M + R$,

$$\begin{aligned} \big\| RF[u, X; v, Y] \big\|^\star_{p^\circ, M^\circ} &\le \int_0^1 \big\| Df[u^\sigma, X^\sigma] - Df[u, X] \big\|^\star_{r,R}\, d\sigma \\ &\qquad \times \Big(\|v-u\|_E + \big\| Y - X \big\|^\star_{p,M} \Big) \\ &= o\Big(\|v-u\|_E + \big\| Y - X \big\|^\star_{p,M} \Big) \end{aligned}$$

on the grounds that $\big| Df[u^\sigma, X^\sigma] - Df[u, X] \big|^\star_{T^\lambda -} \to 0$

as $\|v-u\|_E + \big\| Y-X \big\|^\star_{p,M} \to 0$, pointwise and boundedly for every λ, σ, and $\omega \in \Omega$: the weak uniform differentiability follows. ∎

Exercise 5.3.3 Extend this to randomly autologous coefficients $f[\omega, u, x.]$: Assume that the family $\{f[\omega, u, x.] : \omega \in \Omega\}$ of autologous coefficients is equidifferentiable and has $\sup_\omega \big|Df[\omega, u, x.]\big|^\star_t \le L$. Show that then $F[u, X].(\omega) \stackrel{\text{def}}{=} f[\omega, u, X.(\omega)]$ is weakly differentiable on $\mathfrak{S}^{\star n}_{p,M}$ with

$$DF[u, X] : \begin{pmatrix} \xi \\ \Xi \end{pmatrix} \mapsto Df[\omega, u, X(\omega)]\cdot \begin{pmatrix} \xi \\ \Xi.(\omega) \end{pmatrix} .$$

Example 5.3.4 Example 5.2.8 exhibits a coupling coefficient $\mathfrak{S}^{\star n}_{p,M} \to \mathfrak{S}^{\star n}_{p,M}$ that is linear and bounded for all (p, M) and *ipso facto* uniformly Fréchet differentiable. We leave to the reader the chore of finding the conditions that make examples 5.2.10–5.2.11 weakly differentiable. Suppose the maps $(u, X) \mapsto F[u, X], G[u, X]$ are both weakly differentiable as functions from $U \times \mathfrak{S}^{\star n}_{p,M}$ to $\mathfrak{S}^{\star n}_{p,M}$. Then so is $(u, X) \mapsto F\big[u, G[u, X]\big]$.

Example 5.3.5 Let $\mathcal{T}$ be an adapted partition of $[0, \infty) \times \mathscr{D}^n$. The scalæfication map $x. \mapsto x^{\mathcal{T}}_.$ is a differentiable autologous Lipschitz coupling coefficient. If F is another, then the coupling coefficient F' of equation (5.4.7) on page 312, defined by $F' : Y \mapsto F[Y^{\mathcal{T}}]^{\mathcal{T}}$, is yet another.

---

[24] To see this, reduce to the scalar case by applying a continuous linear functional.

The Derivative of the Solution

Let us then stipulate for the remainder of this section that in the stochastic differential equation (5.3.1) the initial condition $u \mapsto C[u]$ is weakly differentiable as a map from U to $\mathfrak{S}^{\star n}_{p,M}$, and that the coupling coefficients

$$F_\eta : U \times \mathfrak{S}^{\star n}_{p,M} \to \mathfrak{S}^{n}_{p,M} \ , \ F_\eta : (u, X) \mapsto F_\eta[u, X] \ ,$$

are weakly differentiable as well. What is needed at this point is a candidate $DX[u]$ for the derivative of $v \mapsto X[v]$ at $u \in U$. In order to get an idea what it might be, let us assume that X is in fact weakly differentiable. With

$$v = u + \tau\xi \ , \ \ \|\xi\|_E = 1 \ , \ \ \tau = \|v - u\|_E \ ,$$

this can be written as

$$X[v] = X[u] + \tau DX[u]{\cdot}\xi + RX[u; v] \ , \tag{5.3.4}$$

where $\quad \big\|RX[u;v]\big\|_{p^\circ, M^\circ} = o(\tau) \qquad$ for $p^\circ < p \ , \ M^\circ > M$.

On the other hand,

$$\begin{aligned} X[v] &= X[u] + C[v] - C[u] + \Big\{\boldsymbol{F}[v, X[v]] - \boldsymbol{F}[u, X[u]]\Big\}_{\!\leftarrow}{*}\boldsymbol{Z} \\ &= X[u] + DC[u]{\cdot}(v-u) + RC[u;v] \\ &\quad + \left\{D\boldsymbol{F}[u, X[u]]{\cdot}\begin{pmatrix} v-u \\ X[v]-X[u] \end{pmatrix}\right\}_{\!\leftarrow}{*}\boldsymbol{Z} \\ &\quad + R\boldsymbol{F}[u, X[u]; v, X[v]]_{\leftarrow}{*}\boldsymbol{Z} \ , \end{aligned}$$

which translates to

$$X[v] = X[u] + \tau DC[u]{\cdot}\xi + \tau\left\{D\boldsymbol{F}[u, X[u]]{\cdot}\begin{pmatrix} \xi \\ DX[u]{\cdot}\xi \end{pmatrix}\right\}_{\!\leftarrow}{*}\boldsymbol{Z} \tag{5.3.5}$$

$$+ RC + R\boldsymbol{F}_{\leftarrow}{*}\boldsymbol{Z} + (D_2\boldsymbol{F}{\cdot}RX)_{\leftarrow}{*}\boldsymbol{Z} \ . \tag{5.3.6}$$

In the previous line the arguments $[\cdot]$ of RC, RX, and $R\boldsymbol{F}$ are omitted from the display. In view of inequality (5.2.5) the entries in line (5.3.6) are $o(\tau)$, as is the last term in (5.3.4). Thus both equations (5.3.4) and (5.3.5) for $X[v]$ are of the form $X[u] + \mathit{const}\cdot\tau + o(\tau)$ when measured in the pertinent Picard norm, which here is $\|\ \|^{\star}_{p^\circ, M^\circ}$. Clearly,[25] then, the coefficient of τ on both sides must be the same. We arrive at

$$\begin{aligned} DX[u]{\cdot}\xi = DC[u]{\cdot}\xi &+ \big(D_1\boldsymbol{F}[u, X[u]]{\cdot}\xi\big)_{\leftarrow}{*}\boldsymbol{Z} \\ &+ \big\{D_2\boldsymbol{F}[u, X[u]]\cdot(DX[u]{\cdot}\xi)\big\}_{\leftarrow}{*}\boldsymbol{Z} \ . \end{aligned} \tag{5.3.7}$$

[25] To enhance the clarity apply a linear functional that is bounded in the pertinent norm, thus reducing this to the case of real-valued functions. The Hahn–Banach theorem A.2.25 provides the generality stated.

For every fixed $u \in U$ and $\xi \in E$ we see here a linear stochastic differential equation for a process[2] $DX[u]{\cdot}\xi \in \mathfrak{D}^n$, whose initial condition is the sum $DC[u]{\cdot}\xi + (D_1\boldsymbol{F}[u, X[u]]{\cdot}\xi)_{.-}{*}\boldsymbol{Z}$ and whose coupling coefficient is given by $\Xi \mapsto D_2\boldsymbol{F}[u, X[u]]{\cdot}\Xi$, which by exercise A.2.50 (c) has a Lipschitz constant less than L. Its solution depends linearly on ξ, lies in $\mathfrak{S}^{\star n}_{p,M}$ for all pairs (p, M) as in (5.3.3), and there it has size (see inequality (5.2.23) on page 290)

$$\left\|DX[u]{\cdot}\xi\right\|^{\star}_{p,M} \le \frac{1}{1-\gamma}\Big(\left\|DC[u]{\cdot}\xi\right\|^{\star}_{p,M} + \frac{\gamma}{L}{\cdot}\left\|D_1\boldsymbol{F}[u, X[u]]{\cdot}\xi\right\|^{\star}_{p,M}\Big)$$
$$\le \mathit{const} \cdot \|\xi\|_E < \infty \,.$$

Thus if $X[\,\cdot\,]$ is differentiable, then its derivative $DX[u]$ must be given by (5.3.7). Therefore $DX[u]$ is henceforth *defined* as the linear map from E to $\mathfrak{S}^{\star}_{p,M}$ that sends ξ to the unique solution of (5.3.7). It is necessarily our candidate for the derivative of $X[\,\cdot\,]$ at u, and it does in fact do the job:

Theorem 5.3.6 *If $v \mapsto C[v]$ and the $(v, X) \mapsto F_\eta[v, X]$, $\eta = 1, \ldots, d$, are weakly (equi)differentiable maps into $\mathfrak{S}^{\star n}_{p,M}$, then $v \mapsto X[v] \in \mathfrak{S}^{\star n}_{p,M}$ is weakly (equi)differentiable, and its derivative at a point $u \in U$ is the linear map $DX[u]$ defined by (5.3.7).*

Proof. It has to be shown that, for every $p^\circ \in [2, p)$ and $M^\circ > M$,

$$RX[u; v] \stackrel{\text{def}}{=} X[v] - X[u] - DX[u]{\cdot}(v-u)$$

has
$$\left\|RX[u; v]\right\|^{\star}_{p^\circ,M^\circ} = o(\|v-u\|_E) \,.$$

Now it is a simple matter of comparing equalities (5.3.4) and (5.3.5) to see that, in view of (5.3.7), $RX[u; v]$ satisfies the stochastic differential equation

$$RX[u; v] = RC[u; v] + R\boldsymbol{F}[u, X[u]; v, X[v]]_{.-}{*}\boldsymbol{Z}$$
$$+ \big\{D_2\boldsymbol{F}[u, X[u]]{\cdot}RX[u; v]\big\}_{.-}{*}\boldsymbol{Z} \,,$$

whose Lipschitz constant is L. According to (5.2.23) on page 290, therefore,

$$\left\|RX[u; v]\right\|^{\star}_{p^\circ,M^\circ} \le \frac{1}{1-\gamma} \cdot \left\|RC[u; v] + R\boldsymbol{F}[u, X[u]; v, X[v]]_{.-}{*}\boldsymbol{Z}\right\|^{\star}_{p^\circ,M^\circ} \,.$$

Since $\left\|RC[u; v]\right\|_{p^\circ,M^\circ} = o(\|v-u\|_E)$ as $v \to u$, all that needs showing is

$$\left\|R\boldsymbol{F}[u, X[u]; v, X[v]]_{.-}{*}\boldsymbol{Z}\right\|^{\star}_{p^\circ,M^\circ} = o(\|v-u\|_E) \quad \text{as} \quad v \to u \,,$$

and this follows via inequality (5.2.5) on page 284 from

$$\left\|RF_\eta[u, X[u]; v, X[v]]\right\|_{p^\circ,M^\circ} = o\big(\|v-u\|_E + \left\|X[v] - X[u]\right\|^{\star}_{p,M}\big)$$

by A.2.50 (c) and (5.2.40):
$$= o(\|v-u\|_E) \quad \text{as} \quad v \to u \,, \qquad \eta = 1, \ldots d \,.$$

If $C, \boldsymbol{F}$ are weakly uniformly differentiable, then the estimates above are independent of u, and $X[u]$ is in fact uniformly differentiable in u.

Exercise 5.3.7 Taking $E \stackrel{\text{def}}{=} \mathfrak{S}^{\star n}_{p,M}$, show that the coupling coefficient of remark 5.2.20 is differentiable.

Pathwise Differentiability

Consider now the difference $\boldsymbol{D} \stackrel{\text{def}}{=} DX[v] - DX[u]$ of the derivatives at two different points u, v of the parameter domain U, applied to an element ξ of E_1.[26] According to equation (5.3.7) and inequality (5.2.34), $\boldsymbol{D}$ satisfies the estimate

$$\left\|\boldsymbol{D}\cdot\xi\right\|^{\star}_{p,M} \le \frac{1}{1-\gamma}\cdot\Big(\left\|(DC[v]-DC[u])\cdot\xi\right\|^{\star}_{p,M} + \frac{\gamma}{L}\cdot\left\|\left|(D\boldsymbol{F}[v,X[v]] - D\boldsymbol{F}[u,X[u]])\cdot\xi\right|_{\infty}\right\|^{\star}_{p,M} + \frac{\gamma}{L}\cdot\left\|(D_2\boldsymbol{F}[v,X[v]] - D_2\boldsymbol{F}[u,X[u]])\cdot DX[u]\cdot\xi\right\|^{\star}_{p,M}\Big).$$

Let us now assume that $v \mapsto DC[v]$ and $(v, Y) \mapsto D\boldsymbol{F}[v, Y]$ are Lipschitz with constant L', in the sense that for all pairs (p, M) as in (5.3.3), and all $\xi \in E_1$ [26]

$$\left\|(DC[v]-DC[u])\cdot\xi\right\|^{\star}_{p,M} \le L'\cdot\|v-u\|_E\cdot\|\xi\|_E \tag{5.3.8}$$

and

$$\left\|(D\boldsymbol{F}[v,X[u]]-D\boldsymbol{F}[u,X[u]])\cdot\xi\right\|^{\star}_{p,M} \le L'\cdot\|v-u\|_E\cdot\|\xi\|_E\,. \tag{5.3.9}$$

Then an application of proposition 5.2.22 on page 295 produces

$$\left\|DX[v] - DX[u]\right\| \le const\cdot\|v-u\|_E\,, \tag{5.3.10}$$

where $\|\ \|$ denotes the operator norm on $DX[u] : E \to \mathfrak{S}^{\star n}_{p,M}$.

Let us specialize to the situation that $E = \mathbb{R}^k$. Then, by letting ξ run through the usual basis, we see that $DX[u]$ can be identified with an $n\times k$-matrix-valued process in $\mathfrak{D}^{n\times k}$. At this juncture it is necessary to ***assume that***[27] $\boldsymbol{q > p > k}$. Corollary 5.2.23 then puts us in the following situation:

5.3.8 *For every $\omega \in \Omega$, $u \mapsto DX[u].(\omega)$ is a continuous map*[21] *from U to $\mathscr{D}^{n\times k}$.*

Consider now a curve $\gamma : [0,1] \to U$ that is piecewise[28] of class[10] C^1. Then the integral

$$\int_\gamma DX[u]\,d\tau \stackrel{\text{def}}{=} \int_0^1 DX[\gamma(\tau)]\cdot\gamma'(\tau)\,d\tau$$

---

26 E_1 is the unit ball of E.

27 See however theorem 5.3.10 below.

28 That is to say, there exists a càglàd function $\gamma' : [0,1] \to E$ with finitely many discontinuities so that $\gamma(t) = \int_0^t \gamma'(\tau)\,d\tau \in U \quad \forall t \in [0,1]$.

can be understood in two ways: as the Riemann integral[29] of the piecewise continuous curve $t \mapsto DX[\gamma(\tau)]\cdot\gamma'(\tau)$ in $\mathfrak{S}^{\star n}_{p,M}$, yielding an element of $\mathfrak{S}^{\star n}_{p,M}$ that is unique up to evanescence; or else as the Riemann integral of the piecewise continuous curve $\tau \mapsto DX[\gamma(\tau)].(\omega)\cdot\gamma'(\tau)$, one for every $\omega \in \Omega$, and yielding for every $\omega \in \Omega$ an element of path space $\mathscr{D}^n$. Looking at Riemann sums that approximate the integrals will convince the reader that the integral understood in the latter sense is but one of the (many nearly equal) processes that constitute the integral of the former sense, which by the Fundamental Theorem of Calculus equals $X[\gamma(1)]. - X[\gamma(0)].$. In particular, if γ is a closed curve, then the integral in the first sense is evanescent; and this implies that for nearly every $\omega \in \Omega$ the Riemann integral in the second sense,

$$\oint_\gamma DX[u].(\omega)\, d\tau \tag{$*$}$$

is the zero path in $\mathscr{D}^n$.

Now let Γ denote the collection of all closed polygonal paths in $U \subseteq \mathbb{R}^k$ whose corners are rational points. Clearly Γ is countable. For every $\gamma \in \Gamma$ the set of $\omega \in \Omega$ where the integral $(*)$ is non-zero is nearly empty, and so is the union of these sets. Let us remove it from Ω. That puts us in the position that $\oint_\gamma DX[u](\omega)\, d\tau = 0$ for all $\gamma \in \Gamma$ and all $\omega \in \Omega$. Now for every curve that is piecewise[28] of class C^1 there is a sequence of curves $\gamma_n \in \Gamma$ such that both $\gamma_n(\tau) \to \gamma_n(\tau)$ and $\gamma'_n(\tau) \to \gamma'_n(\tau)$ uniformly in $\tau \in [0,1]$. From this it is plain that the integral $(*)$ vanishes for every closed curve γ that is piecewise of class C^1, on every $\omega \in \Omega$.

To bring all of this to fruition, let us pick in every component U_0 of U a base point u_0 and set, for every $\omega \in \Omega$,

$$X[u].(\omega) \stackrel{\text{def}}{=} \int_\gamma DX[\gamma(\tau)].(\omega)\cdot\gamma'(\tau)\, d\tau ,$$

where γ is some C^1-path joining u_0 to $u \in U_0$. This element of $\mathscr{D}^n$ does not depend on γ, and $X[u].$ is one of the (many nearly equal) solutions of our stochastic differential equation (5.3.1). The upshot:

Proposition 5.3.9 *Assume that the initial condition and the coupling coefficients of equation (5.3.1) on page 298 have weak derivatives $DC[u]$ and $DF_\eta[u,X]$ in $\mathfrak{S}^{\star n}_{p,M}$ that are Lipschitz in their argument u, in the sense of (5.3.8) and (5.3.9), respectively. Assume further that $\boldsymbol{Z}$ is a local L^q-integrator for some $q > \dim U$. Then there exists a particular solution $X[u].(\omega)$ that is, for nearly every $\omega \in \Omega$, differentiable as a map from U to path space[21] $\mathscr{D}^n$.*

Using theorem 5.2.24 on page 295, it suffices to assume that $\boldsymbol{Z}$ is an L^0-integrator when $\boldsymbol{F}$ is an autologous coupling coefficient:

---

[29] See exercise A.3.16 on page 401.

Theorem 5.3.10 *Suppose that the F_η are differentiable in the sense of example 5.3.2, their derivatives being Lipschitz in $u \in U \subseteq \mathbb{R}^k$. Then there exists a particular solution $X[u].(\omega)$ that is, for every $\omega \in \Omega$, differentiable as a map from U to path space $\mathscr{D}^n$.*[17]

Higher Order Derivatives

Again let $(E, \| \ \|_E)$ and $(S, \| \ \|_S)$ be seminormed spaces and let $\mathcal{U} \subseteq E$ be open and convex. To paraphrase definition A.2.49, a function $F : \mathcal{U} \to S$ is differentiable at $u \in \mathcal{U}$ if "*it can be approximated at u by an affine function strictly better than linearly.*" We can paraphrase Taylor's formula A.2.42 similarly: a function on $U \subseteq \mathbb{R}^k$ with continuous derivatives up to order l at u "*can be approximated at u by a polynomial of degree l to an order strictly better than l.*" In fact, Taylor's formula is the main merit of having higher order differentiability. It is convenient to use this behavior as the *definition* of differentiability to higher orders. It essentially agrees with the usual recursive definition (exercise 5.3.18 on page 310).

Definition 5.3.11 *Let $\| \ \|_S^\circ \le \| \ \|_S$ be a seminorm on S that satisfies $\|x\|_S = 0 \Leftrightarrow \|x\|_S^\circ = 0 \ \forall x \in S$. The map $F : \mathcal{U} \to S$ is **l-times $\| \ \|_S^\circ$-weakly differentiable** at $u \in \mathcal{U}$ if there exist continuous symmetric λ-forms*[30]

$$D^\lambda F[u] : \underbrace{E \otimes \cdots \otimes E}_{\lambda \text{ factors}} \to S \,, \qquad \lambda = 1, \ldots, l \,,$$

such that

$$F[v] = \sum_{0 \le \lambda \le l} \frac{1}{\lambda!} D^\lambda F[u]\cdot(v-u)^{\otimes\lambda} + R^l F[u;v] \,, \qquad (5.3.11)$$

where

$$\left\| R^l F[u;v] \right\|_S^\circ = o\left(\|v-u\|_E^l\right) \quad \text{as } v \to u \,.$$

*$D^\lambda F[u]$ is the $\boldsymbol{\lambda}^{th}$ **derivative** of F at u; and the first sum on the right in (5.3.11) is the **Taylor polynomial of degree l** of F at u, denoted $T^l F[u] : v \mapsto T^l F[u](v)$. If*

$$\sup\left\{ \frac{\|R^l F[u;v]\|_S^\circ}{\delta^l} : u, v \in U \,,\ \|v-u\|_E < \delta \right\} \xrightarrow[\delta \to 0]{} 0 \,,$$

*then F is **l-times $\| \ \|_S^\circ$-weakly uniformly differentiable**. If the target space of F is $\mathfrak{S}^{\star n}_{p,M}$, we say that F is l-times weakly (uniformly) differentiable provided it is l-times $\| \ \|^\star_{p^\circ,M^\circ}$-weakly (uniformly) differentiable in the sense above for every Picard norm*[11] *$\| \ \|^\star_{p^\circ,M^\circ}$ with $p^\circ < p$ and $M^\circ > M$.*

---

[30] A λ-form on E is a function D of λ arguments in E that is linear in each of its arguments separately. It is ***symmetric*** if it equals its ***symmetrization***, which at $(\xi_1, \cdots, \xi_\lambda) \in E^\lambda$ has the value $\frac{1}{\lambda!} \sum D\cdot\xi_{\pi_1} \otimes \cdots \otimes \xi_{\pi_\lambda}$, the sum being taken over all permutations π of $\{1, \cdots, \lambda\}$.

In (5.3.11) we write $D^\lambda F[u]\cdot\xi_1\otimes\cdots\otimes\xi_\lambda$ for the value of the form $D^\lambda F[u]$ at the argument $(\xi_1,\ldots,\xi_\lambda)$ and abbreviate this to $D^\lambda F[u]\cdot\xi^{\otimes\lambda}$ if $\xi_1=\cdots=\xi_\lambda=\xi$. For $\lambda=0$, $D^0F[u]\cdot(v-u)^{\otimes 0}$ stands as usual for the constant $F[u]\in S$. $D^\lambda F[u]\cdot\xi_1\otimes\cdots\otimes\xi_\lambda$ can be constructed from the values $D^\lambda F[u]\cdot\xi^{\otimes\lambda}$, $\xi\in E$: it is the coefficient of $\tau_1\cdots\tau_\lambda$ in $D^\lambda F[u]\cdot(\tau_1\xi_1+\cdots+\tau_\lambda\xi_\lambda)^{\otimes\lambda}/\lambda!$. To say that $D^\lambda F[u]$ is continuous means that

$$\begin{aligned}\left\|D^\lambda F[u]\right\| &\stackrel{\text{def}}{=} \sup\left\{\left\|D^\lambda F[u]\cdot\xi_1\otimes\cdots\otimes\xi_\lambda\right\|_S : \|\xi_1\|_E\le 1,\ldots,\|\xi_\lambda\|_E\le 1\right\}\\ &\le \lambda^{\lambda/2}\sup\left\{\left\|D^\lambda F[u]\cdot\xi^{\otimes\lambda}\right\|_S : \|\xi\|_E\le 1\right\}\end{aligned}\tag{5.3.12}$$

is finite (inequality (5.3.12) is left to the reader to prove). $D^\lambda F[u]$ does not depend on l; indeed, the last $l-\lambda$ terms of the sum in (5.3.11) are $o(\|v-u\|_E^\lambda)$ if measured with $\|\ \|_S^\circ$. In particular, D^1F is the weak derivative DF of definition A.2.49.

Example 5.3.12 — Trouble Consider a function $f:\mathbb{R}\to\mathbb{R}$ that has l continuous bounded derivatives, *vis.* $f(x)=\cos x$. One hopes that composition with f, which takes ϕ to $F[\phi]\stackrel{\text{def}}{=} f\circ\phi$, might define an l-times weakly[31] differentiable map from $L^p(\mathbb{P})$ to itself. Alas, it does not. Indeed, if it did, then $D^\lambda F[\phi]\cdot\psi^{\otimes\lambda}$ would have to be multiplication of the λth derivative $f^{(\lambda)}(\phi)$ with ψ^λ. For $\psi\in L^p$ this product can be expected to lie in $L^{p/\lambda}$, but not generally in L^p. However:

If $f:\mathbb{R}^n\to\mathbb{R}^n$ has continuous bounded partial derivatives of all orders $\lambda\le l$, then $F:\phi\to F[\phi]\stackrel{\text{def}}{=} f\circ\phi$ is weakly differentiable as a map from $L^p_{\mathbb{R}^n}$ to $L^{p/\lambda}_{\mathbb{R}^n}$, for $1\le\lambda\le l$, and[1]

$$D^\lambda F[\phi]\cdot\psi_1\otimes\cdots\otimes\psi_\lambda=\frac{\partial^\lambda f(\phi)}{\partial x^{\nu_\lambda}\cdots\partial x^{\nu_1}}\times\psi_1^{\nu_1}\cdots\psi_\lambda^{\nu_\lambda}\,.$$

These observations lead to a more modest notion of higher order differentiability, which, though technical and useful only for functions that take values in L^p or in $\mathfrak{S}^{\star n}_{p,M}$, has the merit of being pertinent to the problem at hand:

Definition 5.3.13 *(i) A map $F:\mathcal{U}\to\mathfrak{S}^{\star n}_{p,M}$* ***has l tiered weak derivatives*** *if for every $\lambda\le l$ it is λ-times weakly differentiable as a map from $\mathcal{U}$ to $\mathfrak{S}^{\star n}_{p/\lambda,M\lambda}$.*

(ii) A parameter-dependent coupling coefficient $F:U\times\mathfrak{S}^{\star n}_{p,M}\to\mathfrak{S}^{\star n}_{p,M}$ with l tiered weak derivatives ***has l bounded tiered weak derivatives*** *if*

$$\left\|D^\lambda F_\eta[u,X]\cdot\binom{\xi_1}{\Xi_1}\otimes\cdots\otimes\binom{\xi_\lambda}{\Xi_\lambda}\right\|^\star_{p/\lambda,M\lambda}\le C\prod_{1\le j\le\lambda}\left(\|\xi_j\|_E+\left\|\Xi_j\right\|^\star_{p/i_j,Mi_j}\right)$$

for some constant C whenever $i_1,\ldots,i_\lambda\in\mathbb{N}$ have $i_1+\cdots+i_\lambda\le\lambda$.

[31] That F is not Fréchet differentiable, not even when $l=1$, we know from example A.2.48.

Example 5.3.14 The markovian parameter-dependent coupling coefficient $(u, X) \mapsto F[u, X] \stackrel{\text{def}}{=} f(u, X)$ of example 5.3.1 on page 299 has l bounded tiered weak derivatives provided the function f has bounded continuous derivatives of all orders $\leq l$. This is immediate from Taylor's formula A.2.42.

Example 5.3.15 Example 5.3.2 on page 299 has an extension as well. Assume that the map $f : U \times \mathscr{D}^n \to \mathscr{D}^n$ has l continuous bounded Fréchet derivatives. This is to mean that for every $t < \infty$ the restriction of f to $U \times \mathscr{D}^{nt}$, $\mathscr{D}^{nt}$ the Banach space of paths stopped at t and equipped with the topology of uniform convergence, is l-times continuously Fréchet differentiable, with the norm of the λth derivative being bounded in t. Then $F[u, X]_{.}(\omega) \stackrel{\text{def}}{=} f[u, X_{.}(\omega)]$ again defines a parameter-dependent coupling coefficient that is l-times weakly uniformly differentiable with bounded tiered derivatives.

Theorem 5.3.16 *Assume that in equation (5.3.1) on page 298 the initial value $C[u]$ has l tiered weak derivatives on U and that the coupling coefficients $F_\eta[u, X]$ have l bounded tiered weak derivatives.*

Then the solution $X[u]$ has l tiered weak derivatives on U as well, and $DX^l[u]$ is given by equation (5.3.18) below.

Proof. By theorem 5.3.6 on page 302 this is true when $l = 1$ – a good start for an induction. In order to get an idea what the derivatives $D^\lambda X[u]$ might be when $1 < \lambda \leq l$, let us assume that X does in fact have l tiered weak derivatives. With

$$v = u + \tau\xi\,, \quad \|\xi\|_E = 1\,, \quad \tau = \|v - u\|_E\,,$$

we write this as

$$X[v] - X[u] = \sum_{1 \leq \lambda \leq l} \frac{\tau^\lambda}{\lambda!} D^\lambda X[u] \cdot \xi^{\otimes\lambda} + R^l X[u; v]\,, \tag{5.3.13}$$

where[5]
$$\left\| R^l X[u; v] \right\|_{p^\circ/l, M^\circ l} = o(\tau^l) \qquad \text{for } p^\circ < p\,,\ M^\circ > M\,.$$

On the other hand,

$$\begin{aligned} X[v] - X[u] &= C[v] - C[u] + \Big\{ \boldsymbol{F}[v, X[v]] - \boldsymbol{F}[u, X[u]] \Big\}_{\leftarrow} * \boldsymbol{Z} \\ &= \sum_{1 \leq \lambda \leq l} \frac{\tau^\lambda}{\lambda!} D^\lambda C[u] \cdot \xi^{\otimes\lambda} + R^l C[u; v] \\ &\quad + \sum_{1 \leq \lambda \leq l} \frac{1}{\lambda!} \left\{ D^\lambda \boldsymbol{F}[u, X[u]] \cdot \begin{pmatrix} v - u \\ X[v] - X[u] \end{pmatrix}^{\otimes\lambda} \right\}_{\leftarrow} * \boldsymbol{Z} \\ &\qquad + R^l \boldsymbol{F}[u, X[u]; v, X[v]]_{\leftarrow} * \boldsymbol{Z} \end{aligned} \tag{5.3.14}$$

$$= \sum_{1\le\lambda\le l} \frac{\tau^\lambda}{\lambda!} D^\lambda C[u]\cdot\xi^{\otimes\lambda} + R^l C[u;v] \tag{5.3.15}$$

$$+ \sum_{1\le\lambda\le l} \frac{1}{\lambda!}\Big\{D^\lambda \boldsymbol{F}\big[u, X[u]\big]\cdot\Delta^\lambda[\tau]\Big\}_{\leftarrow} *\boldsymbol{Z}$$

$$+ R^l \boldsymbol{F}\big[u, X[u]; v, X[v]\big]_{\leftarrow} *\boldsymbol{Z} \,, \tag{5.3.16}$$

where by the multinomial formula[32]

$$\Delta^\lambda[\tau] \overset{\text{def}}{=} \begin{pmatrix} v-u \\ X[v]-X[u]\end{pmatrix}^{\otimes\lambda} = \begin{pmatrix} \tau\xi \\ \sum_{1\le\rho\le l} \frac{\tau^\rho D^\rho X[u]\cdot\xi^{\otimes\rho}}{\rho!} + R^l X[u;v]\end{pmatrix}^{\otimes\lambda}$$

$$= \sum_{\lambda_0+\cdots+\lambda_{l+1}=\lambda} \binom{\lambda}{\lambda_0\ldots\lambda_{l+1}} \times \frac{\tau^{\lambda_0+1\lambda_1+\cdots+l\lambda_l}}{1!\cdots l!} \times$$

$$\times \begin{pmatrix}\xi\\0\end{pmatrix}^{\otimes\lambda_0} \otimes \begin{pmatrix}0\\ D^1X[u]\cdot\xi^{\otimes 1}\end{pmatrix}^{\otimes\lambda_1} \otimes\cdots\otimes \begin{pmatrix}0\\ D^lX[u]\cdot\xi^{\otimes l}\end{pmatrix}^{\otimes\lambda_l} \otimes \begin{pmatrix}0\\ R^lX[u;v]\end{pmatrix}^{\otimes\lambda_{l+1}}$$

and where

$$\big\|R^l C\big\|_{p^\circ/l, M^\circ l} = o(\tau^l) = \big\|R^l \boldsymbol{F}\big\|_{p^\circ/l, M^\circ l} \qquad \text{for } p^\circ{<}p\,,\ M^\circ{>}M$$

(the arguments of R^lC and $R^l\boldsymbol{F}$ are not displayed). Line (5.3.13), and lines (5.3.15)–(5.3.16) together, each are of the form "*a polynomial in τ plus terms that are $o(\tau^l)$*" when measured in the pertinent Picard norm, which here is $\| \ \|_{p^\circ/l, M^\circ l}$. Clearly,[25] then, the coefficient of τ^l in the two polynomials must be the same:[32]

$$\frac{D^lX[u]\cdot\xi^{\otimes l}}{l!} = \frac{D^lC[u]\cdot\xi^{\otimes l}}{l!} + \Bigg\{ \sum_{1\le\lambda\le l} \frac{1}{\lambda!} D^\lambda \boldsymbol{F}\big[u, X[u]\big]$$

$$\cdot \sum_{\substack{\lambda_0+\lambda_1+\cdots+\lambda_l=\lambda \\ \lambda_0+1\lambda_1+\cdots+l\lambda_l=l}} \binom{\lambda}{\lambda_0\ldots\lambda_l} \times \frac{1}{1!\cdots l!} \times \tag{5.3.17}$$

$$\times \begin{pmatrix}\xi\\0\end{pmatrix}^{\otimes\lambda_0} \otimes \begin{pmatrix}0\\ D^1X[u]\cdot\xi^{\otimes 1}\end{pmatrix}^{\otimes\lambda_1} \otimes\cdots\otimes \begin{pmatrix}0\\ D^lX[u]\cdot\xi^{\otimes l}\end{pmatrix}^{\otimes\lambda_l} \Bigg\}_{\leftarrow} *\boldsymbol{Z}\,.$$

The term $D^lX[u]\cdot\xi^{\otimes l}$ occurs precisely once on the right-hand side, namely when $\lambda_l = 1$ and then $\lambda = 1$. Therefore the previous equation can be rewritten as a stochastic differential equation for $D^lX[u]\cdot\xi^{\otimes l}$:

$$D^lX[u]\cdot\xi^{\otimes l} = \Big(D^lC[u]\cdot\xi^{\otimes l} + (\overline{C}^l[u]\cdot\xi^{\otimes l})_{\leftarrow} *\boldsymbol{Z}\Big)$$

$$+ \Big(D_2\boldsymbol{F}\big[u, X[u]\big]\cdot D^lX[u]\cdot\xi^{\otimes l}\Big)_{\leftarrow} *\boldsymbol{Z}\,, \tag{5.3.18}$$

[32] Use $\binom{\xi}{D} = \binom{\xi}{0} + \binom{0}{D}$. It is understood that a term of the form $(\cdots)^{\otimes 0}$ is to be omitted.

where $D_2\boldsymbol{F}$ is the partial derivative in the X-direction (see A.2.50) and

$$\overline{C}^l[u]\cdot\xi^{\otimes l} \stackrel{\text{def}}{=} \sum_{1\le\lambda\le l} \frac{D^\lambda \boldsymbol{F}\big[u,X[u]\big]}{\lambda!}\cdot \sum_{\substack{\lambda_0+\lambda_1+\cdots+\lambda_{l-1}=\lambda\\ \lambda_0+1\lambda_1+\cdots+(l-1)\lambda_{l-1}=l}} \binom{\lambda}{\lambda_0\ldots\lambda_{l-1}} \times \frac{1}{1!\cdots(l-1)!} \times$$

$$\times \begin{pmatrix}\xi\\0\end{pmatrix}^{\otimes\lambda_0} \otimes \begin{pmatrix}0\\D^1X[u]\cdot\xi^{\otimes 1}\end{pmatrix}^{\otimes\lambda_1} \otimes\cdots\otimes \begin{pmatrix}0\\D^{l-1}X[u]\cdot\xi^{\otimes(l-1)}\end{pmatrix}^{\otimes\lambda_{l-1}} .$$

Now by induction hypothesis, $D^iX\cdot\xi^{\otimes i}$ stays bounded in $\mathfrak{S}^{\star n}_{p/i,Mi}$ as ξ ranges over the unit ball of E and $1\le i\le l-1$. Therefore $D^\lambda\boldsymbol{F}\big[u,X[u]\big]$ applied to any of the summands stays bounded in $\mathfrak{S}^{\star n}_{p/l,Ml}$, and then so does $\overline{C}^l[u]\cdot\xi^{\otimes l}$. Since the coupling coefficient of (5.3.18) has Lipschitz constant L, we conclude that $D^lX[u]\cdot\xi^{\otimes l}$, *defined by* (5.3.18), stays bounded in $\mathfrak{S}^{\star n}_{p/l,Ml}$ as ξ ranges over E_1.

There is a little problem here, in that (5.3.18) defines $D^lX[u]$ as an l-homogeneous map on E, but not immediately as a l-linear map on $\bigotimes_l E$. To overcome this observe that $\overline{C}[u]\cdot\xi^{\otimes l}$ is in an obvious fashion the value at $\xi^{\otimes l}$ of an l-linear map

$$\vec{\xi}^{\otimes l} \stackrel{\text{def}}{=} \xi_1\otimes\cdots\otimes\xi_l \;\mapsto\; \overline{C}[u]\cdot\vec{\xi}^{\otimes l} .$$

Replacing every $\xi^{\otimes l}$ in (5.3.18) by $\vec{\xi}^{\otimes l}$ produces a stochastic differential equation for an n-vector $D^lX[u]\cdot\vec{\xi}^{\otimes l}\in\mathfrak{S}^{\star n}_{p/l,Ml}$, whose solution defines an l-linear form that at $\vec{\xi}^{\otimes l}=\xi^{\otimes l}$ agrees with the $D^lX[u]$ of equation (5.3.18). The lth derivative $D^lX[u]$ is redefined as the symmetrization [30] of this l-form. It clearly satisfies equation (5.3.18) and is the only symmetric l-linear map that does.

It is left to be shown that for $l>1$ the difference $R^l[u;v]$ of $X[v]-X[u]$ and the Taylor polynomial $T^lX[u](\tau\xi)$ is $o(\tau^l)$ if measured in $\mathfrak{S}^{\star n}_{p^\circ/l,M^\circ l}$. Now by induction hypothesis, $R^{l-1}[u;v]=D^lX[u](v-u)^{\otimes l}/l!+R^l[u;v]$ is $o(\tau^{l-1})$; hence clearly so is $R^l[u;v]$. Subtracting the defining equations (5.3.17) for $l=1,2,\ldots$ from (5.3.13) and (5.3.15)–(5.3.16) leaves us with this equation for the remainder $R^lX[u;v]$:

$$R^lX[u;v]=R^lC[u;v]+\sum_{1\le\lambda\le l}\frac{1}{\lambda!}\Big\{D^\lambda\boldsymbol{F}\big[u,X[u]\big]\cdot\overline{\Delta}^\lambda[\tau]\Big\}_{-}{*}\boldsymbol{Z}$$

$$+R^l\boldsymbol{F}[u,X[u];v,X[v]]_{-}{*}\boldsymbol{Z}\,, \tag{5.3.19}$$

where

$$\overline{\Delta}^\lambda[\tau]\stackrel{\text{def}}{=}\sum_{\substack{\lambda_0+\lambda_1+\cdots+\lambda_l=\lambda\\ \lambda_0+1\lambda_1+\cdots+l\lambda_l>l}}\binom{\lambda}{\lambda_0\ldots\lambda_l}\times\frac{\tau^{\lambda_0+1\lambda_1+\cdots+l\lambda_l}}{1!\cdots l!}\times$$

$$\times \binom{\xi}{0}^{\otimes\lambda_0} \otimes \begin{pmatrix} 0 \\ D^1X[u]\cdot\xi^{\otimes 1}\end{pmatrix}^{\otimes\lambda_1} \otimes\cdots\otimes \begin{pmatrix} 0 \\ D^lX[u]\cdot\xi^{\otimes l}\end{pmatrix}^{\otimes\lambda_l}$$

$$+ \sum_{\substack{\lambda_0+\lambda_1+\cdots+\lambda_{l+1}=\lambda \\ \lambda_{l+1}>0}} \binom{\lambda}{\lambda_0\ldots\lambda_{l+1}} \times \frac{\tau^{\lambda_0+1\lambda_1+\cdots+l\lambda_l}}{1!\cdots l!} \times$$

$$\times \binom{\xi}{0}^{\otimes\lambda_0} \otimes \begin{pmatrix} 0 \\ D^1X[u]\cdot\xi^{\otimes 1}\end{pmatrix}^{\otimes\lambda_1} \otimes\cdots\otimes \begin{pmatrix} 0 \\ D^lX[u]\cdot\xi^{\otimes l}\end{pmatrix}^{\otimes\lambda_l} \otimes \begin{pmatrix} 0 \\ R^lX[u;v]\end{pmatrix}^{\otimes\lambda_{l+1}} .$$

The terms in the first sum all are $o(\tau^l)$. So are all of the terms of the second sum, except the one that arises when $\lambda_{l+1} = 1$ and $\lambda_0 + 1\lambda_1 + \cdots + l\lambda_l = 0$ and then $\lambda = 1$. That term is $\begin{pmatrix} 0 \\ R^lX[u;v]/l! \end{pmatrix}$. Lastly, $R^l\boldsymbol{F}[u, X[u]; v, X[v]]$ is easily seen to be $o(\tau^l)$ as well. Therefore equation (5.3.19) boils down to a stochastic differential equation for $R^lX[u;v]$:

$$R^lX[u;v] = \Big\{R^lC[u;v] + o(\tau^l)_{.-}{*}\boldsymbol{Z}\Big\} + \big(D_2F[u, X[u]]\cdot R^lX[u;v]\big)_{.-} {*}\boldsymbol{Z} .$$

According to inequalities (5.2.23) on page 290 and (5.2.5) on page 284, we have $R^lX[u;v] = o(\|u-v\|_E^l)$, as desired. ∎

Exercise 5.3.17 If in addition C and $\boldsymbol{F}$ are weakly uniformly differentiable, then so is X.

Exercise 5.3.18 Suppose $F : U \to S$ is l-times weakly uniformly differentiable with bounded derivatives:

$$\sup_{\lambda\le l,\, u\in U} \Big\| D^\lambda F[u] \Big\| < \infty \qquad \text{(see inequality (5.3.12))}.$$

Then, for $\lambda < l$, $D^\lambda F$ is weakly uniformly differentiable, and its derivative is $D^{\lambda+1}F$.

Problem 5.3.19 Generalize the pathwise differentiability result 5.3.9 to higher order derivatives.

5.4 Pathwise Computation of the Solution

We return to the stochastic differential equation (5.1.3), driven by a vector $\boldsymbol{Z}$ of integrators:

$$X = C + F_\eta[X]_{.-}{*}Z^\eta = C + \boldsymbol{F}[X]_{.-}{*}\boldsymbol{Z} .$$

Under mild conditions on the coupling coefficients F_η there exists an algorithm that computes the path $X_.(\omega)$ of the solution from the input paths $C_.(\omega), \boldsymbol{Z}_.(\omega)$. It is a slight variant of the well-known adaptive[33] Euler–Peano

[33] Adaptive: the step size is not fixed in advance but is adapted to the situation at every step – see page 281.

scheme of little straight steps, a variant in which the next computation is carried out not after a fixed time has elapsed but when the effect of the noise $\boldsymbol{Z}$ has changed by a fixed threshold – compare this with exercise 3.7.24. There exists an algorithm that takes the $n+d$ paths $t \mapsto C_t^\nu(\omega)$ and $t \mapsto Z_t^\eta(\omega)$ and computes from them a path $t \mapsto {}^\delta X_t(\omega)$, which, when δ is taken through a summable sequence, converges ω–by–ω uniformly on bounded time-intervals to the path $t \mapsto X_t(\omega)$ of the exact solution, irrespective of $\mathbb{P} \in \mathfrak{P}[\boldsymbol{Z}]$. This is shown in theorems 5.4.2 and 5.4.5 below.

The Case of Markovian Coupling Coefficients

One cannot of course expect such an algorithm to exist unless the coupling coefficients F_η are endogenous. This is certainly guaranteed when the coupling coefficients are markovian[8], case treated first. That is to say, we assume here that there are ordinary vector fields $f_\eta : \mathbb{R}^n \to \mathbb{R}^n$ such that

$$F_\eta[X]_t = f_\eta \circ X_t \quad ; \text{and} \quad |f_\eta(y) - f_\eta(x)| \le L \cdot |y - x| \tag{5.4.1}$$

ensures the Lipschitz condition (5.2.8) and with it the existence of a unique solution of equation (5.2.1), which takes the following form:[1]

$$X_t = C_t + \int_0^t f_\eta(X)_{s-}\, dZ_s^\eta \,. \tag{5.4.2}$$

The adaptive[33] Euler–Peano algorithm computing the approximate X' for a fixed threshold[34] $\delta > 0$ works as follows: define $T_0 \overset{\text{def}}{=} 0$, $X'_0 \overset{\text{def}}{=} C_0$ and continue recursively: when the stopping times $T_0 \le T_1 \le \ldots \le T_k$ and the function $X' : [\![0, T_k]\!] \to \mathbb{R}$ have been defined so that $X'_{T_k} \in \mathcal{F}_{T_k}$, then set[1]

$${}^0\Xi'_t \overset{\text{def}}{=} C_t - C_{T_k} + f_\eta(X'_{T_k}) \cdot (Z_t^\eta - Z_{T_k}^\eta) \tag{5.4.3}$$

and
$$T_{k+1} \overset{\text{def}}{=} \inf\left\{t > T_k : |{}^0\Xi'_t| > \delta\right\}, \tag{5.4.4}$$

and extend X':
$$X'_t \overset{\text{def}}{=} X'_{T_k} + {}^0\Xi'_t \quad \text{for } T_k \le t \le T_{k+1} \,. \tag{5.4.5}$$

In other words, the prescription is to wait after time T_k not until some fixed time has elapsed but until random input plus effect of the drivers together have changed sufficiently to warrant a new computation; then extend X' "linearly" to the interval that just passed, and start over. It is obvious how to write a little loop for a computer that will compute the path $X'_.(\omega)$ of the ***Euler–Peano approximate*** $X'_.(\omega)$ as it receives the input paths $C_.(\omega)$ and $\boldsymbol{Z}_.(\omega)$. The scheme (5.4.5) expresses quite intuitively the meaning of the differential equation $dX = \boldsymbol{f}(X)\, d\boldsymbol{Z}$. If one can show that it converges, one should be satisfied that the limit is for all intents and purposes a solution of the differential equation (5.4.2).

[34] Visualize δ as a ***step size*** on the dependent variables' axis.

One can, and it is. An easy induction shows that [1,35]

for $t < T_\infty \overset{\text{def}}{=} \sup_{k<\infty} T_k$

we have $$X'_t = C_t + \sum_{k=0}^{\infty} f_\eta(X'_{T_k}) \cdot (Z^\eta_{T_{k+1}\wedge t} - Z^\eta_{T_k \wedge t})$$

by 3.5.2: $$= C_t + \int_0^t \sum_{0\le k<\infty} f_\eta(X'_{T_k}) \cdot ((T_k, T_{k+1}]]\, dZ^\eta \tag{5.4.6}$$

and exhibits $X' : [[0, T_\infty)) \to \mathbb{R}$ as an adapted process that is right-continuous with left limits on $[[0, T_\infty))$ – but for all we know so far not necessarily at T_∞.

On the way to proving that the Euler–Peano approximate X' is close to the exact solution X, the first order of business is to show that $T_\infty = \infty$. In order to do this recall the scalæfication of processes in definition 3.7.22 that is attached to the random partition[36] $\mathcal{T} \overset{\text{def}}{=} \{0 = T_0 \le T_1 \le \ldots \le T_\infty \le \infty\}$:

$$\text{for } Y \in \mathfrak{D}^n, \quad Y^{\mathcal{T}} \overset{\text{def}}{=} \sum_{0\le k\le\infty} Y_{T_k} \cdot [[T_k, T_{k+1})) \in \mathfrak{D}^n .$$

Attach now a new coupling coefficient $\boldsymbol{F}'$ to $\boldsymbol{F}$ and the partition $\mathcal{T}$ by

$$F'_\eta[Y] \overset{\text{def}}{=} F_\eta[Y^{\mathcal{T}}]^{\mathcal{T}} \quad \text{for } Y \in \mathfrak{D}^n \text{ and } \eta = 1, \ldots, d, \tag{5.4.7}$$

and consider the stochastic differential equation[1]

$$Y = C + F'_\eta[Y]_{.-} {*} Z^\eta , \tag{5.4.8}$$

which reads[35] $$Y_t = C_t + \int_0^t \sum_{0\le k\le\infty} f_\eta(Y_{T_k}) \cdot ((T_k, T_{k+1}]]\, dZ^\eta \tag{5.4.9}$$

in the present markovian case. The coupling coefficient $\boldsymbol{F}'$ evidently satisfies the Lipschitz condition (5.2.8) with Lipschitz constant L from (5.4.1). There is therefore a *unique global* solution Y to equation (5.4.8), and a comparison of (5.4.9) with equation (5.4.6) reveals that $X' = Y$ on $[[0, T_\infty))$. On the set $[T_\infty < \infty]$, $Y \in \mathfrak{D}^n$ has almost surely no oscillatory discontinuity, but X' surely does, since the values of this process at T_k and T_{k+1} differ by at least δ, yet $\sup_k |X'|_{T_k}$ is bounded by $|Y|^\star_{T_\infty} < \infty$. The set $[T_\infty < \infty]$ is therefore negligible, even nearly empty, and thus the T_k increase without bound, nearly.

---

[35] In accordance with convention A.1.5 on page 364, sets are identified with their (idempotent) indicator functions. A stochastic interval $((S, T]]$, for instance, has at the instant s the value $((S,T]]_s = [S < s \le T] = \begin{cases} 1 & \text{if } S(\omega) < s \le T(\omega) \\ 0 & \text{elsewhere} \end{cases}$.

[36] A partition $\mathcal{T}$ is assumed to contain $T_\infty \overset{\text{def}}{=} \sup_{k<\infty} T_k$, and the convention $T_{\infty+1} \overset{\text{def}}{=} \infty$ simplifies formulas.

We now run Picard's iterative scheme, starting with the scalæfication[35]

$$X^{(0)} \overset{\text{def}}{=} X'^{\mathcal{T}} = \sum_{k=0}^{\infty} X'_{T_k} \cdot [\![T_k, T_{k+1})\!) \,.$$

Then $$X^{(1)} \overset{\text{def}}{=} \mathfrak{U}[X^{(0)}] = C + f_\eta(X^{(0)})_{.-} {*} Z^\eta$$

in view of (5.4.6) equals X' and differs from $X^{(0)}$ by less than δ uniformly on $\boldsymbol{B}$. Therefore $X^{(0)}$ and $X^{(1)}$ differ by less than δ in any of the norms $\|\ \|^\star_{p,M}$. The argument of item 5.1.5 immediately provides the estimate (i) below. (Another way to arrive at inequality (5.4.10) is to observe that, on the solution X' of (5.4.8), the F_η and F'_η differ by less than $\delta{\cdot}L$, and to invoke inequality (5.2.35).)

Proposition 5.4.1 *Assume that $\boldsymbol{Z}$ is a local L^q-integrator for some $q \geq 2$ and that equation (5.4.2) with markovian coupling coefficients as in (5.4.1) has its exact global solution X inside $\mathfrak{S}^{\star n}_{p,M}$ for some[23] $p \in [2, q]$ and some[23] $M > M^{\diamond(5.2.20)}_{p,L}$. Then, with $\gamma \overset{\text{def}}{=} M^{\diamond(5.2.20)}_{p,L}/M$,*

(i) $$\|X - X'\|^\star_{p,M} \leq \frac{\gamma}{1-\gamma} \cdot \delta\,; \tag{5.4.10}$$

(ii) and $$|X - X^{(n)}|^\star_t \xrightarrow[n\to\infty]{} 0 \quad \textit{almost surely}$$

at any nearly finite stopping time T, whenever the $X^{(n)}$ are the Euler–Peano approximates constructed via (5.4.5) from a summable sequence of thresholds $\delta_n > 0$. In other words, for nearly every $\omega \in \Omega$ we have $X^{(n)}_t \xrightarrow[n\to\infty]{} X_t$, uniformly in $t \in [0, T(\omega)]$.

Statement (ii) does not change if $\mathbb{P}$ is replaced with an equivalent probability $\mathbb{P}'$ in the manner of the proof of theorem 5.2.15; thus it holds without assuming more about $\boldsymbol{Z}$ than that it be an L^0-integrator:

Theorem 5.4.2 *Let X denote the strong global solution of the markovian system (5.4.1), (5.4.2), driven by the L^0-integrator $\boldsymbol{Z}$. Fix any summable sequence of strictly positive reals δ_n and let $X^{(n)}$ be defined as the Euler–Peano approximates of (5.4.5) for $\delta = \delta_n$. Then $X^{(n)} \xrightarrow[n\to\infty]{} X$ uniformly on bounded time-intervals, nearly.* ∎

Proof of Proposition 5.4.1 (ii). This is a standard application of the Borel–Cantelli lemma. Namely, suppose that T is one of the T^λ, say $T = T^\mu$. Then

$$\|X - X^{(n)}\|^\star_{p,M} \leq \delta_n \cdot \frac{\gamma}{1-\gamma}$$

implies $$\Big\| |X - X^{(n)}|^\star_{T^\mu -} \Big\|_{L^p} \leq \delta_n \cdot \frac{e^{M\mu}\gamma}{1-\gamma}\,,$$

and $$\mathbb{P}\Big[|X - X^{(n)}|^\star_{T^\mu -} > \sqrt{\delta_n}\Big] \leq \Big(\frac{\delta_n \times e^{M\mu}\gamma}{\sqrt{\delta_n}(1-\gamma)}\Big)^p = const \cdot \delta_n^{p/2}\,,$$

which is summable over n, since $p \geq 2$. Therefore

$$\mathbb{P}\Big[\limsup_n |X - X^{(n)}|^\star_{T^\mu -} > 0\Big] = 0 .$$

For arbitrary almost surely finite T, the set $[\limsup_n |X - X^{(n)}|^\star_t > 0]$ is therefore almost surely a subset of $[T \geq T^\mu]$ and is negligible since T^μ can be made arbitrarily large by the choice of μ. ∎

Remark 5.4.3 In the adaptive[33] Euler–Peano algorithm (5.4.5) on page 311, any stochastic partition $\mathcal{T}$ can replace the specific partition (5.4.4), as long as[36] $T_\infty = \infty$ and the quantity ${}^0\Xi'_t$ does not change by more than δ over its intervals. Suppose for instance that C is constant and the f_η are bounded, say[4] $|f_\eta(x)| \leq K$. Then the partition defined recursively by $0 = T_0$, $T_{k+1} \stackrel{\text{def}}{=} \inf\{t > T_k : \sum_\eta |Z^\eta_t - Z^\eta_{T_k}| > \delta/K\}$ will do.

The Case of Endogenous Coupling Coefficients

For any algorithm similar to (5.4.5) and intended to apply in more general situations than the markovian one treated above, the coupling coefficients F_η still must be special. Namely, given *any* input path $(x_., z_.)$, F_η must return an output path. That is to say, the F_η must be endogenous Lipschitz coefficients as in example 5.2.12 on page 289. If they are, then in terms of the f_η the system (5.1.5) reads

$$X_t = C_t + \sum_\eta \int_0^t f_\eta[\boldsymbol{Z}_., X_.]_{s-}\, dZ^\eta_s \tag{5.4.11}$$

or, equivalently, $X = C + \sum_\eta f_\eta[\boldsymbol{Z}_., X_.]_{.-} * Z^\eta = C + \boldsymbol{f}[\boldsymbol{Z}_., X_.]_{.-} * \boldsymbol{Z}$. (5.4.12)

The adaptive[33] Euler–Peano algorithm (5.4.5) needs to be changed a little. Again we fix a strictly positive threshold δ, set $T_0 \stackrel{\text{def}}{=} 0$, $X'_0 \stackrel{\text{def}}{=} C_0$, and continue recursively: when the stopping times $T_0 \leq T_1 \leq \ldots \leq T_k$ and the function $X' : [\![0, T_k]\!] \to \mathbb{R}$ have been defined so that $X'_{T_k} \in \mathcal{F}_{T_k}$, then set[1,3]

$$ {}^0\boldsymbol{f}_t \stackrel{\text{def}}{=} \sup_{\eta,\nu} \left| f^\nu_\eta[\boldsymbol{Z}^{T_k}, X'^{T_k}]_t - f^\nu_\eta[\boldsymbol{Z}^{T_k}, X'^{T_k}]_{T_k} \right| , \quad t \geq T_k ;$$

$$ {}^0\Xi'_t \stackrel{\text{def}}{=} C_t - C_{T_k} + f_\eta\big[\boldsymbol{Z}^{T_k}, X'^{T_k}\big]_{T_k} \cdot \big(Z^\eta_t - Z^\eta_{T_k}\big) , t \geq T_k ;$$

and $$T_{k+1} \stackrel{\text{def}}{=} \inf\{t > T_k : {}^0\boldsymbol{f}_t > \delta \text{ or } |{}^0\Xi'_t| > \delta\} ;$$

and then extend $X' X'_t \stackrel{\text{def}}{=} X'_{T_k} + {}^0\Xi'_t$ for $T_k \leq t \leq T_{k+1}$. (5.4.13)

The spirit is that of (5.4.5), the stopping times T_k are possibly "a bit closer together than there," to make sure that $\boldsymbol{f}[\boldsymbol{Z}, X']_.$ does not vary too much on the intervals of the partition[36] $\mathcal{T} \stackrel{\text{def}}{=} \{T_0 \le T_1 \le \ldots \le \infty\}$. An induction shows as before that

for $\qquad t < T_\infty \stackrel{\text{def}}{=} \sup_{k<\infty} T_k$

we have
$$X'_t = C_t + \sum_{k=0}^{\infty} f_\eta[\boldsymbol{Z}, X']_{T_k} \cdot (Z^\eta_{T_{k+1}\wedge t} - Z^\eta_{T_k \wedge t})$$
$$= C_t + \int_0^t F'_\eta[\boldsymbol{Z}, X']_{.-} dZ^\eta ,$$

where the "$\mathcal{T}$-scalæfied" coupling coefficient F'_η is defined as in (5.4.7), for the present partition $\mathcal{T}$ of course. This exhibits $X' : [\![0, T_\infty)\!) \to \mathbb{R}$ as an adapted process that is right-continuous with left limits on $[\![0, T_\infty)\!)$ – but for all we know so far not necessarily at T_∞. Again we consider the stochastic differential equation (5.4.8):

$$Y = C + F'_\eta[Y]_{.-} {*} Z^\eta$$

and see that X' agrees with its unique global solution Y on $[\![0, T_\infty)\!)$. As in the markovian case we conclude from this that X' has no oscillatory discontinuities. Clearly $f_\eta[\boldsymbol{Z}^{\mathcal{T}}, X'^{\mathcal{T}}]$, which agrees with $f_\eta[\boldsymbol{Z}^{\mathcal{T}}, Y^{\mathcal{T}}]$ on $[\![0, T_\infty)\!)$, has no oscillatory discontinuities either. On the other hand, the very definition of T_∞ implies that one or the other of these processes surely must have a discontinuity on $[T_\infty < \infty]$. This set therefore is negligible, and $T_\infty = \infty$ almost surely.

Let us define $X^{(0)} \stackrel{\text{def}}{=} X'^{\mathcal{T}}$ and $X^{(1)} \stackrel{\text{def}}{=} \mathfrak{U}[X^{(0)}]$. Then

$$X^{(1)}_t = C_t + \int_0^t f_\eta[\boldsymbol{Z}, X^{(0)}]_{.-} \, dZ^\eta$$

and
$$X'_t = C_t + \int_0^t f_\eta[\boldsymbol{Z}, X^{(0)}]^{\mathcal{T}}_{.-} \, dZ^\eta$$

differ by less than $\delta C^\diamond_p / eM$ when measured with the norm $\| \ \|^\star_{p,M}$ (see exercise 5.2.2). X' and $X^{(0)}$ differ uniformly, and therefore in $\| \ \|^\star_{p,M}$, by less than δ. Therefore

$$\|X^{(1)} - X^{(0)}\|^\star_{p,M} \le \delta + \delta C^\diamond_p / eM ,$$

and in view of item 5.1.4 on page 276 the approximate $X^{(1)}$ differs little from the exact solution X of (5.2.1); in fact,

$$\|X - X^{(1)}\|^\star_{p,M} \le \delta \cdot \frac{M + C^\diamond_p/e}{M - M^\diamond} ,$$

and so
$$\|X - X'\|^\star_{p,M} \le \delta \cdot \frac{M + 2C^\diamond_p/e}{M - M^\diamond} . \tag{5.4.14}$$

We have recovered proposition 5.4.1 in the present non-markovian setting:

Proposition 5.4.4 *Assume that* $\boldsymbol{Z}$ *is a local* L^q*-integrator for some* $q \geq 2$*, and pick a* $p \in [2, q]$ *and an* $M > M_{p,L}^{\diamond(5.2.20)}$*,* L *being the Lipschitz constant of the endogenous coefficient* $\boldsymbol{f}$*. Then the global solution* X *of the Lipschitz system (5.4.11) lies in* $\mathfrak{S}_{p,M}^{\star n}$*, and the Euler–Peano approximate* X' *defined in equation (5.4.13) satisfies inequality (5.4.14).*

Even if $\boldsymbol{Z}$ is merely an L^0-integrator, this implies as in theorem 5.4.2 *the*

Theorem 5.4.5 *Fix any summable sequence of strictly positive reals* δ_n *and let* $X^{(n)}$ *be the Euler–Peano approximates of (5.4.13) for* $\delta = \delta_n$*. Then at any almost surely finite stopping time* T *and for almost all* $\omega \in \Omega$ *the sequence* $X_t^{(n)}(\omega)$ *converges to the exact solution* $X_t(\omega)$ *of the Lipschitz system (5.4.11) with endogenous coefficients, uniformly for* $t \in [0, T(\omega)]$*.*

Corollary 5.4.6 *Let* $\boldsymbol{Z}, \boldsymbol{Z}'$ *be* L^0*-integrators and* X, X' *solutions of the Lipschitz systems*

$$X = C + \boldsymbol{f}[\boldsymbol{Z}_., X_.]_{-} * \boldsymbol{Z} \quad \text{and} \quad X' = C' + \boldsymbol{f}[\boldsymbol{Z}_., X'_.]_{-} * \boldsymbol{Z}'$$

with endogenous coefficients, respectively. Let Ω_0 *be a subset of* Ω *and* $T : \Omega \to \mathbb{R}_+$ *a time, neither of them necessarily measurable. If* $C = C'$ *and* $\boldsymbol{Z} = \boldsymbol{Z}'$ *up to and including (excluding) time* T *on* Ω_0*, then* $X = X'$ *up to and including (excluding) time* T *on* Ω_0*, except possibly on an evanescent set.*

The Universal Solution

Consider again the endogenous system (5.4.12), reproduced here as

$$X = C + \boldsymbol{f}[X_., \boldsymbol{Z}_.]_{-} * \boldsymbol{Z} . \tag{5.4.15}$$

In view of Items 2.3.8–2.3.11, the solution can be computed on canonical path space. Here is how. Identify the process $R_t \stackrel{\text{def}}{=} (C_t, \boldsymbol{Z}_t) : \Omega \to \mathbb{R}^{n+d}$ with a representation $\underline{R}$ of $(\Omega, \mathcal{F}_.)$ on the canonical path space $\overline{\Omega} \stackrel{\text{def}}{=} \mathscr{D}^{n+d}$ equipped with its natural filtration $\overline{\mathcal{F}}_. \stackrel{\text{def}}{=} \mathcal{F}_.[\mathscr{D}^{n+d}]$. For consistency's sake let us denote the evaluation processes on $\overline{\Omega}$ by $\overline{\boldsymbol{Z}}$ and $\overline{C}$; to be precise, $\overline{\boldsymbol{Z}}_t(c_., \boldsymbol{z}_.) \stackrel{\text{def}}{=} \boldsymbol{z}_t$ and $\overline{C}_t(c_., \boldsymbol{z}_.) \stackrel{\text{def}}{=} c_t$. We contemplate the stochastic differential equation

$$\overline{X} = \overline{C} + \boldsymbol{f}[\overline{X}_., \overline{\boldsymbol{Z}}_.]_{-} * \overline{\boldsymbol{Z}} \tag{5.4.16}$$

or – see (2.3.10) $\quad \overline{X}_t(c_., \boldsymbol{z}_.) = c_t + \displaystyle\int_0^t f_\eta[\overline{X}_., \boldsymbol{z}_.]_{s-} \, dz_s^\eta .$

We produce a particularly pleasant solution $\overline{X}$ of equation (5.4.16) by applying the Euler–Peano scheme (5.4.13) to it, with $\delta = 2^{-n}$. The corresponding Euler–Peano approximate $\overline{X}^n$ in (5.4.13) is clearly adapted to the natural filtration $\mathcal{F}_.[\mathscr{D}^{n+d}]$ on path space. Next we set $\overline{X} \stackrel{\text{def}}{=} \lim \overline{X}^n$ where this limit exists and $\overline{X} \stackrel{\text{def}}{=} 0$ elsewhere. Note that no probability enters the definition of $\overline{X}$. Yet the process $\overline{X}$ we arrive at solves the stochastic differential equation (5.4.16) in the sense of any of the probabilities in $\mathfrak{P}[\overline{Z}]$. According to equation (2.3.11), $X_t \stackrel{\text{def}}{=} \overline{X}_t \circ \underline{R} = \overline{X}_t(C_., \boldsymbol{Z}_.)$ solves (5.4.15) in the sense of any of the probabilities in $\mathfrak{P}[\boldsymbol{Z}]$.

Summary 5.4.7 *The process $\overline{X}$ is càdlàg and adapted to $\mathcal{F}[\mathscr{D}^{d+n}]$, and it solves (5.4.16). Considered as a map from $\mathscr{D}^{d+n}$ to $\mathscr{D}^n$, it is adapted to the filtrations $\mathcal{F}_.[\mathscr{D}^{n+d}]$ and $\mathcal{F}^0_{.+}[\mathscr{D}^n]$ on these spaces. Since the solution X of (5.4.15) is given by $X_t = \overline{X}_t(C_., \boldsymbol{Z}_.)$, no matter which of the $\mathbb{P} \in \mathfrak{P}[\boldsymbol{Z}]$ prevails at the moment, $\overline{X}$ deserves the name* ***universal solution.***

A Non-Adaptive Scheme

It is natural to ask whether perhaps the stopping times T_k in the Euler–Peano scheme on page 311 can be chosen in advance, without the employ of an "infimum–detector" as in definition (5.4.4). In other words, we ask whether there is a non-adaptive scheme[33] doing the same job.

Consider again the markovian differential equation (5.4.2) on page 311:

$$X_t = C_t + \int_0^t f_\eta(X_{s-})\, dZ^\eta_s \tag{5.4.17}$$

for a vector $X \in \mathbb{R}^n$. We assume here without loss of generality that $\boldsymbol{f}(0) = 0$, replacing C with ${}^0C \stackrel{\text{def}}{=} C + \boldsymbol{f}(0)*\boldsymbol{Z}$ if necessary (see page 272). This has the effect that the Lipschitz condition (5.2.13):

$$|f_\eta(y) - f_\eta(x)| \le L \cdot |y - x| \quad \text{implies} \quad |\boldsymbol{f}(x)| \le L \cdot |x| \,. \tag{5.4.18}$$

To simplify life a little, let us also assume that $\boldsymbol{Z}$ is quasi-left-continuous. Then the intrinsic time Λ, and with it THE time transformation $T^{\bullet}$, can and will be chosen strictly increasing and continuous (see remark 4.5.2).

Remark 5.4.8 Let us see what can be said if we simply define the T_k as usual in calculus by $T_k \stackrel{\text{def}}{=} k\delta$, $k = 0, 1, \ldots$, $\delta > 0$ being the step size. Let us denote by $\mathcal{T} = \mathcal{T}^{(\delta)}$ the (sure) partition so obtained. Then the Euler–Peano approximate X' of (5.4.5) or (5.4.6), defined by $X'_0 \stackrel{\text{def}}{=} C_0$ and recursively for $t \in ((T_k, T_{k+1}]]$ by[1]

$$X'_t \stackrel{\text{def}}{=} X'_{T_k} + \big(C_t - C_{T_k}\big) + f_\eta\big(X'_{T_k}\big)\cdot\big(Z^\eta_t - Z^\eta_{T_k}\big)\,, \tag{5.4.19}$$

is again the solution of the stochastic differential equation (5.4.8). Namely,

$$X' = C + F'_\eta[X']_{.-}*Z^\eta\,,$$

with F'_η as in (5.4.7), to wit,

$$F'_\eta[Y] \stackrel{\text{def}}{=} F_\eta[Y^{\mathcal{T}}]^{\mathcal{T}} = \sum_{0 \le k < \infty} f_\eta(Y_{T_k})\cdot[[T_k, T_{k+1}))\,.$$

The stability estimate (5.2.34) on page 294 says that

$$\|X' - X\|^{\star}_{p,M} \le \frac{\gamma}{L(1-\gamma)} \cdot \left\| |F'[X]-F[X]|_{\infty} \right\|_{p,M} \tag{5.4.20}$$

for any choice of $\gamma \in (0,1)$ and any $M > M^{(5.2.26)}_{L:\gamma}$. A straightforward application of the Dominated Convergence Theorem to the right-hand side of (5.4.20) shows that $\|X' - X\|^{\star}_{p,M} \to 0$ as $\delta \to 0$. Thus X' is an approximate solution, and the path of X', which is an *algebraic construct* of the path of (C, Z), converges uniformly on bounded time-intervals to the path of the exact solution X as $\delta \to 0$, *in probability.*

However, the Dominated Convergence Theorem provides no control of the speed of the convergence, and this line of argument cannot rule out the possibility that *convergence* $X'_.(\omega) \xrightarrow[\delta\to 0]{} X_.(\omega)$ *may obtain for no single course-of-history* $\omega \in \Omega$. True, by exercise A.8.1 (iv) there exists some sequence (δ_n) along which convergence occurs almost surely, but it cannot generally be specified in advance. ∎

A small refinement of the argument in remark 5.4.8 does however result in the desired approximation scheme. The idea is to use equal spacing on the *intrinsic time* λ rather than the external time t (see, however, example 5.4.10 below). Accordingly, fix a step size $\delta > 0$ and set

$$\lambda_k \stackrel{\text{def}}{=} k\delta \quad \text{and} \quad T_k = T_k^{(\delta)} \stackrel{\text{def}}{=} T^{\lambda_k}\ ,\ k = 0, 1, \dots\ . \tag{5.4.21}$$

This produces a stochastic partition $\mathcal{T} = \mathcal{T}^{(\delta)}$ whose mesh tends to zero as $\delta \to 0$. For our purpose it is convenient to estimate the right-hand side of (5.2.35), which reads

$$\|X'-X\|^{\star}_{p,M} \le \frac{\gamma}{L(1-\gamma)} \cdot \left\|F[X'] - F'[X']\right\|_{p,M} .$$

Namely, $\quad \Delta_t \stackrel{\text{def}}{=} F[X']_t - F'[X']_t = f(X'_t) - f(X'^{\mathcal{T}}_t)$

equals[1] $\quad f\big(X'_{T_k} + f_\eta(X'_{T_k})\cdot(Z^\eta_t - Z^\eta_{T_k})\big) - f(X'_{T_k})$

for $T_k \le t < T_{k+1}$, and there satisfies the estimate[35]

$$\begin{aligned} |\Delta^\nu_t| &\le L \cdot \left|f^\nu_\eta(X'_{T_k})\cdot(Z^\eta_t - Z^\eta_{T_k})\right| , \qquad \nu = 1, \dots, n , \\ &\le L \cdot \left|\{f^\nu(X'_{T_k})(\!(T_k, T_{k+1}]\!]\}*Z\right|^{\star}_t . \end{aligned}$$

Thus $\quad |\Delta^\nu_t| \le L \cdot \sum_{0\le k} [\![T_k, T_{k+1})\!)_t \cdot \left|f^\nu(X'_{T_k})(\!(T_k, T_{k+1}]\!]*Z\right|^{\star}_t$

for *all* $t \ge 0$ and, since THE time transformation is strictly increasing,

$$\begin{aligned} \left\|\Delta^\nu_{T^\mu-}\right\|_{L^p} &\le L \cdot \textstyle\sum_k [k\delta < \mu \le (k+1)\delta] \times \\ &\times \left\| \left|f^\nu(X'_{T_k})(\!(T_k, T_{k+1}]\!]*Z\right|^{\star}_{T^\mu-} \right\|_{L^p} \end{aligned}$$

(which is a sum with only one non-vanishing term)

by (4.5.1) and 2.4.7: $$\leq LC_p^\diamond \cdot \textstyle\sum_k [k\delta < \mu \leq (k{+}1)\delta] \times$$
$$\times \max_{\rho=1^\diamond, p^\diamond} \left\| \left(\int_{k\delta}^{\mu} |f^\nu(X'_{T_k})|_\infty^\rho \, d\lambda \right)^{1/\rho} \right\|_{L^p}$$

for $\delta < 1$: $$\leq LC_p^\diamond \cdot \textstyle\sum_k [k\delta < \mu \leq (k{+}1)\delta] \cdot \delta^{1/p^\diamond} \left\| |f^\nu(X'_{T_k})|_\infty \right\|_{L^p} .$$

Therefore, applying $|\ |_p$, Fubini's theorem, and inequality (5.4.18),

$$\|\Delta_{T^\mu -}\|_{L^p} \leq \delta^{1/p^\diamond} \cdot L^2 C_p^\diamond \cdot \left\| X'^\star_{T_k} \right\|_{L^p} \leq \delta^{1/p^\diamond} \cdot L^2 C_p^\diamond \cdot \left\| X'^\star_{T^\mu} \right\|_{L^p} .$$

Multiplying by $e^{-M\mu}$, taking the supremum over μ, and using (5.2.23) results in inequality (5.4.22) below:

Theorem 5.4.9 *Suppose that Z is a quasi-left-continuous local $L^q(\mathbb{P})$-integrator for some $q \geq 2$, let $p \in [2, q]$, $0 < \gamma < 1$, and suppose that the markovian stochastic differential equation (5.4.17) of Lipschitz constant L has its unique global solution in $\mathfrak{S}^{\star n}_{p,M}$, $M = M^{(5.2.26)}_{L:\gamma}$. Then the non-adaptive Euler–Peano approximate X' defined in equation (5.4.19) for $\delta > 0$ satisfies*

$$\left\| X' - X \right\|^\star_{p,M} \leq \delta^{1/p^\diamond} \cdot \frac{C_p^\diamond L\gamma \cdot \|{}^0\!C\|^\star_{p,M}}{(1-\gamma)^2} . \tag{5.4.22}$$

Consequently, if δ runs through a sequence δ_n such that $\sum_n \delta_n^{1/p^\diamond}$ converges, then the corresponding non-adaptive Euler–Peano approximates converge uniformly on bounded time-intervals to the exact solution, nearly. ∎

Example 5.4.10 Suppose Z is a Lévy process whose Lévy measure has q^{th} moments away from zero and therefore is an L^q-integrator (see proposition 4.6.16 on page 267). Then its previsible controller is a multiple of time (ibidem), and $T^\lambda = c\lambda$ for some constant c. In that case the classical subdivision into equal time-intervals coincides with the intrinsic one above, and we get the pathwise convergence of the classical Euler–Peano approximates (5.4.19) under the condition $\sum_n \delta_n^{1/p^\diamond} < \infty$. If in particular Z has no jumps and so is a Wiener process, or if $p = 2$ was chosen, then $p^\diamond = 2$, which implies that square-root summability of the sequence of step sizes suffices for pathwise convergence of the non-adaptive Euler–Peano approximates.

Remark 5.4.11 So why not forget the adaptive algorithm (5.4.3)–(5.4.5) and use the non-adaptive scheme (5.4.19) exclusively?

Well, the former algorithm has order[37] 1 (see (5.4.10)), while the latter has only order 1/2 – or worse if there are jumps, see (5.4.22). (It should be

[37] Roughly speaking, an approximation scheme is of ***order*** r if its ***global error*** is bounded by a multiple of the rth power of the step size. For precise definitions see pages 281 and 324.

pointed out in all fairness that the expected number of computations needed to reach a given final time grows as $1/\delta^2$ in the first algorithm and as $1/\delta$ in the second, when a Wiener process is driving. In other words, the adaptive Euler algorithm essentially has order $1/2$ as well.)

Next, the algorithm (5.4.19) above can (so far) only be shown to make sense and to converge when the driver $\boldsymbol{Z}$ is at least an L^2-integrator. A reduction of the general case to this one by factorization does not seem to offer any practical prospects. Namely, change to another probability in $\mathfrak{P}[\boldsymbol{Z}]$ alters THE time transformation and with it the algorithm: there is no universality property as in summary 5.4.7.

Third, there is the generalization of the adaptive algorithm to general endogenous coupling coefficients (theorem 5.4.5), but not to my knowledge of the non-adaptive one. ∎

The Stratonovich Equation

In this subsection we assume that the drivers Z^η are continuous and the coupling coefficients markovian.[8] On page 271 the original ill-put stochastic differential equation (5.1.1) was replaced by the Itô equation (5.1.2), so as to have its integrands previsible and therefore integrable in the Itô sense. Another approach is to read (5.1.1) as a ***Stratonovich equation***:[38]

$$X = C + f_\eta(X)\circ Z^\eta \stackrel{\text{def}}{=} C + f_\eta(X)*Z^\eta + \frac{1}{2}\big[f_\eta(X), Z^\eta\big] . \tag{5.4.23}$$

Now, in the presence of sufficient smoothness of $\boldsymbol{f}$, there is by Itô's formula a continuous finite variation process V such that[38,16]

$$f_\eta(X) = f_{\eta;\nu}(X)*X^\nu + V .$$

Hence $\big[f_\eta(X), Z^\eta\big] = f_{\eta;\nu}(X)*\big[X^\nu, Z^\eta\big] = f_{\eta;\nu}(X)f_\theta^\nu(X)*\big[Z^\theta, Z^\eta\big]$,

which exhibits the Stratonovich equation (5.4.23) as equivalent with the Itô equation

$$X = C + f_\eta(X)*Z^\eta + \frac{(f_{\eta;\nu}f_\theta^\nu)(X)}{2}*\big[Z^\theta, Z^\eta\big] : \tag{5.4.24}$$

X solves (5.4.23) if and only if it solves (5.4.24). Since the Stratonovich integral has no decent limit properties, the existence and uniqueness of solutions to equation (5.4.23) cannot be established by a contractivity argument. Instead we must read it as the Itô equation (5.4.24); Lipschitz conditions on *both* the f_η and the $f_{\eta;\nu}f_\theta^\nu$ will then produce a unique global solution.

---

[38] Recall that X, C, f_η take values in $\mathbb{R}^n$. For example, $\boldsymbol{f} = \{\boldsymbol{f}^\nu\} = \{f_\eta^\nu\}$. The indices η, θ, ι usually run from 1 to d and the indices $\mu, \nu, \rho \dots$ from 1 to n. Einstein's convention, adopted, implies summation over the same indices in opposite positions.

Exercise 5.4.12 (Coordinate Invariance of the Stratonovich Equation) Let $\Phi : \mathbb{R}^n \to \mathbb{R}^n$ be invertible and twice continuously differentiable. Set $f_\eta^\Phi(y) \stackrel{\text{def}}{=} \Phi(f_\eta(\Phi^{-1}(y)))$. Then $Y \stackrel{\text{def}}{=} \Phi(X)$ is the unique solution of

$$Y = \Phi(C) + f_\eta^\Phi(Y) \circ Z^\eta ,$$

if and only if X is the solution of equation (5.4.23). In other words, the Stratonovich equation behaves like an ordinary differential equation under coordinate transformations – the Itô equation generally does not. This feature, together with theorem 3.9.24 and application 5.4.25, makes the Stratonovich integral very attractive in modeling.

Higher Order Approximation: Obstructions

Approximation schemes of global order 1/2 as offered in theorem 5.4.9 seem unsatisfactory. From ordinary differential equations we are after all accustomed to Taylor or Runge–Kutta schemes of arbitrarily high order.[37] Let us discuss what might be expected in the stochastic case, at the example of the Stratonovich equation (5.4.23) and its equivalent (5.4.24), reproduced here as

$$X = C + \boldsymbol{f}(X) \circ \boldsymbol{Z} \tag{5.4.25}$$

or[38]

$$X = C + f_\eta(X) {*} Z^\eta + \frac{(f_{\eta;\nu} f_\theta^\nu)(X)}{2} {*} [Z^\theta, Z^\eta] \tag{5.4.26}$$

or, equivalently,

$$X = \mathfrak{U}[X] \stackrel{\text{def}}{=} C + \overline{F}_\iota(X) {*} \overline{Z}^\iota , \tag{5.4.27}$$

where $\overline{F}_\iota \stackrel{\text{def}}{=} f_\eta$ and $\overline{Z}^\iota \stackrel{\text{def}}{=} Z^\eta$ when $\iota = \eta \in \{1, \ldots, d\}$

and $\overline{F}_\iota \stackrel{\text{def}}{=} f_{\eta;\nu} f_\theta^\nu$ and $\overline{Z}^\iota \stackrel{\text{def}}{=} [Z^\eta, Z^\theta]$ when $\iota = \eta\theta \in \{11, \ldots, dd\}$.

In order to simplify and to fix ideas we work with the following

Assumption 5.4.13 *The initial condition $C \in \mathcal{F}_0$ is constant in time. $\boldsymbol{Z}$ is continuous with $\boldsymbol{Z}_0 = 0$ – then $\overline{\boldsymbol{Z}}_0 = 0$. The markovian*[8] *coupling coefficient $\overline{\boldsymbol{F}}$ is differentiable and Lipschitz.*

We are then sure that there is a unique solution X of (5.4.27), which also solves (5.4.25) and lies in $\mathfrak{S}_{p,M}^{\star n}$ for any $p \ge 2$ and $M > M_{p,L}^{\diamond(5.2.20)}$ (see proposition 5.2.14).

We want to compare the effect of various step sizes $\delta > 0$ on the accuracy of a given non-adaptive approximation scheme. For every $\delta > 0$ picked, T_k shall denote the ***intrinsically δ-spaced*** stopping times of equation (5.4.21): $T_k \stackrel{\text{def}}{=} T^{k\delta}$.

Surprisingly much – of, alas, a disappointing nature – can be derived from a rather general discussion of single-step approximation methods. We start with the following "metaobservation:" A straightforward[39] generalization of a classical single-step scheme as described on page 280 will result in a method of the following description:

---

[39] and, as it turns out, a bit naive – see notes 5.4.33.

Condition 5.4.14 *The method provides a function* $\Xi' : \mathbb{R}^n \times \mathbb{R}^d \to \mathbb{R}^n$,

$$(x, \boldsymbol{z}) \mapsto \Xi'[x, \boldsymbol{z}] = \Xi'[x, \boldsymbol{z}; \boldsymbol{f}] ,$$

whose role is this: when after k *steps the method has constructed an approximate solution* X'_t *for times* t *up to the* k^{th} *stopping time* T_k, *then* Ξ' *is employed to extend* X' *up to the next time* T_{k+1} *via*

$$X'_t \stackrel{\text{def}}{=} \Xi'[X'_{T_k}, \boldsymbol{Z}_t - \boldsymbol{Z}^{T_k}] \quad \text{for } T_k \le t \le T_{k+1} . \tag{5.4.28}$$

$\boldsymbol{Z} - \boldsymbol{Z}^{T_k}$ *is the upcoming stretch of the driver. The function* Ξ' *is specific for the method at hand, and is constructed from the coupling coefficient* $\boldsymbol{f}$ *and possibly (in Taylor methods) from a number of its derivatives.*

If the approximation scheme meets this description, then we talk about ***the method*** Ξ'.

In an adaptive[33] scheme, Ξ' might also enter the definition of the next stopping time T_{k+1} – see for instance (5.4.4). The function Ξ' should be reasonably simple; the more complex Ξ' is to evaluate the poorer a choice it is, evidently, for an approximation scheme, unless greatly enhanced accuracy pays for the complexity. In the usual single-step methods $\Xi'[x, \boldsymbol{z}; \boldsymbol{f}]$ is an algebraic expression in various derivatives of $\boldsymbol{f}$ evaluated at algebraic expressions made from x and $\boldsymbol{z}$.

Examples 5.4.15 In the Euler–Peano method of theorem 5.4.9

$$\Xi'[x, \boldsymbol{z}; \boldsymbol{f}] = x + f_\eta(x) z^\eta .$$

The classical improved Euler or Heun method generalizes to[1]

$$\Xi'[x, \boldsymbol{z}; \boldsymbol{f}] \stackrel{\text{def}}{=} x + \frac{f_\eta(x) + f_\eta(x + f_\theta(x) z^\theta)}{2} z^\eta .$$

The straightforward[39] generalization of the Taylor method of order 2 is given by

$$\Xi'[x, \boldsymbol{z}; \boldsymbol{f}] \stackrel{\text{def}}{=} x + f_\eta(x) z^\eta + (f_{\eta;\nu} f_\theta^\nu)(x) z^\eta z^\theta / 2 .$$

The classical Runge–Kutta method of global order 4 has the obvious generalization

$$k_1 \stackrel{\text{def}}{=} f_\eta(x) z^\eta , \quad k_2 \stackrel{\text{def}}{=} f_\eta(x + k_1/2) z^\eta , \quad k_3 \stackrel{\text{def}}{=} f_\eta(x + k_2/2) z^\eta , \quad k_4 \stackrel{\text{def}}{=} f_\eta(x + k_3/2) z^\eta$$

and $$\Xi'[x, \boldsymbol{z}; \boldsymbol{f}] \stackrel{\text{def}}{=} x + \frac{k_1 + 2k_2 + 2k_3 + k_4}{6} .$$

The methods Ξ' in this example have a structure in common that is most easily discussed in terms of the following notion. Let us say that the map $\Phi : \mathbb{R}^n \times \mathbb{R}^d \to \mathbb{R}^n$ is ***polynomially bounded in*** $\boldsymbol{z}$ if there is a polynomial P so that

$$|\Phi[x, \boldsymbol{z}]| \le P(|\boldsymbol{z}|) , \qquad (x, \boldsymbol{z}) \in \mathbb{R}^n \times \mathbb{R}^d .$$

The functions polynomially bounded in $\boldsymbol{z}$ evidently form an algebra $\mathcal{BP}$ that is closed under composition: $\Psi\big[\Phi[\,\cdot\,,\cdot\,], \cdot\,\big] \in \mathcal{BP}$ for $\Phi, \Psi \in \mathcal{BP}$. The functions $\Phi \in \mathcal{BP} \cap C^k$ whose first k partials belong to $\mathcal{BP}$ as well form the class $\mathcal{PU}^k$. This is easily seen to form again an algebra closed under composition. For simplicity's sake assume now that $\boldsymbol{f}$ is of class[10] C_b^∞. Then in the examples above, and in fact in all straightforward extensions of the classical single step methods, $\Xi'[\,\cdot\,,\cdot\,;\boldsymbol{f}]$ has all of its partial derivatives in $\mathcal{BP}^\infty$. For the following discussion only this much is needed:

Condition 5.4.16 *Ξ' has partial derivatives of orders 1 and 2 in $\mathcal{BP}$.*

Now, from definition (5.4.28) on page 322 and theorem 3.9.24 on page 170 we get $\Xi'[x,0] = x$ and[40]

$$\Xi'[X_{T_k}, \boldsymbol{Z}-\boldsymbol{Z}^{T_k}] = X_{T_k} + \Xi'_{;\eta}[X_{T_k}, \boldsymbol{Z}-\boldsymbol{Z}^{T_k}]\circ Z^\eta \quad \text{on} \quad [\![T_k, T_{k+1}]\!] \,,$$

so that X' can be viewed as the solution of the Stratonovich equation

$$X' = C + F'_\eta[X']\circ Z^\eta \,, \tag{5.4.29}$$

with Itô equivalent (compare with equation (5.4.27) on page 321)

$$X' = \mathfrak{U}'[X'] \stackrel{\text{def}}{=} C + \overline{F}'_\iota[X']{*}\overline{Z}^\iota \,, \tag{5.4.30}$$

where[35]
$$\overline{F}'_\iota = F'_\eta \stackrel{\text{def}}{=} \textstyle\sum_k [\![T_k, T_{k+1})\!) \cdot \Xi'_{;\eta}[X'_{T_k}, \boldsymbol{Z}-\boldsymbol{Z}^{T_k}] \qquad \text{when } \iota = \eta$$

and
$$\overline{F}'_\iota \stackrel{\text{def}}{=} \textstyle\sum_k [\![T_k, T_{k+1})\!) \cdot \{\Xi'_{;\eta\nu}\Xi'^{\nu}_{;\theta}\}[X'_{T_k}, \boldsymbol{Z}-\boldsymbol{Z}^{T_k}] \qquad \text{for } \iota = \eta\theta \,.$$

Note that $\overline{\boldsymbol{F}}'$ is generally not markovian, in view of the explicit presence of $\boldsymbol{Z} - \boldsymbol{Z}^{T_k}$ in $\Xi'_{;\eta}[x, \boldsymbol{Z}-\boldsymbol{Z}^{T_k}]$.

Exercise 5.4.17 (i) Condition 5.4.16 ensures that $\overline{\boldsymbol{F}}'$ satisfies the Lipschitz condition (5.2.11). Therefore both maps $\mathfrak{U}$ of (5.4.27) and $\mathfrak{U}'$ of (5.4.30) will be strictly contractive in $\mathfrak{S}^{\star n}_{p,M}$ for all $p \ge 2$ and suitably large $M = M(p)$. (ii) Furthermore, there exist constants D', L', M' such that for $0 \le \kappa < \lambda$

$$\Big\| |\Xi'[C, \boldsymbol{Z}_.-\boldsymbol{Z}^{T^\kappa}; \boldsymbol{f}]|^\star_{T^\lambda} \Big\|_{L^p} \le D' \cdot \|C\|_{L^p} \cdot e^{M'(\lambda-\kappa)} \tag{5.4.31}$$

and
$$\Big\| |\Xi'[C', \boldsymbol{Z}_.-\boldsymbol{Z}^{T^\kappa}] - \Xi'[C, \boldsymbol{Z}_.-\boldsymbol{Z}^{T^\kappa}]|^\star_{T^\lambda} \Big\|_{L^p} \le \|C'-C\|_{L^p} \cdot e^{L'(\lambda-\kappa)} \,. \tag{5.4.32}$$

Recall that we are after a method Ξ' of order strictly larger than $1/2$. That is to say, we want it to produce an estimate of the form $X' - X = o(\sqrt{\delta})$ for the difference of the exact solution $X = \Xi[C, \boldsymbol{Z}; \boldsymbol{f}]$ of (5.4.25) from its Ξ'-approximate X' made with step size δ via (5.4.28). The question arises how to measure this difference. We opt[41] for a generalization of the classical notions of order from page 281, replacing time t with intrinsic time λ:

[40] We write $\Xi'_{;\eta} \stackrel{\text{def}}{=} \partial\Xi'/\partial z^\eta$ and $\Xi'_{;\eta\nu} \stackrel{\text{def}}{=} \partial\Xi'_{;\eta}/\partial x^\nu$, etc.[38]

[41] There are less stringent notions; see notes 5.4.33.

Definition 5.4.18 *We say that* Ξ' *has* **local order** r *on the coupling coefficient* $\boldsymbol{f}$ *if there exists a constant* $\underline{M}$ *such that*[4] *for all* $\lambda > \kappa \geq 0$ *and all* $C \in L^p(\mathcal{F}_{T^\lambda})$

$$\left\| \left|\Xi'[C, \boldsymbol{Z}_.-\boldsymbol{Z}^{T^\kappa}; \boldsymbol{f}] - \Xi[C, \boldsymbol{Z}_.-\boldsymbol{Z}^{T^\kappa}; \boldsymbol{f}]\right|_{T^\lambda}^\star \right\|_{L^p} \tag{5.4.33}$$
$$\leq (\|C\|_{L^p}+1) \times \left(\underline{M}(\lambda-\kappa)\right)^r e^{\underline{M}(\lambda-\kappa)} .$$

The least such $\underline{M}$ *is denoted by* $\underline{M}[\boldsymbol{f}]$. *We say* Ξ' *has* **global order** r *on* $\boldsymbol{f}$ *if the difference* $X'-X$ *satisfies an estimate*

$$\left\|X'_. - X_.\right\|_{p,\overline{M}}^\star = \left(\left\|C\right\|_{p,\overline{M}}^\star + 1\right) \cdot O(\delta^r)$$

for some $\overline{M} = \overline{M}[\boldsymbol{f}; \Xi']$. *This amounts to the existence of a* $\overline{B} = \overline{B}[\boldsymbol{f}; \Xi']$

such that $$\left\| \left|X' - \Xi[C, \boldsymbol{Z}; \boldsymbol{f}]\right|_{T^\lambda}^\star \right\|_{L^p} \leq \overline{B}\cdot(\|C\|_{L^p}+1) \times \delta^r e^{\overline{M}\lambda} \tag{5.4.34}$$

for sufficiently small $\delta > 0$ *and all* $\lambda \geq 0$ *and* $C \in L^p(\mathcal{F}_0)$.

Criterion 5.4.19 *(Compare with criterion 5.1.11.) Assume condition 5.4.16.*

(i) If $\left\| \left|\Xi'[C, \boldsymbol{Z}_.-\boldsymbol{Z}^{T^\kappa}; \boldsymbol{f}] - \Xi[C, \boldsymbol{Z}_.-\boldsymbol{Z}^{T^\kappa}; \boldsymbol{f}]\right|_{T^\lambda}^\star \right\|_{L^p} = (\|C\|_{L^p}+1)\cdot O((\lambda-\kappa)^r)$, *then* Ξ' *has local order* r *on* $\boldsymbol{f}$.

(ii) If Ξ' *has local order* r, *then it has global order* $r-1$.

Recall again that we are after a method Ξ' of order strictly larger than $1/2$. In other words, we want it to produce

$$\left\|X' - X\right\|_{p,\overline{M}}^\star = o(\sqrt{\delta}) \tag{5.4.35}$$

for some $p \geq 2$ and some $\overline{M}$. Let us write ${}^0\Xi'(t) \stackrel{\text{def}}{=} \Xi'[C, \boldsymbol{Z}_t] - C$ and ${}^0\Xi'_{;\eta}(t) \stackrel{\text{def}}{=} \Xi'_{;\eta}[C, \boldsymbol{Z}_t]$ for short.[40] According to inequality (5.2.35) on page 294,

(5.4.35) will follow from $$\left\|\overline{\boldsymbol{F}}[X'] - \overline{\boldsymbol{F}}'[X']\right\|_{p,\overline{M}} = o(\sqrt{\delta}) ,$$

which requires $$\left\|f_\eta(C + {}^0\Xi'(t)) - {}^0\Xi'_{;\eta}(t)\right\|_{L^p} = o(\sqrt{t}) . \tag{5.4.36}$$

It is hard to see how (5.4.35) could hold without (5.4.36); at the same time, it is also hard to establish that it implies (5.4.36). We will content ourselves with this much:

Exercise 5.4.20 If Ξ' is to have order > 1 in all circumstances, in particular whenever the driver $\boldsymbol{Z}$ is a standard Wiener process, then equation (5.4.36) must hold.

Letting $\delta \to 0$ in (5.4.36) we see that the method Ξ' must satisfy $\Xi'_{;\eta}[C, 0] = f_\eta(C)$. This can be had in all generality only if[40]

$$\Xi'_{;\eta}[x, 0] = f_\eta(x) \quad \forall\, x \in \mathbb{R}^n . \tag{5.4.37}$$

Then[5]
$$f_\eta(C + {}^0\Xi'(t)) = f_\eta(C) + f_{\eta;\nu}(C)\,{}^0\Xi'^\nu(t) + O(|{}^0\Xi'(t)|^2)$$
$$= f_\eta(C) + f_{\eta;\nu}(C)\,\Xi'^\nu_{;\theta}[C,0]Z^\theta_t + O(|\boldsymbol{Z}_t|^2)$$

by (5.4.37):
$$= f_\eta(C) + f_{\eta;\nu}(C) f^\nu_\theta(C)\,Z^\theta_t + O(|\boldsymbol{Z}_t|^2)\,. \tag{5.4.38}$$

Also,
$$\Xi'_{;\eta}[C, \boldsymbol{Z}_t] = f_\eta(C) + \Xi'_{;\eta\theta}[C,0]Z^\theta_t + O(|\boldsymbol{Z}_t|^2)\,. \tag{5.4.39}$$

Equations (5.4.36), (5.4.38), and (5.4.39) imply that for $t \le T^\delta$

$$\left\| \left\{ (f_{\eta;\nu} f^\nu_\theta)(C) - \Xi'_{;\eta\theta}[C,0] \right\} Z^\theta_t \right\|_{L^p} = o(\sqrt{\delta}) + \left\| O(|\boldsymbol{Z}_t|^2) \right\|_{L^p}. \tag{5.4.40}$$

This condition on Ξ' can be had, of course, if Ξ' is chosen so that

$$M_{\eta\theta}(x) \stackrel{\text{def}}{=} (f_{\eta;\nu} f^\nu_\theta)(x) - \Xi'_{;\eta\theta}[x,0] = 0 \quad \forall\, x \in \mathbb{R}^n\,, \tag{5.4.41}$$

and in general only with this choice. Namely, suppose $\boldsymbol{Z}$ is a standard d-dimensional Wiener process. Then, for $k = 0$, the size in L^p of the martingale $M_t \stackrel{\text{def}}{=} M^\mu_{\eta\theta}(x) Z^\theta_t$ at $t = \delta$ is, by theorem 2.5.19 and inequality (4.2.4), bounded below by a multiple of

$$\|S_\delta[M_.]\|_{L^p} = \left\| \left(\textstyle\sum_\theta |M^\mu_{\eta\theta}(x)|^2 \right)^{1/2} \right\|_{L^p} \cdot \sqrt{\delta}\,,$$

while
$$\|O(|\boldsymbol{Z}_\delta|^2)\|_{L^p} \le const \times \delta = o(\sqrt{\delta})\,.$$

In the presence of equation (5.4.40), therefore,

$$\left\| \left(\textstyle\sum_\theta |M^\mu_{\eta\theta}(x)|^2 \right)^{1/2} \right\|_{L^p} \le \frac{o(\sqrt{\delta})}{\sqrt{\delta}} \xrightarrow[\delta \to 0]{} 0\,.$$

This implies $M_. = 0$ and with it (5.4.41), i.e., $\Xi'_{;\eta\theta}[x,0] = (f_{\eta;\nu} f^\nu_\theta)(x)$ for all $x \in \mathbb{R}^n$. Notice now that $\Xi'_{;\eta\theta}[x,0]$ is symmetric in η, θ. This equality therefore implies that the Lie brackets $[f_\eta, f_\theta] \stackrel{\text{def}}{=} f_{\eta;\nu} f^\nu_\theta - f_{\theta;\nu} f^\nu_\eta$ must vanish:

Condition 5.4.21 *The vector fields $f_1, \ldots, f_d$ commute.*

The following summary of these arguments does not quite deserve to be called a theorem, since the definition of a method and the choice of the norms $\| \;\|_{p,M}$, etc., are not canonical and (5.4.36) was not established rigorously.

Scholium 5.4.22 *We cannot expect a method Ξ' satisfying conditions 5.4.14 and 5.4.16 to provide approximation in the sense of definition 5.4.18 to an order strictly better than 1/2 for all drivers and all initial conditions, unless the coefficient vector fields commute.*

Higher Order Approximation: Results

We seek approximation schemes of an order better than 1/2. We continue to investigate the Stratonovich equation (5.4.25) under assumption 5.4.13, adding condition 5.4.21. This condition, forced by scholium 5.4.22, is a severe restriction on the system (5.4.25). The least one might expect in a just world is that in its presence there are good approximation schemes. Are there? In a certain sense, the answer is affirmative and optimal. Namely, from the change-of-variable formula (3.9.11) on page 171 for the Stratonovich integral, this much is immediate:

Theorem 5.4.23 *Assuming condition 5.4.21, let $\Xi^{\boldsymbol{f}}$ be the action of $\mathbb{R}^d$ on $\mathbb{R}^n$ generated by $\boldsymbol{f}$ (see proposition 5.1.10 on page 279). Then the solution of equation (5.4.25) is given by*

$$X_t = \Xi^{\boldsymbol{f}}[C, \boldsymbol{Z}_t] \, .$$

Examples 5.4.24 (i) Let W be a standard Wiener process. The Stratonovich equation $\mathscr{E} = 1 + \mathscr{E} \circ W$ has the solution e^W, on the grounds that $e^{\cdot}$ solves the corresponding ordinary differential equation $e^t = 1 + \int_0^t e^s \, ds$.

(ii) The vector fields $f_1(x) = x$ and $f_2(x) = -x/2$ on $\mathbb{R}$ commute. Their flows are $\xi[x, t; f_1] = xe^t$ and $\xi[x, t; f_2] = xe^{-t/2}$, respectively, and so the action $\boldsymbol{f} = (f_1, f_2)$ generates is $\Xi^{\boldsymbol{f}}(x, (z_1, z_2)) = x \times e^{z_1} \times e^{-z_2/2}$. Therefore the solution of the Itô equation $\mathscr{E}_t = 1 + \int_0^t \mathscr{E}_s \, dW_s$, which is the same as the Stratonovich equation $\mathscr{E}_t = 1 + \int_0^t \mathscr{E}_s \, \delta W_s - 1/2 \int_0^t \mathscr{E}_s \, ds = 1 + \int_0^t f_1(\mathscr{E}_s) \, \delta W_s + \int_0^t f_2(\mathscr{E}_s) \, ds$, is $\mathscr{E}_t = e^{W_t - t/2}$, which the reader recognizes from proposition 3.9.2 as the Doléans–Dade exponential of W.

(iii) The previous example is about linear stochastic differential equations. It has the following generalization. Suppose $A_1, \ldots, A_d$ are *commuting* $n \times n$-matrices. The vector fields $f_\eta(x) \stackrel{\text{def}}{=} A_\eta x$ then commute. The linear Stratonovich equation $X = C + A_\eta X \circ Z^\eta$ then has the explicit solution $X_t = C \cdot e^{A_\eta Z_t^\eta}$. The corresponding Itô equation $X = C + A_\eta X * Z^\eta$, equivalent with $X = C + A_\eta X \circ Z^\eta - \frac{1}{2} A_\eta A_\theta X \circ [Z^\eta, Z^\theta]$, is solved explicitely by $X_t = C \cdot e^{A_\eta Z_t^\eta - \frac{1}{2} A_\eta A_\theta [Z^\eta, Z^\theta]_t}$.

Application 5.4.25 (Approximating the Stratonovich Equation by an ODE) Let us continue the assumptions of theorem 5.4.23. For $n \in \mathbb{N}$ let $\boldsymbol{Z}^{(n)}$ be that continuous and piecewise linear process which at the times k/n equals $\boldsymbol{Z}$, $k = 1, 2, \ldots$. Then $\boldsymbol{Z}^{(n)}$ has finite variation but is generally not adapted; the solution of the ordinary differential equation $X_t^{(n)} = C + \int_0^t \boldsymbol{f}(X_s^{(n)}) \, d\boldsymbol{Z}_s^{(n)}$ (which depends of course on the parameter $\omega \in \Omega$) converges uniformly on bounded time-intervals to the solution X of the Stratonovich equation (5.4.25), for every $\omega \in \Omega$. This is simply because $\boldsymbol{Z}_t^{(n)}(\omega) \to \boldsymbol{Z}_t(\omega)$ uniformly on bounded intervals and $X_t^{(n)}(\omega) = \Xi^{\boldsymbol{f}}[C, \boldsymbol{Z}_t^{(n)}(\omega)]$. This feature, together with theorem 3.9.24 and exercise 5.4.12, makes the Stratonovich integral very attractive in modeling. ∎

One way of reading theorem 5.4.23 is that $(x, \boldsymbol{z}) \mapsto \Xi^{\boldsymbol{f}}[x, \boldsymbol{z}]$ is a method of infinite order: there is no error. Another, that in order to solve the stochastic differential equation (5.4.25), one merely needs to solve d ordinary differential equations, producing Ξ^f, and then evaluate Ξ^f at Z. All of this looks very satisfactory, until one realizes that Ξ^f is not at all a simple function to evaluate and that it does not lend itself to run time approximation of X.

5.4.26 A Method of Order r An obvious remedy leaps to the mind: approximate the action Ξ^f by some less complex function Ξ'; then $\Xi'[x, Z_t]$ should be an approximation of X_t. This simple idea can in fact be made to work. For starters, observe that one needs to solve *only one* ordinary differential equation in order to compute $X_t(\omega) = \Xi^f[C(\omega), Z_t(\omega)]$ for any given $\omega \in \Omega$. Indeed, by proposition 5.1.10 (iii), $X_t(\omega)$ is the value x_{τ_t} at $\tau_t \stackrel{\text{def}}{=} |Z_t(\omega)|$ of the solution $x_.$ to the ODE

$$x_. = C(\omega) + \int_0^. f(x_\sigma)\,d\sigma \quad \text{, where } \; f(x) \stackrel{\text{def}}{=} \textstyle\sum_\eta f_\eta(x)\, Z_t^\eta(\omega)/\tau_t\,. \tag{5.4.42}$$

Note that knowledge of the whole path of Z is not needed, only of its value $Z_t(\omega)$. We may now use any classical method to approximate $x_{\tau_t} = X_t(\omega)$. Here is a suggestion: given an r, choose a classical method ξ' of global order r, for instance a suitable Runge–Kutta or Taylor method, and use it with step size δ to produce an approximate solution $x'_. = x'_.[c;\delta, f]$ to (5.4.42). According to page 324, to say that the method ξ' chosen has global order r means that there are constants $\overline{b} = \overline{b}[c; f, \xi']$ and $\overline{m} = \overline{m}[f;\xi']$ so that for sufficiently small $\delta > 0$

$$|x_\sigma - x'_\sigma| \le \overline{b}\cdot\delta^r \times e^{\overline{m}\sigma}\,, \qquad \sigma \ge 0\,.$$

Now set
$$b \stackrel{\text{def}}{=} \sup\{\overline{b}[f_\eta z^\eta;\xi'] : |z| \le 1\} \tag{5.4.43}$$
and
$$m \stackrel{\text{def}}{=} \sup\{\overline{m}[f_\eta z^\eta;\xi'] : |z| \le 1\}\,. \tag{5.4.44}$$
Then
$$\left|X_t(\omega) - x'_{\tau_t}\right| \le b\cdot\delta^r \times e^{m\tau_t}\,. \tag{5.4.45}$$

Hidden in (5.4.43), (5.4.44) is another assumption on the method ξ':

Condition 5.4.27 *If ξ' has global order r on $f_1,\ldots,f_d$, then the suprema in equations (5.4.43) and (5.4.44) can be had finite.* (If, as is often the case, $\overline{b}[f;\xi']$ and $\overline{m}[f;\xi']$ can be estimated by polynomials in the uniform bounds of various derivatives of f, then the present condition is easily verified.) ∎

In order to match (5.4.47) with our general definition 5.4.14 of a single-step method, let us define the function $\Xi' : \mathbb{R}^n \times \mathbb{R}^d \to \mathbb{R}^n$ by

$$\Xi'[x, z] \stackrel{\text{def}}{=} x'_\tau\,, \tag{5.4.46}$$

where $\tau \stackrel{\text{def}}{=} |z|$ and $x'_.$ is the ξ'-approximate to $x_. = x + \int_0^. f_\eta(x_\sigma) z^\eta/\tau\, d\sigma$. Then the corresponding Ξ'-approximate for (5.4.25) is

$$X'_t(\omega) = \Xi'[c, Z_t(\omega)] \stackrel{\text{def}}{=} x'_{\tau_t(\omega)}$$

and by (5.4.45) has
$$\left|X(\omega) - X'(\omega)\right|_t^\star \le b\cdot\delta^r \times e^{m|Z|_t^\star(\omega)} \tag{5.4.47}$$

when ξ' is carried out with step size δ. The method Ξ' is still not very simple, requiring as it does $\lceil \tau_t(\omega)/\delta \rceil$ iterations of the classical method ξ' that defines it; but given today's fast computers, one might be able to live with this much complexity. Here is another mitigating observation: if one is interested only in approximating $X_t(\omega)$ at one finite time t, then Ξ' is actually evaluated only once: it is a *one-single-step method.*

Suppose now that $\boldsymbol{Z}$ is in particular of the following ubiquitous form:

Condition 5.4.28 $$Z_t^\eta = \begin{cases} t & \text{for } \eta = 1 \\ W_t^\eta & \text{for } \eta = 2, \ldots, d, \end{cases}$$

where $\boldsymbol{W}$ is a standard $d{-}1$-dimensional Wiener process. ∎

Then the previsible controller becomes $\Lambda_t = d{\cdot}t$ (exercise 4.5.19), THE time transformation is given by $T^\lambda = \lambda/d$, and the Stratonovich equation (5.4.25) reads

$$X = C + \boldsymbol{f}(X){\circ}\boldsymbol{Z} \tag{5.4.48}$$

or, equivalently, $$X = C + \overline{f}_\eta(X){*}Z^\eta \ ,$$

where $$\overline{f}_\eta \stackrel{\text{def}}{=} \begin{cases} f_1 + \dfrac{1}{2}\displaystyle\sum_{\theta>1} f_{\theta;\nu} f_\theta^\nu & \text{for } \eta = 1, \\ f_\eta & \text{for } \eta > 1. \end{cases}$$

$$\sup_{\eta \geq 1} \left| \overline{f}_\eta(x) - \overline{f}_\eta(y) \right| \leq L \cdot |x - y| \tag{5.4.49}$$

is the requisite Lipschitz condition from 5.4.13, which guarantees the existence of a unique solution to (5.4.48), which lies in $\mathfrak{S}_{p,M}^{\star n}$ for any $p \geq 2$ and $M > M_{p,L}^{\diamond(5.2.20)}$. Furthermore, $\boldsymbol{Z}$ is of the form discussed in exercise 5.2.18 (ii) on page 292, and inequality (5.4.47) together with inequality (5.2.30) leads to the existence of constants $B' = B'[b, d, p, r]$ and $M' = M'[d, m, p, r]$ such that

$$\left\| |X' - X|_t^\star \right\|_{L^p} \leq \delta^r \cdot B' e^{M't} \ , \qquad t \geq 0 \, .$$

We have established the following result:

Proposition 5.4.29 *Suppose that the driver $\boldsymbol{Z}$ satisfies condition 5.4.28, the coefficients $\overline{f}_1, \ldots, \overline{f}_d$ are Lipschitz, and the coefficients $f_1, \ldots, f_d$ commute. If ξ' is any classical single-step approximation method of global order r for ordinary differential equations in $\mathbb{R}^n$ (page 280) that satisfies condition 5.4.27, then the one-single-step method Ξ' defined from it in (5.4.46) is again of global order r, in this weak sense: at any fixed time t the difference of the exact solution $X_t = \Xi[C, \boldsymbol{Z}_t]$ of (5.4.25) and its Ξ'-approximate X_t' made with step size δ can be estimated as follows: there exist constants B, M, B_1, M_1 that depend only on $d, \boldsymbol{f}, p > 1, \xi'$ such that*

$$\left| X_t'(\omega) - X_t(\omega)) \right| \leq B{\cdot}\delta^r \times e^{M|\boldsymbol{Z}_t(\omega)|} \qquad \forall\, \omega \in \Omega$$

and $$\left\| |X' - X|_t^\star \right\|_{L^p} \leq B_1{\cdot}\delta^r \times e^{M_1 t} \, . \tag{5.4.50}$$

Discussion 5.4.30 This result apparently has two related shortcomings: the method Ξ' computes an approximation to the value of $X_t(\omega)$ only at the final time t of interest, not to the whole path $X_.(\omega)$, and it waits until that time before commencing the computation – no information from the signal $\boldsymbol{Z}_.$ is processed until the final time t has arrived. In order to approximate K points on the solution path $X_.$ the method Ξ' has to be run K times, each time using $|\boldsymbol{Z}_{T_k}|/\delta$ iterations of the classical method ξ'. In the figure above[42] one expects to perform as many calculations as there are dashes, in order to compute approximations at the K dots.

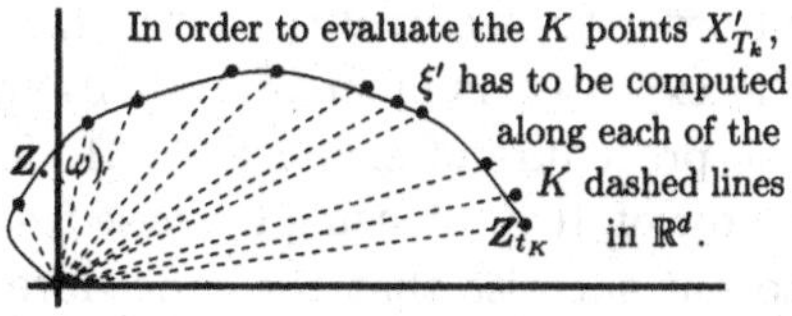
In order to evaluate the K points X'_{T_k}, ξ' has to be computed along each of the K dashed lines in $\mathbb{R}^d$.
$Z_.(\omega)$
Z_{t_K}
Figure 5.13

Exercise 5.4.31 Suppose one wants to compute approximations $X'_{k\delta}$ at the K points $\delta, 2\delta, \ldots, K\delta = t$ via proposition 5.4.29. Then the expected number of evaluations of ξ' is $N_1 \approx B_1(t)/\delta^2$; in terms of the mean error $E \overset{\text{def}}{=} \| |X' - X|_t^\star \|_{L^2}$

$$N_1 \approx C_1(t)/E^{2/r} ,$$

$B_1(t)$, $C_1(t)$ being functions of at most exponential growth that depend only on ξ'.

Figure 5.13 suggests that one should look for a method that at the $k+1^{\text{th}}$ step uses the previous computations, or at least the previously computed value X'_{T_k}. The simplest thing to do here is evidently to apply the classical method ξ' at the kth point to the ordinary differential equation

$$x_\tau = X'_{T_k} + \int_{T_k}^{\tau} (\boldsymbol{f}\cdot(\boldsymbol{Z}_t - \boldsymbol{Z}_t^{T_k}))(x_\sigma)\, d\sigma ,$$

whose exact solution at $\tau = 1$ is $x_1 = \Xi^{\boldsymbol{f}}[X'_{T_k}, \boldsymbol{Z}_t - \boldsymbol{Z}_t^{T_k}]$, so as to obtain $X'_t \overset{\text{def}}{=} x'_1$; apply it in one "giant" step of size 1. In figure 5.13 this propels us from one dot to the next. This prescription defines a single-step method Ξ' in the sense of 5.4.14:

$$\Xi'[x, \boldsymbol{z}; \boldsymbol{f}] \overset{\text{def}}{=} \xi'[x, 1; \boldsymbol{f}\cdot\boldsymbol{z}] ;$$

and $\quad X'_t = \Xi'[X'_{T_k}, \boldsymbol{Z}_t - \boldsymbol{Z}_t^{T_k}; \boldsymbol{f}] = \xi'[X'_{T_k}, 1; \boldsymbol{f}\cdot(\boldsymbol{Z}_t - \boldsymbol{Z}_t^{T_k})] , \quad T_k \le t \le T_{k+1} ,$

is the corresponding approximate as in definition (5.4.28).

Exercise 5.4.32 Continue to consider the Stratonovich equation (5.4.48), assuming conditions 5.4.21, 5.4.28, and inequality (5.4.49). Assume that the classical method ξ' is *scale-invariant* (see note 5.1.12) and has local order $r+1$ – by criterion 5.1.11 on page 281 it has global order r. Show that then Ξ' has global order $r/2 - 1/2$ in the sense of (5.4.34), so that, for suitable constants B_2, M_2, the Ξ'-approximate X' satisfies

$$E \overset{\text{def}}{=} \| |X' - X|_t^\star \|_{L^2} \le B_2(t)\delta^{r/2-1/2} .$$

Consequently the number $N_2 = t/\delta$ of evaluations of ξ' needed as in 5.4.31 is

$$N_2 \approx C_2(t)/E^{2/(r-1)} .$$

[42] It is highly stylized, not showing the wild gyrations the path $\boldsymbol{Z}_.$ will usually perform.

In order to decrease the error E by a factor of $10^{r/2}$, we have to increase the expected number of evaluations of the method ξ' by a factor of 10 in the procedure of exercise 5.4.31. The number of evaluations increases by a factor of $10^{r/r-1}$ using exercise 5.4.32 with the estimate given there. We see to our surprise that the procedure of exercise 5.4.31 is better than that of exercise 5.4.32, at least according to the estimates we were able to establish.

Notes 5.4.33 (i) The adaptive Euler method of theorem 5.4.2 is from [7]. It, its generalization 5.4.5, and its non-adaptive version 5.4.9 have global order 1/2 in the sense of definition 5.4.18. Protter and Talay show in [92] that the latter method has order 1 when the driver is a suitable Lévy process, the coupling coefficients are suitably smooth, and the deviation of the approximate X' from the exact solution X is measured by $\mathbb{E}[g \circ X_t - g \circ X'_t]$ for suitably (rather) smooth g.

(ii) That the coupling coefficients should commute surely is rare. The reaction to scholium 5.4.22 nevertheless should not be despair. Rather, we might distance ourselves from the definition 5.4.14 of a method and possibly entertain less stringent definitions of order than the one adopted in definition 5.4.18. We refer the reader to [85] and [86].

5.5 Weak Solutions

Example 5.5.1 (Tanaka) Let W be a standard Wiener process on its own natural filtration $\mathcal{F}_.[W]$, and consider the stochastic differential equation

$$X = \mathrm{sign}X{*}W \ . \tag{5.5.1}$$

The coupling coefficient $\quad \mathrm{sign}x \stackrel{\mathrm{def}}{=} \begin{cases} 1 & \text{for } x \geq 0 \\ -1 & \text{for } x < 0 \end{cases}$

is of course more general than the ones contemplated so far; it is, in particular, not Lipschitz and returns a previsible rather than a left-continuous process upon being fed $X \in \mathfrak{C}$. Let us show that (5.5.1) cannot have a strong solution in the sense of page 273. By way of contradiction assume that X solves this equation. Then X is a continuous martingale with square function $[X,X]_t = \Lambda_t \stackrel{\mathrm{def}}{=} t$ and $X_0 = 0$, so it is a standard Wiener process (corollary 3.9.5). Then

$$|X|^2 = X^2 = 2X{*}X + \Lambda = 2X\mathrm{sign}X{*}W + \Lambda \ ,$$

and so $$\frac{1}{|X|+\epsilon}{*}|X|^2 = \frac{2|X|}{|X|+\epsilon}{*}W + \frac{1}{|X|+\epsilon}{*}\Lambda \ , \qquad \epsilon > 0 \ .$$

Then $$W = \lim_{\epsilon \to 0} \frac{|X|}{|X|+\epsilon}{*}W = \lim_{\epsilon \to 0} \frac{1}{|X|+\epsilon}{*}(|X|^2 - \Lambda)/2$$

is adapted to the filtration generated by $|X|$: $\mathcal{F}_t[X] \subseteq \mathcal{F}_t[W] \subseteq \mathcal{F}_t[|X|] \quad \forall t$ – this would make X a Wiener process adapted to the filtration generated by its absolute value $|X|$, what nonsense. Thus (5.5.1) has no strong solution. Yet it has a solution in some sense: start with a Wiener process X on its own natural filtration $\mathcal{F}_.[X]$, and set $W \overset{\text{def}}{=} \text{sign}X{*}X$. Again by corollary 3.9.5, W is a standard Wiener process on $\mathcal{F}_.[X]$ (!), and equation (5.5.1) is satisfied. In fact, there is more than one solution, $-X$ being another one. What is going on? In short: the natural filtration of the driver W of (5.5.1) was too small to sustain a solution of (5.5.1).

Example 5.5.1 gives rise to the notion of a weak solution. To set the stage consider the stochastic differential equation

$$X = C + f_\eta[X, \boldsymbol{Z}]_{.-}{*}Z^\eta = C + \boldsymbol{f}[X, \boldsymbol{Z}]_{.-}{*}\boldsymbol{Z} \,. \tag{5.5.2}$$

Here $\boldsymbol{Z}$ is our usual vector of integrators on a measured filtration $(\mathcal{F}_., \mathbb{P})$. The coupling coefficients f_η are assumed to be endogenous and to act in a non-anticipating fashion – see (5.1.4):

$$f_\eta\big[x_., \boldsymbol{z}_.\big]_t = f_\eta\big[x_.^t, \boldsymbol{z}_.^t\big]_t \qquad \forall\, x_. \in \mathscr{D}^n,\ \boldsymbol{z}_. \in \mathscr{D}^d,\ t \ge 0 \,.$$

Definition 5.5.2 *A* ***weak solution*** Ξ' *of equation (5.5.2) is a filtered probability space* $(\Omega', \mathcal{F}'_., \mathbb{P}')$ *together with* $\mathcal{F}'_.$*-adapted processes* $C', \boldsymbol{Z}', X'$ *such that the law of* $(C', \boldsymbol{Z}')$ *on* $\mathscr{D}^{n+d}$ *is the same as that of* $(C, \boldsymbol{Z})$*, and such that (5.5.2) is satisfied:*

$$X' = C' + \boldsymbol{f}[X', \boldsymbol{Z}']_{.-}{*}\boldsymbol{Z}' \,.$$

The problem (5.5.2) is said to have a ***unique weak solution*** *if for any other weak solution* $\Xi'' = (\Omega'', \mathcal{F}''_., \mathbb{P}'', C'', \boldsymbol{Z}'', X'')$ *the laws of* X' *and* X'' *agree, that is to say* $X'[\mathbb{P}'] = X''[\mathbb{P}'']$.

Let us fix $f_\eta[X, \boldsymbol{Z}]_{.-}{*}Z^\eta$ to be the universal integral $f_\eta[X, \boldsymbol{Z}]_{.-} \circledast Z^\eta$ of remarks 3.7.27 and represent $(C_., X_., \boldsymbol{Z}_.)$ on canonical path space $\mathscr{D}^{2n+d}$ in the manner of item 2.3.11. The image of $\mathbb{P}'$ under the representation is then a probability $\overline{\mathbb{P}}'$ on $\mathscr{D}^{2n+d}$ that is carried by the "universal solution set"

$$S \overset{\text{def}}{=} \big\{(c_., x_., \boldsymbol{z}_.) : \ x_. = c_. + \big(\boldsymbol{f}[x_., \boldsymbol{z}_.]_{.-} \circledast \boldsymbol{z}\big)_.\big\} \tag{5.5.3}$$

and whose projection on the " $(C, \boldsymbol{Z})$-component" $\mathscr{D}^{n+d}$ is the law $\mathbb{L}$ of $(C, \boldsymbol{Z})$. Doing this to another weak solution Ξ'' will only change the measure from $\overline{\mathbb{P}}'$ to $\overline{\mathbb{P}}''$. The uniqueness problem turns into the question of whether the solution set S supports different probabilities whose projection on $\mathscr{D}^{n+d}$ is $\mathbb{L}$. Our equation will have a strong solution precisely if there is an *adapted* cross section $\mathscr{D}^{n+d} \to S$. We shall henceforth adopt this picture but write the evaluation processes as $\boldsymbol{Z}_t(c_., x_., \boldsymbol{z}_.) = \boldsymbol{z}_t$, etc., without overbars.

We shall show below that there exist weak solutions to (5.5.2) when $\boldsymbol{Z}$ is continuous and $\boldsymbol{f}$ is endogenous and continuous and has at most linear growth (see theorem 5.5.4 on page 333). This is accomplished by generalizing to the stochastic case the usual proof involving Peano's method of little straight steps. The uniqueness is rather more difficult to treat and has been established only in much more restricted circumstances – when the driver has the special form of condition 5.4.28 and the coupling coefficient is markovian and suitably nondegenerate; below we give two proofs (theorem 5.5.10 and exercise 5.5.14). For more we refer the reader to the literature ([104], [33], [53]).

The Size of the Solution

We continue to assume that $\boldsymbol{Z} = (Z^1, \ldots, Z^d)$ is a local L^q-integrator for some $q \geq 2$ and pick a $p \in [2, q]$. For a suitable choice of M (see (5.2.26)), the arguments of items 5.1.4 and 5.1.5 that led to the inequalities (5.1.16) and (5.1.17) on page 276 provide the a priori estimates (5.2.23) and (5.2.24) of the size of the solution X. They were established using the Lipschitz nature of the coupling coefficient $\boldsymbol{F}$ in an essential way. We shall now prove an a priori growth estimate that assumes no Lipschitz property, merely ***linear growth:*** there exist constants A, B such that up to evanescence

$$\left|\boldsymbol{F}[X]\right|_{\infty p} \leq A + B \cdot \left|X^\star\right|_p . \tag{5.5.4}$$

This implies $\left|\boldsymbol{F}[X]_T\right|_{\infty p} \leq A + B \cdot \left|X_T^\star\right|_p$

for all stopping times T, and in particular

$$\left|\boldsymbol{F}[X]_{T^\lambda -}\right|_{\infty p} \leq A + B \cdot \left|X_{T^\lambda -}^\star\right|_p$$

for the stopping times T^λ of THE time transformation, which in turn implies

$$\Big\|\left|\boldsymbol{F}[X]_{T^\lambda -}\right|_{\infty p}\Big\|_{L^p} \leq A + B \cdot \Big\|\left|X_{T^\lambda -}^\star\right|_p\Big\|_{L^p} \quad \forall \lambda > 0 . \tag{5.5.5}$$

This last is the form in which the assumption of linear growth enters the arguments. We will discuss this in the context of the general equation (5.2.18) on page 289:

$$X = C + F_\eta[X]_{.-} * Z^\eta . \tag{5.5.6}$$

Lemma 5.5.3 *Assume that X is a solution of (5.5.6), that the coupling coefficient $\boldsymbol{F}$ satisfies the linear-growth condition (5.5.5), and that*[43]

$$\Big\|\left|C_{T^\lambda -}^\star\right|_p\Big\|_{L^p} < \infty \quad \textit{and} \quad \Big\|\left|X_{T^\lambda -}^\star\right|_p\Big\|_{L^p} < \infty \tag{5.5.7}$$

for all $\lambda > 0$. Then there exists a constant $M = M_{p;B}$ such that

$$\|X\|_{p,M}^\star \leq 2\Big(A/B + \sup_{\lambda > 0} \Big\|\left|C_{T^\lambda -}^\star\right|_p\Big\|_{L^p}\Big) . \tag{5.5.8}$$

---

[43] If (5.5.4) holds, then inequality (5.5.7) can of course always be had provided we are willing to trade the given probability for a suitable equivalent one and to argue only up to some finite stopping time (see theorem 4.1.2).

Proof. Set $\Delta \stackrel{\text{def}}{=} X - C$ and let $0 \le \kappa < \mu$. Let S be a stopping time with $T^\kappa \le S < T^\mu$ on $[T^\kappa < T^\mu]$. Such S exist arbitrarily close to T^μ due to the predictability of that stopping time. Then

$$\left\|\left|(\Delta - \Delta^{T^\kappa})^\star_S\right|_p\right\|_{L^p} \le \left\|\left|((T^\kappa, S] \cdot \boldsymbol{F}[X]_{.-} * \boldsymbol{Z})^\star_S\right|_p\right\|_{L^p} \le C_p^{\diamond(4.5.1)} \cdot \left|Q\right|_p,$$

where $$Q^\nu \stackrel{\text{def}}{=} \max_{\rho=1^\diamond, p^\diamond} \left\| \left(\int_{T^\kappa}^S \sup_\eta \left|F^\nu_\eta[X]\right|^\rho_{s-} \, d\Lambda_s \right)^{1/\rho} \right\|_{L^p}$$

$$\le \max_{\rho=1^\diamond, p^\diamond} \left\| \left(\int_\kappa^\mu \sup_\eta \left|F^\nu_\eta[X]\right|^\rho_{T^\lambda-} \, d\lambda \right)^{1/\rho} \right\|_{L^p}.$$

Thus $$\left|Q\right|_p \le \max_{\rho=1^\diamond, p^\diamond} \left\| \left| \left(\int_\kappa^\mu \sup_\eta \left|F_\eta[X]\right|^\rho_{T^\lambda-} \, d\lambda \right)^{1/\rho} \right|_p \right\|_{L^p},$$

using A.3.29: $$\le \max_{\rho=1^\diamond, p^\diamond} \left(\int_\kappa^\mu \left\| \left| \sup_\eta \left|F_\eta[X]\right|_{T^\lambda-} \right|_p \right\|^\rho_{L^p} d\lambda \right)^{1/\rho}$$

using (5.5.5): $$\le \max_{\rho=1^\diamond, p^\diamond} \left(\int_\kappa^\mu \left(A + B \left\| \left|X^\star_{T^\lambda-}\right|_p \right\|_{L^p}\right)^\rho d\lambda \right)^{1/\rho}.$$

Taking S through a sequence announcing T^μ gives

$$\left\|\left|(\Delta - \Delta^{T^\kappa})^\star_{T^\mu-}\right|_p\right\|_{L^p} \le C_p^\diamond \max_{\rho=1^\diamond, p^\diamond} \left(\int_\kappa^\mu \left(A+B \left\| \left|X^\star_{T^\lambda-}\right|_p \right\|_{L^p}\right)^\rho d\lambda \right)^{1/\rho}. \tag{5.5.9}$$

For $\kappa = 0$, we have $T^\kappa = 0$, $X_0 = C_0$, and $\Delta_0 = 0$, so (5.5.9) implies

$$\left\| \left|X^\star_{T^\mu-}\right|_p \right\|_{L^p} \le \left\| \left|C^\star_{T^\mu-}\right|_p \right\|_{L^p} + C_p^\diamond \max_{\rho=1^\diamond, p^\diamond} \left(\int_0^\mu \left(A+B \left\| \left|X^\star_{T^\lambda-}\right|_p \right\|_{L^p}\right)^\rho d\lambda \right)^{\frac{1}{\rho}}.$$

Gronwall's lemma in the form of exercise A.2.36 on page 384 now produces the desired inequality (5.5.8). ∎

Existence of Weak Solutions

Theorem 5.5.4 *Assume the driver $\boldsymbol{Z}$ is continuous; the coupling coefficient $\boldsymbol{f}$ is endogenous (p. 289) and non-anticipating, continuous,*[21] *and has at most linear growth; and the initial condition C is constant in time.*

Then the stochastic differential equation $X = C + \boldsymbol{f}[X, \boldsymbol{Z}] * \boldsymbol{Z}$ has a weak solution.

The proof requires several steps. The continuity of the driver entrains that THE previsible controller $\Lambda = \Lambda^{\langle q \rangle}[\boldsymbol{Z}]$ and the solution X of equation (5.5.6) are continuous as well. Both Λ and THE time transformation associated with it are now strictly increasing and continuous. Also, $X^\star_{T^\lambda-} = X^\star_{T^\lambda}$ for all λ, and $p^\diamond = 2$. Using inequality (5.5.8) and carrying out the λ-integral in (5.5.9) provides the inequality

$$\left\| \left| (X - X^{T^\kappa})^\star_{T^\lambda} \right|_p \right\|_{L^p} \le c_\mu \cdot |\kappa - \lambda|^{1/2}, \qquad \kappa, \lambda \in [0, \mu], \ |\kappa - \lambda| < 1,$$

where $c_\mu = c_{\mu;A,B}$ is a constant that grows exponentially with μ and depends only on the indicated quantities $\mu; A, B$. We raise this to the pth power and obtain

$$\mathbb{E}\Big[|X_{T^\kappa} - X_{T^\lambda}|_p^p\Big] \le c_\mu \cdot |\kappa - \lambda|^{p/2} . \qquad (*_1)$$

The driver clearly satisfies a similar inequality:

$$\mathbb{E}\Big[|\boldsymbol{Z}_{T^\kappa} - \boldsymbol{Z}_{T^\lambda}|_p^p\Big] \le c'_\mu \cdot |\kappa - \lambda|^{p/2} . \qquad (*_2)$$

We choose $p > 2$ and invoke Kolmogorov's lemma A.2.37 (ii) to establish as a first step toward the proof of theorem 5.5.4 the

Lemma 5.5.5 *Denote by $\mathbb{X}_{AB}$ the collection of all those solutions of equation (5.5.6) whose coupling coefficient satisfies inequality (5.5.5).*

(i) For every $\alpha < 1$ there exists a set C_α of paths in $\mathscr{C}^{n+d}$, compact[21] and therefore uniformly equicontinuous on every bounded time-interval, such that

$$\mathbb{P}\big[(X_., \boldsymbol{Z}_.) \in C_\alpha\big] > 1 - \alpha , \qquad X_. \in \mathbb{X}_{AB} .$$

(ii) Therefore the set $\big\{(X_., \boldsymbol{Z}_.)[\mathbb{P}] : X_. \in \mathbb{X}_{AB}\big\}$ of laws on $\mathscr{C}^{n+d}$ is uniformly tight and thus is relatively compact.[44]

Proof. Fix an instant u. There exists a $\mu > 0$ so that $\Omega^\Lambda \overset{\text{def}}{=} [\Lambda_u \le \mu] = [T^\mu \ge u]$ has $\mathbb{P}[\Omega^\Lambda] > 1 - \alpha/2$. As in the arguments of pages 14–15 we regard Λ as a $\mathbb{P}^*$-measurable map on $[\Lambda_u < \mu]$ whose codomain is $\mathscr{C}[0, u]$ equipped with the uniform topology. According to the definition 3.4.2 of measurability or Lusin's theorem, there exists a subset $\Omega^\Lambda_\alpha \subset \Omega^\Lambda$ with $\mathbb{P}[\Omega^\Lambda_\alpha] > 1 - \alpha/2$ on which Λ is uniformly continuous in the uniformity generated by the (idempotent) functions in $\mathcal{F}_u$, a uniformity whose completion is compact. Hence the collection $\Lambda_.(\Omega^\Lambda_\alpha)$ of increasing functions has compact closure $\underline{C}^\Lambda_\alpha$ in $\mathscr{C}[0, u]$.

For $X_. \in \mathbb{X}_{AB}$ consider the paths $\lambda \mapsto (\underline{X}_\lambda, \underline{\boldsymbol{Z}}_\lambda) \overset{\text{def}}{=} (X_{T^\lambda}, \boldsymbol{Z}_{T^\lambda})$ on $[0, \mu]$. Kolmogorov's lemma A.2.37 in conjunction with $(*_1)$ and $(*_2)$ provides a compact set $\underline{C}^{AB}_\alpha$ of continuous paths $\lambda \mapsto (\underline{x}_\lambda, \underline{z}_\lambda)$, $0 \le \lambda \le \mu$, such that the set $\Omega^X_\alpha \overset{\text{def}}{=} [(\underline{X}^\mu_., \underline{Z}^\mu_.) \in \underline{C}^{AB}_\alpha]$ has $\mathbb{P}[\Omega^X_\alpha] > 1 - \alpha/2$ simultaneously for every $X_.$ in $\mathbb{X}_{AB}$. Since the paths of $\underline{C}^{AB}_\alpha$ are uniformly equicontinuous (exercise A.2.38), the composition map $\circ$ on $\underline{C}^{AB}_\alpha \times \underline{C}^\Lambda_\alpha$, which sends $((\underline{x}_., \underline{z}_.), \lambda_.)$ to $t \mapsto (\underline{x}_{\lambda_t}, \underline{z}_{\lambda_t})$, is continuous and thus has compact image $C_\alpha \overset{\text{def}}{=} \underline{C}^{AB}_\alpha \circ \underline{C}^\Lambda_\alpha \subset \mathscr{C}^{n+d}[0, u]$. Indeed, let $\epsilon > 0$. There is a $\delta > 0$ so that $|\lambda' - \lambda| < \delta$ implies $|(x_{\lambda'}, z_{\lambda'}) - (x_\lambda, z_\lambda)| < \epsilon/2$ for all $(x_., z_.) \in \underline{C}^{AB}_\alpha$ and all $\lambda, \lambda' \in [0, \mu]$. If $|\lambda' - \lambda| < \delta$ in C^Λ_α and $|(x'_., z'_.) - (x_., z_.)| < \epsilon/2$,

[44] The pertinent topology on the space of probabilities on path spaces is the topology of weak convergence of measures; see section A.4.

then $|(x'_{\lambda'_t}, z'_{\lambda'_t}) - (x_{\lambda_t}, z_{\lambda_t})| \le |(x'_{\lambda'_t}, z'_{\lambda'_t}) - (x_{\lambda'_t}, z_{\lambda'_t})| |(x_{\lambda'_t}, z_{\lambda'_t}) - (x_{\lambda_t}, z_{\lambda_t})| < \epsilon + \epsilon = 2\epsilon$; taking the supremum over $t \in [0, u]$ yields the claimed continuity. Now on $\Omega_\alpha \overset{\text{def}}{=} \Omega_\alpha^\Lambda \cap \Omega_\alpha^X$ we have clearly $(\underline{X}_{\Lambda_t}, \underline{Z}_{\Lambda_t}) = (X_t, Z_t)$, $0 \le t \le u$. That is to say, $(X., Z.)$ maps the set Ω_α, which has $\mathbb{P}[\Omega_\alpha] > 1 - \alpha$, into the compact set $C_\alpha \subset \mathscr{C}^{n+d}[0, u]$, a set that was manufactured from Z and A, B alone. Since $\alpha < 1$ was arbitrary, the set of laws $\{(X., Z.)[\mathbb{P}] : X. \in \mathbb{X}_{AB}\}$ is uniformly tight and thus (proposition A.4.6) is relatively compact.[44]

Actually, so far we have shown only that the projections on $\mathscr{C}^{n+d}[0, u]$ of these laws form a relatively weakly compact set, for any instant u. The fact that they form a relatively compact[44] set of probabilities on $\mathscr{C}^{n+d}[0, \infty)$ and are uniformly tight is left as an exercise. ∎

Proof of Theorem 5.5.4. For $n \in \mathbb{N}$ let $\mathcal{S}^{(n)}$ be the partition $\{k2^{-n} : k \in \mathbb{N}\}$ of time, define the coupling coefficient $F^{(n)}$ as the $\mathcal{S}^{(n)}$-scalæfication of f, and consider the corresponding stochastic differential equation

$$X_t^{(n)} = C + \int_0^t F_s^{(n)}[X^{(n)}]\, dZ_s = C + \sum_{0 \le k} f[X^{(n)}, Z]_{k2^{-n}} \cdot (Z_t - Z_{k2^{-n} \wedge t}) .$$

It has a unique solution, obtained recursively by $X_0^{(n)} = C$ and

$$X_t^{(n)} = X_{k2^{-n}}^{(n)} + f[X_.^{(n)k2^{-n}}, Z_.^{k2^{-n}}]_{k2^{-n}} \cdot (Z_t - Z_{k2^{-n}}) \qquad (5.5.10)$$
$$\text{for } k2^{-n} \le t \le (k{+}1)2^{-n} , \qquad k = 0, 1, \ldots .$$

For later use note here that the map $Z. \mapsto X_.^{(n)}$ is evidently continuous.[21] Also, the linear-growth assumption $|f[x, z]_t| \le A + B \cdot x_t^\star$ implies that the $F^{(n)}$ all satisfy the linear-growth condition (5.5.4). The laws $\mathbb{L}^{(n)}$ of the $(X_.^{(n)}, Z.)$ on path space $\mathcal{C}^{n+d}[0, \infty)$ form, in view of lemma 5.5.5, a relatively compact[44] set of probabilities. Extracting a subsequence and renaming it to $(\mathbb{L}^{(n)})$ we may assume that this sequence converges[44] to a probability $\mathbb{L}'$ on $\mathcal{C}^{n+d}[0, \infty)$. We set $\Omega' \overset{\text{def}}{=} \mathbb{R}^n \times \mathcal{C}^{n+d}[0, \infty)$ and $\mathbb{P}' \overset{\text{def}}{=} C[\mathbb{P}] \times \mathbb{L}'$ and equip Ω' with its canonical filtration $\mathcal{F}'$. On it there live the natural processes $X'_., Z'_.$ defined by

$$X'_t((c, x., z.)) \overset{\text{def}}{=} x_t \quad \text{and} \quad Z'_t((c, x., z.)) \overset{\text{def}}{=} z_t , \qquad t \ge 0 ,$$

and the random variable $C' : (c, x., z.) \mapsto c$. If we can show that, under $\mathbb{P}'$, $X' = C + [\mathbf{X}', Z'] * Z'$, then the theorem will be proved; that $(C', Z'_.)$ has the same distribution under $\mathbb{P}'$ as $(C, Z.)$ has under $\mathbb{P}$, that much is plain.

Let us denote by $\mathbb{E}'$ and $\mathbb{E}^{(n)}$ the expectations with respect to $\mathbb{P}'$ and $\mathbb{P}^{(n)} \overset{\text{def}}{=} C[\mathbb{P}] \times \mathbb{L}^{(n)}$, respectively. Below we will need to know that Z' is a $\mathbb{P}'$-integrator:

$$\lceil Z' \rceil_{\mathcal{I}^p[\mathbb{P}']} \le \lceil Z \rceil_{\mathcal{I}^p[\mathbb{P}]} . \qquad (5.5.11)$$

To see this let $\mathcal{A}_t$ denote the algebra of bounded continuous functions $f : \Omega' \to \mathbb{R}$ that depend only on the values of the path at finitely many instants s prior to t; such f is the composition of a continuous bounded function ϕ on $\mathbb{R}^{(2n+d)\times k}$ with a vector $(c, x_., z_.) \mapsto \big((c, x_{s_i}, z_{s_i}) : 1 \le i \le k\big)$. Evidently $\mathcal{A}_t$ is an algebra and vector lattice closed under chopping that generates $\mathcal{F}_t$. To see (5.5.11) consider an elementary integrand $\boldsymbol{X}'$ on $\mathcal{F}'$ whose d components X'_η are as in equation (2.1.1) on page 46, but special in the sense that the random variables $X'_{\eta s}$ belong to $\mathcal{A}_s$, at all instants s. Consider only such $\boldsymbol{X}'$ that vanish after time t and are bounded in absolute value by 1. An inspection of equation (2.2.2) on page 56 shows that, for such $\boldsymbol{X}'$, $\int \boldsymbol{X}' d\boldsymbol{Z}'$ is a continuous function on Ω'. The composition of $\boldsymbol{X}'$ with $(X_., \boldsymbol{Z}_.)$ is a previsible process $\boldsymbol{X}$ on $(\Omega, \mathcal{F})$ with $|\boldsymbol{X}| \le [\![0, t]\!]$, and

$$\mathbb{E}'\Big[\Big|\int \boldsymbol{X}' d\boldsymbol{Z}'\Big|^p \wedge K\Big] = \lim \mathbb{E}^{(n)}\Big[\Big|\int \boldsymbol{X}' d\boldsymbol{Z}'\Big|^p \wedge K\Big]$$

$$= \lim \mathbb{E}\Big[\Big|\int \boldsymbol{X}\, d\boldsymbol{Z}\Big|^p \wedge K\Big] \le \lceil\!\lceil \boldsymbol{Z}^t \rceil\!\rceil^p_{\mathcal{I}^p[\mathbb{P}]}. \tag{5.5.12}$$

We take the supremum over $K \in \mathbb{N}$ and apply exercise 3.3.3 on page 109 to obtain inequality (5.5.11).

Next let $t \ge 0$, $\alpha \in (0,1)$, and $\epsilon > 0$ be given. There exists a compact[21] subset $C_\alpha \in \mathcal{F}'_t$ such that $\mathbb{P}'[C_\alpha] > 1 - \alpha$ and $\mathbb{P}^{(n)}[C_\alpha] > 1 - \alpha \quad \forall n \in \mathbb{N}$. Then

$$\mathbb{E}'\Big[\big|X' - (C' + \boldsymbol{f}[X', \boldsymbol{Z}']{*}\boldsymbol{Z}')\big|^\star_t \wedge 1\Big]$$

$$\begin{aligned}
&\le \mathbb{E}'\Big[\Big|(\boldsymbol{f}[X', \boldsymbol{Z}'] - \boldsymbol{f}^{(n)}[X', \boldsymbol{Z}']){*}\boldsymbol{Z}'\Big|^\star_t \wedge 1\Big] \\
&\quad + \mathbb{E}'\Big[\Big|X' - \Big(C' + \boldsymbol{f}^{(n)}[X', \boldsymbol{Z}']{*}\boldsymbol{Z}'\Big)\Big|^\star_t \wedge 1\Big] \\
&\le \mathbb{E}'\Big[\Big|(\boldsymbol{f}[X', \boldsymbol{Z}'] - \boldsymbol{f}^{(n)}[X', \boldsymbol{Z}']){*}\boldsymbol{Z}'\Big|^\star_t \wedge 1\Big] \\
&\quad + \Big(\mathbb{E}' - \mathbb{E}^{(m)}\Big)\Big[\Big|X' - \Big(C' + \boldsymbol{f}^{(n)}[X', \boldsymbol{Z}']{*}\boldsymbol{Z}'\Big)\Big|^\star_t \wedge 1\Big] \\
&\quad + \mathbb{E}^{(m)}\Big[\Big|X' - \Big(C' + \boldsymbol{f}^{(n)}[X', \boldsymbol{Z}']{*}\boldsymbol{Z}'\Big)\Big|^\star_t \wedge 1\Big]
\end{aligned}$$

Since $\mathbb{E}^{(m)}\big[\big|X' - (C' + \boldsymbol{f}^{(m)}[X', \boldsymbol{Z}']{*}\boldsymbol{Z}')\big|^\star_t \wedge 1\big] = 0$:

$$\begin{aligned}
&\le \mathbb{E}'\Big[\Big|(\boldsymbol{f}[X', \boldsymbol{Z}'] - \boldsymbol{f}^{(n)}[X', \boldsymbol{Z}']){*}\boldsymbol{Z}'\Big|^\star_t \wedge 1\Big] \\
&\quad + \Big(\mathbb{E}' - \mathbb{E}^{(m)}\Big)\Big[\Big|X' - \Big(C' + \boldsymbol{f}^{(n)}[X', \boldsymbol{Z}']{*}\boldsymbol{Z}'\Big)\Big|^\star_t \wedge 1\Big] \\
&\quad + \mathbb{E}^{(m)}\Big[\Big|(\boldsymbol{f}^{(m)}[X', \boldsymbol{Z}'] - \boldsymbol{f}^{(n)}[X', \boldsymbol{Z}']){*}\boldsymbol{Z}'\Big|^\star_t \wedge 1\Big] \\
&\le 2\alpha + \mathbb{E}'\Big[\Big|(\boldsymbol{f}[X', \boldsymbol{Z}'] - \boldsymbol{f}^{(n)}[X', \boldsymbol{Z}']){*}\boldsymbol{Z}'\Big|^\star_t \cdot C_\alpha\Big]
\end{aligned} \tag{5.5.13}$$

$$+\left(\mathbb{E}-\mathbb{E}^{(m)}\right)\left[\left|X'-\left(C'+\boldsymbol{f}^{(n)}[X',\boldsymbol{Z}']*\boldsymbol{Z}'\right)\right|_t^\star \wedge 1\right] \tag{5.5.14}$$

$$+\,\mathbb{E}^{(m)}\left[\left|\left(\boldsymbol{f}^{(m)}[X',\boldsymbol{Z}']-\boldsymbol{f}^{(n)}[X',\boldsymbol{Z}']\right)*\boldsymbol{Z}'\right|_t^\star \cdot C_\alpha\right]. \tag{5.5.15}$$

Now the image under $\boldsymbol{f}$ of the compact set C_α is compact, on acount of the stipulated continuity of $\boldsymbol{f}$, and thus is uniformly equicontinuous (exercise A.2.38). There is an index N such that $|\boldsymbol{f}((x_., \boldsymbol{z}_.)) - \boldsymbol{f}^{(n)}((x_., \boldsymbol{z}_.))| \le \epsilon$ for all $n \ge N$ and all $(x_., \boldsymbol{z}_.) \in C_\alpha$. Since $\boldsymbol{f}$ is non-anticipating, $\boldsymbol{f}_.[X_., \boldsymbol{Z}_.]$ is a continuous adapted process and so is predictable. So is $\boldsymbol{f}^{(n)}_.[X_., \boldsymbol{Z}_.]$. Therefore $|\boldsymbol{f} - \boldsymbol{f}^{(n)}| \le \epsilon$ on the predictable envelope $\widehat{C}_\alpha$ of C_α. We conclude with exercise 3.7.16 on page 137 that $(\boldsymbol{f}[X', \boldsymbol{Z}'] - \boldsymbol{f}^{(n)}[X', \boldsymbol{Z}'])*\boldsymbol{Z}'$ and $((\boldsymbol{f}[X', \boldsymbol{Z}'] - \boldsymbol{f}^{(n)}[X', \boldsymbol{Z}']) \cdot \widehat{C}_\alpha)*\boldsymbol{Z}'$ agree on C_α. Now the integrand of the previous indefinite integral is uniformly less than ϵ, so the maximal inequality (2.3.5) furnishes the inequality

$$\mathbb{E}'\left[\left|\left(\boldsymbol{f}[X',\boldsymbol{Z}']-\boldsymbol{f}^{(n)}[X',\boldsymbol{Z}']\right)*\boldsymbol{Z}'\right|_t^\star \cdot C_\alpha\right] \le \epsilon \cdot C_1^\star \lceil \boldsymbol{Z}'^t \rceil_{\mathcal{I}^1[\mathbb{P}']}$$

by inequality (5.5.11):
$$\le \epsilon \cdot C_1^\star \lceil \boldsymbol{Z}^t \rceil_{\mathcal{I}^1[\mathbb{P}]} .$$

The term in (5.5.15) can be estimated similarly, so that we arrive at

$$\mathbb{E}'\left[\left|X'-\left(C'+\boldsymbol{f}[X',\boldsymbol{Z}']*\boldsymbol{Z}'\right)_t^\star\right| \wedge 1\right] \le 2\alpha + 2\epsilon \cdot C_1^\star \lceil \boldsymbol{Z}^t \rceil_{\mathcal{I}^1[\mathbb{P}]}$$
$$+\left(\mathbb{E}-\mathbb{E}^{(m)}\right)\left[\left|X'-\left(C'+\boldsymbol{f}^{(n)}[X',\boldsymbol{Z}']*\boldsymbol{Z}'\right)\right|_t^\star \wedge 1\right].$$

Now the expression inside the brackets $[\,]$ of the previous line is a continuous bounded function on $\mathscr{C}^{n+d}$ (see equation (5.5.10)); by the choice of a sufficiently large $m \ge N$ it can be made arbitrarily small. In view of the arbitrariness of α and ϵ, this boils down to $\mathbb{E}'\left[\left|X' - (C'+\boldsymbol{f}[X',\boldsymbol{Z}']*\boldsymbol{Z}')\right|_t^\star \wedge 1\right] = 0$. ∎

Problem 5.5.6 *Find a generalization to càdlàg drivers* $\boldsymbol{Z}$.

Uniqueness

The known uniqueness results for weak solutions cover mainly what might be called the "Classical Stochastic Differential Equation"

$$X_t = x + \int_0^t f_0(s, X_s)\, ds + \sum_{\eta=1}^{d} \int_0^t f_\eta(s, X_s)\, dW_s^\eta . \tag{5.5.16}$$

Here the driver is as in condition 5.4.28. They all require the ***uniform ellipticity***[7] of the symmetric matrix

$$a^{\mu\nu}(t,x) \stackrel{\text{def}}{=} \frac{1}{2}\sum_{\eta=1}^{d} f_\eta^\mu(t,x) f_\eta^\nu(t,x) ,$$

namely,
$$a^{\mu\nu}(t,x)\xi_\mu\xi_\nu \ge \beta^2 \cdot |\xi|^2 \qquad \forall\, \xi, x \in \mathbb{R}^n , \quad \forall\, t \in \mathbb{R}_+ \tag{5.5.17}$$

for some $\beta > 0$. We refer the reader to [104] for the most general results. Here we will deal only with the "Classical Time-Homogeneous Stochastic Differential Equation"

$$X_t = x + \int_0^t f_0(X_s)\, ds + \sum_{\eta=1}^{d} \int_0^t f_\eta(X_s)\, dW_s^\eta \tag{5.5.18}$$

under stronger than necessary assumptions on the coefficients f_η. We give two uniqueness proofs, mostly to exhibit the connection that stochastic differential equations (SDEs) have with elliptic and parabolic partial differential equations (PDEs) of order 2.

The uniform ellipticity can of course be had only if the dimension d of $\boldsymbol{W}$ exceeds the dimension n of the state space; it is really no loss of generality to assume that $n = d$. Then the matrix $f_\eta^\nu(x)$ is invertible at all $x \in \mathbb{R}^n$, with a uniformly bounded inverse named $F(x) \stackrel{\text{def}}{=} f^{-1}(x)$.

We shall also assume that the f_η^ν are continuous and bounded. For ease of thinking let us use the canonical representation of page 331 to shift the whole situation to the path space $\mathscr{C}^n$. Accordingly, the value of X_t at a path $\omega = x_\cdot \in \mathscr{C}^n$ is x_t. Because of the identity

$$W_t^\eta = \textstyle\sum_\nu \displaystyle\int_0^t F_\nu^\eta(X_s)\,(dX_s^\nu - f_0^\nu(X_s)ds)\,, \tag{5.5.19}$$

$\boldsymbol{W}$ is adapted to the natural filtration on path space. In this situation the problem becomes this: denoting by $\mathfrak{P}$ the collection of all probabilities on $\mathscr{C}^n$ under which the process $\boldsymbol{W}_t$ of (5.5.19) is a standard Wiener process, show that $\mathfrak{P}$ – which we know from theorem 5.5.4 to be non-void – is in fact a singleton.

There is no loss of generality in assuming that $f_0 = 0$, so that equation (5.5.18) turns into

$$X_t = x + \sum_{\eta=1}^{d} \int_0^t f_\eta(X_s)\, dW_s^\eta\,. \tag{5.5.20}$$

Exercise 5.5.7 Indeed, one can use Girsanov's theorem 3.9.19 to show the following: If the law of any process X satisfying (5.5.20) is unique, then so is the law of any process X satisfying (5.5.18).

All of the known uniqueness proofs also have in common the need for some input in the form of hard estimates from Fourier analysis or PDE. The proof given below may have some little whimsical appeal in that it does not refer to the martingale problem ([104], [33], and [53]) but uses the existence of solutions of the Dirichlet problem for its outside input. A second slightly simpler proof is outlined in exercises 5.5.13–5.5.14.

5.5.8 The Dirichlet Problem in its form pertinent to the problem at hand is to find for a given domain B of $\mathbb{R}^n$ a *continuous* function $u : \overline{B} \to \mathbb{R}$ with two *continuous* derivatives in the interior $\mathring{B}$ that solves the PDE[40]

$$\mathcal{A}u(x) \stackrel{\text{def}}{=} \frac{a^{\mu\nu}(x)}{2} u_{;\mu\nu}(x) = 0 \quad \forall\, x \in \mathring{B} \tag{5.5.21}$$

and satisfies the boundary condition

$$u(x) = g(x) \quad \forall\, x \in \partial B \stackrel{\text{def}}{=} \overline{B} \backslash \mathring{B} \,. \tag{5.5.22}$$

If a is the identity matrix, then this is the classical Dirichlet problem asking for a function u harmonic inside B, continuous on $\overline{B}$, and taking the prescribed value g on the boundary. This problem has a unique solution if B is a box and g is continuous; it can be constructed with the time-honored method of separation of variables, which the reader has seen in third-semester calculus. The solution of the classical problem can be parlayed into a solution of (5.5.21)–(5.5.22) when the coefficient matrix $a(x)$ is continuous, the domain B is a box, and the boundary value g is smooth ([36]). For the sake of accountability we put this result as an assumption on $\boldsymbol{f}$:

Assumption 5.5.9 *The coefficients $f^\mu_\eta(x)$ are continuous and bounded and (i) the matrix $a^{\mu\nu}(x) \stackrel{\text{def}}{=} \sum_\eta f^\mu_\eta(x) f^\nu_\eta(x)$ satisfies the strict ellipticity (5.5.17); (ii) the Dirichlet problem (5.5.21)–(5.5.22) with smooth boundary data g has a solution of class $C^2(\mathring{B}) \cap C^0(\overline{B})$ on every box B in $\mathbb{R}^n$ whose sides are perpendicular to the axes.*

The connection of our uniqueness quest with the Dirichlet problem is made through the following observation. Suppose that X^x is a weak solution of the stochastic differential equation (5.5.20). Let u be a solution of the Dirichlet problem above, with B being some relatively compact domain in $\mathbb{R}^n$ containing the point x in its interior. By exercise 3.9.10, the first time T at which X^x hits the boundary of the domain is almost surely finite, and Itô's formula gives

$$\begin{aligned} u(X^x_T) &= u(X^x_0) + \int_0^T u_{;\nu}(X^x_s)\, dX^{x\nu}_s + \frac{1}{2}\int_0^T u_{;\mu\nu}(X^x_s)\, d[X^{x\mu}, X^{x,\nu}]_s \\ &= u(x) + \int_0^T u_{;\nu}(X^x_s) f^\nu_\eta(X^x_s)\, dW^\eta \,, \end{aligned}$$

since $\mathcal{A}u = 0$ and thus $u_{;\mu\nu}(X^x_s) d[X^{x\mu}, X^{x\nu}]_s = 2\mathcal{A}u(X^x_s) ds = 0$ on $[s < T]$. Now u, being continuous on $\overline{B}$, is bounded there. This exhibits the right-hand side as the value at T of a bounded local martingale. Finally, since $u = g$ on ∂B,

$$u(x) = \mathbb{E}\big[g(X^x_T)\big] \,. \tag{5.5.23}$$

This equality provides two uniqueness statements: the solution u of the Dirichlet problem, for whose existence we relied on the literature, is unique;

indeed, the equality expresses $u(x)$ as a construct of the vector fields f_η and the boundary function g. We can also read off the ***maximum principle***: u takes its maximum and its minimum on the boundary ∂B. The uniqueness of the solution implies at the same time that the map $g \mapsto u(x)$ is linear. Since it satisfies $|u(x)| \le \sup\{|g(x)| : x \in \partial B\}$ on the algebra of functions g that are restrictions to ∂B of smooth functions, an algebra that is uniformly dense in $C(\partial B)$ by theorem A.2.2 (iii), it has a unique extension to a continuous linear map on $C(\partial B)$, a Radon measure. This is called the ***harmonic measure*** for the problem (5.5.21)–(5.5.22) and is denoted by $\eta^x_{\partial B}(d\sigma)$.

The second uniqueness result concerns any probability $\mathbb{P}$ under which X is a weak solution of equation (5.5.20). Namely, (5.5.23) also says that the ***hitting distribution*** of X^x on the boundary ∂B, by which we mean the law of the process X^x at the first time T it hits the boundary, or the distribution $\lambda^x_{\partial B}(d\sigma) \stackrel{\text{def}}{=} X^x_T[\mathbb{P}]$ of the ∂B-valued random variable X^x_T, is determined by the matrix $a^{\mu\nu}(x)$ alone. In fact it is harmonic measure:

$$\lambda^x_{\partial B} \stackrel{\text{def}}{=} X^x_T[\mathbb{P}] = \eta^x_{\partial B} \qquad \forall\, \mathbb{P} \in \mathfrak{P}\,.$$

Look at things this way: varying B but so that $x \in \mathring{B}$ will produce lots of hitting distributions $\lambda^x_{\partial B}$ that are all images of $\mathbb{P}$ under various maps X^x_T but do actually not depend on $\mathbb{P}$. Any other $\mathbb{P}'$ under which X^x solves equation (5.5.20) will give rise to exactly the same hitting distributions $\lambda^x_{\partial B}$. Our goal is to parlay this observation into the uniqueness $\mathbb{P} = \mathbb{P}'$:

Theorem 5.5.10 *Under assumption 5.5.9, equation (5.5.18) has a unique weak solution.*

Proof. Only the uniqueness is left to be established. Let $H^\nu_{\ell,k}$ be the hyperplane in $\mathbb{R}^n$ with equation $x^\nu = k2^{-n}$, $\nu = 1,\ldots,n$, $1 \le \ell \in \mathbb{N}$, $k \in \mathbb{Z}$. According to exercise 3.9.10 on page 162, we may remove from $\mathscr{C}^n$ a $\mathfrak{P}$-nearly empty set N such that on the remainder $\mathscr{C}^n \backslash N$ the stopping times $S^\nu_{\ell,k} \stackrel{\text{def}}{=} \inf\{t : X_t \in H^\nu_{\ell,k}\}$ are continuous.[21] The random variables $X_{S^\nu_{\ell,k}}$ will then be continuous as well. Then we may remove a further $\mathfrak{P}$-nearly empty set N' such that the stopping times

$$S^{\nu,\nu'}_{\ell,\ell',k,k'} \stackrel{\text{def}}{=} \inf\{t > S^\nu_{\ell,k} : X_t \in H^{\nu'}_{\ell',k'}\}\,,$$

too, are continuous on $\Omega \stackrel{\text{def}}{=} \mathscr{C}^n \setminus (N \cup N')$, and so on. With this in place let us define for every $\ell \in \mathbb{N}$ the stopping times $T^\ell_0 = 0$,

$$T^\ell_\nu \stackrel{\text{def}}{=} \inf\{t > T^{\ell,\nu} : X_t \in \bigcup\nolimits_k H^\nu_{\ell,k}\}$$

and

$$T^\ell_{k+1} \stackrel{\text{def}}{=} \inf\{T^{\ell,\nu} : T^{\ell,\nu} > T^\ell_k\}\,, \qquad k = 0,1,\ldots\,.$$

T^ℓ_{k+1} is the first time after T^ℓ_k that the path leaves the smallest box with sides in the $H^\nu_{\ell,k}$ that contains $X_{T^\ell_k}$ in its interior. The T^ℓ_k are continuous on Ω, and so are the maps $\omega \mapsto X_{T_k^\ell}(\omega)$. In this way we obtain for every $\ell \in \mathbb{N}$ and $\omega = x_. \in \Omega$ a discrete path $\boldsymbol{x}_.^{(\ell)} : \mathbb{N} \ni k \mapsto X_{T_k^\ell}(\omega)$ in $\ell^0_{\mathbb{R}^n}$. The map $\omega \to \boldsymbol{x}_.^{(\ell)}$ is clearly continuous from Ω to $\ell^0_{\mathbb{R}^n}$. Let us identify $\boldsymbol{x}_.^{(\ell)}$ with the path $x'_. \in \mathscr{C}^n$ that at time T_k^ℓ has the value $\boldsymbol{x}_k^{(\ell)} = X_{T_k^\ell}$ and is linear between T_k^ℓ and T_{k+1}^ℓ. The map $\boldsymbol{x}_.^{(\ell)} \mapsto x'_.$ is evidently continuous from $\ell^0_{\mathbb{R}^n}$ to $\mathscr{C}^n$. We leave it to the reader to check that for $\jmath \le \ell$ the times $T_k^\jmath$ agree on $x_.$ and $x'_.$, and that therefore $x_{T_k^\jmath} = x'_{T_k^\jmath}$, $\jmath \le \ell$, $1 \le k < \infty$. The point: for $\jmath \le \ell$

$$\boldsymbol{x}_.^{(\jmath)} \in \ell^0_{\mathbb{R}^n} \quad \text{is a continuous function of } \boldsymbol{x}_.^{(\ell)} \in \ell^0_{\mathbb{R}^n} . \tag{5.5.24}$$

Next let $\mathcal{A}^{(\ell)}$ denote the algebra of functions on Ω of the form $x_. \mapsto \phi(\boldsymbol{x}_.^{(\ell)})$, where $\phi : \ell^0_{\mathbb{R}^n} \to \mathbb{R}$ is bounded and continuous. Equation (5.5.24) shows that $\mathcal{A}^{(\jmath)} \subset \mathcal{A}^{(\ell)}$ for $\jmath \le \ell$. Therefore $\mathcal{A} \overset{\text{def}}{=} \bigcup_\ell \mathcal{A}^{(\ell)}$ is an algebra of bounded continuous functions on Ω.

Lemma 5.5.11 *(i) If $x_.$ and $x'_.$ are two paths in Ω on which every function of $\mathcal{A}$ agrees, then $x_.$ and $x'_.$ describe the same arc (see definition 3.8.17). (ii) In fact, after removal from Ω of another $\mathfrak{P}$-nearly empty set $\mathcal{A}$ separates the points of Ω.*

Proof. (i) First observe that $x_0 = x'_0$. Otherwise there would exist a continuous bounded function ϕ on $\mathbb{R}^n$ that separates these two points. The function $x_. \mapsto \phi(\boldsymbol{x}_{T_0^\ell})$ of $\mathcal{A}$ would take different values on $x_.$ and on $x'_.$. An induction in k using the same argument shows that $x_{T_k^\ell} = x'_{T_k^\ell}$ for all $\ell, k \in \mathbb{N}$. Given a $t > 0$ we now set

$$t' \overset{\text{def}}{=} \sup\{T_k^\ell(x'_.) : T_k^\ell(x_.) \le t\} .$$

Clearly $x_.$ and $x'_.$ describe the same arc via $t \mapsto t'$.

(ii) Using exercise 3.8.18 we adjust Ω so that whenever ω and ω' describe the same arc via $t \mapsto t'$ then, in view of equation (5.5.19), $\boldsymbol{W}_.(\omega)$ and $\boldsymbol{W}_.(\omega')$ also describe the same arc via $t \mapsto t'$, which forces $t = t'$ $\forall t$: any two paths of $X_.$ on which all the functions of $\mathcal{A}$ agree not only describe the same arc, they are actually identical. It is at this point that the differential equation (5.5.18) is used, through its consequence (5.5.19). ∎

Since every probability on the polish space $\mathscr{C}^n$ is tight, the uniqueness claim is immediate from proposition A.3.12 on page 399 once the following is established:

Lemma 5.5.12 *Any two probabilities in $\mathfrak{P}$ agree on $\mathcal{A}$.*

Proof. Let $\mathbb{P}, \mathbb{P}' \in \mathfrak{P}$, with corresponding expectations $\mathbb{E}, \mathbb{E}'$. We shall prove by induction in k the following: $\mathbb{E}$ and $\mathbb{E}'$ agree on functions in $\mathcal{A}^\ell$ of the form

$$\phi_0\big(X_{T_0^\ell}\big) \cdots \phi_k\big(X_{T_k^\ell}\big) , \tag{$*$}$$

$\phi_\kappa \in C_b(\mathbb{R}^n)$. This is clear if $k = 0$: $\phi_0(X_{T_0^\ell}) = \phi_0(x)$. We preface the induction step with a remark: $X_{T_k^\ell}$ is contained in a finite number of $n-1$-dimensional "squares" S^i of side length $2^{-\ell}$. About each of these there is a minimal box B^i containing S^i in its interior, and $X_{T_{k+1}^\ell}$ will lie in the union $\bigcup_i \partial B^i$ of their boundaries. Let u_{k+1}^i denote the solution of equation (5.5.21) on B^i that equals ϕ_{k+1} on ∂B^i. Then[35]

$$\phi_{k+1}(X_{T_{k+1}^\ell}) \cdot S^i \circ X_{T_k^\ell} = u_{k+1}^i(X_{T_{k+1}^\ell}) \cdot S^i \circ X_{T_k^\ell}$$

$$= u_{k+1}^i(X_{T_k^\ell}) \cdot S^i \circ X_{T_k^\ell} + \int_{T_k^\ell}^{T_{k+1}^\ell} u_{k+1;\nu}^i(X)\, dX^\nu$$

has the conditional expectation

$$\mathbb{E}\Big[\phi_{k+1}(X_{T_{k+1}^\ell}) \cdot S^i \circ X_{T_k^\ell} \big| \mathcal{F}_{T_k^\ell}\Big] = u_{k+1}^i(X_{T_k^\ell}) \cdot S^i \circ X_{T_k^\ell} ,$$

whence

$$\mathbb{E}\Big[\phi_{k+1}(X_{T_{k+1}^\ell}) \big| \mathcal{F}_{T_k^\ell}\Big] = \textstyle\sum_i u_{k+1}^i(X_{T_k^\ell}) .$$

Therefore, after conditioning on $\mathcal{F}_{T_k^\ell}$,

$$\mathbb{E}\Big[\phi_0(X_{T_0^\ell}) \cdots \phi_{k+1}(X_{T_{k+1}^\ell})\Big] = \mathbb{E}\Big[\phi_0(X_{T_0^\ell}) \cdots \big(\phi_k \textstyle\sum_i u_{k+1}^i\big)(X_{T_k^\ell})\Big] .$$

By the same token

$$\mathbb{E}'\Big[\phi_0(X_{T_0^\ell}) \cdots \phi_{k+1}(X_{T_{k+1}^\ell})\Big] = \mathbb{E}'\Big[\phi_0(X_{T_0^\ell}) \cdots \big(\phi_k \textstyle\sum_i u_{k+1}^i\big)(X_{T_k^\ell})\Big] .$$

By the induction hypothesis the two right-hand sides agree. The induction is complete. Since the functions of the form (*), $k \in \mathbb{N}$, form a multiplicative class generating $\mathcal{A}$, $\mathbb{E} = \mathbb{E}'$ on $\mathcal{A}$. The proof of the lemma is complete, and with it that of theorem 5.5.10. ∎

The next two exercise comprise another proof of the uniqueness theorem.

Exercise 5.5.13 The ***initial value problem*** for the differential operator $\mathcal{A}$ is the problem of finding, for every $\phi \in C_0(\mathbb{R}^n)$, a function $u(t, x)$ that is twice continuously differentiable in x and bounded on every strip $(0, t') \times \mathbb{R}^n$ and satisfies the ***evolution equation*** $\dot{u} = \mathcal{A}u$ ($\dot{u}$ denotes the t-partial $\partial u/\partial t$) and the initial condition $u(0, x) = \phi(x)$. Suppose X^x solves equation (5.5.20) under $\mathbb{P}$ and u solves the initial value problem. Then $[0, t'] \ni t \mapsto u(t' - t, X_t)$ is a martingale under $\mathbb{P}$.

Exercise 5.5.14 Retain assumption 5.5.9 (i) and assume that the initial value problem of exercise 5.5.13 has a solution in C^2 for every $\phi \in C_b^\infty(\mathbb{R}^n)$ (this holds if the coefficient matrix a is Hölder continuous, for example). Then again equation (5.5.18) has a unique weak solution.

5.6 Stochastic Flows

Consider now the situation that the coupling coefficient $\boldsymbol{F}$ is strongly Lipschitz with constant L, and the initial condition a (constant) point $x \in \mathbb{R}^n$. We want to investigate how the solution X^x of

$$X_t^x = x + \int_0^t \boldsymbol{F}[X^x]_{s-}\, d\boldsymbol{Z}_s \tag{5.6.1}$$

depends on x. Considering x as the parameter in $U \stackrel{\text{def}}{=} \mathbb{R}^n$ and applying theorem 5.2.24, we may assume that $x \mapsto X_.^x(\omega)$ is continuous from $\mathbb{R}^n$ to $\mathscr{D}^n$, for every $\omega \in \Omega$. In particular, the maps $\Xi_t = \Xi_t^\omega : x \mapsto X_t^x(\omega)$, one for every $\omega \in \Omega$ and every $t \geq 0$, map $\mathbb{R}^n$ continuously into itself. They constitute the ***stochastic flow*** that comes with (5.6.1). We shall now consider several special circumstances.

Stochastic Flows with a Continuous Driver

Theorem 5.6.1 *Suppose that the driver $\boldsymbol{Z}$ of equation (5.6.1) is continuous. Then for nearly every $\omega \in \Omega$ all of the Ξ_t^ω, $t \geq 0$, are homeomorphisms of $\mathbb{R}^n$ onto itself.*

Proof [91]. The hard part is to show that Ξ_t^ω *is injective.* Let us replace $\boldsymbol{F}$ by $\boldsymbol{F}/2L$ and $\boldsymbol{Z}$ by $2L\boldsymbol{Z}$. This does not change the differential equation (5.6.1) nor the solutions, but has the effect that now $L \leq 1/2$. Λ shall be the previsible controller for the adjusted driver $\boldsymbol{Z}$. $|\ |$ denotes the euclidean norm on $\mathbb{R}^n$ and $\langle\ |\ \rangle$ the inner product. Let $x, y \in \mathbb{R}^n$ and set $\Delta = \Delta^{x,y} \stackrel{\text{def}}{=} X^x - X^y$. According to equation (5.2.32), Δ satisfies the stochastic differential equation

$$\Delta = (x-y) + G_\eta[\Delta]{*}Z^\eta ,$$

where $$G_\eta[\Delta] \stackrel{\text{def}}{=} F_\eta[\Delta + X^y] - F_\eta[X^y] \qquad ,\eta = 1,\ldots,d ,$$

has $G_\eta[0] = 0$ and is strongly Lipschitz with constant $\leq 1/2$. Clearly

$$\begin{aligned} |\Delta|^2 &= |x-y|^2 + 2\langle\Delta \,{*}\, \Delta\rangle + \textstyle\sum_{\nu=1}^n [\Delta^\nu, \Delta^\nu] \\ &= |x-y|^2 + 2\langle\Delta|G_\eta[\Delta]\rangle{*}Z^\eta + \langle G_\eta[\Delta]|G_\theta[\Delta]\rangle{*}[Z^\eta, Z^\theta] , \end{aligned}$$

and $$\big[|\Delta|^2, |\Delta|^2\big] = 4\langle\Delta|G_\eta[\Delta]\rangle\langle\Delta|G_\theta[\Delta]\rangle{*}[Z^\eta, Z^\theta] .$$

For $\epsilon \geq 0$ set $$|{}^\epsilon\Delta| \stackrel{\text{def}}{=} \sqrt{|\Delta|^2 + \epsilon} .$$

If $\epsilon > 0$, then Itô's formula applies to $|{}^\epsilon\Delta|^{-1} = \big(|\Delta|^2 + \epsilon\big)^{-1/2}$ and gives

$$|{}^\epsilon\Delta|^{-1} = |{}^\epsilon\Delta_0|^{-1} + |{}^\epsilon\Delta|^{-1}{*}{}^\epsilon Y , \tag{5.6.2}$$

where $${}^\epsilon Y = {}^\epsilon Y[x,y] \stackrel{\text{def}}{=} {}^\epsilon J_\eta{*}Z^\eta + {}^\epsilon K_{\eta\theta}{*}[Z^\eta, Z^\theta] ,$$

$$ {}^\epsilon\!J_\eta = {}^\epsilon\!J_\eta[x,y] \stackrel{\text{def}}{=} -\frac{\langle\Delta|G_\eta[\Delta]\rangle}{|\Delta|^2+\epsilon}\,, $$

and $$ {}^\epsilon\!K_{\eta\theta} = {}^\epsilon\!K_{\eta\theta}[x,y] \stackrel{\text{def}}{=} -\frac{\langle G_\eta[\Delta]|G_\theta[\Delta]\rangle}{2|\Delta|^2+\epsilon} + \frac{3\langle\Delta|G_\eta[\Delta]\rangle\langle\Delta|G_\theta[\Delta]\rangle}{2(|\Delta|^2+\epsilon)^2}\,. $$

In view of exercise 3.9.4 the solution of the linear equation (5.6.2) can be expressed in terms of the Doléans–Dade exponential $\mathscr{E}[{}^\epsilon Y]$ as

$$ |{}^\epsilon\Delta|^{-1} = |{}^\epsilon\Delta_0|^{-1}\cdot\mathscr{E}[{}^\epsilon Y] = \big(|x-y|^2+\epsilon\big)^{-1/2}\cdot e^{{}^\epsilon Y-[{}^\epsilon Y,{}^\epsilon Y]/2}\,. $$

Now ${}^\epsilon\!J_\eta$ and ${}^\epsilon\!K_{\eta\theta}$ are bounded in absolute value by 1, independently of ϵ, and converge to $J_\eta \stackrel{\text{def}}{=} {}^0\!J_\eta$ and $K_{\eta\theta} \stackrel{\text{def}}{=} {}^0\!K_{\eta\theta}$, respectively, where we have ${}^0\!J_{\eta\theta} = {}^0\!K_{\eta\theta} \stackrel{\text{def}}{=} 0$ on the (evanescent as we shall see) set $[|\Delta|=0]$. By the DCT, ${}^\epsilon Y \xrightarrow[\epsilon\to 0]{} Y \stackrel{\text{def}}{=} {}^0Y$ and $[{}^\epsilon Y,{}^\epsilon Y] \xrightarrow[\epsilon\to 0]{} [Y,Y]$ as integrators and therefore uniformly on bounded time-intervals, nearly (see exercise 3.8.10). Therefore the limit

$$ |\Delta|^{-1} = \lim_{\epsilon\to 0} |{}^\epsilon\Delta|^{-1} $$

exists and $$ |\Delta|^{-1} = |x-y|^{-1}\cdot e^{Y-[Y,Y]/2} = |x-y|^{-1}\cdot\mathscr{E}[Y]\,. $$

Now $\Lambda[Z]$ is a controller for Y, so by (5.2.23) and for $\gamma \le 10p/\sqrt{M}$

$$ \|Y\|^\star_{p,M} \le \frac{1}{1-\gamma} $$

and $$ \Big\||X^x-X^y|^{-1}\Big\|^\star_{p,M} \le |x-y|^{-1}\cdot\frac{1}{1-\gamma}\,. $$

Clearly, therefore, Δ is bounded away from zero on any bounded time-interval, nearly. Note also the obvious fact that $\Lambda[\boldsymbol{Z}]$ is a previsible controller for Y, so that $|\Delta|^{-1} \in \mathfrak{S}^\star_{p,M}$ for all $p \ge 2$ and all sufficiently large $M = M(p)$, say for $M > M^{\diamond(5.2.20)}_{p,1}$.

We would like to show at this point that the maps $(x,y) \mapsto J_\eta[x,y]$ and $(x,y) \mapsto K_{\eta\theta}[x,y]$ are Lipschitz from $D \stackrel{\text{def}}{=} \big\{(x,y)\in\mathbb{R}^n\times\mathbb{R}^n : x\ne y\big\}$ to $\mathfrak{S}^{\star n}_{p,M}$. By inequality (5.2.5) on page 284 then so are the $\mathscr{C}^n$-valued maps $D \ni (x,y) \mapsto Y[x,y]$ and $D \ni (x,y) \mapsto [Y,Y]$, and another invocation of Kolmogoroff's lemma shows that versions can be chosen so that, for all $\omega\in\Omega$, $D \ni (x,y) \mapsto Y_.[x,y](\omega)$ and $D \ni (x,y) \mapsto \big[Y[x,y],Y[x,y]\big]_{\bullet}(\omega)$ are continuous. This then implies that, simultaneously for all $(x,y)\in D$,

$$ |\Delta^{x,y}_{\bullet}(\omega)|^{-1} = |x-y|^{-1}\cdot e^{Y_.[x,y](\omega)-\big[Y[x,y],Y[x,y]\big]_{\bullet}(\omega)/2} \tag{5.6.3} $$

for nearly all $\omega\in\Omega$, and that $|X^x_t - X^y_t| > 0$ at all times and all pairs $x \ne y$, except possibly in a nearly empty set.

To this end note that $J_\eta[x,y]$ and $K_{\eta\theta}[x,y]$ both are but (inner) products of factors such as $H_1 \stackrel{\text{def}}{=} G_\eta[\Delta^{x,y}]/|\Delta^{x,y}|$ and $H_2 \stackrel{\text{def}}{=} \Delta^{x,y}/|\Delta^{x,y}|$, which are bounded in absolute value by 1. To show that J_η or $K_{\eta\theta}$ are Lipschitz from D to $\mathfrak{S}^{\star n}_{p,M}$ it evidently suffices to show this for the H_i. Let then $(\overline{x},\overline{y})$ be another pair in D and write

$$\Delta \stackrel{\text{def}}{=} \Delta^{x,y} = X^x - X^y \quad\text{and}\quad \overline{\Delta} \stackrel{\text{def}}{=} \overline{\Delta}^{x,y} = X^{\overline{x}} - X^{\overline{y}} ,$$

$$G[\Delta] \stackrel{\text{def}}{=} F_\eta[\Delta + X^y] - F_\eta[X^y] \quad\text{and}\quad \overline{G}[\overline{\Delta}] \stackrel{\text{def}}{=} F_\eta[\overline{\Delta} + X^{\overline{y}}] - F_\eta[X^{\overline{y}}] .$$

Then
$$\left|\frac{\overline{G}[\overline{\Delta}]}{|\overline{\Delta}|} - \frac{G[\Delta]}{|\Delta|}\right| = \left|\frac{|\Delta|\cdot\overline{G}[\overline{\Delta}] - |\overline{\Delta}|\cdot G[\Delta]}{|\overline{\Delta}||\Delta|}\right|$$

$$= \left|\frac{(|\Delta| - |\overline{\Delta}|)\cdot\overline{G}[\overline{\Delta}] + |\overline{\Delta}|\cdot(\overline{G}[\overline{\Delta}] - G[\overline{\Delta}]) + |\overline{\Delta}|\cdot(G[\overline{\Delta}] - G[\Delta])}{|\overline{\Delta}||\Delta|}\right|$$

$$\le \frac{\big||\Delta| - |\overline{\Delta}|\big|}{|\Delta|} + \frac{|\overline{G}[\overline{\Delta}] - G[\overline{\Delta}]|}{|\Delta|} + \frac{\big||\overline{\Delta}| - |\Delta|\big|}{|\Delta|}$$

$$\le \frac{2|\Delta - \overline{\Delta}|}{|\Delta|} + \frac{2|X^{\overline{y}} - X^y|}{|\Delta|} \le 4\frac{|X^{\overline{x}} - X^x| + |X^{\overline{y}} - X^y|}{|\Delta|}$$

and therefore

$$\left\|\frac{\overline{G}[\overline{\Delta}]}{|\overline{\Delta}|} - \frac{G[\Delta]}{|\Delta|}\right\|^\star_{p,M}$$

$$\le 4\left\||\Delta|^{-1}\right\|^\star_{2p,\frac{M}{2}} \times \left(\left\|X^x - X^{\overline{x}}\right\|^\star_{2p,\frac{M}{2}} + \left\|X^y - X^{\overline{y}}\right\|^\star_{2p,\frac{M}{2}}\right)$$

by (5.2.40):
$$\le 4\,|x-y|^{-1}\left\|\mathscr{E}[Y]\right\|^\star_{2p,\frac{M}{2}} \times (|x - \overline{x}| + |y - \overline{y}|) \times \frac{1+\gamma}{1-\gamma}$$

$$\le \frac{const}{|x-y|}\cdot|(x,y) - (\overline{x},\overline{y})| .$$

This shows that H_1 is Lipschitz, but only on $D_k \stackrel{\text{def}}{=} \{(x,y)\in D : |x-y| \ge 1/k\}$. The estimate for H_2 is similar but easier. We can conclude now that (5.6.3) holds for all $(x,y) \in D_k$ simultaneously, whence $\inf_{s\le t}|X^x_s - X^y_s| > 0$ for all such pairs, except in a nearly empty set $N_k \subset \Omega$. Of course, we then have $\inf_{s\le t}|X^x_s - X^y_s| > 0$ for all $x \ne y$, except possibly on the nearly empty set $\bigcup_k N_k$. After removal of this set from Ω we are in the situation that $x \mapsto X^x_t(\omega)$ is continuous and injective for all $\omega \in \Omega$. *Ξ^ω_t is indeed injective.*

To see the surjectivity define

$$Y^x \stackrel{\text{def}}{=} \begin{cases} \left|X^{x/|x|^2} - X^0\right|^{-1} & \text{for } x \ne 0 \\ \left|X^0\right|^{-1} & \text{for } x = 0. \end{cases}$$

Then
$$\left|Y^x - Y^y\right| \le \frac{\left|X^{x/|x|^2} - X^{y/|y|^2}\right|}{\left|X^{x/|x|^2} - X^0\right|\left|X^{y/|y|^2} - X^0\right|}$$

and $$\left\|\left|Y^x - Y^y\right|\right\|^\star_{p,M} \le \frac{1}{(1-\gamma)^3} \cdot |x||y| \left| \frac{x}{|x|^2} - \frac{y}{|y|^2} \right| = \frac{1}{(1-\gamma)^3} \cdot |x - y| \,.$$

This shows that $x \mapsto X^{x/|x|^2}$ is continuous at 0 and means that $x \mapsto X^x_t$ can be viewed as an injection of the n-sphere into itself that maps the north pole (the point at infinity of $\mathbb{R}^n$) to itself. Such a map is surjective. Indeed, because of the compactness of S_n it is a homeomorphism onto its image; and if it did miss as much as a single point, then the image would be contractible, and then so would be S_n.

Drivers with Small Jumps

Suppose Z is an integrator that has a jump of size -1 at the stopping time T. By exercise 3.9.4 the solutions of the exponential equation $X^x = x + X^x_{.-}{*}Z$ will all vanish at and after time T. So we cannot expect the result of theorem 5.6.1 to hold in general. It is shown in [91] that it persists in some cases where the driver $\boldsymbol{Z}$ has sufficiently small or suitably distributed jumps. Here is some information that becomes available when the coupling coefficients are differentiable:

Example 5.6.2 Consider again the stochastic differential equation (5.6.1):

$$X_t = x + \boldsymbol{F}[X]_{.-}{*}\boldsymbol{Z} \,, \qquad x \in \mathbb{R}^n \,.$$

Its driver $\boldsymbol{Z}$ is now an arbitrary L^0-integrator. Its unique solution is a process $X^x_.$ that starts at x: $X^x_0 = x$. We consider $\mathbb{R}^n$ the parameter domain and assume that $\boldsymbol{F} : \mathfrak{S}^{\star n}_{p,M} \to \mathfrak{S}^{\star n}_{p,M}$ has a Lipschitz weak derivative and $\boldsymbol{Z}$ is an L^q-integrator with $q > n$, or, if $\boldsymbol{Z}$ is merely an L^0-integrator, that $\boldsymbol{F}$ is a differentiable randomly autologous coefficient with a Lipschitz derivative as in exercise 5.3.3. Then there exists a particular solution $X^x_.$ such that $x \mapsto X^x_.(\omega)$ is of class[21, 10] C^1 for every $\omega \in \Omega$. In particular, the maps $\Xi_t : x \mapsto X^x_t(\omega)$, which constitute the stochastic flow, each are differentiable from $\mathbb{R}^n$ to itself. One says that the solution is a ***stochastic flow of class C^1***.

The differential equation (5.3.7) for DX^x reads

$$DX^x_t = I + \int_0^t DF_\eta[X^x]DX^x_{s-}\, dZ^\eta_s = I + \int_0^t dY_s\, DX^x_{s-} \,,$$

where $Y_s \overset{\text{def}}{=} F_\eta[X^x]_{.-}{*}Z^\eta$ (see exercise 5.6.3). Clearly $|\Delta Y| \le L \cdot \sum_\eta |\Delta Z^\eta|$.

Assume henceforth that $L \cdot \sup_{\varpi} \sum_\eta |\Delta Z^\eta| < 1$.

Then $DX^x_t(\omega)$ is invertible for all $\varpi = (t, \omega) \in \boldsymbol{B}$ (*ibidem, mutatis mutandis*). From this it is easy to see that every member $\Xi^\omega_t : \mathbb{R}^n \to \mathbb{R}^n$ of the flow is locally a homeomorphism. Since its image is closed, it must be all of $\mathbb{R}^n$, and Ξ_t^ω is a (finite) covering of $\mathbb{R}^n$. Since $\mathbb{R}^n$ is simply connected, Ξ_t^ω is a homeomorphism of $\mathbb{R}^n$ onto itself. The inverse function theorem now permits us to conclude that *Ξ_t^ω is a diffeomorphism for all $(t,\omega) \in \boldsymbol{B}$.*

This example persists *mutatis mutandis* when ζ is a spatially bounded L^q-random measure for some $q > n$, or when $n = 1$. ∎

Exercise 5.6.3 Let $Y = Y_\nu^\mu$ be an $n \times n$-matrix of L^q-integrators, $q \ge 2$. Consider $Y_t(\omega)$ as a linear operator from euclidean space $\mathbb{R}^n$ to itself, with operator norm $\|Y\|$. Its jump is the matrix

$$\Delta Y_s \stackrel{\text{def}}{=} (\Delta Y^\mu_{\nu s})^{\mu=1\ldots n}_{\nu=1\ldots n} ;$$

its square function is[1]

$$[Y,Y] = ([Y,Y])^\mu_\nu \stackrel{\text{def}}{=} [Y^\mu_\rho, Y^\rho_\nu] ,$$

which by theorems 3.8.4 and 3.8.9 is an $L^{q/2}$-integrator. Set

$$\overline{Y}_t \stackrel{\text{def}}{=} -Y_t + {}^c[Y,Y]_t + \sum_{0<s\le t} (I + \Delta Y_s)^{-1}(\Delta Y_s)^2 .$$

(i) Assume that $\overline{Y}$ is an L^0-integrator. Then the solutions of

$$D_t = I + \int_0^t D_{s-}\, dY_s \quad \text{and} \quad \overline{D}_t = I + \int_0^t d\overline{Y}\, \overline{D}_{s-}$$

are inverse to each other:

$$D_t \overline{D}_t = I \qquad \forall t \ge 0 .$$

Here[1]

$$(d\overline{Y}\, \overline{D}_{.-})^\mu_\nu \stackrel{\text{def}}{=} D^\rho_{\nu .-}\, d\overline{Y}^\mu_\rho .$$

(ii) If $\sup_{(s,\omega)\in\boldsymbol{B}} \|\Delta Y_s(\omega)\| < 1$, then $(I+\Delta Y_s)^{-1}$ is bounded. If $(I+\Delta Y_s)^{-1}$ is bounded, then

$$J_t \stackrel{\text{def}}{=} \sum_{0<s\le t} (I + \Delta Y_s)^{-1}(\Delta Y_s)^2$$

is an $L^{q/2}$-integrator, and so is $\overline{Y}$.

Markovian Stochastic Flows

If the coupling coefficient of (5.6.1) is markovian, that equation becomes

$$X_t^x = x + \int_0^t \boldsymbol{f}(X_{s-}^x)\, d\boldsymbol{Z}_s . \tag{5.6.4}$$

The f_η comprising $\boldsymbol{f}$ are assumed Lipschitz with constant L, and the driver $\boldsymbol{Z}$ is an arbitrary L^0-integrator with $\boldsymbol{Z}_0 = 0$, at least for a while. Summary 5.4.7 on page 317 provides the universal solution $\overline{X}(x, \boldsymbol{z}_.)$ of

$$\overline{X}_t = x + \int_0^t \boldsymbol{f}(\overline{X}_{s-})\, d\overline{\boldsymbol{Z}}_s . \tag{5.6.5}$$

Since at this point we are interested only in constant initial conditions, we restrict $\overline{X}$ to $\mathbb{R}^n \times \mathscr{D}^d$, so that it becomes a map from $\mathbb{R}^n \times \mathscr{D}^d$ to $\mathscr{D}^n$, adapted to the filtrations $\mathcal{B}^\bullet(\mathbb{R}^n) \otimes \mathcal{F}_.[\mathscr{D}^d]$ and $\mathcal{F}_.[\mathscr{D}^n]$ (see item 2.3.8). $\overline{Z}$ is the process representing Z on the path space $\mathbb{R}^n \times \mathscr{D}^d$: $\overline{Z}_t(x, z_.) = z_t$, and the assumption $Z_0 = 0$ has the effect that any of the probabilities of $\overline{\mathfrak{P}} \stackrel{\text{def}}{=} Z[\mathfrak{P}] \subset \mathfrak{P}[\overline{Z}]$ is carried by the set $[\overline{Z}_0 = 0] = \{(x, z_.) : z_0 = 0\}$. Taking for the parameter domain U of theorem 5.2.24 the space $\mathbb{R}^n$ itself, we see that $\overline{X}$ can be modified on a $\mathfrak{P}[\overline{Z}]$-nearly empty set in such a way that both

$$z_. \mapsto \overline{X}_.(x, z_.) \text{ is adapted to } \mathcal{F}_.[\mathscr{D}^d] \text{ and } \mathcal{F}_.[\mathscr{D}^n] \quad \forall x \in \mathbb{R}^n \tag{5.6.6}$$

and $$x \mapsto \overline{X}_.(x, z_.) \text{ is continuous}^{21} \text{ from } \mathbb{R}^n \text{ to } \mathscr{D}^n \quad \forall z_. \in \mathscr{D}^d. \tag{5.6.7}$$

Note that $\overline{X}$ is a construct made from the coupling coefficient f alone; in particular, the prevailing probability does not enter its definition. $X^x_. \stackrel{\text{def}}{=} \overline{X}_.(x, Z_.)$ is a solution of (5.6.4), and any other solution that depends continuously on x differs from $X^x_.$ in a nearly empty set that does not depend on x.

Here is a straightforward observation. Let S be a stopping time and $t \geq 0$. Then

$$\begin{aligned} X^x_{S+t} &= X^x_S + \int_S^{S+t} f(X^x_{\sigma-})\, d(Z_\sigma - Z_S) \\ &= X^x_S + \int_0^t f(X^x_{(S+\tau)-})\, d(Z_{S+\tau} - Z_S)\,. \end{aligned}$$

This says that the process $X_{S+.}$ satisfies the same differential equation (5.6.4) as does $X_.$, except that the initial condition is X_S and the driver $Z_. - Z_S$. Now upon representing this driver on $\mathscr{D}^d$, every $\mathbb{P} \in \mathfrak{P}[Z]$ turns into a probability in $\mathfrak{P}[\overline{Z}]$, inasmuch as $Z_. - Z_S$ is a $\mathbb{P}$-integrator. Therefore this stochastic differential equation is solved by

$$X^x_{S+t} = \overline{X}_t(X^x_S, Z_{S+.} - Z_S)\,. \tag{5.6.8}$$

Applying this very argument to equation (5.6.5) on $\mathbb{R}^n \times \mathscr{D}^d$ results in

$$\overline{X}_{S+t}(x, z_.) = \overline{X}_t\big(\overline{X}_S(x, z_.), z_{S+.} - z_S\big)\,. \tag{5.6.9}$$

For any $\mathcal{F}_.[\mathscr{D}^d]$-stopping time S, any $t \geq 0$, and any $x \in \mathbb{R}^n$, this equation holds a priori only $\mathfrak{P}[\overline{Z}]$-nearly, of course. At a fixed stopping time S we may assume, by throwing out a $\mathfrak{P}[\overline{Z}]$-nearly empty set, that (5.6.9) holds for all rational instants t and all rational points $x \in \mathbb{R}^n$. Since $\overline{X}$ is continuous in its first argument, though, and $t \mapsto \overline{X}_t(x, z_.)$ is right-continuous, we get

Proposition 5.6.4 *The universal solution for equation (5.6.4) satisfies (5.6.6) and (5.6.7); and for any stopping time S there exists a nearly empty set outside which equation (5.6.9) holds simultaneously for all $x \in \mathbb{R}^n$ and all $t \geq 0$.*

Problem 5.6.5 Can (5.6.9) be made to hold identically for all $s \geq 0$? ∎

Markovian Stochastic Flows Driven by a Lévy Process

The situation and notations are the same as in the previous subsection, except that we now investigate the case that the driver $\boldsymbol{Z}$ is a Lévy process. In this case THE previsible controller $\Lambda_t[\boldsymbol{Z}]$ is just a constant multiple of time t (equation (4.6.30)) and THE time transformation $T^{\cdot}$ is also simply a linear scaling of time. Let us define the positive linear operator T_t on $C_b(\mathbb{R}^n)$ by

$$T_t\phi(x) \stackrel{\text{def}}{=} \mathbb{E}[\phi(X_t^x)] = \mathbb{E}[\phi \circ \overline{X}_t(x, \boldsymbol{Z}_.)] . \tag{5.6.10}$$

It follows from inequality (5.2.36) that $T_t\phi$ is continuous; and it is not too hard to see with the help of inequality (5.2.25) that T_t maps $C_0(\mathbb{R}^n)$ into $C_0(\mathbb{R}^n)$ when $\boldsymbol{f}$ is bounded. Now fix an $\mathcal{F}_.$-stopping time S. Since $\boldsymbol{Z}_{.+S} - \boldsymbol{Z}_S$ is independent of $\mathcal{F}_S$ and has the same law as $\boldsymbol{Z}_.$ (exercise 4.6.1),

$$\begin{aligned}\mathbb{E}[\phi \circ \overline{X}_t(x, \boldsymbol{Z}_{.+S} - \boldsymbol{Z}_S)|\mathcal{F}_S] &= \mathbb{E}[\phi \circ \overline{X}_t(x, \boldsymbol{Z}_{.+S} - \boldsymbol{Z}_S)] \\ &= \mathbb{E}[\phi \circ \overline{X}_t(x, \boldsymbol{Z}_.)] = T_t\phi(x) ,\end{aligned}$$

whence

$$\mathbb{E}\big[\phi \circ \overline{X}_t(X_S^x, \boldsymbol{Z}_{.+S} - \boldsymbol{Z}_S)|\mathcal{F}_S\big] = T_t\phi \circ X_S^x ,$$

thanks to exercise A.3.26 on page 408. With equation (5.6.8) this gives

$$\mathbb{E}[\phi \circ X_{S+t}^x|\mathcal{F}_S] = T_t\phi \circ X_S^x \tag{5.6.11}$$

and, taking the expectation,

$$T_{s+t}\phi(x) = T_s[T_t\phi](x) \tag{5.6.12}$$

for $s, t \in \mathbb{R}_+$ and $x \in \mathbb{R}^n$. The remainder of this subsection is given over to a discussion of these two equalities.

5.6.6 (The Markov Property) Taking in (5.6.11) the conditional expectation under X_S^x (see page 407) gives

$$\mathbb{E}\big[\phi \circ X_{S+t}^x|\mathcal{F}_S\big] = \mathbb{E}\big[\phi \circ X_{S+t}^x|X_S^x\big] \circ X_S^x . \tag{5.6.13}$$

That is to say, as far as the distribution of X_{S+t}^x is concerned, knowledge of the whole path up to time S provides no better clue than knowing the position X_S^x at that time: "the process continually forgets its past." A process satisfying (5.6.13) is said to have the ***strong Markov property***. If (5.6.13) can be ascertained only at deterministic instants S, then the process has the "plain" ***Markov property***. Equation (5.6.11) is actually stronger than the strong Markov property, in this sense: the function $T_t\phi$ is a common conditional expectation of $\phi(X_{S+t}^x)$ under X_S^x, one that depends neither on S nor on x. We leave it to the reader to parlay equation (5.6.11) into the following. Let $F : \mathscr{D}^n \to \mathbb{R}$ be bounded and measurable on the Baire σ-algebra for the pointwise topology, that is, on $\mathcal{F}_\infty^0[\mathscr{D}^n]$. Then

$$\mathbb{E}\big[F(X_{S+.}^x)|\mathcal{F}_S\big] = \mathbb{E}\big[F(X_{S+.}^x)|X_S^x\big] \circ X_S^x . \tag{5.6.14}$$

To paraphrase: "in order to predict at time S anything[45] about the future of $X^x_.$, knowledge of the whole history $\mathcal{F}_S$ of the world up to that time is not more helpful than knowing merely the position X^x_S at that time."

Exercise 5.6.7 Let us wean the processes X^x, $x \in \mathbb{R}^n$, of their provenance, by representing each on $\mathscr{D}^n$ (see items 2.3.8–2.3.11). The image under $\underline{X}^x$ of the given probability is a probability on $\mathcal{F}_\infty[\mathscr{D}^n]$, which shall be written $\mathbb{P}^x$; the corresponding expectation is $\mathbb{E}^x$. The evaluation process $\underline{(t, x.) \mapsto} x_t$ will be written $\overline{X}$, as often before. Show the following. (i) Under $\mathbb{P}^x$, $\overline{X}$ starts at x: $\mathbb{P}^x[\overline{X}_0 = x] = 1$. (ii) For $F \in L^\infty(\mathcal{F}_\infty[\mathscr{D}^n])$, $x \mapsto \mathbb{E}^x[F]$ is universally measurable. (iii) For all $\phi \in C_0(E)$, all stopping times S on $\mathcal{F}.[\mathscr{D}^n]$, all $t \geq 0$, and all $x \in E$ we have $\mathbb{P}^x$-almost surely $\mathbb{E}^x[\phi(X_{S+t})|\mathcal{F}_S] = T_t\phi \circ X_S$. (iv) $\overline{X}$ is quasi-left-continuous.

5.6.8 (The Feller Semigroup of the Flow) Equation (5.6.12) says that the operators T_t form a semigroup under composition: $T_{s+t} = T_s \circ T_t$ for $s, t \geq 0$. Since evidently $\sup\{T_t\phi(x) : \phi \in C_0(E)\,,\ 0 \leq \phi \leq 1\} = \mathbb{E}[1] = 1$, we have

Proposition 5.6.9 *$\{T_t\}_{t\geq 0}$ forms a conservative Feller semigroup on $C_0(\mathbb{R}^n)$.* ∎

Let us go after the generator of this semigroup. Itô's formula applied to $X^x_.$ and a function ϕ on $\mathbb{R}^n$ of class[10] C^2_b gives[1]

$$\begin{aligned}
\phi(X^x_t) &= \phi(X^x_0) + \int_0^t \phi_{;\nu}(X^x_{s-})\, dX^x_s + \frac{1}{2}\int_0^t \phi_{;\mu\nu}(X^x_s) d^c[X^{x\mu}, X^{x\nu}] \\
&\quad + \textstyle\sum_{0\leq s\leq t} \phi(X^x_{s-} + \Delta X^x_s) - \phi(X^x_{s-}) - \phi_{;\nu}(X^x_{s-})\Delta X^{x\nu}_s \\
&= \phi(x) + \int_0^t (\phi_{;\nu} f^\nu_\eta)(X^x_{s-})\, dZ^\eta_s + \frac{1}{2}\int_0^t (\phi_{;\mu\nu} f^\mu_\eta f^\nu_\theta)(X^x_s)\, d^c[Z^\eta, Z^\theta]_s \\
&\quad + \textstyle\sum_{0\leq s\leq t} \phi(X^x_{s-} + f_\eta(X^x_{s-})\Delta Z^\eta_s) - \phi(X^x_{s-}) - (\phi_{;\nu} f^\nu_\eta)(X^x_{s-})\Delta Z^\eta_s \\
&= \phi(x) + \int_0^t (\phi_{;\nu} f^\nu_\eta)(X^x_{s-})\, \big(dZ^\eta_s - y^\eta[|\boldsymbol{y}| > 1]\, \jmath_Z(d\boldsymbol{y}, ds)\big) \\
&\quad + \frac{1}{2}\int_0^t (\phi_{;\mu\nu} f^\mu_\eta f^\nu_\theta)(X^x_s)\, d^c[Z^\eta, Z^\theta]_s \\
&\quad + \int_0^t \phi(X^x_{s-} + f_\eta(X^x_{s-})y^\eta) - \phi(X^x_{s-}) - (\phi_{;\nu} f^\nu_\eta)(X^x_{s-})y^\eta[|\boldsymbol{y}|\leq 1]\, \jmath_Z(d\boldsymbol{y}, ds) \\
&= Mart_t + \phi(x) + \int_0^t (\phi_{;\nu} f^\nu_\eta)(X^x_s) A^\eta\, ds + \frac{1}{2}\int_0^t (\phi_{;\mu\nu} f^\mu_\eta f^\nu_\theta)(X^x_s) B^{\eta\theta}\, ds \\
&\quad + \int_0^t \int \phi(X^x_s + f_\eta(X^x_s)y^\eta) - \phi(X^x_s) - (\phi_{;\mu} f^\mu_\eta)(X^x_s)y^\eta[|\boldsymbol{y}|\leq 1]\, \nu(d\boldsymbol{y})\, ds
\end{aligned}$$

In the penultimate equality the large jumps were shifted into the first term as in equation (4.6.32). We take the expectation and differentiate in t at $t = 0$ to obtain

---

45 Well, at least anything that depends Baire measurably on the path $t \mapsto X^x_{S+t}$.

Proposition 5.6.10 *The generator $\mathcal{A}$ of $\{T_t\}_{t\geq 0}$ acts on a function $\phi \in C_0^2$ by*[1]

$$\mathcal{A}\phi(x) = A^\eta f_\eta^\nu(x)\frac{\partial\phi}{\partial x^\eta}(x) + \frac{B^{\eta\theta}}{2} f_\eta^\mu(x) f_\theta^\nu(x)\frac{\partial^2\phi}{\partial x^\eta \partial x^\theta}(x)$$
$$+ \int \Big(\phi(x + f_\eta(x)y^\eta) - \phi(x) - \frac{\partial\phi}{\partial x^\mu}(x)\, f_\eta^\mu(x)\, y^\eta [|\boldsymbol{y}| \leq 1]\Big)\nu(dy) .$$

Differentiating at $t \neq 0$ gives

$$\frac{dT_t\phi}{dt} = T_t\mathcal{A}\phi = \overline{\mathcal{A}}T_t\phi ,$$

where $\overline{\mathcal{A}}$ is the closure of $\mathcal{A}$. Suppose the coefficients $\boldsymbol{f}$ have two continuous bounded derivatives. Then, by theorem 5.3.16, $x \mapsto X_t^x$ is twice continuously differentiable as a map from $\mathbb{R}^n$ to any of the L^p, $p < \infty$, and hence $x \mapsto T_t\phi(x) = \mathbb{E}[\phi(X_t^x)]$ is twice continuously differentiable. On this function $\overline{\mathcal{A}}$ agrees with $\mathcal{A}$:

Corollary 5.6.11 *If $\boldsymbol{f} \in C_b^2$ and $\phi \in C_0^2$, then $u(t,x) \stackrel{\text{def}}{=} \mathbb{E}[\phi(X_t^x)]$ is continuous and twice continuously differentiable in x and satisfies the evolution equation*

$$\frac{du(t,x)}{dt} = \mathcal{A}u(t,x) .$$

5.7 Semigroups, Markov Processes, and PDE

We have encountered several occasion where a Feller semigroup arose from a process (page 268) or a stochastic differential equation (item 5.6.8) and where a PDE appeared in such contexts (item 5.5.8, exercise 5.5.13, corollary 5.6.11). This section contains some rudimentary discussions of these connections.

Stochastic Representation of Feller Semigroups

Not only do some processes give rise to Feller semigroups, every Feller semigroup comes this way:

Definition 5.7.1 *A **stochastic representation** of the conservative Feller semigroup $T_.$ consists of a filtered measurable space $(\Omega, \mathcal{F}_.)$ together with an E-valued adapted*[46] *process $X_.$ and a slew $\{\mathbb{P}^x\}$ of probabilities on $\mathcal{F}_\infty$, one for every point $x \in E$, satisfying the following description:*
(i) for every $x \in E$, X starts at x under $\mathbb{P}^x$: $\mathbb{P}^x[X_0 = x] = 1$;
(ii) $x \mapsto \mathbb{E}^x[F] \stackrel{\text{def}}{=} \int F\, d\mathbb{P}^x$ is universally measurable, for every $F \in L^\infty(\mathcal{F}_\infty)$;
(iii) for all $\phi \in C_0(E)$, $0 \leq s < t < \infty$, and $x \in E$ we have $\mathbb{P}^x$-almost surely

$$\mathbb{E}^x\big[\phi(X_t)|\mathcal{F}_s\big] = T_{t-s}\phi \circ X_s . \tag{5.7.1}$$

[46] X_t is $\mathcal{F}_t/\mathcal{B}^*(E)$-measurable for $0 \leq t < \infty$ — see page 391.

5.7.2 Here are some easily verified consequences. (i) With $s = 0$ equation (5.7.1) yields

$$\mathbb{E}^x\left[\phi(X_t)\right] = T_t\phi\,(x)\ .$$

(ii) For any $\phi \in C_0(E)$ and $x \in E$, $t \mapsto \phi(X_t)$ is a uniformly continuous curve in the complete seminormed space $\mathcal{L}^2(\mathbb{P}^x)$. So is the curve $t \mapsto T_{u-t}\phi\,(X_t)$ for $0 \le t \le u$. (iii) For any $\alpha > 0$, $x \in E$, and positive $\gamma \in C_0(E)$,

$$t \mapsto Z_t^{\alpha,\gamma} \stackrel{\text{def}}{=} e^{-\alpha t} \cdot U_\alpha\gamma \circ X_t$$

is a positive bounded $\mathbb{P}^x$-supermartingale on $(\Omega, \mathcal{F}_.)$. Here $U_.$ is the resolvent, see page 463.

Theorem 5.7.3 *(i) Every conservative Feller semigroup $T_.$ has a stochastic representation. (ii) In fact, there is a stochastic representation $(\Omega, \mathcal{F}_., X_., \mathfrak{P})$ in which $\mathcal{F}_.$ satisfies the natural conditions*[47] *with respect to $\mathfrak{P} \stackrel{\text{def}}{=} \{\mathbb{P}^x\}_{x\in E}$ and in which the paths of $X_.$ are right-continuous with left limits and stay in compact subsets of E during any finite interval. Such we call a* ***regular stochastic representation.***

(iii) A regular stochastic representation has these additional properties: ***The Strong Markov Property****: for any finite $\mathcal{F}_.$-stopping time S, number $\sigma \ge 0$, bounded Baire function f, and $x \in E$ we have $\mathbb{P}^x$-almost surely*

$$\mathbb{E}^x\left[f(X_{S+\sigma})|\mathcal{F}_S\right] = \int_E T_\sigma(X_S, dy)\, f(y) = \mathbb{E}^{X_S}\left[f(X_\sigma)\right]\ . \qquad (5.7.2)$$

Quasi-Left-Continuity*: for any strictly increasing sequence T_n of $\mathcal{F}_.$-stopping times with finite supremum T and all $\mathbb{P}^x \in \mathfrak{P}$*

$$\lim X_{T_n} = X_T \qquad \mathbb{P}^x\text{-nearly.}$$

Blumenthal's Zero-One Law*: the $\mathfrak{P}$-regularization of the basic filtration of $X_.$ is right-continuous.*

Remarks 5.7.4 Equation (5.7.2) has this consequence:[48] under $\mathbb{P} = \mathbb{P}^x$

$$\mathbb{E}^{\mathbb{P}}[f \circ X_{S+\sigma}|\mathcal{F}_S] = \mathbb{E}^{\mathbb{P}}[f \circ X_{S+\sigma}|X_S] \circ X_S\ . \qquad (5.7.3)$$

This says that as far as the distribution of $X_{S+\sigma}$ is concerned, knowledge of the whole path up to time S provides no better clue than knowing the position X_S at that time: "the process continually forgets its past." An adapted process X on $(\Omega, \mathcal{F}_., \mathbb{P})$ obeying (5.7.3) for all finite stopping times S and $\sigma \ge 0$ is said to have the ***strong Markov property.*** It has the "plain" ***Markov property*** as long as (5.7.3) can be ascertained for sure times S.

---

[47] See definition 1.3.38 and warning 1.3.39 on page 39.
[48] See theorem A.3.24 for the conditional expectation under the map X_S. $\mathbb{E}[f \circ X_T|X_S]$ is a function on the range of X_S and is unique up to sets negligible for the law of X_S.

Actually, equation (5.7.2) gives much more than merely the strong Markov property of $X.$ on $(\Omega, \mathcal{F}_., \mathbb{P}^x)$ for every starting point x. Namely, it says that the Baire function $x' \mapsto \int T_\sigma(x', dy)\, f(y)$ serves as a common member of every one of the classes[48] $\mathbb{E}^x\big[f(X_{S+\sigma})|X_S\big]$. Note that this function depends neither on S nor on x.

Consider the case that S is an instant s and apply (5.7.2) to a function $\phi \in C_0(E)$ and a Borel set B in E.[49] Then

$$\mathbb{E}^x\big[f \circ X_{s+\sigma}\big|\mathcal{F}_s\big] = T_\sigma\phi \circ X_s$$

and
$$\mathbb{P}^x\big[X_{s+\sigma} \in B\big|\mathcal{F}_s\big] = T_\sigma(X_s, B)\,. \tag{5.7.4}$$

Visualize X as the position of a meandering particle. Then (5.7.4) says that the probability of finding it in B at time $s+\sigma$ given the whole history up to time s is the same as the probability that during σ it made its way from its position at time s to B, *no matter how it got to that position.* ∎

Exercise 5.7.5 [11, page 50] If for every compact $K \subset E$ and neighborhood G of K

$$\lim_{t\to 0} \sup_{x\in K} \frac{T_t(x, E\setminus G)}{t} = 0\,,$$

then the paths of $X.$ are $\mathbb{P}^x$-almost surely continuous, for all $x \in E$.

Exercise 5.7.6 For every $x \in E$ and every function ψ in the domain $\breve{\mathcal{D}}$ of the natural extension $\breve{A}$ of the generator (see pages 467–468)

$$M_t^\psi \stackrel{\text{def}}{=} \psi(X_t) - \int_0^t \breve{A}\psi \circ X_s\, ds$$

is a $\mathbb{P}^x$-martingale. Conversely, if there exists a $\phi \in \breve{C}$ so that for all $x \in E$ $M_t^\psi \stackrel{\text{def}}{=} \psi(X_t) - \int_0^t \phi(X_s)\, ds$ is a $\mathbb{P}^x$-martingale, then $\psi \in \breve{\mathcal{D}}$ and $\breve{A}\psi = \phi$.

Proof of Theorem 5.7.3. To start with, let us assume that E is compact.

(i) Prepare, for every $t \in \mathbb{R}_+$, a copy E_t of E and take for Ω the product

$$E^{[0,\infty)} = \prod_{t\in[0,\infty)} E_t$$

of all $[0,\infty)$-tuples $(x_t)_{0\le t<\infty}$ with entry x_t in E_t. Ω is compact in the product topology. X_t is simply the tth coordinate: $X_t\big((x_s)_{0\le s<\infty}\big) = x_t$. Next let $\mathbb{T}$ denote the collection of all finite ordered subsets $\tau \subset [0,\infty)$ and $\mathbb{T}_0$ those that contain $t_0 = 0$; both $\mathbb{T}$ and $\mathbb{T}_0$ are ordered by inclusion. For any $\tau = \{t_1 \ldots t_k\} \in \mathbb{T}$ (with $t_i < t_{i+1}$) write

$$E_\tau = E_{t_1\ldots t_k} \stackrel{\text{def}}{=} E_{t_1} \times \cdots \times E_{t_k}\,.$$

There are the natural projections

$$\begin{aligned} X_\tau = X_{t_1\ldots t_k} :\ & E^{[0,\infty)} \ \to\ E_\tau \\ & : (x_t)_{0\le t<\infty} \mapsto (x_{t_1}, \ldots, x_{t_k})\,. \end{aligned}$$

[49] See convention A.1.5 on page 364.

"X_τ forgets the coordinates not listed in τ." If τ is a singleton $\{t\}$, then X_τ is the tth coordinate function X_t, so that we may also write

$$X_{t_1 \ldots t_k} = \left(X_{t_1}, \ldots, X_{t_k}\right) .$$

We shall define the expectations $\mathbb{E}^x$ first on a special class of functions, the ***cylinder functions***. $F : E^{[0,\infty)} \to \mathbb{R}$ is a cylinder function based on $\tau = \{t_0 t_1 \ldots t_k\} \in \mathbb{T}_0$ if there exists $f : E_\tau \to \mathbb{R}$ with $F = f \circ X_\tau$, i.e.,

$$F = f(X_{t_0 t_1 \ldots t_k}) = f(X_{t_0}, \ldots, X_{t_k}) , \qquad t_0 = 0 .$$

We say that F is Borel or continuous, etc., if f is. If F is based on τ, it is also based on any τ' containing τ; in the representation $F = f \circ X_{\tau'}$, f simply does not depend on the coordinates that are not listed in τ. For instance, if F is based on X_τ and G on $X_{\tau'}$, then both are based on X_υ, where $\upsilon \overset{\text{def}}{=} \tau \cup \tau'$, and can be written $F = f \circ X_\upsilon$ and $G = g \circ X_\upsilon$; thus $F + G = (f + g) \circ X_\upsilon$ is again a cylinder function. Repeating this with $\cdot, \wedge, \vee$ replacing $+$ we see that the Borel or continuous cylinder functions form both an algebra and a vector lattice.

We are ready to define $\mathbb{P}^x$. Let $F = f \circ X_{t_0 t_1 \ldots t_k}$ be a bounded Borel cylinder function. Define inductively $f^{(k)} = f$, $\sigma_i = t_i - t_{i-1}$, and

$$f^{(i-1)}(x_0, x_1, \ldots, x_{k-1}) \overset{\text{def}}{=} \int T_{\sigma_i}(x_{i-1}, dx_i)\, f^{(i)}(x_0, x_1, \ldots, x_{i-1}, x_i) ,$$

as long as $i \geq 1$, and finally set $\mathbb{E}^x[F] \overset{\text{def}}{=} f^{(0)}(x)$. In other words,

$$\mathbb{E}^x[F] \overset{\text{def}}{=} \int T_{t_1}(x, dx_1) \int \cdots \int T_{t_{k-1}-t_{k-2}}(x_{k-2}, dx_{k-1}) \int T_{t_k - t_{k-1}}(x_{k-1}, dx_k) \\ f(x, x_1, \ldots, x_{k-2}, x_{k-1}, x_k) . \qquad (5.7.5)$$

To see that this makes sense assume for the moment that $f \in C(E_\tau)$. We see by inspection[50] that then $f^{(i)}$ belongs to $C(E_{t_0 \ldots t_i})$, $i = k, k-1, \ldots, 0$. Thus

$$x \mapsto \mathbb{E}^x[F] \text{ is continuous for } F \in C(E_\tau) \circ X_\tau .$$

Next consider the class of bounded Borel functions f on $E_{t_0 \ldots t_k}$ such that $f^{(i)}$ is a bounded Borel function on $E_{t_0 \ldots t_i}$ for all i. It contains $C(E_\tau)$ and is closed under bounded sequential limits, so it contains all bounded Borel functions: the iterated integral (5.7.5) makes sense.

Suppose that f does not depend on the coordinate x_j, for some j between 1 and k. The Chapman–Kolmogorov equation (A.9.9) applied with $s = \sigma_j$, $t = \sigma_{j+1}$, and $y = x_{j+1}$ to $\phi(y) = f^{(j+1)}(x_0, \ldots, x_{j-1}, y)$ shows that for

[50] This is particularly obvious when f is of the form $f(x_0, \ldots, x_k) = \prod_i \phi_i(x_i)$, $\phi_i \in C$, or is a linear combination of such functions. The latter form an algebra uniformly dense in $C(E_\tau)$ (theorem A.2.2).

$i < j$ we get the same functions $f^{(i)}$ whether we consider F as a cylinder function based on X_τ or on $X_{\tau'}$, where τ' is τ with t_j removed. Using this observation it is easy to see that $\mathbb{E}^x$ is well-defined on Borel cylinder functions and is linear. It is also evidently positive and has $\mathbb{E}^x[1] = 1$. It is a positive linear functional of mass one defined on the continuous cylinder functions, which are uniformly dense[50] in C: $\mathbb{E}^x$ is a Radon measure. Also, $\mathbb{E}^x[\phi(X_0)] = \phi(x)$. Let $\mathcal{F}^0_.$ denote the basic filtration of $X_.$. Its final σ-algebra $\mathcal{F}^0_\infty$ is clearly nothing but the Baire σ-algebra of Ω. The collection of all bounded Baire functions F on Ω for which $x \mapsto \mathbb{E}^x[F]$ is Baire measurable on E contains the cylinder functions and is sequentially closed. Thus it contains all Baire functions on Ω.

Only equation (5.7.1) is left to be proved. Fix $0 \le s < t$, and let F be a Borel cylinder function based on a partition $\tau \subset [0, s]$ in $\mathbb{T}_0$ and let $\phi \in C_0(E)$. The very definition of $\mathbb{E}^x$ gives

$$\mathbb{E}^x\big[F \cdot \phi(X_t)\big] = \mathbb{E}^x\big[F \cdot T_{t-s}\phi(X_s)\big] .$$

Since the functions F generate $\mathcal{F}^0_s$, equation (5.7.1) follows. Part (i) of theorem 5.7.3 is proved.

(ii) We continue to assume that E is compact, and we employ the stochastic representation constructed above. Its filtration is the basic filtration of $X_.$. Let $\alpha_n > 0$ in $\mathbb{R}$ and $\gamma_n \ge 0$ in C be such that the sequence $(U_{\alpha_n}\gamma_n)$ is dense in C_{0+}. Let $Z^n_.$ be the process

$$t \mapsto e^{-\alpha_n t} \cdot U_{\alpha_n}\gamma_n \circ X_t$$

of item 5.7.2 (ii). It is a global $L^2(\mathbb{P}^x)$-integrator for every single $x \in E$ (lemma 2.5.27). The set Osc of points $\omega \in \Omega$ where $\mathbb{Q} \ni q \mapsto Z^n_q(\omega)$ oscillates anywhere for any n belongs to $\mathcal{F}_\infty$ and is $\mathbb{P}$-nearly empty for every probability $\mathbb{P}$ with respect to which the Z^n are integrators (lemma 2.3.1), in particular for every one of the $\mathbb{P}^x$. Since the differences of the $U_{\alpha_n}\gamma_n$ are dense in C, we are assured that for all $\omega \in \Omega' \stackrel{\text{def}}{=} \Omega \setminus Osc$ and all $\phi \in C_0(E)$

$$\mathbb{Q} \ni q \mapsto \phi\big(X_q(\omega)\big) \text{ has left and right limits at all finite instants.}$$

Fix an instant t and an $\omega \in \Omega'$ and denote by $L_t(\omega)$ the intersection over n of the closures of the sets $\{X_q(\omega) : t \le q \le t + 1/n\} \subset E$. This set contains at least one point, by compactness, but not any more. Indeed, if there were two, we would choose a $\phi \in C$ that separates them; the path $q \mapsto \phi\big(X_q(\omega)\big)$ would have two limit points as $q \downarrow t$, which is impossible. Therefore

$$X'_t(\omega) \stackrel{\text{def}}{=} \lim_{\mathbb{Q} \ni q \downarrow t} X_q(\omega)$$

exists for every $t \ge 0$ and $\omega \in \Omega'$ and defines a right-continuous E-valued process $X'_.$ on Ω'. A similar argument shows that $X'_.$ has a unique left limit in E at all instants. The set $\big[X_t \ne X'_t\big]$ equals the union of the sets

$[Z_t^n \neq Z_{t+}^n] \in \mathcal{F}_{t+1}^0$, which are $\mathfrak{P}$-nearly empty in view of item 5.7.2 (i). Thus $X'_.$ is adapted to the $\mathfrak{P}$-regularization of $\mathcal{F}_.^0$. It is clearly also adapted to that filtration's right-continuous version $\mathcal{F}'_.$. Moreover, equation (5.7.1) stays when $\mathcal{F}_.^0$ is replaced by $\mathcal{F}'_.$. Namely, if $\mathbb{Q} \ni q_n \downarrow s$, then the left-hand side of

$$\mathbb{E}^x\left[\phi(X_t)|\mathcal{F}_{q_n}\right] = T_{t-q_n}\phi(X_{q_n})$$

converges in $\mathbb{E}^x$-mean to $\mathbb{E}^x[\phi(X_t)|\mathcal{F}'_s]$, while the right-hand side converges to $T_{t-s}\phi(X_s)$ by a slight extension of item 5.7.2 (i). Part (ii) is proved: the primed representation $(\Omega', \mathcal{F}'_., X'_., \mathfrak{P})$ meets the description.

Let us then drop the primes and address the case of noncompact E. We let E^Δ denote the one-point compactification $E \cup \{\Delta\}$ of E, and consider on E^Δ the Feller semigroup $T_.^\Delta$ of remark A.9.6 on page 465. Functions in $C(E^\Delta)$ that vanish at Δ are identified with functions of $C_0(E)$. Let $\mathfrak{X}^\Delta \overset{\text{def}}{=} (\Omega, \mathcal{F}_., X_.^\Delta, \mathfrak{P}^\Delta)$ be the corresponding stochastic representation provided by the proof above, with $\mathfrak{P}^\Delta = \{\mathbb{P}^x : x \in E^\Delta\}$. Note that $T_t^\Delta(x, \{\Delta\}) = 0$, which has the effect that $X_t^\Delta \neq \Delta$ $\mathbb{P}^x$-almost surely for all $x \in E$. Pick a function $\gamma^\Delta \in C(E^\Delta)$ that is strictly positive on E and vanishes at Δ and note that $U_1^\Delta\gamma^\Delta$ is of the same description. Then $t \mapsto Z_t^\Delta = e^{-t} \cdot U_1\gamma^\Delta \circ X_t^\Delta$ is a positive bounded right-continuous supermartingale on $\mathcal{F}_.$ with left limits and equals zero at time t if and only if $X_t^\Delta = \Delta$. Due to exercise 2.5.32, the path of Z^Δ is bounded away from zero on finite time-intervals, which means that $X_.^\Delta$ stays in a compact subset of E during any finite time-interval, except possibly in a $\mathfrak{P}$-nearly empty set N. Removing N from Ω leaves us with a stochastic representation of $T_.$ on E. It clearly satisfies (ii) of theorem 5.7.3.

Proof of Theorem 5.7.3 (iii). We start with the strong Markov property. It clearly suffices to prove equation (5.7.2) for the case $f \in C_0(E)$. The general case will follow from an application of the monotone class theorem. Let then $A \in \mathcal{F}_S$ and fix $\mathbb{P}^x \in \mathfrak{P}$. To start with, assume S that takes only countably many values. Then

$$\int A \cdot f(X_{S+\sigma})\, d\mathbb{P}^x = \sum_{s\geq 0} \int [S=s] \cap A \cdot f(X_{S+\sigma})\, d\mathbb{P}^x$$

since $[S=s]\cap A \in \mathcal{F}_s$:
$$= \sum_{s\geq 0} [S=s] \cap A \cdot \mathbb{E}^x\Big[f(X_{S+\sigma})\big|\mathcal{F}_s\Big]\, d\mathbb{P}^x$$

by (5.7.1):
$$= \sum_{s\geq 0} \int [S=s] \cap A \cdot \mathbb{E}^{X_s}\left[f(X_\sigma)\right] d\mathbb{P}^x$$

as $X_s = X_S$ on $[S=s]$:
$$= \sum_{s\geq 0} \int [S=s] \cap A \cdot \mathbb{E}^{X_S}\left[f(X_\sigma)\right] d\mathbb{P}^x$$

$$= \int A \cdot \mathbb{E}^{X_S}\Big[f(X_\sigma)\Big]\, d\mathbb{P}^x .$$

In the general case let $S^{(n)}$ be the approximating stopping times of exercise 1.3.20. Since $x \mapsto \mathbb{E}^x[f(X_\sigma)]$ is continuous and bounded, taking the limit in

$$\int A \cdot f(X_{S^{(n)}+\sigma})\, d\mathbb{P}^x = \int A \cdot \mathbb{E}^{X_{S^{(n)}}}\Big[f(X_\sigma)\Big]\, d\mathbb{P}^x$$

produces the desired equality

$$\int A \cdot f(X_{S+\sigma})\, d\mathbb{P}^x = \int A \cdot \mathbb{E}^{X_S}\Big[f(X_\sigma)\Big]\, d\mathbb{P}^x .$$

This proves the strong Markov property.

Now to the quasi-left-continuity. Let $\phi, \psi \in C_0(E)$. Thanks to the right-continuity of the paths

$$X_T = \lim_{t \downarrow 0} \lim_{n \to \infty} X_{T_n + t} ,$$

and therefore

$$\begin{aligned}
\mathbb{E}^x\Big[\phi(X_{T-}) \cdot \psi(X_T)\Big] &= \lim_{t \downarrow 0} \lim_{n \to \infty} \mathbb{E}^x\Big[\phi(X_{T_n}) \cdot \psi(X_{T_n+t})\Big] \\
&= \lim_{t \downarrow 0} \lim_{n \to \infty} \mathbb{E}^x\Big[\phi(X_{T_n}) \cdot (T_t\psi)(X_{T_n})\Big] \\
&= \lim_{t \downarrow 0} \mathbb{E}^x\Big[\phi(X_{T-}) \cdot (T_t\psi)(X_{T-})\Big] \\
&= \mathbb{E}^x\Big[\phi(X_{T-}) \cdot \psi(X_{T-})\Big] .
\end{aligned}$$

The equality

$$\mathbb{E}^x\Big[f(X_{T-}, X_T)\Big] = \mathbb{E}^x\Big[f(X_{T-}, X_{T-})\Big]$$

therefore holds for functions on $E \times E$ of the form $(x,y) \mapsto \sum_i \phi_i(x)\psi_i(y)$, which form a multiplicative class generating the Borels of $E \times E$. Thus $\mathbb{E}^x[h(X_{T-} - X_T)] = 0$ for all Borel functions h on E, which implies that $[X_{T-} \ne X_T] = \bigcup_n [X_{(T\wedge n)-} \ne X_{T\wedge n}]$ is $\mathbb{P}^x$-nearly empty: the quasi-left-continuity follows.

Finally let us address the regularity. Let us denote by $\mathcal{X}^0_. \subseteq \mathcal{F}_.$ the basic filtration of $X_.$. Fix a $\mathbb{P}^x \in \mathfrak{P}$, a $t \ge 0$, and a bounded $\mathcal{X}^0_{t+}$-measurable function F. Set $G \stackrel{\text{def}}{=} F - \mathbb{E}^x[F|\mathcal{X}^0_t]$. This function is measurable on $\mathcal{X}^0_\tau$ for all $\tau > t$. Now let $f : \mathscr{D}_b(E) \to \mathbb{R}$ be a bounded function of the form

$$f(\omega) = f^t(\omega) \cdot \phi(X_u(\omega)) , \tag{$*$}$$

where $f^t \in \mathcal{X}^0_t$, $\phi \in C_0(E)$, and $u > t$, and consider

$$I^f \stackrel{\text{def}}{=} \int f \cdot G\, d\mathbb{P}^x = \int f^t \cdot G \cdot \phi(X_u)\, d\mathbb{P}^x . \tag{$**$}$$

Pick a $\tau \in (t,u)$ and condition on $\mathcal{X}^0_\tau$. Then

$$I^f = \int f^t \cdot G \cdot T_{u-\tau}\phi(X_\tau)\, d\mathbb{P}^x .$$

Due to item 5.7.2 (i), taking $\tau \downarrow t$ produces

$$I^f = \int f^t \cdot G \cdot T_{u-t}\phi(X_t)\, d\mathbb{P}^x .$$

The factor of G in this integral is measurable on $\mathcal{X}^0_t$, so the very definition of G results in $I^f = 0$.

Now let $\overline{f}(\omega) = \overline{f}^t(\omega) \cdot \overline{\phi}(X_v(\omega))$ be a second function of the form (*), where $v \ge u$, say. Conditioning on $\mathcal{X}^0_u$ produces

$$I^{f\overline{f}} = \int f^t \overline{f}^t \cdot G \cdot \phi(X_u)\overline{\phi}(X_v)\, d\mathbb{P}^x = \int f^t \overline{f}^t \cdot G \cdot \big(\phi \cdot T_{v-u}\overline{\phi}\big)(X_u)\, d\mathbb{P}^x .$$

This is an integral of the form (**) and therefore vanishes. That is to say, $\int f \cdot G\, d\mathbb{P}^x{=}0$ for all f in the algebra generated by functions of the form (*). Now this algebra generates $\mathcal{X}^0_\infty$. We conlude that $[G \ne 0]$ is $\mathbb{P}^x$-negligible. Since this set belongs to $\mathcal{X}^0_{t+1}$, it is even $\mathbb{P}^x$-nearly empty, and thus $\mathbb{P}^x$-nearly $F = \mathbb{E}^x[F|\mathcal{X}^0_t]$. To summarize: if $F \in \mathcal{X}^0_{t+}$, then F differs $\mathbb{P}^x$-nearly from some $\mathcal{X}^0_t$-measurable function $F^{\mathbb{P}^x} = \mathbb{E}^x[F|\mathcal{X}^0_t]$, and this holds for all $\mathbb{P}^x \in \mathfrak{P}$. This says that F belongs to the regularization $\mathcal{X}^{\mathfrak{P}}_t$. Thus $\mathcal{X}^0_{t+} \subset \mathcal{X}^{\mathfrak{P}}_t$ and then $\mathcal{X}^{\mathfrak{P}}_{t+} = \mathcal{X}^{\mathfrak{P}}_t \quad \forall\, t \ge 0$. Theorem 5.7.3 is proved in its entirety. ∎

Exercise 5.7.7 Let $\mathfrak{X}^{\vdash} = (\Omega, \mathcal{F}_., X^{\vdash}_., \{\mathbb{P}^x\})$ be a regular stochastic representation of $T^{\vdash}_.$. Describe $\mathfrak{X}^{\vdash}$ and the projection $X_. \overset{\text{def}}{=} \pi_E \circ X^{\vdash}_.$ on the second factor.

Exercise 5.7.8 To appreciate the designation "Zero-One Law" of theorem 5.7.3 (iiic) show the following: for any $x \in E$ and $A \in \mathcal{F}^{\mathfrak{P}}_{0+}[X_.]$, $\mathbb{P}^x[A]$ is either zero or one. In particular, for $B \in \mathcal{B}^\bullet(E)$,

$$T^B \overset{\text{def}}{=} \inf\{t > 0 : X_t \in B\}$$

is $\mathbb{P}^x$-almost surely either strictly positive or identically zero.

Exercise 5.7.9 (The Canonical Representation of $T_.$) Let $\mathfrak{X} = (\Omega, \mathcal{F}_., X_., \{\mathbb{P}^x\})$ be a regular stochastic representation of the conservative Feller semigroup $T_.$. It gives rise to a map $\rho : \Omega \to \mathscr{D}_E$, space of right-continuous paths $x_. : [0,\infty) \to E$ with left limits, via $\rho(\omega)_t = X_t(\omega)$. Equip $\mathscr{D}_E$ with its basic filtration $\mathcal{F}^0_.[\mathscr{D}_E]$, the filtration generated by the evaluations $x_. \mapsto x_t$, $t \in [0,\infty)$, which we denote again by X_t. Then ρ is $\mathcal{F}^0_\infty[\mathscr{D}_E]/\mathcal{F}_\infty$-measurable, and we may define ***the laws of*** $\mathfrak{X}$ as the images under ρ of the $\mathbb{P}^x$ and denote them again by $\mathbb{P}^x$. They depend only on the semigroup $T_.$, not on its representation $\mathfrak{X}$. We now replace $\mathcal{F}^0_.[\mathscr{D}_E]$ on $\mathscr{D}_E$ by the natural enlargement $\mathcal{F}^{\mathfrak{P}}_{.+}[\mathscr{D}_E]$, where $\mathfrak{P} = \{\mathbb{P}^x : x \in E\}$, and then rename $\mathcal{F}^{\mathfrak{P}}_{.+}[\mathscr{D}_E]$ to $\mathcal{F}_.$. The regular stochastic representation $(\mathscr{D}_E, \mathcal{F}_., X_., \{\mathbb{P}^x\})$ is the canonical representation of $T_.$.

Exercise 5.7.10 (Continuation) Let us denote the typical path in $\mathscr{D}_E$ by ω, and let $\theta_s : \mathscr{D}_b(E) \to \mathscr{D}_b(E)$ be the ***time shift operator*** on paths defined by

$$(\theta_s(\omega))_t = \omega_{s+t} , \qquad s,t \geq 0 .$$

Then $\theta_s \circ \theta_t = \theta_{s+t}$, $X_t \circ \theta_s = X_{t+s}$, $\theta_s \in \mathcal{F}_{s+t}/\mathcal{F}_t$, and $\theta_s \in \mathcal{F}^{\mathfrak{P}}_{s+t}/\mathcal{F}^{\mathfrak{P}}_t$ for all $s,t \geq 0$, and for any finite $\mathcal{F}$-stopping time S and bounded $\mathcal{F}$-measurable random variable F

$$\mathbb{E}^x\left[F \circ \theta_S | \mathcal{F}_S\right] = \mathbb{E}^{X_S}\left[F\right] :$$

"the semigroup $T_.$ is represented by the flow $\theta_.$."

Exercise 5.7.11 Let E be $\mathbb{N}$ equipped with the discrete topology and define the ***Poisson semigroup*** T_t by

$$(T_t\phi)(k) = e^{-t} \sum_{i=0}^{\infty} \phi(k+i) \frac{t^i}{i!} , \qquad \phi \in C_0(\mathbb{N}) .$$

This is a Feller semigroup whose generator $A\phi : n \mapsto \phi(n+1) - \phi(n)$ is defined for all $\phi \in C_0(\mathbb{N})$. Any regular process representing this semigroup is ***Poisson process***.

Exercise 5.7.12 Fix a $t > 0$, and consider a bounded continuous function defined on *all* bounded paths $\omega : [0,\infty) \to E$ that is continuous in the topology of pointwise convergence of paths and depends on the path prior to t only; that is to say, if the stopped paths ω^t and ω'^t agree, then $F(\omega) = F(\omega')$. (i) There exists a countable set $\tau \in [0,t]$ such that F is a cylinder function based on τ; in other words, there is a function f defined on all bounded paths $\xi : \tau \to E$ and continuous in the product topology of E_τ such that $F = f(X_\tau)$. (ii) The function $x \mapsto \mathbb{E}^x[F]$ is continuous. (iii) Let $T_{n,\cdot}$ be a sequence of Feller semigroups converging to $T_.$ in the sense that $T_{n,t}\phi(x) \to T_t\phi(x)$ for all $\phi \in C_0(E)$ and all $x \in E$. Then $\mathbb{E}^x_n[F] \to \mathbb{E}^x[F]$.

Exercise 5.7.13 For every $x \in E$, $t > 0$, and $\epsilon > 0$ there exist a compact set K such that

$$\mathbb{P}^x[X_s \in K \quad \forall\, s \in [0,t]] > 1 - \epsilon . \tag{5.7.6}$$

Exercise 5.7.14 Assume the semigroup $T_.$ is compact; that is to say, the image under T_t of the unit ball of $C_0(E)$ is compact, for arbitrarily small, and then all, $t > 0$. Then T_t maps bounded Borel functions to continuous functions and $x \mapsto \mathbb{E}^x[F \circ \theta_t]$ is continuous for bounded $F \in \mathcal{F}_\infty$, provided $t > 0$. Equation (5.7.6) holds for all x in an open set.

Theorem 5.7.15 (A Feynman–Kac Formula) *Let $T_{s,t}$ be a conservative family of Feller transition probabilities, with corresponding infinitesimal generators A_t, and let $\mathfrak{X}^{\vdash} = \left(\Omega, \mathcal{F}_., X^{\vdash}_., \{\mathbb{P}^x\}\right)$ be a regular stochastic representation of its time-rectification $T^{\vdash}_.$. Denote by $X_.$ its trace on E, so that $X^{\vdash}_t = (t, X_t)$.*

*Suppose that $\Phi \in \operatorname{dom}(\breve{A}^{\vdash})$ satisfies on $[0,u] \times E$ the **backward equation***

$$\frac{\partial \Phi}{\partial t}(t,x) + A_t\Phi\,(t,x) = \big(q{\cdot}\Phi - g\big)(t,x) \tag{5.7.7}$$

and the final condition $\qquad \Phi(u,x) = f(x)$,

where $q, g : [0, u] \times E \to \mathbb{R}$ and $f : E \to \mathbb{R}$ are continuous. Then $\Phi(t, x)$ has the following stochastic representation:

$$\Phi(t,x) = \mathbb{E}^{t,x}\big[Q_u f(X_u)\big] + \mathbb{E}^{t,x}\Big[\int_t^u Q_\tau g(X_\tau^{\vdash})\, d\tau\Big], \tag{5.7.8}$$

where $$Q_\tau \stackrel{\text{def}}{=} \exp\Big(-\int_t^\tau q(X_s^{\vdash})\, ds\Big),$$

provided** (a) q is bounded below, (b) $g \in \check{C}$ or $g \ge 0$, (c) $f \in \check{C}$ or $f \ge 0$, **and

(d) $X_.^{\vdash}$ *has continuous paths* *or*

(d') $\int_{E^{\vdash}} |\Phi(s,y)|^p\, T_t^{\vdash}\big((x,t), ds \times dy\big)$ *is finite for some $p > 1$.*

Proof. Let $S \le u$ be a stopping time and set $G_v = \int_t^v Q_\tau g(X_\tau^{\vdash})\, d\tau$. Itô's formula gives $\mathbb{P}^{t,x}$-almost surely

$$\begin{aligned}
G_S + Q_S\Phi(X_S^{\vdash}) - \Phi(t,x) &= G_S + Q_S\Phi(X_S^{\vdash}) - Q_t\Phi(X_t^{\vdash}) \\
&= G_S + \int_{t+}^S \Phi(X_\tau^{\vdash})\, dQ_\tau + \int_{t+}^S Q_\tau\, d\Phi(X_\tau^{\vdash}) \\
&= G_S - \int_t^S Q_\tau \cdot q\Phi{\circ}X_\tau^{\vdash}\, d\tau \\
\text{by 5.7.6:}\qquad &\quad + \int_t^S Q_\tau \cdot \breve{A}^{\vdash}\Phi{\circ}X_\tau^{\vdash}\, d\tau + \int_{t+}^S Q_\tau\, dM_\tau^\Phi \\
&= \int_t^S Q_\tau \cdot (g - q\Phi){\circ}X_\tau^{\vdash}\, d\tau \\
\text{by A.9.15 and (5.7.7):}\qquad &\quad + \int_t^S Q_\tau \cdot (q\Phi - g){\circ}X_\tau^{\vdash}\, d\tau + \int_{t+}^S Q_\tau\, dM_\tau^\Phi \\
&= \int_{t+}^S Q_\tau\, dM_\tau^\Phi .
\end{aligned}$$

Since the paths of $X_.^{\vdash}$ stay in compact sets $\mathbb{P}^{t,x}$-almost surely and all functions appearing are continuous, the maximal function of every integrand above is $\mathbb{P}^{t,x}$-almost surely finite at time S, in the $d\Phi(X_\tau^{\vdash})$- and dM_τ^Φ-integrals. Thus every integral makes sense (theorem 3.7.17 on page 137) and the computation is kosher. Therefore

$$\begin{aligned}
\Phi(t,x) &= Q_S\Phi(S, X_S) + \int_t^S Q_\tau g(X_\tau^{\vdash})\, d\tau - \int_{t+}^S Q_\tau\, dM_\tau^\Phi \\
&= \mathbb{E}^{t,x}\big[Q_S\Phi(S,X_S)\big] + \mathbb{E}^{t,x}\Big[\int_t^S Q_\tau g(X_\tau^{\vdash})\, d\tau\Big] - \mathbb{E}^{t,x}\Big[\int_{t+}^S Q_\tau\, dM_\tau^\Phi\Big],
\end{aligned}$$

provided the random variables have finite $\mathbb{P}^{t,x}$-expectation. The proviso in the statement of the theorem is designed to achieve this and to have the last expectation vanish. The assumption that q be bounded below has the effect that $Q_u^\star$ is bounded. If $g \geq 0$, then the second expectation exists at time u.

The desired equality equation (5.7.8) now follows upon application of $\mathbb{E}^{t,x}$, and it is to make this expectation applicable that assumptions (a)–(d) are needed. Namely, since q is bounded below, Q is bounded above. The solidity of $\check{C}$ together with assumptions (b) and (c) make sure that the expectation of the first two integrals exists. If (d') is satisfied, then $M^\Phi_.$ is an L^1-integrator (theorem 2.5.30 on page 85) and the expectation of $\int_t^u Q_\tau \, dM_\tau^\Phi$ vanishes. If (d) is satisfied, then $X \in K_{n+1}$ up to and incuding time S_n, so that M^Φ stopped at time S_n is a bounded martingale: we take the expectation at time S_n, getting zero for the martingale integral, and then let $n \to \infty$. ∎

Repeated Footnotes: 271[1] 272[2] 273[3] 274[4] 277[5] 278[7] 280[8] 281[10] 282[11] 282[12] 287[16] 288[17] 293[20] 295[21] 297[23] 301[25] 303[26] 303[28] 305[30] 308[32] 310[33] 312[35] 312[36] 319[37] 320[38] 321[39] 323[40] 334[44] 352[48] 354[50]

Appendix A

Complements to Topology and Measure Theory

We review here the facts about topology and measure theory that are used in the main body. Those that are covered in every graduate course on integration are stated without proof. Some facts that might be considered as going beyond the very basics are proved, or at least a reference is given. The presentation is not linear – the two indexes will help the reader navigate.

A.1 Notations and Conventions

Convention A.1.1 *The reals are denoted by* $\mathbb{R}$, *the complex numbers by* $\mathbb{C}$, *the rationals by* $\mathbb{Q}$. $\mathbb{R}^d_*$ *is* ***punctured d-space*** $\mathbb{R}^d \setminus \{0\}$.

A real number a *will be called* ***positive*** *if* $a \ge 0$, *and* $\mathbb{R}_+$ *denotes the set of positive reals. Similarly, a real-valued function* f *is positive if* $f(x) \ge 0$ *for all points* x *in its domain* $\operatorname{dom}(f)$. *If we want to emphasize that* f *is* ***strictly positive:*** $f(x) > 0 \quad \forall x \in \operatorname{dom}(f)$, *we shall say so. It is clear what the words "**negative**" and "**strictly negative**" mean. If* $\mathcal{F}$ *is a collection of functions, then* $\mathcal{F}_+$ *will denote the positive functions in* $\mathcal{F}$, *etc. Note that a positive function may be zero on a large set, in fact, everywhere. The statements "b **exceeds** a," "b **is bigger than** a," and "a **is less than** b" all mean "$a \le b$;" modified by the word "strictly" they mean "$a < b$."*

A.1.2 The Extended Real Line $\overline{\mathbb{R}}$ is the real line augmented by the two symbols $-\infty$ and $\infty = +\infty$:

$$\overline{\mathbb{R}} = \{-\infty\} \cup \mathbb{R} \cup \{+\infty\}.$$

We adopt the usual conventions concerning the arithmetic and order structure of the extended reals $\overline{\mathbb{R}}$:

$$-\infty < r < +\infty \quad \forall r \in \mathbb{R}; \quad |\pm\infty| = +\infty;$$
$$-\infty \wedge r = -\infty, \quad \infty \vee r = \infty \quad \forall r \in \mathbb{R};$$
$$-\infty + r = -\infty, \quad +\infty + r = +\infty \quad \forall r \in \mathbb{R};$$
$$r \cdot \pm\infty = \begin{cases} \pm\infty & \text{for } r > 0, \\ 0 & \text{for } r = 0, \\ \mp\infty & \text{for } r < 0; \end{cases} \qquad \pm\infty^p = \begin{cases} \pm\infty & \text{for } p > 0, \\ 1 & \text{for } p = 0, \\ 0 & \text{for } p < 0. \end{cases}$$

The symbols $\infty - \infty$ and $0/0$ are not defined; there is no way to do this without confounding the order or the previous conventions.

A function whose values may include $\pm\infty$ is often called a ***numerical function***. The ***extended reals*** $\overline{\mathbb{R}}$ form a complete metric space under the ***arctan metric***

$$\overline{\rho}(r,s) \stackrel{\text{def}}{=} \left|\arctan(r) - \arctan(s)\right| , \qquad r,s \in \overline{\mathbb{R}} .$$

Here $\arctan(\pm\infty) \stackrel{\text{def}}{=} \pm\,\pi/2$. $\overline{\mathbb{R}}$ is compact in the topology $\overline{\tau}$ of $\overline{\rho}$. The natural injection $\mathbb{R} \hookrightarrow \overline{\mathbb{R}}$ is a ***homeomorphism***; that is to say, $\overline{\tau}$ agrees on $\mathbb{R}$ with the usual topology.

$a \vee b$ ($a \wedge b$) is the larger (smaller) of a and b. If f,g are numerical functions, then $f \vee g$ ($f \wedge g$) denote the pointwise maximum (minimum) of f,g. When a set $S \subset \overline{\mathbb{R}}$ is unbounded above we write $\sup S = \infty$, and $\inf S = -\infty$ when it is not bounded below. It is convenient to define the infimum of the empty set $\emptyset$ to be $+\infty$ and to write $\sup\emptyset = -\infty$.

A.1.3 Different length measurements of vectors and sequences come in handy in different places. For $0 < p < \infty$ the ***ℓ^p-length*** of $x = (x^1, \ldots, x^n) \in \mathbb{R}^n$ or of a sequence $(x^1, x^2, \ldots)$ is written variously as

$$|x|_p = \|x\|_{\ell^p} \stackrel{\text{def}}{=} \left(\textstyle\sum_\nu |x^\nu|^p\right)^{1/p}, \quad \text{while} \quad |z|_\infty = \|z\|_{\ell^\infty} \stackrel{\text{def}}{=} \sup_\eta |z^\eta|$$

denotes the ℓ^∞-length of a d-tuple $z = (z^1, z^2, \ldots, z^d)$ or a sequence $z = (z^0, z^1, \ldots)$. The vector space of all scalar sequences is a Fréchet space (which see) under the topology of pointwise convergence and is denoted by ℓ^0. The sequences x having $|x|_p < \infty$ form a Banach space under $|\ |_p$, $1 \le p < \infty$, which is denoted by ℓ^p. For $0 < p < q$ we have

$$|z|_q \le |z|_p \; ; \quad \text{and} \quad |z|_p \le d^{1/(q-p)} \cdot |z|_q \tag{A.1.1}$$

on sequences z of finite length d. $|\ |$ stands not only for the ordinary absolute value on $\mathbb{R}$ or $\mathbb{C}$ but for any of the norms $|\ |_p$ on $\mathbb{R}^n$ when $p \in [0,\infty]$ need not be displayed.

Next some notation and conventions concerning sets and functions, which will simplify the typography considerably:

Notation A.1.4 (Knuth [56]) *A statement enclosed in rectangular brackets denotes the set of points where it is true. For instance, the symbol* $[f = 1]$ *is short for* $\{x \in \operatorname{dom}(f) : f(x) = 1\}$. *Similarly,* $[f > r]$ *is the set of points* x *where* $f(x) > r$, $[f_n \not\to]$ *is the set of points where the sequence* (f_n) *fails to converge, etc.*

Convention A.1.5 (Knuth [56]) *Occasionally we shall use the same name or symbol for a set* A *and its indicator function:* A *is also the function that returns* 1 *when its argument lies in* A, 0 *otherwise. For instance,* $[f > r]$ *denotes not only the set of points where* f *strictly exceeds* r *but also the function that returns* 1 *where* $f > r$ *and* 0 *elsewhere.*

Remark A.1.6 The indicator function of A is written 1_A by most mathematicians, $\imath_A$, χ_A, or I_A or even $1A$ by others, and A by a select few. There is a considerable typographical advantage in writing it as A: $[T_n^k < r]$ or $\left[U_S^{[a,b]} \ge n\right]$ are rather easier on the eye than $1_{[T_n^k<r]}$ or $1_{\left[U_S^{[a,b]}\ge n\right]}$, respectively. When functions z are written $s \mapsto z_s$, as is common in stochastic analysis, the indicator of the interval $(a,b]$ has under this convention at s the value $(a,b]_s$ rather than $1_{(a,b]s}$.

In deference to prevailing custom we shall use this swell convention only sparingly, however, and write 1_A when possible. We do invite the aficionado of the 1_A-notation to compare how much eye strain and verbiage is saved on the occasions when we employ Knuth's nifty convention.

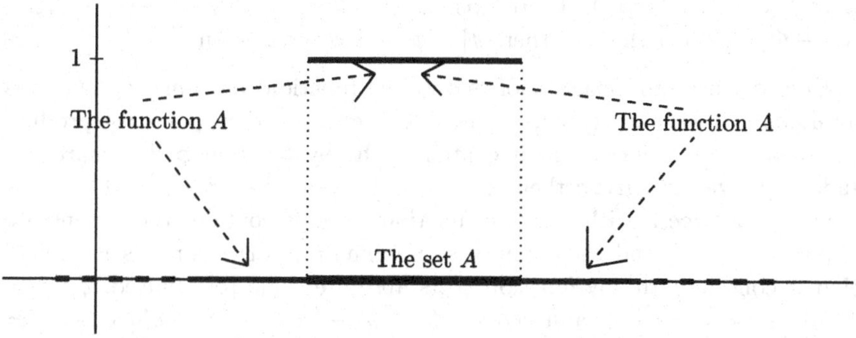

Figure A.14 A set and its indicator function have the same name

Exercise A.1.7 Here is a little practice to get used to Knuth's convention: (i) Let f be an ***idempotent function***: $f^2 = f$. Then $f = [f = 1] = [f \ne 0]$. (ii) Let (f_n) be a sequence of functions. Then the sequence $([f_n \to] \cdot f_n)$ converges everywhere. (iii) $\int (0,1](x) \cdot x^2\,dx = 1/3$. (iv) For $f_n(x) \stackrel{\text{def}}{=} \sin(nx)$ compute $[f_n \not\to]$. (v) Let $A_1, A_2, \ldots$ be sets. Then $A_1^c = 1 - A_1$, $\bigcup_n A_n = \sup_n A_n = \bigvee_n A_n$, $\bigcap_n A_n = \inf_n A_n = \bigwedge_n A_n$. (vi) Every real-valued function is the pointwise limit of the simple step functions

$$f_n = \sum_{k=-4^n}^{4^n} k2^{-n} \cdot \left[k2^{-n} \le f < (k+1)2^{-n}\right].$$

(vii) The sets in an algebra of functions form an algebra of sets. (viii) A family of idempotent functions is a σ-algebra if and only if it is closed under $f \mapsto 1-f$ and under finite infima and countable suprema. (ix) Let $f : X \to Y$ be a map and $S \subset Y$. Then $f^{-1}(S) = S \circ f \subset X$.

A.2 Topological Miscellanea

The Theorem of Stone–Weierstraß

All measures appearing in nature owe their σ-additivity to the following very simple result or some variant of it; it is also used in the proof of theorem A.2.2.

Lemma A.2.1 (Dini's Theorem) *Let $\boldsymbol{B}$ be a topological space and Φ a collection of positive continuous functions on $\boldsymbol{B}$ that vanish at infinity.*[1] *Assume Φ is decreasingly directed*[2,3] *and the pointwise infimum of Φ is zero. Then $\Phi \to 0$ uniformly; that is to say, for every $\epsilon > 0$ there is a $\psi \in \Phi$ with $\psi(x) \le \epsilon$ for all $x \in \boldsymbol{B}$ and therefore $\phi \le \epsilon$ uniformly for all $\phi \in \Phi$ with $\phi \le \psi$.*

Proof. The sets $[\phi \ge \epsilon]$, $\phi \in \Phi$, are compact and have void intersection. There are finitely many of them, say $[\phi_i \ge \epsilon]$, $i = 1, \ldots, n$, whose intersection is void (exercise A.2.12). There exists a $\psi \in \Phi$ smaller than[3] $\phi_1 \wedge \cdots \wedge \phi_n$. If $\phi \in \Phi$ is smaller than ψ, then $|\phi| = \phi < \epsilon$ everywhere on $\boldsymbol{B}$. ∎

Consider a vector space $\mathcal{E}$ of real-valued functions on some set $\boldsymbol{B}$. It is an ***algebra*** if with any two functions ϕ, ψ it contains their pointwise product $\phi\psi$. For this it suffices that it contain with any function ϕ its square ϕ^2. Indeed, by ***polarization*** then $\phi\psi = 1/2\big((\phi+\psi)^2 - \phi^2 - \psi^2\big) \in \mathcal{E}$. $\mathcal{E}$ is a ***vector lattice*** if with any two functions ϕ, ψ it contains their pointwise maximum $\phi \vee \psi$ and their pointwise minimum $\phi \wedge \psi$. For this it suffices that it contain with any function ϕ its absolute value $|\phi|$. Indeed, $\phi \vee \psi = 1/2\big(|\phi - \psi| + (\phi + \psi)\big)$, and $\phi \wedge \psi = (\phi + \psi) - (\phi \vee \psi)$. $\mathcal{E}$ is ***closed under chopping*** if with any function ϕ it contains the ***chopped function*** $\phi \wedge 1$. It then contains $f \wedge q = q(f/q \wedge 1)$ for any strictly positive scalar q. A ***lattice algebra*** is a vector space of functions that is both an algebra and a lattice under pointwise operations.

Theorem A.2.2 (Stone–Weierstraß) *Let $\mathcal{E}$ be an algebra* OR *a vector lattice closed under chopping, of bounded real-valued functions on some set $\boldsymbol{B}$. We denote by Z the set $\{x \in \boldsymbol{B} : \phi(x) = 0 \;\; \forall\, \phi \in \mathcal{E}\}$ of common zeroes of $\mathcal{E}$, and identify a function of $\mathcal{E}$ in the obvious fashion with its restriction to $\boldsymbol{B}_0 \overset{\text{def}}{=} \boldsymbol{B} \setminus Z$.*

(i) The uniform closure $\overline{\mathcal{E}}$ of $\mathcal{E}$ is both an algebra AND *a vector lattice closed under chopping. Furthermore, if $\Phi : \mathbb{R} \to \mathbb{R}$ is continuous with $\Phi(0) = 0$, then $\Phi \circ \overline{\phi} \in \overline{\mathcal{E}}$ for any $\overline{\phi} \in \overline{\mathcal{E}}$.*

---

[1] A set is ***relatively compact*** if its closure is compact. ϕ vanishes at ∞ if its carrier $[|\phi| \ge \epsilon]$ is relatively compact for every $\epsilon > 0$. The collection of continuous functions vanishing at infinity is denoted by $C_0(\boldsymbol{B})$ and is given the topology of uniform convergence. $C_0(\boldsymbol{B})$ is identified in the obvious way with the collection of continuous functions on the one-point compactification $\boldsymbol{B}^\Delta \overset{\text{def}}{=} \boldsymbol{B} \cup \{\Delta\}$ (see page 374) that vanish at Δ.

[2] That is to say, for any two $\phi_1, \phi_2 \in \Phi$ there is a $\phi \in \Phi$ less than both ϕ_1 and ϕ_2. Φ is ***increasingly directed*** if for any two $\phi_1, \phi_2 \in \Phi$ there is a $\phi \in \Phi$ with $\phi \ge \phi_1 \vee \phi_2$.

[3] See convention A.1.1 on page 363 about language concerning order relations.

(ii) There exist a locally compact Hausdorff space $\widehat{\boldsymbol{B}}$ and a map $j : \boldsymbol{B}_0 \to \widehat{\boldsymbol{B}}$ with dense image such that $\widehat{\phi} \mapsto \widehat{\phi} \circ j$ is an algebraic and order isomorphism of $C_0(\widehat{\boldsymbol{B}})$ with $\overline{\mathcal{E}} \simeq \overline{\mathcal{E}}_{|\boldsymbol{B}_\circ}$. We call $\widehat{\boldsymbol{B}}$ the **spectrum** *of $\mathcal{E}$ and $j : \boldsymbol{B}_0 \to \widehat{\boldsymbol{B}}$ the* **local $\mathcal{E}$-compactification** *of $\boldsymbol{B}$. $\widehat{\boldsymbol{B}}$ is compact if and only if $\mathcal{E}$ contains a function that is bounded away from zero.*[4] *If $\mathcal{E}$* **separates the points.**[5] *of $\boldsymbol{B}_0$, then j is injective. If $\mathcal{E}$ is countably generated,*[6] *then $\widehat{\boldsymbol{B}}$ is separable and metrizable.*

(iii) Suppose that there is a locally compact Hausdorff topology τ on $\boldsymbol{B}$ and $\mathcal{E} \subset C_0(\boldsymbol{B}, \tau)$, and assume that $\mathcal{E}$ separates the points of $\boldsymbol{B}_0 \stackrel{\text{def}}{=} \boldsymbol{B} \setminus Z$. Then $\overline{\mathcal{E}}$ equals the algebra of all continuous functions that vanish at infinity and on Z.[7]

Proof. (i) There are several steps. (a) If $\mathcal{E}$ is an algebra OR a vector lattice closed under chopping, then its uniform closure $\overline{\mathcal{E}}$ is clearly an algebra OR a vector lattice closed under chopping, respectively.

(b) Assume that $\mathcal{E}$ is an algebra and let us show that then $\overline{\mathcal{E}}$ is a vector lattice closed under chopping. To this end define polynomials $p_n(t)$ on $[-1, 1]$ inductively by $p_0 = 0$, $p_{n+1}(t) = 1/2\big(t^2 + 2p_n(t) - (p_n(t))^2\big)$. Then $p_n(t)$ is a polynomial in t^2 with zero constant term. Two easy manipulations result in

$$2\big(|t| - p_{n+1}(t)\big) = \big(2 - |t|\big)|t| - \big(2 - p_n(t)\big)p_n(t)$$

and
$$2\big(p_{n+1}(t) - p_n(t)\big) = t^2 - \big(p_n(t)\big)^2 .$$

Now $(2 - x)x = 2x - x^2$ is increasing on $[0, 1]$. If, by induction hypothesis, $0 \le p_n(t) \le |t|$ for $|t| \le 1$, then $p_{n+1}(t)$ will satisfy the same inequality; as it is true for p_0, it holds for all the p_n. The second equation shows that $p_n(t)$ increases with n for $t \in [-1, 1]$. As this sequence is also bounded, it has a limit $p(t) \ge 0$. $p(t)$ must satisfy $0 = t^2 - (p(t))^2$ and thus equals $|t|$. Due to Dini's theorem A.2.1, $|t| - p_n(t)$ decreases uniformly on $[-1, 1]$ to 0. Given a $\overline{\phi} \in \overline{\mathcal{E}}$, set $M = \|\overline{\phi}\|_\infty \vee 1$. Then $P_n(t) \stackrel{\text{def}}{=} M p_n(t/M)$ converges to $|t|$ uniformly on $[-M, M]$, and consequently $|f| = \lim P_n(f)$ belongs to $\overline{\overline{\mathcal{E}}} = \overline{\mathcal{E}}$. To see that $\overline{\mathcal{E}}$ is closed under chopping consider the polynomials $Q'_n(t) \stackrel{\text{def}}{=} 1/2\big(t + 1 - P_n(t - 1)\big)$. They converge uniformly on $[-M, M]$ to $1/2\big(t + 1 - |t - 1|\big) = t \wedge 1$. So do the polynomials $Q_n(t) = Q'_n(t) - Q'_n(0)$, which have the virtue of vanishing at zero, so that $Q_n \circ \overline{\phi} \in \overline{\mathcal{E}}$. Therefore $\overline{\phi} \wedge 1 = \lim Q_n \circ \overline{\phi}$ belongs to $\overline{\overline{\mathcal{E}}} = \overline{\mathcal{E}}$.

(c) Next assume that $\mathcal{E}$ is a vector lattice closed under chopping, and let us show that then $\overline{\mathcal{E}}$ is an algebra. Given $\overline{\phi} \in \overline{\mathcal{E}}$ and $\epsilon \in (0, 1)$, again set

---

[4] ϕ is ***bounded away from zero*** if $\inf\{|\phi(x)| : x \in \boldsymbol{B}\} > 0$.

[5] That is to say, for any $x \ne y$ in $\boldsymbol{B}_0$ there is a $\phi \in \mathcal{E}$ with $\phi(x) \ne \phi(y)$.

[6] That is to say, there is a countable set $\mathcal{E}_0 \subset \mathcal{E}$ such that $\mathcal{E}$ is contained in the smallest uniformly closed algebra containing $\mathcal{E}_0$.

[7] If $Z = \emptyset$, this means $\overline{\mathcal{E}} = C_0(\boldsymbol{B})$; if in addition τ is compact, then this means $\overline{\mathcal{E}} = C(\boldsymbol{B})$.

$M = \|\overline{\phi}\|_\infty + 1$. For $k \in \mathbb{Z} \cap [-M/\epsilon, M/\epsilon]$ let $\ell_k(t) = 2k\epsilon t - k^2\epsilon^2$ denote the tangent to the function $t \mapsto t^2$ at $t = k\epsilon$. Since $\ell_k \vee 0 = (2k\epsilon t - k^2\epsilon^2) \vee 0 = 2k\epsilon t - (2k\epsilon t \wedge k^2\epsilon^2)$, we have

$$\begin{aligned}\Phi_\epsilon(t) &\stackrel{\text{def}}{=} \bigvee \{2k\epsilon t - k^2\epsilon^2 : q\, k \in \mathbb{Z},\, |k| < M/\epsilon\} \\ &= \bigvee \{2k\epsilon t - (2k\epsilon t \wedge k^2\epsilon^2) :\; k \in \mathbb{Z},\, |k| < M/\epsilon\} .\end{aligned}$$

Now clearly $t^2 - \epsilon \le \Phi_\epsilon(t) \le t^2$ on $[-M, M]$, and the second line above shows that $\Phi_e \circ \overline{\phi} \in \overline{\mathcal{E}}$. We conclude that $\overline{\phi}^2 = \lim_{\epsilon\to 0} \Phi_e \circ \overline{\phi} \in \overline{\overline{\mathcal{E}}} = \overline{\mathcal{E}}$.

We turn to (iii), assuming to start with that τ is compact. Let $\overline{\mathcal{E}} \oplus \mathbb{R} \stackrel{\text{def}}{=} \{\phi + r : \phi \in \overline{\mathcal{E}},\, r \in \mathbb{R}\}$. This is an algebra AND a vector lattice[8] over $\mathbb{R}$ of bounded τ-continuous functions. It is uniformly closed and contains the constants. Consider a continuous function f that is constant on Z, and let $\epsilon > 0$ be given. For any two different points $s, t \in \boldsymbol{B}$, not both in Z, there is a function $\overline{\psi}_{s,t}$ in $\overline{\mathcal{E}}$ with $\overline{\psi}_{s,t}(s) \ne \overline{\psi}_{s,t}(t)$. Set

$$\overline{\phi}_{s,t}(\tau) = f(s) + \left(\frac{f(t) - f(s)}{\overline{\psi}_{s,t}(t) - \overline{\psi}_{s,t}(s)}\right) \cdot (\overline{\psi}_{s,t}(\tau) - \overline{\psi}_{s,t}(s)) .$$

If $s = t$ or $s, t \in Z$, set $\overline{\phi}_{s,t}(\tau) = f(t)$. Then $\overline{\phi}_{s,t}$ belongs to $\overline{\mathcal{E}} \oplus \mathbb{R}$ and takes at s and t the same values as f. Fix $t \in \boldsymbol{B}$ and consider the sets $U_s^t = [\overline{\phi}_{s,t} > f - \epsilon]$. They are open, and they cover $\boldsymbol{B}$ as s ranges over $\boldsymbol{B}$; indeed, the point $s \in \boldsymbol{B}$ belongs to U_s^t. Since $\boldsymbol{B}$ is compact, there is a finite subcover $\{U_{s_i}^t : 1 \le i \le n\}$. Set $\overline{\phi}^t = \bigvee_{i=1}^n \overline{\phi}_{s_i,t}$. This function belongs to $\overline{\mathcal{E}} \oplus \mathbb{R}$, is everywhere bigger than $f - \epsilon$, and coincides with f at t. Next consider the open cover $\{[\overline{\phi}^t < f + \epsilon] : t \in \boldsymbol{B}\}$. It has a finite subcover $\{[\overline{\phi}^{t_i} < f + \epsilon] : 1 \le i \le k\}$, and the function $\phi \stackrel{\text{def}}{=} \bigwedge_{i=1}^k \overline{\phi}^{t_i} \in \overline{\mathcal{E}} \oplus \mathbb{R}$ is clearly uniformly as close as ϵ to f. In other words, there is a sequence $\overline{\phi}_n + r_n \in \overline{\mathcal{E}} \oplus \mathbb{R}$ that converges uniformly to f. Now if Z is non-void and f vanishes on Z, then $r_n \to 0$ and $\overline{\phi}_n \in \overline{\mathcal{E}}$ converges uniformly to f. If $Z = \emptyset$, then there is, for every $s \in \boldsymbol{B}$, a $\overline{\phi}_s \in \overline{\mathcal{E}}$ with $\overline{\phi}_s(s) > 1$. By compactness there will be finitely many of the $\overline{\phi}_s$, say $\overline{\phi}_{s_1}, \ldots, \overline{\phi}_{s_n}$, with $\overline{\phi} \stackrel{\text{def}}{=} \bigvee_i \overline{\phi}_{s_i} > 1$. Then $1 = \overline{\phi} \wedge 1 \in \overline{\mathcal{E}}$ and consequently $\overline{\mathcal{E}} \oplus \mathbb{R} = \overline{\mathcal{E}}$. In both cases $f \in \overline{\overline{\mathcal{E}}} = \overline{\mathcal{E}}$. If τ is not compact, we view $\overline{\mathcal{E}} \oplus \mathbb{R}$ as a uniformly closed algebra of bounded continuous functions on the one-point compactification $\boldsymbol{B}^\Delta = \boldsymbol{B} \,\dot\cup\, \{\Delta\}$ and an $f \in C_0(\boldsymbol{B})$ that vanishes on Z as a continuous bounded function on $\boldsymbol{B}^\Delta$ that vanishes on $Z \cup \{\Delta\}$, the common zeroes of $\mathcal{E}$ on $\boldsymbol{B}^\Delta$, and apply the above: if $\overline{\mathcal{E}} \oplus \mathbb{R} \ni \phi_n + r_n \to f$ uniformly on $\boldsymbol{B}^\Delta$, then $r_n \to 0$, and $f \in \overline{\overline{\mathcal{E}}} = \overline{\mathcal{E}}$.

---

[8] To see that $\overline{\mathcal{E}} \oplus \mathbb{R}$ is closed under pointwise infima write $(\overline{\phi} + r) \wedge (\overline{\psi} + s) = (\overline{\phi} - \overline{\psi}) \wedge (s - r) + \overline{\psi} + r$. Since without loss of generality $r \le s$, the right-hand side belongs to $\overline{\mathcal{E}} \oplus \mathbb{R}$.

(d) Of (i) only the last claim remains to be proved. Now thanks to (iii) there is a sequence of polynomials q_n that vanish at zero and converge uniformly on the compact set $[-\|\overline{\phi}\|_\infty, \|\overline{\phi}\|_\infty]$ to Φ. Then

$$\Phi \circ \overline{\phi} = \lim q_n \circ \overline{\phi} \in \overline{\overline{\mathcal{E}}} = \overline{\mathcal{E}} .$$

(ii) Let $\mathcal{E}_0$ be a subset of $\mathcal{E}$ that generates $\mathcal{E}$ in the sense that $\mathcal{E}$ is contained in the smallest uniformly closed algebra containing $\mathcal{E}_0$. Set

$$\Pi = \prod_{\psi \in \mathcal{E}_0} \left[-\|\psi\|_u, +\|\psi\|_u\right] .$$

This product of compact intervals is a compact Hausdorff space in the product topology (exercise A.2.13), metrizable if $\mathcal{E}_0$ is countable. Its typical element is an "$\mathcal{E}_0$-tuple" $(\xi_\psi)_{\psi\in\mathcal{E}_0}$ with $\xi_\psi \in [-\|\psi\|_u, +\|\psi\|_u]$. There is a natural map $j : \boldsymbol{B} \to \Pi$ given by $x \mapsto (\psi(x))_{\psi\in\mathcal{E}_0}$. Let $\overline{\boldsymbol{B}}$ denote the closure of $j(\boldsymbol{B})$ in Π, the ***$\mathcal{E}$-completion*** of $\boldsymbol{B}$ (see lemma A.2.16). The finite linear combinations of finite products of coordinate functions $\widehat{\phi} : (\xi_\psi)_{\psi\in\mathcal{E}_0} \mapsto \xi_\phi$, $\phi \in \mathcal{E}_0$, form an algebra $\mathcal{A} \subset C(\overline{\boldsymbol{B}})$ that separates the points. Now set $\overline{Z} \stackrel{\text{def}}{=} \{\widehat{z} \in \overline{\boldsymbol{B}} : \widehat{\phi}(\widehat{z}) = 0 \quad \forall \phi \in \mathcal{A}\}$. This set is either empty or contains one point, $(0, 0, \ldots)$, and j maps $\boldsymbol{B}_0 \stackrel{\text{def}}{=} \boldsymbol{B} \backslash Z$ into $\widehat{\boldsymbol{B}} \stackrel{\text{def}}{=} \overline{\boldsymbol{B}} \backslash \overline{Z}$. View $\mathcal{A}$ as a subalgebra of $C_0(\widehat{\boldsymbol{B}})$ that separates the points of $\boldsymbol{B}$. The linear multiplicative map $\widehat{\phi} \mapsto \widehat{\phi} \circ j$ evidently takes $\mathcal{A}$ to the smallest algebra containing $\mathcal{E}_0$ and preserves the uniform norm. It extends therefore to a linear isometry of $\overline{\mathcal{A}}$ – which by (iii) coincides with $C_0(\widehat{\boldsymbol{B}})$ – with $\overline{\mathcal{E}}$; it is evidently linear and multiplicative and preserves the order. Finally, if $\phi \in \mathcal{E}$ separates the points $x, y \in \boldsymbol{B}$, then the function $\widehat{\phi} \in \overline{\mathcal{A}}$ that has $\phi = \widehat{\phi} \circ j$ separates $j(x), j(y)$, so when $\mathcal{E}$ separates the points then j is injective. ∎

Exercise A.2.3 Let $\boldsymbol{A}$ be any subset of $\boldsymbol{B}$. (i) A function f can be approximated uniformly on $\boldsymbol{A}$ by functions in $\mathcal{E}$ if and only if it is the restriction to $\boldsymbol{A}$ of a function in $\overline{\mathcal{E}}$. (ii) If $f_1, f_2 : \boldsymbol{B} \to \overline{\mathbb{R}}$ can be approximated uniformly on $\boldsymbol{A}$ by functions in $\mathcal{E}$ (in the arctan metric $\overline{\rho}$; see item A.1.2), then $\overline{\rho}(f_1, f_2)$ is the restriction to $\boldsymbol{A}$ of a function in $\overline{\mathcal{E}}$.

All spaces of elementary integrands that we meet in this book are self-confined in the following sense.

Definition A.2.4 *A subset $S \subset \boldsymbol{B}$ is called* ***$\mathcal{E}$-confined*** *if there is a function $\phi \in \mathcal{E}$ that is greater than 1 on S: $\phi \geq 1_S$. A function $f : \boldsymbol{B} \to \mathbb{R}$ is $\mathcal{E}$-confined if its carrier*[9] *$[f \neq 0]$ is $\mathcal{E}$-confined; the collection of $\mathcal{E}$-confined functions in $\mathcal{E}$ is denoted by $\mathcal{E}_{00}$. A sequence of functions f_n on $\boldsymbol{B}$ is $\mathcal{E}$-confined if the f_n all vanish outside the same $\mathcal{E}$-confined set; and $\mathcal{E}$ is* ***self-confined*** *if all of its members are $\mathcal{E}$-confined, i.e., if $\mathcal{E} = \mathcal{E}_{00}$. A function f is the* ***$\mathcal{E}$-confined uniform limit*** *of the sequence (f_n) if (f_n) is $\mathcal{E}$-confined*

---

[9] The ***carrier of a function*** ϕ is the set $[\phi \neq 0]$.

and converges uniformly to f. The typical examples of self-confined lattice algebras are the step functions over a ring of sets and the space $C_{00}(\boldsymbol{B})$ of continuous functions with compact support on $\boldsymbol{B}$. The product $\mathcal{E}_1 \otimes \mathcal{E}_2$ of two self-confined algebras or vector lattices closed under chopping is clearly self-confined.

A.2.5 The notion of a confined uniform limit is a topological notion: for every $\mathcal{E}$-confined set A let $\mathcal{F}_A$ denote the algebra of bounded functions confined by A. Its natural topology is the topology of uniform convergence. The natural topology on the vector space $\mathcal{F}_{\mathcal{E}}$ of bounded $\mathcal{E}$-confined functions, union of the $\mathcal{F}_A$, the ***topology of*** $\mathcal{E}$***-confined uniform convergence*** is the finest topology on bounded $\mathcal{E}$-confined functions that agrees on every $\mathcal{F}_A$ with the topology of uniform convergence. It makes the bounded $\mathcal{E}$-confined functions, the union of the $\mathcal{F}_A$, into a topological vector space. Now let $\mathcal{I}$ be a linear map from $\mathcal{F}_{\mathcal{E}}$ to a topological vector space and show that the following are equivalent: (i) $\mathcal{I}$ is continuous in this topology; (ii) the restriction of $\mathcal{I}$ to any of the $\mathcal{F}_A$ is continuous; (iii) $\mathcal{I}$ maps order-bounded subsets of $\mathcal{F}_{\mathcal{E}}$ to bounded subsets of the target space.

Exercise A.2.6 Show: if $\mathcal{E}$ is a self-confined algebra or vector lattice closed under chopping, then a uniform limit $\overline{\phi} \in \overline{\mathcal{E}}$ is $\mathcal{E}$-confined if and only if it is the uniform limit of a $\mathcal{E}$-confined sequence in $\mathcal{E}$; we then say "$\overline{\phi}$ is the ***confined uniform limit***" of a sequence in $\mathcal{E}$. Therefore the "confined uniform closure $\overline{\mathcal{E}}_{00}$ of $\mathcal{E}$" is a self-confined algebra and a vector lattice closed under chopping.

The next two corollaries to Weierstraß' theorem employ the local $\mathcal{E}$-compactification $j : \boldsymbol{B}_0 \to \widehat{\boldsymbol{B}}$ to establish results that are crucial for the integration theory of integrators and random measures (see proposition 3.3.2 and lemma 3.10.2). In order to ease their statements and the arguments to prove them we introduce the following notation: for every $X \in \overline{\mathcal{E}}$ the unique continuous function $\widehat{X}$ on $\widehat{\boldsymbol{B}}$ that has $\widehat{X} \circ j = X$ will be called the ***Gelfand transform*** of X; next, given any functional $\mathcal{I}$ on $\mathcal{E}$ we define $\widehat{\mathcal{I}}$ on $\widehat{\mathcal{E}}$ by $\widehat{\mathcal{I}}(\widehat{X}) = \mathcal{I}(\widehat{X} \circ j)$ and call it the ***Gelfand transform*** of $\mathcal{I}$.

For simplicity's sake we assume in the remainder of this subsection that $\mathcal{E}$ is a self-confined algebra and a vector lattice closed under chopping, of bounded functions on some set $\boldsymbol{B}$.

Corollary A.2.7 *Let* (L, τ) *be a topological vector space and* $\tau_0 \subset \tau$ *a weaker Hausdorff topology on* L. *If* $\mathcal{I} : \mathcal{E} \to L$ *is a linear map whose Gelfand transform* $\widehat{\mathcal{I}}$ *has an extension satisfying the Dominated Convergence Theorem, and if* $\mathcal{I}$ *is* σ*-continuous in the topology* τ_0, *then it is in fact* σ*-additive in the topology* τ.

Proof. Let $\mathcal{E} \ni X_n \downarrow 0$. Then the sequence $(\widehat{X}_n)$ of Gelfand transforms decreases on $\widehat{\boldsymbol{B}}$ and has a pointwise infimum $\widehat{K} : \widehat{\boldsymbol{B}} \to \mathbb{R}$. By the DCT, the sequence $\big(\widehat{\mathcal{I}}(\widehat{X}_n)\big)$ has a τ-limit f in L, the value of the extension at $\widehat{K}$. Clearly $f = \tau\text{-}\lim \widehat{\mathcal{I}}(\widehat{X}_n) = \tau\text{-}\lim \mathcal{I}(X_n) = \tau_0\text{-}\lim \mathcal{I}(X_n) = 0$. Since

τ_0 is Hausdorff, $f = 0$. The σ-continuity of $\mathcal{I}$ is established, and exercise 3.1.5 on page 90 produces the σ-additivity. This argument repeats that of proposition 3.3.2 on page 108. ∎

Corollary A.2.8 *Let $\boldsymbol{H}$ be a locally compact space equipped with the algebra $\mathcal{H} \stackrel{\text{def}}{=} C_{00}(\boldsymbol{H})$ of continuous functions of compact support. The cartesian product $\check{\boldsymbol{B}} \stackrel{\text{def}}{=} \boldsymbol{H} \times \boldsymbol{B}$ is equipped with the algebra $\check{\mathcal{E}} \stackrel{\text{def}}{=} \mathcal{H} \otimes \mathcal{E}$ of functions*

$$(\eta, \varpi) \mapsto \sum_i H_i(\eta) X_i(\varpi) , \qquad H_i \in \mathcal{H}, X_i \in \mathcal{E}, \text{ the sum finite.}$$

Suppose θ is a real-valued linear functional on $\check{\mathcal{E}}$ that maps order-bounded sets of $\check{\mathcal{E}}$ to bounded sets of reals and that is ***marginally σ-additive*** *on $\mathcal{E}$; that is to say, the measure $X \mapsto \theta(H \otimes X)$ on $\mathcal{E}$ is σ-additive for every $H \in \mathcal{H}$. Then θ is in fact σ-additive.*[10]

Proof. First observe easily that $\mathcal{E} \ni X_n \downarrow 0$ implies $\theta(\check{H} \cdot X_n) \to 0$ for every $\check{H} \in \check{\mathcal{E}}$. Another way of saying this is that for every $\check{H} \in \check{\mathcal{E}}$ the measure $X \mapsto \theta^{\check{H}}(X) \stackrel{\text{def}}{=} \theta(\check{H} \cdot X)$ on $\mathcal{E}$ is σ-additive.

From this let us deduce that the variation $|\theta|$ has the same property. To this end let $\jmath : \boldsymbol{B}_0 \to \widehat{\boldsymbol{B}}$ denote the local $\mathcal{E}$-compactification of $\boldsymbol{B}$; the local $\mathcal{H}$-compactification of $\boldsymbol{H}$ clearly is the identity map $id : \boldsymbol{H} \to \boldsymbol{H}$. The spectrum of $\check{\mathcal{E}}$ is $\widehat{\check{\boldsymbol{B}}} = \boldsymbol{H} \times \widehat{\boldsymbol{B}}$ with local $\check{\mathcal{E}}$-compactification $\check{\jmath} \stackrel{\text{def}}{=} id \otimes \jmath$. The Gelfand transform $\widehat{\theta}$ is a σ-additive measure on $\widehat{\check{\mathcal{E}}}$ of finite variation $|\widehat{\theta}| = \widehat{|\theta|}$; in fact, $|\widehat{\theta}|$ is a positive Radon measure on $\widehat{\check{\boldsymbol{B}}}$. There exists a locally $|\widehat{\theta}|$-integrable function $\widehat{\Gamma}$ with $\widehat{\Gamma}^2 = 1$ and $\widehat{\theta} = \widehat{\Gamma} \cdot |\widehat{\theta}|$ on $\widehat{\check{\boldsymbol{B}}}$, to wit, the Radon–Nikodym derivative $d\widehat{\theta}/d|\widehat{\theta}|$. With these notations in place pick an $H \in \mathcal{H}_+$ with compact carrier K and let (X_n) be a sequence in $\mathcal{E}$ that decreases pointwise to 0. There is no loss of generality in assuming that both $X_1 < 1$ and $H < 1$. Given an $\epsilon > 0$, let $\widehat{E}$ be the closure in $\widehat{\boldsymbol{B}}$ of $\jmath([X_1 > 0])$ and find an $\widehat{\check{X}} \in \widehat{\check{\mathcal{E}}}$ with

$$\int_{K \times E} |\widehat{\Gamma} - \widehat{\check{X}}| \, d|\widehat{\theta}| < \epsilon .$$

Then
$$\begin{aligned} |\widehat{\theta}|(H \otimes \widehat{X}_n) &= \int_{K \times \widehat{E}} H(\eta) \widehat{X}_n(\widehat{\varpi}) \widehat{\Gamma}(\eta, \widehat{\varpi}) \, \widehat{\theta}(d\eta, d\widehat{\varpi}) \\ &\le \int_{K \times \widehat{E}} H(\eta) \widehat{X}_n(\widehat{\varpi}) \widehat{\check{X}}(\eta, \widehat{\varpi}) \, \widehat{\theta}(d\eta, d\widehat{\varpi}) + \epsilon \\ &= \int_{\boldsymbol{H} \times \widehat{\boldsymbol{B}}} H(\eta) \widehat{X}_n(\widehat{\varpi}) \widehat{\check{X}}(\eta, \widehat{\varpi}) \, \widehat{\theta}(d\eta, d\widehat{\varpi}) + \epsilon \\ &= \theta^{H\check{X}}(X_n) + \epsilon \end{aligned}$$

[10] Actually, it suffices to assume that $\boldsymbol{H}$ is Suslin, that the vector lattice $\mathcal{H} \subset \mathcal{B}^*(\boldsymbol{H})$ generates $\mathcal{B}^*(\boldsymbol{H})$, and that θ is also marginally σ-additive on $\mathcal{H}$ – see [93].

has limit less than ϵ by the very first observation above. Therefore

$$|\theta|(H \otimes X_n) = |\widehat{\theta}|(H \otimes \widehat{X}_n) \downarrow 0 .$$

Now let $\check{\mathcal{E}} \ni \check{X}_n \downarrow 0$. There are a compact set $K \subset \boldsymbol{H}$ so that $\check{X}_1(\eta, \varpi) = 0$ whenever $\eta \notin K$, and an $H \in \mathcal{H}$ equal to 1 on K. The functions $X_n : \varpi \mapsto \max_{\eta \in \boldsymbol{H}} \check{X}(\eta, \varpi)$ belong to $\mathcal{E}$, thanks to the compactness of K, and decrease pointwise to zero on $\boldsymbol{B}$ as $n \to \infty$. Since $\check{X}_n \le H \otimes X_n$, $|\theta|(\check{X}_n) \le |\theta|(H \otimes X_n) \xrightarrow[n\to\infty]{} 0$: $|\theta|$ and with it θ is indeed σ-additive. ∎

Exercise A.2.9 Let $\theta : \mathcal{E} \to \mathbb{R}$ be a linear functional of finite variation. Then its Gelfand transform $\widehat{\theta} : \widehat{\mathcal{E}} \to \mathbb{R}$ is σ-additive due to Dini's theorem A.2.1 and has the usual integral extension featuring the Dominated Convergence Theorem (see pages 395–398). Show: θ is σ-additive if and only if $\int \widehat{k}\, d\widehat{\theta} = 0$ for every function $\widehat{k}$ on the spectrum $\widehat{\boldsymbol{B}}$ that is the pointwise infimum of a sequence in $\widehat{\mathcal{E}}$ and vanishes on $j(\boldsymbol{B})$.

Exercise A.2.10 Consider a linear map $\mathcal{I}$ on $\mathcal{E}$ with values in a space $L^p(\mu)$, $1 \le p < \infty$, that maps order intervals of $\mathcal{E}$ to bounded subsets of L^p. Show: if $\mathcal{I}$ is weakly σ-additive, then it is σ-additive in in the norm topology L^p.

Weierstraß' original proof of his approximation theorem applies to functions on $\mathbb{R}^d$ and employs the heat kernel γ_{tI} (exercise A.3.48 on page 420). It yields in its less general setting an approximation in a finer topology than the uniform one. We give a sketch of the result and its proof, since it is used in Itô's theorem, for instance. Consider an open subset D of $\mathbb{R}^d$. For $0 \le k \in \mathbb{N}$ denote by $C^k(D)$ the algebra of real-valued functions on D that have continuous partial derivatives of orders $1, \ldots, k$. The natural topology of $C^k(D)$ is the topology of uniform convergence on compact subsets of D, of functions and all of their partials up to and including order k. In order to describe this topology with seminorms let $\mathfrak{D}^k$ denote the collection of all partial derivative operators of order not exceeding k; $\mathfrak{D}^k$ contains by convention in particular the zeroeth-order partial derivative $\Phi \mapsto \Phi$. Then set, for any compact subset $K \subset D$ and $\Phi \in C^k(D)$,

$$\|\Phi\|_{k,K} \stackrel{\text{def}}{=} \sup_{x \in K} \sup_{\partial \in \mathfrak{D}^k} |\partial\Phi(x)| .$$

These seminorms, one for every compact $K \subset D$, actually make for a metrizable topology: there is a sequence K_n of compact sets with $K_n \subset \mathring{K}_{n+1}$ whose interiors $\mathring{K}_n$ exhaust D; and

$$\rho(\Phi, \Psi) \stackrel{\text{def}}{=} \sum_n 2^{-n} \left(1 \wedge \|\Phi - \Psi\|_{n,K_n}\right)$$

is a metric defining the natural topology of $C^k(D)$, which is clearly much finer than the topology of uniform convergence on compacta.

Proposition A.2.11 *The polynomials are dense in $C^k(D)$ in this topology.*

Here is a terse sketch of the proof. Let K be a compact subset of D. There exists a compact $K' \subset D$ whose interior $\mathring{K}'$ contains K. Given $\Phi \in C^k(D)$, denote by Φ_σ the convolution of the heat kernel γ_{tI} with the product[11] $\mathring{K}' \cdot \Phi$. Since Φ and its partials are bounded on $\mathring{K}'$, the integral defining the convolution exists and defines a real-analytic function Φ_t. Some easy but space-consuming estimates show that all partials of Φ_t converge uniformly on K to the corresponding partials of Φ as $t \downarrow 0$: the real-analytic functions are dense in $C^k(D)$. Then of course so are the polynomials.

Topologies, Filters, Uniformities

A topology on a space S is a collection $\mathfrak{t}$ of subsets that contains the whole space S and the empty set $\emptyset$ and that is closed under taking finite intersections and arbitrary unions. The sets of $\mathfrak{t}$ are called the ***open sets*** or $\mathfrak{t}$-***open sets***. Their complements are the ***closed sets***. Every subset $A \subseteq S$ contains a largest open set, denoted by $\mathring{A}$ and called the $\mathfrak{t}$-***interior*** of A; and every $A \subseteq S$ is contained in a smallest closed set, denoted by $\bar{A}$ and called the $\mathfrak{t}$-***closure*** of. A subset $A \subset S$ is given the ***induced topology*** $\mathfrak{t}_A \stackrel{\text{def}}{=} \{A \cap U : U \in \mathfrak{t}\}$. For details see [55] and [34]. A ***filter*** on S is a collection $\mathfrak{F}$ of non-void subsets of S that is closed under taking finite intersections and arbitrary supersets. The ***tail filter*** of a sequence (x_n) is the collection of all sets that contain a whole tail $\{x_n : n \geq N\}$ of the sequence. The ***neighborhood filter*** $\mathfrak{V}(x)$ of a point $x \in S$ for the topology $\mathfrak{t}$ is the filter of all subsets that contain a $\mathfrak{t}$-open set containing x. The filter $\mathfrak{F}$ ***converges*** to x if $\mathfrak{F}$ ***refines*** $\mathfrak{V}(x)$, that is to say if $\mathfrak{F} \supset \mathfrak{V}(x)$. Clearly a sequence converges if and only if its tail filter does. By Zorn's lemma, every filter is contained in (refined by) an ***ultrafilter***, that is to say, in a filter that has no proper refinement.

Let $(S, \mathfrak{t}_S)$ and $T, \mathfrak{t}_T)$ be topological spaces. A map $f : S \to T$ is ***continuous*** if the inverse image of every set in $\mathfrak{t}_T$ belongs to $\mathfrak{t}_S$. This is the case if and only if $\mathfrak{V}(x)$ refines $f^{-1}(\mathfrak{V}(f(x))$ at all $x \in S$.

The topology $\mathfrak{t}$ is ***Hausdorff*** if any two distinct points $x, x' \in S$ have non-intersecting neighborhoods V, V', respectively. It is completely regular if given $x \in E$ and $C \subset E$ closed one can find a continuous function that is zero on C and non-zero at x.

If the closure $\bar{U}$ is the whole ambient set S, then U is called $\mathfrak{t}$-***dense***. The topology $\mathfrak{t}$ is ***separable*** if S contains a countable $\mathfrak{t}$-dense set.

Exercise A.2.12 A filter $\mathfrak{U}$ on S is an ultrafilter if and only if for every $A \subseteq S$ either A or its complement A^c belongs to $\mathfrak{U}$. The following are equivalent: (i) every cover of S by open sets has a finite subcover; (ii) every collection of closed subsets with void intersection contains a finite subcollection whose intersection is void; (iii) every ultrafilter in S converges. In this case the topology is called ***compact***.

---

[11] $\mathring{K}'$ denotes both the set $\mathring{K}'$ and its indicator function – see convention A.1.5 on page 364.

Exercise A.2.13 (Tychonoff's Theorem) Let $E_\alpha, \mathfrak{t}_\alpha$, $\alpha \in A$, be topological spaces. The product topology $\mathfrak{t}$ on $E = \prod E_\alpha$ is the coarsest topology with respect to which all of the projections onto the factors E_α are continuous. The projection of an ultrafilter on E onto any of the factors is an ultrafilter there. Use this to prove Tychonoff's theorem: if the $\mathfrak{t}_\alpha$ are all compact, then so is $\mathfrak{t}$.

Exercise A.2.14 If $f : S \to T$ is continuous and $A \subset S$ is compact (in the induced topology, of course), then the forward image $f(A)$ is compact.

A topological space $(S, \mathfrak{t})$ is ***locally compact*** if every point has a basis of compact neighborhoods, that is to say, if every neighborhood of every point contains a compact neighborhood of that point. The ***one-point compactification*** S^Δ of $(S, \mathfrak{t})$ is obtained by adjoining one point, often denoted by Δ and called the ***point at infinity*** or the ***grave***, and declaring its neighborhood system to consist of the complements of the compact subsets of S. If S is already compact, then Δ is evidently an isolated point of $S^\Delta = S \dot\cup \{\Delta\}$.

A ***pseudometric*** on a set E is a function $d : E \times E \to \mathbb{R}_+$ that has $d(x,x) = 0$; is symmetric: $d(x,y) = d(y,x)$; and obeys the ***triangle inequality***: $d(x,z) \le d(x,y) + d(y,z)$. If $d(x,y) = 0$ implies that $x = y$, then d is a ***metric***. Let $\mathfrak{u}$ be a collection of pseudometrics on E. Another pseudometric d' is ***uniformly continuous*** with respect to $\mathfrak{u}$ if for every $\epsilon > 0$ there are $d_1, \ldots, d_k \in \mathfrak{u}$ and $\delta > 0$ such that

$$d_1(x,y) < \delta, \ldots, d_k(x,y) < \delta \implies d'(x,y) < \epsilon , \qquad \forall\, x,y \in E .$$

The ***saturation*** of $\mathfrak{u}$ consists of all pseudometrics that are uniformly continuous with respect to $\mathfrak{u}$. It contains in particular the pointwise sum and maximum of any two pseudometrics in $\mathfrak{u}$, and any positive scalar multiple of any pseudometric in $\mathfrak{u}$. A ***uniformity*** on E is simply a collection $\mathfrak{u}$ of pseudometrics that is saturated; a ***basis*** of $\mathfrak{u}$ is any subcollection $\mathfrak{u}_0 \subset \mathfrak{u}$ whose saturation equals $\mathfrak{u}$. The topology of $\mathfrak{u}$ is the topology $\mathfrak{t}_\mathfrak{u}$ generated by the open "pseudoballs" $B_{d,\epsilon}(x_0) \stackrel{\text{def}}{=} \{x \in E : d(x,x_0) < \epsilon\}$, $d \in \mathfrak{u}$, $\epsilon > 0$.

A map $f : E \to E'$ between uniform spaces $(E, \mathfrak{u})$ and $(E', \mathfrak{u}')$ is ***uniformly continuous*** if the pseudometric $(x,y) \mapsto d'\big(f(x), f(y)\big)$ belongs to $\mathfrak{u}$, for every $d' \in \mathfrak{u}'$. The composition of two uniformly continuous functions is obviously uniformly continuous again. The restrictions of the pseudometrics in $\mathfrak{u}$ to a fixed subset A of S clearly generate a uniformity on A, the ***induced uniformity***. A function on S is ***uniformly continuous on A*** if its restriction to A is uniformly continuous in this induced uniformity.

The filter $\mathfrak{F}$ on E is ***Cauchy*** if it contains arbitrarily small sets; that is to say, for every pseudometric $d \in \mathfrak{u}$ and every $\epsilon > 0$ there is an $F \in \mathfrak{F}$ with d-diam$(F) \stackrel{\text{def}}{=} \sup\{d(x,y) : x,y \in F\} < \epsilon$. The uniform space $(E, \mathfrak{u})$ is ***complete*** if every Cauchy filter $\mathfrak{F}$ converges. Every uniform space $(E, \mathfrak{u})$ has a ***Hausdorff completion***. This is a complete uniform space $(\overline{E}, \overline{\mathfrak{u}})$ whose topology $\mathfrak{t}_{\overline{\mathfrak{u}}}$ is Hausdorff, together with a uniformly continuous map $j : E \to \overline{E}$ such that the following holds: whenever $f : E \to Y$ is a uniformly continuous map into a Hausdorff complete uniform space Y, then there exists a unique uniformly continuous map $\overline{f} : \overline{E} \to Y$ such that $f = \overline{f} \circ j$. If a topology $\mathfrak{t}$ can be generated by some uniformity $\mathfrak{u}$, then it is ***uniformizable***; if $\mathfrak{u}$ has a basis consisting of a singleton d, then $\mathfrak{t}$ is ***pseudometrizable*** and ***metrizable*** if d is a metric; if $\mathfrak{u}$ and d can be chosen complete, then $\mathfrak{t}$ is ***completely (pseudo)metrizable***.

Exercise A.2.15 A Cauchy filter $\mathfrak{F}$ that has a convergent refinement converges. Therefore, if the topology of the uniformity $\mathfrak{u}$ is compact, then $\mathfrak{u}$ is complete.

A *compact* topology is generated by a unique uniformity: it is uniformizable in a unique way; if its topology has a countable basis, then it is completely pseudometrizable and completely metrizable if and only if it is also Hausdorff. A continuous function on a compact space and with values in a uniform space is uniformly continuous.

In this book two types of uniformity play a role. First there is the case that $\mathfrak{u}$ has a basis consisting of a single element d, usually a metric. The second instance is this: suppose $\mathcal{E}$ is a collection of real-valued functions on E. The ***$\mathcal{E}$-uniformity*** on E is the saturation of the collection of pseudometrics d_ϕ defined by

$$d_\phi(x, y) = \big|\phi(x) - \phi(y)\big| , \qquad \phi \in \mathcal{E} , \; x, y \in E .$$

It is also called the uniformity generated by $\mathcal{E}$ and is denoted by $\mathfrak{u}[\mathcal{E}]$. We leave to the reader the following facts:

Lemma A.2.16 *Assume that $\mathcal{E}$ consists of bounded functions on some set E.*

(i) The uniformities generated by $\mathcal{E}$ and the smallest uniformly closed algebra containing $\mathcal{E}$ and the constants coincide.

(ii) If $\mathcal{E}$ contains a countable uniformly dense set, then $\mathfrak{u}[\mathcal{E}]$ is pseudometrizable: it has a basis consisting of a single pseudometric d. If in addition $\mathcal{E}$ separates the points of E, then d is a metric and $\mathfrak{t}_{\mathfrak{u}[\mathcal{E}]}$ is Hausdorff.

(iii) The Hausdorff completion of $(E, \mathfrak{u}[\mathcal{E}])$ is the space $\overline{E}$ of the proof of theorem A.2.2 equipped with the uniformity generated by its continuous functions: if $\mathcal{E}$ contains the constants, it equals $\widehat{E}$; otherwise it is the one-point compactification of $\widehat{E}$. In any case, it is compact.

(iv) Let $A \subset E$ and let $f : A \to E'$ be a uniformly continuous[12] map to a complete uniform space $(E', \mathfrak{u}')$. Then $f(A)$ is relatively compact in $(E', \mathfrak{t}_{\mathfrak{u}'})$. Suppose $\mathcal{E}$ is an algebra or a vector lattice closed under chopping; then a real-valued function on A is uniformly continuous[12] if and only if it can be approximated uniformly on A by functions in $\mathcal{E} \oplus \mathbb{R}$, and an $\overline{\mathbb{R}}$-valued function is uniformly continuous if and only if it is the uniform limit (under the arctan metric $\overline{\rho}$!) of functions in $\mathcal{E} \oplus \mathbb{R}$.

[12] The uniformity of A is of course the one induced from $\mathfrak{u}[\mathcal{E}]$: it has the basis of pseudometrics $(x, y) \mapsto d_\phi(x, y) = \big|\phi(x) - \phi(y)\big|$, $d_\phi \in \mathfrak{u}[\mathcal{E}]$, $x, y \in A$, and is therefore the uniformity generated by the restrictions of the $\phi \in \mathcal{E}$ to A. The uniformity on $\mathbb{R}$ is of course given by the usual metric $\rho(r, s) \stackrel{\text{def}}{=} |r - s|$, the uniformity of the extended reals by the arctan metric $\overline{\rho}(r, s)$ – see item A.1.2.

Exercise A.2.17 A subset of a uniform space is called ***precompact*** if its image in the completion is relatively compact. A precompact subset of a complete uniform space is relatively compact.

Exercise A.2.18 Let (D,d) be a metric space. The distance of a point $x \in D$ from a set $F \subset D$ is $d(x,F) \stackrel{\text{def}}{=} \inf\{d(x,x') : x' \in F\}$. The ϵ-neighborhood of F is the set of all points whose distance from F is strictly less than ϵ; it evidently equals the union of all ϵ-balls with centers in F. A subset $K \subset D$ is called ***totally bounded*** if for every $\epsilon > 0$ there is a finite set $F_\epsilon \subset D$ whose ϵ-neighborhood contains K. Show that a subset $K \subseteq D$ is precompact if and only if it is totally bounded.

Semicontinuity

Let E be a topological space. The collection of bounded continuous real-valued functions on E is denoted by $C_b(E)$. It is a lattice algebra containing the constants. A real-valued function f on E is ***lower semicontinuous*** at $x \in E$ if $\liminf_{y\to x} f(y) \geq f(x)$; it is called ***upper semicontinuous*** at $x \in E$ if $\limsup_{y\to x} f(y) \leq f(x)$. f is simply lower (upper) semicontinuous if it is lower (upper) semicontinuous at every point of E. For example, an open set is a lower semicontinuous function, and a closed set is an upper semicontinuous function.[13]

Lemma A.2.19 *Assume that the topological space E is completely regular.*
(a) For a bounded function f the following are equivalent:

(i) f is lower (upper) semicontinuous;
(ii) f is the pointwise supremum of the continuous functions $\phi \leq f$ (f is the pointwise infimum of the continuous functions $\phi \geq f$);
(iii) $-f$ is upper (lower) semicontinuous;
(iv) for every $r \in \mathbb{R}$ the set $[f > r]$ (the set $[f < r]$) is open.

(b) Let $\mathcal{A}$ be a vector lattice of bounded continuous functions that contains the constants and generates the topology.[14] Then:

(i) If $U \subset E$ is open and $K \subset U$ compact, then there is a function $\phi \in \mathcal{A}$ with values in $[0,1]$ that equals 1 on K and vanishes outside U.
(ii) Every bounded lower semicontinuous function h is the pointwise supremum of an increasingly directed subfamily $\mathcal{A}^h$ of $\mathcal{A}$.

Proof. We leave (a) to the reader. (b) The sets of the form $[\phi > r]$, $\phi \in \mathcal{A}$, $r > 0$, clearly form a subbasis of the topology generated by $\mathcal{A}$. Since $[\phi > r] = [(\phi/r - 1) \vee 0 > 0]$, so do the sets of the form $[\phi > 0]$, $0 \leq \phi \in \mathcal{A}$. A finite intersection of such sets is again of this form: $\bigcap_i [\phi_i > 0]$ equals $[\bigvee_i \phi_i > 0]$.

---

[13] S denotes both the set S and its indicator function – see convention A.1.5 on page 364.
[14] The ***topology generated by a collection*** Γ ***of functions*** is the coarsest topology with respect to which every $\gamma \in \Gamma$ is continuous. A net (x_α) converges to x in this topology if and only if $\gamma(x_\alpha) \to \gamma(x)$ for all $\gamma \in \Gamma$. Γ is said to define the given topology τ if the topology it generates coincides with τ; if τ is metrizable, this is the same as saying that a sequence x_n converges to x if and only if $\gamma(x_n) \to \gamma(x)$ for all $\gamma \in \Gamma$.

The sets of the form $[\phi > 0]$, $\phi \in \mathcal{A}_+$, thus form a basis of the topology generated by $\mathcal{A}$.

(i) Since K is compact, there is a finite collection $\{\phi_i\} \subset \mathcal{A}_+$ such that $K \subset \bigcup_i [\phi_i > 0] \subset U$. Then $\psi \stackrel{\text{def}}{=} \bigvee \phi_i$ vanishes outside U and is strictly positive on K. Let $r > 0$ be its minimum on K. The function $\phi \stackrel{\text{def}}{=} (\psi/r) \wedge 1$ of $\mathcal{A}_+$ meets the description of (i).

(ii) We start with the case that the lower semicontinuous function h is positive. For every $q > 0$ and $x \in [h > q]$ let $\phi_x^q \in \mathcal{A}$ be as provided by (i): $\phi_x^q(x) = 1$, and $\phi_x^q(x') = 0$ where $h(x') \le q$. Clearly $q \cdot \phi_x^q < h$. The finite suprema of the functions $q \cdot \phi_x^q \in \mathcal{A}$ form an increasingly directed collection $\mathcal{A}^h \subset \mathcal{A}$ whose pointwise supremum evidently is h. If h is not positive, we apply the foregoing to $h + \| h \|_\infty$. ∎

Separable Metric Spaces

Recall that a topological space E is ***metrizable*** if there exists a metric d that defines the topology in the sense that the neighborhood filter $\mathcal{V}(x)$ of every point $x \in E$ has a basis of ***d-balls*** $B_r(x) \stackrel{\text{def}}{=} \{x' : d(x, x') < r\}$ – then there are in general many metrics doing this. The next two results facilitate the measure theory on separable and metrizable spaces.

Lemma A.2.20 *Assume that E is separable and metrizable.*

(i) There exists a countably generated[6] *uniformly closed lattice algebra $\mathcal{U}[E]$ of bounded uniformly continuous functions that contains the constants, generates the topology,*[14] *and has in addition the property that every bounded lower semicontinuous function is the pointwise supremum of an increasing sequence in $\mathcal{U}[E]$, and every bounded upper semicontinuous function is the pointwise infimum of a decreasing sequence in $\mathcal{U}[E]$.*

(ii) Any increasingly (decreasingly) directed[2] *subset Φ of $C_b(E)$ contains a sequence that has the same pointwise supremum (infimum) as Φ.*

Proof. (i) Let d be a metric for E and $D = \{x_1, x_2, \ldots\}$ a countable dense subset. The collection Γ of bounded uniformly continuous functions $\gamma_{k,n} : x \mapsto kd(x, x_n) \wedge 1$, $x \in E$, $k, n \in \mathbb{N}$, is countable and generates the topology; indeed the open balls $[\gamma_{k,n} < 1/2]$ evidently form a basis of the topology. Let $\mathcal{A}$ denote the collection of finite $\mathbb{Q}$-linear combinations of 1 and finite products of functions in Γ. This is a countable algebra over $\mathbb{Q}$ containing the scalars whose uniform closure $\mathcal{U}[E]$ is both an algebra and a vector lattice (theorem A.2.2).

Let h be a lower semicontinuous function. Lemma A.2.19 provides an increasingly directed family $\mathcal{U}^h \subset \mathcal{U}[E]$ whose pointwise supremum is h; that it can be chosen countable follows from (ii).

(ii) Assume Φ is increasingly directed and has bounded pointwise supremum h. For every $\phi \in \Phi$, $x \in E$, and $n \in \mathbb{N}$ let $\psi_{\phi,x,n}$ be an element of $\mathcal{A}$ with $\psi_{\phi,x,n} \le \phi$ and $\psi_{\phi,x,n}(x) > \phi(x) - 1/n$. The collection $\mathcal{A}^h$ of these

$\psi_{\phi,x,n}$ is at most countable: $\mathcal{A}^h = \{\psi_1, \psi_2, \ldots\}$, and its pointwise supremum is h. For every n select a $\phi'_n \in \Phi$ with $\psi_n \le \phi'_n$. Then set $\phi_1 = \phi'_1$, and when $\phi_1, \ldots, \phi_n \in \Phi$ have been defined let ϕ_{n+1} be an element of Φ that exceeds $\phi_1, \ldots, \phi_n, \phi'_{n+1}$. Clearly $\phi_n \uparrow h$. ∎

Lemma A.2.21 *(a) Let $\mathcal{X}, \mathcal{Y}$ be metric spaces, $\mathcal{Y}$ compact, and suppose that $K \subset \mathcal{X} \times \mathcal{Y}$ is σ-compact and non-void. Then there is a Borel cross section; that is to say, there is a Borel map $\gamma : \mathcal{X} \to \mathcal{Y}$ "whose graph lies in K when it can:" when $x \in \pi_{\mathcal{X}}(K)$ then $(x, \gamma(x)) \in K$ – see figure A.15.*

(b) Let $\mathcal{X}, \mathcal{Y}$ be separable metric spaces, $\mathcal{X}$ locally compact and $\mathcal{Y}$ compact, and suppose $G : \mathcal{X} \times \mathcal{Y} \to \mathbb{R}$ is a continuous function. There exists a Borel function $\gamma : \mathcal{X} \to \mathcal{Y}$ such that for all $x \in \mathcal{X}$

$$\sup\Big\{G(x,y) : y \in \mathcal{Y}\Big\} = G\big(x, \gamma(x)\big) .$$

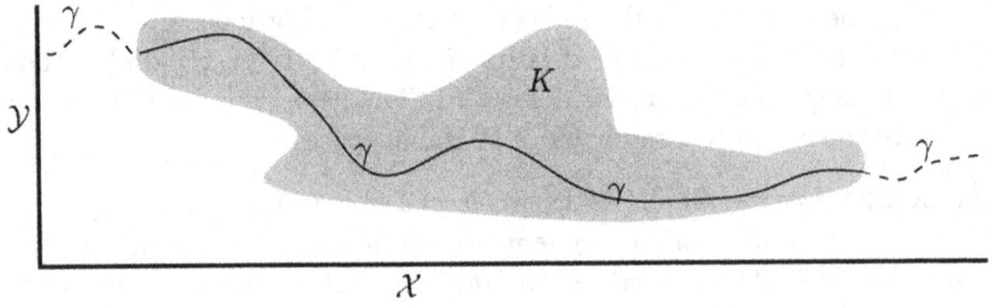

Figure A.15 The Cross Section Lemma

Proof. (a) To start with, consider the case that $\mathcal{Y}$ is the unit interval I and that K is compact. Then $\gamma^K(x) \stackrel{\text{def}}{=} \inf\{t : (x,t) \in K\} \wedge 1$ defines an upper semicontinuous function from $\mathcal{X}$ to I with $\big(x, \gamma(x)\big) \in K$ when $x \in \pi_{\mathcal{X}}(K)$. If K is σ-compact, then there is an increasing sequence $\big(K_n\big)$ of compacta with union K. The cross sections γ^{K_n} give rise to the decreasing and ultimately constant sequence (γ_n) defined inductively by $\gamma_1 \stackrel{\text{def}}{=} \gamma^{K_1}$,

$$\gamma_{n+1} \stackrel{\text{def}}{=} \begin{cases} \gamma_n & \text{on } [\gamma_n < 1], \\ \gamma^{K_{n+1}} & \text{on } [\gamma_n = 1]. \end{cases}$$

Clearly $\gamma \stackrel{\text{def}}{=} \inf \gamma_n$ is Borel, and $\big(x, \gamma(x)\big) \in K$ when $x \in \pi_{\mathcal{X}}(K)$. If $\mathcal{Y}$ is not the unit interval, then we use the universality A.2.22 of the Cantor set $C \subset I$: it provides a continuous surjection $\phi : C \to \mathcal{Y}$. Then $K' \stackrel{\text{def}}{=} (\phi \times id_{\mathcal{X}})^{-1}(K)$ is a σ-compact subset of $I \times \mathcal{X}$, there is a Borel function $\gamma^{K'} : \mathcal{X} \to C$ whose restriction to $\pi_{\mathcal{X}}(K') = \pi_{\mathcal{X}}(K)$ has its graph in K', and $\gamma \stackrel{\text{def}}{=} \phi \circ \gamma^{K'}$ is the desired Borel cross section.

(b) Set $\sigma(x) \stackrel{\text{def}}{=} \sup\big\{\{G(x,y) : y \in \mathcal{Y}\big\}$. Because of the compactness of $\mathcal{Y}$, σ is a continuous function on $\mathcal{X}$ and $K \stackrel{\text{def}}{=} \{(x,y) : G(x,y) = \sigma(x)\}$ is a σ-compact subset of $\mathcal{X} \times \mathcal{Y}$ with $\mathcal{X}$-projection $\mathcal{X}$. Part (a) furnishes γ. ∎

Exercise A.2.22 (Universality of the Cantor Set) For every compact metric space $\mathcal{Y}$ there exists a continuous map from the Cantor set onto $\mathcal{Y}$.

Exercise A.2.23 Let F be a Hausdorff space and E a subset whose induced topology can be defined by a complete metric ρ. Then E is a $\mathcal{G}_\delta$-set; that is to say, there is a sequence of open subsets of F whose intersection is E.

Exercise A.2.24 Let (P, d) be a separable complete metric space. There exists a compact metric space $\widehat{P}$ and a homeomorphism j of P onto a subset of $\widehat{P}$. j can be chosen so that $j(P)$ is a dense $\mathcal{G}_\delta$-set and a $\mathcal{K}_{\sigma\delta}$-set of $\widehat{P}$.

Topological Vector Spaces

A real vector space $\mathcal{V}$ together with a topology on it is a ***topological vector space*** if the linear and topological structures are compatible in this sense: the maps $(f, g) \mapsto f + g$ from $\mathcal{V} \times \mathcal{V}$ to $\mathcal{V}$ and $(r, f) \mapsto r \cdot f$ from $\mathbb{R} \times \mathcal{V}$ to $\mathcal{V}$ are continuous.

A subset B of the topological vector space $\mathcal{V}$ is ***bounded*** if it is ***absorbed*** by any neighborhood V of zero; this means that there exists a scalar λ so that $B \subset \lambda V \stackrel{\text{def}}{=} \{\lambda v : v \in V\}$.

The main examples of topological vector spaces concerning us in this book are the spaces $\mathcal{L}^p$ and L^p for $0 \le p \le \infty$ and the spaces $C_0(E)$ and $C(E)$ of continuous functions. We recall now a few common notions that should help the reader navigate their topologies.

A set $V \subset \mathcal{V}$ is ***convex*** if for any two scalars λ_1, λ_2 with absolute value less than 1 and sum 1 and for any two points $v_1, v_2 \in V$ we have $\lambda_1 v_1 + \lambda_2 v_2 \in V$. A topological vector space $\mathcal{V}$ is ***locally convex*** if the neighborhood filter at zero (and then at any point) has a basis of convex sets. The examples above all have this feature, except the spaces $\mathcal{L}^p$ and L^p when $0 \le p < 1$.

One of the most useful facts about topological vector spaces is doubtlessly the

Theorem A.2.25 (Hahn–Banach) *Let C, K be two convex sets in a topological vector space $\mathcal{V}$, C closed and K compact. There exists a continuous linear functional $x^* : \mathcal{V} \to \mathbb{R}$ so that $x^*(k) < 1$ for all $k \in K$ and $x^*(x) \ge 1$ for all $x \in C$. In other words, C lies on one side of the **hyperplane** $[x^* = 1]$ and K strictly on the other. Therefore a closed convex subset of $\mathcal{V}$ is weakly closed (see item A.2.32).*

A.2.26 Gauges It is easy to see that a topological vector space admits a collection Γ of gauges $\lceil\!\lceil\ \rceil\!\rceil : \mathcal{V} \to \mathbb{R}_+$ that define the topology in the sense that $f_n \to f$ if and only if $\lceil\!\lceil f - f_n \rceil\!\rceil \to 0$ for all $\lceil\!\lceil\ \rceil\!\rceil \in \Gamma$. This is the same as saying that the "balls"

$$B_\epsilon(0) \stackrel{\text{def}}{=} \{f : \lceil\!\lceil f \rceil\!\rceil < \epsilon\}, \qquad \lceil\!\lceil\ \rceil\!\rceil \in \Gamma, \ \epsilon > 0,$$

form a basis of the neighborhood system at 0 and implies that $\lceil\!\lceil rf \rceil\!\rceil \xrightarrow[r \to 0]{} 0$ for all $f \in \mathcal{V}$ and all $\lceil\!\lceil\ \rceil\!\rceil \in \Gamma$. There are always many such gauges. Namely, let $\{V_n\}$ be a decreasing sequence of neighborhoods of 0 with $V_0 = \mathcal{V}$. Then

$$\lceil\!\lceil f \rceil\!\rceil \stackrel{\text{def}}{=} \Big(\inf\{n : f \in V_n\}\Big)^{-1}$$

will be a gauge. If the v_n form basis of neighborhoods at zero, then Γ can be taken to be the singleton $\{\lceil\!\lceil\ \rceil\!\rceil\}$ above. With a little more effort it can be shown that there are continuous gauges defining the topology that are ***sub-additive***: $\lceil\!\lceil f+g \rceil\!\rceil \le \lceil\!\lceil f \rceil\!\rceil + \lceil\!\lceil g \rceil\!\rceil$. For such a gauge, $dist(f,g) \stackrel{\text{def}}{=} \lceil\!\lceil f-g \rceil\!\rceil$ defines a translation-invariant pseudometric, a metric if and only if $\mathcal{V}$ is Hausdorff. From now on the word ***gauge*** will mean a continuous subadditive gauge.

A locally convex topological vector space whose topology can be defined by a complete metric is a ***Fréchet space***. Here are two examples that recur throughout the text:

Examples A.2.27 (i) Suppose E is a locally compact separable metric space and F is a Fréchet space with translation-invariant metric ρ (visualize $\mathbb{R}$). Let $C_F(E)$ denote the vector space of all continuous functions from E to F. The ***topology of uniform convergence on compacta*** on $C_F(E)$ is given by the following collection of gauges, one for every compact set $K \subset E$,

$$\lceil\!\lceil \phi \rceil\!\rceil_K \stackrel{\text{def}}{=} \sup\big\{\rho(\phi(x)) : x \in K\big\}\,,\quad \phi : E \to F\,. \tag{A.2.1}$$

It is Fréchet. Indeed, a cover by compacta K_n with $K_n \subset \mathring{K}_{n+1}$ gives rise to the gauge

$$\phi \mapsto \textstyle\sum_n \lceil\!\lceil \phi \rceil\!\rceil_{K_n} \wedge 2^{-n}\,, \tag{A.2.2}$$

which in turn gives rise to a complete metric for the topology of $C_F(E)$. If F is separable, then so is $C_F(E)$.

(ii) Suppose that $E = \mathbb{R}_+$, but consider the space $\mathscr{D}_F$, the ***path space***, of functions $\phi : \mathbb{R}_+ \to F$ that are right-continuous and have a left limit at every instant $t \in \mathbb{R}_+$. Inasmuch as such a càdlàg path is bounded on every bounded interval, the supremum in (A.2.1) is finite, and (A.2.2) again describes the Fréchet topology of uniform convergence on compacta. But now this topology is not separable in general, even when F is as simple as $\mathbb{R}$. The indicator functions $\phi_t \stackrel{\text{def}}{=} 1_{[0,t)}$, $0 < t < 1$, have $\lceil\!\lceil \phi_s - \phi_t \rceil\!\rceil_{[0,1]} = 1$, yet they are uncountable in number. ∎

With every convex neighborhood V of zero there comes the ***Minkowski functional*** $\|f\| \stackrel{\text{def}}{=} \inf\{|r| : rf \in V\}$. This continuous gauge is both sub-additive and ***absolute-homogeneous***: $\|r\cdot f\| = |r|\cdot\|f\|$ for $f \in \mathcal{V}$ and $r \in \mathbb{R}$. An absolute-homogeneous subadditive gauge is a ***seminorm***. If $\mathcal{V}$ is locally convex, then their collection defines the topology. Prime examples of spaces whose topology is defined by a single seminorm are the spaces $\mathcal{L}^p$ and L^p for $1 \le p \le \infty$, and $C_0(E)$.

Exercise A.2.28 Suppose that $\mathcal{V}$ has a countable basis at 0. Then $B \subset \mathcal{V}$ is bounded if and only if for one, and then every, continuous gauge $\lceil\!\lceil\ \rceil\!\rceil$ on $\mathcal{V}$ that defines the topology

$$\sup\{\lceil\!\lceil \lambda \cdot f \rceil\!\rceil : f \in B\} \xrightarrow[\lambda \to 0]{} 0 .$$

Exercise A.2.29 Let $\mathcal{V}$ be a topological vector space with a countable base at 0, and $\lceil\!\lceil\ \rceil\!\rceil$ and $\lceil\!\lceil\ \rceil\!\rceil'$ two gauges on $\mathcal{V}$ that define the topology – they need not be subadditive nor continuous except at 0. There exists an increasing right-continuous function $\Phi : \mathbb{R}_+ \to \overline{\mathbb{R}}_+$ with $\Phi(r) \xrightarrow[r \to 0]{} 0$ such that $\lceil\!\lceil f \rceil\!\rceil' \le \Phi(\lceil\!\lceil f \rceil\!\rceil)$ for all $f \in \mathcal{V}$.

A.2.30 Quasinormed Spaces In some contexts it is more convenient to use the homogeneity of the $\|\ \|_{L^p}$ on L^p rather than the subadditivity of the $\lceil\!\lceil\ \rceil\!\rceil_{L^p}$. In order to treat Banach spaces and spaces L^p simultaneously one uses the notion of a ***quasinorm*** on a vector space E. This is a function $\|\ \| : E \to \mathbb{R}_+$ such that

$$\|x\| = 0 \Longleftrightarrow x = 0 \text{ and } \|r\cdot x\| = |r| \cdot \|x\| \qquad \forall\, r \in \mathbb{R}, x \in E .$$

A topological vector space is ***quasinormed*** if it is equipped with a quasinorm $\|\ \|$ that defines the topology, i.e., such that $x_n \xrightarrow[n\to\infty]{} x$ if and only if $\|x_n - x\| \xrightarrow[n\to\infty]{} 0$. If $(E, \|\ \|_E)$ and $(F, \|\ \|_F)$ are quasinormed topological vector spaces and $u : E \to F$ is a continuous linear map between them, then the ***size of u*** is naturally measured by the number

$$\|u\| = \|u\|_{L(E,F)} \stackrel{\text{def}}{=} \sup\big\{ \|u(x)\|_F : x \in E, \|x\|_E \le 1 \big\} .$$

A subadditive quasinorm clearly is a seminorm; so is an absolute-homogeneous gauge.

Exercise A.2.31 Let $\mathcal{V}$ be a vector space equipped with a seminorm $\|\ \|$. The set $N \stackrel{\text{def}}{=} \{x \in \mathcal{V} : \|x\| = 0\}$ is a vector subspace and coincides with the closure of $\{0\}$. On the quotient $\dot{\mathcal{V}} \stackrel{\text{def}}{=} \mathcal{V}/N$ set $\|\dot{x}\| \stackrel{\text{def}}{=} \|x\|$. This does not depend on the representative x in the equivalence class $\dot{x} \in \dot{\mathcal{V}}$ and makes $(\dot{\mathcal{V}}, \|\ \|)$ a normed space. The transition from $(\mathcal{V}, \|\ \|)$ to $(\dot{\mathcal{V}}, \|\ \|)$ is such a standard operation that it is sometimes not mentioned, that $\mathcal{V}$ and $\dot{\mathcal{V}}$ are identified, and that reference is made to "the norm" $\|\ \|$ on $\mathcal{V}$.

A.2.32 Weak Topologies Let $\mathcal{V}$ be a vector space and $\mathcal{M}$ a collection of linear functionals $\mu : \mathcal{V} \to \mathbb{R}$. This gives rise to two topologies. One is the topology $\sigma(\mathcal{V}, \mathcal{M})$ on $\mathcal{V}$, the coarsest topology with respect to which every functional $\mu \in \mathcal{M}$ is continuous; it makes $\mathcal{V}$ into a locally convex topological vector space. The other is $\sigma(\mathcal{M}, \mathcal{V})$, the topology on $\mathcal{M}$ of pointwise convergence on $\mathcal{V}$. For an example assume that $\mathcal{V}$ is already a topological vector space under some topology τ and $\mathcal{M}$ consists of all τ-continuous linear functionals on $\mathcal{V}$, a vector space usually called the ***dual of*** $\mathcal{V}$ and denoted by $\mathcal{V}^*$. Then $\sigma(\mathcal{V}, \mathcal{V}^*)$ is called by analysts the ***weak topology*** on $\mathcal{V}$ and $\sigma(\mathcal{V}^*, \mathcal{V})$ the ***weak* topology*** on $\mathcal{V}^*$. When $\mathcal{V} = C_0(E)$ and $\mathcal{M} = \mathfrak{P}^* \subset C_0(E)^*$ probabilists like to call the latter the topology of weak convergence – as though life weren't confusing enough already!

Exercise A.2.33 If $\mathcal{V}$ is given the topology $\sigma(\mathcal{V},\mathcal{M})$, then the dual of $\mathcal{V}$ coincides with the vector space generated by $\mathcal{M}$.

The Minimax Theorem, Lemmas of Gronwall and Kolmogoroff

Lemma A.2.34 (Ky–Fan) *Let K be a compact convex subset of a topological vector space and $\mathcal{H}$ a family of upper semicontinuous concave numerical functions on K. Assume that the functions of $\mathcal{H}$ do not take the value $+\infty$ and that any convex combination of any two functions in $\mathcal{H}$ majorizes another function of $\mathcal{H}$. If every function $h \in \mathcal{H}$ is nonnegative at some point $k_h \in K$, then there is a common point $k \in K$ at which all of the functions $h \in \mathcal{H}$ take a nonnegative value.*

Proof. We argue by contradiction and assume that the conclusion fails. Then the convex compact sets $[h \geq 0]$, $h \in \mathcal{H}$, have void intersection, and there will be finitely many $h \in \mathcal{H}$, say $h_1, \ldots, h_N$, with

$$\bigcap_{n=1}^{N} [h_n \geq 0] = \emptyset\,. \tag{A.2.3}$$

Let the collection $\{h_1, \ldots, h_N\}$ be chosen so that N is minimal. Since $[h \geq 0] \neq \emptyset$ for every $h \in \mathcal{H}$, we must have $N \geq 2$. The compact convex set

$$K' \stackrel{\text{def}}{=} \bigcap_{n=3}^{N} [h_n \geq 0] \subset K$$

is contained in $[h_1 < 0] \cup [h_2 < 0]$ (if $N = 2$ it equals K). Both h_1 and h_2 take nonnegative values on K'; indeed, if h_1 did not, then h_2 could be struck from the collection, and vice versa, in contradiction to the minimality of N.

Let us see how to proceed in a very simple situation: suppose K is the unit interval $I = [0,1]$ and $\mathcal{H}$ consists of affine functions. Then K' is a closed subinterval I' of I, and h_1 and h_2 take their positive maxima at one of the endpoints of it, evidently not in the same one. In particular, I' is not degenerate. Since the open sets $[h_1 < 0]$ and $[h_2 < 0]$ together cover the interval I', but neither does by itself, there is a point $\xi \in I'$ at which both h_1 and h_2 are strictly negative; ξ evidently lies in the interior of I'. Let $\eta = \max\{h_1(\xi), h_2(\xi)\}$. Any convex combination $h' = r_1 h_1 + r_2 h_2$ of h_1, h_2 will at ξ have a value less than $\eta < 0$. It is clearly possible to choose $r_1, r_2 \geq 0$ with sum 1 so that h' has at the left endpoint of I' a value in $(-\eta/2, 0)$. The affine function h' is then evidently strictly less than zero on all of I'. There exists by assumption a function $h \in \mathcal{H}$ with $h \leq h'$; it can replace the pair $\{h_1, h_2\}$ in equation (A.2.3), which is in contradiction to the minimality of N. The desired result is established in the simple case that K is the unit interval and $\mathcal{H}$ consists of affine functions.

The general case follows readily from this. First note that the set $[h_i > -\infty]$ is convex, as the increasing union of the convex sets $\bigcup_{k\in\mathbb{N}}[h_i \ge -k]$, $i = 1, 2$. Thus

$$K_0' \stackrel{\text{def}}{=} K' \cap [h_1 > -\infty] \cap [h_2 > -\infty]$$

is convex. Next observe that there is an $\epsilon > 0$ such that the open set $[h_1 + \epsilon < 0] \cup [h_2 + \epsilon < 0]$ still covers K'. For every $k \in K_0'$ consider the affine function

$$a_k : t \mapsto -\Big(t \cdot \big(h_1(k) + \epsilon\big) + (1-t) \cdot \big(h_2(k) + \epsilon\big)\Big),$$

i.e.,
$$a_k(t) \stackrel{\text{def}}{=} -\Big(t \cdot h_1(k) + (1-t) \cdot h_2(k)\Big) - \epsilon,$$

on the unit interval I. Every one of them is nonnegative at some point of I; for instance, if $k \in [h_1 + \epsilon < 0]$, then $\lim_{t\to 1} a_k(t) = -\big(h_1(k) + \epsilon\big) > 0$. An easy calculation using the concavity of h_i shows that a convex combination $ra_k + (1-r)a_{k'}$ majorizes $a_{rk+(1-r)k'}$. We can apply the first part of the proof and conclude that there exists a $\tau \in I$ at which every one of the functions a_k is nonnegative. This reads

$$h'(k) \stackrel{\text{def}}{=} \tau \cdot h_1(k) + (1-\tau) \cdot h_2(k) \le -\epsilon < 0 \qquad k \in K_0'.$$

Now τ is not the right endpoint 1; if it were, then we would have $h_1 < -\epsilon$ on K_0', and a suitable convex combination $rh_1 + (1-r)h_2$ would majorize a function $h \in \mathcal{H}$ that is strictly negative on K; this then could replace the pair $\{h_1, h_2\}$ in equation (A.2.3). By the same token $\tau \ne 0$. But then h' is strictly negative on all of K' and there is an $h \in \mathcal{H}$ majorized by h', which can then replace the pair $\{h_1, h_2\}$. In all cases we arrive at a contradiction to the minimality of N. ∎

Lemma A.2.35 (Gronwall's Lemma) *Let $\phi : [0,\infty] \to [0,\infty)$ be an increasing function satisfying*

$$\phi(t) \le A(t) + \int_0^t \phi(s)\,\eta(s)\,ds \qquad t \ge 0,$$

where $\eta : [0,\infty) \to \mathbb{R}$ is positive and Borel, and $A : [0,\infty) \to [0,\infty)$ is increasing. Then

$$\phi(t) \le A(t) \cdot \exp\left(\int_0^t \eta(s)\,ds\right), \qquad t \ge 0.$$

Proof. To start with, assume that ϕ is right-continuous and A constant.

Set $\quad H(t) \stackrel{\text{def}}{=} \exp\left(\int_0^t \eta(s)\,ds\right)$, fix an $\epsilon > 0$,

and set $\quad t_0 \stackrel{\text{def}}{=} \inf\big\{s : \phi(s) \ge (A+\epsilon) \cdot H(s)\big\}$.

Then
$$\begin{aligned}\phi(t_0) &\le A + (A+\epsilon)\int_0^{t_0} H(s)\,\eta(s)ds \\ &= A + (A+\epsilon)\big(H(t_0) - H(0)\big) = (A+\epsilon)H(t_0) - \epsilon \\ &< (A+\epsilon)H(t_0)\,.\end{aligned}$$

Since this is a strict inequality and ϕ is right-continuous, $\phi(t_0') \le (A+\epsilon)H(t_0')$ for some $t_0' > t_0$. Thus $t_0 = \infty$, and $\phi(t) \le (A+\epsilon)H(t)$ for all $t \ge 0$. Since $\epsilon > 0$ was arbitrary, $\phi(t) \le AH(t)$. In the general case set

$$\psi(s) = \inf\{\phi(\tau \wedge t) : \tau > s\}\,.$$

ψ is right-continuous and satisfies $\psi(\tau) \le A(t) + \int_0^\tau \psi(s)\,\eta(s)\,ds$ for $\tau \ge 0$. The first part of the proof applies and yields $\phi(t) \le \psi(t) \le A(t)\cdot H(t)$. ∎

Exercise A.2.36 Let $x : [0,\infty] \to [0,\infty)$ be an increasing function satisfying

$$x_\mu \le C + \max_{\rho=p,q}\Big(\int_0^\mu (A + Bx_\lambda)^\rho\,d\lambda\Big)^{1/\rho}, \qquad \mu \ge 0\,,$$

for some $1 \le p \le q < \infty$ and some constants $A, B > 0$. Then there exist constants $\alpha \le 2(A/B + C)$ and $\beta \le \max_{\rho=p,q}(2B)^\rho/\rho$ such that $x_\lambda \le \alpha e^{\beta\lambda}$ for all $\lambda > 0$.

Lemma A.2.37 (Kolmogorov) *Let U be an open subset of $\mathbb{R}^d$, and let*

$$\{X_u \,:\, u \in U\}$$

be a family of functions,[15] *all defined on the same probability space $(\Omega, \mathcal{F}, \mathbb{P})$ and having values in the same complete metric space (E, ρ). Assume that $\omega \mapsto \rho(X_u(\omega), X_v(\omega))$ is measurable for any two $u, v \in U$ and that there exist constants $p, \beta > 0$ and $C < \infty$ so that*

$$\mathbb{E}\left[\rho(X_u, X_v)^p\right] \le C\cdot|u - v|^{d+\beta} \quad \textit{for } u, v \in U\,. \tag{A.2.4}$$

Then there exists a family $\{X'_u \,:\, u \in U\}$ of the same description which in addition has the following properties: (i) $X'_.$ is a modification of $X_.$, meaning that $\mathbb{P}[X_u \ne X'_u] = 0$ for every $u \in U$; and (ii) for every single $\omega \in \Omega$ the map $u \mapsto X_u(\omega)$ from U to E is continuous. In fact there exists, for every $\alpha > 0$, a subset $\Omega_\alpha \in \mathcal{F}$ with $\mathbb{P}[\Omega_\alpha] > 1 - \alpha$ such that the family

$$\{u \mapsto X'_u(\omega) \,:\, \omega \in \Omega_\alpha\}$$

of E-valued functions is equicontinuous on U and uniformly equicontinuous on every compact subset K of U; that is to say, for every $\epsilon > 0$ there is a $\delta > 0$ independent of $\omega \in \Omega_\alpha$ such that $|u - v| < \delta$ implies $\rho(X'_u(\omega), X'_v(\omega)) < \epsilon$ for all $u, v \in K$ and all $\omega \in \Omega_\alpha$. (In fact, $\delta = \delta_{K;\alpha,p,C,\beta}(\epsilon)$ depends only on the indicated quantities.)

---

[15] Not necessarily measurable for the Borels of (E, ρ).

Exercise A.2.38 (Ascoli–Arzelà) Let $K \subseteq U$, $\epsilon \mapsto \delta(\epsilon)$ an increasing function from $(0,1)$ to $(0,1)$, and $C \subseteq E$ compact. The collection $\mathcal{K}(\delta(\cdot))$ of paths $x : K \to C$ satisfying

$$|u-v| \le \delta(\epsilon) \implies \rho(x_u, x_v) \le \epsilon$$

is compact in the topology of uniform convergence of paths; conversely, a compact set of continuous paths is uniformly equicontinuous and the union of their ranges is relatively compact. Therefore, if E happens to be compact, then the set of paths $\{X'_.(\omega) : \omega \in \Omega_\alpha\}$ of lemma A.2.37 is relatively compact in the topology of uniform convergence on compacta.

Proof of A.2.37. Instead of the customary euclidean norm $|\ |_2$ we may and shall employ the sup-norm $|\ |_\infty$. For $n \in \mathbb{N}$ let U_n be the collection of vectors in U whose coordinates are of the form $k2^{-n}$, with $k \in \mathbb{Z}$ and $|k2^{-n}| < n$. Then set $U_\infty = \bigcup_n U_n$. This is the set of ***dyadic-rational points*** in U and is clearly in U. To start with we investigate the random variables[15] X_u, $u \in U_\infty$.

Let $0 < \lambda < \beta/p$.[16] If $u, v \in U_n$ are nearest neighbors, that is to say if $(|u-v| = 2^{-n})$, then Chebyscheff's inequality and (A.2.4) give

$$\begin{aligned}\mathbb{P}\Big(\big[\rho(X_u, X_v) > 2^{-\lambda n}\big]\Big) &\le 2^{p\lambda n} \cdot \mathbb{E}\left[\rho(X_u, X_v)^p\right] \\ &\le C \cdot 2^{p\lambda n} \cdot 2^{-n(d+\beta)} = C \cdot 2^{(p\lambda-\beta-d)n}.\end{aligned}$$

Now a point $u \in U_n$ has less than 3^d nearest neighbors v in U_n, and there are less than $(2n2^n)^d$ points in U_n. Consequently

$$\mathbb{P}\Big[\bigcup_{\substack{u,v \in U_n \\ |u-v| = 2^{-n}}} \big[\rho(X_u, X_v) > 2^{-\lambda n}\big]\Big]$$

$$\le C \cdot 2^{(p\lambda-\beta-d)n} \cdot (6n)^d \cdot 2^{nd} = C \cdot (6n)^d 2^{-(\beta-p\lambda)n}.$$

Since $2^{-(\beta-p\lambda)} < 1$, these numbers are summable over n. Given $\alpha > 0$, we can find an integer N_α depending only[16] on C, β, p such that the set

$$\mathcal{N}_\alpha = \bigcup_{n \ge N_\alpha} \bigcup_{\substack{u,v \in U_n \\ |u-v| = 2^{-n}}} \big[\rho(X_u, X_v) > 2^{-\lambda n}\big]$$

has $\mathbb{P}[\mathcal{N}_\alpha] < \alpha$. Its complement $\Omega_\alpha = \Omega \setminus \mathcal{N}_\alpha$ has measure

$$\mathbb{P}[\Omega_\alpha] > 1 - \alpha.$$

A point $\omega \in \Omega_\alpha$ has the property that whenever $n > N_\alpha$ and $u, v \in U_n$ have distance $|u-v| = 2^{-n}$ then

$$\rho\big(X_u(\omega), X_v(\omega)\big) \le 2^{-\lambda n}. \tag{$*$}$$

---

[16] For instance, $\lambda = \beta/2p$.

Let K be a compact subset of U, and let us start on the last claim by showing that $\{u \mapsto X_u(\omega) : \omega \in \Omega_\alpha\}$ *is uniformly equicontinuous on* $U_\infty \cap K$. To this end let $\epsilon > 0$ be given. There is an $n_0 > N_\alpha$ such that

$$2^{-\lambda n_0} < \epsilon \cdot (1 - 2^{-\lambda}) \cdot 2^{\lambda - 1} .$$

Note that this number depends only on α, ϵ, and the constants of inequality (A.2.4). Next let n_1 be so large that 2^{-n_1} is smaller than the distance of K from the complement of U, and let n_2 be so large that K is contained in the centered ball (the shape of a box) of diameter (side) $2n_2$. We respond to ϵ by setting

$$n = n_0 \vee n_1 \vee n_2 \quad \text{and} \quad \delta = 2^{-n} .$$

Clearly δ was manufactured from $\epsilon, K\alpha, p, C, \beta$ alone. We shall show that $|u - v| < \delta$ implies $\rho\big(X_u(\omega), X_v(\omega)\big) \le \epsilon$ for all $\omega \in \Omega_\alpha$ and $u, v \in K \cap U_\infty$. Now if $u, v \in U_\infty$, then there is a "mesh-size" $m \ge n$ such that both u and v belong to U_m. Write $u = u_m$ and $v = v_m$. There exist $u_{m-1}, v_{m-1} \in U_{m-1}$

with $$|u_m - u_{m-1}|_\infty \le 2^{-m} , \quad |v_m - v_{m-1}| \le 2^{-m} ,$$

and $$|u_{m-1} - v_{m-1}| \le |u_m - v_m| .$$

Namely, if $u = (k_1 2^{-m}, \ldots, k_d 2^{-m})$ and $v = (\ell_1 2^{-m}, \ldots, \ell_d 2^{-m})$, say, we add or subtract 1 from an odd k_δ according as $k_\delta - \ell_\delta$ is strictly positive or negative; and if k_δ is even or if $k_\delta - \ell_\delta = 0$, we do nothing. Then we go through the same procedure with v. Since $\delta \le 2^{n_1}$, the (box-shaped) balls with radius 2^{-n} about u_m, v_m lie entirely inside U, and then so do the points u_{m-1}, v_{m-1}. Since $\delta \le 2^{-n_2}$, they actually belong to U_{m-1}. By the same token there exist $u_{m-2}, v_{m-2} \in U_{m-2}$

with $$|u_{m-1} - u_{m-2}| \le 2^{-m-1} , \quad |v_{m-1} - v_{m-2}| \le 2^{-m-1} ,$$

and $$|u_{m-2} - v_{m-2}| \le |u_{m-1} - v_{m-1}| .$$

Continue on. Clearly $u_n = v_n$. In view of $(*)$ we have, for $\omega \in \Omega_\alpha$,

$$\begin{aligned}
\rho\big(X_u(\omega), X_v(\omega)\big) &\le \rho\big(X_u(\omega), X_{u_{m-1}}(\omega)\big) + \ldots + \rho\big(X_{u_{n+1}}(\omega), X_{u_n}(\omega)\big) \\
&\quad + \rho\big(X_{u_n}(\omega), X_{v_n}(\omega)\big) \\
&\quad + \rho\big(X_{v_n}(\omega), X_{v_{n+1}}(\omega)\big) + \ldots + \rho\big(X_{v_{m-1}}(\omega), X_v(\omega)\big) \\
&\le 2^{-m\lambda} + 2^{-(m-1)\lambda} + \ldots + 2^{-(n+1)\lambda} \\
&\quad + 0 \\
&\quad + 2^{-(n+1)\lambda} + \ldots + 2^{-(m-1)\lambda} + 2^{-m\lambda} \\
&\le 2 \sum_{i=n+1}^{\infty} (2^{-\lambda})^i = 2 \cdot 2^{-\lambda n} \cdot \big(2^{-\lambda}/(1 - 2^{-\lambda})\big) \le \epsilon .
\end{aligned}$$

To summarize: the family $\{u \mapsto X_u(\omega) : \omega \in \Omega_\alpha\}$ of E-valued functions is uniformly equicontinuous on every relatively compact subset K of U_∞.

Now set $\Omega_0 = \bigcup_n \Omega_{1/n}$. For every $\omega \in \Omega_0$, the map $u \mapsto X_u(\omega)$ is uniformly continuous on relatively compact subsets of U_∞ and thus has a unique continuous extension to all of U. Namely, for arbitrary $u \in U$ we set

$$X'_u(\omega) \stackrel{\text{def}}{=} \lim \left\{X_q(\omega) : U_\infty \ni q \to u\right\}, \qquad \omega \in \Omega_0 .$$

This limit exists, since $\left\{X_q(\omega) : U_\infty \ni q \to u\right\}$ is Cauchy and E is complete. In the points ω of the negligible set $\mathcal{N} = \Omega_0^c = \bigcap_\alpha \mathcal{N}_\alpha$ we set X'_u equal to some fixed point $x_0 \in E$. From inequality (A.2.4) it is plain that $X'_u = X_u$ almost surely. The resulting selection meets the description; it is, for instance, an easy exercise to check that the δ given above as a response to K and ϵ serves as well to show the uniform equicontinuity of the family $\left\{u \mapsto X'_u(\omega) : \omega \in \Omega_\alpha\right\}$ of functions on K. ∎

Exercise A.2.39 The proof above shows that there is a negligible set $\mathcal{N}$ such that, for every $\omega \notin \mathcal{N}$, $q \mapsto X_q(\omega)$ is uniformly continuous on every bounded set of dyadic rationals in U.

Exercise A.2.40 Assume that the set U, while possibly not open, is contained in the closure of its interior. Assume further that the family $\{X_u : u \in U\}$ satisfies merely, for some fixed $p > 0, \beta > 0$:

$$\limsup_{U \ni v, v' \to u} \frac{\mathbb{E}\left[\rho(X_v, X_{v'})^p\right]}{\left|v - v'\right|^{d+\beta}} < \infty \qquad \forall\, u \in U .$$

Again a modification can be found that is continuous in $u \in U$ for all $\omega \in \Omega$.

Exercise A.2.41 Any two continuous modifications X'_u, X''_u are indistinguishable in the sense that the set $\{\omega : \exists u \in U \text{ with } X'_u(\omega) \neq X''_u(\omega)\}$ is negligible.

Lemma A.2.42 (Taylor's Formula) *Suppose $D \subset \mathbb{R}^d$ is open and convex and $\Phi : D \to \mathbb{R}$ is n-times continuously differentiable. Then*[17]

$$\begin{aligned}\Phi(\boldsymbol{z} + \boldsymbol{\Delta}) - \Phi(\boldsymbol{z}) &= \Phi_{;\eta}(\boldsymbol{z}) \cdot \Delta^\eta + \int_0^1 (1-\lambda)\, \Phi_{;\eta\theta}\big(\boldsymbol{z}+\lambda\boldsymbol{\Delta}\big)\, d\lambda \cdot \Delta^\eta \Delta^\theta \\ &= \sum_{\nu=1}^{n-1} \frac{1}{\nu!} \Phi_{;\eta_1\ldots\eta_\nu}(\boldsymbol{z}) \cdot \Delta^{\eta_1} \cdots \Delta^{\eta_\nu} \\ &\quad + \int_0^1 \frac{(1-\lambda)^{n-1}}{(n-1)!}\, \Phi_{;\eta_1\ldots\eta_n}\big(\boldsymbol{z}+\lambda\boldsymbol{\Delta}\big)\, d\lambda \cdot \Delta^{\eta_1} \cdots \Delta^{\eta_n}\end{aligned}$$

for any two points $\boldsymbol{z}, \boldsymbol{z} + \boldsymbol{\Delta} \in D$.

---

[17] Subscripts after semicolons denote partial derivatives, e.g.,

$$\Phi_{;\eta} \stackrel{\text{def}}{=} \frac{\partial \Phi}{\partial x^\eta} \quad \text{and} \quad \Phi_{;\eta\theta} \stackrel{\text{def}}{=} \frac{\partial^2 \Phi}{\partial x^\eta \partial x^\theta} .$$

Summation over repeated indices in opposite positions is implied by Einstein's convention, which is adopted throughout.

A.2.43 Let $1 \le p < \infty$, and set $n = [p]$ and $\epsilon = p - n$. Then for $z, \delta \in \mathbb{R}$

$$|z+\delta|^p = |z|^p + \sum_{\nu=1}^{n-1} \binom{p}{\nu} |z|^{p-\nu} (\operatorname{sgn} z)^\nu \cdot \delta^\nu$$
$$+ \int_0^1 n(1-\lambda)^{n-1} \binom{p}{n} |z+\lambda\delta|^\epsilon \big(\operatorname{sgn}(z+\lambda\delta)\big)^n \, d\lambda \cdot \delta^n ,$$

where $\binom{p}{\nu} \stackrel{\text{def}}{=} \frac{p(p-1)\cdots(p-\nu+1)}{\nu!}$ and $\operatorname{sgn} z \stackrel{\text{def}}{=} \begin{cases} 1 & \text{if } z > 0 \\ 0 & \text{if } z = 0 \\ -1 & \text{if } z < 0 . \end{cases}$

Differentiation

Definition A.2.44 (Big O and Little o) *Let N, D, s be real-valued functions depending on the same arguments $u, v, \ldots$. One says*

$$\text{“}N = O(D) \text{ as } s \to 0\text{” if } \lim_{\delta \to 0} \sup \left\{ \frac{N(u,v,\ldots)}{D(u,v,\ldots)} : s(u,v,\ldots) \le \delta \right\} < \infty ,$$

$$\text{“}N = o(D) \text{ as } s \to 0\text{” if } \lim_{\delta \to 0} \sup \left\{ \frac{N(u,v,\ldots)}{D(u,v,\ldots)} : s(u,v,\ldots) \le \delta \right\} = 0 .$$

If $D = s$, one simply says “$N = O(D)$” or “$N = o(D)$,” respectively.

This nifty convention eases many arguments, including the usual definition of differentiability, also called ***Fréchet differentiability***:

Definition A.2.45 *Let F be a map from an open subset U of a seminormed space E to another seminormed space S. F is **differentiable at a point** $u \in U$ if there exists a bounded*[18] *linear operator $DF[u] : E \to S$, written $\eta \mapsto DF[u]\cdot\eta$ and called the **derivative of** F **at** u, such that the **remainder** RF, defined by*

$$F(v) - F(u) = DF[u]\cdot(v-u) + RF[v;u] ,$$

has $\quad \|RF[v;u]\|_S = o(\|v-u\|_E) \quad$ *as* $v \to u$.

*If F is differentiable at all points of U, it is called differentiable on U or simply **differentiable**; if in that case $u \mapsto DF[u]$ is continuous in the operator norm, then F is **continuously differentiable**; if in that case $\| RF[v;u] \|_S = o(\| v-u \|_E)$,*[19] *$F$ is **uniformly differentiable**.*

*Next let $\mathfrak{F}$ be a whole family of maps from U to S, all differentiable at $u \in U$. Then $\mathfrak{F}$ is **equidifferentiable** at u if* $\sup\{\| RF[v;u] \|_S : F \in \mathfrak{F}\}$ *is*

---

18 This means that the operator norm $\| DF[u] \|_{E \to S} \stackrel{\text{def}}{=} \sup\{\| DF[u]\cdot\eta \|_S : \|\eta\|_E \le 1\}$ is finite.

19 That is to say $\sup\{\| RF[v;u] \|_S / \| v-u \|_E : \| v-u \|_E \le \delta\} \xrightarrow[\delta \to 0]{} 0$, which explains the word “uniformly.”

$o(\|v-u\|_E)$ *as* $v \to u$, *and* ***uniformly equidifferentiable*** *if the previous supremum is* $o(\|v-u\|_E)$ *as* $\|v-u\|_E \to 0$.

Exercise A.2.46 (i) Establish the usual rules of differentiation. (ii) If F is differentiable at u, then $\|F(v)-F(u)\|_S = O(\|v-u\|_E)$ as $v \to u$. (iii) Suppose now that U is open and convex and F is differentiable on U. Then F is Lipschitz with constant L if and only if $\|DF[u]\|_{E\to S}$ is bounded; and in that case

$$L = \sup_u \|DF[u]\|_{E\to S} .$$

(iv) If F is continuously differentiable on U, then it is uniformly differentiable on every relatively compact subset of U; furthermore, there is this representation of the remainder:

$$RF[v;u] = \int_0^1 (DF[u+\lambda(v-u)] - DF[u])\cdot(v-u)\, d\lambda .$$

Exercise A.2.47 For differentiable $f : \mathbb{R} \to \mathbb{R}$, $Df[x]$ is multiplication by $f'(x)$.

Now suppose $F = F[u,x]$ is a differentiable function of two variables, $u \in U$ and $x \in V \subset X$, X being another seminormed space. This means of course that F is differentiable on $U \times V \subset E \times X$. Then $DF[u,x]$ has the form

$$\begin{aligned} DF[u,x]\cdot\begin{pmatrix}\eta\\ \xi\end{pmatrix} &= \Big(D_1F[u,x], D_2F[u,x]\Big)\cdot\begin{pmatrix}\eta\\ \xi\end{pmatrix} \\ &= D_1F[u,x]\cdot\eta + D_2F[u,x]\cdot\xi , \qquad \eta \in E, \xi \in X , \end{aligned}$$

where $D_1F[u,x]$ is the ***partial in the u-direction*** and $D_2F[u,x]$ the ***partial in the x-direction.*** In particular, when the arguments u, x are real we often write $F_{;1} = F_{;u} \stackrel{\text{def}}{=} D_1F$, $F_{;2} = F_{;x} \stackrel{\text{def}}{=} D_2F$, etc.

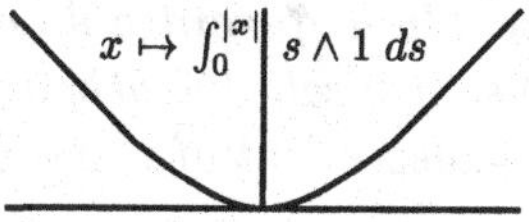

Example A.2.48 — of Trouble Consider a differentiable function f on the line of not more than linear growth, for example $f(x) \stackrel{\text{def}}{=} \int_0^{|x|} s \wedge 1\, ds$. One hopes that composition with f, which takes ϕ to $F[\phi] \stackrel{\text{def}}{=} f \circ \phi$, might define a Fréchet differentiable map F from $L^p(\mathbb{P})$ to itself. Alas, it does not. Namely, if $DF[0]$ exists, it must equal multiplication by $f'(0)$, which in the example above equals zero – but then $RF(0,\phi) = F[\phi] - F[0] - DF[0]\cdot\phi = F[\phi] = f \circ \phi$ does not go to zero faster in $L^p(\mathbb{P})$-mean $\|\ \|_p$ than does $\|\phi - 0\|_p$ – simply take ϕ through a sequence of indicator functions converging to zero in $L^p(\mathbb{P})$-mean.

F is, however, differentiable, even uniformly so, as a map from $L^p(\mathbb{P})$ to $L^{p^\circ}(\mathbb{P})$ for any p° *strictly smaller* than p, whenever the derivative f' is continuous and bounded. Indeed, by Taylor's formula of order one (see lemma A.2.42 on page 387)

$$\begin{aligned} F[\psi] = F[\phi] &+ f'(\phi)\cdot(\psi-\phi) \\ &+ \int_0^1 \Big[f'(\phi+\sigma(\psi-\phi)) - f'(\phi)\Big]\, d\sigma \cdot (\psi-\phi) , \end{aligned}$$

whence, with Hölder's inequality and $1/p^\circ = 1/p + 1/r$ defining r,

$$\|RF[\psi;\phi]\|_{p^\circ} \le \left\| \int_0^1 \Big[f'(\phi + \sigma(\psi-\phi)) - f'(\phi) \Big]\, d\sigma \right\|_r \cdot \|\psi-\phi\|_p \,.$$

The first factor tends to zero as $\|\psi-\phi\|_p \to 0$, due to theorem A.8.6, so that $\|RF[\psi;\phi]\|_{p^\circ} = o(\|\psi-\phi\|_p)$. Thus F is uniformly differentiable as a map from $L^p(\mathbb{P})$ to $L^{p^\circ}(\mathbb{P})$, with[20] $DF[\phi] = f' \circ \phi$.

Note the following phenomenon: the derivative $\xi \mapsto DF[\phi]\cdot\xi$ is actually a continuous linear map from $L^p(\mathbb{P})$ to itself whose operator norm is bounded independently of ϕ by $\|f'\| \overset{\text{def}}{=} \sup_x |f'(x)|$. It is just that the remainder $RF[\psi;\phi]$ is $o(\|\psi-\phi\|_p)$ only if it is measured with the weaker seminorm $\|\ \|_{p^\circ}$. The example gives rise to the following notion:

Definition A.2.49 *(i) Let $(S, \|\ \|_S)$ be a seminormed space and $\|\ \|_S^\circ \le \|\ \|_S$ a weaker seminorm. A map F from an open subset $\mathcal{U}$ of a seminormed space $(E, \|\ \|_E)$ to S is $\|\ \|_S^\circ$-weakly differentiable at $u \in \mathcal{U}$ if there exists a bounded*[18] *linear map $DF[u] : E \to S$ such that*

$$F[v] = F[u] + DF[u]\cdot(v-u) + RF[u;v] \qquad \forall\, v \in \mathcal{U}\,,$$

with $\|RF[u;v]\|_S^\circ = o(\|v-u\|_E)$ *as* $v \to u$ *, i.e.,* $$\frac{\|RF[u;v]\|_S^\circ}{\|v-u\|_E} \xrightarrow[v\to u]{} 0\,.$$

(ii) Suppose that S comes equipped with a family $\mathfrak{N}^\circ$ of seminorms $\|\ \|_S^\circ \le \|\ \|_S$ such that $\|x\|_S = \sup\{\|x\|_S^\circ : \|\ \|_S^\circ \in \mathfrak{N}^\circ\}$ $\forall\, x \in S$. If F is $\|\ \|_S^\circ$-weakly differentiable at $u \in \mathcal{U}$ for every $\|\ \|_S^\circ \in \mathfrak{N}^\circ$, then we call F ***weakly differentiable at*** *u. If F is weakly differentiable at every $u \in \mathcal{U}$, it is simply called weakly differentiable; if, moreover, the decay of the remainder is independent of $u, v \in U$:*

$$\sup\left\{ \frac{\|RF[u;v]\|_S^\circ}{\delta} : u, v \in U\,,\ \|v-u\|_E < \delta \right\} \xrightarrow[\delta\to 0]{} 0$$

for every $\|\ \|_S^\circ \in \mathfrak{N}^\circ$, then F is ***uniformly weakly differentiable on*** *U.*

Here is a reprise of the calculus for this notion:

Exercise A.2.50 (a) The linear operator $DF[u]$ of (i) if extant is unique, and $F \to DF[u]$ is linear. To say that F is weakly differentiable means that F is, for every $\|\ \|_S^\circ \in \mathfrak{N}^\circ$, Fréchet differentiable as a map from $(E, \|\ \|_E)$ to $(S, \|\ \|_S^\circ)$ and has a derivative that is continuous as a linear operator from $(E, \|\ \|_E)$ to $(S, \|\ \|_S)$. (b) Formulate and prove the product rule and the chain rule for weak differentiability. (c) Show that if F is $\|\ \|_S^\circ$-weakly differentiable, then for all $u, v \in E$

$$\|F[v] - F[u]\|_S^\circ \le \sup\{ \|DF[u]\| : u \in E \} \cdot \|v-u\|_E \,.$$

[20] $DF[\phi]\cdot\xi = f' \circ \phi \cdot \xi$. In other words, $DF[\phi]$ is multiplication by $f' \circ \phi$.

A.3 Measure and Integration

σ-Algebras

A ***measurable space*** is a set F equipped with a σ-algebra $\mathcal{F}$ of subsets of F. A ***random variable*** is a map f whose domain is a measurable space $(F,\mathcal{F})$ and which takes values in another measurable space $(G,\mathcal{G})$. It is understood that a random variable f is ***measurable***: the inverse image $f^{-1}(G_0)$ of every set $G_0 \in \mathcal{G}$ belongs to $\mathcal{F}$. If there is need to specify which σ-algebra on the domain is meant, we say "***f is measurable on $\mathcal{F}$***" and write $f \in \mathcal{F}$. If we want to specify both σ-algebras involved, we say "f is $\mathcal{F}/\mathcal{G}$-measurable" and write $f \in \mathcal{F}/\mathcal{G}$. If $G = \mathbb{R}$ or $G = \mathbb{R}^n$, then it is understood that $\mathcal{G}$ is the σ-algebra of Borel sets (see below). A random variable is ***simple*** if it takes only finitely many different values.

The intersection of any collection of σ-algebras is a σ-algebra. Given some property P of σ-algebras, we may therefore talk about the ***σ-algebra generated by*** P: it is the intersection of all σ-algebras having P. We assume here that there is at least one σ-algebra having P, so that the collection whose intersection is taken is not empty – the σ-algebra of all subsets will usually do. Given a collection Φ of functions on F with values in measurable spaces, the ***σ-algebra generated by Φ*** is the smallest σ-algebra on which every function $\phi \in \Phi$ is measurable. For instance, if F is a topological space, there are the σ-algebra $\mathcal{B}^*(F)$ of ***Baire sets*** and the σ-algebra $\mathcal{B}^\bullet(F)$ of ***Borel sets***. The former is the smallest σ-algebra on which all continuous real-valued functions are measurable, and the latter is the generally larger σ-algebra generated by the open sets.[13] Functions measurable on $\mathcal{B}^*(F)$ or $\mathcal{B}^\bullet(F)$ are called ***Baire functions*** or ***Borel functions***, respectively.

Exercise A.3.1 On a metrizable space the Baire and Borel functions coincide. In particular, on $\mathbb{R}^n$ and on the path spaces $\mathscr{C}^n$ or the Skorohod spaces $\mathscr{D}^n$, $n = 1, 2, \ldots$, there is no distinction between the two.

Sequential Closure

Consider random variables $f_n : (F,\mathcal{F}) \to (G,\mathcal{G})$, where $\mathcal{G}$ is the Borel σ-algebra of some topology on G. If $f_n \to f$ pointwise on F, then f is again a random variable. This is because for any open set $U \subset G$, $f^{-1}(U) = \bigcup_N \bigcap_{n>N} f_n^{-1}(U) \in \mathcal{F}$ – since $\{B : f^{-1}(B) \in \mathcal{F}\}$ is a σ-algebra containing the open sets, it contains the Borels. A similar argument applies when $\mathcal{G}$ is the Baire σ-algebra $\mathcal{B}^*(G)$. Inasmuch as this permanence property under pointwise limits of sequences is the main merit of the notions of σ-algebra and $\mathcal{F}/\mathcal{G}$-measurability, it deserves a bit of study of its own:

A collection $\mathcal{B}$ of functions defined on some set E and having values in a topological space is called ***sequentially closed*** if the limit of any pointwise convergent sequence in $\mathcal{B}$ belongs to $\mathcal{B}$ as well. In most of the applications of this notion the functions in $\mathcal{B}$ are considered ***numerical***, i.e., they are allowed to take values in the extended reals $\overline{\mathbb{R}}$. For example, the collection of $\mathcal{F}/\mathcal{G}$-measurable random variables above, the collection of $[\![\ ]\!]^*$-measurable processes, and the collection of $[\![\ ]\!]^*$-measurable *sets* each are sequentially closed.

The intersection of any family of sequentially closed collections of functions on E plainly is sequentially closed. If $\mathcal{E}$ is any collection of functions, then there is thus a smallest sequentially closed collection $\mathcal{E}^\sigma$ of functions containing $\mathcal{E}$, to wit, the intersection of *all* sequentially closed collections containing $\mathcal{E}$. $\mathcal{E}^\sigma$ can be constructed by transfinite induction as follows. Set $\mathcal{E}_0 \stackrel{\text{def}}{=} \mathcal{E}$. Suppose that $\mathcal{E}_\alpha$ has been defined for all ordinals $\alpha < \beta$. If β is the successor of α, then define $\mathcal{E}_\beta$ to be the set of all functions that are limits of a sequence in $\mathcal{E}_\alpha$; if β is not a successor, then set $\mathcal{E}_\beta \stackrel{\text{def}}{=} \bigcup_{\alpha<\beta} \mathcal{E}_\alpha$. Then $\mathcal{E}_\beta = \mathcal{E}^\sigma$ for all β that exceed the first uncountable ordinal $\aleph_1$.

It is reasonable to call $\mathcal{E}^\sigma$ the ***sequential closure*** or ***sequential span*** of $\mathcal{E}$. If $\mathcal{E}$ is considered as a collection of numerical (real-valued) functions, and if this point must be emphasized, we shall denote the sequential closure by $\mathcal{E}^\sigma_{\overline{\mathbb{R}}}$ ($\mathcal{E}^\sigma_{\mathbb{R}}$). It will generally be clear from the context which is meant.

Exercise A.3.2 (i) Every $f \in \mathcal{E}^\sigma$ is contained in the sequential closure of a countable subcollection of $\mathcal{E}$. (ii) $\mathcal{E}$ is called ***σ-finite*** if it contains a countable subcollection whose pointwise supremum is everywhere strictly positive. Show: if $\mathcal{E}$ is a ring of sets or a vector lattice closed under chopping or an algebra of bounded functions, then $\mathcal{E}$ is σ-finite if and only if $1 \in \mathcal{E}^\sigma$. (iii) The collection of real-valued functions in $\mathcal{E}^\sigma_{\overline{\mathbb{R}}}$ coincides with $\mathcal{E}^\sigma_{\mathbb{R}}$.

Lemma A.3.3 *(i) If $\mathcal{E}$ is a vector space, algebra, or vector lattice of real-valued functions, then so is its sequential closure $\mathcal{E}^\sigma_{\mathbb{R}}$. (ii) If $\mathcal{E}$ is an algebra of bounded functions or a vector lattice closed under chopping, then $\mathcal{E}^\sigma_{\mathbb{R}}$ is both. Furthermore, if $\mathcal{E}$ is σ-finite, then the collection $\mathcal{E}^\sigma_e$ of sets in $\mathcal{E}^\sigma$ is the σ-algebra generated by $\mathcal{E}$, and $\mathcal{E}^\sigma$ consists precisely of the functions measurable on $\mathcal{E}^\sigma_e$.*

Proof. (i) Let $*$ stand for $+, -, \cdot, \vee, \wedge$, etc. Suppose $\mathcal{E}$ is closed under $*$.

The collection $$\mathcal{E}^* \stackrel{\text{def}}{=} \{f : f * \phi \in \mathcal{E}^\sigma \quad \forall \phi \in \mathcal{E}\} \tag{$*$}$$
then contains $\mathcal{E}$, and it is sequentially closed. Thus it contains $\mathcal{E}^\sigma$. This shows that

the collection $$\mathcal{E}^{**} \stackrel{\text{def}}{=} \{g : f * g \in \mathcal{E}^\sigma \quad \forall f \in \mathcal{E}^\sigma\} \tag{$**$}$$

contains $\mathcal{E}$. This collection is evidently also sequentially closed, so it contains $\mathcal{E}^\sigma$. That is to say, $\mathcal{E}^\sigma$ is closed under $*$ as well.

(ii) The constant 1 belongs to $\mathcal{E}^\sigma$ (exercise A.3.2). Therefore $\mathcal{E}^\sigma_e$ is not merely a σ-ring but a σ-algebra. Let $f \in \mathcal{E}^\sigma$. The set[13] $[f > 0] = \lim_{n\to\infty} 0 \vee \big((n \cdot f) \wedge 1\big)$, being the limit of an increasing bounded sequence, belongs to $\mathcal{E}^\sigma_e$. We conclude that for every $r \in \mathbb{R}$ and $f \in \mathcal{E}^\sigma$ the set[13]

$[f > r] = [f - r > 0]$ belongs to $\mathcal{E}_e^\sigma$: f is measurable on $\mathcal{E}_e^\sigma$. Conversely, if f is measurable on $\mathcal{E}_e^\sigma$, then it is the limit of the functions

$$\sum_{|\nu|\le 2^n} \nu 2^{-n}\big[\nu 2^{-n} < f \le (\nu+1)2^{-n}\big]$$

in $\mathcal{E}^\sigma$ and thus belongs to $\mathcal{E}^\sigma$. Lastly, since every $\phi \in \mathcal{E}$ is measurable on $\mathcal{E}_e^\sigma$, $\mathcal{E}_e^\sigma$ contains the σ-algebra $\mathcal{E}^\Sigma$ generated by $\mathcal{E}$; and since the $\mathcal{E}^\Sigma$-measurable functions form a sequentially closed collection containing $\mathcal{E}$, $\mathcal{E}_e^\sigma \subset \mathcal{E}^\Sigma$. ∎

Theorem A.3.4 (The Monotone Class Theorem) *Let $\mathcal{V}$ be a collection of real-valued functions on some set that is closed under pointwise limits of increasing or decreasing sequences – this makes it a* **monotone class**. *Assume further that $\mathcal{V}$ forms a real vector space and contains the constants. With any subcollection $\mathcal{M}$ of bounded functions that is closed under multiplication – a* **multiplicative class** *– $\mathcal{V}$ then contains every real-valued function measurable on the σ-algebra $\mathcal{M}^\Sigma$ generated by $\mathcal{M}$.*

Proof. The family $\mathcal{E}$ of all finite linear combinations of functions in $\mathcal{M}\cup\{1\}$ is an algebra of bounded functions and is contained in $\mathcal{V}$. Its uniform closure $\overline{\mathcal{E}}$ is contained in $\mathcal{V}$ as well. For if $\mathcal{E} \ni f_n \to \overline{f}$ uniformly, we may without loss of generality assume that $\|\overline{f} - f_n\|_\infty < 2^{-n}/4$. The sequence $f_n - 2^{-n} \in \mathcal{E}$ then converges increasingly to $\overline{f}$. $\overline{\mathcal{E}}$ is a vector lattice (theorem A.2.2). Let $\mathcal{E}^{\uparrow\downarrow}$ denote the smallest collection of functions that contains $\mathcal{E}$ and is closed under pointwise limits of monotone sequences; it is evidently contained in $\mathcal{V}$. We see as in $(*)$ and $(**)$ above that $\mathcal{E}^{\uparrow\downarrow}$ is a vector lattice; namely, the collections $\mathcal{E}^*$ and $\mathcal{E}^{**}$ from the proof of lemma A.3.3 are closed under limits of monotone sequences. Since $\lim f_n = \sup_N \inf_{n>N} f_n$, $\mathcal{E}^{\uparrow\downarrow}$ is sequentially closed. If f is measurable on $\mathcal{M}^\Sigma$, it is evidently measurable on $\mathcal{E}^\Sigma = \mathcal{E}_e^\sigma$ and thus belongs to $\mathcal{E}^\sigma \subset \mathcal{E}^{\uparrow\downarrow} \subset \mathcal{V}$ (lemma A.3.3). ∎

Exercise A.3.5 (The Complex Bounded Class Theorem) Let $\mathcal{V}$ be a complex vector space of complex-valued functions on some set, and assume that $\mathcal{V}$ contains the constants and is closed under taking limits of bounded pointwise convergent sequences. With any subfamily $\mathcal{M} \subset \mathcal{V}$ that is closed under multiplication and complex conjugation – a ***complex multiplicative class*** – $\mathcal{V}$ contains every bounded complex-valued function that is measurable on the σ-algebra $\mathcal{M}^\Sigma$ generated by $\mathcal{M}$. In consequence, if two σ-additive measures of totally finite variation agree on the functions of $\mathcal{M}$, then they agree on $\mathcal{M}^\Sigma$.

Exercise A.3.6 On a topological space E the class of Baire functions is the sequential closure of the class $C_b(E)$ of bounded continuous functions. If E is completely regular, then the class of Borel functions is the sequential closure of the set of differences of lower semicontinuous functions.

Exercise A.3.7 Suppose that $\mathcal{E}$ is a self-confined vector lattice closed under chopping or an algebra of bounded functions on some set E (see exercise A.2.6). Let us denote by $\mathcal{E}_{00}^\sigma$ the smallest collection of functions on f that is closed under taking pointwise limits of bounded $\mathcal{E}$-confined sequences. Show: (i) $f \in \mathcal{E}_{00}^\sigma$ if and only if $f \in \mathcal{E}^\sigma$ is bounded and $\mathcal{E}$-confined; (ii) $\mathcal{E}_{00}^\sigma$ is both a vector lattice closed under chopping and an algebra.

Measures and Integrals

A σ-additive measure on the σ-algebra $\mathcal{F}$ is a function $\mu : \mathcal{F} \to \mathbb{R}$[21] that satisfies $\mu(\bigcup_n A_n) = \sum_n \mu(A_n)$ for every disjoint sequence (A_n) in $\mathcal{F}$. Σ-algebras have no *raison d'être* but for the σ-additive measures that live on them. However, rare is the instance that a measure appears on a σ-algebra. Rather, measures come naturally as linear functionals on some small space $\mathcal{E}$ of functions (Radon measures, Haar measure) or as set functions on a ring $\mathcal{A}$ of sets (Lebesgue measure, probabilities). They still have to undergo a lengthy extension procedure before their domain contains the σ-algebra generated by $\mathcal{E}$ or $\mathcal{A}$ and before they can integrate functions measurable on that.

Set Functions A ***ring of sets*** on a set F is a collection $\mathcal{A}$ of subsets of F that is closed under taking relative complements and finite unions, and then under taking finite intersections. A ring is an ***algebra*** if it contains the whole ambient set F, and a ***δ-ring*** if it is closed under taking countable intersections (if both, it is a ***σ-algebra*** or ***σ-field***). A *measure* on the ring $\mathcal{A}$ is a σ-additive function $\mu : \mathcal{A} \to \mathbb{R}$ of finite variation. The additivity means that $\mu(A + A') = \mu(A) + \mu(A')$ for[13] $A, A', A + A' \in \mathcal{A}$. The σ-additivity means that $\mu(\bigcup_n A_n) = \sum_n \mu(A_n)$ for every disjoint sequence (A_n) of sets in $\mathcal{A}$ whose union A happens to belong to $\mathcal{A}$. In the presence of finite additivity this is equivalent with σ-continuity: $\mu(A_n) \to 0$ for every decreasing sequence (A_n) in $\mathcal{A}$ that has void intersection. The additive set function $\mu : \mathcal{A} \to \mathbb{R}$ has ***finite variation*** on $A \subset F$ if

$$\left\lvert\mu\right\rvert(A) \stackrel{\text{def}}{=} \sup\{\mu(A') - \mu(A'') : A', A'' \in \mathcal{A},\ A' + A'' \le A\}$$

is finite. To say that μ has finite variation means that $\lvert\mu\rvert(A) < \infty$ for all $A \in \mathcal{A}$. The function $\lvert\mu\rvert : \mathcal{A} \to \overline{\mathbb{R}}_+$ then is a positive σ-additive measure on $\mathcal{A}$, called the ***variation*** of μ. μ has ***totally finite variation*** if $\lvert\mu\rvert(F) < \infty$. A σ-additive set function on a σ-algebra automatically has totally finite variation. Lebesgue measure on the finite unions of intervals $(a, b]$ is an example of a measure that appears naturally as a set function on a ring of sets.

Radon measures are examples of measures that appear naturally as linear functionals on a space of functions. Let E be a locally compact Hausdorff space and $C_{00}(E)$ the set of continuous functions with compact support. A Radon measure is simply a linear functional $\mu : C_{00}(E) \to \mathbb{R}$ that is bounded on order-bounded (confined) sets.

Elementary Integrals The previous two instances of measures look so disparate that they are often treated quite differently. Yet a little teleological thinking reveals that they fit into a common pattern. Namely, while measuring sets is a pleasurable pursuit, integrating functions surely is what measure

[21] For numerical measures, i.e., measures that are allowed to take their values in the extended reals $\overline{\mathbb{R}}$, see exercise A.3.27.

theory ultimately is all about. So, given a measure μ on a ring $\mathcal{A}$ we immediately extend it by linearity to the linear combinations of the sets in $\mathcal{A}$, thus obtaining a linear functional on functions. Call their collection $\mathcal{E}[\mathcal{A}]$. This is the family of step functions ϕ over $\mathcal{A}$, and the linear extension is the natural one: $\mu(\phi)$ is the sum of the products height–of–step times μ-size–of–step.

In both instances we now face a linear functional $\mu : \mathcal{E} \to \mathbb{R}$. If μ was a Radon measure, then $\mathcal{E} = C_{00}(E)$; if μ came as a set function on $\mathcal{A}$, then $\mathcal{E} = \mathcal{E}[\mathcal{A}]$ and μ is replaced by its linear extension. In both cases the pair $(\mathcal{E}, \mu)$ has the following properties:

(i) *$\mathcal{E}$ is an algebra and vector lattice closed under chopping.* The functions in $\mathcal{E}$ are called ***elementary integrands***.

(ii) *μ is **σ-continuous***: $\mathcal{E} \ni \phi_n \downarrow 0$ pointwise implies $\mu(\phi_n) \to 0$.

(iii) *μ has **finite variation***: for all $\phi \ge 0$ in $\mathcal{E}$

$$\left|\mu\right|(\phi) \stackrel{\text{def}}{=} \sup\left\{|\mu(\psi)| : \psi \in \mathcal{E} \,,\ |\psi| \le \phi\right\}$$

is finite; in fact, $\left|\mu\right|$ extends to a σ-continuous positive[22] linear functional on $\mathcal{E}$, the ***variation*** of μ. We shall call such a pair $(\mathcal{E}, \mu)$ an elementary integral.

(iv) Actually, all elementary integrals that we meet in this book have a ***σ-finite domain*** $\mathcal{E}$ (exercise A.3.2). This property facilitates a number of arguments. We shall therefore subsume the requirement of σ-finiteness on $\mathcal{E}$ in the definition of an ***elementary integral*** $(\mathcal{E}, \mu)$.

Extension of Measures and Integration The reader is no doubt familiar with the way Lebesgue succeeded in 1905[23] to extend the length function on the ring of finite unions of intervals to many more sets, and with Caratheodory's generalization to positive σ-additive set functions μ on arbitrary rings of sets. The main tools are the inner and outer measures μ_* and μ^*.

Once the measure is extended there is still quite a bit to do before it can integrate functions. In 1918 the French mathematician Daniell noticed that many of the arguments used in the extension procedure for the set function and again in the integration theory of the extension are the same. He discovered a way of melding the extension of a measure and its integration theory into one procedure. This saves labor and has the additional advantage of being applicable in more general circumstances, such as the stochastic integral. We give here a short overview. This will furnish both notation and motivation for the main body of the book. For detailed treatments see for example [9] and [12]. The reader not conversant with Daniell's extension procedure can actually find it in all detail in chapter 3, if he takes Ω to consist of a single point.

Daniell's idea is really rather simple: get to the main point right away, the main point being the integration of functions. Accordingly, when given a

---

[22] A linear functional is called ***positive*** if it maps positive functions to positive numbers.
[23] A fruitful year – see page 9.

measure μ on the ring $\mathcal{A}$ of sets, extend it right away to the step functions $\mathcal{E}[\mathcal{A}]$ as above. In other words, in whichever form the elementary data appear, keep them as, or turn them into, an elementary integral.

Daniell saw further that Lebesgue's expand–contract construction of the outer measure of sets has a perfectly simple analog in an up–down procedure that produces an upper integral for functions. Here is how it works. Given a *positive* elementary integral $(\mathcal{E}, \mu)$, let $\mathcal{E}^{\uparrow}$ denote the collection of functions h on F that are pointwise suprema of some sequence in $\mathcal{E}$:

$$\mathcal{E}^{\uparrow} = \{h : \exists \phi_1, \phi_2, \ldots \text{ in } \mathcal{E} \text{ with } h = \sup_n \phi_n \} .$$

Since $\mathcal{E}$ is a lattice, the sequence (ϕ_n) can be chosen increasing, simply by replacing ϕ_n with $\phi_1 \vee \cdots \vee \phi_n$. $\mathcal{E}^{\uparrow}$ corresponds to Lebesgue's collection of open sets, which are countable suprema[13] of intervals. For $h \in \mathcal{E}^{\uparrow}$ set

$$\int^* h \, d\mu \stackrel{\text{def}}{=} \sup\left\{\int \phi \, d\mu : \mathcal{E} \ni \phi \leq h\right\} . \tag{A.3.1}$$

Similarly, let $\mathcal{E}_{\downarrow}$ denote the collection of functions k on the ambient set F that are pointwise infima of some sequence in $\mathcal{E}$, and set

$$\int_* k \, d\mu \stackrel{\text{def}}{=} \inf\left\{\int \phi \, d\mu : \mathcal{E} \ni \phi \geq k\right\} .$$

Due to the σ-continuity of μ, $\int^* d\mu$ and $\int_* d\mu$ are σ-continuous on $\mathcal{E}^{\uparrow}$ and $\mathcal{E}_{\downarrow}$, respectively, in this sense: $\mathcal{E}^{\uparrow} \ni h_n \uparrow h$ implies $\int^* h_n \, d\mu \to \int^* h \, d\mu$ and $\mathcal{E}_{\downarrow} \ni k_n \downarrow k$ implies $\int_* k_n \, d\mu \to \int_* k \, d\mu$. Then set for arbitrary functions $f : F \to \overline{\mathbb{R}}$

$$\int^* f \, d\mu \stackrel{\text{def}}{=} \inf\left\{\int^* h \, d\mu : h \in \mathcal{E}^{\uparrow}, h \geq f\right\} \quad \text{and}$$

$$\int_* f \, d\mu \stackrel{\text{def}}{=} \sup\left\{\int_* k \, d\mu : k \in \mathcal{E}_{\downarrow}, k \leq f\right\} \quad \left(= -\int^* -f \, d\mu \leq \int^* f \, d\mu\right).$$

$\int^* d\mu$ and $\int_* d\mu$ are called the ***upper integral*** and ***lower integral*** associated with μ, respectively. Their restrictions to sets are precisely the ***outer and inner measures*** μ^* and μ_* of Lebesgue–Caratheodory. The upper integral is ***countably subadditive***,[24] and the lower integral is countably superadditive. A function f on F is called μ-integrable if $\int^* f \, d\mu = \int_* f \, d\mu$, and the common value is the integral $\int f \, d\mu$. The idea is of course that on the integrable functions the integral is countably additive. The all-important Dominated Convergence Theorem follows from the countable additivity with little effort.

The procedure outlined is intuitively just as appealing as Lebesgue's, and much faster. Its real benefit lies in a slight variant, though, which is based on

---

[24] $\int^* \sum_{n=1}^{\infty} f_n \leq \sum_{n=1}^{\infty} \int^* f_n$.

the easy observation that a function f is ***μ-integrable*** if and only if there is a sequence (ϕ_n) of elementary integrands with $\int^* |f - \phi_n| \, d\mu \to 0$, and then $\int f \, d\mu = \lim \int \phi_n \, d\mu$. So we might as well *define* integrability and the integral this way: the integrable functions are the closure of $\mathcal{E}$ under the seminorm $f \mapsto \| f \|_\mu^* \stackrel{\text{def}}{=} \int^* |f| \, d\mu$, and the integral is the extension by continuity. One now does not even have to introduce the lower integral, saving labor, and the proofs of the main results speed up some more.

Let us rewrite the definition of the ***Daniell mean*** $\| \ \|_\mu^*$:

$$\| f \|_\mu^* = \inf_{|f| \le h \in \mathcal{E}^\uparrow} \ \sup_{\phi \in \mathcal{E}, |\phi| \le h} \left| \int \phi \, d\mu \right| . \tag{D}$$

As it stands, this makes sense even if μ is not positive. It must merely have finite variation, in order that $\| \ \|_\mu^*$ be finite on $\mathcal{E}$. Again the integral can be defined simply as the extension by continuity of the elementary integral. The famous limit theorems are all consequences of two properties of the mean: it is countably subadditive on positive functions and additive on $\mathcal{E}_+$, as it agrees with the variation $|\mu|$ there.

As it stands, (D) even makes sense for measures μ that take values in some Banach space F, or even some space more general than that; one only needs to replace the absolute value in (D) by the norm or quasinorm of F. Under very mild assumptions on F, ordinary integration theory with its beautiful limit results can be established simply by repeating the classical arguments. In chapter 3 we go this route to do stochastic integration.

The main theorems of integration theory use only the properties of the mean $\| \ \|_\mu^*$ listed in definition 3.2.1. The proofs given in section 3.2 apply of course *a fortiori* in the classical case and produce the Monotone and Dominated Convergence Theorems, etc. Functions and sets that are $\| \ \|_\mu^*$-negligible or $\| \ \|_\mu^*$-measurable[25] are usually called ***μ-negligible*** or ***μ-measurable***, respectively. Their permanence properties are the ones established in sections 3.2 and 3.4. The integrability criterion 3.4.10 characterizes μ-integrable functions in terms of their local structure: μ-measurability, and their $\| \ \|_\mu^*$-size.

Let $\mathcal{E}^\sigma$ denote the sequential closure of $\mathcal{E}$. The sets in $\mathcal{E}^\sigma$ form a σ-algebra[26] $\mathcal{E}_e^\sigma$, and $\mathcal{E}^\sigma$ consists precisely of the functions measurable on $\mathcal{E}_e^\sigma$. In the case of a Radon measure, $\mathcal{E}^\sigma$ are the Baire functions. In the case that the starting point was a ring $\mathcal{A}$ of sets, $\mathcal{E}_e^\sigma$ is the σ-algebra generated by $\mathcal{A}$. The functions in $\mathcal{E}^\sigma$ are by Egoroff's theorem 3.4.4 μ-measurable for every measure μ on $\mathcal{E}$, but their collection is in general much smaller than the collection of μ-measurable functions, even in cardinality. Proposition 3.6.6, on the other hand, supplies μ-envelopes and for every μ-measurable function an equivalent one in $\mathcal{E}^\sigma$.

---

[25] See definitions 3.2.3 and 3.4.2

[26] The assumption that $\mathcal{E}$ be σ-finite is used here – see lemma A.3.3.

For the proof of the general Fubini theorem A.3.18 below it is worth stating that $\| \ \|_\mu^*$ is ***maximal***: any other mean that agrees with $\| \ \|_\mu^*$ on $\mathcal{E}_+$ is less than $\| \ \|_\mu^*$ (see 3.6.1); and for applications of capacity theory that it is ***continuous along arbitrary increasing sequences*** (see 3.6.5):

$$0 \le f_n \uparrow f \quad \text{pointwise implies} \quad \| f_n \|_\mu^* \uparrow \| f \|_\mu^* . \tag{A.3.2}$$

Exercise A.3.8 (Regularity) Let Ω be a set, $\mathcal{A}$ a σ-finite ring of subsets, and μ a positive σ-additive measure on the σ-algebra $\mathcal{A}^\sigma$ generated by $\mathcal{A}$. Then μ coincides with the Daniell extension of its restriction to $\mathcal{A}$.

(i) For any μ-integrable set A, $\mu(A) = \sup\{\mu(K) : K \in \mathcal{A}_\delta, K \subset A\}$.

(ii) Any subset Ω' of Ω has a ***measurable envelope***. This is a subset $\widetilde{\Omega'} \in \mathcal{A}^\sigma$ that contains Ω' and has the same outer measure. Any two measurable envelopes differ μ^*-negligibly.

Order-Continuous and Tight Elementary Integrals

Order-Continuity A positive Radon measure $(C_{00}(E), \mu)$ has a continuity property stronger than mere σ-continuity. Namely, if Φ is a decreasingly directed[2] subset of $C_0(E)$ with pointwise infimum zero, not necessarily countable, then $\inf\{\mu(\phi) : \phi \in \Phi\} = 0$. This is called order-continuity and is easily established using Dini's theorem A.2.1. Order-continuity occurs in the absence of local compactness as well: for instance, Dirac measure or, more generally, any measure that is carried by a countable number of points is order-continuous.

Definition A.3.9 *Let $\mathcal{E}$ be an algebra or vector lattice closed under chopping, of bounded functions on some set E. A positive linear functional $\mu : \mathcal{E} \to \mathbb{R}$ is **order-continuous** if*

$$\inf\{\mu(\phi) : \phi \in \Phi\} = 0$$

for any decreasingly directed family $\Phi \subset \mathcal{E}$ whose pointwise infimum is zero.

Sometimes it is useful to rephrase order-continuity this way: $\mu(\sup \Phi) = \sup \mu(\Phi)$ for any increasingly directed subset $\Phi \subset \mathcal{E}_+$ with pointwise supremum $\sup \Phi$ in $\mathcal{E}$.

Exercise A.3.10 If E is separable and metrizable, then any positive σ-continuous linear functional μ on $C_b(E)$ is automatically order-continuous.

In the presence of order-continuity a slightly improved integration theory is available: let $\mathcal{E}^\Uparrow$ denote the family of all functions that are pointwise suprema of arbitrary – not only the countable – subcollections of $\mathcal{E}$, and set as in (D)

$$\| f \|_\mu^{\bullet} = \inf_{|f| \le h \in \mathcal{E}^\Uparrow} \ \sup_{\phi \in \mathcal{E}, |\phi| \le h} \left| \int \phi \, d\mu \right| .$$

The functional $\| \, \|_\mu^\bullet$ is a mean[27] that agrees with $\| \, \|_\mu^*$ on $\mathcal{E}_+$, so thanks to the maximality of the latter it is smaller than $\| \, \|_\mu^*$ and consequently has more integrable functions. It is order-continuous[2] in the sense that $\|\sup H\|_\mu^\bullet = \sup \|H\|_\mu^\bullet$ for any increasingly directed subset $H \subset \mathcal{E}_+^\Uparrow$, and among all order-continuous means that agree with μ on $\mathcal{E}_+$ it is the maximal one.[27] The elements of $\mathcal{E}^\Uparrow$ are $\| \, \|_\mu^\bullet$-measurable; in fact,[27] assume that $H \subset \mathcal{E}^\Uparrow$ is increasingly directed with pointwise supremum h'. If $\|h'\|_\mu^\bullet < \infty$, then[27] h' is integrable and $H \to h'$ in $\| \, \|_\mu^\bullet$-mean:

$$\inf \left\{ \left\| h' - h \right\|_\mu^\bullet : h \in H \right\} = 0 . \tag{A.3.3}$$

For an example most pertinent in the sequel consider a completely regular space and let μ be an order-continuous positive linear functional on the lattice algebra $\mathcal{E} = C_b(E)$. Then $\mathcal{E}^\Uparrow$ contains all bounded lower semicontinuous functions, in particular all open sets (lemma A.2.19). The unique extension under $\| \, \|_\mu^\bullet$ integrates all bounded semicontinuous functions, and all Borel functions – not merely the Baire functions – are $\| \, \|_\mu^\bullet$-measurable. Of course, if E is separable and metrizable, then $\mathcal{E}^\uparrow = \mathcal{E}^\Uparrow$, $\| \, \|_\mu^* = \| \, \|_\mu^\bullet$ for any σ-continuous $\mu : C_b(E) \to \mathbb{R}$, and the two integral extensions coincide.

Tightness If E is locally compact, and in fact in most cases where a positive order-continuous measure μ appears naturally, μ is tight in this sense:

Definition A.3.11 *Let E be a completely regular space. A positive order-continuous functional $\mu : C_b(E) \to \mathbb{R}$ is **tight** and is called a **tight measure** on E if its integral extension with respect to $\| \, \|_\mu^\bullet$ satisfies*

$$\mu(E) = \sup\{\mu(K) : \ K \text{ compact } \} .$$

Tight measures are easily distinguished from each other:

Proposition A.3.12 *Let $\mathcal{M} \subset C_b(E;\mathbb{C})$ be a multiplicative class that is closed under complex conjugation, separates the points[5] of E, and has no common zeroes. Any two tight measures μ, ν that agree on $\mathcal{M}$ agree on $C_b(E)$.*

Proof. μ and ν are of course extended in the obvious complex-linear way to complex-valued bounded continuous functions. Clearly μ and ν agree on the set $\mathcal{A}^\mathbb{C}$ of complex-linear combinations of functions in $\mathcal{M}$ and then on the collection $\mathcal{A}^\mathbb{R}$ of real-valued functions in $\mathcal{A}^\mathbb{C}$. $\mathcal{A}^\mathbb{R}$ is a real algebra of real-valued functions in $C_b(E)$, and so is its uniform closure $\mathcal{A}[\mathcal{M}]$. In fact, $\mathcal{A}[\mathcal{M}]$ is also a vector lattice (theorem A.2.2), still separates the points, and $\mu = \nu$ on $\mathcal{A}[\mathcal{M}]$. There is no loss of generality in assuming that $\mu(1) = \nu(1) = 1$.

Let $f \in C_b(E)$ and $\epsilon > 0$ be given, and set $M = \|f\|_\infty$. The tightness of μ, ν provides a compact set K with $\mu(K) > 1 - \epsilon/M$ and $\nu(K) > 1 - \epsilon/M$.

[27] This is left as an exercise.

The restriction $f_{|K}$ of f to K can be approximated uniformly on K to within ϵ by a function $\phi \in \mathcal{A}[\mathcal{M}]$ (ibidem). Replacing ϕ by $-M \vee \phi \wedge M$ makes sure that ϕ is not too large. Now

$$|\mu(f) - \mu(\phi)| \le \int_K |f - \phi| \, d\mu + \int_{K^c} |f - \phi| \, d\mu \le \epsilon + 2M\mu(K^c) \le 3\epsilon \,.$$

The same inequality holds for ν, and as $\mu(\phi) = \nu(\phi)$, $|\mu(f) - \nu(f)| \le 6\epsilon$. This is true for all $\epsilon > 0$, and hence $\mu(f) = \nu(f)$. ∎

Exercise A.3.13 Let E be a completely regular space and $\mu : C_b(E) \to \mathbb{R}$ a positive order-continuous measure. Then $U_0 \overset{\text{def}}{=} \sup\{\phi \in C_b(E) : 0 \le \phi \le 1,\ \mu(\phi) = 0\}$ is integrable. It is the largest open μ-negligible set, and its complement U_0^c, the "smallest closed set of full measure," is called the ***support*** of μ.

Exercise A.3.14 An order-continuous tight measure μ on $C_b(E)$ is ***inner regular***; that is to say, its $\| \ \|_\mu^\bullet$-extension to the Borel sets satisfies

$$\mu(B) = \sup\{\mu(K) : \ K \subset B,\, K \ \text{compact}\,\}$$

for any Borel set B, in fact for every $\| \ \|_\mu^\bullet$-integrable set B. Conversely, the Daniell $\| \ \|_\mu^*$-extension of a positive σ-additive inner regular set function on the Borels of a completely regular space E is order-continuous on $C_b(E)$, and the extension of the resulting linear functional on $C_b(E)$ agrees on $\mathcal{B}^\bullet(E)$ with μ. (If E is polish or Suslin, then any σ-continuous positive measure on $C_b(E)$ is inner regular – see proposition A.6.2.)

A.3.15 The Bochner Integral Suppose $(\mathcal{E}, \mu)$ is a σ-additive positive elementary integral on E and $\mathcal{V}$ is a Fréchet space equipped with a distinguished subadditive continuous gauge $\lceil \ \rceil_{\mathcal{V}}$. Denote by $\mathcal{E} \otimes \mathcal{V}$ the collection of all functions $f : E \to \mathcal{V}$ that are finite sums of the form

$$f(x) = \sum_i v_i \, \phi_i(x) \,, \qquad v_i \in \mathcal{V} \,,\ \phi_i \in \mathcal{E} \,,$$

and define
$$\int_E f(x)\, \mu(dx) \overset{\text{def}}{=} \sum_i v_i \int_E \phi_i \, d\mu \qquad \text{for such } f \,.$$

For any $f : E \to \mathcal{V}$ set
$$\lceil f \rceil_{\mathcal{V},\mu}^* \overset{\text{def}}{=} \lceil \lceil f \rceil_{\mathcal{V}} \rceil_\mu^* = \int^* \lceil f \rceil_{\mathcal{V}} \, d\mu$$

and let
$$\mathfrak{F}[\lceil \ \rceil_{\mathcal{V},\mu}^*] \overset{\text{def}}{=} \{f : E \to \mathcal{V} : \lceil \lambda f \rceil_{\mathcal{V},\mu}^* \xrightarrow[\lambda \to 0]{} 0\} \,.$$

The ***elementary*** $\mathcal{V}$***-valued integral*** in the second line is a linear map from $\mathcal{E} \otimes \mathcal{V}$ to $\mathcal{V}$ majorized by the gauge $\lceil \ \rceil_{\mathcal{V},\mu}^*$ on $\mathfrak{F}[\lceil \ \rceil_{\mathcal{V},\mu}^*]$. Let us call a function $f : E \to \mathcal{V}$ ***Bochner*** μ***-integrable*** if it belongs to the closure of $\mathcal{E} \otimes \mathcal{V}$ in $\mathfrak{F}[\lceil \ \rceil_{\mathcal{V},\mu}^*]$. Their collection $\mathcal{L}_{\mathcal{V}}^1(\mu)$ forms a Fréchet space with gauge $\lceil \ \rceil_{\mathcal{V},\mu}^*$, and the elementary integral has a unique continuous linear extension to this space. This extension is called the ***Bochner integral.*** Neither $\mathcal{L}_{\mathcal{V}}^1(\mu)$ nor the integral extension depend on the choice of the gauge $\lceil \ \rceil_{\mathcal{V}}$. The

Dominated Convergence Theorem holds: if $\mathcal{L}^1_{\mathcal{V}}(\mu) \ni f_n \to f$ pointwise and $\lceil\!\lceil f_n \rceil\!\rceil_{\mathcal{V}} \le g \in \mathfrak{F}[\lceil\!\lceil\ \rceil\!\rceil^*_\mu]$ $\forall n$, then $f_n \to f$ in $\lceil\!\lceil\ \rceil\!\rceil^*_{\mathcal{V},\mu}$-mean, etc.

Exercise A.3.16 Suppose that $(E, \mathcal{E}, \mu)$ is the Lebesgue integral $(\mathbb{R}_+, \mathcal{E}(\mathbb{R}_+), \lambda)$. Then the Fundamental Theorem of Calculus holds: if $f : \mathbb{R}_+ \to \mathcal{V}$ is *continuous*, then the function $F : t \mapsto \int_0^t f(s)\ \lambda(ds)$ is differentiable on $[0, \infty)$ and has the derivative $f(t)$ at t; conversely, if $F : [0, \infty) \to \mathcal{V}$ has a continuous derivative F' on $[0, \infty)$, then $F(t) = F(0) + \int_0^t F'\, d\lambda$.

Projective Systems of Measures

Let $\mathbb{T}$ be an increasingly directed index set. For every $\tau \in \mathbb{T}$ let $(E_\tau, \mathcal{E}_\tau, \mathbb{P}_\tau)$ be a triple consisting of a set E_τ, an algebra and/or vector lattice $\mathcal{E}_\tau$ of bounded elementary integrands on E_τ that contains the constants, and a σ-continuous probability $\mathbb{P}_\tau$ on $\mathcal{E}_\tau$. Suppose further that there are given surjections $\pi^\tau_\sigma : E_\tau \to E_\sigma$ such that

$$\phi \circ \pi^\tau_\sigma \in \mathcal{E}_\tau$$

and

$$\int \phi \circ \pi^\tau_\sigma\, d\mathbb{P}_\tau = \int \phi\, d\mathbb{P}_\sigma$$

for $\sigma \le \tau$ and $\phi \in \mathcal{E}_\sigma$. The data $\big((E_\tau, \mathcal{E}_\tau, \mathbb{P}_\tau, \pi^\tau_\sigma) : \sigma \le \tau \in \mathbb{T}\big)$ are called a ***consistent family*** or ***projective system of probabilities***.

Let us call a ***thread*** on a subset $S \subset \mathbb{T}$ any element $(x_\sigma)_{\sigma \in S}$ of $\prod_{\sigma \in S} E_\sigma$ with $\pi^\tau_\sigma(x_\tau) = x_\sigma$ for $\sigma < \tau$ in S and denote by $E_{\mathbb{T}} = \varprojlim E_\tau$ the set of all threads on $\mathbb{T}$.[28] For every $\tau \in \mathbb{T}$ define the map

$$\pi_\tau : E_{\mathbb{T}} \to E_\tau \text{ by } \pi_\tau\big((x_\sigma)_{\sigma \in \mathbb{T}}\big) = x_\tau\ .$$

Clearly

$$\pi^\tau_\sigma \circ \pi_\tau = \pi_\sigma\ , \qquad \sigma < \tau \in \mathbb{T}\ .$$

A function $f : E_{\mathbb{T}} \to \mathbb{R}$ of the form $f = \phi \circ \pi_\tau$, $\phi \in \mathcal{E}_\tau$, is called a ***cylinder function*** based on π_τ. We denote their collection by

$$\mathcal{E}_{\mathbb{T}} = \bigcup_{\tau \in \mathbb{T}} \mathcal{E}_\tau \circ \pi_\tau\ .$$

Let $f = \phi \circ \pi_\sigma$ and $g = \psi \circ \pi_\tau$ be cylinder functions based on σ, τ, respectively. Then, assuming without loss of generality that $\sigma \le \tau$, $f + g = (\phi \circ \pi^\tau_\sigma + \psi) \circ \pi_\tau$ belongs to $\mathcal{E}_{\mathbb{T}}$. Similarly one sees that $\mathcal{E}_{\mathbb{T}}$ is closed under multiplication, finite infima, etc: $\mathcal{E}_{\mathbb{T}}$ is an algebra and/or vector lattice of bounded functions on $E_{\mathbb{T}}$. If the function $f \in \mathcal{E}_{\mathbb{T}}$ is written as $f = \phi \circ \pi_\sigma = \psi \circ \pi_\tau$ with $\phi \in C_b(E_\sigma), \psi \in C_b(E_\tau)$, then there is an $\upsilon > \sigma, \tau$ in $\mathbb{T}$, and with $\rho \overset{\text{def}}{=} \phi \circ \pi^\upsilon_\sigma = \psi \circ \pi^\upsilon_\tau$, $\mathbb{P}_\sigma(\phi) = \mathbb{P}_\upsilon(\rho) = \mathbb{P}_\tau(\psi)$ due to the consistency. We may thus define unequivocally for $f \in \mathcal{E}_{\mathbb{T}}$, say $f = \phi \circ \pi_\sigma$,

$$\mathbb{P}(f) \overset{\text{def}}{=} \mathbb{P}_\sigma(\phi)\ .$$

[28] It may well be empty or at least rather small.

Clearly $\mathbb{P} : \mathcal{E}_{\mathbb{T}} \to \mathbb{R}$ is a positive linear map with $\sup\{\mathbb{P}(f) : |f| \le 1\} = 1$. It is denoted by $\varprojlim \mathbb{P}_\tau$ and is called the ***projective limit*** of the $\mathbb{P}_\tau$. We also call $(E_T, \mathcal{E}_T, \mathbb{P})$ the ***projective limit*** of the elementary integrals $(E_\tau, \mathcal{E}_\tau, \mathbb{P}_\tau, \pi^\tau_\sigma)$ and denote it by $= \varprojlim(E_\tau, \mathcal{E}_\tau, \mathbb{P}_\tau, \pi^\tau_\sigma)$. $\mathbb{P}$ will not in general be σ-additive. The following theorem identifies sufficient conditions under which it is. To facilitate its statement let us call the projective system ***full*** if every thread on any subset of indices can be extended to a thread on all of $\mathbb{T}$. For instance, when $\mathbb{T}$ has a countable cofinal subset then the system is full.

Theorem A.3.17 (Kolmogorov) *Assume that*

(i) the projective system $\big((E_\tau, \mathcal{E}_\tau, \mathbb{P}_\tau, \pi^\tau_\sigma) : \sigma \le \tau \in \mathbb{T}\big)$ *is full;*
(ii) every $\mathbb{P}_\tau$ *is tight under the topology generated by* $\mathcal{E}_\tau$.

Then the projective limit $\mathbb{P} = \varprojlim \mathbb{P}_\tau$ *is* σ*-additive.*

Proof. Suppose the sequence of functions $f_n = \phi_n \circ \pi_{\tau_n} \in \mathcal{E}_{\mathbb{T}}$ decreases pointwise to zero. We have to show that $\mathbb{P}(f_n) \to 0$. By way of contradiction assume there is an $\epsilon > 0$ with $\mathbb{P}(f_n) > 2\epsilon \ \forall n$. There is no loss of generality in assuming that the τ_n increase with n and that $f_1 \le 1$. Let K_n be a compact subset of E_{τ_n} with $\mathbb{P}_{\tau_n}(K_n) > 1 - \epsilon 2^{-n}$, and set $K^n = \bigcap_{N \ge n} \pi^{\tau_N}_{\tau_n}(K_N)$. Then $\mathbb{P}_{\tau_n}(K^n) \ge 1 - \epsilon$, and thus $\int_{K^n} \phi_n \, d\mathbb{P}_{\tau_n} > \epsilon$ for all n. The compact sets $\overline{K}^n \overset{\text{def}}{=} K^n \cap [\phi_n \ge \epsilon]$ are non-void and have $\pi^{\tau_n}_{\tau_m}(\overline{K}^n) \supset \overline{K}^m$ for $m \le n$, so there is a thread $(x_{\tau_1}, x_{\tau_2}, \ldots)$ with $\phi_n(x_{\tau_n}) \ge \epsilon$. This thread can be extended to a thread θ on all of $\mathbb{T}$, and clearly $f_n(\theta) \ge \epsilon \ \forall n$. This contradiction establishes the claim. ∎

Products of Elementary Integrals

Let $(E, \mathcal{E}_E, \mu)$ and $(F, \mathcal{E}_F, \nu)$ be positive elementary integrals. Extending μ and ν as usual, we may assume that $\mathcal{E}_E$ and $\mathcal{E}_F$ are the step functions over the σ-algebras $\mathcal{A}_E, \mathcal{A}_F$, respectively. The ***product σ-algebra*** $\mathcal{A}_E \otimes \mathcal{A}_F$ is the σ-algebra on the cartesian product $G \overset{\text{def}}{=} E \times F$ generated by the product paving of rectangles

$$\mathcal{A}_E \times \mathcal{A}_F \overset{\text{def}}{=} \{A \times B : \ A \in \mathcal{A}_E, B \in \mathcal{A}_F\} .$$

Let $\mathcal{E}_G$ be the collection of functions on G of the form

$$\phi(x,y) = \sum_{k=1}^{K} \phi_k(x)\psi_k(y) , \quad K \in \mathbb{N}, \phi_k \in \mathcal{E}_E, \psi_k \in \mathcal{E}_F . \tag{A.3.4}$$

Clearly[13] $\mathcal{A}_E \otimes \mathcal{A}_F$ is the σ-algebra generated by $\mathcal{E}_G$. Define the product measure $\gamma = \mu \times \nu$ on a function as in equation (A.3.4) by

$$\begin{aligned} \int_G \phi(x,y)\, \gamma(dx,dy) &\overset{\text{def}}{=} \sum_k \int_E \phi_k(x)\, \mu(dx) \cdot \int_F \psi_k(y)\, \nu(dy) \\ &= \int_F \Big(\int_E \phi(x,y)\, \mu(dx) \Big)\, \nu(dy) . \end{aligned} \tag{A.3.5}$$

The first line shows that this definition is symmetric in x, y, the second that it is independent of the particular representation (A.3.4) and that γ is σ-continuous, that is to say, $\phi_n(x,y) \downarrow 0$ implies $\int \phi_n \, d\gamma \to 0$. This is evident since the inner integral in equation (A.3.5) belongs to $\mathcal{E}_F$ and decreases pointwise to zero. We can now extend the integral to all $\mathcal{A}_E \otimes \mathcal{A}_F$-measurable functions with finite upper γ-integral, etc.

Fubini's Theorem says that the integral $\int f \, d\gamma$ can be evaluated iteratively as $\int \big(\int f(x,y) \, \mu(dx) \big) \nu(dy)$ for γ-integrable f. In several instances we need a generalization, one that refers to a slightly more general setup.

Suppose that we are given for every $y \in F$ not the fixed measure $\phi(x,y) \mapsto \int_E f(x,y) \, d\mu(x)$ on $\mathcal{E}_G$ but a measure μ_y that varies with $y \in F$, but so that $y \mapsto \int \phi(x,y) \, \mu_y(dx)$ is ν-integrable for all $\phi \in \mathcal{E}_G$. We can then define a measure $\gamma = \int \mu_y \, \nu(dy)$ on $\mathcal{E}_G$ via iterated integration:

$$\int \phi \, d\gamma \stackrel{\text{def}}{=} \int \Big(\int \phi(x,y) \, \mu_y(dx) \Big) \nu(dy) \,, \qquad \phi \in \mathcal{E}_G \,.$$

If $\mathcal{E}_G \ni \phi_n \downarrow 0$, then $\mathcal{E}_F \ni \int \phi_n(x,y) \, \mu_y(dx) \downarrow 0$ and consequently $\int \phi_n \, d\gamma \to 0$: γ is σ-continuous. Fubini's theorem can be generalized to say that the γ-integral can be evaluated as an iterated integral:

Theorem A.3.18 (Fubini) *If f is γ-integrable, then $\int f(x,y) \, \mu_y(dx)$ exists for ν-almost all $y \in Y$ and is a ν-integrable function of y, and*

$$\int f \, d\gamma = \int \Big(\int f(x,y) \, \mu_y(dx) \Big) \, \nu(dy) \,.$$

Proof. The assignment $f \mapsto \int^* \big(\int^* |f(x,y)| \, \mu_y(dx) \big) \, \nu(dy)$ is a mean that coincides with the usual Daniell mean $\| \ \|_\gamma^*$ on $\mathcal{E}$, and the maximality of Daniell's mean gives

$$\int^* \Big(\int^* |f(x,y)| \, \mu_y(dx) \Big) \, \nu(dy) \le \| f \|_\gamma^* \tag{$*$}$$

for all $f : G \to \overline{\mathbb{R}}$. Given the γ-integrable function f, find a sequence (ϕ_n) of functions in $\mathcal{E}_G$ with $\sum \| \phi_n \|_\gamma^* < \infty$ and such that $f = \sum \phi_n$ both in $\| \ \|_\gamma^*$-mean and γ-almost surely. Applying $(*)$ to the set of points $(x,y) \in G$ where $\sum \phi_n(x,y) \ne f(x,y)$, we see that the set N_1 of points $y \in F$ where not $\sum \phi_n(\cdot,y) = f(\cdot,y)$ μ_y-almost surely is ν-negligible. Since

$$\Big\| \Big\| \sum |\phi_n(\cdot,y)| \Big\|_{\mu_y}^* \Big\|_\nu^* \le \Big\| \sum \| \phi_n(\cdot,y) \|_{\mu_y}^* \Big\|_\nu^* \le \sum \| \phi_n \|_\gamma^* < \infty \,,$$

the sum $g(y) \stackrel{\text{def}}{=} \| \sum |\phi_n(\cdot,y)| \|_{\mu_y}^* = \sum \int |\phi_n(x,y)| \, \mu_y(dx)$ is ν-measurable and finite in ν-mean, so it is ν-integrable. It is, in particular, finite ν-almost surely (proposition 3.2.7). Set $N_2 = [g = \infty]$ and fix a $y \notin N_1 \cup N_2$. Then $\overline{f}(\cdot,y) \stackrel{\text{def}}{=} \sum |\phi_n(\cdot,y)|$ is μ_y-almost surely finite (ibidem). Hence $\sum \phi_n(\cdot,y)$

converges μ_y-almost surely absolutely. In fact, since $y \notin N_1$, the sum is $f(\cdot,y)$. The partial sums are dominated by $\overline{f}(\cdot,y) \in \mathfrak{L}^1(\mu_y)$. Thus $f(\cdot,y)$ is μ_y-integrable with integral

$$I(y) \stackrel{\text{def}}{=} \int f(x,y)\, \mu_y(dx) = \lim_n \int \sum_{\nu \le n} \phi_\nu(x,y)\mu_y(dx) \ .$$

I is ν-almost surely defined and ν-measurable, with $|I| \le g$ having $\|I\|_\nu^* < \infty$; it is thus ν-integrable with integral

$$\int I(y)\, \nu(dy) = \lim \int \int \sum_{\nu \le n} \phi_\nu(x,y)\mu_y(dx)\, \nu(dy) = \int f\, d\gamma \ .$$

The interchange of limit and integral here is justified by the observation that $|\int \sum_{\nu \le n} \phi_\nu(x,y)\mu_y(dx)| \le \sum_{\nu \le n} \int |\phi_\nu(x,y)|\mu_y(dx) \le g(y)$ for all n. ❑

Infinite Products of Elementary Integrals

Suppose for every t in an index set $\mathcal{T}$ the triple $(E_t, \mathcal{E}_t, \mathbb{P}_t)$ is a positive σ-additive elementary integral of total mass 1. For any finite subset $\tau \subset \mathcal{T}$ let $(E_\tau, \mathcal{E}_\tau, \mathbb{P}_\tau)$ be the product of the elementary integrals $(E_t, \mathcal{E}_t, \mathbb{P}_t)$, $t \in \tau$. For $\sigma \subset \tau$ there is the obvious projection $\pi_\sigma^\tau : E_\tau \to E_\sigma$ that "forgets the components not in σ," and the projective limit (see page 401)

$$(E, \mathcal{E}, \mathbb{P}) \stackrel{\text{def}}{=} \prod_{t \in \mathcal{T}} (E_t, \mathcal{E}_t, \mathbb{P}_t) \stackrel{\text{def}}{=} \varprojlim_{\tau \subset \mathcal{T}} (E_\tau, \mathcal{E}_\tau, \mathbb{P}_\tau, \pi_\sigma^\tau)$$

of this system is the ***product of the elementary integrals*** $(E_t, \mathcal{E}_t, \mathbb{P}_t)$, $t \in \mathcal{T}$. It has for its underlying set the cartesian product $E = \prod_{t \in \mathcal{T}} E_t$. The cylinder functions of $\mathcal{E}$ are finite sums of functions of the form

$$(e_t)_{t \in \mathcal{T}} \mapsto \phi_1(e_{t_1}) \cdot \phi_2(e_{t_2}) \cdots \phi_j(e_{t_j}) \ , \qquad \phi_i \in \mathcal{E}_{t_i} \ ,$$

"that depend only on finitely many components." The projective limit $\mathbb{P} = \varprojlim \mathbb{P}_\tau$ clearly has mass 1.

Exercise A.3.19 Suppose $\mathcal{T}$ is the disjoint union of two non-void subsets $\mathcal{T}_1, \mathcal{T}_2$. Set $(E_i, \mathcal{E}_i, \mathbb{P}_i) \stackrel{\text{def}}{=} \prod_{t \in \mathcal{T}_i} (E_t, \mathcal{E}_t, \mathbb{P}_t)$, $i = 1, 2$. Then in a canonical way

$$\prod_{t \in \mathcal{T}} (E_t, \mathcal{E}_t, \mathbb{P}_t) = (E_1, \mathcal{E}_1, \mathbb{P}_1) \times (E_2, \mathcal{E}_2, \mathbb{P}_2) \ , \qquad \text{so that,}$$

$$\text{for } \phi \in \mathcal{E}, \quad \int \phi(e_1, e_2)\mathbb{P}(de_1, de_2) = \int \Big(\int \phi(e_1, e_2)\mathbb{P}_2(de_2) \Big) \mathbb{P}_1(de_1) \ . \qquad \text{(A.3.6)}$$

The present projective system is clearly full, in fact so much so that no tightness is needed to deduce the σ-additivity of $\mathbb{P} = \varprojlim \mathbb{P}_\tau$ from that of the factors $\mathbb{P}_t$:

Lemma A.3.20 *If the $\mathbb{P}_t$ are σ-additive, then so is $\mathbb{P}$.*

Proof. Let (ϕ_n) be a pointwise decreasing sequence in $\mathcal{E}_+$ and assume that for all n

$$\int \phi_n(e)\,\mathbb{P}(de) \geq a > 0\,.$$

There is a countable collection $\mathcal{T}_0 = \{t_1, t_2, \ldots\} \subset \mathcal{T}$ so that every ϕ_n depends only on coordinates in $\mathcal{T}_0$. Set $\mathcal{T}_1 \stackrel{\text{def}}{=} \{t_1\}$, $\mathcal{T}_2 \stackrel{\text{def}}{=} \{t_2, t_3, \ldots\}$. By (A.3.6),

$$\int \Big(\int \phi_n(e_1, e_2)\mathbb{P}_2(de_2) \Big) \mathbb{P}_{t_1}(de_1) > a$$

for all n. This can only be if the integrands of $\mathbb{P}_{t_1}$, which form a pointwise decreasing sequence of functions in $\mathcal{E}_{t_1}$, exceed a at some common point $e'_1 \in E_{t_1}$: for all n

$$\int \phi_n(e'_1, e_2)\,\mathbb{P}_2(de_2) \geq a\,.$$

Similarly we deduce that there is a point $e'_2 \in E_{t_2}$ so that for all n

$$\int \phi_n(e'_1, e'_2, e_3)\,\mathbb{P}_3(de_3) \geq a\,,$$

where $e_3 \in E_3 \stackrel{\text{def}}{=} E_{t_3} \times E_{t_4} \times \cdots$. There is a point $e' = (e_t) \in E$ with $e'_{t_i} = e'_i$ for $i = 1, 2, \ldots$, and clearly $\phi_n(e') \geq a$ for all n. ∎

So our product measure is σ-additive, and we can effect the usual extension upon it (see page 395 ff.).

Exercise A.3.21 State and prove Fubini's theorem for a $\mathbb{P}$-integrable function $f : E \to \mathbb{R}$.

Images, Law, and Distribution

Let $(X, \mathcal{E}_X)$ and $(Y, \mathcal{E}_Y)$ be two spaces, each equipped with an algebra of bounded elementary integrands, and let $\mu : \mathcal{E}_X \to \mathbb{R}$ be a positive σ-continuous measure on $(X, \mathcal{E}_X)$. A map $\Phi : X \to Y$ is called ***μ-measurable***[29] if $\psi \circ \Phi$ is μ-integrable for every $\psi \in \mathcal{E}_Y$. In this case the ***image of μ under Φ*** is the measure $\nu = \Phi[\mu]$ on $\mathcal{E}_Y$ defined by

$$\nu(\psi) = \int \psi \circ \Phi\, d\mu\,, \qquad \psi \in \mathcal{E}_Y\,.$$

Some authors write $\mu \circ \Phi^{-1}$ for $\Phi[\mu]$. ν is also called the ***distribution*** or ***law*** of Φ under μ. For every $x \in X$ let λ_x be the Dirac measure at $\Phi(x)$. Then clearly

$$\int_Y \psi(y)\,\nu(dy) = \int_X \int_Y \psi(y)\lambda_x(dy)\,\mu(dx)$$

[29] The "right" definition is actually this: Φ is μ-measurable if it is ***largely uniformly continuous*** in the sense of definition 3.4.2 on page 110, where of course X, Y are given the uniformities generated by $\mathcal{E}_X, \mathcal{E}_Y$, respectively.

for $\psi \in \mathcal{E}_Y$, and Fubini's theorem A.3.18 says that this equality extends to all ν-integrable functions. This fact can be read to say: if h is ν-integrable, then $h \circ \Phi$ is μ-integrable and

$$\int_Y h\, d\nu = \int_X h \circ \Phi\, d\mu\,. \tag{A.3.7}$$

We leave it to the reader to convince herself that this definition and conclusion stay *mutatis mutandis* when both μ and ν are σ-finite.

Suppose $\mathcal{X}$ and $\mathcal{Y}$ are (the step functions over) σ-algebras. If μ is a probability $\mathbb{P}$, then the law of Φ is evidently a probability as well and is given by

$$\Phi[\mathbb{P}](B) \stackrel{\text{def}}{=} \mathbb{P}([\Phi \in B])\,, \quad \forall\, B \in \mathcal{Y}\,. \tag{A.3.8}$$

Suppose Φ is real-valued. Then the ***cumulative distribution function***[30] of (the law of) Φ is the function $t \mapsto F_\Phi(t) = \mathbb{P}[\Phi \le t] = \Phi[\mathbb{P}]\big((-\infty, t]\big)$. Theorem A.3.18 applied to $(y, \lambda) \mapsto \phi'(\lambda)[\Phi(y) > \lambda]$ yields

$$\begin{aligned} \int \phi\, d\Phi[\mathbb{P}] &= \int \phi \circ \Phi\, d\mathbb{P} \\ &= \int_{-\infty}^{+\infty} \phi'(\lambda)\mathbb{P}[\Phi > \lambda]\, d\lambda = \int_{-\infty}^{+\infty} \phi'(\lambda)\big(1 - F_\Phi(\lambda)\big)\, d\lambda \end{aligned} \tag{A.3.9}$$

for any differentiable function ϕ that vanishes at $-\infty$. One defines the ***cumulative distribution function*** $F = F_\mu$ for any measure μ on the line or half-line by $F(t) = \mu((-\infty, t])$, and then denotes μ by dF and the variation $\mid\!\mu\!\mid$ variously by $|dF|$ or by $d\!\mid\! F\!\mid$.

The Vector Lattice of All Measures

Let $\mathcal{E}$ be a σ-finite algebra and vector lattice closed under chopping, of bounded functions on some set F. We denote by $\mathfrak{M}^*[\mathcal{E}]$ the set of all measures – i.e., σ-continuous elementary integrals of finite variation – on $\mathcal{E}$. This is a vector space under the usual addition and scalar multiplication of functions. Defining an order by saying that $\mu \le \nu$ is to mean that $\nu - \mu$ is a positive measure[22] makes $\mathfrak{M}^*[\mathcal{E}]$ into a vector lattice. That is to say, for every two measures $\mu, \nu \in \mathfrak{M}^*[\mathcal{E}]$ there is a least measure $\mu \vee \nu$ greater than both μ and ν and a greatest measure $\mu \wedge \nu$ less than both. In these terms the variation $\mid\!\mu\!\mid$ is nothing but $\mu \vee (-\mu)$. In fact, $\mathfrak{M}^*[\mathcal{E}]$ is order-complete: suppose $\mathcal{M} \subset \mathfrak{M}^*[\mathcal{E}]$ is ***order-bounded*** from above, i.e., there is a $\nu \in \mathfrak{M}^*[\mathcal{E}]$ greater than every element of $\mathcal{M}$; then there is a least upper order bound $\bigvee \mathcal{M}$ [5].

Let $\mathcal{E}_0^\sigma = \{\phi \in \mathcal{E}^\sigma : |\phi| \le \psi \text{ for some } \psi \in \mathcal{E}\}$, and for every $\mu \in \mathfrak{M}^*[\mathcal{E}]$ let μ^σ denote the restriction of the extension $\int d\mu$ to $\mathcal{E}_0^\sigma$. The map $\mu \mapsto \mu^\sigma$ is an order-preserving linear isomorphism of $\mathfrak{M}^*[\mathcal{E}]$ onto $\mathfrak{M}^*(\mathcal{E}_0^\sigma)$.

---

30 A ***distribution function*** of a measure μ on the line is any function $F : (-\infty, \infty) \to \mathbb{R}$ that has $\mu((a, b]) = F(b) - F(a)$ for $a < b$ in $(-\infty, \infty)$. Any two differ by a constant. The cumulative distribution function is thus that distribution function which has $F(-\infty) = 0$.

Every $\mu \in \mathfrak{M}^*[\mathcal{E}]$ has an extension whose σ-algebra of μ-measurable sets includes $\mathcal{E}_e^\sigma$ but is generally cardinalities bigger. The ***universal completion*** of $\mathcal{E}$ is the collection of all sets that are μ-measurable for every single $\mu \in \mathfrak{M}^*[\mathcal{E}]$. It is denoted by $\mathcal{E}_e^*$. It is clearly a σ-algebra containing $\mathcal{E}_e^\sigma$. A function f measurable on $\mathcal{E}_e^*$ is called ***universally measurable***. This is of course the same as saying that f is μ-measurable for every $\mu \in \mathfrak{M}^*[\mathcal{E}]$.

Theorem A.3.22 (Radon–Nikodym) *Let $\mu, \nu \in \mathfrak{M}^*[\mathcal{E}]$, with $\mathcal{E}$ σ-finite. The following are equivalent:*[26]

(i) $|\mu| = \bigvee_{k\in\mathbb{N}} |\mu| \wedge (k|\nu|)$.

(ii) For every decreasing sequence $\phi_n \in \mathcal{E}_+$, $\nu(\phi_n) \to 0 \implies \mu(\phi_n) \to 0$.

(iii) For every decreasing sequence $\phi_n \in \mathcal{E}_{0+}^\sigma$, $\nu^\sigma(\phi_n) \to 0 \implies \mu^\sigma(\phi_n) \to 0$.

(iv) For $\phi \in \mathcal{E}_0^\sigma$, $\nu^\sigma(\phi) = 0$ implies $\mu^\sigma(\phi) = 0$.

(v) A ν-negligible set is μ-negligible.

(vi) There exists a function $g \in \mathcal{E}^\sigma$ such that $\mu(\phi) = \nu^\sigma(g\phi)$ for all $\phi \in \mathcal{E}$.

In this case μ is called ***absolutely continuous*** *with respect to ν and we write $\mu \ll \nu$; furthermore, then $\int f\,d\mu = \int fg\,d\nu$ whenever either side makes sense. The function g is the* ***Radon–Nikodym derivative*** *or* ***density*** *of μ with respect to ν, and it is customary to write $\mu = g\nu$, that is to say, for $\phi \in \mathcal{E}$ we have $(g\nu)(\phi) = \nu^\sigma(g\phi)$.*

If $\mu \ll \rho$ for all $\mu \in \mathcal{M} \subset \mathfrak{M}^*[\mathcal{E}]$, then $\bigvee \mathcal{M} \ll \rho$.

Exercise A.3.23 Let $\mu, \nu : C_b(E) \to \mathbb{R}$ be σ-additive with $\mu \ll \nu$. If ν is order-continuous and tight, then so is μ.

Conditional Expectation

Let $\Phi : (\Omega, \mathcal{F}) \to (Y, \mathcal{Y})$ be a measurable map of measurable spaces and μ a positive finite measure on $\mathcal{F}$ with image $\nu \stackrel{\text{def}}{=} \Phi[\mu]$ on $\mathcal{Y}$.

Theorem A.3.24 *(i) For every μ-integrable function $f : \Omega \to \mathbb{R}$ there exists a ν-integrable $\mathcal{Y}$-measurable function $\mathbb{E}[f|\Phi] = \mathbb{E}^\mu[f|\Phi] : Y \to \mathbb{R}$, called the* ***conditional expectation of f given Φ****, such that*

$$\int_\Omega f \cdot h \circ \Phi \, d\mu = \int_Y \mathbb{E}[f|\Phi] \cdot h \, d\nu$$

for all bounded $\mathcal{Y}$-measurable $h : Y \to \mathbb{R}$. Any two conditional expectations differ at most in a ν-negligible set of $\mathcal{Y}$ and depend only on the class of f.

(ii) The map $f \mapsto \mathbb{E}^\mu[f|\Phi]$ is linear and positive, maps 1 to 1, and is contractive[31] *from $L^p(\mu)$ to $L^p(\nu)$ when $1 \le p \le \infty$.*

[31] A linear map $\Phi : E \to S$ between seminormed spaces is ***contractive*** if there exists a $\gamma \le 1$ such that $\|\Phi(x)\|_S \le \gamma \cdot \|x\|_E$ for all $x \in E$; the least γ satisfying this inequality is the ***modulus of contractivity*** of Φ. If the contractivity modulus is strictly less than 1, then Φ is ***strictly contractive***.

(iii) Assume $\Gamma:\mathbb{R}\to\mathbb{R}$ *is convex*[32] *and* $f:\Omega\to\mathbb{R}$ *is* $\mathcal{F}$*-measurable and such that* $\Gamma\circ f$ *is* ν*-integrable. Then if* $\mu(1)\geq 1$, *we have* ν*-almost surely*

$$\Gamma(\mathbb{E}^{\mu}[f|\Phi])\leq\mathbb{E}^{\mu}[\Gamma(f)|\Phi]\,. \tag{A.3.10}$$

Proof. (i) Consider the measure $f\mu: B\mapsto\int_B f\,d\mu$, $B\in\mathcal{F}$, and its image $\nu'=\Phi[f\mu]$. This is a measure on the σ-algebra $\mathcal{Y}$, absolutely continuous with respect to ν. The Radon–Nikodym theorem provides a derivative $d\nu'/d\nu$, which we may call $\mathbb{E}^{\mu}[f|\Phi]$. If f is changed μ-negligibly, then the measure ν' and thus the (class of the) derivative do not change.

(ii) The linearity and positivity are evident. The contractivity follows from (iii) and the observation that $x\mapsto|x|^p$ is convex when $1\leq p<\infty$.

(iii) There is a countable collection of linear functions $\ell_n(x)=\alpha_n+\beta_n x$ such that $\Gamma(x)=\sup_n\ell_n(x)$ at every point $x\in\mathbb{R}$. Linearity and positivity give

$$\ell_n(\mathbb{E}^{\mu}[f|\Phi])=\mathbb{E}^{\mu}\left[\ell_n(f)|\Phi\right]\leq\mathbb{E}^{\mu}[\Gamma(f)|\Phi]\qquad\text{a.s.}\quad\forall n\in\mathbb{N}\,.$$

Upon taking the supremum over n, ***Jensen's inequality*** (A.3.10) follows. ∎

Frequently the situation is this: Given is not a map Φ but a sub-σ-algebra $\mathcal{Y}$ of $\mathcal{F}$. In that case we understand Φ to be the identity $(\Omega,\mathcal{F})\to(\Omega,\mathcal{Y})$. Then $\mathbb{E}^{\mu}[f|\Phi]$ is usually denoted by $\mathbb{E}[f|\mathcal{Y}]$ or $\mathbb{E}^{\mu}[f|\mathcal{Y}]$ and is called the ***conditional expectation of f given*** $\mathcal{Y}$. It is thus defined by the identity

$$\int f\cdot H\,d\mu=\int\mathbb{E}^{\mu}[f|\mathcal{Y}]\cdot H\,d\mu\,,\qquad H\in\mathcal{Y}_b\,,$$

and (i)–(iii) continue to hold, *mutatis mutandis.*

Exercise A.3.25 Let μ be a subprobability ($\mu(1)\leq 1$) and $\phi:\mathbb{R}_+\to\mathbb{R}_+$ a concave function. Then for all μ-integrable functions z

$$\int\phi(|z|)\,d\mu\leq\phi\Big(\int|z|\,d\mu\Big)\,.$$

Exercise A.3.26 On the probability triple $(\Omega,\mathcal{G},\mathbb{P})$ let $\mathcal{F}$ be a sub-σ-algebra of $\mathcal{G}$, X an $\mathcal{F}/\mathcal{X}$-measurable map from Ω to some measurable space $(\Xi,\mathcal{X})$, and Φ a bounded $\mathcal{X}\otimes\mathcal{G}$-measurable function. For every $x\in\Xi$ set $\overline{\Phi}(x,\omega)\overset{\text{def}}{=}\mathbb{E}[\Phi(x,\cdot)|\mathcal{F}](\omega)$. Then $\mathbb{E}[\Phi(X(\cdot),\cdot)|\mathcal{F}](\omega)=\overline{\Phi}(X(\omega),\omega)$ $\mathbb{P}$-almost surely.

Numerical and σ-Finite Measures

Many authors define a measure to be a triple $(\Omega,\mathcal{F},\mu)$, where $\mathcal{F}$ is a σ-algebra on Ω and $\mu:\mathcal{F}\to\overline{\mathbb{R}}_+$ is ***numerical***, i.e., is allowed to take values in the extended reals $\overline{\mathbb{R}}$, with suitable conventions about the meaning of $r+\infty$, etc.

---

[32] Γ is ***convex*** if $\Gamma(\lambda x+(1-\lambda)x')\leq\lambda\Gamma(x)+(1-\lambda)\Gamma(x')$ for $x,x'\in\operatorname{dom}\Gamma$ and $0\leq\lambda\leq 1$; it is ***strictly convex*** if $\Gamma(\lambda x+(1-\lambda)x')<\lambda\Gamma(x)+(1-\lambda)\Gamma(x')$ for $x\neq x'\in\operatorname{dom}\Gamma$ and $0<\lambda<1$.

(see A.1.2). μ is σ-additive if it satisfies $\mu(\bigcup F_n) = \sum \mu(F_n)$ for mutually disjoint sets $F_n \in \mathcal{F}$. Unless the δ-ring $\mathcal{D}_\mu \stackrel{\text{def}}{=} \{F \in \mathcal{F} : \mu(F) < \infty\}$ generates the σ-algebra $\mathcal{F}$, examples of quite unnatural behavior can be manufactured [110]. If this requirement is made, however, then any reasonable integration theory of the measure space $(\Omega, \mathcal{F}, \mu)$ is essentially the same as the integration theory of $(\Omega, \mathcal{D}_\mu, \mu)$ explained above.

μ is called ***σ-finite*** if $\mathcal{D}_\mu$ is a σ-finite class of sets (exercise A.3.2), i.e., if there is a countable family of sets $F_n \in \mathcal{F}$ with $\mu(F_n) < \infty$ and $\bigcup_n F_n = \Omega$; in that case the requirement is met and $(\Omega, \mathcal{F}, \mu)$ is also called a ***σ-finite measure space***.

Exercise A.3.27 Consider again a measurable map $\Phi : (\Omega, \mathcal{F}) \to (Y, \mathcal{Y})$ of measurable spaces and a measure μ on $\mathcal{F}$ with image ν on $\mathcal{Y}$, and assume that both μ and ν are σ-finite on their domains.

(i) With μ_0 denoting the restriction of μ to $\mathcal{D}_\mu$, $\mu = \mu_0^*$ on $\mathcal{F}$.
(ii) Theorem A.3.24 stays, including Jensen's inequality (A.3.10).
(iii) If Γ is strictly convex, then equality holds in inequality (A.3.10) if and only if f is almost surely equal to a function of the form $f' \circ \Phi$, f' $\mathcal{Y}$-measurable.
(iv) For $h \in \mathcal{Y}_b$, $\mathbb{E}[fh \circ \Phi|\Phi] = h \cdot \mathbb{E}[f|\Phi]$ provided both sides make sense.
(v) Let $\Psi : (Y, \mathcal{Y}) \to (Z, \mathcal{Z})$ be measurable, and assume $\Psi[\nu]$ is σ-finite. Then

$$\mathbb{E}^\mu[f|\Psi \circ \Phi] = \mathbb{E}^\nu[\mathbb{E}^\mu[f|\Phi]|\Psi] , \text{ and } \mathbb{E}[f|\mathcal{Z}] = \mathbb{E}[\mathbb{E}[f|\mathcal{Y}]|\mathcal{Z}]$$

when $\Omega = Y = Z$ and $\mathcal{Z} \subset \mathcal{Y} \subset \mathcal{F}$.
(vi) If $\mathbb{E}[f \cdot b] = \mathbb{E}[f \cdot \mathbb{E}[b|\mathcal{Y}]]$ for all $b \in L^\infty(\mathcal{Y})$, then f is measurable on $\mathcal{Y}$.

Exercise A.3.28 The argument in the proof of Jensen's inequality theorem A.3.24 can be used in a slightly different context. Let E be a Banach space, ν a signed measure with σ-finite variation $|\nu|$, and $f \in L^1_E(\nu)$ (see item A.3.15). Then

$$\left\| \int f \, d\nu \right\|_E \le \int \|f\|_E \, d|\nu| .$$

Exercise A.3.29 Yet another variant of the same argument can be used to establish the following inequality, which is used repeatedly in chapter 5. Let $(F, \mathcal{F}, \mu)$ and $(G, \mathcal{G}, \nu)$ be σ-finite measure spaces. Let f be a function measurable on the product σ-algebra $\mathcal{F} \otimes \mathcal{G}$ on $F \times G$. Then

$$\Big\| \|f\|_{L^p(\mu)} \Big\|_{L^q(\nu)} \le \Big\| \|f\|_{L^q(\nu)} \Big\|_{L^p(\mu)}$$

for $0 < p \le q \le \infty$.

Characteristic Functions

It is often difficult to get at the law of a random variable $\Phi : (F, \mathcal{F}) \to (G, \mathcal{G})$ through its definition (A.3.8). There is a recurring situation when the powerful tool of characteristic functions can be applied. Namely, let us suppose that $\mathcal{G}$ is generated by a vector space Γ of real-valued functions. Now, inasmuch as $\gamma = -i \lim_{n\to\infty} n\big(e^{i\gamma/n} - e^{i0}\big)$, $\mathcal{G}$ is also generated by the functions

$$y \mapsto e^{i\gamma(y)} , \qquad \gamma \in \Gamma .$$

These functions evidently form a complex multiplicative class $e^{i\Gamma}$, and in view of exercise A.3.5 any σ-additive measure μ of totally finite variation on $\mathcal{G}$ is determined by its values

$$\widehat{\mu}(\gamma) = \int_G e^{i\gamma(y)}\, \mu(dy) \tag{A.3.11}$$

on them. $\widehat{\mu}$ is called the ***characteristic function*** of μ. We also write $\widehat{\mu}^\Gamma$ when it is necessary to indicate that this notion depends on the generating vector space Γ, and then talk about the characteristic function of μ for Γ. If μ is the law of $\Phi : (F, \mathcal{F}) \to (G, \mathcal{G})$ under $\mathbb{P}$, then (A.3.7) allows us to rewrite equation (A.3.11) as

$$\widehat{\Phi[\mathbb{P}]}(\gamma) = \int_G e^{i\gamma(y)}\, \Phi[\mathbb{P}](dy) = \int_F e^{i\gamma\circ\Phi}\, d\mathbb{P}\,, \qquad \gamma \in \Gamma\,.$$

$\widehat{\Phi[\mathbb{P}]} = \widehat{\Phi[\mathbb{P}]}^\Gamma$ is also called the ***characteristic function*** of Φ.

Example A.3.30 Let $G = \mathbb{R}^n$, equipped of course with its Borel σ-algebra. The vector space Γ of linear functions $x \mapsto \langle \xi | x \rangle$, one for every $\xi \in \mathbb{R}^n$, generates the topology of $\mathbb{R}^n$ and therefore also generates $\mathcal{B}^\bullet(\mathbb{R}^n)$. Thus any measure μ of finite variation on $\mathbb{R}^n$ is determined by its characteristic function for Γ

$$\mathfrak{F}[\mu(dx)](\xi) = \widehat{\mu}(\xi) = \int_{\mathbb{R}^n} e^{i\langle \xi | x \rangle}\, \mu(dx)\,, \qquad \xi \in \mathbb{R}^n\,.$$

$\widehat{\mu}$ is a bounded uniformly continuous complex-valued function on the dual $\mathbb{R}^n$ of $\mathbb{R}^n$.

Suppose that μ has a density g with respect to Lebesgue measure λ; that is to say, $\mu(dx) = g(x)\lambda(dx)$. μ has totally finite variation if and only if g is Lebesgue integrable, and in fact $\rvert\mu\lvert = |g|\lambda$. It is customary to write $\widehat{g}$ or $\mathfrak{F}[g(x)]$ for $\widehat{\mu}$ and to call this function the ***Fourier transform***[33] of g (and of μ). The Riemann–Lebesgue lemma says that $\widehat{g} \in C_0(\mathbb{R}^n)$. As g runs through $\mathcal{L}^1(\lambda)$, the $\widehat{g}$ form a subalgebra of $C_0(\mathbb{R}^n)$ that is practically impossible to characterize. It does however contain the ***Schwartz space*** $\mathcal{S}$ of infinitely differentiable functions that together with their partials of any order decay at ∞ faster than $|\xi|^{-k}$ for any $k \in \mathbb{N}$. By theorem A.2.2 this algebra is dense in $C_0(\mathbb{R}^n)$. For $g, h \in \mathcal{S}$ (and whenever both sides make sense) and $1 \le \nu \le n$

$$\mathfrak{F}[ix^\nu g(x)](\xi) = \frac{\partial}{\partial \xi^\nu}\, \mathfrak{F}[g(x)](\xi) \quad \text{and} \quad \mathfrak{F}\Big[\frac{\partial g(x)}{\partial x^\nu}\Big](\xi) = -i\xi^\nu \cdot \widehat{g}(\xi)$$

$$\widehat{g\star h} = \widehat{g} \cdot \widehat{h}\,,\ \widehat{g \cdot h} = \widehat{g} \star \widehat{h} \ \text{ and}^{34} \quad \widehat{\overset{*}{\mu}} = \overset{*}{\widehat{\mu}}\,. \tag{A.3.12}$$

---

33 Actually it is the widespread custom among analysts to take for Γ the space of linear functionals $y \mapsto 2\pi\langle \xi | x \rangle$, $\xi \in \mathbb{R}^n$, and to call the resulting characteristic function the Fourier transform. This simplifies the Fourier inversion formula (A.3.13) to $g(x) = \int e^{-2\pi i \langle \xi | x \rangle} \widehat{g}(\xi)\, d\xi$.

34 $\overset{*}{\phi}(x) \overset{\text{def}}{=} \phi(-x)$ and $\overset{*}{\mu}(\phi) \overset{\text{def}}{=} \mu(\overset{*}{\phi})$ define the ***reflections through the origin*** $\overset{*}{\phi}$ and $\overset{*}{\mu}$. Note the perhaps somewhat unexpected equality $g \cdot \overset{*}{\lambda} = (-1)^n \cdot \overset{*}{g} \cdot \lambda$.

Roughly: the Fourier transform turns partial differentiation into multiplication with $-i$ times the corresponding coordinate function, and vice versa; it turns convolution into the pointwise product, and vice versa. It commutes with ***reflection*** $\mu \mapsto \overset{*}{\mu}$ ***through the origin.*** g can be recovered from its Fourier transform $\widehat{g}$ by the ***Fourier inversion formula***

$$g(x) = \mathfrak{F}^{-1}[\widehat{g}](x) = \frac{1}{(2\pi)^n} \int_{\mathbb{R}^n} e^{-i\langle\xi|x\rangle} \widehat{g}(\xi)\, d\xi \,. \tag{A.3.13}$$

Example A.3.31 Next let $(G, \mathcal{G})$ be the path space $\mathscr{C}^n$, equipped with its Borel σ-algebra. $\mathcal{G} = \mathcal{B}^\bullet(\mathscr{C}^n)$ is generated by the functions $w \mapsto \langle\alpha|w\rangle_t$ (see page 15). These do not form a vector space, however, so we emulate example A.3.30. Namely, every continuous linear functional on $\mathscr{C}^n$ is of the form

$$w \mapsto \langle w|\gamma\rangle \overset{\text{def}}{=} \int_0^\infty \sum_{\nu=1}^n w_t^\nu \, d\gamma_t^\nu \,,$$

where $\gamma = (\gamma^\nu)_{\nu=1}^n$ is an n-tupel of functions of finite variation and of compact support on the half-line. The continuous linear functionals do form a vector space $\Gamma = \mathscr{C}^{n*}$ that generates $\mathcal{B}^\bullet(\mathscr{C}^n)$ (ibidem). Any law $\mathbb{L}$ on $\mathscr{C}^n$ is therefore determined by its characteristic function

$$\widehat{\mathbb{L}}(\gamma) = \int_{\mathscr{C}^n} e^{i\langle w|\gamma\rangle} \, \mathbb{L}(dw) \,.$$

An aside: the topology generated[14] by Γ is the weak topology $\sigma(\mathscr{C}^n, \mathscr{C}^{n*})$ on $\mathscr{C}^n$ (item A.2.32) and is distinctly weaker than the topology of uniform convergence on compacta.

Example A.3.32 Let $\mathcal{H}$ be a countable index set, and equip the "sequence space" $\mathbb{R}^{\mathcal{H}}$ with the topology of pointwise convergence. This makes $\mathbb{R}^{\mathcal{H}}$ into a Fréchet space. The stochastic analysis of random measures leads to the space $\mathscr{D}_{\mathbb{R}^{\mathcal{H}}}$ of all càdlàg paths $[0, \infty) \to \mathbb{R}^{\mathcal{H}}$ (see page 175). This is a polish space under the Skorohod topology; it is also a vector space, but topology and linear structure do not cooperate to make it into a topological vector space. Yet it is most desirable to have the tool of characteristic functions at one's disposal, since laws on $\mathscr{D}_{\mathbb{R}^{\mathcal{H}}}$ do arise (ibidem). Here is how this can be accomplished. Let Γ denote the vector space of all functions of compact support on $[0, \infty)$ that are continuously differentiable, say. View each $\gamma \in \Gamma$ as the cumulative distribution function of a measure $d\gamma_t = \dot{\gamma}_t dt$ of compact support. Let $\Gamma_0^{\mathcal{H}}$ denote the vector space of all $\mathcal{H}$-tuples $\gamma = \{\gamma^h : h \in \mathcal{H}\}$ of elements of Γ all but finitely many of which are zero. Each $\gamma \in \Gamma_0^{\mathcal{H}}$ is naturally a linear functional on $\mathscr{D}_{\mathbb{R}^{\mathcal{H}}}$, via

$$\mathscr{D}_{\mathbb{R}^{\mathcal{H}}} \ni z. \mapsto \langle z.|\gamma\rangle \overset{\text{def}}{=} \sum_{h\in\mathcal{H}} \int_0^\infty z_t^h \, d\gamma_t^h \,,$$

a finite sum. In fact, the $\langle\cdot|\gamma\rangle$ are continuous in the Skorohod topology and separate the points of $\mathscr{D}_{\mathbb{R}^{\mathcal{H}}}$; they form a linear space $\boldsymbol{\Gamma}$ of continuous linear functionals on $\mathscr{D}_{\mathbb{R}^{\mathcal{H}}}$ that separates the points. Therefore, for one good thing, the weak topology $\sigma(\mathscr{D}_{\mathbb{R}^{\mathcal{H}}}, \boldsymbol{\Gamma})$ is a Lusin topology on $\mathscr{D}_{\mathbb{R}^{\mathcal{H}}}$, for which every σ-additive probability is tight, and whose Borels agree with those of the Skorohod topology; and for another, we can define the characteristic function of any probability $\mathbb{P}$ on $\mathscr{D}_{\mathbb{R}^{\mathcal{H}}}$ by

$$\widehat{\mathbb{P}}(\gamma) \stackrel{\text{def}}{=} \mathbb{E}\left[e^{i\langle\cdot|\gamma\rangle}\right] .$$

To amplify on examples A.3.31 and A.3.32 and to prepare the way for an easy proof of the Central Limit Theorem A.4.4 we provide here a simple result:

Lemma A.3.33 *Let Γ be a real vector space of real-valued functions on some set E. The topologies generated*[14] *by Γ and by the collection $e^{i\Gamma}$ of functions $x \mapsto e^{i\gamma(x)}$, $\gamma \in \Gamma$, have the same convergent sequences.*

Proof. It is evident that the topology generated by $e^{i\Gamma}$ is coarser than the one generated by Γ. For the converse, let (x_n) be a sequence that converges to $x \in E$ in the former topology, i.e., so that $e^{i\gamma(x_n)} \to e^{i\gamma(x)}$ for all $\gamma \in \Gamma$. Set $\delta_n = \gamma(x_n) - \gamma(x)$. Then $e^{it\delta_n} \to 1$ for all t. Now

$$\begin{aligned}\frac{1}{K}\int_{-K}^{K} 1 - e^{it\delta_n}\,dt &= 2\left(1 - \frac{1}{K}\int_0^K \cos(t\delta_n)\,dt\right)\\ &= 2\left(1 - \frac{\sin(\delta_n K)}{\delta_n K}\right) \geq 2\left(1 - \frac{1}{|\delta_n K|}\right) .\end{aligned}$$

For sufficiently large indices $n \geq n(K)$ the left-hand side can be made smaller than 1, which implies $1/|\delta_n K| \geq 1/2$ and $|\delta_n| \leq 2/K$: $\delta_n \to 0$ as desired. ∎

The conclusion may fail if Γ is merely a vector space over the rationals $\mathbb{Q}$: consider the $\mathbb{Q}$-vector space Γ of rational linear functions $x \mapsto qx$ on $\mathbb{R}$. The sequence $(2\pi n!)$ converges to zero in the topology generated by $e^{i\Gamma}$, but not in the topology generated by Γ, which is the usual one. On subsets of E that are precompact in the Γ-topology, the Γ-topology and the $e^{i\Gamma}$-topology coincide, of course, whatever Γ. However,

Exercise A.3.34 A sequence (x_n) in $\mathbb{R}^d$ converges if and only if $(e^{i\langle t|x_n\rangle})$ converges for almost all $t \in \mathbb{R}^d$.

Exercise A.3.35 If $\widehat{\mathbb{L}_1}(\gamma) = \widehat{\mathbb{L}_2}(\gamma)$ for all γ in the real vector space Γ, then $\mathbb{L}_1$ and $\mathbb{L}_2$ agree on the σ-algebra generated by Γ.

Independence On a probability space $(\Omega, \mathcal{F}, \mathbb{P})$ consider n $\mathbb{P}$-measurable maps $\Phi_\nu : \Omega \to E_\nu$, where E_ν is equipped with the algebra $\mathcal{E}_\nu$ of elementary integrands. If the law of the product map $(\Phi_1, \ldots, \Phi_n) : \Omega \to E_1 \times \cdots \times E_n$ – which is clearly $\mathbb{P}$-measurable if $E_1 \times \cdots \times E_n$ is equipped with $\mathcal{E}_1 \otimes \cdots \otimes \mathcal{E}_n$ –

happens to coincide with the product of the laws $\Phi_1[\mathbb{P}], \ldots, \Phi_n[\mathbb{P}]$, then one says that the family $\{\Phi_1, \ldots, \Phi_n\}$ is ***independent*** under $\mathbb{P}$. This definition generalizes in an obvious way to countable collections $\{\Phi_1, \Phi_2, \ldots\}$ (page 404).

Suppose $\mathcal{F}_1, \mathcal{F}_2, \ldots$ are sub-σ-algebras of $\mathcal{F}$. With each goes the (trivially measurable) identity map $\Phi_n : (\Omega, \mathcal{F}) \to (\Omega, \mathcal{F}_n)$. The σ-algebras $\mathcal{F}_n$ are called independent if the Φ_n are.

Exercise A.3.36 Suppose that the sequential closure of $\mathcal{E}_\nu$ is generated by the vector space Γ_ν of real-valued functions on E_ν. Write Φ for the product map $\prod_{\nu=1}^n \Phi_\nu$ from Ω to $\prod_\nu E_\nu$. Then $\Gamma \stackrel{\text{def}}{=} \bigotimes_{\nu=1}^n \Gamma_\nu$ generates the sequential closure of $\bigotimes_{\nu=1}^n \mathcal{E}_\nu$, and $\{\Phi_1, \ldots, \Phi_n\}$ is independent if and only if

$$\widehat{\Phi[\mathbb{P}]}^{\Gamma}(\gamma_1 \otimes \cdots \otimes \gamma_n) = \prod_{1 \le \nu \le n} \widehat{\Phi_\nu[\mathbb{P}]}^{\Gamma_\nu}(\gamma_\nu) .$$

Convolution

Fix a commutative locally compact group G whose topology has a countable basis. The group operation is denoted by $+$ or by juxtaposition, and it is understood that group operation and topology are compatible in the sense that "the subtraction map" $- : G \times G \to G$, $(g, g') \mapsto g - g'$, is continuous. In the instances that occur in the main body $(G, +)$ is either $\mathbb{R}^n$ with its usual addition or $\{-1, 1\}^n$ under pointwise multiplication. On such a group there is an essentially unique translation-invariant[35] Radon measure η called ***Haar measure***. In the case $G = \mathbb{R}^n$, Haar measure is taken to be Lebesgue measure, so that the mass of the unit box is unity; in the second example it is the normalized counting measure, which gives every point equal mass 2^{-n} and makes it a probability.

Let μ_1 and μ_2 be two Radon measures on G that have bounded variation: $\|\mu_i\| \stackrel{\text{def}}{=} \sup\{\mu_i(\phi) : \phi \in C_{00}(G)\ ,\ |\phi| \le 1\} < \infty$. Their ***convolution*** $\mu_1{\star}\mu_2$ is defined by

$$\mu_1{\star}\mu_2(\phi) = \int_{G \times G} \phi(g_1 + g_2)\ \mu_1(dg_1)\mu_2(dg_2) . \qquad \text{(A.3.14)}$$

In other words, apply the product $\mu_1 \times \mu_2$ to the particular class of functions $(g_1, g_2) \mapsto \phi(g_1 + g_2)$, $\phi \in C_{00}(G)$. It is easily seen that $\mu_1{\star}\mu_2$ is a Radon measure on $C_{00}(G)$ of total variation $\|\mu_1{\star}\mu_2\| \le \|\mu_1\| \cdot \|\mu_2\|$, and that convolution is associative and commutative. The usual sequential closure argument shows that equation (A.3.14) persists if ϕ is a bounded Baire function on G.

Suppose μ_1 has a Radon–Nikodym derivative h_1 with respect to Haar measure: $\mu_1 = h_1\eta$ with $h_1 \in \mathcal{L}^1[\eta]$. If ϕ in equation (A.3.14) is negligible for Haar measure, then $\mu_1{\star}\mu_2$ vanishes on ϕ by Fubini's theorem A.3.18.

[35] This means that $\int \phi(x + g)\ \eta(dx) = \int \phi(x)\ \eta(dx)$ for all $g \in G$ and $\phi \in C_{00}(G)$ and persists for η-integrable functions ϕ. If η is translation-invariant, then so is $c\eta$ for $c \in \mathbb{R}$, but this is the only ambiguity in the definition of Haar measure.

Therefore $\mu_1{\star}\mu_2$ is absolutely continuous with respect to Haar measure. Its density is then denoted by $h_1{\star}\mu_2$ and can be calculated easily:

$$(h_1{\star}\mu_1)(g) = \int_G h_1(g - g_2)\ \mu_2(dg_2)\ . \tag{A.3.15}$$

Indeed, repeated applications of Fubini's theorem give

$$\begin{aligned}
\int_G \phi(g)\ \mu_1{\star}\mu_2(dg) &= \int_{G\times G} \phi(g_1 + g_2)\ h_1(g_1)\ \eta(dg_1)\,\mu_2(dg_2) \\
\text{by translation-invariance:} \quad &= \int_{G\times G} \phi((g_1{-}g_2) + g_2))\ h_1(g_1{-}g_2)\ \eta(dg_1)\,\mu_2(dg_2) \\
\text{with } g = g_1: \quad &= \int_{G\times G} \phi(g) h_1(g - g_2)\ \eta(dg)\,\mu_2(dg_2) \\
&= \int_G \phi(g)\Big(\int_G h_1(g - g_2)\ \mu_2(dg_2)\Big)\ \eta(dg)\ ,
\end{aligned}$$

which exhibits $\int h_1(g - g_2)\ \mu_2(dg_2)$ as the density of $\mu_1{\star}\mu_2$ and yields equation (A.3.15).

Exercise A.3.37 (i) If $h_1 \in C_0(G)$, then $h_1{\star}\mu_2 \in C_0(G)$ as well. (ii) If μ_2, too, has a density $h_2 \in \mathcal{L}^1[\eta]$ with respect to Haar measure η, then the density of $\mu_1{\star}\mu_2$ is commonly denoted by $h_1{\star}h_2$ and is given by

$$(h_1{\star}h_2)(g) = \int h_1(g_1)\, h_2(g - g_1)\ \eta(dg_1)\ .$$

Let us compute the characteristic function of $\mu_1{\star}\mu_2$ in the case $G = \mathbb{R}^n$:

$$\begin{aligned}
\widehat{\mu_1{\star}\mu_2}(\zeta) &= \int_{\mathbb{R}^n} e^{i\langle\zeta|z_1+z_2\rangle}\ \mu_1(dz_1)\mu_2(dz_2) \\
&= \int e^{i\langle\zeta|z_1\rangle}\ \mu_1(dz_1)\cdot \int e^{i\langle\zeta|z_2\rangle}\ \mu_2(dz_2) \\
&= \widehat{\mu_1}(\zeta)\cdot\widehat{\mu_2}(\zeta)\ .
\end{aligned} \tag{A.3.16}$$

Exercise A.3.38 Convolution commutes with reflection through the origin[34]: if $\mu = \mu_1{\star}\mu_2$, then $\overset{*}{\mu} = \overset{*}{\mu_1}{\star}\overset{*}{\mu_2}$.

Liftings, Disintegration of Measures

For the following fix a σ-algebra $\mathcal{F}$ on a set F and a positive σ-finite measure μ on $\mathcal{F}$ (exercise A.3.27). We assume that $(\mathcal{F}, \mu)$ is complete, i.e., that $\mathcal{F}$ equals the ***μ-completion*** $\mathcal{F}^\mu$, the σ-algebra generated by $\mathcal{F}$ and all subsets of μ-negligible sets in $\mathcal{F}$. We distinguish carefully between a function $f \in \mathcal{L}^\infty$[36] and its class modulo negligible functions $\dot f \in L^\infty$, writing $f \dot\le g$ to mean that $\dot f \le \dot g$, i.e., that $\dot f, \dot g$ contain representatives f', g' with $f'(x) \le g'(x)$ at all points $x \in F$, etc.

---

[36] f is $\mathcal{F}$-measurable and bounded.

Definition A.3.39 *(i) A* ***density*** *on* $(F, \mathcal{F}, \mu)$ *is a map* $\theta : \mathcal{F} \to \mathcal{F}$ *with the following properties:*
a) $\theta(\emptyset) = \emptyset$ *and* $\theta(F) = F$; *b)* $A \dot{\subseteq} B \Longrightarrow \theta(A) \subseteq \theta(B) \in \dot{B} \quad \forall A, B \in \mathcal{F}$;
c) $A_1 \cap \ldots \cap A_k \dot{=} \emptyset \Longrightarrow \theta(A_1) \cap \ldots \cap \theta(A_k) = \emptyset \quad \forall k \in \mathbb{N}, A_1, \ldots, A_k \in \mathcal{F}$.

(ii) A ***dense topology*** *is a topology* $\tau \subset \mathcal{F}$ *with the following properties: a) a negligible set in* τ *is void; and b) every set of* $\mathcal{F}$ *contains a* τ*-open set from which it differs negligibly.*

(iii) A ***lifting*** *is an algebra homomorphism* $T : \mathcal{L}^\infty \to \mathcal{L}^\infty$ *that takes the constant function* 1 *to itself and obeys* $f \dot{=} g \Longrightarrow Tf = Tg \in \dot{g}$.

Viewed as a map $T : L^\infty \to \mathcal{L}^\infty$, a lifting T is a linear multiplicative inverse of the natural quotient map $\dot{} : f \mapsto \dot{f}$ from $\mathcal{L}^\infty$ to L^∞. A lifting T is positive; for if $0 \le f \in \mathcal{L}^\infty$, then f is the square of some function g and thus $Tf = (Tg)^2 \ge 0$. A lifting T is also contractive; for if $\|f\|_\infty \le a$, then $-a \le f \le a \Longrightarrow -a \le Tf \le a \Longrightarrow \|Tf\|_\infty \le a$.

Lemma A.3.40 *Let* $(F, \mathcal{F}, \mu)$ *be a complete totally finite measure space.*

(i) If $(F, \mathcal{F}, \mu)$ *admits a density* θ, *then it has a dense topology* τ_θ *that contains the family* $\{\theta(A) : A \in \mathcal{F}\}$.

(ii) Suppose $(F, \mathcal{F}, \mu)$ *admits a dense topology* τ. *Then every function* $f \in \mathcal{L}^\infty$ *is* μ*-almost surely* τ*-continuous, and there exists a lifting* T_τ *such that* $T_\tau f(x) = f(x)$ *at all* τ*-continuity points* x *of* f.

(iii) If $(F, \mathcal{F}, \mu)$ *admits a lifting, then it admits a density.*

Proof. (i) Given a density θ, let τ_θ be the topology generated by the sets $\theta(A) \setminus N$, $A \in \mathcal{F}$, $\mu(N) = 0$. It has the basis τ_0 of sets of the form

$$\bigcap_{i=1}^{I} \theta(A_i) \setminus N_i \,, \qquad\qquad I \in \mathbb{N},\ A_i \in \mathcal{F},\ \mu(N_i) = 0 \,.$$

If such a set is negligible, it must be void by A.3.39 (ic). Also, any set $A \in \mathcal{F}$ is μ-almost surely equal to its τ_θ-open subset $\theta(A) \cap A$. The only thing perhaps not quite obvious is that $\tau_\theta \subset \mathcal{F}$. To see this, let $U \in \tau_\theta$. There is a subfamily $\mathcal{U} \subset \tau_0 \subset \mathcal{F}$ with union U. The family $\mathcal{U}^{\cup f} \subset \mathcal{F}$ of finite unions of sets in $\mathcal{U}$ also has union U. Set $u = \sup\{\mu(B) : B \in \mathcal{U}^{\cup f}\}$, let $\{B_n\}$ be a countable subset of $\mathcal{U}^{\cup f}$ with $u = \sup_n \mu(B_n)$, and set $B = \bigcup_n B_n$ and $C = \theta(F \setminus B)$. Thanks to A.3.39(ic), $B \cap C = \emptyset$ for all $B \in \mathcal{U}$, and thus $B \subset U \subset C^c \dot{=} B$. Since $\mathcal{F}$ is μ-complete, $U \in \mathcal{F}$.

(ii) A set $A \in \mathcal{F}$ is evidently continuous on its τ-interior and on the τ-interior of its complement A^c; since these two sets add up almost everywhere to F, A is almost everywhere τ-continuous. A linear combination of sets in $\mathcal{F}$ is then clearly also almost everywhere continuous, and then so is the uniform limit of such. That is to say, every function in $\mathcal{L}^\infty$ is μ-almost everywhere τ-continuous.

By theorem A.2.2 there exists a map j from F into a compact space $\widehat{F}$ such that $\widehat{f} \mapsto \widehat{f} \circ j$ is an isometric algebra isomorphism of $C(\widehat{F})$ with $\mathcal{L}^\infty$.

Fix a point $x \in F$ and consider the set $\mathcal{I}_x$ of functions $f \in \mathcal{L}^\infty$ that differ negligibly from a function $f' \in \mathcal{L}^\infty$ that is zero and τ-continuous at x. This is clearly an ideal of $\mathcal{L}^\infty$. Let $\widehat{\mathcal{I}}_x$ denote the corresponding ideal of $C(\widehat{F})$. Its zero-set $\widehat{Z}_x$ is not void. Indeed, if it were, then there would be, for every $\widehat{y} \in \widehat{F}$, a function $\widehat{f}_y \in \widehat{\mathcal{I}}_x$ with $\widehat{f}_y(\widehat{y}) \neq 0$. Compactness would produce a finite subfamily $\{\widehat{f}_{y^i}\}$ with $\widehat{f} \stackrel{\text{def}}{=} \sum \widehat{f}^2_{y^i} \in \widehat{\mathcal{I}}_x$ bounded away from zero. The corresponding function $f = \widehat{f} \circ j \in \mathcal{I}_x$ would also be bounded away from zero, say $f > \epsilon > 0$. For a function $f' \doteq f$ continuous at x, $[f' < \epsilon]$ would be a negligible τ-neighborhood of x, necessarily void. This contradiction shows that $\widehat{Z}_x \neq \emptyset$.

Now pick for every $x \in F$ a point $\widehat{x} \in \widehat{Z}_x$ and set

$$T_\tau f(x) \stackrel{\text{def}}{=} \widehat{f}(\widehat{x}), \qquad f \in \mathcal{L}^\infty .$$

This is the desired lifting. Clearly T_τ is linear and multiplicative. If $f \in \mathcal{L}^\infty$ is τ-continuous at x, then $g \stackrel{\text{def}}{=} f - f(x) \in \mathcal{I}_x$, $\widehat{g} \in \widehat{\mathcal{I}}_x$, and $\widehat{g}(\widehat{x}) = 0$, which signifies that $Tg(x) = 0$ and thus $T_\tau f(x) = f(x)$: the function $T_\tau f$ differs negligibly from f, namely, at most in the discontinuity points of f. If f, g differ negligibly, then $f - g$ differs negligibly from the function zero, which is τ-continuous at all points. Therefore $f - g \in \mathcal{I}_x \;\; \forall x$ and thus $T(f - g) = 0$ and $Tf = Tg$.

(iii) Finally, if T is a lifting, then its restriction to the sets of $\mathcal{F}$ (see convention A.1.5) is plainly a density. ∎

Theorem A.3.41 *Let $(F, \mathcal{F}, \mu)$ be a σ-finite measure space (exercise A.3.27) and denote by $\mathcal{F}^\mu$ the μ-completion of $\mathcal{F}$.*

(i) There exists a lifting T for $(F, \mathcal{F}^\mu, \mu)$.

(ii) Let $\mathcal{C}$ be a countable collection of bounded $\mathcal{F}$-measurable functions. There exists a set $G \in \mathcal{F}$ with $\mu(G^c) = 0$ such that $G \cdot Tf = G \cdot f$ for all f that lie in the algebra generated by $\mathcal{C}$ or in its uniform closure, in fact for all bounded f that are continuous in the topology generated by $\mathcal{C}$.

Proof. (i) We assume to start with that $\mu \geq 0$ is finite. Consider the set $\mathfrak{L}$ of all pairs $(\mathcal{A}, T^{\mathcal{A}})$, where $\mathcal{A}$ is a sub-σ-algebra of $\mathcal{F}^\mu$ that contains all μ-negligible subsets of $\mathcal{F}^\mu$ and $T^{\mathcal{A}}$ is a lifting on $(E, \mathcal{A}, \mu)$. $\mathfrak{L}$ is not void: simply take for $\mathcal{A}$ the collection of negligible sets and their complements and set $T^{\mathcal{A}} f = \int f \, d\mu$. We order $\mathfrak{L}$ by saying $(\mathcal{A}, T^{\mathcal{A}}) \ll (\mathcal{B}, T^{\mathcal{B}})$ if $\mathcal{A} \subset \mathcal{B}$ and the restriction of $T^{\mathcal{B}}$ to $\mathcal{L}^\infty(\mathcal{A})$ is $T^{\mathcal{A}}$. The proof of the theorem consists in showing that this order is inductive and that a maximal element has σ-algebra $\mathcal{F}^\mu$.

Let then $\mathfrak{C} = \{(\mathcal{A}_\sigma, T^{\mathcal{A}_\sigma}) : \sigma \in \Sigma\}$ be a chain for the order $\ll$. If the index set Σ has no countable cofinal subset, then it is easy to find an upper bound for $\mathfrak{C}$: $\mathcal{A} \stackrel{\text{def}}{=} \bigcup_{\sigma \in \Sigma} \mathcal{A}^\sigma$ is a σ-algebra and and $T^{\mathcal{A}}$, defined to coincide with $T^{\mathcal{A}_\sigma}$ on $\mathcal{A}_\sigma$ for $\sigma \in \Sigma$, is a lifting on $\mathcal{L}^\infty(\mathcal{A})$. Assume then that Σ has a countable cofinal subset – that is to say, there exists a countable subset $\Sigma_0 \subset \Sigma$ such that every $\sigma \in \Sigma$ is exceeded by some index in Σ_0. We may then assume as well that $\Sigma = \mathbb{N}$. Letting $\mathcal{B}$ denote the σ-algebra generated by $\bigcup_n \mathcal{A}_n$ we define a density θ on $\mathcal{B}$ as follows:

$$\theta(B) \stackrel{\text{def}}{=} \left[\lim_{n\to\infty} T^{\mathcal{A}_n}\mathbb{E}^\mu[B|\mathcal{A}_n] = 1 \right] , \qquad B \in \mathcal{B} .$$

The uniformly integrable martingale $\mathbb{E}^\mu[B|\mathcal{A}_n]$ converges μ-almost everywhere to B (page 75), so that $\theta(B) = B$ μ-almost everywhere. Properties a) and b) of a density are evident; as for c), observe that $B_1 \cap \ldots \cap B_k \dot{=} \emptyset$ implies $B_1 + \cdots + B_k \dot{\le} k-1$, so that due to the linearity of the $\mathbb{E}[\cdot|\mathcal{A}_n]$ and $T^{\mathcal{A}_n}$ not all of the $\theta(B_i)$ can equal 1 at any one point: $\theta(B_1) \cap \ldots \cap \theta(B_k) = \emptyset$. Now let τ denote the dense topology τ_θ provided by lemma A.3.40 and $T^{\mathcal{B}}$ the lifting T_τ (ibidem). If $A \in \mathcal{A}_n$, then $\theta(A) = T^{\mathcal{A}_n}A$ is τ-open, and so is $T^{\mathcal{A}_n}A^c = 1 - T^{\mathcal{A}_n}A$. This means that $T^{\mathcal{A}_n}A$ is τ-continuous at all points, and therefore $T^{\mathcal{A}}A = T^{\mathcal{A}_n}A$: $T^{\mathcal{B}}$ extends $T^{\mathcal{A}_n}$ and $(\mathcal{B}, T^{\mathcal{B}})$ is an upper bound for our chain. Zorn's lemma now provides a maximal element $(\mathcal{M}, T^{\mathcal{M}})$ of $\mathfrak{L}$.

It is left to be shown that $\mathcal{M} = \mathcal{F}$. By way of contradiction assume that there exists a set $F \in \mathcal{F}$ that does not belong to $\mathcal{M}$. Let $\tau^{\mathcal{M}}$ be the dense topology that comes with $T^{\mathcal{M}}$ considered as a density. Let $\underline{\mathring{F}} \stackrel{\text{def}}{=} \bigcup\{U \in \tau^{\mathcal{M}} : U \dot{\subset} F\}$ denote the *essential interior* of F, and replace F by the equivalent set $(F \cup \underline{\mathring{F}}) \setminus \underline{\mathring{F^c}}$. Let $\mathcal{N}$ be the σ-algebra generated by $\mathcal{M}$ and F,

$$\mathcal{N} = \{(M \cap F) \cup (M' \cap F^c) : M, M' \in \mathcal{M}\} ,$$

and

$$\tau^{\mathcal{N}} = \{ (U \cap F) \cup (U' \cap F^c) : U, U' \in \tau^{\mathcal{M}} \}$$

the topology generated by $\tau^{\mathcal{M}}, F, F^c$. A little set algebra shows that $\tau^{\mathcal{N}}$ is a dense topology for $\mathcal{N}$ and that the lifting $T^{\mathcal{N}}$ provided by lemma A.3.40 (ii) extends $T^{\mathcal{M}}$. $(\mathcal{N}, T^{\mathcal{N}})$ strictly exceeds $(\mathcal{M}, T^{\mathcal{M}})$ in the order $\ll$, which is the desired contradiction.

If μ is merely σ-finite, then there is a countable collection $\{F_1, F_2, \ldots\}$ of mutually disjoint sets of finite measure in $\mathcal{F}$ whose union is F. There are liftings T_n for μ on the restriction of $\mathcal{F}$ to F_n. We glue them together: $T : f \mapsto \sum T_n(F_n \cdot f)$ is a lifting for $(F, \mathcal{F}, \mu)$.

(ii) $\bigcup_{f\in\mathcal{C}}[Tf \ne f]$ is contained in a μ-negligible subset $B \in \mathcal{F}$ (see A.3.8). Set $G = B^c$. The $f \in \mathcal{L}^\infty$ with $GTf = Gf \in \mathcal{F}$ form a uniformly closed algebra that contains the algebra $\mathcal{A}$ generated by $\mathcal{C}$ and its uniform closure $\overline{\mathcal{A}}$, which is a vector lattice (theorem A.2.2) generating the same topology $\tau_{\mathcal{C}}$ as $\mathcal{C}$. Let h be bounded and continuous in that topology. There exists an increasingly directed family $\overline{\mathcal{A}}^h \subset \overline{\mathcal{A}}$ whose pointwise supremum is h (lemma A.2.19). Let $G' = G \cap TG$. Then $G'h = \sup G'\overline{\mathcal{A}}^h = \sup G'T\overline{\mathcal{A}}^h$ is lower semicontinuous in the dense topology τ_T of T. Applying this to $-h$ shows that $G'h$ is upper semicontinuous as well, so it is τ_T-continuous and therefore μ-measurable. Now $GTh \ge \sup GT\overline{\mathcal{A}}^h = \sup G\overline{\mathcal{A}}^h = Gh$. Applying this to $-h$ shows that $GTh \le Gh$ as well, so that $GTh = Gh \in \mathcal{F}$. ∎

Corollary A.3.42 (Disintegration of Measures) *Let $\boldsymbol{H}$ be a locally compact space with a countable basis for its topology and equip it with the algebra $\mathcal{H} = C_{00}(\boldsymbol{H})$ of continuous functions of compact support; let $\boldsymbol{B}$ be a set equipped with $\mathcal{E}$, a σ-finite algebra or vector lattice closed under chopping of bounded functions, and let θ be a positive σ-additive measure on $\mathcal{H} \otimes \mathcal{E}$.*

There exist a positive measure μ on $\mathcal{E}$ and a slew $\varpi \mapsto \nu_\varpi$ of positive Radon measures, one for every $\varpi \in \boldsymbol{B}$, having the following two properties: (i) for every $\phi \in \mathcal{H} \otimes \mathcal{E}$ the function $\varpi \mapsto \int \phi(\eta, \varpi)\, \nu_\varpi(d\eta)$ is measurable on the σ-algebra $\mathcal{P}$ generated by $\mathcal{E}$; (ii) for every θ-integrable function $f : \boldsymbol{H} \times \boldsymbol{B} \to \mathbb{R}$, $f(\cdot, \varpi)$ is ν_ϖ-integrable for μ-almost all $\varpi \in \boldsymbol{B}$, the function $\varpi \mapsto \int f(\eta, \varpi)\, \nu_\varpi(d\eta)$ is μ-integrable, and

$$\int_{\boldsymbol{H}\times\boldsymbol{B}} f(\eta, \varpi)\, \theta(d\eta, d\varpi) = \int_{\boldsymbol{B}} \int_{\boldsymbol{H}} f(\eta, \varpi) \nu_\varpi(d\eta)\, \mu(d\varpi) .$$

If $\theta(1_{\boldsymbol{H}} \otimes X) < \infty$ for all $X \in \mathcal{E}$, then the ν_ϖ can be chosen to be probabilities.

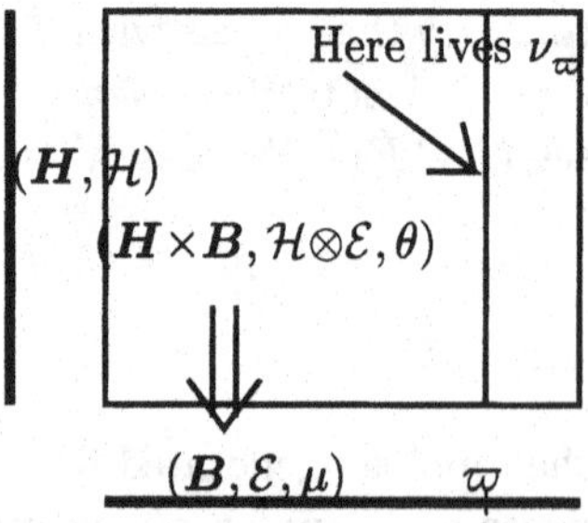

Proof. There is an increasing sequence of functions $X_i \in \mathcal{E}$ with pointwise supremum 1. The sets $P_i \stackrel{\text{def}}{=} [X_i > 1 - 1/i]$ belong to the sequential closure $\mathcal{E}^\sigma$ of $\mathcal{E}$ and increase to $\boldsymbol{B}$. Let $\mathcal{E}_0^\sigma$ denote the collection of those bounded functions in $\mathcal{E}^\sigma$ that vanish off one of the P_i. There is an obvious extension of θ to $\mathcal{H} \otimes \mathcal{E}_0^\sigma$. We shall denote it again by θ and prove the corollary with $(\mathcal{E}, \theta)$ replaced by $(\mathcal{E}_0^\sigma, \theta)$. The original claim is then immediate from the observation that every function in $\phi \in \mathcal{E}$ is the dominated limit of the sequence $\phi \cdot P_i \in \mathcal{E}_0^\sigma$. There is also an increasing sequence of compacta $K_i \subset \boldsymbol{H}$ whose interiors cover $\boldsymbol{H}$.

For an $h \in \mathcal{H}$ consider the map $\mu^h : \mathcal{E}_0^\sigma \to \mathbb{R}$ defined by

$$\mu^h(X) = \int h \cdot X\, d\theta , \qquad X \in \mathcal{E}_0^\sigma$$

and set

$$\mu = \sum_i a_i \mu^{K_i} ,$$

where the $a_i > 0$ are chosen so that $\sum a_i \mu^{K_i}(1) < \infty$. Then μ^h is a measure whose variation $\mu^{|h|}$ is majorized by a multiple of μ. Indeed, there is an index i such that K_i contains the support of h, and then $\mu^{|h|} \le a_i^{-1} \| h \|_\infty \cdot \mu$. There exists a bounded Radon–Nikodym derivative $\dot{g}^h = d\mu^h / d\mu$. The map $\mathcal{H} \ni h \mapsto \dot{g}^h \in L^\infty(\mu)$ is clearly positive and linear. Fix a lifting $T : L^\infty(\mu) \to \mathcal{L}^\infty(\mu)$, producing the set C of theorem A.3.41 (ii) by picking for every h in a uniformly dense countable subcollection of $\mathcal{H}$ a representative $g^h \in \dot{g}^h$ that is measurable on $\mathcal{P}$. There then exists a set $G \in \mathcal{P}$ of

μ-negligible complement such that $G{\cdot}g^h = G{\cdot}T\dot{g}^h$ for all $h \in \mathcal{H}$. We define now the maps $\nu_\varpi : \mathcal{H} \to \mathbb{R}$, one for every $\varpi \in \boldsymbol{B}$, by

$$\nu_\varpi(h) = G(\varpi) \cdot T_\varpi(\dot{g}^h) .$$

As positive linear functionals on $\mathcal{H}$ they are Radon measures, and for every $h \in \mathcal{H}$, $\varpi \mapsto \nu_\varpi(h)$ is $\mathcal{P}$-measurable. Let $\phi = \sum_k h_k X_k \in \mathcal{H} \otimes \mathcal{E}_0^\sigma$. Then

$$\begin{aligned}\int \phi(\eta, \varpi)\, \theta(d\eta, d\varpi) &= \sum_k \int X_k(\varpi)\, \mu^{h_k}(d\varpi) = \sum_k \int X_k \dot{g}^{h_k}\, d\mu \\ &= \sum_k \int X_k \int h_k(\eta)\, \nu_\varpi(d\eta)\, \mu(d\varpi) \\ &= \int \int \phi(\eta, \varpi)\, \nu_\varpi(d\eta)\, \mu(d\varpi) .\end{aligned}$$

The functions ϕ for which the left-hand side and the ultimate right-hand side agree and for which $\varpi \mapsto \int \phi(\eta, \varpi)\, \nu_\varpi(d\eta)$ is measurable on $\mathcal{P}$ form a collection closed under $\mathcal{H} \otimes \mathcal{E}$-dominated sequential limits and thus contains $\left(\mathcal{H} \otimes \mathcal{E}\right)_{00}^\sigma$. This proves (i). Theorem A.3.18 on page 403 yields (ii). The last claim is left to the reader. ∎

Exercise A.3.43 Let $(\Omega, \mathcal{F}, \mu)$ be a measure space, $\mathcal{C}$ a countable collection of μ-measurable functions, and τ the topology generated by $\mathcal{C}$. Every τ-continuous function is μ-measurable.

Exercise A.3.44 Let E be a separable metric space and $\mu : C_b(E) \to \mathbb{R}$ a σ-continuous positive measure. There exists a ***strong lifting***, that is to say, a lifting $T : \mathcal{L}^\infty(\mu) \to \mathcal{L}^\infty(\mu)$ such that $T\phi(x) = \phi(x)$ for all $\phi \in C_b(E)$ and all x in the support of μ.

Gaussian and Poisson Random Variables

The ***centered Gaussian distribution*** with ***variance*** t is denoted by γ_t:

$$\gamma_t(dx) = \frac{1}{\sqrt{2\pi t}} e^{-x^2/2t}\, dx .$$

A real-valued random variable X whose law is γ_t is also said to be $N(0,t)$, pronounced "normal zero–t." The ***standard deviation*** of such X or of γ_t by definition is $\sqrt{t}$; if it equals 1, then X is said to be a ***normalized Gaussian***. Here are a few elementary integrals involving Gaussian distributions. They are used on various occasions in the main text. $|\boldsymbol{a}|$ stands for the Euclidean norm $|\boldsymbol{a}|_2$.

Exercise A.3.45 A $N(0,t)$-random variable X has expectation $\mathbb{E}[X] = 0$, variance $\mathbb{E}[X^2] = t$, and its characteristic function is

$$\mathbb{E}\Big[e^{i\xi X}\Big] = \int_{\mathbb{R}} e^{i\xi x} \gamma_t(dx) = e^{-t\xi^2/2} . \tag{A.3.17}$$

Exercise A.3.46 The ***Gamma function*** Γ, defined for complex z with a strictly positive real part by

$$\Gamma(z) = \int_0^\infty u^{z-1} e^{-u}\, du ,$$

is convex on $(0,\infty)$ and satisfies $\Gamma(1/2) = \sqrt{\pi}$, $\Gamma(1) = 1$, $\Gamma(z+1) = z \cdot \Gamma(z)$, and $\Gamma(n+1) = n!$ for $n \in \mathbb{N}$.

Exercise A.3.47 The Gauß kernel has moments

$$\int_{-\infty}^{+\infty} |x|^p\, \gamma_t(dx) = \frac{(2t)^{p/2}}{\sqrt{\pi}} \cdot \Gamma\Big(\frac{p+1}{2}\Big) , \qquad (p > -1).$$

Now let $X_1, \ldots, X_n$ be independent Gaussians distributed $N(0,t)$. Then the distribution of the vector $\boldsymbol{X} = (X_1, \ldots, X_n) \in \mathbb{R}^n$ is $\gamma_{tI}(\boldsymbol{x})d\boldsymbol{x}$, where

$$\gamma_{tI}(\boldsymbol{x}) \stackrel{\text{def}}{=} \frac{1}{(\sqrt{2\pi t})^n}\, e^{-|\boldsymbol{x}|^2/2t}$$

is the n-dimensional ***Gauß kernel*** or ***heat kernel***. I indicates the identity matrix.

Exercise A.3.48 The characteristic function of the vector $\boldsymbol{X}$ (or of γ_{tI}) is $e^{-t|\boldsymbol{\xi}|^2/2}$. Consequently, the law of $\boldsymbol{\xi}$ is invariant under rotations of $\mathbb{R}^n$. Next let $0 < p < \infty$ and assume that $t = 1$. Then

$$\frac{1}{(\sqrt{2\pi})^n} \int_{\mathbb{R}^n} |\boldsymbol{x}|^p\, e^{-|\boldsymbol{x}|^2/2}\, dx_1\, dx_2 \ldots dx_n = \frac{2^{p/2} \cdot \Gamma\left(\frac{n+p}{2}\right)}{\Gamma(\frac{n}{2})} ,$$

and for any vector $\boldsymbol{a} \in \mathbb{R}^n$

$$\frac{1}{(\sqrt{2\pi})^n} \int_{\mathbb{R}^n} \Big|\sum_{\nu=1}^n x_\nu \cdot a_\nu\Big|^p e^{-|\boldsymbol{x}|^2/2}\, dx_1 \ldots dx_n = \frac{|\boldsymbol{a}|^p \cdot 2^{p/2}\Gamma\left(\frac{p+1}{2}\right)}{\sqrt{\pi}} .$$

Consider next a symmetric ***positive semidefinite*** $d \times d$-matrix B; that is to say, $x_\eta x_\theta B^{\eta\theta} \geq 0$ for every $\boldsymbol{x} \in \mathbb{R}^d$.

Exercise A.3.49 There exists a matrix U that depends continuously on B such that $B^{\eta\theta} = \sum_{\iota=1}^n U_\iota^\eta U_\iota^\theta$.

Definition A.3.50 *The* ***Gaussian with covariance matrix B*** *or* ***centered normal distribution with covariance matrix B*** *is the image of the heat kernel γ_I under the linear map $U : \mathbb{R}^d \to \mathbb{R}^d$.*

Exercise A.3.51 The name is justified by these facts: the ***covariance matrix*** $\int_{\mathbb{R}^d} x^\eta x^\theta\ \gamma_B(dx)$ equals $B^{\eta\theta}$; for any $t > 0$ the characteristic function of γ_{tB} is given by

$$\widehat{\gamma_{tB}}(\boldsymbol{\xi}) = e^{-t\xi_\eta \xi_\theta B^{\eta\theta}/2} .$$

Changing topics: a random variable N that takes only positive integer values is ***Poisson*** with mean $\lambda > 0$ if

$$\mathbb{P}[N = n] = e^{-\lambda}\frac{\lambda^n}{n!} , \qquad n = 0, 1, 2 \ldots .$$

Exercise A.3.52 Its characteristic function $\widehat{N}$ is given by

$$\widehat{N}(\alpha) \stackrel{\text{def}}{=} \mathbb{E}\Big[e^{i\alpha N}\Big] = e^{\lambda(e^{i\alpha}-1)} .$$

The sum of independent Poisson random variables N_i with means λ_i is Poisson with mean $\sum \lambda_i$.

A.4 Weak Convergence of Measures

In this section we fix a completely regular space E and consider σ-continuous measures μ of finite total variation $|\mu|(1) = \|\mu\|$ on the lattice algebra $C_b(E)$. Their collection is $\mathfrak{M}^*(E)$. Each has an extension that integrates all bounded Baire functions and more. The order-continuous elements of $\mathfrak{M}^*(E)$ form the collection $\mathfrak{M}^{\cdot}(E)$. The positive[22] σ-continuous measures of total mass 1 are the ***probabilities*** on E, and their collection is denoted by $\mathfrak{M}^*_{1,+}(E)$ or $\mathfrak{P}^*(E)$. We shall be concerned mostly with the order-continuous probabilities on E; their collection is denoted by $\mathfrak{M}^{\cdot}_{1,+}(E)$ or $\mathfrak{P}^{\cdot}(E)$. Recall from exercise A.3.10 that $\mathfrak{P}^*(E) = \mathfrak{P}^{\cdot}(E)$ when E is separable and metrizable. The advantage that the order-continuity of a measure conveys is that every Borel set, in particular every compact set, is integrable with respect to it under the integral extension discussed on pages 398–400.

Equipped with the uniform norm $C_b(E)$ is a Banach space, and $\mathfrak{M}^*(E)$ is a subset of the dual $C_b^*(E)$ of $C_b(E)$. The pertinent topology on $\mathfrak{M}^*(E)$ is the trace of the weak*-topology on $C_b^*(E)$; unfortunately, probabilists call the corresponding notion of convergence ***weak convergence***,[37] and so *nolens volens* will we: a sequence[38] (μ_n) in $\mathfrak{M}^*(E)$ converges weakly to $\mu \in \mathfrak{P}^*(E)$, written $\mu_n \Rightarrow \mu$, if

$$\mu_n(\phi) \xrightarrow[n\to\infty]{} \mu(\phi) \qquad \forall\, \phi \in C_b(E)\,.$$

In the typical application made in the main body E is a path space $\mathscr{C}$ or $\mathscr{D}$, and μ_n, μ are the laws of processes $X^{(n)}, X$ considered as E-valued random variables on probability spaces $\big(\Omega^{(n)}, \mathcal{F}^{(n)}, \mathbb{P}^{(n)}\big)$, which may change with n. In this case one also writes $X^{(n)} \Rightarrow X$ and says that $X^{(n)}$ ***converges to*** X ***in law*** or ***in distribution***.

It is generally hard to verify the convergence of μ_n to μ on every single function of $C_b(E)$. Our first objective is to reduce the verification to fewer functions.

Proposition A.4.1 *Let $\mathcal{M} \subset C_b(E;\mathbb{C})$ be a multiplicative class that is closed under complex conjugation and generates the topology,*[14] *and let μ_n, μ belong to $\mathfrak{P}^{\cdot}(E)$.*[39] *If $\mu_n(\phi) \to \mu(\phi)$ for all $\phi \in \mathcal{M}$, then $\mu_n \Rightarrow \mu$*[38]*; moreover,*

(i) For h bounded and lower semicontinuous, $\int h\,d\mu \le \liminf_{n\to\infty} \int h\,d\mu_n$; and for k bounded and upper semicontinuous, $\int k\,d\mu \ge \limsup_{n\to\infty} \int k\,d\mu_n$.

(ii) If f is a bounded function that is integrable for every one of the μ_n and is μ-almost everywhere continuous, then still $\int f\,d\mu = \lim_{n\to\infty} \int f\,d\mu_n$.

---

[37] Sometimes called "strict convergence," "convergence étroite" in French. In the parlance of functional analysts, weak convergence of measures is convergence for the trace of the weak*-topology (!) $\sigma(C_b^*(E), C_b(E))$ on $\mathfrak{P}^*(E)$; they reserve the words "weak convergence" for the weak topology $\sigma(C_b(E), C_b^*(E))$ on $C_b(E)$. See item A.2.32 on page 381.

[38] Everything said applies to nets and filters as well.

[39] The proof shows that it suffices to check that the μ_n are σ-continuous and that μ is order-continuous on the real part of the algebra generated by $\mathcal{M}$.

Proof. Since $\mu_n(1) = \mu(1) = 1$, we may assume that $1 \in \mathcal{M}$. It is easy to see that the lattice algebra $\mathcal{A}[\mathcal{M}]$ constructed in the proof of proposition A.3.12 on page 399 still generates the topology and that $\mu_n(\phi) \to \mu(\phi)$ for all of its functions ϕ. In other words, we may assume that $\mathcal{M}$ is a lattice algebra.

(i) We know from lemma A.2.19 on page 376 that there is an increasingly directed set $\mathcal{M}^h \subset \mathcal{M}$ whose pointwise supremum is h. If $a < \int h\,d\mu$,[40] there is due to the order-continuity of μ a function $\phi \in \mathcal{M}$ with $\phi \le h$ and $a < \mu(\phi)$. Then $a < \liminf \mu_n(\phi) \le \liminf \int h\,d\mu_n$. Consequently $\int h\,d\mu \le \liminf \int h\,d\mu_n$. Applying this to $-k$ gives the second claim of (i).

(ii) Set $k(x) \stackrel{\text{def}}{=} \limsup_{y\to x} f(y)$ and $h(x) \stackrel{\text{def}}{=} \liminf_{y\to x} f(y)$. Then k is upper semicontinuous and h lower semicontinuous, both bounded. Due to (i),

$$\limsup \int f\,d\mu_n \le \limsup \int k\,d\mu_n$$

as $h = f = k$ μ-a.e.: $$\le \int k\,d\mu = \int f\,d\mu = \int h\,d\mu$$

$$\le \liminf \int h\,d\mu_n \le \liminf \int f\,d\mu_n\ :$$

equality must hold throughout. *A fortiori*, $\mu_n \Rightarrow \mu$. ∎

For an application consider the case that E is separable and metrizable. Then every σ-continuous measure on $C_b(E)$ is automatically order-continuous (see exercise A.3.10). If $\mu_n \to \mu$ on *uniformly* continuous bounded functions, then $\mu_n \Rightarrow \mu$ and the conclusions (i) and (ii) persist. Proposition A.4.1 not only reduces the need to check $\mu_n(\phi) \to \mu(\phi)$ to fewer functions ϕ, it can also be used to deduce $\mu_n(f) \to \mu(f)$ for more than μ-almost surely continuous functions f once $\mu_n \Rightarrow \mu$ is established:

Corollary A.4.2 *Let E be a completely regular space and (μ_n) a sequence*[38] *of order-continuous probabilities on E that converges weakly to $\mu \in \mathfrak{P}^{\bullet}(E)$. Let F be a subset of E, not necessarily measurable, that has full measure for every μ_n and for μ (i.e., $\int^{\bullet} F\,d\mu = \int^{\bullet} F\,d\mu_n = 1 \quad \forall n$). Then $\int f\,d\mu_n \to \int f\,d\mu$ for every bounded function f that is integrable for every μ_n and for μ and whose restriction to F is μ-almost everywhere continuous.*

Proof. Let $\mathcal{E}$ denote the collection of restrictions $\phi_{|F}$ to F of functions ϕ in $C_b(E)$. This is a lattice algebra of bounded functions on F and generates the induced topology. Let us define a positive linear functional $\mu_{|F}$ on $\mathcal{E}$ by

$$\mu_{|F}(\phi_{|F}) \stackrel{\text{def}}{=} \mu(\phi)\,, \qquad \phi \in C_b(E)\,.$$

$\mu_{|F}$ is well-defined; for if $\phi, \phi' \in C_b(E)$ have the same restriction to F, then $F \subset [\phi = \phi']$, so that the Baire set $[\phi \ne \phi']$ is μ-negligible and consequently

[40] This is of course the μ-integral under the extension discussed on pages 398–400.

$\mu(\phi) = \mu(\phi')$. $\mu_{|F}$ is also order-continuous on $\mathcal{E}$. For if $\Phi \subset \mathcal{E}$ is decreasingly directed with pointwise infimum zero on F, without loss of generality consisting of the restrictions to F of a decreasingly directed family $\Psi \subset C_b(E)$, then $[\inf \Psi = 0]$ is a Borel set of E containing F and thus has μ-negligible complement: $\inf \mu_{|F}(\Phi) = \inf \mu(\Psi) = \mu(\inf \Psi) = 0$. The extension of $\mu_{|F}$ discussed on pages 398–400 integrates all bounded functions of $\mathcal{E}^{\Uparrow}$, among which are the bounded continuous functions of F (lemma A.2.19), and the integral is order-continuous on $C_b(F)$ (equation (A.3.3)). We might as well identify $\mu_{|F}$ with the probability in $\mathfrak{P}^{\bullet}(F)$ so obtained. The order-continuous mean $f \to \| f_{|F} \|^{\bullet}_{\mu_{|F}}$ is the same whether built with $\mathcal{E}$ or $C_b(F)$ as the elementary integrands,[27] agrees with $\| \ \|^{\bullet}_{\mu}$ on $C_{b+}(E)$, and is thus smaller than the latter. From this observation it is easy to see that if $f : E \to \mathbb{R}$ is μ-integrable, then its restriction to F is $\mu_{|F}$-integrable and

$$\int f_{|F} \, d\mu_{|F} = \int f \, d\mu \,. \tag{A.4.1}$$

The same remarks apply to the $\mu_{n|F}$. We are in the situation of proposition A.4.1: $\mu_n \Rightarrow \mu$ clearly implies $\mu_{n|F}(\psi_{|F}) \to \mu_{|F}(\psi_{|F})$ for all $\psi_{|F}$ in the multiplicative class $\mathcal{E}$ that generates the topology, and therefore $\mu_{n|F}(\phi) \to \mu_{|F}(\phi)$ for all bounded functions ϕ on F that are $\mu_{|F}$-almost everywhere continuous.

This translates easily into the claim. Namely, the set of points in F where $f_{|F}$ is discontinuous is by assumption μ-negligible, so by (A.4.1) it is $\mu_{|F}$-negligible: $f_{|F}$ is $\mu_{|F}$-almost everywhere continuous. Therefore

$$\int f \, d\mu_n = \int f_{|F} \, d\mu_{n|F} \to \int f_{|F} \, d\mu_{|F} = \int f \, d\mu \,. \qquad \blacksquare$$

Proposition A.4.1 also yields the Continuity Theorem on $\mathbb{R}^d$ without further ado. Namely, since the complex multiplicative class $\{x \mapsto e^{i\langle x|\alpha\rangle} : \alpha \in \mathbb{R}^d\}$ generates the topology of $\mathbb{R}^d$ (lemma A.3.33), the following is immediate:

Corollary A.4.3 (The Continuity Theorem) *Let μ_n be a sequence of probabilities on $\mathbb{R}^d$ and assume that their characteristic functions $\widehat{\mu}_n$ converge pointwise to the characteristic function $\widehat{\mu}$ of a probability μ. Then $\mu_n \Rightarrow \mu$, and the conclusions (i) and (ii) of proposition A.4.1 continue to hold.*

Theorem A.4.4 (Central Limit Theorem with Lindeberg Criteria) *For $n \in \mathbb{N}$ let $X_n^1, \ldots, X_n^{r_n}$ be independent random variables, defined on probability spaces $(\Omega_n, \mathcal{F}_n, \mathbb{P}_n)$. Assume that $\mathbb{E}_n[X_n^k = 0]$ and $(\sigma_n^k)^2 \overset{\text{def}}{=} \mathbb{E}_n[|X_n^k|^2] < \infty$ for all $n \in \mathbb{N}$ and $k \in \{1, \ldots, r_n\}$, set $S_n = \sum_{k=1}^{r_n} X_n^k$ and $s_n^2 = \mathrm{var}(S_n) = \sum_{k=1}^{r_n} |\sigma_n^k|^2$, and assume the* ***Lindeberg condition***

$$\frac{1}{s_n^2} \sum_{k=1}^{r_n} \int_{|X_n^k| > \epsilon s_n} |X_n^k|^2 \, d\mathbb{P}_n \xrightarrow[n\to\infty]{} 0 \quad \textit{for all } \epsilon > 0 \,. \tag{A.4.2}$$

Then S_n / s_n converges in law to a normalized Gaussian random variable.

Proof. Corollary A.4.3 reduces the problem to showing that the characteristic functions $\widehat{S_n}(\xi)$ of S_n converge to $e^{-\xi^2/2}$ (see A.3.45). This is a standard estimate [10]: replacing X_n^k by X_n^k/s_n we may assume that $s_n = 1$. The inequality[27]

$$\left|e^{i\xi x} - (1 + i\xi x - \xi^2 x^2/2)\right| \le |\xi x|^2 \wedge |\xi x|^3$$

results in the inequality

$$\begin{aligned}\left|\widehat{X_n^k}(\xi) - (1 - \xi^2|\sigma_n^k|^2/2)\right| &\le \mathbb{E}_n\left[\left|\xi X_n^k\right|^2 \wedge \left|\xi X_n^k\right|^3\right] \\ &\le \xi^2 \int_{|X_n^k|\ge\epsilon} \left|X_n^k\right|^2 d\mathbb{P}_n + \int_{|X_n^k|<\epsilon} \left|\xi X_n^k\right|^3 d\mathbb{P}_n \\ &\le \xi^2 \int_{|X_n^k|\ge\epsilon} \left|X_n^k\right|^2 d\mathbb{P}_n + \epsilon|\xi|^3|\sigma_n^k|^2 \qquad \forall\, \epsilon > 0\end{aligned}$$

for the characteristic function of X_n^k. Since the $|\sigma_n^k|^2$ sum to 1 and $\epsilon > 0$ is arbitrary, Lindeberg's condition produces

$$\sum_{k=1}^{r_n} \left|\widehat{X_n^k}(\xi) - (1 - \xi^2|\sigma_n^k|^2/2)\right| \xrightarrow[n\to\infty]{} 0 \tag{A.4.3}$$

for any fixed ξ. Now for $\epsilon > 0$, $|\sigma_n^k|^2 \le \epsilon^2 + \int_{|X_n^k|\ge\epsilon} \left|X_n^k\right|^2 d\mathbb{P}_n$, so Lindeberg's condition also gives $\max_{k=1}^{r_n} \sigma_n^k \xrightarrow[n\to\infty]{} 0$. Henceforth we fix a ξ and consider only indices n large enough to ensure that $\left|1 - \xi^2|\sigma_n^k|^2/2\right| \le 1$ for $1 \le k \le r_n$. Now if $z_1, \dots, z_m$ and $w_1, \dots, w_m$ are complex numbers of absolute value less than or equal to one, then

$$\left|\prod_{k=1}^m z_k - \prod_{k=1}^m w_k\right| \le \sum_{k=1}^m \left|z_k - w_k\right| , \tag{A.4.4}$$

so (A.4.3) results in

$$\widehat{S_n}(\xi) = \prod_{k=1}^{r_n} \widehat{X_n^k}(\xi) = \prod_{k=1}^{r_n} (1 - \xi^2|\sigma_n^k|^2/2) + R_n , \tag{A.4.5}$$

where $R_n \xrightarrow[n\to\infty]{} 0$: it suffices to show that the product on the right converges to $e^{-\xi^2/2} = \prod_{k=1}^{r_n} e^{-\xi^2|\sigma_n^k|^2/2}$. Now (A.4.4) also implies that

$$\left|\prod_{k=1}^{r_n} e^{-\xi^2|\sigma_n^k|^2/2} - \prod_{k=1}^{r_n} (1 - \xi^2|\sigma_n^k|^2/2)\right| \le \sum_{k=1}^{r_n} \left|e^{-\xi^2|\sigma_n^k|^2/2} - (1 - \xi^2|\sigma_n^k|^2/2)\right| .$$

Since $|e^{-x} - (1-x)| \le x^2$ for $x \in \mathbb{R}_+$,[27] the left-hand side above is majorized by

$$\xi^4 \sum_{k=1}^{r_n} |\sigma_n^k|^4 \le \xi^4 \max_{k=1}^{r_n} |\sigma_n^k|^2 \times \sum_{k=1}^{r_n} |\sigma_n^k|^2 \xrightarrow[n\to\infty]{} 0 .$$

This in conjunction with (A.4.5) yields the claim. ∎

Uniform Tightness

Unless the underlying completely regular space E is $\mathbb{R}^d$, as in corollary A.4.3, or the topology of E is rather weak, it is hard to find multiplicative classes of bounded functions that define the topology, and proposition A.4.1 loses its utility. There is another criterion for the weak convergence $\mu_n \Rightarrow \mu$, though, one that can be verified in many interesting cases, to wit that the family $\{\mu_n\}$ be uniformly tight and converge on a multiplicative class that separates the points.

Definition A.4.5 *The set $\mathfrak{M}$ of measures in $\mathfrak{M}^{\cdot}(E)$ is* ***uniformly tight*** *if $M \overset{\text{def}}{=} \sup\{\,|\mu|(1) : \mu \in \mathfrak{M}\}$ is finite and if for every $\alpha > 0$ there is a compact subset $K_\alpha \subset E$ such that $\sup\{\,|\mu|(K_\alpha^c)$[40] $: \mu \in \mathfrak{M}\} < \alpha$.*

A set $\mathfrak{P} \subset \mathfrak{P}^{\cdot}(E)$ clearly is uniformly tight if and only if for every $\alpha < 1$ there is a compact set K_α such that $\mu(K_\alpha) \ge 1 - \alpha$ for all $\mu \in \mathfrak{P}$.

Proposition A.4.6 (Prokhoroff) *A uniformly tight collection $\mathfrak{M} \subset \mathfrak{M}^{\cdot}(E)$ is relatively compact in the topology of weak convergence of measures*[37]*; the closure of $\mathfrak{M}$ belongs to $\mathfrak{M}^{\cdot}(E)$ and is uniformly tight as well.*

Proof. The theorem of Alaoglu, a simple consequence of Tychonoff's theorem, shows that the closure of $\mathfrak{M}$ in the topology $\sigma\big(C_b^*(E), C_b(E)\big)$ consists of linear functionals on $C_b(E)$ of total variation less than M. What may not be entirely obvious is that a limit point μ' of $\mathfrak{M}$ is order-continuous. This is rather easy to see, though. Namely, let $\Phi \subset C_b(E)$ be decreasingly directed with pointwise infimum zero. Pick a $\phi_0 \in \Phi$. Given an $\alpha > 0$, find a compact set K_α as in definition A.4.5. Thanks to Dini's theorem A.2.1 there is a $\phi_\alpha \le \phi_0$ in Φ smaller than α on all of K_α. For any $\phi \in \Phi$ with $\phi \le \phi_\alpha$

$$|\mu(\phi)| \le \alpha |\mu|(K_\alpha) + \int_{K_\alpha^c} \phi_\alpha \, d|\mu| \le \alpha(M + \|\phi_0\|_\infty) \qquad \forall \mu \in \mathfrak{M}.$$

This inequality will also hold for the limit point μ'. That is to say, $\mu'(\Phi) \to 0$: μ' is order-continuous.

If ϕ is any continuous function less than[13] K_α^c, then $|\mu(\phi)| \le \alpha$ for all $\mu \in \mathfrak{M}$ and so $|\mu'(\phi)| \le \alpha$. Taking the supremum over such ϕ gives $|\mu'|(K_\alpha^c) \le \alpha$: the closure of $\mathfrak{M}$ is "just as uniformly tight as $\mathfrak{M}$ itself." ∎

Corollary A.4.7 *Let (μ_n) be a uniformly tight sequence*[38] *in $\mathfrak{P}^{\cdot}(E)$ and assume that $\mu(\phi) = \lim \mu_n(\phi)$ exists for all ϕ in a complex multiplicative class $\mathcal{M}$ of bounded continuous functions that separates the points. Then (μ_n) converges weakly*[37] *to an order-continuous tight measure that agrees with μ on $\mathcal{M}$. Denoting this limit again by μ we also have the conclusions (i) and (ii) of proposition A.4.1.*

Proof. All limit points of $\{\mu_n\}$ agree on $\mathcal{M}$ and are therefore identical (see proposition A.3.12). ∎

Exercise A.4.8 There exists a partial converse of proposition A.4.6, which is used in section 5.5 below: if E is polish, then a relatively compact subset $\mathfrak{P}$ of $\mathfrak{P}^{\bullet}(E)$ is uniformly tight.

Application: Donsker's Theorem

Recall the normalized random walk

$$Z_t^{(n)} = \frac{1}{\sqrt{n}} \sum_{k \le tn} X_k^{(n)} = \frac{1}{\sqrt{n}} \sum_{k \le [tn]} X_k^{(n)} , \qquad t \ge 0 ,$$

of example 2.5.26. The $X_k^{(n)}$ are independent Bernoulli random variables with $\mathbb{P}^{(n)}[X_k^{(n)} = 1] = 1/2$; they may be living on probability spaces $(\Omega^{(n)}, \mathcal{F}^{(n)}, \mathbb{P}^{(n)})$ that vary with n. The Central Limit Theorem easily shows[27] that, for every fixed instant t, $Z_t^{(n)}$ converges in law to a Gaussian random variable with expectation zero and variance t. Donsker's theorem extends this to the whole path: viewed as a random variable with values in the path space $\mathscr{D}$, $Z^{(n)}$ converges in law to a standard Wiener process W. The pertinent topology on $\mathscr{D}$ is the topology of uniform convergence on compacta; it is defined by, and complete under, the metric

$$\rho(z, z') = \sum_{u \in \mathbb{N}} 2^{-u} \wedge \sup_{0 \le s \le u} \bigl| z(s) - z'(s) \bigr| , \qquad z, z' \in \mathscr{D} .$$

What we mean by $Z^{(n)} \Rightarrow W$ is this: for all continuous bounded functions ϕ on $\mathscr{D}$,

$$\mathbb{E}^{(n)}\bigl[\phi(Z_.^{(n)})\bigr] \xrightarrow[n\to\infty]{} \mathbb{E}[\phi(W_.)] . \tag{A.4.6}$$

It is necessary to spell this out, since a priori the law $\mathbb{W}$ of a Wiener process is a measure on $\mathscr{C}$, while the $Z^{(n)}$ take values in $\mathscr{D}$ – so how then can the law $\mathbb{Z}^{(n)}$ of $Z^{(n)}$, which lives on $\mathscr{D}$, converge to $\mathbb{W}$? Equation (A.4.6) says how: read Wiener measure as the probability

$$\overline{\mathbb{W}} : \phi \mapsto \int \phi_{|\mathscr{C}} \, d\mathbb{W} , \qquad \phi \in C_b(\mathscr{D}) ,$$

on $\mathscr{D}$. Since the restrictions $\phi_{|\mathscr{C}}$, $\phi \in C_b(\mathscr{D})$, belong to $C_b(\mathscr{C})$, $\overline{\mathbb{W}}$ is actually order-continuous (exercise A.3.10). Now $\mathscr{C}$ is a Borel set in $\mathscr{D}$ (exercise A.2.23) that carries $\overline{\mathbb{W}}$,[27] so we shall henceforth identify $\mathbb{W}$ with $\overline{\mathbb{W}}$ and simply write $\mathbb{W}$ for both.

The left-hand side of (A.4.6) raises a question as well: what is the meaning of $\mathbb{E}^{(n)}\bigl[\phi(Z^{(n)})\bigr]$? Observe that $Z^{(n)}$ takes values in the subspace $\mathscr{D}^{(n)} \subset \mathscr{D}$ of paths that are constant on intervals of the form $[k/n, (k+1)/n)$, $k \in \mathbb{N}$, and take values in the discrete set $\mathbb{N}/\sqrt{n}$. One sees as in exercise 1.2.4 that $\mathscr{D}^{(n)}$ is separable and complete under the metric ρ and that the evaluations

$z \mapsto z_t$, $t \geq 0$, generate the Borel σ-algebra of this space. We see as above for Wiener measure that

$$\mathbb{Z}^{(n)} : \phi \mapsto \mathbb{E}^{(n)}\big[\phi(Z^{(n)}_{.})\big] , \qquad \phi \in C_b(\mathscr{D}) ,$$

defines an order-continuous probability on $\mathscr{D}$. It makes sense to state the

Theorem A.4.9 (Donsker) $\mathbb{Z}^{(n)} \Rightarrow \mathbb{W}$. *In other words, the $Z^{(n)}$ converge in law to a standard Wiener process.*

We want to show this using corollary A.4.7, so there are two things to prove: **1)** the laws $\mathbb{Z}^{(n)}$ form a uniformly tight family of probabilities on $C_b(\mathscr{D})$, and **2)** there is a multiplicative class $\mathcal{M} = \overline{\mathcal{M}} \subset C_b(\mathscr{D};\mathbb{C})$ separating the points so that

$$\mathbb{E}^{\mathbb{Z}^{(n)}}[\phi] \xrightarrow[n\to\infty]{} \mathbb{E}^{\mathbb{W}}[\phi] \qquad \forall\, \phi \in \mathcal{M} .$$

We start with point 2). Let Γ denote the vector space of all functions $\gamma : [0,\infty) \to \mathbb{R}$ of compact support that have a continuous derivative $\dot\gamma$. We view $\gamma \in \Gamma$ as the cumulative distribution function of the measure $d\gamma_t = \dot\gamma_t dt$ and also as the functional $z. \mapsto \langle z.|\gamma\rangle \stackrel{\text{def}}{=} \int_0^\infty z_t \; d\gamma_t$. We set $\mathcal{M} = e^{i\Gamma} \stackrel{\text{def}}{=} \{e^{i\langle\cdot|\gamma\rangle} : \gamma \in \Gamma\}$ as on page 410. Clearly $\mathcal{M}$ is a multiplicative class closed under complex conjugation and separating the points; for if $e^{i\int_0^\infty z_t\, d\gamma_t} = e^{i\int_0^\infty z'_t\, d\gamma_t}$ for all $\gamma \in \Gamma$, then the two right-continuous paths $z, z' \in \mathscr{D}$ must coincide.

Lemma A.4.10 $\mathbb{E}^{(n)}\left[e^{i\int_0^\infty Z_t^{(n)}\, d\gamma_t}\right] \xrightarrow[n\to\infty]{} e^{-\frac{1}{2}\int_0^\infty \gamma_t^2\, dt}$.

Proof. Repeated applications of l'Hospital's rule show that $(\tan x - x)/x^3$ has a finite limit as $x \to 0$, so that $\tan x = x + O(x^3)$ at $x = 0$. Integration gives $\ln\cos x = -x^2/2 + O(x^4)$. Since γ is continuous and bounded and vanishes after some finite instant, therefore,

$$\sum_{k=1}^{\infty} \ln\cos\left(\frac{\gamma_{k/n}}{\sqrt{n}}\right) = -\frac{1}{2}\sum_{k=1}^{\infty}\frac{\gamma_{k/n}^2}{n} + O(1/n) \xrightarrow[n\to\infty]{} -\frac{1}{2}\int_0^\infty \gamma_t^2\, dt ,$$

and so
$$\prod_{k=1}^{\infty} \cos\left(\frac{\gamma_{k/n}}{\sqrt{n}}\right) \xrightarrow[n\to\infty]{} e^{-\frac{1}{2}\int_0^\infty \gamma_t^2\, dt} . \qquad (*)$$

Now
$$\int_0^\infty Z_t^{(n)}\, d\gamma_t = -\int_0^\infty \gamma_t\, dZ_t^{(n)} = -\sum_{k=1}^{\infty}\frac{\gamma_{k/n}}{\sqrt{n}} \cdot X_k^{(n)} ,$$

and so
$$\mathbb{E}^{(n)}\left[e^{i\int_0^\infty Z_t^{(n)}\, d\gamma_t}\right] = \mathbb{E}^{(n)}\left[e^{-i\sum_{k=1}^\infty \frac{\gamma_{k/n}}{\sqrt{n}}\cdot X_k^{(n)}}\right]$$
$$= \prod_{k=1}^{\infty}\cos\left(\frac{\gamma_{k/n}}{\sqrt{n}}\right) \xrightarrow[n\to\infty]{} e^{-\frac{1}{2}\int_0^\infty \gamma_t^2\, dt} .$$

∎

Now to point 1), the tightness of the $\mathbb{Z}^{(n)}$. To start with we need a criterion for compactness in $\mathscr{D}$. There is an easy generalization of the Ascoli–Arzela theorem A.2.38:

Lemma A.4.11 *A subset $\mathcal{K} \subset \mathscr{D}$ is relatively compact if and only if the following two conditions are satisfied:*

(a) For every $u \in \mathbb{N}$ there is a constant M^u such that $|z_t| \le M^u$ for all $t \in [0, u]$ and all $z \in \mathcal{K}$.

(b) For every $u \in \mathbb{N}$ and every $\epsilon > 0$ there exists a finite collection $\mathcal{T}^{u,\epsilon} = \{0 = t_0^{u,\epsilon} < t_1^{u,\epsilon} < \ldots < t_{N(\epsilon)}^{u,\epsilon} = u\}$ of instants such that for all $z \in \mathcal{K}$

$$\sup\{|z_s - z_t| : s, t \in [t_{n-1}^{u,\epsilon}, t_n^{u,\epsilon})\} \le \epsilon\,, \qquad 1 \le n \le N(\epsilon)\,. \tag{$*$}$$

Proof. We shall need, and therefore prove, only the sufficiency of these two conditions. Assume then that they are satisfied and let $\mathfrak{F}$ be a filter on $\mathcal{K}$. Tychonoff's theorem A.2.13 in conjunction with (a) provides a refinement $\mathfrak{F}'$ that converges pointwise to some path z. Clearly z is again bounded by M^u on $[0, u]$ and satisfies $(*)$. A path $z' \in \mathcal{K}$ that differs from z in the points $t_n^{u,\epsilon}$ by less than ϵ is uniformly as close as 3ϵ to z on $[0, u]$. Indeed, for $t \in [t_{n-1}^{u,\epsilon}, t_n^{u,\epsilon})$

$$\left|z_t - z'_t\right| \le \left|z_t - z_{t_{n-1}^{u,\epsilon}}\right| + \left|z_{t_{n-1}^{u,\epsilon}} - z'_{t_{n-1}^{u,\epsilon}}\right| + \left|z'_{t_{n-1}^{u,\epsilon}} - z'_t\right| < 3\epsilon\,.$$

That is to say, the refinement $\mathfrak{F}'$ converges uniformly $[0, u]$ to z. This holds for all $u \in \mathbb{N}$, so $\mathfrak{F}' \to z \in \mathscr{D}$ uniformly on compacta. ∎

We use this to prove the tightness of the $\mathbb{Z}^{(n)}$:

Lemma A.4.12 *For every $\alpha > 0$ there exists a compact set $\mathcal{K}_\alpha \subset \mathscr{D}$ with the following property: for every $n \in \mathbb{N}$ there is a set $\Omega_\alpha^{(n)} \in \mathcal{F}^{(n)}$ such that*

$$\mathbb{P}^{(n)}\left[\Omega_\alpha^{(n)}\right] > 1 - \alpha \quad \textit{and} \quad Z^{(n)}\left(\Omega_\alpha^{(n)}\right) \subset \mathcal{K}_\alpha\,.$$

Consequently the laws $\mathbb{Z}^{(n)}$ form a uniformly tight family.

Proof. For $u \in \mathbb{N}$, let

$$M_\alpha^u \stackrel{\text{def}}{=} \sqrt{u}\, 2^{u+1}/\alpha$$

and set
$$\Omega_{\alpha,1}^{(n)} \stackrel{\text{def}}{=} \bigcap_{u \in \mathbb{N}} \left[\left|Z^{(n)}\right|_u^* \le M_\alpha^u\right].$$

Now $Z^{(n)}$ is a martingale that at the instant u has square expectation u, so Doob's maximal lemma 2.5.18 and a summation give

$$\mathbb{P}^{(n)}\left[\left|Z^{(n)}\right|_u^* > M_\alpha^u\right] < \sqrt{u}/M_\alpha^u \quad \text{and} \quad \mathbb{P}\left[\Omega_{\alpha,1}^{(n)}\right] > 1 - \alpha/2\,.$$

For $\omega \in \Omega_{\alpha,1}^{(n)}$, $Z_.^{(n)}(\omega)$ is bounded by M_α^u on $[0, u]$.

The construction of a large set $\Omega^{(n)}_{\alpha,2}$ on which the paths of $Z^{(n)}$ satisfy (b) of lemma A.4.11 is slightly more complicated. Let $0 \le s \le \tau \le t$. $Z^{(n)}_\tau - Z^{(n)}_s = \sum_{[sn]<k\le[\tau n]} X^{(n)}_k$ has the same distribution as $M^{(n)}_\tau \overset{\text{def}}{=} \sum_{k\le[\tau n]-[sn]} X^{(n)}_k$. Now $M^{(n)}_t$ has fourth moment

$$\mathbb{E}\Big[\big|M^{(n)}_t\big|^4\Big] = 3\big([tn]-[sn]\big)^2/n^2 \le 3\big(t-s+1/n\big)^2 .$$

Since $M^{(n)}$ is a martingale, Doob's maximal theorem 2.5.19 gives

$$\mathbb{E}\Big[\sup_{s\le\tau\le t} \big|Z^{(n)}_\tau - Z^{(n)}_s\big|^4\Big] \le \left(\frac{4}{3}\right)^4 \cdot 3(t-s+1/n)^2 < 10(t-s+1/n)^2 \quad (*)$$

for all $n \in \mathbb{N}$. Since $\sum N2^{-N/2} < \infty$, there is an index N_α such that

$$40 \sum_{N\ge N_\alpha} N2^{-N/2} < \alpha/2 .$$

If $n \le 2^{N_\alpha}$, we set $\Omega^{(n)}_{\alpha,2} = \Omega^{(n)}$. For $n > 2^{N_\alpha}$ let $\mathcal{N}^{(n)}$ denote the set of integers $N \ge N_\alpha$ with $2^N \le n$. For every one of them $(*)$ and Chebyscheff's inequality produce

$$\mathbb{P}\Big[\sup_{k2^{-N}\le\tau\le(k+1)2^{-N}} \big|Z^{(n)}_\tau - Z^{(n)}_{k2^{-N}}\big| > 2^{-N/8}\Big]$$
$$\le 2^{N/2}\cdot 10\big(2^{-N}+1/n\big)^2 \le 40\cdot 2^{-3N/2} , \qquad k = 0,1,2,\ldots .$$

Hence $$\bigcup_{N\in\mathcal{N}^{(n)}} \bigcup_{0\le k<N2^N} \Big[\sup_{k2^{-N}\le\tau\le(k+1)2^{-N}} \big|Z^{(n)}_\tau - Z^{(n)}_{k2^{-N}}\big| > 2^{-N/8}\Big]$$

has measure less than $\alpha/2$. We let $\Omega^{(n)}_{\alpha,2}$ denote its complement and set

$$\Omega^{(n)}_\alpha = \Omega^{(n)}_{\alpha,1} \cap \Omega^{(n)}_{\alpha,2} .$$

This is a set of $\mathbb{P}^{(n)}$-measure greater than $1-\alpha$.

For $N \in \mathbb{N}$, let $\mathcal{T}^N$ be the set of instants that are of the form k/l, $k \in \mathbb{N}, l \le 2^{N_\alpha}$, or of the form $k2^{-N}$, $k \in \mathbb{N}$. For the set $\mathcal{K}_\alpha$ we take the collection of paths z that satisfy the following description: for every $u \in \mathbb{N}$ z is bounded on $[0,u]$ by M^u_α and varies by less than $2^{-N/8}$ on any interval $[s,t)$ whose endpoints s,t are consecutive points of $\mathcal{T}^N$. Since $\mathcal{T}^N \cap [0,u]$ is finite, $\mathcal{K}_\alpha$ is compact (lemma A.4.11).

It is left to be shown that $Z^{(n)}_.(\omega) \in \mathcal{K}_\alpha$ for $\omega \in \Omega^{(n)}_\alpha$. This is easy when $n \le 2^{N_\alpha}$: the path $Z^{(n)}_.(\omega)$ is actually constant on $[s,t)$, whatever $\omega \in \Omega^{(n)}$. If $n > 2^{N_\alpha}$ and s,t are consecutive points in $\mathcal{T}^N$, then $[s,t)$ lies in an interval of the form $\big[k2^{-N},(k+1)2^{-N}\big)$, $N \in \mathcal{N}^{(n)}$, and $Z^{(n)}_.(\omega)$ varies by less than $2^{-N/8}$ on $[s,t)$ as long as $\omega \in \Omega^{(n)}_\alpha$.

Thanks to equation (A.3.7),

$$\mathbb{Z}^{(n)}(\mathcal{K}_\alpha) = \mathbb{E}\Big[\mathcal{K}_\alpha \circ Z^{(n)}\Big] \geq \mathbb{P}^{(n)}\big[\Omega^{(n)}\big] \geq 1-\alpha\,, \qquad n \in \mathbb{N}\,.$$

The family $\{\mathbb{Z}^{(n)} : n \in \mathbb{N}\}$ is thus uniformly tight. ∎

Proof of Theorem A.4.9. Lemmas A.4.10 and A.4.12 in conjuction with criterion A.4.7 allow us to conclude that $(\mathbb{Z}^{(n)})$ converges weakly to an order-continuous tight (proposition A.4.6) probability $\mathbb{Z}$ on $\mathscr{D}$ whose characteristic function $\widehat{\mathbb{Z}}^{\Gamma}$ is that of Wiener measure (corollary 3.9.5). By proposition A.3.12, $\mathbb{Z} = \mathbb{W}$. ∎

Example A.4.13 Let $\delta_1, \delta_2, \ldots$ be strictly positive numbers. On $\mathscr{D}$ define by induction the functions $\tau_0 = \tau_0^+ = 0$,

$$\tau_{k+1}(z) = \inf\Big\{t : \big|z_t - z_{\tau_k(z)}\big| \geq \delta_{k+1}\Big\}$$

and

$$\tau_{k+1}^+(z) = \inf\Big\{t : \big|z_t - z_{\tau_k^+(z)}\big| > \delta_{k+1}\Big\}\,.$$

Let $\Phi = \Phi(t_1, \zeta_1, t_2, \zeta_2, \ldots)$ be a bounded continuous function on $\mathbb{R}^{\mathbb{N}}$ and set

$$\phi(z) = \Phi\big(\tau_1(z), z_{\tau_1(z)}, \tau_2(z), z_{\tau_2(z)}, \ldots\big)\,, \qquad z \in \mathscr{D}\,.$$

Then

$$\mathbb{E}^{\mathbb{Z}^{(n)}}[\phi] \xrightarrow[n\to\infty]{} \mathbb{E}^{\mathbb{W}}[\phi]\,.$$

Proof. The processes $Z^{(n)}, W$ take their values in the Borel[41] subset

$$\mathscr{D}^{\simeq} \overset{\text{def}}{=} \bigcup_n \mathscr{D}^{(n)} \cup \mathscr{C}$$

of $\mathscr{D}$. $\mathscr{D}^{\simeq}$ therefore[27] has full measure for their laws $\mathbb{Z}^{(n)}, \mathbb{W}$. Henceforth we consider τ_k, τ_k^+ as functions on this set. At a point $z^0 \in \mathscr{D}^{(n)}$ the functions τ_k, τ_k^+, $z \mapsto z_{\tau_k(z)}$, and $z \mapsto z_{\tau_k^+(z)}$ are continuous. Indeed, pick an instant of the form p/n, where p, n are relatively prime. A path $z \in \mathscr{D}^{\simeq}$ closer than $1/(6\sqrt{n})$ to z^0 uniformly on $[0, 2p/n]$ must jump at p/n by at least $2/(3\sqrt{n})$, and no path in $\mathscr{D}^{\simeq}$ other than z^0 itself does that. In other words, every point of $\bigcup_n \mathscr{D}^{(n)}$ is an isolated point in $\mathscr{D}^{\simeq}$, so that every function is continuous at it: we have to worry about the continuity of the functions above only at points $w \in \mathscr{C}$. Several steps are required.

a) If $z \to z_{\tau_k(z)}$ is continuous on $E_k \overset{\text{def}}{=} \mathscr{C} \cap [\tau_k = \tau_k^+ < \infty]$, then (a1) τ_{k+1} is lower semicontinuous and (a2) τ_{k+1}^+ is upper semicontinuous on this set.

To see (a1) let $w \in E_k$, set $s = \tau_k(w)$, and pick $t < \tau_{k+1}(w)$. Then $\alpha \overset{\text{def}}{=} \delta_{k+1} - \sup_{s\leq\sigma\leq t} |w_\sigma - w_s| > 0$. If $z \in \mathscr{D}^{\simeq}$ is so close to w uniformly

---

[41] See exercise A.2.23 on page 379.

on a suitable interval containing $[0,t+1]$ that $|w_s - z_{\tau_k(z)}| < \alpha/2$, and is uniformly as close as $\alpha/2$ to w there, then

$$\begin{aligned}\left|z_\sigma - z_{\tau_k(z)}\right| &\le \left|z_\sigma - w_\sigma\right| + \left|w_\sigma - w_s\right| + \left|w_s - z_{\tau_k(z)}\right| \\ &< \alpha/2 + (\delta_{k+1} - \alpha) + \alpha/2 = \delta_{k+1}\end{aligned}$$

for all $\sigma \in [s,t]$ and consequently $\tau_{k+1}(z) > t$. Therefore we have as desired $\liminf_{z\to w} \tau_{k+1}(z) \ge \tau_{k+1}(w)$.

To see (a2) consider a point $w \in \left[\tau^+_{k+1} < u\right] \cap E_k$. Set $s = \tau^+_k(w)$. There is an instant $t \in (s,u)$ at which $\alpha \stackrel{\text{def}}{=} |w_t - w_s| - \delta_{k+1} > 0$. If $z \in \mathscr{D}^{\simeq}$ is sufficiently close to w uniformly on some interval containing $[0,u]$, then $|z_t - w_t| < \alpha/2$ and $|w_s - z_{\tau_k(z)}| < \alpha/2$ and therefore

$$\begin{aligned}\left|z_t - z_{\tau_k(z)}\right| &\ge -\left|z_t - w_t\right| + \left|w_t - w_s\right| - \left|w_s - z_{\tau_k(z)}\right| \\ &> -\alpha/2 + (\delta_{k+1} + \alpha) - \alpha/2 = \delta_{k+1}\ .\end{aligned}$$

That is to say, $\tau^+_{k+1} < u$ in a whole neighborhood of w in $\mathscr{D}^{\simeq}$, wherefore as desired $\limsup_{z\to w} \tau^+_{k+1}(z) \le \tau^+_{k+1}(w)$.

b) $z \to z_{\tau_k(z)}$ *is continuous on* E_k *for all* $k \in \mathbb{N}$. This is trivially true for $k = 0$. Asssume it for k. By a) τ_{k+1} and τ^+_{k+1}, which on E_{k+1} agree and are finite, are continuous there. Then so is $z \to z_{\tau_k(z)}$.

c) $\mathbb{W}[E_k] = 1$, *for* $k = 1, 2, \ldots$. This is plain for $k = 0$. Assuming it for $1, \ldots, k$, set $E^k = \bigcap_{\kappa \le k} E_\kappa$. This is then a Borel subset of $\mathscr{C} \subset \mathscr{D}$ of Wiener measure 1 on which plainly $\tau_{k+1} \le \tau^+_{k+1}$. Let $\delta = \delta_1 + \cdots + \delta_{k+1}$. The stopping time τ^+_{k+1} occurs before $T \stackrel{\text{def}}{=} \inf\{t : |w_t| > \delta\}$, which is integrable (exercise 4.2.21). The continuity of the paths results in

$$\delta^2_{k+1} = \left(w_{\tau_{k+1}} - w_{\tau_k}\right)^2 = \left(w_{\tau^+_{k+1}} - w_{\tau^+_k}\right)^2 .$$

$$\begin{aligned}\text{Thus}\quad \delta^2_{k+1} &= \mathbb{E}^{\mathbb{W}}\left[\left(w_{\tau_{k+1}} - w_{\tau_k}\right)^2\right] = \mathbb{E}^{\mathbb{W}}\left[w^2_{\tau_{k+1}} - w^2_{\tau_k}\right] \\ &= \mathbb{E}^{\mathbb{W}}\left[2\int_{\tau_k}^{\tau_{k+1}} w_s\, dw_s + \left(\tau_{k+1} - \tau_k\right)\right] = \mathbb{E}^{\mathbb{W}}\left[\tau_{k+1} - \tau_k\right] .\end{aligned}$$

The same calculation can be made for τ^+_{k+1}, so that $\mathbb{E}^{\mathbb{W}}[\tau^+_{k+1} - \tau_{k+1}] = 0$ and consequently $\tau^+_{k+1} = \tau_{k+1}$ $\mathbb{W}$-almost surely on E^k: we have $\mathbb{W}[E_{k+1}] = 1$, as desired.

Let $E = \bigcup_n \mathscr{D}^{(n)} \cup \bigcap_k E^k$. This is a Borel subset of $\mathscr{D}$ with $\mathbb{W}[E] = \mathbb{Z}^{(n)}[E] = 1$ $\forall n$. The restriction of ϕ to it is continuous. Corollary A.4.2 applies and gives the claim. ∎

Exercise A.4.14 Assume the coupling coefficient f of the markovian SDE (5.6.4), which reads

$$X_t = x + \int_0^t f(X^x_{s-})\, dZ_s\ , \tag{A.4.7}$$

is a bounded Lipschitz vector field. As Z runs through the sequence $Z^{(n)}$ the solutions $X^{(n)}$ converge in law to the solution of (A.4.7) driven by Wiener process.

A.5 Analytic Sets and Capacity

The *preimages* under continuous maps of open sets are open; the *preimages* under measurable maps of measurable sets are measurable. Nothing can be said in general about *direct* or *forward* images, with one exception: the continuous image of a compact set is compact (exercise A.2.14). (Even Lebesgue himself made a mistake here, thinking that the projection of a Borel set would be Borel.) This dearth is alleviated slightly by the following abbreviated theory of analytic sets, initiated by Lusin and his pupil Suslin. The presentation follows [20] – see also [17]. The class of *analytic sets* is designed to be invariant under direct images of certain simple maps, projections. Their theory implicitly uses the fact that continuous direct images of compact sets are compact.

Let F be a set. Any collection $\mathcal{F}$ of subsets of F is called a ***paving*** of F, and the pair $(F, \mathcal{F})$ is a ***paved set.*** $\mathcal{F}_\sigma$ denotes the collection of subsets of F that can be written as countable unions of sets in $\mathcal{F}$, and $\mathcal{F}_\delta$ denotes the collection of subsets of F that can be written as countable intersections of members of $\mathcal{F}$. Accordingly $\mathcal{F}_{\sigma\delta}$ is the collection of sets that are countable intersections of sets each of which is a countable union of sets in $\mathcal{F}$, etc. If $(K, \mathcal{K})$ is another paved set, then the ***product paving*** $\mathcal{K} \times \mathcal{F}$ consists of the "rectangles" $A \times B$, $A \in \mathcal{K}, B \in \mathcal{F}$. The family $\mathcal{K}$ of subsets of K constitutes a ***compact paving*** if it has the *finite intersection property:* whenever a subfamily $\mathcal{K}' \subset \mathcal{K}$ has void intersection there exists a finite subfamily $\mathcal{K}'_0 \subset \mathcal{K}'$ that already has void intersection. We also say that K is ***compactly paved*** by $\mathcal{K}$.

Definition A.5.1 (Analytic Sets) *Let $(F, \mathcal{F})$ be a paved set. A set $A \subset F$ is called **$\mathcal{F}$-analytic** if there exist an auxiliary set K equipped with a compact paving $\mathcal{K}$ and a set $B \in (\mathcal{K} \times \mathcal{F})_{\sigma\delta}$ such that A is the projection of B on F:*

$$A = \pi_F(B) .$$

Here $\pi_F = \pi_F^{K \times F}$ is the natural projection of $K \times F$ onto its second factor F – see figure A.16. The collection of $\mathcal{F}$-analytic sets is denoted by $\mathcal{A}[\mathcal{F}]$.

Theorem A.5.2 *The sets of $\mathcal{F}$ are $\mathcal{F}$-analytic. The intersection and the union of countably many $\mathcal{F}$-analytic sets are $\mathcal{F}$-analytic.*

Proof. The first statement is obvious. For the second, let $\{A_n : n = 1, 2 \ldots\}$ be a countable collection of $\mathcal{F}$-analytic sets. There are auxiliary spaces K_n equipped with compact pavings $\mathcal{K}_n$ and $(\mathcal{K}_n \times \mathcal{F})_{\sigma\delta}$-sets $B_n \subset K_n \times F$ whose projection onto F is A_n. Each B_n is the countable intersection of sets $B_n^j \in (\mathcal{K}_n \times \mathcal{F})_\sigma$.

To see that $\bigcap A_n$ is $\mathcal{F}$-analytic, consider the product $K = \prod_{n=1}^\infty K_n$. Its paving $\mathcal{K}$ is the product paving, consisting of sets $C = \prod_{n=1}^\infty C_n$, where

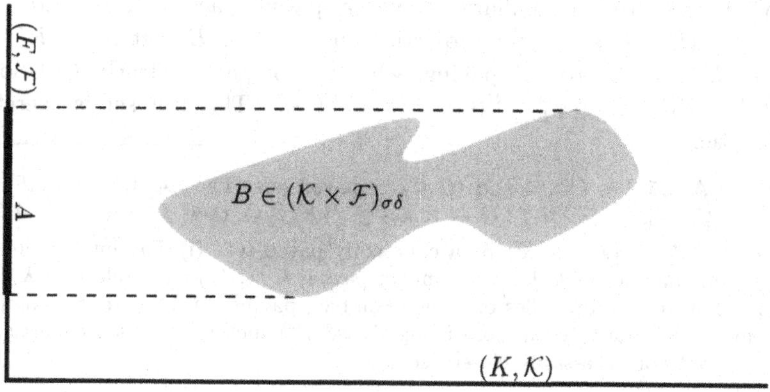

Figure A.16 An $\mathcal{F}$-analytic set A

$C_n = K_n$ for all but finitely many indices n and $C_n \in \mathcal{K}_n$ for the finitely many exceptions. $\mathcal{K}$ is compact. For if $\{C^\alpha\} = \{\prod_n C_n^\alpha : \alpha \in A\} \subset \mathcal{K}$ has void intersection, then one of the collections $\{C_n^\alpha : \alpha\}$, say C_1^α, must have void intersection, otherwise $\prod_n \bigcap_\alpha C_n^\alpha \neq \emptyset$ would be contained in $\bigcap_\alpha C^\alpha$. There are then $\alpha_1, \ldots, \alpha_k$ with $\bigcap_i C_1^{\alpha_i} = \emptyset$, and thus $\bigcap_i C^{\alpha_i} = \emptyset$. Let

$$B'_n = \prod_{m \neq n} K_m \times B_n = \prod_{m \neq n} K_m \times \bigcup_{j=1}^{\infty} B_n^j \subset F \times K .$$

Clearly $B = \bigcap B'_n$ belongs to $(\mathcal{K} \times \mathcal{F})_{\sigma\delta}$ and has projection $\bigcap A_n$ onto F. Thus $\bigcap A_n$ is $\mathcal{F}$-analytic.

For the union consider instead the disjoint union $K = \biguplus_n K_n$ of the K_n. For its paving $\mathcal{K}$ we take the direct sum of the $\mathcal{K}_n$: $C \subset K$ belongs to $\mathcal{K}$ if and only if $C \cap K_n$ is void for all but finitely many indices n and a member of $\mathcal{K}_n$ for the exceptions. $\mathcal{K}$ is clearly compact. The set $B \stackrel{\text{def}}{=} \biguplus_n B_n$ equals $\bigcap_{j=1}^{\infty} \biguplus_n B_n^j$ and has projection $\bigcup A_n$. Thus $\bigcup A_n$ is $\mathcal{F}$-analytic. ∎

Corollary A.5.3 *$\mathcal{A}[\mathcal{F}]$ contains the σ-algebra generated by $\mathcal{F}$ if and only if the complement of every set in $\mathcal{F}$ is $\mathcal{F}$-analytic. In particular, if the complement of every set in $\mathcal{F}$ is the countable union of sets in $\mathcal{F}$, then $\mathcal{A}[\mathcal{F}]$ contains the σ-algebra generated by $\mathcal{F}$.*

Proof. Under the hypotheses the collection of sets $A \subset F$ such that both A and its complement A^c are $\mathcal{F}$-analytic contains $\mathcal{F}$ and is a σ-algebra. ∎

The *direct* or *forward* image of an analytic set is analytic, under certain projections. The precise statement is this:

Proposition A.5.4 *Let $(K, \mathcal{K})$ and $(F, \mathcal{F})$ be paved sets, with $\mathcal{K}$ compact. The projection of a $\mathcal{K} \times \mathcal{F}$-analytic subset B of $K \times F$ onto F is $\mathcal{F}$-analytic.*

Proof. There exist an auxiliary compactly paved space $(K', \mathcal{K}')$ and a set $C \in \big(\mathcal{K}' \times (\mathcal{K} \times \mathcal{F})\big)_{\sigma\delta}$ whose projection on $K \times F$ is B. Set $K'' = K' \times K$ and let $\mathcal{K}''$ be its product paving, which is compact. Clearly C belongs to $(\mathcal{K}'' \times \mathcal{F})_{\sigma\delta}$, and $\pi_F^{K'' \times F}(C) = \pi_F^{K' \times F}(B)$. This last set is therefore $\mathcal{F}$-analytic. ∎

Exercise A.5.5 Let $(F, \mathcal{F})$ and $(G, \mathcal{G})$ be paved sets. Then $\mathcal{A}[\mathcal{A}[\mathcal{F}]] = \mathcal{A}[\mathcal{F}]$ and $\mathcal{A}[\mathcal{F}] \times \mathcal{A}[\mathcal{G}] \subset \mathcal{A}[\mathcal{F} \times \mathcal{G}]$. If $f : G \to F$ has $f^{-1}(\mathcal{F}) \subset \mathcal{G}$, then $f^{-1}(\mathcal{A}[\mathcal{F}]) \subset \mathcal{A}[\mathcal{G}]$.

Exercise A.5.6 Let $(K, \mathcal{K})$ be a compactly paved set. (i) The intersections of arbitrary subfamilies of $\mathcal{K}$ form a compact paving $\mathcal{K}^{\cap_a}$. (ii) The collection $\mathcal{K}^{\cup_f}$ of all unions of finite subfamilies of $\mathcal{K}$ is a compact paving. (iii) There is a compact topology on K (possibly far from being Hausdorff) such that $\mathcal{K}$ is a collection of compact (but not necessarily closed) sets.

Definition A.5.7 (Capacities and Capacitability) *Let $(F, \mathcal{F})$ be a paved set. (i) An $\mathcal{F}$-**capacity** is a positive numerical set function I that is defined on all subsets of F and is increasing: $A \subset B \implies I(A) \le I(B)$; is continuous along arbitrary increasing sequences: $F \supset A_n \uparrow A \implies I(A_n) \uparrow I(A)$; and is continuous along decreasing sequences of $\mathcal{F}$: $\mathcal{F} \ni F_n \downarrow F \implies I(F_n) \downarrow I(F)$. (ii) A subset C of F is called $(\mathcal{F}, I)$-capacitable, or **capacitable** for short, if*

$$I(C) = \sup\{I(K) : K \subset C\,,\ K \in \mathcal{F}_\delta\}\,.$$

The point of the compactness that is required of the auxiliary paving is the following consequence:

Lemma A.5.8 *Let $(K, \mathcal{K})$ and $(F, \mathcal{F})$ be paved sets, with $\mathcal{K}$ compact. Denote by $\mathcal{K} \otimes \mathcal{F}$ the paving $(\mathcal{K} \times \mathcal{F})^{\cup_f}$ of finite unions of rectangles from $\mathcal{K} \times \mathcal{F}$.*

(i) For any sequence (C_n) in $\mathcal{K} \otimes \mathcal{F}$, $\pi_F\big(\bigcap_n C_n\big) = \bigcap_n \pi_F(C_n)$.

(ii) If I is an $\mathcal{F}$-capacity, then

$$I \circ \pi_F : A \mapsto I\big(\pi_F(A)\big)\,, \qquad A \subset K \times F\,,$$

is a $\mathcal{K} \otimes \mathcal{F}$-capacity.

Proof. (i) Let $x \in \bigcap_n \pi_F(C_n)$. The sets $K_n^x \stackrel{\text{def}}{=} \{k \in K : (k, x) \in C_n\}$ belong to $\mathcal{K}^{\cup_f}$ and are non-void. Exercise A.5.6 furnishes a point k in their intersection, and clearly (k, x) is a point in $\bigcap_n C_n$ whose projection on F is x. Thus $\bigcap_n \pi_F(C_n) \subset \pi_F\big(\bigcap_n C_n\big)$. The reverse inequality is obvious.

Here is a direct proof that avoids ultrafilters. Let x be a point in $\bigcap_n \pi_F(C_n)$ and let us show that it belongs to $\pi_F\big(\bigcap_n C_n\big)$. Now the sets $K_n^x = \{k \in K : (k, x) \in C_n\}$ are not void. K_n^x is a finite union of sets in $\mathcal{K}$, say $K_n^x = \bigcup_{i=1}^{I(n)} K_{n,i}^x$. For at least one index $i = n(1)$, $K_{1,n(1)}^x$ must intersect all of the subsequent sets K_n^x, $n > 1$, in a non-void set. Replacing K_n^x by $K_{1,n(1)}^x \cap K_n^x$ for $n = 1, 2 \ldots$ reduces the situation to $K_1^x \in \mathcal{K}^{\cap_f}$. For at least one index $i = n(2)$, $K_{2,n(2)}^x$ must intersect all of the subsequent sets K_n^x, $n > 2$, in a non-void set. Replacing K_n^x by $K_{2,n(2)}^x \cap K_n^x$ for $n = 2, 3 \ldots$

reduces the situation to $K_2^x \in \mathcal{K}^{\cap f}$. Continue on. The K_n^x so obtained belong to $\mathcal{K}^{\cap f}$, still decrease with n, and are non-void. There is thus a point $k \in \bigcap_n K_n^x$. The point $(k,x) \in \bigcap_n C_n$ evidently has $\pi_F(k,x) = x$, as desired.

(ii) First, it is evident that $I \circ \pi_F$ is increasing and continuous along arbitrary sequences; indeed, $K \times F \supset A_n \uparrow A$ implies $\pi_F(A_n) \uparrow \pi_F(A)$, whence $I \circ \pi_F(A_n) \uparrow I \circ \pi_F(A_n)$. Next, if C_n is a decreasing sequence in $\mathcal{K} \otimes \mathcal{F}$, then $\pi_F(C_n)$ is a decreasing sequence of $\mathcal{F}$, and by (i) $I \circ \pi_F(C_n) = I\big(\pi_F(C_n)\big)$ decreases to $I\big(\bigcap_n \pi_F(C_n)\big) = I \circ \pi_F\big(\bigcap_n C_n\big)$: the continuity along decreasing sequences of $\mathcal{K} \otimes \mathcal{F}$ is established as well. ∎

Theorem A.5.9 (Choquet's Capacitability Theorem) *Let $\mathcal{F}$ be a paving that is closed under finite unions and finite intersections, and let I be an $\mathcal{F}$-capacity. Then every $\mathcal{F}$-analytic set A is $(\mathcal{F}, I)$-capacitable.*

Proof. To start with, let $A \in \mathcal{F}_{\sigma\delta}$. There is a sequence of sets $F_n^\sigma \in \mathcal{F}_\sigma$ whose intersection is A. Every one of the F_n^σ is the union of a countable family $\{F_n^j : j \in \mathbb{N}\} \subset \mathcal{F}$. Since $\mathcal{F}$ is closed under finite unions, we may replace F_n^j by $\bigcup_{i=1}^j F_n^i$ and thus assume that F_n^j increases with j: $F_n^j \uparrow F_n^\sigma$. Suppose $I(A) > r$. We shall construct by induction a sequence (F'_n) in $\mathcal{F}$

such that $$F'_n \subset F_n^\sigma \quad \text{and} \quad I(A \cap F'_1 \cap \ldots \cap F'_n) > r\,.$$

Since $$I(A) = I(A \cap F_1^\sigma) = \sup_j I(A \cap F_1^j) > r\,,$$

we may choose for F'_1 an F_1^j with sufficiently high index j. If $F'_1, \ldots, F'_n$ in $\mathcal{F}_{\sigma\delta}$ have been found, we note that

$$\begin{aligned} I(A \cap F'_1 \ldots \cap F'_n) &= I(A \cap F'_1 \cap \ldots \cap F'_n \cap F_{n+1}^\sigma) \\ &= \sup_j I(A \cap F'_1 \cap \ldots \cap F'_n \cap F_{n+1}^j) > r\,; \end{aligned}$$

for F'_{n+1} we choose F_{n+1}^j with j sufficiently large. The construction of the F'_n is complete. Now $F^\delta \overset{\text{def}}{=} \bigcap_{n=1}^\infty F'_n$ is an $\mathcal{F}_\delta$-set and is contained in A, inasmuch as it is contained in every one of the F_n^σ. The continuity along decreasing sequences of $\mathcal{F}$ gives $I(F^\delta) \ge r$. The claim is proved for $A \in \mathcal{F}_{\sigma\delta}$.

Now let A be a general $\mathcal{F}$-analytic set and $r < I(A)$. There are an auxiliary compactly paved set $(K, \mathcal{K})$ and an $(\mathcal{K} \times \mathcal{F})_{\sigma\delta}$-set $B \subset K \times F$ whose projection on F is A. We may assume that $\mathcal{K}$ is closed under taking finite intersections by the simple expedient of adjoining to $\mathcal{K}$ the intersections of its finite subcollections (exercise A.5.6). The paving $\mathcal{K} \otimes \mathcal{F}$ of $K \times F$ is then closed under both finite unions and finite intersections, and B still belongs to $(\mathcal{K} \otimes \mathcal{F})_{\sigma\delta}$. Due to lemma A.5.8 (ii), $I \circ \pi_F$ is a $\mathcal{K} \otimes \mathcal{F}$-capacity with $r < I \circ \pi_F(B)$, so the above provides a set $C \subset B$ in $(\mathcal{K} \otimes \mathcal{F})_\delta$ with $r < I\big(\pi_F(C)\big)$. Clearly $F_r \overset{\text{def}}{=} \pi_F(C)$ is a subset of A with $r < I(F_r)$. Now C is the intersection of a decreasing family $C_n \in \mathcal{K} \otimes \mathcal{F}$, each of which has $\pi_F(C_n) \in \mathcal{F}$, so by lemma A.5.8 (i) $F_r = \bigcap_n \pi_F(C_n) \in \mathcal{F}_\delta$. Since $r < I(A)$ was arbitrary, A is $(\mathcal{F}, I)$-capacitable. ∎

Applications to Stochastic Analysis

Theorem A.5.10 (The Measurable Section Theorem) *Let* $(\Omega, \mathcal{F})$ *be a measurable space and* $B \subset \mathbb{R}_+ \times \Omega$ *measurable on* $\mathcal{B}^\bullet(\mathbb{R}_+) \otimes \mathcal{F}$.

(i) For every $\mathcal{F}$*-capacity*[42] I *and* $\epsilon > 0$ *there is an* $\mathcal{F}$*-measurable function* $R : \Omega \to \overline{\mathbb{R}}_+$, *"an* $\mathcal{F}$*-measurable random time," whose graph is contained in* B *and such that*

$$I[R < \infty] > I[\pi_\Omega(B)] - \epsilon\,.$$

(ii) $\pi_\Omega(B)$ *is measurable on the universal completion* $\mathcal{F}^*$.

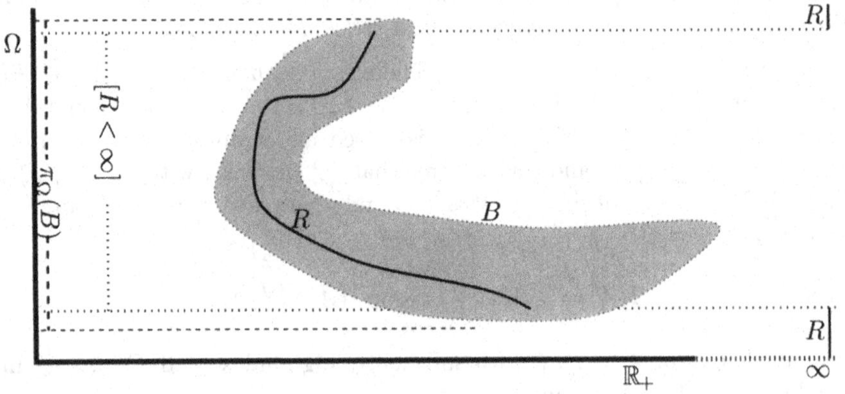

Figure A.17 The Measurable Section Theorem

Proof. (i) π_Ω denotes, of course, the natural projection of $\boldsymbol{B}$ onto Ω. We equip $\mathbb{R}_+$ with the paving $\mathcal{K}$ of compact intervals. On $\Omega \times \mathbb{R}_+$ consider the pavings

$$\mathcal{K} \times \mathcal{F} \quad \text{and} \quad \mathcal{K} \otimes \mathcal{F}\,.$$

The latter is closed under finite unions and intersections and generates the σ-algebra $\mathcal{B}^\bullet(\mathbb{R}_+) \otimes \mathcal{F}$. For every set $M = \bigcup_i [s_i, t_i] \times A_i$ in $\mathcal{K} \otimes \mathcal{F}$ and every $\omega \in \Omega$ the path $M_.(\omega) = \bigcup_{i: A_i(\omega) \neq \emptyset} [s_i, t_i]$ is a compact subset of $\mathbb{R}_+$. Inasmuch as the complement of every set in $\mathcal{K} \times \mathcal{F}$ is the countable union of sets in $\mathcal{K} \times \mathcal{F}$, the paving of $\mathbb{R}_+ \times \Omega$ which generates the σ-algebra $\mathcal{B}^\bullet(\mathbb{R}_+) \otimes \mathcal{F}$, every set of $\mathcal{B}^\bullet(\mathbb{R}_+) \otimes \mathcal{F}$, in particular B, is $\mathcal{K} \times \mathcal{F}$-analytic (corollary A.5.3) and *a fortiori* $\mathcal{K} \otimes \mathcal{F}$-analytic. Next consider the set function

$$F \mapsto J(F) \stackrel{\text{def}}{=} I[\pi(F)] = I \circ \pi_\Omega(F)\,, \qquad F \subset \boldsymbol{B}.$$

According to lemma A.5.8, J is a $\mathcal{K} \otimes \mathcal{F}$-capacity. Choquet's theorem provides a set $K \in (\mathcal{K} \otimes \mathcal{F})_\delta$, the intersection of a decreasing countable family

[42] In most applications I is the outer measure $\mathbb{P}^*$ of a probability $\mathbb{P}$ on $\mathcal{F}$, which by equation (A.3.2) is a capacity.

$\{C_n\} \subset \mathcal{K} \otimes \mathcal{F}$, that is contained in B and has $J(K) > J(B) - \epsilon$. The "left edges" $R_n(\omega) \stackrel{\text{def}}{=} \inf\{t : (t, \omega) \in C_n\}$ are simple $\mathcal{F}$-measurable random variables, with $R_n(\omega) \in C_m(\omega)$ for $n \geq m$ at points ω where $R_n(\omega) < \infty$. Therefore $R \stackrel{\text{def}}{=} \sup_n R_n$ is $\mathcal{F}$-measurable, and thus $R(\omega) \in \bigcap_m C_m(\omega) = K(\omega) \subset B(\omega)$ where $R(\omega) < \infty$. Clearly $[R < \infty] = \pi_\Omega[K] \in \mathcal{F}$ has $I[R < \infty] > I[\pi_\Omega(B)] - \epsilon$.

(ii) To say that the filtration $\mathcal{F}_.$ is ***universally complete*** means of course that $\mathcal{F}_t$ is universally complete for all $t \in [0, \infty]$ ($\mathcal{F}_t = \mathcal{F}_t^*$; see page 407); and this is certainly the case if $\mathcal{F}_.$ is $\mathfrak{P}$-regular, no matter what the collection $\mathfrak{P}$ of pertinent probabilities. Let then $\mathbb{P}$ be a probability on $\mathcal{F}$, and R_n $\mathcal{F}$-measurable random times whose graphs are contained in B and that have $\mathbb{P}^*[\pi_\Omega(B)] < \mathbb{P}[R_n < \infty] + 1/n$. Then $A \stackrel{\text{def}}{=} \bigcup_n [R_n < \infty] \in \mathcal{F}$ is contained in $\pi_\Omega(B)$ and has $\mathbb{P}[A] = \mathbb{P}^*[\pi_\Omega(B)]$: the inner and outer measures of $\pi_\Omega(B)$ agree, and so $\pi_\Omega(B)$ is $\mathbb{P}$-measurable. This is true for every probability $\mathbb{P}$ on $\mathcal{F}$, so $\pi_\Omega(B)$ is universally measurable. ∎

A slight refinement of the argument gives further information:

Corollary A.5.11 *Suppose that the filtration $\mathcal{F}_.$ is universally complete, and let T be a stopping time. Then the projection $\pi_\Omega[B]$ of a progressively measurable set $B \subset [\![0, T]\!]$ is measurable on $\mathcal{F}_T$.*

Proof. Fix an instant $t < \infty$. We have to show that $\pi_\Omega[B] \cap [T \leq t] \in \mathcal{F}_t$. Now this set equals the intersection of $\pi_\Omega[B^t]$ with $[T \leq t]$, so as $[T \leq t] \in \mathcal{F}_T$ it suffices to show that $\pi_\Omega[B^t] \in \mathcal{F}_t$. But this is immediate from theorem A.5.10 (ii) with $\mathcal{F} = \mathcal{F}_t$, since the stopped process B^t is measurable on $\mathcal{B}^\bullet(\mathbb{R}_+) \otimes \mathcal{F}_t$ by the very definition of progressive measurability. ∎

Corollary A.5.12 (First Hitting Times Are Stopping Times) *If the filtration $\mathcal{F}_.$ is right-continuous and universally complete, in particular if it satisfies the natural conditions, then the* ***debut***

$$D_B(\omega) \stackrel{\text{def}}{=} \inf\{t : (t, \omega) \in B\}$$

of a progressively measurable set $B \subset \boldsymbol{B}$ is a stopping time.

Proof. Let $0 \leq t < \infty$. The set $B \cap [\![0, t)\!)$ is progressively measurable and contained in $[\![0, t]\!]$, and its projection on Ω is $[D_B < t]$. Due to the universal completeness of $\mathcal{F}_.$, $[D_B < t]$ belongs to $\mathcal{F}_t$ (corollary A.5.11). Due to the right-continuity of $\mathcal{F}_.$, D_B is a stopping time (exercise 1.3.30 (i)). ∎

Corollary A.5.12 is a pretty result. Consider for example a progressively measurable process Z with values in some measurable state space $(S, \mathcal{S})$, and let $A \in \mathcal{S}$. Then $T_A \stackrel{\text{def}}{=} \inf\{t : Z_t \in A\}$ is the debut of the progressively measurable set $B \stackrel{\text{def}}{=} [Z \in A]$ and is therefore a stopping time. T_A is the "first time Z hits A," or better "the last time Z has not touched A." We can of course not claim that Z is in A at that time. If Z is right-continuous, though, and A is closed, then B is left-closed and $Z_{T_A} \in A$.

Corollary A.5.13 (The Progressive Measurability of the Maximal Process) *If the filtration $\mathcal{F}_.$ is universally complete, then the maximal process $X^\star$ of a progressively measurable process X is progressively measurable.*

Proof. Let $0 \le t < \infty$ and $a > 0$. The set $[X_t^\star > a]$ is the projection on Ω of the $\mathcal{B}^\bullet[0,\infty) \otimes \mathcal{F}_t$-measurable set $[|X^t| > a]$ and is by theorem A.5.10 measurable on $\mathcal{F}_t = \mathcal{F}_t^*$: $X^\star$ is adapted. Next let T be the debut of $[X^\star > a]$. It is identical with the debut of $[|X| > a]$, a progressively measurable set, and so is a stopping time on the right-continuous version $\mathcal{F}_{.+}$ (corollary A.5.12). So is its reduction S to $[|X|_T > a] \in \mathcal{F}_{T+}$ (proposition 1.3.9). Now clearly $[X^\star > a] = [\![S]\!] \cup (\!(T,\infty)\!)$. This union is progressively measurable for the filtration $\mathcal{F}_{.+}$. This is obvious for $(\!(T,\infty)\!)$ (proposition 1.3.5 and exercise 1.3.17) and also for the set $[\![S]\!] = \bigcap [\![S, S+1/n)\!)$ (ibidem). Since this holds for all $a > 0$, $X^\star$ is progressively measurable for $\mathcal{F}_{.+}$. Now apply exercise 1.3.30 (v). ∎

Theorem A.5.14 (Predictable Sections) *Let $(\Omega, \mathcal{F}_.)$ be a filtered measurable space and $B \subset \boldsymbol{B}$ a predictable set. For every $\mathcal{F}_\infty$-capacity I[42] and every $\epsilon > 0$ there exists a predictable stopping time R whose graph is contained in B and that satisfies (see figure A.17)*

$$I[R < \infty] > I[\pi_\Omega(B)] - \epsilon\,.$$

Proof. Consider the collection $\mathcal{M}$ of finite unions of stochastic intervals of the form $[\![S,T]\!]$, where S,T are predictable stopping times. The arbitrary left-continuous stochastic intervals

$$(\!(S,T]\!] = \bigcup_n \bigcap_k [\![S+1/n, T+1/k]\!] \in \mathcal{M}_{\delta\sigma}\,,$$

with S,T arbitrary stopping times, generate the predictable σ-algebra $\mathcal{P}$ (exercise 2.1.6), and then so does $\mathcal{M}$. Let $[\![S,T]\!]$, S,T predictable, be an element of $\mathcal{M}$. Its complement is $[\![0,S)\!) \cup (\!(T,\infty)\!)$. Now $(\!(T,\infty)\!) = \bigcup [\![T+1/n, n]\!]$ is $\mathcal{M}$-analytic as a member of $\mathcal{M}_\sigma$ (corollary A.5.3), and so is $[\![0,S)\!)$. Namely, if S_n is a sequence of stopping times announcing S, then $[\![0,S)\!) = \bigcup [\![0,S_n]\!]$ belongs to $\mathcal{M}_\sigma$. Thus every predictable set, in particular B, is $\mathcal{M}$-analytic.

Consider next the set function

$$F \mapsto J(F) \stackrel{\text{def}}{=} I[\pi_\Omega(F)]\,, \qquad F \subset \boldsymbol{B}.$$

We see as in the proof of theorem A.5.10 that J is an $\mathcal{M}$-capacity. Choquet's theorem provides a set $K \in \mathcal{M}_\delta$, the intersection of a decreasing countable family $\{M_n\} \subset \mathcal{M}$, that is contained in B and has $J(K) > J(B) - \epsilon$. The "left edges" $R_n(\omega) \stackrel{\text{def}}{=} \inf\{t : (t,\omega) \in M_n\}$ are predictable stopping times with $R_n(\omega) \in M_m(\omega)$ for $n \ge m$. Therefore $R \stackrel{\text{def}}{=} \sup_n R_m$ is a predictable stopping time (exercise 3.5.10). Also, $R(\omega) \in \bigcap_m M_m(\omega) = K(\omega) \subset B(\omega)$ where $R(\omega) < \infty$. Evidently $[R < \infty] = \pi_\Omega[K]$, and therefore $I[R < \infty] > I[\pi_\Omega(B)] - \epsilon$. ∎

Corollary A.5.15 (The Predictable Projection) *Let X be a bounded measurable process. For every probability $\mathbb{P}$ on $\mathcal{F}_\infty$ there exists a predictable process $X^{\mathcal{P},\mathbb{P}}$ such that for all predictable stopping times T*

$$\mathbb{E}\big[X_T[T<\infty]\big]=\mathbb{E}\big[X_T^{\mathcal{P},\mathbb{P}}[T<\infty]\big]\ .$$

$X^{\mathcal{P},\mathbb{P}}$ is called ***a predictable $\mathbb{P}$-projection*** *of X. Any two predictable $\mathbb{P}$-projections of X cannot be distinguished with $\mathbb{P}$.*

Proof. Let us start with the uniqueness. If $X^{\mathcal{P},\mathbb{P}}$ and $\overline{X}^{\mathcal{P},\mathbb{P}}$ are predictable projections of X, then $N \stackrel{\text{def}}{=} \big[X^{\mathcal{P},\mathbb{P}} > \overline{X}^{\mathcal{P},\mathbb{P}}\big]$ is a predictable set. It is $\mathbb{P}$-evanescent. Indeed, if it were not, then there would exist a predictable stopping time T with its graph contained in N and $\mathbb{P}[T<\infty]>0$; then we would have $\mathbb{E}\big[X_T^{\mathcal{P},\mathbb{P}}\big]>\mathbb{E}\big[\overline{X}_T^{\mathcal{P},\mathbb{P}}\big]$, a plain impossibility. The same argument shows that if $X\le Y$, then $X^{\mathcal{P},\mathbb{P}}\le Y^{\mathcal{P},\mathbb{P}}$ except in a $\mathbb{P}$-evanescent set.

Now to the existence. The family $\mathcal{M}$ of bounded processes that have a predictable projection is clearly a vector space containing the constants, and a monotone class. For if X^n have predictable projections $X^{n\mathcal{P},\mathbb{P}}$ and, say, increase to X, then $\limsup X^{n\mathcal{P},\mathbb{P}}$ is evidently a predictable projection of X. $\mathcal{M}$ contains the processes of the form $X=(t,\infty)\times g$, $g\in L^\infty(\mathcal{F}^*_\infty)$, which generate the measurable σ-algebra. Indeed, a predictable projection of such a process is $M^g_{.-}\cdot(t,\infty)$. Here M^g is the right-continuous martingale $M^g_t=\mathbb{E}[g|\mathcal{F}_t]$ (proposition 2.5.13) and $M^g_{.-}$ its left-continuous version. For let T be a predictable stopping time, announced by (T_n), and recall from lemma 3.5.15 (ii) that the strict past of T is $\bigvee \mathcal{F}_{T_n}$ and contains $[T>t]$.

Thus $\quad \mathbb{E}[g|\mathcal{F}_{T-}]=\mathbb{E}\big[g|\bigvee\mathcal{F}_{T_n}\big]$

by exercise 2.5.5: $\quad =\lim\mathbb{E}\big[g|\mathcal{F}_{T_n}\big]$

by theorem 2.5.22: $\quad =\lim M^g_{T_n}=M^g_{T-}$

$\mathbb{P}$-almost surely, and therefore

$$\begin{aligned}\mathbb{E}\big[X_T\cdot[T<\infty]\big]&=\mathbb{E}\big[g\cdot[T>t]\big]\\&=\mathbb{E}\big[\mathbb{E}[g|\mathcal{F}_{T-}]\cdot[T>t]\big]=\mathbb{E}\big[M^g_{T-}\cdot[T>t]\big]\\&=\mathbb{E}\big[\big(M^g_{.-}\cdot(t,\infty)\big)_T\cdot[T<\infty]\big]\ .\end{aligned}$$

This argument has a flaw: M^g is generally adapted only to the natural enlargement $\mathcal{F}^{\mathbb{P}}_{+}$ and $M^g_{.-}$ only to the $\mathbb{P}$-regularization $\mathcal{F}^{\mathbb{P}}$. It can be fixed as follows. For every dyadic rational q let $\overline{M}^g_q$ be an $\mathcal{F}_q$-measurable random variable $\mathbb{P}$-nearly equal to M^g_{q-} (exercise 1.3.33) and set

$$\overline{M}^{g,n}\stackrel{\text{def}}{=}\sum_k \overline{M}_{k2^{-n}}\big[k2^{-n},(k+1)2^{-n}\big)\ .$$

This is a predictable process, and so is $\overline{M}^g \stackrel{\text{def}}{=} \limsup_n \overline{M}^{g,n}$. Now the paths of $\overline{M}^g$ differ from those of $M^g_{\leftarrow}$ only in the $\mathbb{P}$-nearly empty set $\bigcup_q [M^g_{q-} \neq \overline{M}^g_q]$. So $\overline{M}^g \cdot (t,\infty)$ is a predictable projection of $X = (t,\infty) \times g$. An application of the monotone class theorem A.3.4 finishes the proof. ❑

Exercise A.5.16 For any predictable right-continuous increasing process I

$$\mathbb{E}\Big[\int X\,dI\Big] = \mathbb{E}\Big[\int X^{\mathcal{P},\mathbb{P}}\,dI\Big].$$

Supplements and Additional Exercises

Definition A.5.17 (Optional or Well-Measurable Processes) *The σ-algebra generated by the càdlàg adapted processes is called the σ-algebra of* ***optional*** *or* ***well-measurable*** *sets on* $\boldsymbol{B}$ *and is denoted by $\mathcal{O}$. A function measurable on $\mathcal{O}$ is an* ***optional*** *or* ***well-measurable process****.*

Exercise A.5.18 The optional σ-algebra $\mathcal{O}$ is generated by the right-continuous stochastic intervals $[\![S,T)\!)$, contains the previsible σ-algebra $\mathcal{P}$, and is contained in the σ-algebra of progressively measurable sets. For every optional process X there exist a predictable process X' and a countable family $\{T^n\}$ of stopping times such that $[X \neq X']$ is contained in the union $\bigcup_n [\![T^n]\!]$ of their graphs.

Corollary A.5.19 (The Optional Section Theorem) *Suppose that the filtration $\mathcal{F}_.$ is right-continuous and universally complete, let $\mathcal{F} = \mathcal{F}_\infty$, and let $B \subset \boldsymbol{B}$ be an optional set. For every $\mathcal{F}$-capacity I* [42] *and every $\epsilon > 0$ there exists a stopping time R whose graph is contained in B and which satisfies (see figure A.17)*

$$I[R < \infty] > I[\pi_\Omega(B)] - \epsilon .$$

[Hint: Emulate the proof of theorem A.5.14, replacing $\mathcal{M}$ by the finite unions of arbitrary right-continuous stochastic intervals $[\![S,T)\!)$.]

Exercise A.5.20 (The Optional Projection) Let X be a measurable process. For every probability $\mathbb{P}$ on $\mathcal{F}_\infty$ there exists a process $X^{\mathcal{O},\mathbb{P}}$ that is measurable on the optional σ-algebra of the natural enlargement $\mathcal{F}^{\mathbb{P}}_{.+}$ and such that for all stopping times

$$\mathbb{E}[X_T[T < \infty]] = \mathbb{E}[X^{\mathcal{O},\mathbb{P}}_T[T < \infty]] .$$

$X^{\mathcal{O},\mathbb{P}}$ is called ***an optional $\mathbb{P}$-projection*** of X. Any two optional $\mathbb{P}$-projections of X are indistinguishable with $\mathbb{P}$.

Exercise A.5.21 (The Optional Modification) Assume that the measured filtration is right-continuous and regular. Then an adapted measurable process X has an optional modification.

A.6 Suslin Spaces and Tightness of Measures

Polish and Suslin Spaces

A topological space is ***polish*** if it is Hausdorff and separable and if its topology can be defined by a metric under which it is complete. Exercise 1.2.4 on page 15 amounts to saying that the path space $\mathscr{C}$ is polish. The name seems to have arisen this way: the Poles decided that, being a small nation, they should concentrate their mathematical efforts in one area and do it well rather than spread themselves thin. They chose analysis, extending the achievements of the great Pole Banach. They excelled. The theory of analytic spaces, which are the continuous images of polish spaces, is essentially due to them. A Hausdorff topological space F is called a ***Suslin space*** if there exists a polish space P and a continuous surjection $p : P \to F$. A Suslin space is evidently separable.[43] If a continuous injective surjection $p : P \to F$ can be found, then F is a ***Lusin space***. A subset of a Hausdorff space is a ***Suslin set*** or a ***Lusin set***, of course, if it is Suslin or Lusin in the induced topology. The attraction of Suslin spaces in the context of measure theory is this: they contain an abundance of large compact sets: every σ-additive measure on their Borels is tight.

Exercise A.6.1 (i) If P is polish, then there exists a compact metric space $\widehat{P}$ and a homeomorphism j of P onto a dense subset of $\widehat{P}$ that is both a $\mathcal{G}_\delta$-set and a $\mathcal{K}_{\sigma\delta}$-set of $\widehat{P}$.

(ii) A closed subset and a continuous Hausdorff image of a Suslin set are Suslin. This fact is the clue to everthing that follows.

(iii) The union and intersection of countably many Suslin sets are Suslin.

(iv) In a metric Suslin space every Borel set is Suslin.

Proposition A.6.2 *Let F be a Suslin space and $\mathcal{E}$ be an algebra of continuous bounded functions that separates the points of F, e.g., $\mathcal{E} = C_b(F)$.*

(i) Every Borel subset of F is $\mathcal{K}$-analytic.

(ii) $\mathcal{E}$ contains a countable algebra $\mathcal{E}_0$ over $\mathbb{Q}$ that still separates the points of F. The topology generated by $\mathcal{E}_0$ is metrizable, Suslin, and weaker than the given one (but not necessarily strictly weaker).

(iii) Let $m : \mathcal{E} \to \mathbb{R}$ be a positive σ-continuous linear functional with $\|m\| \stackrel{\text{def}}{=} \sup\{m(\phi) : \phi \in \mathcal{E}, |\phi| \le 1\} < \infty$. Then the Daniell extension of m integrates all bounded Borel functions. There exists a unique σ-additive measure μ on $\mathcal{B}^\bullet(F)$ that represents m:

$$m(\phi) = \int \phi \, d\mu \,, \qquad \phi \in \mathcal{E} .$$

This measure is tight and inner regular, and order-continuous on $C_b(F)$.

Proof. Scaling reduces the situation to the case that $\|m\| = 1$. Also, the Daniell extension of m certainly integrates any function in the uniform closure of $\mathcal{E}$ and is σ-continuous thereon. We may thus assume without loss of generality that $\mathcal{E}$ is uniformly closed and thus is both an algebra and a vector lattice (theorem A.2.2). Fix a polish space P and a continuous surjection $p : P \to F$. There are several steps.

(ii) There is a countable subset $\Phi \subset \mathcal{E}$ that still separates the points of F. To see this note that $P \times P$ is again separable and metrizable, and

[43] In the literature Suslin and Lusin spaces are often metrizable by definition. We don't require this, so we don't have to check a topology for metrizability in an application.

let $\mathcal{U}_0[P\times P]$ be a countable uniformly dense subset of $\mathcal{U}[P\times P]$ (see lemma A.2.20). For every $\phi \in \mathcal{E}$ set $g_\phi(x',y') \stackrel{\text{def}}{=} |\phi(p(x')) - \phi(p(y'))|$, and let $\mathcal{U}^{\mathcal{E}}$ denote the countable collection

$$\{f \in \mathcal{U}_0[P\times P] : \exists \phi \in \mathcal{E} \text{ with } f \le g_\phi\} .$$

For every $f \in \mathcal{U}^{\mathcal{E}}$ select one particular $\phi_f \in \mathcal{E}$ with $f \le g_{\phi_f}$, thus obtaining a countable subcollection Φ of $\mathcal{E}$. If $x = p(x') \ne p(y') = y$ in F, then there are a $\phi \in \mathcal{E}$ with $0 < |\phi(x) - \phi(y)| = g_\phi(x',y')$ and an $f \in \mathcal{U}[P \times P]$ with $f \le g_\phi$ and $f(x',y') > 0$. The function $\phi_f \in \Phi$ has $g_{\phi_f}(x',y') > 0$, which signifies that $\phi_f(x) \ne \phi_f(y)$: $\Phi \subset \mathcal{E}$ still separates the points of F. The finite $\mathbb{Q}$-linear combinations of finite products of functions in Φ form a countable $\mathbb{Q}$-algebra $\mathcal{E}_0$.

Henceforth m' denotes the restriction m to the uniform closure $\mathcal{E}' \stackrel{\text{def}}{=} \overline{\mathcal{E}}_0$ in $\mathcal{E} = \overline{\mathcal{E}}$. Clearly $\mathcal{E}'$ is a vector lattice and algebra and m' is a positive linear σ-continuous functional on it.

Let $j : F \to \widehat{F}$ denote the local $\mathcal{E}_0$-compactification of F provided by theorem A.2.2 (ii). $\widehat{F}$ is metrizable and j is injective, so that we may identify F with the dense subset $j(F)$ of $\widehat{F}$. Note however that the Suslin topology of F is a priori finer than the topology induced by $\widehat{F}$. Every $\phi \in \mathcal{E}'$ has a unique extension $\widehat{\phi} \in C_0(\widehat{F})$ that agrees with ϕ on F, and $\widehat{\mathcal{E}'} = C_0(\widehat{F})$. Let us define $\widehat{m}' : C(\widehat{F}) \to \mathbb{R}$ by $\widehat{m}'(\widehat{\phi}) = m'(\phi)$. This is a Radon measure on $\widehat{F}$. Thanks to Dini's theorem A.2.1, $\widehat{m}'$ is automatically σ-continuous. We convince ourselves next that F has upper integral (= outer measure) 1. Indeed, if this were not so, then the inner measure $\widehat{m}'_*(\widehat{F} - F) = 1 - \widehat{m}'^*(F)$ of its complement would be strictly positive. There would be a function $k \in C_\downarrow(\widehat{F})$, pointwise limit of a decreasing sequence $\widehat{\phi}_n$ in $C(\widehat{F})$, with $k \le \widehat{F} - F$ and $0 < \widehat{m}'_*(k) = \lim \widehat{m}'(\widehat{\phi}_n) = \lim m'(\phi_n)$. Now (ϕ_n) decreases to zero on F and therefore the last limit is zero, a plain contradiction.

So far we do not know that *F is $\widehat{m}'$-measurable in $\widehat{F}$*. To prove that it is it will suffice to show that F is $\mathcal{K}$-analytic in $\widehat{F}$: since $\widehat{m}'^*$ is a $\mathcal{K}$-capacity, Choquet's theorem will then provide compact sets $K_n \subset F$ with $\sup \widehat{m}'(K_n) = \sup \widehat{m}'^*(K_n) = 1$, showing that the inner measure of F equals 1 as well, so that F is measurable. To show the analyticity of F we embed P homeomorphically as a $\mathcal{G}_\delta$ in a compact metric space $\widehat{P}$, as in exercise A.6.1. Actually, we shall view P as a $\mathcal{G}_\delta$ in $\Pi \stackrel{\text{def}}{=} \widehat{F} \times \widehat{P}$, embedded via $x \mapsto (p(x), x)$. The projection π of Π on its first factor $\widehat{F}$ coincides with p on P.

Now the topologies of both $\widehat{F}$ and $\widehat{P}$ have a countable basis of relatively compact open balls, and the rectangles made from them constitute a countable basis for the topology of Π. Therefore every open set of Π is the countable union of compact rectangles and is thus a $\mathcal{K}^\times_\sigma$-set for the paving $\mathcal{K}^\times \stackrel{\text{def}}{=} \mathcal{K}[\widehat{F}] \times \mathcal{K}[\widehat{P}]$. The complement of a set in $\mathcal{K}^\times$ is the countable union of sets in $\mathcal{K}^\times$, so every Borel set of Π is $\mathcal{K}^\times$-analytic (corollary A.5.3). In particular,

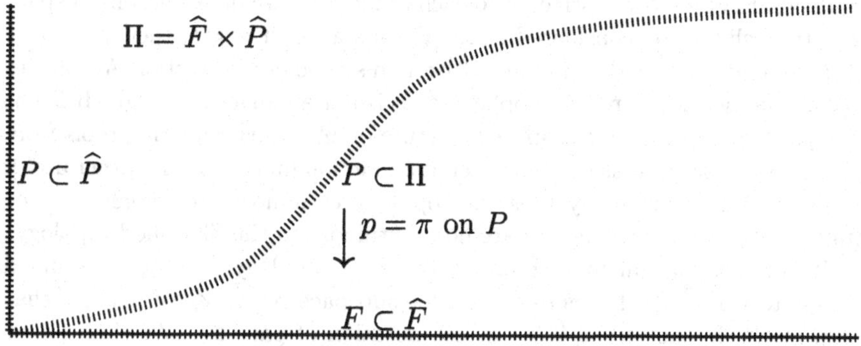

Figure A.18 Compactifying the situation

P, which is the countable intersection of open subsets of Π (exercise A.6.1), is $\mathcal{K}^\times$-analytic. By proposition A.5.4, $F = \pi(P)$ is $\mathcal{K}[\widehat{F}]$-analytic and a *fortiori* $\mathcal{K}[F]$-analytic.

This argument also shows that *every Borel set B of F is analytic.* Namely, $B^\uparrow \stackrel{\text{def}}{=} p^{-1}(B)$ is a Borel set of P and then of Π, and its image $B = \pi(B^\uparrow)$ is $\mathcal{K}[\widehat{F}]$-analytic in F and in $\widehat{F}$. As above we see that therefore $\widehat{m}'^*(B) = \sup\{\int K\, d\widehat{m}' : K \subset B, K \in \mathcal{K}[\widehat{F}]\}$: $B \mapsto \mu(B) \stackrel{\text{def}}{=} \int B\, d\widehat{m}'$ is an inner regular measure representing m.

Only the uniqueness of μ remains to be shown. Assume then that μ' is another measure on the Borel sets of F that agrees with m on $\mathcal{E}$. Taking $C_b(F)$ instead of $\mathcal{E}$ in the previous arguments shows that μ' is inner regular. By theorem A.3.4, μ and μ' agree on the sets of $\mathcal{K}[\widehat{F}]$ in F, by the inner regularity on all Borel sets. ∎

A.7 The Skorohod Topology

In this section (E, ρ) is a fixed complete separable metric space. Replacing if necessary ρ by $\rho \wedge 1$, we may and shall assume that $\rho \le 1$. We consider the set $\mathscr{D}_E$ of all paths $z : [0, \infty) \to E$ that are right-continuous and have left limits $z_{t-} \stackrel{\text{def}}{=} \lim_{s \uparrow t} z_s \in E$ at all finite instants $t > 0$. Integrators and solutions of stochastic differential equations are random variables with values in $\mathscr{D}_{\mathbb{R}^n}$. In the stochastic analysis of them the maximal process plays an important role. It is finite at finite times (lemma 2.3.2), which seems to indicate that the topology of uniform convergence on bounded intervals is the appropriate topology on $\mathscr{D}$. This topology is not Suslin, though, as it is not separable: the functions $[0, t)$, $t \in \mathbb{R}_+$, are uncountable in number, but any two of them have uniform distance 1 from each other. Results invoking tightness, as for instance proposition A.6.2 and theorem A.3.17, are not applicable.

Skorohod has given a polish topology on $\mathscr{D}_E$. It rests on the idea that temporal as well as spatial measurements are subject to errors, and that paths that can be transformed into each other by small deformations of space and time should be considered close. For instance, if $s \approx t$, then $[0, s)$ and $[0, t)$ should be considered close. Skorohod's topology of section A.7 makes the above-mentioned results applicable. It is not a panacea, though. It is not compatible with the vector space structure of $\mathscr{D}_{\mathbb{R}}$, thus rendering tools from Fourier analysis such as the characteristic function unusable; the rather useful example A.4.13 genuinely uses the topology of uniform convergence – the functions ϕ appearing therein are not continuous in the Skorohod topology.

It is most convenient to study Skorohod's topology first on a bounded time-interval $[0, u]$, that is to say, on the subspace $\mathscr{D}_E^u \subset \mathscr{D}_E$ of paths z that stop at the instant u: $z = z^u$ in the notation of page 23. We shall follow rather closely the presentation in Billingsley [10]. There are two equivalent metrics for the Skorohod topology whose convenience of employ depends on the situation. Let Λ denote the collection of all strictly increasing functions from $[0, \infty)$ onto itself, the "time transformations." The first Skorohod metric $d^{(0)}$ on $\mathscr{D}_E^u$ is defined as follows: for $z, y \in \mathscr{D}_E^u$, $d^{(0)}(z, y)$ is the infimum of the numbers $\epsilon > 0$ for which there exists a $\lambda \in \Lambda$ with

$$\|\lambda\|^{(0)} \stackrel{\text{def}}{=} \sup_{0 \le t < \infty} |\lambda(t) - t| < \epsilon$$

and

$$\sup_{0 \le t < \infty} \rho\big(z_t, y_{\lambda(t)}\big) < \epsilon .$$

It is left to the reader to check that

$$\|\lambda\|^{(0)} = \|\lambda^{-1}\|^{(0)} \quad \text{and} \quad \|\lambda \circ \mu\|^{(0)} \le \|\lambda\|^{(0)} + \|\mu\|^{(0)} , \qquad \lambda, \mu \in \Lambda ,$$

and that $d^{(0)}$ is a metric satisfying $d^{(0)}(z, y) \le \sup_t \rho(z_t, y_t)$. The topology of $d^{(0)}$ is called the Skorohod topology on $\mathscr{D}_E^u$. It is coarser than the uniform topology. A sequence $\big(z^{(n)}\big)$ of $\mathscr{D}_E^u$ converges to $z \in \mathscr{D}_E^u$ if and only if there exist $\lambda_n \in \Lambda$ with $\|\lambda_n\|^{(0)} \to 0$ and $z^{(n)}_{\lambda_n(t)} \to z_t$ uniformly in t.

We now need a couple of tools. The ***oscillation*** of any stopped path $z : [0, \infty) \to E$ on an interval $I \subset \mathbb{R}_+$ is $o_I[z] \stackrel{\text{def}}{=} \sup\{\rho(z_s, z_t) : s, t \in I\}$, and there are two pertinent ***moduli of continuity***:

$$\gamma^\delta[z] \stackrel{\text{def}}{=} \sup_{0 \le t < \infty} o_{[t, t+\delta]}[z]$$

$$\text{and} \quad \gamma^\delta_{[)}[z] \stackrel{\text{def}}{=} \inf\Big\{ \sup_i o_{[t_i, t_{i+1})}[z] : \ 0 = t_0 < t_1 < \ldots , \ |t_{i+1} - t_i| > \delta \Big\} .$$

They are related by $\gamma^\delta_{[)}[z] \le \gamma^{2\delta}[z]$. A stopped path $z = z^u : [0, \infty) \to E$ is in $\mathscr{D}_E^u$ if and only if $\gamma^\delta_{[)}[z] \xrightarrow[\delta \to 0]{} 0$ and is continuous if and only if $\gamma^\delta[z] \xrightarrow[\delta \to 0]{} 0$ – we then write $z \in \mathscr{C}_E^u$. Evidently

$$\begin{aligned} \rho\big(z_t, z^{(n)}_t\big) &\le \rho\Big(z_t, z_{\lambda_n^{-1}(t)}\Big) + \rho\Big(z_{\lambda_n^{-1}(t)}, z^{(n)}_t\Big) \le \rho\Big(z_t, z_{\lambda_n^{-1}(t)}\Big) + d^{(0)}\big(z, z^{(n)}\big) \\ &\le \gamma^{\|\lambda_n^{-1}\|}[z] + d^{(0)}\big(z, z^{(n)}\big) . \end{aligned}$$

The first line shows that $d^{(0)}(z, z^{(n)}) \to 0$ implies $z_t^{(n)} \to z_t$ in continuity points t of z, and the second that the convergence is uniform if $z = z^u$ is continuous (and then uniformly continuous). The Skorohod topology therefore coincides on $\mathscr{C}_E^u$ with the topology of uniform convergence.

It is separable. To see this let E_0 be a countable dense subset of E and let $\mathscr{D}^{(k)}$ be the collection of paths in $\mathscr{D}_E^u$ that are constant on intervals of the form $[i/k, (i+1)/k)$, $i \le ku$, and take a value from E_0 there. $\mathscr{D}^{(k)}$ is countable and $\bigcup_k \mathscr{D}^{(k)}$ is $d^{(0)}$-dense in $\mathscr{D}_E^u$.

However, $\mathscr{D}_E^u$ is not complete under this metric – $[u/2 - 1/n, u/2)$ is a $d^{(0)}$-Cauchy sequence that has no limit. There is an equivalent metric, one that defines the same topology, under which $\mathscr{D}_E^u$ is complete; thus $\mathscr{D}_E^u$ is polish in the Skorohod topology. The second Skorohod metric $d^{(1)}$ on $\mathscr{D}_E^u$ is defined as follows: for $z, y \in \mathscr{D}_E^u$, $d^{(1)}(z, y)$ is the infimum of the numbers $\epsilon > 0$ for which there exists a $\lambda \in \Lambda$ with

$$\|\lambda\|^{(1)} \stackrel{\text{def}}{=} \sup_{0 \le s < t < \infty} \ln \left| \frac{\lambda(t) - \lambda(s)}{t - s} \right| < \epsilon \tag{A.7.1}$$

and

$$\sup_{0 \le t < \infty} \rho\big(z_t, y_{\lambda(t)}\big) < \epsilon\,. \tag{A.7.2}$$

Roughly speaking, (A.7.1) restricts the time transformations to those with "slope close to unity." Again it is left to the reader to show that

$$\|\lambda\|^{(1)} = \|\lambda^{-1}\|^{(1)} \quad \text{and} \quad \|\lambda \circ \mu\|^{(1)} \le \|\lambda\|^{(1)} + \|\mu\|^{(1)}\,, \qquad \lambda, \mu \in \Lambda\,,$$

and that $d^{(1)}$ is a metric.

Theorem A.7.1 *(i) $d^{(0)}$ and $d^{(1)}$ define the same topology on $\mathscr{D}_E^u$.*

(ii) $(\mathscr{D}_E^u, d^{(1)})$ is complete. $\mathscr{D}_E$ is separable and complete under the metric

$$d(z, y) \stackrel{\text{def}}{=} \sum_{u \in \mathbb{N}} 2^{-u} \wedge d^{(1)}(z^u, y^u)\,, \qquad z, y \in \mathscr{D}_E\,.$$

The polish topology τ of d on $\mathscr{D}_E$ is called the ***Skorohod topology*** *and coincides on $\mathscr{C}_E$ with the topology of uniform convergence on bounded intervals and on $\mathscr{D}_E^u \subset \mathscr{D}_E$ with the topology of $d^{(0)}$ or $d^{(1)}$. The stopping maps $z \mapsto z^u$ are continuous projections from $(\mathscr{D}_E, d)$ onto $(\mathscr{D}_E^u, d^{(1)})$, $0 < u < \infty$.*

(iii) The Hausdorff topology σ on $\mathscr{D}_{\mathbb{R}^d}$ generated by the linear functionals

$$z \mapsto \langle z | \phi \rangle \stackrel{\text{def}}{=} \int_0^\infty \sum_{a=1}^d z_s^{(a)} \phi_a(s)\, ds\,, \qquad \phi \in C_{00}(\mathbb{R}_+, \mathbb{R}^d)\,,$$

is weaker than τ and makes $\mathscr{D}_{\mathbb{R}^d}$ into a Lusin topological vector space.

(iv) Let $\mathcal{F}_t^0$ denote the σ-algebra generated by the evaluations $z \mapsto z_s$, $0 \le s \le t$, the ***basic filtration of path space****. Then $\mathcal{F}_t^0 \subset \mathcal{B}^\bullet(\mathscr{D}_E^t, \tau)$, and $\mathcal{F}_t^0 = \mathcal{B}^\bullet(\mathscr{D}_E^t, \sigma)$ if $E = \mathbb{R}^d$.*

(v) Suppose $\mathbb{P}_t$ is, for every $t < \infty$, a σ-additive probability on $\mathcal{F}_t^0$, such that the restriction of $\mathbb{P}_t$ to $\mathcal{F}_s^0$ equals $\mathbb{P}_s$ for $s < t$. Then there exists a unique tight probability $\mathbb{P}$ on the Borels of $(\mathscr{D}_E, \tau)$ that equals $\mathbb{P}_t$ on $\mathcal{F}_t$ for all t.

Proof. (i) If $d^{(1)}(z,y) < \epsilon < 1/(4+u)$, then there is a $\lambda \in \Lambda$ with $\|\lambda\|^{(1)} < \epsilon$ satisfying (A.7.2). Since $\lambda(0) = 0$, $\ln(1-2\epsilon) < -\epsilon < \ln\big(\lambda(t)/t\big) < \epsilon < \ln(1+2\epsilon)$, which results in $|\lambda(t) - t| \le 2\epsilon t \le 2\epsilon(u+1) < 1/2$ for $0 \le t \le u+1$. Changing λ on $[u+1,\infty)$ to continue with slope 1 we get $\|\lambda\|^{(0)} \le 2\epsilon(u+1)$ and $d^{(0)}(z,y) \le 2\epsilon(u+1)$. That is to say, $d^{(1)}(z,z^{(n)}) \to 0$ implies $d^{(0)}(z,z^{(n)}) \to 0$. For the converse we establish the following claim: *if $d^{(0)}(z,y) < \delta^2 < 1/4$, then $d^{(1)}(z,y) \le 4\delta + \gamma_{[)}^{\delta}[z]$*. To see this choose instants $0 = t_0 < t_1 \ldots$ with $t_{i+1} - t_i > \delta$ and $o_{[t_i,t_{i+1})}[z] < \gamma_{[)}^{\delta}[z] + \delta$ and $\mu \in \Lambda$ with $\|\mu\|^{(0)} < \delta^2$ and $\sup_t \rho\big(z_{\mu^{-1}(t)}, y_t\big) < \delta^2$. Let λ be that element of Λ which agrees with μ at the instants t_i and is linear in between, $i = 0,1,\ldots$. Clearly $\mu^{-1}\circ\lambda$ maps $[t_i,t_{i+1})$ to itself, and

$$\begin{aligned}\rho\big(z_t, y_{\lambda(t)}\big) &\le \rho\big(z_t, z_{\mu^{-1}\circ\lambda(t)}\big) + \rho\big(z_{\mu^{-1}\circ\lambda(t)}, y_{\lambda(t)}\big)\\ &\le \gamma_{[)}^{\delta}[z] + \delta + \delta^2 < 4\delta + \gamma_{[)}^{\delta}[z]\,.\end{aligned}$$

So if $d^{(0)}(z,z^{(n)}) \to 0$ and $0 < \epsilon < 1/2$ is given, we choose $0 < \delta < \epsilon/8$ so that $\gamma_{[)}^{\delta}[z] < \epsilon/2$ and then N so that $d^{(0)}(z,z^{(n)}) < \delta^2$ for $n > N$. The claim above produces $d^{(1)}(z,z^{(n)}) < \epsilon$ for such n.

(ii) Let $\big(z^{(n)}\big)$ be a $d^{(1)}$-Cauchy sequence in $\mathscr{D}_E^u$. Since it suffices to show that a subsequence converges, we may assume that $d^{(1)}\big(z^{(n)}, z^{(n+1)}\big) < 2^{-n}$. Choose $\mu_n \in \Lambda$ with

$$\|\mu_n\|^{(1)} < 2^{-n} \quad \text{and} \quad \sup_t \rho\big(z_t^{(n)}, z_{\mu_n(t)}^{(n+1)}\big) < 2^{-n}\,.$$

Denote by μ_n^{n+m} the composition $\mu_{n+m}\circ\mu_{n+m-1}\circ\cdots\circ\mu_n$. Clearly

$$\sup_t \Big|\mu_{n+m+1}\circ\mu_n^{n+m}(t) - \mu_n^{n+m}(t)\Big| < 2^{-n-m-1}\,,$$

and by induction
$$\sup_t \Big|\mu_n^{n+m'}(t) - \mu_n^{n+m}(t)\Big| \le 2^{-n-m}\,, \qquad 1 \le m < m'\,.$$

The sequence $\big(\mu_n^{n+m}\big)_{m=1}^{\infty}$ is thus uniformly Cauchy and converges uniformly to some function λ_n that has $\lambda_n(0) = 0$ and is increasing. Now

$$\ln\left|\frac{\mu_n^{n+m}(t) - \mu_n(s)}{t-s}\right| \le \sum_{i=n+1}^{n+m} \|\mu_i\|^{(1)} \le 2^{-n}\,,$$

so that $\|\lambda_n\|^{(1)} \le 2^{-n}$. Therefore λ_n is strictly increasing and belongs to Λ. Also clearly $\lambda_n = \lambda_{n+1}\circ\mu_n$. Thus

$$\sup_t \rho\Big(z_{\lambda_n^{-1}(t)}^{(n)}, z_{\lambda_{n+1}^{-1}(t)}^{(n+1)}\Big) = \sup_t \rho\Big(z_t^{(n)}, z_{\mu^n(t)}^{(n+1)}\Big) \le 2^{-n}\,,$$

and the paths $z_{\lambda_n^{-1}}^{(n)}$ converge uniformly to some right-continuous path z

with left limits. Since $z^{(n)}$ is constant on $[\lambda_n^{-1} \geq u]$, and since $\lambda_n^{-1}(t) \to t$ uniformly on $[0, u+1]$, z is constant on $(u, \infty) \subset \liminf[\lambda_n^{-1} \geq u]$ and by right-continuity belongs to $\mathscr{D}_E^u$. Since $\| \lambda_n \|^{(1)} \xrightarrow[n\to\infty]{} 0$ and $\sup_t \rho(z_t, z^{(n)}_{\lambda_n^{-1}(t)})$ converges to 0, $d^{(1)}(z, z^{(n)}) \xrightarrow[n\to\infty]{} 0$: $(\mathscr{D}_E^u, d^{(1)})$ is indeed complete. The remaining statements of (ii) are left to the reader.

(iii) Let $\boldsymbol{\phi} = (\phi_1, \ldots, \phi_d) \in C_{00}(\mathbb{R}_+, \mathbb{R}^d)$. To see that the linear functional $\boldsymbol{z} \mapsto \langle \boldsymbol{z} | \boldsymbol{\phi} \rangle$ is continuous in the Skorohod topology, let u be the supremum of supp $\boldsymbol{\phi}$. If $\boldsymbol{z}^{(n)} \to \boldsymbol{z}$, then $\sup_{n\in\mathbb{N}, t\leq u} |\boldsymbol{z}_t^{(n)}|$ is finite and $\boldsymbol{z}_t^{(n)} \xrightarrow[n\to\infty]{} \boldsymbol{z}_t$ in all but the countably many points t where $\boldsymbol{z}_t$ jumps. By the DCT the integrals $\langle \boldsymbol{z}^{(n)} | \boldsymbol{\phi} \rangle$ converge to $\langle \boldsymbol{z} | \boldsymbol{\phi} \rangle$. The topology σ is thus weaker than τ. Since $\langle \boldsymbol{z}^{(n)} | \boldsymbol{\phi} \rangle = 0$ for all continuous $\boldsymbol{\phi} : [0, \infty) \to \mathbb{R}^d$ with compact support implies $\boldsymbol{z} \equiv 0$, this evidently linear topology is Lusin. Writing $\langle \boldsymbol{z} | \boldsymbol{\phi} \rangle$ as a limit of Riemann sums shows that the σ-Borels of $\mathscr{D}^t_{\mathbb{R}^d}$ are contained in $\mathcal{F}_t^0$. Conversely, letting $\boldsymbol{\phi}$ run through an ***approximate identity*** $\boldsymbol{\phi}_n$ supported on the right of $s \leq t$ [44] shows that $\boldsymbol{z}_s = \lim \langle \boldsymbol{z} | \boldsymbol{\phi}_n \rangle$ is measurable on $\mathcal{B}^\bullet(\sigma)$, so that there is coincidence.

(iv) In general, when E is just some polish space, one can define a Lusin topology σ weaker than τ as well: it is the topology generated by the τ-continuous functions $z \mapsto \int_0^\infty \psi(z_s) \cdot \phi(s)\, ds$, where $\phi \in C_{00}(\mathbb{R}_+)$ and $\psi : E \to \mathbb{R}$ is uniformly continuous. It follows as above that $\mathcal{F}_t^0 = \mathcal{B}^\bullet(\mathscr{D}_E^t, \sigma)$.

(v) The $\mathbb{P}_t$ viewed as probabilities on the Borels of $(\mathscr{D}_E^t, \sigma)$ form a tight (proposition A.6.2) projective system, evidently full, under the stopping maps $\pi_u^t(z) \stackrel{\text{def}}{=} z^t$ [45]. There is thus on $\bigcup_t C_b(\mathscr{D}_E^t, \sigma) \circ \pi^t$ a σ-additive projective limit $\mathbb{P}$ (theorem A.3.17). This algebra of bounded Skorohod-continuous functions separates the points of $\mathscr{D}_E$, so $\mathbb{P}$ has a unique extension to a tight probability on the Borels of the Skorohod topology τ (proposition A.6.2). ∎

Proposition A.7.2 *A subset $\mathscr{K} \subset \mathscr{D}_E$ is relatively compact if and only if both (i) $\{z_t : z \in \mathscr{K}\}$ is relatively compact in E for every $t \in [0, \infty)$ and (ii) for every instant $u \in [0, \infty)$, $\lim_{\delta\to 0} \gamma_{\mathfrak{D}}^\delta[z^u] = 0$ uniformly in $z \in \mathscr{K}$. In this case the sets $\{z_t : 0 \leq t \leq u, z \in \mathscr{K}\}$, $u < \infty$, are relatively compact in E.*
For a proof see for example Ethier and Kurtz [33, page 123].

Proposition A.7.3 *A family $\mathcal{M}$ of probabilities on $\mathscr{D}_E$ is uniformly tight provided that for every instant $u < \infty$ we have, uniformly in $\mu \in \mathcal{M}$,*

$$\int_{\mathscr{D}_E} \gamma_{\mathfrak{D}}^\delta[z^u] \wedge 1\, \mu(dz) \xrightarrow[\delta\to 0]{} 0$$

$$\textit{or} \quad \sup\left\{ \int_{\mathscr{D}_E} \rho(z_T^u, z_S^u) \wedge 1\, \mu(dz) : S, T \in \mathfrak{T}, 0 \leq S \leq T \leq S + \delta \right\} \xrightarrow[\delta\to 0]{} 0 .$$

Here $\mathfrak{T}$ denotes collection of stopping times for the right-continuous version of the basic filtration on $\mathscr{D}_E$.

---

[44] supp $\boldsymbol{\phi}_n \in [s, s+1/n]$, $\phi_n^\eta \geq 0$, and $\int \phi_n^\eta(r)\, dr = 1$.
[45] The set of threads is identified naturally with $\mathscr{D}_E$.

A.8 The L^p-Spaces

Let $(\Omega, \mathcal{F}, \mathbb{P})$ be a probability space and recall from page 33 the following measurements of the size of a measurable function f:

$$\lceil\!\lceil f \rceil\!\rceil_p = \lceil\!\lceil f \rceil\!\rceil_{L^p(\mathbb{P})} \stackrel{\text{def}}{=} \begin{cases} \|f\|_p = \left(\int |f|^p \, d\mathbb{P}\right)^{1/p} & \text{for } 1 \le p < \infty, \\ \|f\|_p^p = \int |f|^p \, d\mathbb{P} & \text{for } 0 < p \le 1, \\ \inf\left\{\lambda : \mathbb{P}[|f| > \lambda] \le \lambda\right\} & \text{for } p = 0; \end{cases}$$

and

$$\|f\|_{[\alpha]} = \|f\|_{[\alpha;\mathbb{P}]} = \inf\{\lambda > 0 : \mathbb{P}[|f| > \lambda] \le \alpha\} \qquad \text{for } p = 0 \text{ and } \alpha > 0.$$

The space $L^p(\mathbb{P})$ is the collection of all measurable functions f that satisfy $\lceil\!\lceil rf \rceil\!\rceil_{L^p(\mathbb{P})} \xrightarrow[r\to 0]{} 0$. Customarily, the slew of L^p-spaces is extended at $p = \infty$ to include the space $L^\infty = L^\infty(\mathbb{P}) = L^\infty(\mathcal{F}, \mathbb{P})$ of bounded measurable functions equipped with the seminorm

$$\|f\|_\infty = \|f\|_{L^\infty(\mathbb{P})} \stackrel{\text{def}}{=} \inf\{c : \mathbb{P}[|f| > c] = 0\},$$

which we also write $\lceil\!\lceil \ \rceil\!\rceil_\infty$, if we want to stress its subadditivity. L^∞ plays a minor role in this book, bsince it is not suited to be the range of a vector measure such as the stochastic integral.

Exercise A.8.1 (i) $\lceil\!\lceil f \rceil\!\rceil_0 \le a \iff \mathbb{P}[|f| > a] \le a$.

(ii) For $1 \le p \le \infty$, $\lceil\!\lceil \ \rceil\!\rceil_p$ is a seminorm. For $0 \le p < 1$, it is subadditive but not homogeneous.

(iii) Let $0 \le p \le \infty$. A measurable function f is said to be ***finite in p-mean*** if

$$\lim_{r\to 0} \lceil\!\lceil rf \rceil\!\rceil_p = 0.$$

For $0 < p \le \infty$ this means simply that $\lceil\!\lceil f \rceil\!\rceil_p < \infty$. A numerical measurable function f belongs to L^p if and only if it is finite in p-mean.

(iv) Let $0 \le p \le \infty$. The spaces L^p are vector lattices, i.e., are closed under taking finite linear combinations and finite pointwise maxima and minima. They are not in general algebras, except for L^0, which is one. They are complete under the metric $dist_p(f, g) = \lceil\!\lceil f - g \rceil\!\rceil_p$, and every mean-convergent sequence has an almost surely convergent subsequence.

(v) Let $0 \le p < \infty$. The simple measurable functions are p-mean dense. (A measurable function is simple if it takes only finitely many values, all of them finite.)

Exercise A.8.2 For $0 < p < 1$, the homogeneous functionals $\| \ \|_p$ are not subadditive, but there is a substitute for subadditivity:

$$\|f_1 + \ldots + f_n\|_p \le n^{0 \vee (1-p)/p} \cdot \left(\|f_1\|_p + \ldots + \|f_n\|_p\right) \qquad 0 < p \le \infty.$$

Exercise A.8.3 For any set K and $p \in [0, \infty]$, $\lceil\!\lceil K \rceil\!\rceil_{L^p(\mathbb{P})} = (\mathbb{P}[K])^{1 \wedge 1/p}$.

Theorem A.8.4 (Hölder's Inequality) *For any three exponents* $0 < p, q, r \le \infty$ *with* $1/r = 1/p + 1/q$ *and any two measurable functions* f, g,

$$\|fg\|_r \le \|f\|_p \cdot \|g\|_q .$$

If $p, p' \in [1, \infty]$ *are related by* $1/p + 1/p' = 1$, *then* p, p' *are called* ***conjugate exponents****. Then* $p' = p/(p-1)$ *and* $p = p'/(p'-1)$. *For conjugate exponents* p, p'

$$\|f\|_p = \sup\{\int fg : g \in L^{p'}, \|g\|_{p'} \le 1\} .$$

All of this remains true if the underlying measure is merely σ*-finite.*[46]

Proof: Exercise. ∎

Exercise A.8.5 Let μ be a positive σ-additive measure and f a μ-measurable function. The set $I_f \stackrel{\text{def}}{=} \{1/p \in \mathbb{R}_+ : \|f\|_p < \infty\}$ is either empty or is an interval. The function $1/p \mapsto \|f\|_p$ is logarithmically convex and therefore continuous on the interior $\mathring{I}_f$. Consequently, for $0 < p_0 < p_1 < \infty$,

$$\sup_{p_0 < p < p_1} \|f\|_p = \|f\|_{p_0} \vee \|f\|_{p_1} .$$

Uniform Integrability Let $0 < p < \infty$ and let $\mathbb{P}$ be a positive measure of finite total mass. A collection $\mathcal{C}$ of L^p-integrable functions is ***uniformly p-integrable*** if for every $\epsilon > 0$ there is a constant K_ϵ with the following property: for every $f \in \mathcal{C}$ there is an f' with

$$-K_\epsilon \le f' \le K_\epsilon \quad \text{and} \quad \|f - f'\|_{L^p(\mathbb{P})} < \epsilon ,$$

that is to say, so that the L^p-distance of f from the uniformly bounded set

$$\{f' : -K_\epsilon \le f' \le K_\epsilon\}$$

is less than ϵ. $f' = (-K_\epsilon) \vee f \wedge K_\epsilon$ will minimize this distance. Uniform integrability generalizes the domination condition in the DCT in this sense: if there is a $g \in L^p$ with $|f| \le g$ for all $f \in \mathcal{C}$, then $\mathcal{C}$ is uniformly p-integrable. Using this notion one can establish the most general and sharp convergence theorem of Lebesgue's theory – it makes a good exercise:

Theorem A.8.6 (Dominated Convergence Theorem on L^p) *Suppose that the measure has finite total mass and let* $0 \le p < \infty$. *A sequence* (f_n) *in* L^p *converges in p-mean if and only if it is uniformly p-integrable and converges in measure.*

Exercise A.8.7 (Fatou's Lemma) (i) Let (f_n) be a sequence of positive measurable functions. Then

$$\Big\|\liminf_{n\to\infty} f_n\Big\|_{L^p} \le \liminf_{n\to\infty} \|f_n\|_{L^p} , \qquad 0 < p \le \infty;$$

$$\Big\lceil\!\Big\lceil\liminf_{n\to\infty} f_n\Big\rceil\!\Big\rceil_{L^p} \le \liminf_{n\to\infty} \lceil\!\lceil f_n \rceil\!\rceil_{L^p} , \qquad 0 \le p < \infty;$$

$$\Big\|\liminf_{n\to\infty} f_n\Big\|_{[\alpha]} \le \liminf_{n\to\infty} \|f_n\|_{[\alpha]} , \qquad 0 < \alpha < \infty.$$

[46] I.e., $L^1(\mu)$ is σ-finite (exercise A.3.2).

(ii) Let (f_n) be a sequence in L^0 that converges in measure to f. Then

$$\|f\|_{L^p} \le \liminf_{n\to\infty} \|f_n\|_{L^p}\,, \qquad 0<p\le\infty;$$
$$\lceil f \rceil_{L^p} \le \liminf_{n\to\infty} \lceil f_n \rceil_{L^p}\,, \qquad 0\le p<\infty;$$
$$\|f\|_{[\alpha]} \le \liminf_{n\to\infty} \|f_n\|_{[\alpha]}\,, \qquad 0<\alpha<\infty.$$

A.8.8 Convergence in Measure — the Space L^0 The space L^0 of almost surely finite functions plays a large role in this book. Exercise A.8.10 amounts to saying that L^0 is a topological vector space for the topology of convergence in measure, whose neighborhood system at 0 has a countable basis, and which is defined by the gauge $\lceil\ \rceil_0$. Here are a few exercises that are used in the main text.

Exercise A.8.9 The functional $\lceil\ \rceil_0$ is by no means the canonical way with which to gauge the size of a.s. finite measurable functions. Here is another one that serves just as well and is sometimes easier to handle: $\lceil f \rceil_\odot \stackrel{\text{def}}{=} \mathbb{E}[|f|\wedge 1]$, with associated distance

$$dist_\odot(f,g) \stackrel{\text{def}}{=} \lceil f-g \rceil_\odot = \mathbb{E}[|f-g|\wedge 1]\,.$$

Show: (i) $\lceil\ \rceil_\odot$ is subadditive, and $dist_\odot$ is a pseudometric. (ii) A sequence (f_n) in L^0 converges to $f \in L^0$ in measure if and only if $\lceil f - f_n \rceil_\odot \xrightarrow[n\to\infty]{} 0$. In other words, $\lceil\ \rceil_\odot$ is a gauge on $L^0(\mathbb{P})$.

Exercise A.8.10 L^0 is an algebra. The maps $(f,g) \mapsto f+g$ and $(f,g) \mapsto f\cdot g$ from $L^0 \times L^0$ to L^0 and $(r,f) \mapsto r\cdot f$ from $\mathbb{R}\times L^0$ to L^0 are continuous. The neighborhood system at 0 has a basis of "balls" of the form

$$B_r(0) \stackrel{\text{def}}{=} \{f : \lceil f \rceil_0 < r\} \text{ and of the form } B_r(0) \stackrel{\text{def}}{=} \{f : \lceil f \rceil_\odot < r\}\,, \quad r>0\,.$$

Exercise A.8.11 Here is another gauge on $L^0(\mathbb{P})$, which is used in proposition 3.6.20 on page 129 to study the behavior of the stochastic integral under a change of measure. Let $\mathbb{P}'$ be a probability equivalent to $\mathbb{P}$ on $\mathcal{F}$, i.e., a probability having the same negligible sets as $\mathbb{P}$. The Radon–Nikodym theorem A.3.22 on page 407 provides a strictly positive $\mathcal{F}$-measurable function g' such that $\mathbb{P}' = g'\mathbb{P}$. Show that the spaces $L^0(\mathbb{P})$ and $L^0(\mathbb{P}')$ coincide; moreover, the topologies of convergence in $\mathbb{P}$-measure and in $\mathbb{P}'$-measure coincide as well.

The mean $\lceil\ \rceil_{L^0(\mathbb{P}')}$ thus describes the topology of convergence in $\mathbb{P}$-measure just as satisfactorily as does $\lceil\ \rceil_{L^0(\mathbb{P})}$: $\lceil\ \rceil_{L^0(\mathbb{P}')}$ is a gauge on $L^0(\mathbb{P})$.

Exercise A.8.12 Let $(\Omega,\mathcal{F})$ be a measurable space and $\mathbb{P} \ll \mathbb{P}'$ two probabilities on $\mathcal{F}$. There exists an increasing right-continuous function $\Phi : (0,1] \to (0,1]$ with $\lim_{r\to 0}\Phi(r) = 0$ such that $\lceil f \rceil_{L^0(\mathbb{P})} \le \Phi(\lceil f \rceil_{L^0(\mathbb{P}')})$ for all $f \in \mathcal{F}$.

A.8.13 The Homogeneous Gauges $\|\ \|_{[\alpha]}$ on $\mathbf{L^p}$ Recall the definition of the homogeneous gauges on measurable functions f,

$$\|f\|_{[\alpha]} = \|f\|_{[\alpha;\mathbb{P}]} \stackrel{\text{def}}{=} \inf\{\lambda > 0 : \ \mathbb{P}[|f| > \lambda] \le \alpha\}\,.$$

Of course, if $\alpha < 0$, then $\|f\|_{[\alpha]} = \infty$, and if $\alpha \ge 1$, then $\|f\|_{[\alpha]} = 0$. Yet it streamlines some arguments a little to make this definition for all real α.

Exercise A.8.14 (i) $\|r\cdot f\|_{[\alpha]} = |r|\cdot\|f\|_{[\alpha]}$ for any measurable f and any $r\in\mathbb{R}$.

(ii) For any measurable function f and any $\alpha>0$, $\lceil\!\lceil f \rceil\!\rceil_{L^0} < \alpha \iff \|f\|_{[\alpha]} < \alpha$, $\|f\|_{[\alpha]} \le \lambda \iff \mathbb{P}[|f|>\lambda] \le \alpha$, and $\lceil\!\lceil f \rceil\!\rceil_{L^0} = \inf\{\alpha : \|f\|_{[\alpha]} \le \alpha\}$.

(iii) A sequence (f_n) of measurable functions converges to f in measure if and only if $\|f_n - f\|_{[\alpha]} \xrightarrow[n\to\infty]{} 0$ for all $\alpha>0$, i.e., iff $\mathbb{P}[|f-f_n|>\alpha] \xrightarrow[n\to\infty]{} 0 \quad \forall\alpha>0$.

(iv) The function $\alpha\mapsto\|f\|_{[\alpha]}$ is decreasing and right-continuous. Considered as a measurable function on the Lebesgue space $(0,1)$ it has the same distribution as $|f|$. It is thus often called the ***non-increasing rearrangement*** of $|f|$.

Exercise A.8.15 (i) Let f be a measurable function. Then for $0<p<\infty$

$$\big\||f|^p\big\|_{[\alpha]} = \left(\|f\|_{[\alpha]}\right)^p, \quad \|f\|_{[\alpha]} \le \alpha^{-1/p}\cdot\|f\|_{L^p},$$

and

$$\mathbb{E}[|f|^p] = \int_0^1 \|f\|^p_{[\alpha]}\,d\alpha\,.$$

In fact, for any continuous Φ on $\mathbb{R}_+$, $\int\Phi(|f|)\,d\mathbb{P} = \int_0^1\Phi(\|f\|_{[\alpha]})\,d\alpha$.

(ii) $f\mapsto\|f\|_{[\alpha]}$ is not subadditive, but there is a substitute:

$$\|f+g\|_{[\alpha+\beta]} \le \|f\|_{[\alpha]} + \|g\|_{[\beta]}\,.$$

Exercise A.8.16 In the proofs of 2.3.3 and theorems 3.1.6 and 4.1.12 a "Fubini-type estimate" for the gauges $\|\ \|_{[\alpha]}$ is needed. It is this: let $\mathbb{P},\tau$ be probabilities, and $f(\omega,t)$ a $\mathbb{P}\times\tau$-measurable function. Then for $\alpha,\beta,\gamma>0$

$$\Big\|\,\|f\|_{[\beta;\tau]}\Big\|_{[\alpha;\mathbb{P}]} \le \Big\|\,\|f\|_{[\gamma;\mathbb{P}]}\Big\|_{[\alpha\beta-\gamma;\tau]}\,.$$

Exercise A.8.17 Suppose f,g are positive random variables with $\|f\|_{L^r(\mathbb{P}/g)} \le E$, where $r>0$. Then

$$\|f\|_{[\alpha+\beta]} \le E\cdot\left(\frac{\|g\|_{[\alpha]}}{\beta}\right)^{1/r}, \quad \|f\|_{[\alpha]} \le E\cdot\left(\frac{2\,\|g\|_{[\alpha/2]}}{\alpha}\right)^{1/r} \tag{A.8.1}$$

and

$$\|fg\|_{[\alpha]} \le E\cdot\|g\|_{[\alpha/2]}\cdot\left(\frac{2\|g\|_{[\alpha/2]}}{\alpha}\right)^{1/r}\,. \tag{A.8.2}$$

Bounded Subsets of L^p Recall from page 379 that a subset B of a topological vector space $\mathcal{V}$ is ***bounded*** if it can be absorbed by any neighborhood V of zero; that is to say, if for any neighborhood V of zero there is a scalar r such that $C\subset r\cdot V$.

Exercise A.8.18 (Cf. A.2.28) Let $0\le p\le\infty$. A set $\mathcal{C}\subset L^p$ is bounded if and only if

$$\sup\{\lceil\!\lceil\lambda\cdot f\rceil\!\rceil_p : f\in\mathcal{C}\} \xrightarrow[\lambda\to 0]{} 0\,, \tag{A.8.3}$$

which is the same as saying $\sup\{\lceil\!\lceil f\rceil\!\rceil_p : f\in\mathcal{C}\}<\infty$ in the case that p is strictly positive. If $p=0$, then the previous supremum is always less than or equal to 1 and equation (A.8.3) describes boundedness. Namely, for $\mathcal{C}\subset L^0(\mathbb{P})$, the following are equivalent: (i) $\mathcal{C}$ is bounded in $L^0(\mathbb{P})$; (ii) $\sup\{\lceil\!\lceil\lambda\cdot f\rceil\!\rceil_{L^0(\mathbb{P})} : f\in\mathcal{C}\} \xrightarrow[\lambda\to 0]{} 0$; (iii) for every $\alpha>0$ there exists $C_\alpha<\infty$ such that

$$\|f\|_{[\alpha;\mathbb{P}]} \le C_\alpha \quad \forall f\in\mathcal{C}\,.$$

Exercise A.8.19 Let $\mathbb{P}' \ll \mathbb{P}$. Then the natural injection of $L^0(\mathbb{P})$ into $L^0(\mathbb{P}')$ is continuous and thus maps bounded sets of $L^0(\mathbb{P})$ into bounded sets of $L^0(\mathbb{P}')$.

The elementary stochastic integral is a linear map from a space $\mathcal{E}$ of functions to one of the spaces L^p. It is well to study the continuity of such a map. Since both the domain and range are metric spaces, continuity and boundedness coincide – recall that a linear map is ***bounded*** if it maps bounded sets of its domain to bounded sets of its range.

Exercise A.8.20 Let $\mathcal{I}$ be a linear map from the normed linear space $(\mathcal{E}, \| \ \|_{\mathcal{E}})$ to $L^p(\mathbb{P})$, $0 \le p \le \infty$. The following are equivalent: (i) $\mathcal{I}$ is continuous. (ii) $\mathcal{I}$ is continuous at zero. (iii) $\mathcal{I}$ is bounded.

(iv) $$\sup\Big\{ \lceil\!\lceil \mathcal{I}(\lambda\cdot\phi) \rceil\!\rceil_p : \ \phi \in \mathcal{E}, \ \|\phi\|_{\mathcal{E}} \le 1 \Big\} \xrightarrow[\lambda\to 0]{} 0 \,.$$

If $p = 0$, then $\mathcal{I}$ is continuous if and only if for every $\alpha > 0$ the number

$$\|\mathcal{I}\|_{[\alpha;\mathbb{P}]} \stackrel{\text{def}}{=} \sup\Big\{ \|\mathcal{I}(X)\|_{[\alpha;\mathbb{P}]} : \ X \in \mathcal{E}\,, \ \|X\|_{\mathcal{E}} \le 1 \Big\} \quad \text{is finite.}$$

A.8.21 Occasionally one wishes to estimate the size of a function or set without worrying about its measurability. In this case one argues with the upper integral $\int^*$ or the outer measure $\mathbb{P}^*$. The corresponding constructs for $\| \ \|_p$, $\lceil\!\lceil \ \rceil\!\rceil_p$, and $\| \ \|_{[\alpha]}$ are

$$\left.\begin{aligned} \|f\|_p^* = \|f\|_{L^p(\mathbb{P})}^* &\stackrel{\text{def}}{=} \Big(\int^* |f|^p \, d\mathbb{P} \Big)^{1/p} \\ \lceil\!\lceil f \rceil\!\rceil_p^* = \lceil\!\lceil f \rceil\!\rceil_{L^p(\mathbb{P})}^* &\stackrel{\text{def}}{=} \Big(\int^* |f|^p \, d\mathbb{P} \Big)^{1\wedge 1/p} \end{aligned}\right\} \quad \text{for } 0 < p < \infty\,,$$

$$\left.\begin{aligned} \lceil\!\lceil f \rceil\!\rceil_0^* = \lceil\!\lceil f \rceil\!\rceil_{L^0(\mathbb{P})} &\stackrel{\text{def}}{=} \inf\{\lambda : \ \mathbb{P}^*[|f| \ge \lambda] \le \lambda\} \\ \|f\|_{[\alpha]}^* = \|f\|_{[\alpha;\mathbb{P}]}^* &\stackrel{\text{def}}{=} \inf\{\lambda : \ \mathbb{P}^*[|f| \ge \lambda] \le \alpha\} \end{aligned}\right\} \quad \text{for } p = 0\,.$$

Exercise A.8.22 It is well known that $\int^*$ is ***continuous along arbitrary increasing sequences***:

$$0 \le f_n \uparrow f \implies \int^* f_n \uparrow \int^* f \,.$$

Show that the starred constructs above all share this property.

Exercise A.8.23 Set $\|f\|_{1,\infty} = \|f\|_{L^{1,\infty}(\mathbb{P})} = \sup_{\lambda>0} \lambda \cdot \mathbb{P}[|f| > \lambda]$. Then

$$\|f\|_p \le p\Big(\frac{2-p}{1-p}\Big)^{1/p} \cdot \|f\|_{1,\infty} \qquad \text{for } 0 < p < 1\,,$$

$$\|f+g\|_{1,\infty} \le 2\cdot \Big(\|f\|_{1,\infty} + \|g\|_{1,\infty}\Big)\,,$$

and $$\|rf\|_{1,\infty} = |r| \cdot \|f\|_{1,\infty} \,.$$

Marcinkiewicz Interpolation

Interpolation is a large field, and we shall establish here only the one rather elementary result that is needed in the proof of theorem 2.5.30 on page 85. Let $U : L^\infty(\mathbb{P}) \to L^\infty(\mathbb{P})$ be a subadditive map and $1 \le p \le \infty$. U is said to be of ***strong type*** $p{-}p$ if there is a constant A_p with

$$\|U(f)\|_p \le A_p \cdot \|f\|_p ,$$

in other words if it is continuous from $L^p(\mathbb{P})$ to $L^p(\mathbb{P})$. It is said to be of ***weak type*** $p{-}p$ if there is a constant A'_p such that

$$\mathbb{P}[|U(f)| \ge \lambda] \le \Big(\frac{A'_p}{\lambda} \cdot \|f\|_p\Big)^p .$$

"Weak type $\infty{-}\infty$" is to mean "strong type $\infty{-}\infty$." Chebyscheff's inequality shows immediately that a map of strong type $p{-}p$ is of weak type $p{-}p$.

Proposition A.8.24 (An Interpolation Theorem) *If U is of weak types $p_1{-}p_1$ and $p_2{-}p_2$ with constants A'_{p_1}, A'_{p_2}, respectively, then it is of strong type $p{-}p$ for $p_1 < p < p_2$:*

$$\|U(f)\|_p \le A_p \cdot \|f\|_p$$

with constant $$A_p \le p^{1/p} \cdot \Big(\frac{(2A'_{p_1})^{p_1}}{p - p_1} + \frac{(2A'_{p_2})^{p_2}}{p_2 - p}\Big)^{1/p} .$$

Proof. By the subadditivity of U we have for every $\lambda > 0$

$$|U(f)| \le |U(f \cdot [|f| \ge \lambda])| + |U(f \cdot [|f| < \lambda])| ,$$

and consequently

$$\begin{aligned}\mathbb{P}[|U(f)| \ge \lambda] &\le \mathbb{P}[|U(f \cdot [|f| \ge \lambda])| \ge \lambda/2] + \mathbb{P}[|U(f \cdot [|f| < \lambda])| \ge \lambda/2] \\ &\le \Big(\frac{A'_{p_1}}{\lambda/2}\Big)^{p_1} \int_{[|f| \ge \lambda]} |f|^{p_1}\, d\mathbb{P} + \Big(\frac{A'_{p_2}}{\lambda/2}\Big)^{p_2} \int_{[|f| < \lambda]} |f|^{p_2}\, d\mathbb{P} .\end{aligned}$$

We multiply this with λ^{p-1} and integrate against $d\lambda$, using Fubini's theorem A.3.18:

$$
\begin{aligned}
\frac{\mathbb{E}(|U(f)|^p)}{p} &= \int\int_0^{|U(f)|} \lambda^{p-1}\, d\lambda d\mathbb{P} = \int\int_0^{\infty} \mathbb{P}[|U(f)| \ge \lambda]\lambda^{p-1}\, d\lambda \\
&\le \int \left(\frac{A'_{p_1}}{\lambda/2}\right)^{p_1} \int_{[|f|\ge\lambda]} |f|^{p_1}\lambda^{p-1}\, d\lambda d\mathbb{P} \\
&\quad + \int \left(\frac{A'_{p_2}}{\lambda/2}\right)^{p_2} \int_{[|f|<\lambda]} |f|^{p_2}\lambda^{p-1}\, d\lambda d\mathbb{P} \\
&= \left(2A'_{p_1}\right)^{p_1} \int\int_0^{|f|} |f|^{p_1}\lambda^{p-p_1-1}\, d\lambda d\mathbb{P} \\
&\quad + \left(2A'_{p_2}\right)^{p_2} \int\int_{|f|}^{\infty} |f|^{p_2}\lambda^{p-p_2-1}\, d\lambda d\mathbb{P} \\
&= \frac{\left(2A'_{p_1}\right)^{p_1}}{p-p_1} \int |f|^{p_1}|f|^{p-p_1}\, d\mathbb{P} + \frac{\left(2A'_{p_2}\right)^{p_2}}{p_2-p} \int |f|^{p_2}|f|^{p-p_2}\, d\mathbb{P} \\
&= \left(\frac{\left(2A'_{p_1}\right)^{p_1}}{p-p_1} + \frac{\left(2A'_{p_2}\right)^{p_2}}{p_2-p}\right) \cdot \int |f|^p\, d\mathbb{P}\,.
\end{aligned}
$$

We multiply both sides by p and take p^{th} roots; the claim follows. ∎

Note that the constant A_p blows up as $p \downarrow p_1$ or $p \uparrow p_2$. In the one application we make, in the proof of theorem 2.5.30, the map U happens to be ***self-adjoint***, which means that

$$
\mathbb{E}[U(f) \cdot g] = \mathbb{E}[f \cdot U(g)]\,, \qquad f, g \in L^2\,.
$$

Corollary A.8.25 *If U is self-adjoint and of weak types $1\!-\!1$ and $2\!-\!2$, then U is of strong type $p\!-\!p$ for all $p \in (1, \infty)$.*

Proof. Let $2 < p < \infty$. The conjugate exponent $p' = p/(p-1)$ then lies in the open interval $(1, 2)$, and by proposition A.8.24

$$
\mathbb{E}[U(f) \cdot g] = \mathbb{E}[f \cdot U(g)] \le \|f\|_p \cdot \|U(g)\|_{p'} \le \|f\|_p \cdot A_{p'}\|g\|_{p'}\,.
$$

We take the supremum over g with $\|g\|_{p'} \le 1$ and arrive at

$$
\|U(f)\|_p \le A_{p'} \cdot \|f\|_p\,.
$$

The claim is satisfied with $A_p = A_{p'}$. Note that, by interpolating now between $3/2$ and 3, say, the estimate of the constants A_p can be improved so no pole at $p = 2$ appears. (It suffices to assume that U is of weak types $1\!-\!1$ and $(1+\epsilon) - (1+\epsilon)$). ∎

Khintchine's Inequalities

Let T be the product of countably many two-point sets $\{1,-1\}$. Its elements are sequences $t = (t_1, t_2, \ldots)$ with $t_\nu = \pm 1$. Let $\epsilon_\nu : t \mapsto t_\nu$ be the νth coordinate function. If we equip T with the product τ of uniform measure on $\{1,-1\}$, then the ϵ_ν are independent and identically distributed Bernoulli variables that take the values ± 1 with τ-probability $1/2$ each. The ϵ_ν form an orthonormal set in $L^2(\tau)$ that is far from being a basis; in fact, they form so sparse or *lacunary* a collection that on their linear span

$$\mathcal{R} = \Big\{ \sum_{\nu=1}^{n} a_\nu \epsilon_\nu : n \in \mathbb{N},\ a_\nu \in \mathbb{R} \Big\}$$

all of the $L^p(\tau)$-topologies coincide:

Theorem A.8.26 (Khintchine) *Let $0 < p < \infty$. There are universal constants k_p, K_p such that for any natural number n and reals $a_1, \ldots, a_n$*

$$\Big\| \sum_{\nu=1}^{n} a_\nu \epsilon_\nu \Big\|_{L^p(\tau)} \le k_p \cdot \Big(\sum_{\nu=1}^{n} a_\nu^2 \Big)^{1/2} \tag{A.8.4}$$

and

$$\Big(\sum_{\nu=1}^{n} a_\nu^2 \Big)^{1/2} \le K_p \cdot \Big\| \sum_{\nu=1}^{n} a_\nu \epsilon_\nu \Big\|_{L^p(\tau)} . \tag{A.8.5}$$

For $p = 0$, there are universal constants $\kappa_0 > 0$ and $K_0 < \infty$ such that

$$\Big(\sum_{\nu=1}^{n} a_\nu^2 \Big)^{1/2} \le K_0 \cdot \Big\| \sum_{\nu=1}^{n} a_\nu \epsilon_\nu \Big\|_{[\kappa_0;\tau]} . \tag{A.8.6}$$

In particular, a subset of $\mathcal{R}$ that is bounded in the topology induced by any of the $L^p(\tau)$, $0 \le p < \infty$, is bounded in $L^2(\tau)$. The completions of $\mathcal{R}$ in these various topologies being the same, the inequalities above stay if the finite sums are replaced with infinite ones. Bounds for the universal constants k_p, K_p, K_0, κ_0 are discussed in remark A.8.28 and exercise A.8.29 below.

Proof. Let us write $\|f\|_p$ for $\|f\|_{L^p(\tau)}$. Note that for $f = \sum_{\nu=1}^{n} a_\nu \epsilon_\nu \in \mathcal{R}$

$$\|f\|_2 = \Big(\sum_{\nu} a_\nu^2 \Big)^{1/2} .$$

In proving (A.8.4)–(A.8.6) we may by homogeneity assume that $\|f\|_2 = 1$. Let $\lambda > 0$. Since the ϵ_ν are independent and $\cosh x \le e^{x^2/2}$ (as a term-by-term comparison of the power series shows),

$$\int e^{\lambda f(t)}\, \tau(dt) = \prod_{\nu=1}^{N} \cosh(\lambda a_\nu) \le \prod_{\nu=1}^{N} e^{\lambda^2 a_\nu^2/2} = e^{\lambda^2/2} ,$$

and consequently

$$\int e^{\lambda|f(t)|}\,\tau(dt) \le \int e^{\lambda f(t)}\,\tau(dt) + \int e^{-\lambda f(t)}\,\tau(dt) \le 2e^{\lambda^2/2}\,.$$

We apply Chebysheff's inequality to this and obtain

$$\tau([|f| \ge \lambda]) = \tau\Big(\Big[e^{\lambda|f|} \ge e^{\lambda^2}\Big]\Big) \le e^{-\lambda^2}\cdot 2e^{\lambda^2/2} = 2e^{-\lambda^2/2}\,.$$

Therefore, if $p \ge 2$, then

$$\begin{aligned}\int |f(t)|^p\,\tau(dt) &= p\cdot\int_0^\infty \lambda^{p-1}\,\tau([|f|\ge\lambda])\,d\lambda\\ &\le 2p\cdot\int_0^\infty \lambda^{p-1}e^{-\lambda^2/2}\,d\lambda\\ \text{with } z=\lambda^2/2: \quad &= 2p\cdot\int_0^\infty (2z)^{(p-2)/2}e^{-z}\,dz\\ &= 2^{\frac{p}{2}+1}\cdot\frac{p}{2}\Gamma(\frac{p}{2}) = 2^{\frac{p}{2}+1}\Gamma(\frac{p}{2}+1)\,.\end{aligned}$$

We take p^{th} roots and arrive at (A.8.4) with

$$k_p = \begin{cases} 2^{\frac{1}{p}+\frac{1}{2}}(\Gamma(\frac{p}{2}+1))^{1/p} \le \sqrt{2p} & \text{if } p \ge 2\\ 1 & \text{if } p < 2.\end{cases}$$

As to inequality (A.8.5), which is only interesting if $0 < p < 2$, write

$$\begin{aligned}\|f\|_2^2 &= \int |f|^2(t)\,\tau(dt) = \int |f|^{p/2}(t)\cdot|f|^{2-p/2}(t)\,\tau(dt)\\ \text{using Hölder:} \quad &\le \Big(\int |f|^p(t)\,\tau(dt)\Big)^{1/2}\cdot\Big(\int |f|^{4-p}(t)\,\tau(dt)\Big)^{1/2}\\ &= \|f\|_p^{p/2}\cdot\|f\|_{4-p}^{2-p/2} \le \|f\|_p^{p/2}\cdot k_{4-p}^{2-p/2}\|f\|_2^{2-p/2}\,,\end{aligned}$$

and thus $\|f\|_2^{p/2} \le k_{4-p}^{2-p/2}\cdot\|f\|_p^{p/2}$ and $\|f\|_2 \le k_{4-p}^{4/p-1}\cdot\|f\|_p$.

The estimate

$$k_{4-p} \le 2\cdot\Big(\Gamma\Big(\frac{4-p}{2}+1\Big)\Big)^{1/(4-p)} \le 2\cdot(\Gamma(3))^{1/4} = 2^{5/4}$$

leads to (A.8.5) with $K_p \le 2^{5/p-5/4}$. In summary:

$$K_p \le \begin{cases} 1 & \text{if } p \ge 2,\\ 2^{5/p-5/4} < 2^{5/p} & \text{if } 0 < p < 2\\ \le 16 & \text{if } 1 \le p < \infty.\end{cases}$$

Finally let us prove inequality (A.8.6). From

$$\|f\|_1 = \int |f(t)|\,\tau(dt) = \int_{[|f|\ge\|f\|_1/2]} |f(t)|\,\tau(dt) + \int_{[|f|<\|f\|_1/2]} |f(t)|\,\tau(dt)$$
$$\le \|f\|_2 \cdot \Big(\tau\big[|f| \ge \|f\|_1/2\big]\Big)^{1/2} + \|f\|_1/2$$
$$\le K_1 \cdot \|f\|_1 \cdot \Big(\tau\big[|f| \ge \|f\|_1/2\big]\Big)^{1/2} + \|f\|_1/2$$

we deduce that $\qquad 1/2 \le K_1 \cdot \Big(\tau\big[|f| \ge \|f\|_1/2\big]\Big)^{1/2}$

and thus $$\frac{1}{(2K_1)^2} \le \tau\big([|f| \ge \|f\|_1/2]\big) \le \tau\Big[|f| \ge \frac{\|f\|_2}{2K_1}\Big]\,. \tag{$*$}$$

Recalling that $\qquad \|f\|_{[\alpha;\tau]} = \inf\{\lambda : \tau[|f| \ge \lambda] < \alpha\}\,,$

we rewrite $(*)$ as $\qquad \|f\|_2 \le 2K_1 \cdot \|f\|_{[1/(4K_1^2);\tau]}\,,$

which is inequality (A.8.6) with $K_0 = 2K_1$ and $\kappa_0 = 1/(4K_1^2)$. ∎

Exercise A.8.27 $\|f\|_2 \le \big(\sqrt{1/2} - \sqrt{\kappa}\big)^{-1} \cdot \|f\|_{[\kappa]}$ for $0 < \kappa < 1/2$.

Remark A.8.28 Szarek [105] found the smallest possible value for K_1: $K_1 = \sqrt{2}$. Haagerup [38] found the best possible constants k_p, K_p for all $p > 0$:

$$k_p = \begin{cases} 1 & \text{for } 0 < p \le 2, \\ \sqrt{2} \cdot \Big(\dfrac{\Gamma((p+1/2))}{\sqrt{\pi}}\Big)^{1/p} & \text{for } 2 \le p < \infty; \end{cases} \tag{A.8.7}$$

$$K_p = \begin{cases} 2^{-1/2} \cdot \Big(\dfrac{\sqrt{\pi}}{\Gamma((p+1)/2)} \vee 2\Big)^{1/p} & \text{for } 0 < p \le 2, \\ 1 & \text{for } 2 \le p < \infty. \end{cases} \tag{A.8.8}$$

In the text we shall use the following values to estimate the K_p and κ_0 (they are only slightly worse but rather simpler to read):

Exercise A.8.29

$$K_0^{(A.8.6)} \le 2\sqrt{2} \text{ and } \kappa_0^{(A.8.6)} \ge 1/8 \text{ for } p = 0\,;$$
$$K_p^{(A.8.5)} \le \begin{cases} 2^{1/p\,-\,1/2} & \text{for } 0 < p \le 1.8 \\ 1.00037 \cdot 2^{1/p\,-\,1/2} & \text{for } 1.8 < p \le 2 \\ 1 & \text{for } 2 \le p < \infty. \end{cases} \tag{A.8.9}$$

Exercise A.8.30 The values of k_p and K_p in (A.8.7) and (A.8.8) are best possible.

Exercise A.8.31 The *Rademacher functions* r_n are defined on the unit interval by

$$r_n(x) = \operatorname{sgn}\sin(2^n\pi x)\,, \qquad x \in (0,1),\ n = 1,2,\ldots$$

The sequences (ϵ_ν) and (r_ν) have the same distribution.

Stable Type

In the proof of the general factorization theorem 4.1.2 on page 191, other special sequences of random variables are needed, sequences that show a behavior similar to that of the Rademacher functions, in the sense that on their linear span all L^p-topologies coincide. They are the sequences of independent symmetric q-stable random variables.

Symmetric Stable Laws Let $0 < q \le 2$. There exists a random variable $\gamma^{(q)}$ whose characteristic function is

$$\mathbb{E}\Big[e^{i\alpha\gamma^{(q)}}\Big] = e^{-|\alpha|^q} .$$

This can be proved using Bochner's theorem: $\alpha \mapsto |\alpha|^q$ is of negative type, so $\alpha \mapsto e^{-|\alpha|^q}$ is of positive type and thus is the characteristic function of a probability on $\mathbb{R}$. The random variable $\gamma^{(q)}$ is said to have a ***symmetric stable law of order q*** or to be ***symmetric q-stable***. For instance, $\gamma^{(2)}$ is evidently a centered normal random variable with variance 2; a Cauchy random variable, i.e., one that has density $1/\pi(1+x^2)$ on $(-\infty,\infty)$, is symmetric 1-stable. It has moments of orders $0 < p < 1$.

We derive the existence of $\gamma^{(q)}$ without appeal to Bochner's theorem in a computational way, since some estimates of their size are needed anyway. For our purposes it suffices to consider the case $0 < q \le 1$.

The function $\alpha \mapsto e^{-|\alpha|^q}$ is continuous and has very fast decay at $\pm\infty$, so its inverse Fourier transform

$$f^{(q)}(x) \stackrel{\text{def}}{=} \frac{1}{2\pi}\int_{-\infty}^{\infty} e^{-ix\alpha}\cdot e^{-|\alpha|^q}\,d\alpha$$

is a smooth square integrable function with the right Fourier transform $\alpha \mapsto e^{-|\alpha|^q}$. It has to be shown, of course, that $f^{(q)}$ is positive, so that it qualifies as the probability density of a random variable – its integral is automatically 1, as it equals the value of its Fourier transform at $\alpha = 0$.

Lemma A.8.32 *Let $0 < q \le 1$. (i) The function $f^{(q)}$ is positive: it is indeed the density of a probability measure. (ii) If $0 < p < q$, then $\big|\gamma^{(q)}\big|$ has p^{th} moment*

$$\|\gamma^{(q)}\|_p^p = \Gamma\big((q-p)/q\big)\cdot b(p) , \tag{A.8.10}$$

where
$$b(p) \stackrel{\text{def}}{=} \frac{2}{\pi}\int_0^\infty \xi^{p-1}\sin\xi\,d\xi$$

is a strictly decreasing function with $\lim_{p\to 0} = 1$ and

$$(3-p)/\pi \le b(p) \le 1-(1-2/\pi)p \le 1 .$$

Proof (Charlie Friedman). For strictly positive x write

$$f^{(q)}(x) = \frac{1}{2\pi}\int_{-\infty}^{\infty} e^{-ix\alpha}\cdot e^{-\alpha^q}\,d\alpha = \frac{1}{\pi}\int_0^{\infty}\cos(\alpha x)\cdot e^{-\alpha^q}\,d\alpha$$

$$= \frac{1}{\pi}\sum_{k=0}^{\infty}\int_{k\pi/x}^{(k+1)\pi/x}\cos(\alpha x)\cdot e^{-\alpha^q}\,d\alpha$$

with $\alpha x = \xi + k\pi$:

$$= \frac{1}{\pi x}\sum_{k=0}^{\infty}\int_0^{\pi}\cos(\xi + k\pi)\cdot e^{-\left(\frac{\xi+k\pi}{x}\right)^q}\,d\xi$$

$$= \frac{1}{\pi x}\sum_{k=0}^{\infty}(-1)^k\int_0^{\pi}\cos\xi\cdot e^{-\left(\frac{\xi+k\pi}{x}\right)^q}\,d\xi$$

integrating by parts:

$$= \sum_{k=0}^{\infty}\frac{q(-1)^k}{\pi x^{1+q}}\int_0^{\pi}\sin\xi\cdot e^{-\left(\frac{\xi+k\pi}{x}\right)^q}\cdot(\xi + k\pi)^{q-1}\,d\xi$$

$$= \frac{q}{\pi x^{1+q}}\int_0^{\infty}\sin\xi\cdot e^{-(\frac{\xi}{x})^q}\cdot\xi^{q-1}\,d\xi\,.$$

The penultimate line represents $f^{(q)}(x)$ as the sum of an alternating series. In the present case $0 < q \le 1$ it is evident that $e^{-((\xi+k\pi)/x)^q}\,(\xi + k\pi)^{q-1}$ decreases as k increases. The series' terms decrease in absolute value, so the sum converges. Since the first term is positive, so is the sum: $f^{(q)}$ is indeed a probability density.

(ii) To prove the existence of pth moments write $\|\gamma^{(q)}\|_p^p$ as

$$\int_{-\infty}^{\infty}|x|^p\,f^{(q)}(x)\,dx = 2\int_0^{\infty}x^p\,f^{(q)}(x)\,dx$$

$$= \frac{2q}{\pi}\int_0^{\infty}\int_0^{\infty}x^{p-1-q}\sin\xi\,e^{-(\xi/x)^q}\,\xi^{q-1}\,dx\,d\xi$$

with $(\xi/x)^q = y$:

$$= \frac{2}{\pi}\int_0^{\infty}\int_0^{\infty}y^{-p/q}e^{-y}\xi^{p-1}\sin\xi\,dy\,d\xi$$

$$= \Gamma((q-p)/q)\cdot b(p)\,.$$

The estimates of $b(p)$ are left as an exercise. ∎

Lemma A.8.33 *Let $0 < q \le 2$, let $\left(\gamma_1^{(q)}, \gamma_2^{(q)}, \ldots\right)$ be a sequence of independent symmetric q-stable random variables defined on a probability space $(X, \mathcal{X}, dx)$, and let $c_1, c_2, \ldots$ be a sequence of reals. (i) For any $p \in (0, q)$*

$$\Big\|\sum_{\nu=1}^{\infty}c_\nu\,\gamma_\nu^{(q)}\Big\|_{L^p(dx)} = \|\gamma^{(q)}\|_p\cdot\Big(\sum_{\nu=1}^{\infty}|c_\nu|^q\Big)^{1/q}\,. \tag{A.8.11}$$

In other words, the map $\ell^q \ni (c_\nu) \mapsto \sum_\nu c_\nu\gamma_\nu^{(q)}$ is, up to the factor $\|\gamma^{(q)}\|_p$, an isometry from ℓ^q into $L^p(dx)$.

(ii) This map is also a homeomorphism from ℓ^q into $L^0(dx)$. To be precise: for every $\beta \in (0,1)$ there exist universal constants $A_{[\beta],q}$ and $B_{[\beta],q}$ with

$$A_{[\beta],q} \cdot \Big\| \sum_\nu c_\nu \gamma_\nu^{(q)} \Big\|_{[\beta;dx]} \le \| (c_\nu) \|_{\ell^q} \le B_{[\beta],q} \cdot \Big\| \sum_\nu c_\nu \gamma_\nu^{(q)} \Big\|_{[\beta;dx]} . \tag{A.8.12}$$

Proof. (i) The functions $\sum_\nu c_\nu \gamma_\nu^{(q)}$ and $\| (c_\nu) \|_{\ell^q} \cdot \gamma_1^{(q)}$ are easily seen to have the same characteristic function, so their L^p-norms coincide.

(ii) In view of exercise A.8.15 and equation (A.8.11),

$$\| f \|_{[\beta]} \le \beta^{-1/p} \| f \|_{L^p(dx)} = \beta^{-1/p} \|\gamma^{(q)}\|_p \cdot \| (c_\nu) \|_{\ell^q}$$

for any $p \in (0,q)$: $A_{[\beta],q} \stackrel{\text{def}}{=} \sup \Big\{ \beta^{1/p} . \|\gamma^{(q)}\|_p^{-1} : 0 < p < q \Big\}$ answers the left-hand side of inequality (A.8.12).

The constant $B_{[\beta],q}$ is somewhat harder to come by. Let $0 < p_1 < p_2 < q$ and $r > 1$ be so that $r \cdot p_1 = p_2$. Then the conjugate exponent of r is $r' = p_2/(p_2 - p_1)$. For $\lambda > 0$ write

$$\|f\|_{p_1}^{p_1} = \int_{[|f| > \lambda \|f\|_{p_1}]} |f|^{p_1} + \int_{[|f| \le \lambda \|f\|_{p_1}]} |f|^{p_1}$$

using Hölder: $$\le \Big(\int |f|^{p_2} \Big)^{\frac{1}{r}} \cdot \big(dx[|f| > \lambda \|f\|_{p_1}] \big)^{\frac{1}{r'}} + \lambda^{p_1} \|f\|_{p_1}^{p_1} ,$$

and so $$(1 - \lambda^{p_1}) \|f\|_{p_1}^{p_1} \le \|f\|_{p_2}^{p_1} \cdot \big(dx[|f| > \lambda \|f\|_{p_1}] \big)^{\frac{1}{r'}}$$

and $$dx[|f| > \lambda \|f\|_{p_1}] \ge \left(\frac{(1 - \lambda^{p_1}) \|f\|_{p_1}^{p_1}}{\|f\|_{p_2}^{p_1}} \vee 0 \right)^{\frac{p_2}{p_2 - p_1}}$$

by equation (A.8.11): $$= \left(\frac{(1 - \lambda^{p_1}) \|\gamma^{(q)}\|_{p_1}^{p_1}}{\|\gamma^{(q)}\|_{p_2}^{p_1}} \vee 0 \right)^{\frac{p_2}{p_2 - p_1}} .$$

Therefore, setting $\lambda = \big(B \cdot \|\gamma^{(q)}\|_{p_1} \big)^{-1}$ and using equation (A.8.11):

$$dx[|f| > \|(c_\nu)\|_{\ell^q}/B] \ge \left(\frac{\big(\|\gamma^{(q)}\|_{p_1}^{p_1} - B^{-p_1}\big)}{\|\gamma^{(q)}\|_{p_2}^{p_1}} \vee 0 \right)^{\frac{p_2}{p_2 - p_1}} . \tag{$*$}$$

This inequality means

$$\| (c_\nu) \|_{\ell^q} \le B \cdot \| f \|_{[\beta(B,p_1,p_2,q)]} ,$$

where $\beta(B, p_1, p_2, q)$ denotes the right-hand side of $(*)$. The question is whether $\beta(B, p_1, p_2, q)$ can be made larger than the $\beta < 1$ given in the statement, by a suitable choice of B. To see that it can be we solve the inequality $\beta(B,p_1,p_2,q) \ge \beta$ for B:

$$B \ge \left(\left(\|\gamma^{(q)}\|_{p_1}^{p_1} - \beta^{\frac{p_2-p_1}{p_2}} \|\gamma^{(q)}\|_{p_2}^{p_1}\right) \vee 0\right)^{\frac{-1}{p_1}}$$

by (A.8.10):
$$= \left(\left(\left(b(p_1)\cdot\Gamma\left(\frac{q-p_1}{q}\right) - \beta^{\frac{p_2-p_1}{p_2}} \left(b(p_2)\cdot\Gamma\left(\frac{q-p_2}{q}\right)\right)^{\frac{p_1}{p_2}}\right) \vee 0\right)\right)^{\frac{-1}{p_1}} .$$

Fix $p_2 < q$. As $p_1 \to 0$, $b(p_1) \to 1$ and $\Gamma(\frac{q-p_1}{q}) \to 1$, so the expression in the outermost parentheses will eventually be strictly positive, and $B < \infty$ satisfying this inequality can be found. We get the estimate:

$$B_{[\beta],q} \le \inf\left(\left(\|\gamma^{(q)}\|_{p_1} - \beta^{\frac{p_2-p_1}{p_2}} \|\gamma^{(q)}\|_{p_2}^{\frac{p_1}{p_2}}\right) \vee 0\right)^{\frac{-1}{p_1}} , \tag{A.8.13}$$

the infimum being taken over all p_1, p_2 with $0 < p_1 < p_2 < q$. ∎

Maps and Spaces of Type (p, q) One-half of equation (A.8.11) holds even when the c_ν belong to a Banach space E, provided $0 < p < q < 1$: in this range there are universal constants $T_{p,q}$ so that for $x_1, \ldots, x_n \in E$

$$\left\|\sum_{\nu=1}^{n} x_\nu\, \gamma_\nu^{(q)}\right\|_{L^p(dx)} \le T_{p,q} \cdot \left(\sum_{\nu=1}^{n} \|x_\nu\|_E^q\right)^{1/q} .$$

In order to prove this inequality it is convenient to extend it to maps and to make it into a definition:

Definition A.8.34 *Let $u : E \to F$ be a linear map between quasinormed vector spaces and $0 < p < q \le 2$. u is a **map of type (p, q)** if there exists a constant $C < \infty$ such that for every finite collection $\{x_1, \ldots, x_n\} \subset E$*

$$\left\|\left\|\sum_{\nu=1}^{n} u(x_\nu)\, \gamma_\nu^{(q)}\right\|_F\right\|_{L^p(dx)} \le C \cdot \left(\sum_{\nu=1}^{n} \|x_\nu\|_E^q\right)^{1/q} . \tag{A.8.14}$$

*Here $\gamma_1^{(q)}, \ldots, \gamma_n^{(q)}$ are independent symmetric q-stable random variables defined on some probability space $(X, \mathcal{X}, dx)$. The smallest constant C satisfying (A.8.14) is denoted by $T_{p,q}(u)$. A quasinormed space E (see page 381) is said to be a **space of type (p, q)** if its identity map id_E is of type (p, q), and then we write $T_{p,q}(E) = T_{p,q}(id_E)$.*

Exercise A.8.35 The continuous linear maps of type (p, q) form a two-sided operator ideal: if $u : E \to F$ and $v : F \to G$ are continuous linear maps between quasinormed spaces, then $T_{p,q}(v \circ u) \le \|u\| \cdot T_{p,q}(v)$ and $T_{p,q}(v \circ u) \le \|v\| \cdot T_{p,q}(u)$.

Example A.8.36 [66] Let $(Y, \mathcal{Y}, dy)$ be a probability space. Then the natural injection j of $L^q(dy)$ into $L^p(dy)$ has type (p, q) with

$$T_{p,q}(j) \le \|\gamma^{(q)}\|_p . \tag{A.8.15}$$

Indeed, if $f_1, \ldots, f_n \in L^q(dy)$, then

$$\Big\| \Big\| \sum_{\nu=1}^{n} j(f_\nu)\,\gamma_\nu^{(q)} \Big\|_{L^p(dy)} \Big\|_{L^p(dx)} = \Big(\int\!\!\int \Big| \sum_{\nu=1}^{n} f_\nu(y)\,\gamma_\nu^{(q)}(x) \Big|^p \,dx\,dy \Big)^{1/p}$$

by equation (A.8.11):

$$= \|\gamma^{(q)}\|_p \cdot \Big(\int \Big(\sum_{\nu=1}^{n} |f_\nu(y)|^q \Big)^{p/q} dy \Big)^{1/p}$$

$$\le \|\gamma^{(q)}\|_p \cdot \Big(\int \Big(\sum_{\nu=1}^{n} |f_\nu(y)|^q \Big)\, dy \Big)^{1/q}$$

$$= \|\gamma^{(q)}\|_p \cdot \Big(\sum_{\nu=1}^{n} \|f_\nu\|^q_{L^q(dy)} \Big)^{1/q} .$$

Exercise A.8.37 Equality obtains in inequality (A.8.15).

Example A.8.38 [66] For $0 < p < q < 1$,

$$T_{p,q}(\ell^1) \le \|\gamma^{(q)}\|_p \cdot \frac{\|\gamma^{(1)}\|_q}{\|\gamma^{(1)}\|_p} .$$

To see this let $\gamma_1^{(1)}, \gamma_2^{(1)}, \ldots$ be a sequence of independent symmetric 1-stable (i.e., Cauchy) random variables defined on a probability space $(X, \mathcal{X}, dx)$, and consider the map u that associates with $(a_i) \in \ell^1$ the random variable $\sum_i a_i \gamma_i^{(1)}$. According to equation (A.8.11), u has norm $\|\gamma^{(1)}\|_q$ if considered as a map $u_q : \ell^1 \to L^q(dx)$, and has norm $\|\gamma^{(1)}\|_p$ if considered as a map $u_p : \ell^1 \to L^p(dx)$. Let j denote the injection of $L^q(dx)$ into $L^p(dx)$. Then by equation (A.8.11)

$$v \stackrel{\text{def}}{=} \|\gamma^{(1)}\|_p^{-1} \cdot j \circ u_q$$

is an isometry of ℓ^1 onto a subspace of $L^p(dx)$. Consequently

$$T_{p,q}(\ell^1) = T_{p,q}\big(id_{\ell^1}\big) = T_{p,q}(v^{-1} \circ v)$$

by exercise A.8.35:

$$\le T_{p,q}(v) \le \|\gamma^{(1)}\|_p^{-1} \cdot \big\|u_q\big\| \cdot T_{p,q}(j)$$

by example A.8.36:

$$\le \|\gamma^{(1)}\|_p^{-1} \cdot \|\gamma^{(1)}\|_q \cdot \|\gamma^{(q)}\|_p .$$

Proposition A.8.39 [66] *For $0 < p < q < 1$ every normed space E is of type (p, q):*

$$T_{p,q}(E) \le T_{p,q}(\ell^1) \le \|\gamma^{(q)}\|_p \cdot \frac{\|\gamma^{(1)}\|_q}{\|\gamma^{(1)}\|_p} . \tag{A.8.16}$$

Proof. Let $x_1, \ldots, x_n \in E$, and let E_0 be the finite-dimensional subspace spanned by these vectors. Next let $x'_1, x'_2, \ldots \in E_0$ be a sequence dense in the unit ball of E_0 and consider the map $\pi : \ell^1 \to E_0$ defined by

$$\pi\big((a_i)\big) = \sum_{i=1}^{\infty} a_i\, x'_i , \qquad (a_i) \in \ell^1 .$$

It is easily seen to be a contraction: $\|\pi(\boldsymbol{a})\|_{E_0} \le \|\boldsymbol{a}\|_{\ell^1}$. Also, given $\epsilon > 0$, we can find elements $\boldsymbol{a}_\nu \in \ell^1$ with $\pi(\boldsymbol{a}_\nu) = x_\nu$ and $\|\boldsymbol{a}_\nu\|_{\ell^1} \le \|x_\nu\|_E + \epsilon$. Using the independent symmetric q-stable random variables $\gamma_\nu^{(q)}$ from definition A.8.34 we get

$$\Big\|\Big\|\sum_{\nu=1}^n x_\nu\,\gamma_\nu^{(q)}\Big\|_E\Big\|_{L^p(dx)} = \Big\|\Big\|\sum_{\nu=1}^n \pi \circ id_{\ell^1}(\boldsymbol{a}_\nu)\,\gamma_\nu^{(q)}\Big\|_E\Big\|_{L^p(dx)}$$

$$\le \|\pi\| \cdot T_{p,q}(\ell^1) \cdot \Big(\sum_{\nu=1}^n \|\boldsymbol{a}_\nu\|_{\ell^1}^q\Big)^{1/q}$$

$$\le T_{p,q}(\ell^1) \cdot \Big(\sum_{\nu=1}^n \|x_\nu\|_E^q + \epsilon^q\Big)^{1/q}.$$

Since $\epsilon > 0$ was arbitrary, inequality (A.8.16) follows. ∎

A.9 Semigroups of Operators

Definition A.9.1 *A family $\{T_t : 0 \le t < \infty\}$ of bounded linear operators on a Banach space C is a* ***semigroup*** *if $T_{s+t} = T_s \circ T_t$ for $s,t \ge 0$ and T_0 is the* ***identity operator*** *I. We shall need to consider only* ***contraction semigroups****: the* ***operator norm*** *$\|T_t\| \stackrel{\text{def}}{=} \sup\{\|T_t\phi\|_C : \|\phi\|_C \le 1\}$ is bounded by 1 for all $t \in [0,\infty)$. Such $T.$ is* ***(strongly) continuous*** *if $t \mapsto T_t\phi$ is continuous from $[0,\infty)$ to C, for all $\phi \in C$.*

Exercise A.9.2 Then $T.$ is, in fact, uniformly strongly continuous. That is to say, $\|T_t\phi - T_s\phi\|_\infty \xrightarrow[(t-s)\to 0]{} 0$ for every $\phi \in C$.

Resolvent and Generator

The resolvent or Laplace transform of a continuous contraction semigroup $T.$ is the family $U.$ of bounded linear operators U_α, defined on $\phi \in C$ as

$$U_\alpha\phi \stackrel{\text{def}}{=} \int_0^\infty e^{-\alpha t} \cdot T_t\phi\, dt\,, \qquad \alpha > 0\,.$$

This can be read as an improper Riemann integral or a Bochner integral (see A.3.15). U_α is evidently linear and has $\|\alpha U_\alpha\| \le 1$. The ***resolvent, identity***

$$U_\alpha - U_\beta = (\beta - \alpha) U_\alpha U_\beta\,, \tag{A.9.1}$$

is a straightforward consequence of a variable substitution and implies that all of the U_α have the same range $\mathcal{U} \stackrel{\text{def}}{=} U_1 C$. Since evidently $\alpha U_\alpha \phi \xrightarrow[\alpha\to 0]{} \phi$ for all $\phi \in C$, $\mathcal{U}$ is dense in C.

The generator of a continuous contraction semigroup $T.$ is the linear operator $\mathcal{A}$ defined by

$$\mathcal{A}\psi \stackrel{\text{def}}{=} \lim_{t\downarrow 0} \frac{T_t\psi - \psi}{t}\,. \tag{A.9.2}$$

It is not, in general, defined for all $\psi \in C$, so there is need to talk about its domain $\text{dom}(\mathcal{A})$. This is the set of all $\psi \in C$ for which the limit (A.9.2) exists in C. It is very easy to see that T_t maps $\text{dom}(\mathcal{A})$ to itself, and that $\mathcal{A}T_t\psi = T_t\mathcal{A}\psi$ for $\psi \in \text{dom}(\mathcal{A})$ and $t \ge 0$. That is to say, $t \mapsto T_t\psi$ has a continuous right derivative at all $t \ge 0$; it is then actually differentiable at all $t > 0$ ([115, page 237 ff.]). In other words, $u_t \stackrel{\text{def}}{=} T_t\psi$ solves the C-valued initial value problem

$$\frac{du_t}{dt} = \mathcal{A}u_t \ , \ u_0 = \psi \ .$$

For an example pick a $\phi \in C$ and set $\psi \stackrel{\text{def}}{=} \int_0^s T_\sigma\phi \, d\sigma$. Then $\psi \in \text{dom}(\mathcal{A})$ and a simple computation results in

$$\mathcal{A}\psi = T_s\phi - \phi \ . \tag{A.9.3}$$

The Fundamental Theorem of Calculus gives

$$T_t\psi - T_s\psi = \int_s^t T_\tau \, \mathcal{A}\psi \, d\tau \ . \tag{A.9.4}$$

for $\psi \in \text{dom}(\mathcal{A})$ and $0 \le s < t$.

If $\phi \in C$ and $\psi \stackrel{\text{def}}{=} U_\alpha\phi$, then the curve $t \mapsto T_t\psi = e^{\alpha t}\int_t^\infty e^{-\alpha s}T_s\phi \, ds$ is plainly differentiable at every $t \ge 0$, and a simple calculation produces

$$\mathcal{A}[T_t\psi] = T_t[\mathcal{A}\psi] = T_t[\alpha\psi - \phi] \ ,$$

and so at $t = 0$ $\quad \mathcal{A}U_\alpha\phi = \alpha U_\alpha\phi - \phi$,

or, equivalently, $\quad (\alpha I - \mathcal{A})U_\alpha = I$ or $(I - \mathcal{A}/\alpha)^{-1} = \alpha U_\alpha$. $\quad$ (A.9.5)

This implies $\quad \left\|(I - \epsilon\mathcal{A})^{-1}\right\| \le 1$ for all $\epsilon > 0$. $\quad$ (A.9.6)

Exercise A.9.3 From this it is easy to read off these properties of the generator $\mathcal{A}$: (i) The domain of $\mathcal{A}$ contains the common range $\mathcal{U}$ of the resolvent operators U_α. In fact, $\text{dom}(\mathcal{A}) = \mathcal{U}$, and therefore ***the domain of $\mathcal{A}$ is dense*** [53, p. 316]. (ii) Equation (A.9.3) also shows easily that ***$\mathcal{A}$ is a closed operator***, meaning that its graph $G_\mathcal{A} \stackrel{\text{def}}{=} \{(\psi, \mathcal{A}\psi) : \psi \in \text{dom}(\mathcal{A})\}$ is a closed subset of $C \times C$. Namely, if $\text{dom}(\mathcal{A}) \ni \psi_n \to \psi$ and $\mathcal{A}\psi_n \to \phi$, then by equation (A.9.3) $T_s\psi - \psi = \lim \int_0^s T_\sigma\mathcal{A}\psi_n \, d\sigma = \int_0^s T_\sigma\phi \, d\sigma$; dividing by s and letting $s \to 0$ shows that $\psi \in \text{dom}(\mathcal{A})$ and $\phi = \mathcal{A}\psi$. (iii) $\mathcal{A}$ is ***dissipative***. This means that

$$\|(I - \epsilon\mathcal{A})\psi\|_C \ge \|\psi\|_C \tag{A.9.7}$$

for all $\epsilon > 0$ and all $\psi \in \text{dom}(\mathcal{A})$ and follows directly from (A.9.6).

A subset $\mathcal{D}_0 \subset \text{dom}\,\mathcal{A}$ is a ***core*** for $\mathcal{A}$ if the restriction $\mathcal{A}_0$ of $\mathcal{A}$ to $\mathcal{D}_0$ has closure $\mathcal{A}$ (meaning of course that the closure of $G_{\mathcal{A}_0}$ in $C \times C$ is $G_\mathcal{A}$).

Exercise A.9.4 $\mathcal{D}_0 \subset \text{dom}\,\mathcal{A}$ is a core if and only if $(\alpha - \mathcal{A})\mathcal{D}_0$ is dense in C for some, and then for all, $\alpha > 0$. A dense ***invariant***[47] subspace $\mathcal{D}_0 \subset \text{dom}\,\mathcal{A}$ is a core.

[47] I.e., $T_t\mathcal{D}_0 \subseteq \mathcal{D}_0 \quad \forall t \ge 0$.

Feller Semigroups

In this book we are interested only in the case where the Banach space C is the space $C_0(E)$ of continuous functions vanishing at infinity on some separable locally compact space E. This Banach space carries an additional structure, the order, and the semigroups of interest are those that respect it:

Definition A.9.5 *A* ***Feller semigroup*** *on E is a strongly continuous semigroup $T_.$ on $C_0(E)$ of* ***positive***[48] *contractive linear operators T_t from $C_0(E)$ to itself. The Feller semigroup $T_.$ is* ***conservative*** *if for every $x \in E$ and $t \ge 0$*

$$\sup\{T_t\phi\,(x) : \phi \in C_0(E)\ ,\ 0 \le \phi \le 1\} = 1\ .$$

The positivity and contractivity of a Feller semigroup imply that the linear functional $\phi \mapsto T_t\phi(x)$ on $C_0(E)$ is a positive Radon measure of total mass ≤ 1. It extends in any of the usual fashions (see, e.g., page 395) to a subprobability $T_t(x,\cdot)$ on the Borels of E. We may use the measure $T_t(x,\cdot)$ to write, for $\phi \in C_0(E)$,

$$T_t\phi(x) = \int T_t(x,dy)\ \phi(y)\ . \tag{A.9.8}$$

In terms of the ***transition subprobabilities*** $T_t(x,dy)$, the semigroup property of $T_.$ reads

$$\int T_{s+t}(x,dy)\ \phi(y) = \int T_s(x,dy) \int T_t(y,dy')\ \phi(y') \tag{A.9.9}$$

and extends to all bounded Baire functions ϕ by the Monotone Class Theorem A.3.4; (A.9.9) is known as ***the Chapman–Kolmogorov equations***.

Remark A.9.6 Conservativity simply signifies that the $T_t(s,\cdot)$ are all probabilities. The study of general Feller semigroups can be reduced to that of conservative ones with the following little trick. Let us identify $C_0(E)$ with those continuous functions on the one-point compactification E^Δ that vanish at "the grave Δ." On any $\Phi \in C^\Delta \stackrel{\text{def}}{=} C(E^\Delta)$ define the semigroup $T^\Delta_.$ by

$$T^\Delta_t\Phi\,(x) = \begin{cases} \Phi(\Delta) + \int_E T_t(x,dy)\big(\Phi(y) - \Phi(\Delta)\big) & \text{if } x \in E, \\ \Phi(\Delta) & \text{if } x = \Delta. \end{cases} \tag{A.9.10}$$

We leave it to the reader to convince herself that $T^\Delta_.$ is a strongly continuous conservative Feller semigroup on $C(E^\Delta)$, and that "the grave" Δ is ***absorbing***: $T_t(\Delta,\{\Delta\}) = 1$. This terminology comes from the behavior of any process $X_.$ stochastically representing $T^\Delta_.$ (see definition 5.7.1); namely, once $X_.$ has reached the grave it stays there. The compactification $T_. \to T^\Delta_.$ comes in handy even when $T_.$ is conservative but E is not compact.

[48] That is to say, $\phi \ge 0$ implies $T_t\phi \ge 0$.

Examples A.9.7 (i) The simplest example of a conservative Feller semigroup perhaps is this: suppose that $\{\theta_s : 0 \le s < \infty\}$ is a semigroup under composition, of continuous maps $\theta_s : E \to E$ with $\lim_{s\downarrow 0} \theta_s(x) = x = \theta_0(x)$ for all $x \in E$, a ***flow***. Then $T_s\phi = \phi \circ \theta_s$ defines a Feller semigroup $T_.$ on $C_0(E)$, provided that the inverse image $\theta_s^{-1}(K)$ is compact whenever $K \subset E$ is.

(ii) Another example is the Gaussian semigroup of exercise 1.2.13 on $\mathbb{R}^d$:

$$\Gamma_t\phi(x) \stackrel{\text{def}}{=} \frac{1}{(2\pi t)^{d/2}} \int_{\mathbb{R}^d} \phi(x+y)\, e^{-|y|^2/2t}\, dy = \overset{*}{\gamma_t}\star\phi\,(x)\;.$$

(iii) The Poisson semigroup is introduced in exercise 5.7.11, and the semigroup that comes with a Lévy process in equation (4.6.31).

(iv) ***A convolution semigroup of probabilities*** on $\mathbb{R}^n$ is a family $\{\mu_t : t > 0\}$ of probabilities so that $\mu_{s+t} = \mu_s\star\mu_t$ for $s, t > 0$ and $\mu_0 = \delta_0$. Such gives rise to a semigroup of bounded positive linear operators T_t on $C_0(\mathbb{R}^n)$ by the prescription

$$T_t\phi\,(z) \stackrel{\text{def}}{=} \overset{*}{\mu_t}\star\phi\,(x) = \int_{\mathbb{R}^n} \phi(z+z')\,\mu_t(dz')\,, \qquad \phi \in C_0(\mathbb{R}^n)\,,\; z \in \mathbb{R}^n\,.$$

It follows directly from proposition A.4.1 and corollary A.4.3 that the following are equivalent: (a) $\lim_{t\downarrow 0} T_t\phi = \phi$ for all $\phi \in C_0(\mathbb{R}^n)$; (b) $t \mapsto \widehat{\mu_t}(\zeta)$ is continuous on $\mathbb{R}_+$ for all $\zeta \in \mathbb{R}^n$; and (c) $\mu_{t_n} \Rightarrow \mu_t$ weakly as $t_n \to t$. If any and then all of these continuity properties is satisfied, then $\{\mu_t : t > 0\}$ is called a ***conservative Feller convolution semigroup***.

A.9.8 Here are a few observations. They are either readily verified or substantiated in appendix C or are accessible in the concise but detailed presentation of Kallenberg [53, pages 313–326].

(i) The positivity of the T_t causes the resolvent operators to be positive as well. It causes the generator $\mathcal{A}$ to obey the ***positive maximum principle***; that is to say, whenever $\psi \in \text{dom}(\mathcal{A})$ attains a positive maximum at $x \in E$, then $\mathcal{A}\psi\,(x) \le 0$.

(ii) If the semigroup $T_.$ is conservative, then its generator $\mathcal{A}$ is ***conservative*** as well. This means that there exists a sequence $\psi_n \in \text{dom}(\mathcal{A})$ with $\sup_n \|\psi_n\|_\infty < \infty$; $\sup_n \|\mathcal{A}\psi_n\|_\infty < \infty$; and $\psi_n \to 1$, $\mathcal{A}\psi_n \to 0$ pointwise on E.

A.9.9 The Hille–Yosida Theorem states that the closure $\overline{\mathcal{A}}$ of a closable operator[49] $\mathcal{A}$ is the generator of a Feller semigroup – which is then unique – if and only if $\mathcal{A}$ is densely defined and satisfies the positive maximum principle, and $\alpha - \mathcal{A}$ has dense range in $C_0(E)$ for some, and then all, $\alpha > 0$.

---

[49] $\mathcal{A}$ is closable if the closure of its graph $G_{\mathcal{A}}$ in $C_0(E) \times C_0(E)$ is the graph of an operator $\overline{\mathcal{A}}$, which is then called the closure of $\mathcal{A}$. This simply means that the relation $\overline{G_{\mathcal{A}}} \subset C_0(E) \times C_0(E)$ actually is (the graph of) a function, equivalently, that $(0, \phi) \in \overline{G_{\mathcal{A}}}$ implies $\phi = 0$.

For proofs see [53, page 321], [100], and [115]. One idea is to emulate the formula $e^{at} = \lim_{n\to\infty}(1 - ta/n)^{-n}$ for real numbers a by proving that

$$T_t\phi \stackrel{\text{def}}{=} \lim_{n\to\infty} (I - t\overline{\mathcal{A}}/n)^{-n}\phi$$

exists for every $\phi \in C_0(E)$ and defines a contraction semigroup $T_.$ whose generator is $\overline{\mathcal{A}}$. This idea succeeds and we will take this for granted. It is then easy to check that the conservativity of $\mathcal{A}$ implies that of $T_.$.

The Natural Extension of a Feller Semigroup

Consider the second example of A.9.7. The Gaussian semigroup $\Gamma_.$ applies naturally to a much larger class of continuous functions than merely those vanishing at infinity. Namely, if ϕ grows at most exponentially at infinity,[50] then $\Gamma_t\phi$ is easily seen to show the same limited growth. This phenomenon is rather typical and asks for appropriate definitions. In other words, given a strongly continuous semigroup $T_.$, we are looking for a suitable extension $\breve{T}_.$ in the space $C = C(E)$ of continuous functions on E.

The natural topology of C is that of uniform convergence on compacta, which makes it a Fréchet space (examples A.2.27).

Exercise A.9.10 A curve $[0,\infty) \ni t \mapsto \psi_t$ in C is continuous if and only if the map $(s,x) \mapsto \psi_s(x)$ is continuous on every compact subset of $\mathbb{R}_+ \times E$.

Given a Feller semigroup $T_.$ on E and a function $f : E \to \mathbb{R}$, set[51]

$$\|f\|_{t,K}^{\breve{}} \stackrel{\text{def}}{=} \sup\left\{ \int_E^* T_s(x,dy)\,|f(y)| : 0 \le s \le t,\ x \in K \right\} \qquad \text{(A.9.11)}$$

for any $t > 0$ and any compact subset $K \subset E$, and then set

$$\lceil\!\lceil f \rceil\!\rceil^{\breve{}} \stackrel{\text{def}}{=} \sum_\nu 2^{-\nu} \wedge \|f\|_{\nu,K_\nu}^{\breve{}} ,$$

and

$$\begin{aligned} \breve{\mathfrak{F}} &\stackrel{\text{def}}{=} \{f : E \to \overline{\mathbb{R}} : \lceil\!\lceil \lambda f \rceil\!\rceil^{\breve{}} \xrightarrow[\lambda\to 0]{} 0\} \\ &= \{f : E \to \overline{\mathbb{R}} : \|f\|_{t,K}^{\breve{}} < \infty\ \forall t < \infty,\ \forall K \text{ compact}\} . \end{aligned}$$

The $\|\cdot\|_{t,K}^{\breve{}}$ and $\lceil\!\lceil \cdot \rceil\!\rceil^{\breve{}}$ are clearly solid and countably subadditive; therefore $\breve{\mathfrak{F}}$ is complete (theorem 3.2.10 on page 98). Since the $\|\cdot\|_{t,K}^{\breve{}}$ are seminorms, this space is also locally convex. Let us now define the ***natural domain*** $\breve{C}$ of $T_.$ as the $\lceil\!\lceil\ \rceil\!\rceil^{\breve{}}$-closure of $C_{00}(E)$ in $\breve{\mathfrak{F}}$, and the ***natural extension*** $\breve{T}_.$ on $\breve{C}$ by

$$\breve{T}_t\breve{\phi}\,(x) \stackrel{\text{def}}{=} \int_E T_t(x,dy)\,\breve{\phi}(y)$$

---

[50] $|\phi(x)| \le Ce^{c\|x\|}$ for some $C, c > 0$. In fact, for $\phi = e^{c\|x\|}$, $\int \Gamma_t(x,dy)\,\phi(y) = e^{tc^2/2}e^{c\|x\|}$.

[51] $\int^*$ denotes the upper integral – see equation (A.3.1) on page 396.

for $t \geq 0$, $\check{\phi} \in \check{C}$, and $x \in E$. Since the injection $C_{00}(E) \hookrightarrow \check{C}$ is evidently continuous and $C_{00}(E)$ is separable, so is $\check{C}$; since the topology is defined by the *seminorms* (A.9.11), $\check{C}$ is locally convex; since $\check{\mathfrak{F}}$ is complete, so is $\check{C}$: $\check{C}$ is a Fréchet space under the gauge $\lceil\!\lceil \ \rceil\!\rceil$. Since $\lceil\!\lceil \ \rceil\!\rceil$ is solid and $C_{00}(E)$ is a vector lattice, so is $\check{C}$. Here is a reasonably simple membership criterion:

Exercise A.9.11 (i) A continuous function ϕ belongs to $\check{C}$ if and only if for every $t < \infty$ and compact $K \subset E$ there exists a $\psi \in C$ with $|\phi| \leq \psi$ so that the function $(s, x) \mapsto \int_E^* T_s(x, dy)\ \psi(y)$ is finite and continuous on $[0, t] \times K$. In particular, when $T_.$ is conservative then $\check{C}$ contains the bounded continuous functions $C_b = C_b(E)$ and in fact is a module over C_b.

(ii) $\check{T}_.$ is a strongly continuous semigroup of positive continuous linear operators.

A.9.12 The Natural Extension of Resolvent and Generator The Bochner integral[52]

$$\check{U}_\alpha \psi = \int_0^\infty e^{-\alpha t} \cdot \check{T}_t \psi \, dt \tag{A.9.12}$$

may fail to exist for some functions ψ in $\check{C}$ and some $\alpha > 0$.[50] So we introduce the ***natural domains of the extended resolvent***

$$\check{\mathcal{D}}_\alpha = \check{\mathcal{D}}[\check{U}_\alpha] \stackrel{\text{def}}{=} \{\psi \in \check{C} : \text{the integral (A.9.12) exists and belongs to } \check{C}\} ,$$

and on this set define the ***natural extension of the resolvent operator*** $\check{U}_\alpha$ by (A.9.12). Similarly, the ***natural extension of the generator*** is defined by

$$\check{A}\psi \stackrel{\text{def}}{=} \lim_{t \downarrow 0} \frac{\check{T}_t \psi - \psi}{t} \tag{A.9.13}$$

on the subspace $\check{\mathcal{D}} = \check{\mathcal{D}}[\check{A}] \subset \check{C}$ where this limit exists and lies in $\check{C}$. It is convenient and sufficient to understand the limit in (A.9.13) as a pointwise limit.

Exercise A.9.13 $\check{\mathcal{D}}_\alpha$ increases with α and is contained in $\check{\mathcal{D}}$. On $\check{\mathcal{D}}_\alpha$ we have

$$(\alpha I - \check{A})\check{U}_\alpha = I .$$

The requirement that $\check{A}\psi \in \check{C}$ has the effect that $\check{T}_t \check{A}\psi = \check{A}\check{T}_t \psi$ for all $t \geq 0$ and

$$\check{T}_t \psi - \check{T}_s \psi = \int_s^t \check{T}_\sigma \check{A}\psi \, d\sigma , \qquad 0 \leq s \leq t < \infty .$$

A.9.14 A Feller Family of Transition Probabilities is a slew $\{T_{t,s} : 0 \leq s \leq t\}$ of positive contractive linear operators from $C_0(E)$ to itself such that for all $\phi \in C_0(E)$ and all $0 \leq s \leq t \leq u < \infty$

$$T_{u,s}\phi = T_{u,t} \circ T_{t,s}\phi \ \text{ and } \ T_{t,t}\phi = \phi ;$$

$$(s, t) \mapsto T_{t,s}\phi \quad \text{is continuous from } \mathbb{R}_+ \times \mathbb{R}_+ \text{ to } C_0(E).$$

[52] See item A.3.15 on page 400.

It is ***conservative*** if for every pair $s \le t$ and $x \in E$ the positive Radon measure

$$\phi \mapsto T_{t,s}\phi\,(x) \text{ is a probability } T_{t,s}(x,\cdot)\ ;$$

we may then write $\quad T_{t,s}\phi\,(x) = \int_E T_{t,s}(x,dy)\ \phi(y)\ .$

The study of $T_{\cdot,\cdot}$ can be reduced to that of a Feller semigroup with the following little trick: let $E^{\vdash} \overset{\text{def}}{=} \mathbb{R}_+ \times E$, and define $T_t^{\vdash} : C_0^{\vdash} \overset{\text{def}}{=} C_0(E^{\vdash}) \to C_0^{\vdash}$ by

$$T_s^{\vdash}\phi\,(t,x) = \int_E T_{s+t,t}(x,dy)\ \phi(s+t,y)\ ,$$

or $$T_s^{\vdash}\big((t,x), d\tau{\times}d\xi\big) = \delta_{s+t}(d\tau) \times T_{s+t,t}(x,d\xi)\ .$$

Then $T_{\bm{.}}^{\vdash}$ is a Feller semigroup on $E^{\vdash}$. We call it the ***time-rectification*** of $T_{\cdot,\cdot}$. This little procedure can be employed to advantage even if $T_{\cdot,\cdot}$ is already "time-rectified," that is to say, even if $T_{t,s}$ depends only on the elapsed time $t-s$: $T_{t,s} = T_{t-s,0} = T_{t-s}$, where $T_{\bm{.}}$ is some semigroup.

Let us define the operators

$$\mathcal{A}_\tau : \phi \mapsto \lim_{t\downarrow 0} \frac{T_{\tau+t,\tau}\phi - \phi}{t}\ , \qquad \phi \in \operatorname{dom}(\mathcal{A}_\tau)\ ,$$

on the sets $\operatorname{dom}(\mathcal{A}_\tau)$ where these limits exist, and write

$$(\mathcal{A}_\tau\phi)(\tau,x) = \big(\mathcal{A}_\tau\phi(\tau,\cdot)\big)(x)$$

for $\phi(\tau,\cdot) \in \operatorname{dom}(\mathcal{A}_\tau)$. We leave it to the reader to connect these operators with the generator $A^{\vdash}$:

Exercise A.9.15 Describe the generator $A^{\vdash}$ of $T_{\bm{.}}^{\vdash}$ in terms of the operators $\mathcal{A}_\tau$. Identify $\operatorname{dom}(\breve{T}_{\bm{.}}^{\vdash})$, $\mathcal{D}[\breve{U}_{\bm{.}}^{\vdash}]$, $\breve{U}_{\bm{.}}^{\vdash}$, $\operatorname{dom}(\breve{A}^{\vdash})$, and $\breve{A}^{\vdash}$. In particular, for $\psi \in \operatorname{dom}(\breve{A}^{\vdash})$,

$$A^{\vdash}\psi(\tau,x) = \frac{\partial\psi}{\partial t}(\tau,x) + \mathcal{A}_\tau\psi(\tau,x)\ .$$

Repeated Footnotes: 366[1] 366[2] 366[3] 367[5] 367[6] 375[12] 376[13] 376[14] 384[15] 385[16] 388[18] 395[22] 397[26] 399[27] 410[34] 421[37] 421[38] 422[40] 436[42] 467[50]

Appendix B

Answers to Selected Problems

Answers to most of the problems can be found in Appendix C, which is available on the web via `http://www.ma.utexas.edu/users/cup/Answers`.

2.1.6 Define $T_{n+1} = \inf\{t > T_n : X_{t+} - X_t \neq 0\} \wedge t$, where t is an instant past which X vanishes. The T_n are stopping times (why?), and X is a linear combination of $X_0 \cdot [\![0]\!]$ and the intervals $(\!(T_n, T_{n+1}]\!]$.

2.3.6 For $N = 1$ this amounts to $\zeta_1 \leq L_q \zeta_1$, which is evidently true. Assuming that the inequality holds for $N-1$, estimate

$$
\begin{aligned}
\zeta_N^2 &= \zeta_{N-1}^2 + (\zeta_N - \zeta_{N-1})(\zeta_N + \zeta_{N-1}) \\
&\leq L_q^2 \sum_{n=1}^{N} (\zeta_n - \zeta_{n-1})^2 + (\zeta_N - \zeta_{N-1})(\zeta_N + \zeta_{N-1}) \\
&\quad - L_q^2 (\zeta_N - \zeta_{N-1})(\zeta_N - \zeta_{N-1}) \\
&= L_q^2 \sum_{n=1}^{N} (\zeta_n - \zeta_{n-1})^2 + (\zeta_N - \zeta_{N-1})\big((1 - L_q^2)\zeta_N + (L_q^2 + 1)\zeta_{N-1}\big) .
\end{aligned}
$$

Now with $L_q^2 = (q+1)/(q-1)$ we have $1 - L_q^2 = -2/(q-1)$ and $L_q^2 + 1 = 2q/(q-1)$, and therefore

$$
(1 - L_q^2)\zeta_N + (L_q^2 + 1)\zeta_{N-1} = \frac{2}{q-1}\big(- \zeta_N + q\zeta_{N-1}\big) \leq 0 .
$$

2.5.17 Replacing F_n by $F_n - p$, we may assume that $p = 0$. Then $S_n \stackrel{\text{def}}{=} \sum_{\nu=1}^{n} F_\nu$ has expectation zero. Let $\mathcal{F}_n$ denote the σ-algebra generated by $F_1, F_2, \ldots, F_{n-1}$. Clearly S_n is a right-continuous square integrable martingale on the filtration $\{\mathcal{F}_n\}$. More precisely, the fact that $\mathbb{E}[F_\nu | \mathcal{F}_\mu] = 0$ for $\mu < \nu$ implies that

$$
\mathbb{E}\Big[S_n^2\Big] = \mathbb{E}\Big[\big(\textstyle\sum_{\nu=1}^{n} F_\nu\big)^2\Big] = \mathbb{E}\Big[\textstyle\sum_{\nu=1}^{n} F_\nu^2\Big] + 2\textstyle\sum_{1 \leq \mu < \nu \leq n} \mathbb{E}\big[F_\mu \mathbb{E}[F_\nu | \mathcal{F}_\mu]\big] \leq n\sigma^2 .
$$

Therefore $Z_n \stackrel{\text{def}}{=} S_n/n$ has $\|Z_n\|_{L^2} \leq \sigma\sqrt{1/n} \xrightarrow[n\to\infty]{} 0$. (Using Chebyscheff's inequality here we may now deduce the weak law of large numbers.) The strong law says that $Z_n \xrightarrow[n\to\infty]{} 0$ almost surely and evidently follows if we can show that Z_n converges almost surely to something as $n \to \infty$. To see that it does we write

$$
Z_\nu - Z_{\nu-1} = \frac{F_\nu}{\nu} - \frac{1}{\nu(\nu-1)} \sum_{1 \leq i < \nu} F_i , \qquad \nu = 2, 3, \ldots ,
$$

and set
$$\widetilde{Z}_n \stackrel{\text{def}}{=} \sum_{1\le\nu\le n} \frac{F_\nu}{\nu}\,, \qquad n = 1,2,3,\ldots\,,$$

and $\widehat{Z}_1 = 0$,
$$\widehat{Z}_n \stackrel{\text{def}}{=} \sum_{1<\nu\le n} \frac{S_{\nu-1}}{\nu(\nu-1)}\,, \qquad n = 2,3,\ldots\,.$$

Then
$$Z_n = \widetilde{Z}_n - \widehat{Z}_n\,,$$

and it suffices to show that both $\widetilde{Z}_.$ and $\widehat{Z}_.$ converge almost surely. Now $\widetilde{Z}_.$ is an L^2-bounded martingale and so has a limit almost surely. Namely, because of the cancellation as above,
$$\mathbb{E}\Big[\widetilde{Z}_n^2\Big] \le \sum_{1\le\nu\le n} \frac{\mathbb{E}[F_\nu^2]}{\nu^2} < \sum_{\nu<\infty} \frac{\sigma^2}{\nu^2} < \infty\,.$$

As to $\widehat{Z}$, since
$$\lceil \widehat{Z} \rceil_\infty \stackrel{\text{def}}{=} \sum_{2\le\nu<\infty} \frac{|S_{\nu-1}|}{\nu(\nu-1)}$$

has
$$\mathbb{E}[\lceil \widehat{Z} \rceil_\infty] \le \sum_{2\le\nu<\infty} \frac{\|S_{\nu-1}\|_{L^2}}{\nu(\nu-1)} \le \sum_{2\le\nu<\infty} \frac{\sigma\sqrt{\nu-1}}{\nu(\nu-1)} < \infty\,,$$

the sum defining $\widehat{Z}_\infty \stackrel{\text{def}}{=} \lim \widehat{Z}_n$ almost surely converges, even *absolutely*.

3.8.10 (i) Both $(Y,Z) \to [Y,Z](\varpi)$ and $(Y,Z) \to {}^c[Y,Z](\varpi)$ are inner products with associated lengths $S[Z](\varpi), \sigma[Z](\varpi)$. The claims are Minkowski's inequalities and follow from a standard argument.

(ii) Take $U = V = [\![0,T]\!]$ in theorem 3.8.9 and apply Hölder's inequality.

3.8.18 (i) Choose numbers $\alpha_i > 0$ satisfying inequality (3.7.12):
$$\textstyle\sum_{i=1}^\infty \lceil i\alpha_i \cdot X^i \rceil_{\mathcal{I}^0} < \infty$$

(X^i is X stopped at i). With each α_i goes a $\delta_i > 0$ so that $|x - x'| \le \delta_i$ implies $|F(x) - F(x')| \le \alpha_i$. Set $T_0^i = 0$ and $T_{k+1}^i \stackrel{\text{def}}{=} \inf\{t > T_k^i : |Z_t - Z_{T_k^i}| \ge \alpha_i\}$. Then set
$${}^i\!Y_t^\eta \stackrel{\text{def}}{=} \textstyle\sum_{k,\nu} F_\nu^\eta(X_{T_k^i})(X^\nu_{T_{k+1}^i\wedge t} - X^\nu_{T_k^i\wedge t})\,, \qquad \eta = 1,\ldots,d\,.$$

According to theorem 3.7.26, nearly everywhere ${}^i\!Y_. \xrightarrow[i\to\infty]{} (F(X){*}X)_.$ uniformly on bounded time-intervals. Let us toss out the nearly empty set where the limit does not exist and elsewhere select this limit for $F(X){*}X$. Now let ω, ω' be such that $X_.(\omega)$ and $X_.(\omega')$ describe the same arc via $t \mapsto t'$. The times T_k^i may well differ at ω and ω', in fact $T_k^i(\omega') = (T_k^i(\omega))'$, but the values $X_{T_k^i}(\omega)$ and $X_{T_k^i}(\omega')$ clearly agree. Therefore ${}^i\!Y_t(\omega) = {}^i\!Y_{t'}(\omega')$ for all $n \in \mathbb{N}$ and all $t \ge 0$. In the limit $(F(X){*}X)_t(\omega) = (F(X){*}X)_{t'}(\omega')$.

(ii) We start with the case $d = 1$. Apply (i) with $f(x) = 2x$ and toss out in addition the nearly empty set of $\omega \in \Omega$ where $[W,W]_t(\omega) \ne t$ for some t. Now if $W_.(\omega)$ and $W_.(\omega')$ describe the same arc via $t \mapsto t'$, then $[W,W]_.(\omega) = W_.^2(\omega) - 2(W_.{*}W)_.(\omega)$ and $[W,W]_.(\omega') = W_.^2(\omega') - 2(W_.{*}W)_.(\omega')$ describe the same arc also via $t \mapsto t'$. This reads $t = t'$.

In the case $d > 1$ apply the foregoing to the components W^η of $\boldsymbol{W}$ separately.

3.9.10 (i) Set $M_t \stackrel{\text{def}}{=} \langle \xi | X_{T+t} - X_T \rangle$ and assume without loss of generality that $M_0 \le a$. This continuous local martingale has a continuous strictly increasing square function $[M,M]_t = \xi_\mu\xi_\nu([X^\mu,X^\nu]_{T+t} - [X^\mu,X^\nu]_T) \xrightarrow[t\to\infty]{} \infty$. The time transformation $T^\lambda = T^{\lambda+} \stackrel{\text{def}}{=} \inf\{t : [M,M]_t \ge \lambda\}$ of exercise 3.9.8 turns M into a Wiener process. The zero-one law of exercise 1.3.47 on page 41 shows that $T = T^\pm$ almost surely.

(ii) In the previous argument replace T by S. The continuous time transformation $T^\lambda = T^{\lambda+} \stackrel{\text{def}}{=} \inf\{t : [M,M]_t \ge \lambda\}$ turns $M_t \stackrel{\text{def}}{=} \langle \xi | X_{S+t} - X_S \rangle$ into a standard Wiener process, which exceeds any number, including $a \pm \langle \xi | X_S \rangle$, in finite time (exercise 1.3.48 on page 41). Therefore the stopping times T and $T^\pm$ are almost surely finite. Consider the set $\mathcal{X}^-$ of paths $[S(\omega), \infty) \ni t \mapsto X_t(\omega)$ that have $T(\omega) > \tau$. Each of them stays on the same side of H as $X_S(\omega)$ for $t \in [S(\omega), \tau]$, in fact by continuity it stays a strictly positive distance away from H during this interval. Any other path $X_.(\omega')$ sufficiently close to a path in $\mathcal{X}^-$ will not enter H during $[S(\omega'), \tau]$ either and thus will have $T(\omega') > \tau$: the set $\mathcal{X}^-$ is open.

Next consider the set $\mathcal{X}^+$ of paths $[S(\omega), \infty) \ni t \mapsto X_t(\omega)$ that have $T^+(\omega) < \tau$. If $X_.(\omega) \in \mathcal{X}^+$, then there exists a $\sigma < \tau$ with $\langle \xi | X_\sigma \rangle > a$: $X_S(\omega)$ and $X_\sigma(\omega)$ lie on different sides of H. Clearly if ω' is such that the path $X_.(\omega')$ is sufficiently close to that of ω, then $X_\sigma(\omega')$ will also lie on the other side of H: $\mathcal{X}^+$ is open as well. After removal of the nearly empty set $[T \ne T^+]$ we are in the situation that the sets $\{X_.(\omega) : T(\omega) \gtrless \tau\}$ are open for all τ: T depends continuously on the path.

3.10.6 (i) Let $K^1, K^2 \subset \boldsymbol{H}$ be disjoint compact sets. There are $\phi^i_n \in C_{00}[\boldsymbol{H}]$ with $\phi^i_n \downarrow_n K^i$ pointwise and such that ϕ^1_1 and ϕ^2_1 have disjoint support. Then $\phi^i_n * \beta \to K^i * \beta$ as L^2-integrators and therefore uniformly on bounded time-intervals (theorem 2.3.6) and in law. Also, clearly $K^1 * \beta$ and $K^2 * \beta$ are independent, inasmuch as $\phi^1_m * \beta$ and $\phi^2_n * \beta$ are. In the next step exhaust disjoint relatively compact Borel subsets B^1, B^2 of $\boldsymbol{H}$ by compact subsets with respect to the Daniell means $B \mapsto \| B \times [0,n] \|^*_{\beta\text{-}2}$, $n \in \mathbb{N}$, which are $\mathcal{K}$-capacities (proposition 3.6.5), to see that $B^1 * \beta$ and $B^2 * \beta$ are independent Wiener processes. Clearly $\nu(B) \stackrel{\text{def}}{=} \mathbb{E}[(B*\beta)^2_1]$ defines an additive measure on the Borels of $\boldsymbol{H}$. It is σ-continuous, beacause it is majorized by $B \mapsto [\| B \times [0,1] \|^*_{\beta\text{-}2}]^2$, even inner regular.

3.10.9 Let I denote the image in $L^2 \stackrel{\text{def}}{=} L^2(\mathcal{F}^0_\infty[\beta], \mathbb{P})$ of $\mathfrak{L}^1[\beta\text{-}2]$ under the map $\check{X} \mapsto \int_0^\infty \check{X}\, d\beta$, and $\overline{I}$ the algebraic sum $I \oplus \mathbb{R}$. By theorem 3.10.6 (ii), I and then $\overline{I}$ are closed in L^2, and their complexifications $I_\mathbb{C}, \overline{I}_\mathbb{C}$ are closed in $L^2_\mathbb{C}$. By the Dominated Convergence Theorem for the L^2-mean both vector spaces are closed under pointwise limits of bounded sequences.

For $h \in \mathcal{E}[\check{\boldsymbol{H}}] \stackrel{\text{def}}{=} \mathcal{E}[\boldsymbol{H}] \otimes C_{00}(\mathbb{R}_+)$ set $M^h \stackrel{\text{def}}{=} h*\beta$, which is a martingale with square bracket $[M^h, M^h]_t = \int_0^t h^2(\eta, s)\ \nu(d\eta) ds$. Then let $G^h = \exp(iM^h + [M^h, M^h]/2)$ be the Doléans–Dade exponential of iM^h. Clearly $G^h_\infty = 1 + \int_0^\infty iG^h\, dM^h$ belongs to $\overline{I}_\mathbb{C}$, and so does its scalar multiple $\exp(iM^h_\infty)$. The latter form a multiplicative class $\mathcal{M}$ contained in $\overline{I}_\mathbb{C}$ and generating $\mathcal{F}^0_\infty[\beta]$ (page 409). By exercise A.3.5 the vector space $\overline{I}_\mathbb{C}$ contains all bounded $\mathcal{F}^0_\infty[\beta]$-measurable functions. As it is mean closed, it contains all of $L^2_\mathbb{C}$. Thus $\overline{I} = L^2$.

4.3.10 Use exercise 3.7.19, proposition 3.7.33, and exercise 3.7.9.

4.5.8 Inequality (4.5.9) and the homogeneity argument following it show that for any bounded previsible $X \ge 0$

$$\mu^{\langle\sigma\rangle}(X) \le \left(\mu^{\langle\rho\rangle}(X)\right)^{\frac{\tau-\sigma}{\tau-\rho}} \cdot \left(\mu^{\langle\tau\rangle}(X)\right)^{\frac{\sigma-\rho}{\tau-\rho}} \le \left(\mu^{\langle\rho\rangle} \vee \mu^{\langle\tau\rangle}\right)(X) .$$

4.5.9 Since $z \mapsto (z+\zeta)^p - z^p$ increases, inequality (4.5.16) can with the help of theorem 2.3.6 be continued as

$$\| Z_\infty \|^p_{L^p} \le \left(C^\star_p \,\rceil Z \lceil_{\mathcal{I}^p} + \zeta[Z]\right)^p - C^{\star p}_p \,\rceil Z \lceil^p_{\mathcal{I}^p} .$$

Replace Z by $X{*}Z$ and take the supremum over X with $|X| \le 1$ to obtain

$$\lceil Z \rceil_{\mathcal{I}^p}^p \le \left(C_p^\star \lceil Z \rceil_{\mathcal{I}^p} + \zeta[Z]\right)^p - C_p^{\star p} \lceil Z \rceil_{\mathcal{I}^p}^p ,$$

or
$$(1 + C_p^{\star p})^{1/p} \lceil Z \rceil_{\mathcal{I}^p} \le C_p^\star \lceil Z \rceil_{\mathcal{I}^p} + \zeta[Z]$$

and
$$\lceil Z \rceil_{\mathcal{I}^p} \le c_p^\diamond \zeta[Z]$$

with
$$c_p^\diamond \le \left((1 + C_p^{\star p})^{1/p} - C_p^\star\right)^{-1} \le 0.6 \cdot 4^p .$$

4.5.21 Let $T = \inf\{t : A_t \ge a\}$ and $S = \inf\{t : Y_t \ge y\}$. Both are stopping times, and T is predictable (theorem 3.5.13); there is an increasing sequence T_n of finite stopping times announcing T. On the set $[Y_\infty > y, A_\infty < a]$, S is finite, $Y_S \ge y$, and the T_n increase without bound. Therefore

$$[Y_\infty > y, A_\infty < a] \le \frac{1}{y} \cdot \sup_n Y_{S \wedge T_n}$$

and so
$$\mathbb{P}[Y_\infty > y, A_\infty < a] \le \frac{1}{y} \cdot \sup_n \mathbb{E}[Y_{S \wedge T_n}]$$

$$\le \frac{1}{y} \cdot \sup_n \mathbb{E}[A_{S \wedge T_n}] \le \frac{1}{y} \cdot \mathbb{E}[A_\infty \wedge a] .$$

Applying this to sequences $y_n \downarrow y$ and $a_n \uparrow a$ yields inequality (4.5.30). This then implies $\mathbb{P}[Y = \infty, A \le a] = 0$ for all $a < \infty$; then $\mathbb{P}[Y = \infty, A < \infty] = 0$, which is (4.5.31).

4.5.24 Use the characterizations 4.5.12, 4.5.13, and 4.5.14. Consider, for instance, the case of $\boldsymbol{Z}^{\langle q \rangle}$. Let ${}^{\prime}\boldsymbol{X}, {}^{\prime}\check{H}$ be the quantities of exercise 4.5.14 and its answer for ${}'\boldsymbol{Z}$. Then $\check{H}(\boldsymbol{y}, s) \overset{\text{def}}{=} {}^{\prime}\check{H} \circ C(\boldsymbol{y}, s) = {}^{\prime}\check{H}(C\boldsymbol{y}, s) = \langle {}^{\prime}\boldsymbol{X}_s | C\boldsymbol{y} \rangle = \langle C^T {}^{\prime}\boldsymbol{X}_s | \boldsymbol{y} \rangle$, where $C^T : \ell^\infty('d) \to \ell^\infty(d)$ denotes the (again contractive) transpose of C. By exercise 4.5.14, the Doléans–Dade measure ${}'\mu$ of $|\check{H}|^q {*} \jmath_{\boldsymbol{Z}}$ is majorized by that of $\Lambda^{\langle q \rangle}[\boldsymbol{Z}]$. But ${}'\mu$ is the Doléans–Dade measure of $\Lambda^{\langle q \rangle}['\boldsymbol{Z}]$! Indeed, the compensator of $|\check{H}|^q {*} \jmath_{\boldsymbol{Z}} = |{}^{\prime}\check{H} \circ C|^q {*} \jmath_{\boldsymbol{Z}} = |{}^{\prime}\check{H}|^q {*} C[\jmath_{\boldsymbol{Z}}] = |{}^{\prime}\check{H}|^q {*} \jmath_{'\boldsymbol{Z}}$ is $\Lambda^{\langle q \rangle}['\boldsymbol{Z}]$. The other cases are similar but easier.

5.2.2 Let $S < T^\mu$ on $[T^\mu > 0]$. From inequality (4.5.1)

$$\left\| |\boldsymbol{\Delta}{*}\boldsymbol{Z}|_S^\star \right\|_{L^p} \le C_p^\diamond \cdot \max_{\rho=1,p^\diamond} \left\| \left(\int_0^S |\boldsymbol{\Delta}|^\rho \, d\Lambda \right)^{1/\rho} \right\|_{L^p}$$

$$\le C_p^\diamond \cdot \max_{\rho=1,p^\diamond} \left\| \left(\int_0^\mu \delta^\rho \, d\lambda \right)^{1/\rho} \right\|_{L^p} = \delta \cdot C_p^\diamond \max_{\rho=1,p^\diamond} \mu^{1/\rho} .$$

Letting S run through a sequence announcing T^μ, multiplying the resulting inequality $\left\| |\boldsymbol{\Delta}{*}\boldsymbol{Z}|_{T^\mu -}^\star \right\|_{L^p} \le \delta \cdot C_p^\diamond \max_{\rho=1,p^\diamond} \mu^{1/\rho}$ by $e^{-M\mu}$, and taking the supremum over $\mu > 0$ produces the claim after a little calculus.

5.2.18 (i) Since $e^{pmW_t - p^2 m^2 t/2} = \mathscr{E}_t[pmW]$ is a martingale of expectation one

we have
$$|\mathscr{E}_t[mW]|^p = e^{pmW_t - pm^2 t/2} = \mathscr{E}_t[pmW] \cdot e^{(p^2-p)m^2 t/2} ,$$

$$\mathbb{E}[|\mathscr{E}_t[mW]|^p] = e^{(p^2-p)m^2 t/2} , \quad \text{and} \quad \|\mathscr{E}_t[mW]\|_{L^p} = e^{m^2(p-1)t/2} .$$

Next, from $e^{|x|} \le e^x + e^{-x}$ we get

$$e^{|mW_t|} \le e^{mW_t} + e^{-mW_t} = e^{m^2 t/2} \times \left(\mathscr{E}_t[mW] + \mathscr{E}_t[-mW]\right) ,$$

$$e^{|mW|_t^\star} \le e^{m^2t/2} \times \left(\mathscr{E}_t^\star[mW] + \mathscr{E}_t^\star[-mW]\right) ,$$

and $$\left\| e^{|mW|_t^\star} \right\|_{L^p} \le e^{m^2t/2} \times \left(\left\|\mathscr{E}_t^\star[mW]\right\|_{L^p} + \left\|\mathscr{E}_t^\star[-mW]\right\|_{L^p} \right)$$

by theorem 2.5.19: $$\le e^{m^2t/2} \times 2p' \cdot e^{m^2(p-1)t/2} = 2p' \cdot e^{m^2pt/2} .$$

(ii) We do this with $|\ |$ denoting the ℓ^1-norm on $\mathbb{R}^d$. First,

$$\left\| e^{|m\boldsymbol{Z}^\star|_t} \right\|_{L^p} = e^{|m|t} \times \prod_\eta \left\| e^{m|W^{\eta\star}|_t} \right\|_{L^p}$$

by independence of the W^η: $$\le e^{|m|t} \times \left(2p' \cdot e^{m^2pt/2}\right)^{d-1}$$

$$= (2p')^{d-1} \times e^{(|m|+(d-1)m^2p/2)\cdot t} .$$

Thus $$\left\| e^{|m\boldsymbol{Z}^\star|_t} \right\|_{L^p} \le A_{p,d} \times e^{M_{d,m,p}\cdot t} . \tag{1}$$

Next, $$\left\| |\boldsymbol{Z}^\star|_t^r \right\|_{L^p} = \left\| |\boldsymbol{Z}^\star|_t \right\|_{L^{rp}}^r \le \left(t + \sum_\eta \left\| |W^{\eta\star}|_t \right\|_{L^{rp}}\right)^r$$

$$= \left(t + (d-1)\cdot \left\| |W|_t^\star \right\|_{L^{rp}}\right)^r$$

by theorem 2.5.19: $$\le \left(t + (d-1)(rp)' \cdot \left\| |W|_t \right\|_{L^{rp}}\right)^r$$

$$\le 2^{r'}\left(t^r + (d-1)^r(rp)'^r \cdot \left\| |W|_t \times \right\|_{L^{rp}}^r\right)$$

by exercise A.3.47 with $\sigma = \sqrt{t}$: $$= 2^{r'}\left(t^r + (d-1)^r(rp)'^r\Gamma_{p,r} \cdot t^{r/2}\right) .$$

Thus $$\left\| |\boldsymbol{Z}^\star|_t^r \right\|_{L^p} \le B_r t^r + B_{d,r,p}t^{r/2} . \tag{2}$$

Applying Hölder's inequality to (1) and (2), we get

$$\left\| |\boldsymbol{Z}^\star|_t^r \cdot e^{|m\boldsymbol{Z}^\star|_t} \right\|_{L^p} \le \left(B_r t^r + B_{d,r,2p}t^{r/2}\right) \times \left(A_{2p,d}e^{M_{d,m,p}t}\right)$$

$$= t^{r/2}A_{2p,d}\left(B_{d,r,2p} + B_r t^{r/2}\right) \times e^{M_{d,m,2p}t} :$$

we get, for suitable $B' = B'_{d,p,r}$, $M' = M'_{d,m,p,r}$, the desired inequality

$$\left\| |\boldsymbol{Z}^\star|_t^r \cdot e^{|m\boldsymbol{Z}^\star|_t} \right\|_{L^p} \le B' \cdot t^{r/2}\, e^{M't} .$$

5.3.1 $\mathfrak{S}_{p,M}^{\star n}$ is naturally equipped with the collection $\mathfrak{N}^\circ$ of seminorms $\|\ \|_{p^\circ,M^\circ}$, where $2 \le p^\circ < p$ and $M^\circ > M$. $\mathfrak{N}^\circ$ forms an increasing family with pointwise limit $\|\ \|_{p,M}$. For $0 \le \sigma \le 1$ set $u^\sigma \overset{\text{def}}{=} u + \sigma(v-u)$ and $X^\sigma \overset{\text{def}}{=} X + \sigma(Y-X)$. Write F for F_η, etc. Then the remainder $F[v,Y] - F[u,X] - D_1F[u,X]\cdot(v-u) - D_2F[u,X]\cdot(Y-X)$ becomes, as in example A.2.48,

$$RF[u,X;v,Y] = \int_0^1 \left(Df(u^\sigma, X^\sigma) - Df(u,X)\right) \cdot \begin{pmatrix} v-u \\ Y-X \end{pmatrix} d\sigma .$$

With $R^\sigma f \overset{\text{def}}{=} Df(u^\sigma, X^\sigma) - Df(u, X)$, $1/p^\circ = 1/p + 1/r$, and $\|R\|_{pp^\circ}$ denoting the operator norm of a linear operator $R : \ell^p(k+n) \to \ell^{p^\circ}(n)$ we get

$$|RF[u, X; v, Y]^\star_{T^\lambda-}|_{p^\circ} \le C \cdot r_\lambda \cdot \left(|v-u| + ||Y-X|^\star_{T^\lambda-}|_p\right),$$

where
$$r_\lambda \overset{\text{def}}{=} \sup_{t<T^\lambda} \sup_{0\le\sigma\le1} \|R^\sigma f\|_{pp^\circ},$$

and where C is a suitable constant depending only on k, n, p, p°. Now r_λ is a bounded random variable and converges to zero in probability as $|v - u| + \|Y-X\|^\star_{p,M} \to 0$. (Use that the T^λ of definition (5.2.4) on page 283 are bounded; then X^σ_t ranges over a relatively compact set as $[0 \le \sigma \le 1]$ and $0 \le t \le T^\lambda$.) In other words, the uniformly bounded increasing functions $\lambda \mapsto \|r_\lambda\|_r$ converge pointwise – and thus uniformly on compacta – to zero as $|v-u|+\|Y-X\|^\star_{p,M} \to 0$. Therefore the first factor on the right in

$$\|RF[u, X; v, Y]\|_{p^\circ,M^\circ} \le C \sup_\lambda e^{(M-M^\circ)\lambda} \|r_\lambda\|_r \cdot \left(|v-u| + \|Y-X\|^\star_{p,M}\right)$$

converges to zero as $|v-u|+\|Y-X\|^\star_{p,M} \to 0$, which is to say $\|RF[u, X; v, Y]\|_{p^\circ,M^\circ} = o(|v - u| + \|Y-X\|^\star_{p,M})$.

5.4.19 (ii) For fixed μ and δ set $k \overset{\text{def}}{=} \lceil \mu/\delta \rceil$, and $\lambda_i \overset{\text{def}}{=} i\delta$ and $T_i \overset{\text{def}}{=} T^{\lambda_i}$ for $i = 0, 1, \ldots, k$. Then $\lambda_{k-1} < \mu \le \lambda_k$. Let $\Delta^\star_i$ denote the maximal function of the difference of the global solution at T_i, which is $X_{T_i} = \Xi[C, \boldsymbol{Z}]_{T_i}$, from its Ξ'-approximate X'_{T_i}. Consider an $s \in [T^{\lambda_i}, T^{\lambda_{i+1}}]$.

Since
$$\begin{aligned} X'_s - X_s &= \Xi'[X'_{T_i}, \boldsymbol{Z}_s - \boldsymbol{Z}_{T_i}] - \Xi'[X_{T_i}, \boldsymbol{Z}_s - \boldsymbol{Z}_{T_i}] \\ &\quad + \Xi'[X_{T_i}, \boldsymbol{Z}_s - \boldsymbol{Z}_{T_i}] - \Xi[X_{T_i}, \boldsymbol{Z}_s - \boldsymbol{Z}_{T_i}], \end{aligned}$$

5.4.17 gives
$$\|\Delta^\star_{i+1}\|_{L^p} \le \|\Delta^\star_i\|_{L^p} \times e^{L'\delta} + (\||X|^\star_{T_i}+1\|_{L^p}) \times (\underline{M}\delta)^r e^{\underline{M}\delta},$$

which implies
$$\begin{aligned} |\Delta^\star_k| &\le (|X|^\star_{T_k}+1) \times (\underline{M}\delta)^r e^{\underline{M}\delta} \cdot \textstyle\sum_{0\le i<k} e^{iL'\delta} \\
\text{for } X_. \in \mathfrak{S}_{p,M},\ \text{as } \lambda_k = k\delta: \quad &\le (\|X_.\|^\star_M e^{M\lambda_k} + 1) \times (\underline{M}\delta)^r e^{\underline{M}\delta} \cdot \frac{e^{L'k\delta} - 1}{e^{L'\delta} - 1} \\
\text{by (5.2.23), as } \delta \to 0: \quad &\le \frac{2}{1-\gamma}\left(\|{}^0C\|^\star_M + 1\right) e^{M\lambda_k} \times (\underline{M}\delta)^r e^{\underline{M}\delta} \cdot k\, e^{L'\lambda_k} \\
\text{since } k = \lambda_k/\delta: \quad &\le \mathit{const}\,(\|C\|^\star_M + 1)\underline{M}^r e^{\underline{M}\delta} \delta^{r-1} \times \lambda_k \cdot e^{(M+L')\lambda_k} \\
&\le \overline{B} \cdot (\|C\|_{L^p}+1) \times \delta^{r-1} \cdot e^{\overline{M}\lambda_k} \end{aligned}$$

for suitable $\overline{B} = \overline{B}[f; \Xi']$ and $\overline{M} = \overline{M}[f; \Xi'] > M+L'$.

5.5.7 Let $\mathbb{P}, \overline{\mathbb{P}} \in \mathfrak{P}$. Thanks to the uniform ellipticity (5.5.17) there exist bounded functions h^η so that $f_0 = -\sum_\eta f_\eta \cdot h^\eta$. Then $M \overset{\text{def}}{=} \boldsymbol{h}{*}\boldsymbol{W}$ is a martingale under both $\mathbb{P}$ and $\overline{\mathbb{P}}$, and by exercise 3.9.12 so is the Doléans–Dade exponential G' of M. The Girsanov theorem 3.9.19 asserts that $\boldsymbol{W}' \overset{\text{def}}{=} \boldsymbol{W}_t + \int_0^. \boldsymbol{h}_s\, ds$ is a Wiener process under the probabilities $\mathbb{P}'$ and $\overline{\mathbb{P}}'$ that on every $\mathcal{F}_t$ agree with $G'_t\mathbb{P}$ and $G'_t\overline{\mathbb{P}}$, respectively. Now X satisfies equation (5.5.20) with $\boldsymbol{W}'$ replacing $\boldsymbol{W}$, and therefore by assumption $\mathbb{P}' = \overline{\mathbb{P}}'$. This clearly implies $\mathbb{P} = \overline{\mathbb{P}}$.

5.5.13 Let $s \le t \le t'$. Then by Itô's formula

$$u(t'-t, X_t) = u(t'-s, X_s) - \int_s^t \dot{u}(t'-\sigma, X_\sigma)\, d\sigma$$
$$+ \int_s^t u_{;\nu}(t'-\sigma, X_\sigma)\, dX_\sigma^{x\nu} + \int_s^t Au(t'-\sigma, X_\sigma)\, d\sigma$$
$$= u(t'-s, X_s) + \int_s^t u_{;\nu}(t'-\sigma, X_\sigma)\, dX_\sigma^{x\nu} .$$

Taking the conditional expectation under $\mathcal{F}_s$ exhibits $t \mapsto u(t'-t, X_t)$ as a bounded local martingale.

5.5.14 It suffices to consider equation (5.5.20). Recall that $\mathfrak{P}$ is the collection of all probabilities on $\mathscr{C}^n$ under which the process $\boldsymbol{W}_t$ of (5.5.19) is a standard Wiener process. Let $\mathbb{P}, \overline{\mathbb{P}} \in \mathfrak{P}$. From $\phi_0, \phi_1, \ldots, \phi_k \in C_b^\infty(\mathbb{R}^n)$ and $0 = t_0 < t_1 < \cdots < t_k$ make the function $\Phi \overset{\text{def}}{=} \phi_0(X_0^x) \cdot \phi_1(X_{t_1}^x) \cdots \phi_k(X_{t_k}^x)$ on the path space $\mathscr{C}^n$. Their collection forms a multiplicative class that separates the points of $\mathscr{C}^n$. Since path space is polish and consequently every probability on it is tight (proposition A.6.2), or simply because the functions Φ generate the Borel σ-algebra on path space, $\mathbb{P} = \overline{\mathbb{P}}$ will follow if we can show that $\mathbb{E} = \overline{\mathbb{E}}$ on the functions Φ (proposition A.3.12). This we do by induction in k. The case $k = 0$ is trivial. Note that the equality $\mathbb{E}[\Phi] = \overline{\mathbb{E}}[\Phi]$ on Φ made from smooth functions ϕ_i persists on functions Φ made from continuous, even bounded Baire, functions ϕ_i, by the usual sequential closure argument. To propel ourselves from k to $k+1$ let u denote a solution to the initial value problem $\dot{u} = Au$ with $u(0,x) = \phi_{k+1}(x)$ and write $\Phi \overset{\text{def}}{=} \phi_0 \circ X_0^x \cdot \phi_1 \circ X_{t_1}^x \cdots \phi_k \circ X_{t_k}^x$. Then, with $t = t' = t_{k+1}$ and $s = t_k$, exercise 5.5.13 produces

$$\mathbb{E}\Big[\phi_{k+1}(X_{t_{k+1}}^x)|\mathcal{F}_{t_k}\Big] = \mathbb{E}\Big[u(0, X_{t_{k+1}}^x)|\mathcal{F}_{t_k}\Big]$$
$$= \mathbb{E}\big[u(t_{k+1}-t_k, X_{t_k}^x)\big]$$

and so
$$\mathbb{E}[\Phi \cdot \phi_{k+1}(X_{t_{k+1}}^x)] = \mathbb{E}[\Phi \cdot u(t_{k+1}-t_k, X_{t_k}^x)]$$
$$= \overline{\mathbb{E}}[\Phi \cdot u(t_{k+1}-t_k, X_{t_k}^x)] \qquad (*)$$

by the same token:
$$= \overline{\mathbb{E}}[\Phi \cdot \phi_{k+1}(X_{t_{k+1}}^x)] .$$

At $(*)$ we used the fact that the argument of the expectation is a k-fold product of the same form as Φ, so that the induction hypothesis kicks in.

A.2.23 Let U_n be the set of points x in F that have a neighborhood $V_n(x)$ whose intersection with E has ρ-diameter strictly less than $1/n$. The U_n clearly are open and contain E. Their intersection is E. Indeed, if $x \in \bigcap U_n$, then the sets $E \cap V_n(x)$ form a Cauchy filter basis in E whose limit must be x.

A.3.5 The family of all complex finite linear combinations of functions in $\mathcal{M} \cup \{1\}$ is a complex algebra $\mathcal{A}$ of bounded functions in $\mathcal{V}$ that is closed under complex conjugation; the σ-algebra it generates is again $\mathcal{M}^\Sigma$. The real-valued functions in $\mathcal{A}$ form a real algebra $\mathcal{A}_0$ of bounded functions that again generates $\mathcal{M}^\Sigma$. It is a multiplicative class contained in the bounded monotone class $\mathcal{V}_0$ of real-valued functions in $\mathcal{V}$. Now apply theorem A.3.4 suitably.

References

LNM stands for Lecture Notes in Mathematics, Springer, Berlin, Heidelberg, New York

1. R. A. Adams, *Sobolev Spaces,* Academic Press, New York, 1975.
2. L. Arnold, *Stochastic Differential Equations: Theory and Applications*, Wiley, New York, 1974.
3. J. Azéma, "Sur les fermés aléatoires," in: *Séminaire de Probabilités XIX*, LNM **1123**, 1985, pp. 397–495.
4. J. Azéma and M. Yor, "Etude d'une martingale remarquable," in: *Séminaire de Probabilités XXIII,* LNM **1372**, 1989, pp. 88–130.
5. K. Bichteler, *Integration Theory*, LNM **315**, 1973.
6. K. Bichteler, "Stochastic integrators," *Bull. Amer. Math. Soc.* **1** (1979), 761–765.
7. K. Bichteler, "Stochastic integration and L^p–theory of semimartingales," *Annals of Probability* **9** (1981), 49–89.
8. K. Bichteler and J. Jacod, "Random measures and stochastic integration," *Lecture Notes in Control Theory and Information Sciences* **49**, Springer, Berlin, Heidelberg, New York, 1982, pp. 1–18.
9. K. Bichteler, *Integration, a Functional Approach*, Birkhäuser, Basel, 1998.
10. P. Billingsley, *Probability and Measure*, 2nd ed., John Wiley & Sons, New York, 1985.
11. R. M. Blumenthal and R. K. Getoor, *Markov Processes and Potential Theory*, Academic Press, New York, 1968.
12. N. Bourbaki, *Intégration*, Hermann, Paris, 1965–9.
13. J. L. Bretagnolle, "Processus à accroissements indépendants," in: *Ecole d'Eté de Probabilités*, LNM **307**, 1973, pp. 1–26.
14. D. L. Burkholder, "Sharp norm comparison of martingale maximal functions and stochastic integrals," *Proceedings of the Norbert Wiener Centenary Congress* (1994), 343–358.
15. E. Çinlar and J. Jacod, "Representation of semimartingale Markov processes in terms of Wiener processes and Poisson random measures," in: *Seminar on Stochastic Processes*, Progr. Prob. Statist. **1**, Birkhäuser, Boston, 1981, pp. 159–242.

16. P. Courrège, "Intégrale stochastique par rapport à une martingale de carré intégrable," in: *Seminaire Brelot–Choquet–Dény*, 7 année (1962–63), Institut Henri Poincaré, Paris.
17. C. Dellacherie, *Capacités et Processus Stochastiques*, Springer, Berlin, Heidelberg, New York, 1981.
18. C. Dellacherie, "Un survol de la théorie de l'intégrale stochastique," *Stochastic Processes and Their Applications* **10** (1980), 115–144.
19. C. Dellacherie, "Measurabilité des débuts et théorèmes de section," in: *Séminaire de Probabilités XV*, LNM **850**, 1981, pp. 351–360.
20. C. Dellacherie and P. A. Meyer, *Probability and Potential*, North Holland, Amsterdam, New York, 1978.
21. C. Dellacherie and P. A. Meyer, *Probability and Potential B*, North Holland, Amsterdam, New York, 1982.
22. J. Dieudonnée, *Foundations of Modern Analysis*, Academic Press, New York, London, 1964.
23. C. Doléans–Dade, "Quelques applications de la formule de changement de variables pour les semimartingales," *Z. für Wahrscheinlichkeitstheorie* **16** (1970), 181–194.
24. C. Doléans–Dade, "On the existence and unicity of solutions of stochastic differential equations," *Z. für Wahrscheinlichkeitstheorie* **36** (1976), 93–101.
25. C. Doléans–Dade, "Stochastic processes and stochastic differential equations," in: *Stochastic Differential Equations*, Centro Internazionale Matematico Estivo (Cortona), Liguori Editore, Naples, 1981, pp. 5–75.
26. J. L. Doob, *Stochastic Processes*, Wiley, New York, 1953.
27. A. Dvoretsky, P. Erdös, and S. Kakutani "Nonincreasing everywhere of the Brownian motion process," in: 4th *BSMSP* **2** (1961), pp. 103–116.
28. R. J. Elliott, *Stochastic Calculus and Applications*, Springer, Berlin, Heidelberg, New York, 1982.
29. K.D. Elworthy, "Stochastic differential equations on manifolds," in: *Probability towards 2000*, Lecture Notes in Statistics **128**, Springer, New York, 1998, pp. 165–178.
30. M. Emery, "Une topologie sur l'espaces des semimartingales," in: *Séminaire de Probabilités XIII*, LNM **721**, 1979, pp. 260–280.
31. M. Emery, "Equations différentielles stochastiques lipschitziennes: étude de la stabilité," in: *Séminaire de Probabilités XIII*, LNM **721**, 1979, pp. 281–293.
32. M. Emery, "On the Azéma martingales," in: *Séminaire de Probabilités XXIII*, LNM **1372**, 1989, pp. 66–87.
33. S. Ethier and T. G. Kurtz, *Markov Processes: Characterization and Convergence*, Wiley, New York, 1986.
34. W. W. Fairchild and C. Ionescu Tulcea, *Topology*, W. B. Saunders, Philadelphia, London, Toronto, 1971.

35. A. Garsia, *Martingale Inequalities*, Seminar Notes on Recent Progress, Benjamin, New York, 1973.
36. D. Gilbarg and N.S. Trudinger, *Elliptic Partial Differential Equations of Second Order*, Springer, Berlin, Heidelberg, New York, 1977.
37. I. V. Girsanov, "On tranforming a certain class of stochastic processes by absolutely continuous substitutions of measures," *Theory Proba. Appl.* **5** (1960), 285–301.
38. U. Haagerup, "Les meilleurs constantes de l'inegalités de Khintchine," *C. R. Acad. Sci. Paris Sér. A–B* **286/5** (1978), A-259–262.
39. N. Ikeda and S. Watanabe, *StochAstic Differential Equations and Diffusion Processes*, North Holland, Amsterdam, New York, 1981.
40. K. Itô, "Stochastic integral," *Proc. Imp. Acad. Tokyo* **20** (1944), 519–524.
41. K. Itô, "On stochastic integral equations," *Proc. Japan Acad.* **22** (1946), 32–35.
42. K. Itô, "Stochastic differential equations in a differentiable manifold," *Nagoya Math. J.* **1** (1950), 35–47.
43. K. Itô, "On a formula concerning stochastic differentials," *Nagoya Math. J.* **3** (1951), 55–65.
44. K. Itô, *Stochastic Differential Equations*, Memoirs of the American Math. Soc. **4** (1951).
45. K. Itô, "Multiple Wiener integral," *J. Math. Soc. Japan* **3** (1951), 157–169.
46. K. Itô, "Extension of stochastic integrals," *Proceedings of the International Symposium on Stochastic Differential Equations*, Kyoto (1976), 95–109.
47. K. Itô and H. P. McKean, *Diffusion Processes and Their Sample Paths*, Die Grundlehren der mathematischen Wissenschaften **125**, Springer, Berlin-New York, 1974.
48. K. Itô and M. Nisio, "On stationary solutions of stochastic differential equations," *J. Math. Kyoto Univ.* **4** (1964), 1–75.
49. J. Jacod, *Calcul Stochastique et Problème de Martingales*, LNM **714**, 1979.
50. J. Jacod and A. N. Shiryaev, *Limit Theorems for Stochastic Processes*, Springer, Berlin, Heidelberg, New York, 1987.
51. T. Kailath, A. Segal, and M. Zakai, "Fubini-type theorems for stochastic integrals," *Sankhya (Series A)* **40** (1987), 138–143.
52. O. Kallenberg, *Random Measures*, Akademie-Verlag, Berlin, 1983.
53. O. Kallenberg, *Foundations of Modern Probability*, Springer, New York, 1997.
54. I. Karatzas and S. Shreve, *Brownian Motion and Stochastic Calculus*, Springer, Berlin, Heidelberg, New York, 1988.
55. J. L. Kelley, *General Topology*, van Nostrand, 1955.

56. D. E. Knuth, "Two notes on notation," *Amer. Math. Monthly* **99/5** (1992), 403–422.
57. A.N. Kolmogorov, *Foundations of the Theory of Probability*, Chelsea, New York, 1933.
58. H. Kunita, *Lectures on Stochastic Flows and Applications*, Tata Institute, Bombay; Springer, Berlin, Heidelberg, New York, 1986.
59. H. Kunita and S. Watanabe, "On square integrable martingales," *Nagoya Math. J.* **30** (1967), 209–245.
60. T. G. Kurtz and P. Protter, "Weak convergence of stochastic integrals and differential equations I & II," LNM **1627**, 1996, pp. 1–41, 197–285.
61. E. Lenglart, "Semimartingales et intégrales stochastiques en temps continue," *Revu du CETHEDEC–Ondes et Signal* **75** (1983), 91–160.
62. G. Letta, *Martingales et Intégration Stochastique*, Scuola Normale Superiore, Pisa, 1984.
63. P. Lévy, *Processus Stochastiques et Mouvement Brownien*, 2nd ed., Gauthiers–Villards, Paris, 1965.
64. P. Lévy, "Wiener's random function, and other Laplacian random functions," *Proc. of the Second Berkeley Symp. on Math. Stat. and Proba.* (1951), 171–187.
65. R.S Liptser and A. N. Shiryayev, *Statistics of Random Processes*, v. I, Springer, New York, 1977.
66. B. Maurey, "Théorèmes de factorization pour les opérateurs linéaires à valeurs dans les espaces L^p," *Astérisque* **11** (1974), 1–163.
67. B. Maurey and L. Schwartz, *Espaces L^p, Applications Radonifiantes, et Géometrie des Espaces de Banach*, Séminaire Maurey–Schwartz, (1973–1975), École Polytechnique, Paris.
68. H. P. McKean, *Stochastic Integrals*, Academic Press, New York, 1969.
69. E.J. McShane, *Stochastic Calculus and Stochastic Models*, Academic Press, New York, 1974.
70. P. A. Meyer, "A decomposition theorem for supermartingales," *Illinois J. Math.* **6** (1962), 193–205.
71. P. A. Meyer, "Decomposition of supermartingales: the uniqueness theorem," *Illinois J. Math.* **7** (1963), 1–17.
72. P. A. Meyer, *Probability and Potentials*, Blaisdell, Waltham, 1966.
73. P. A. Meyer, "Intégrales stochastiques I–IV" in: *Séminaire de Probabilités I*, LNM **39**, 1967, pp. 72–162.
74. P. A. Meyer, "Un Cours sur les intégrales stochastiques," in: *Séminaire de Probabilités X*, LNM **511**, 1976, pp. 246–400.
75. P. A. Meyer, "Le théorème fondamental sur les martingales locales," in: *Séminaire de Probabilités XI*, LNM **581**, 1976, pp. 463–464.
76. P. A. Meyer, "Inégalités de normes pour les intégrales stochastiques," in: *Séminaire de Probabilités XII*, LNM **649**, 1978, pp. 757–760.

77. P. A. Meyer, "Flot d'une équation différentielle stochastique," in: *Séminaire de Probabilités XV*, LNM **850**, 1981, pp. 103–117.
78. P. A. Meyer, "Géometrie différentielle stochastique," Colloque en l'Honneur de Laurent Schwartz, *Astérisque* **131** (1985), 107–114.
79. P. A. Meyer, "Construction de solutions d'équation de structure," in: *Séminaire de Probabilités XXIII*, LNM **1372**, 1989, pp. 142–145.
80. F. Móricz, "Strong laws of large numbers for orthogonal sequences of random variables," *Limit Theorems in Probability and Statistics*, v.II, North–Holland, 1982, pp. 807–821.
81. A.A. Novikov, "On an identity for stochastic integrals," *Theory Probab. Appl.* **16** (1972) 548–541.
82. E. Nelson, *Dynamical Theories of Brownian Motion*, Mathematical Notes, Princeton University Press, 1967.
83. B. K. Øksendal, *Stochastic Differential Equations: An Introduction with Applications*, 4th ed., Springer, Berlin, 1995.
84. G. Pisier, *Factorization of Linear Operators and Geometry of Banach Spaces*, Conference Series in Mathematics, 60. Published for the Conference Board of the Mathematical Sciences, Washington, DC, by the American Mathematical Society, Providence, RI, 1986.
85. P. Kloeden and E. Platen, *Numerical Solutions of Stochastic Differential Equations*, Applications of Mathematics **23**, Springer, Berlin, Heidelberg, New York, 1994.
86. P. Kloeden, E. Platen and H. Schurz, *Numerical Solutions of SDE through Computer Experiments*, Springer, Berlin, Heidelberg, New York, 1997.
87. P. Protter, "Right-continuous solutions of systems of stochastic integral equations," *J. Multivariate Analysis* **7** (1977), 204–214.
88. P. Protter, "Markov Solutions of stochastic differential equations," *Z. für Wahrscheinlichkeitstheorie* **41** (1977), 39–58.
89. P. Protter, "H^p–stability of solutions of stochastic differential equations," *Z. für Wahrscheinlichkeitstheorie* **44** (1978), 337–352.
90. P. Protter, "A comparison of stochastic integrals," *Annals of Probability* **7** (1979), 176–189.
91. P. Protter, *Stochastic Integration and Differential Equations*, Springer, Berlin, Heidelberg, New York, 1990.
92. M. H. Protter and D. Talay, "The Euler scheme for Lévy driven stochastic differential equations," *Ann. Probab.* **25/1** (1997), 393–423.
93. M. Revesz, *Random Measures*, Dissertation, The University of Texas at Austin, 2000.
94. D. Revuz and M. Yor, *Continuous Martingales and Brownian Motion*, Springer, Berlin, 1991.
95. H. Rosenthal, "On subspaces of L^p," *Annals of Mathematics* **97/2** (1973), 344–373

96. H. L. Royden, *Real Analysis*, 2nd ed., Macmillan, NewYork, 1968.
97. S. Sakai *c^*-algebras and W^*-algebras*, Springer, Berlin, Heidelberg, New York, 1971.
98. L. Schwartz, *Semimartingales and their Stochastic Calculus on Manifolds*, (I. Iscoe, editor), Les Presses de l'Université de Montréal, 1984.
99. M. Sharpe, *General Theory of Markov Processes*, Academic Press, New York, 1988.
100. R. E. Showalter, *Hilbert Space Methods for Partial Differential Equations*, Monographs and Studies in Mathematics, Vol. 1. Pitman, London, San Francisco, Melbourne, 1977. Also available online via http://ejde.math.swt.edu/Monographs/01/abstr.html
101. R. L. Stratonovich, "A new representation for stochastic integrals," *SIAM J. Control* **4** (1966), 362–371.
102. C. Stricker, "Quasimartingales, martingales locales, semimartingales, et filtration naturelles," *Z. für Wahrscheinlichkeitstheorie* **39** (1977), 55–64.
103. C. Stricker and M. Yor, "Calcul stochastique dépendant d'un paramètre," *Z. für Wahrscheinlichkeitstheorie* **45** (1978), 109–134.
104. D. W. Stroock and S. R. S. Varadhan, *Multidimensional Diffusion Processes*, Springer, Berlin, Heidelberg, New York, 1979.
105. S. J. Szarek, "On the best constants in the Khinchin inequality," *Studia Math.* **58/2** (1976), 197–208.
106. M. Talagrand, "Les mesures vectorielles a valeurs dans L^0 sont bornées," *Ann. scient. Éc. Norm. Sup.* **4/14** (1981), 445-452.
107. J. Walsh, "An Introduction to stochastic partial differential equations," in: *LNM* **1180**, 1986, pp. 265–439.
108. D. Williams, *Diffusions, Markov processes, and Martingales, Vol. 1: Foundations*, Wiley, New York, 1979.
109. N. Wiener, "Differential-space," *J. of Mathematics and Physics* **2** (1923), 131–174.
110. G.L. Wise and E.B. Hall, *Counterexamples in Probability and Real Analysis*, Oxford University Press, New York, Oxford, 1993.
111. C. Yoeurp and M. Yor, "Espace orthogonal à une semimartingale; applications," Unpublished (1977).
112. M. Yor, "Un example de processus qui n'est pas une semimartingale," *Temps Locaux*, Asterisque **52-53** (1978), 219–222.
113. M. Yor, "Remarques sur une formule de Paul Lévy," in: *Séminaire de Probabilités XIV*, LNM **784**, 1980, pp. 343–346.
114. M. Yor, "Inégalités entre processus minces et applications," *C. R. Acad. Sci. Paris* **286** (1978), 799–801.
115. K. Yosida, *Functional Analysis*, Springer, Berlin, 1980.

Index of Notations

Symbols

Index

The page where an item is defined appears in boldface.
A full index is at `http://www.ma.utexas.edu/users/cup/Indexes`

Q

R

V

W

Z